III International Scientific Conference: Modernization, Innovations, Progress: Advanced Technologies in Material Science, Mechanical and Automation Engineering (MIP-III 2021)

IOP Conference Series: Materials Science and Engineering
Volume 1155

Krasnoyarsk, Russia
29 - 30 April 2021

Editors:

Igor V. Kovalev
Alena A. Stupina
Hans-Ruediger Fluck
Sergey Ognerubov
Normakhmad Ravshanov
Alekper Aliev
Anna Voroshilova
Oleg Kravets
Karl-Erwin Grosspietsch
Komil D. Astanakulov

ISBN: 978-1-7138-4623-9
ISSN: 1757-8981

Printed from e-media with permission by:

Curran Associates, Inc.
57 Morehouse Lane
Red Hook, NY 12571

Some format issues inherent in the e-media version may also appear in this print version.

Institute of Physics
Dirac House, Temple Back
Bristol BS1 6BE UK

Phone: 44 1 17 929 7481
Fax: 44 1 17 920 0979

techtracking@iop.org

Additional copies of this publication are available from:

Curran Associates, Inc.
57 Morehouse Lane
Red Hook, NY 12571 USA
Phone: 845-758-0400
Fax: 845-758-2633
Email: curran@proceedings.com
Web: www.proceedings.com

TABLE OF CONTENTS

Overview of the III International Conference "MIP -III-2021: Modernization, Innovations and Progress"

Igor V Kovalev[1,2,3], Nikolay A Testoyedov[3,4] and Anna A Voroshilova[1]

[1]Krasnoyarsk Science and Technology City Hall of the Russian Union of Scientific and Engineering Associations, 61 Uritskogo street, Krasnoyarsk, 660049, Russia
[2]Siberian Federal University, 79, Svobodny pr., Krasnoyarsk, 660041, Russia
[3]Reshetnev Siberian State University of Science and Technology, 31, Krasnoyarsky Rabochy Av., Krasnoyarsk, 660037, Russia
[4]JSC "Academician M F Reshetnev Information satellite systems", 52 Lenin street, Zheleznogorsk, Krasnoyarsk region, 662972, Russia

E-mail: krasnio@bk.ru

Abstract. The overview describes the main directions and results of the III International Conference MIP-III-2021 held in Krasnoyarsk in April 29-30, 2021. It gives the details about the participants and the proceedings.

The Volume contains Proceedings of the III International Conference "MIP-III-2021: Modernization, Innovations and Progress" which was held in Krasnoyarsk, Russian Federation in April 29-30, 2021. The Conference took place at the Krasnoyarsk Science and Technology City Hall.

The purpose of the Conference is to share the results and prospects of the achievements in using advanced scientific, innovative and information technologies in materials science, mechanical, aerospace, chemical, ecological and power engineering, metallurgy and design of new materials, economics and organization of mechanical engineering production, automation engineering for industry 4.0, etc. The Conference provided an international forum for the presentation of original results in mathematical modelling for software- and hardware applications in various fields of engineering, natural sciences, cybernetics, computer science, etc.

The Conference was jointly organised by the Russian Union of Scientific and Engineering Associations (Moscow, Russia), Krasnoyarsk Science and Technology City Hall and Siberian Scientific Centre DNIT. Tashkent Institute of Irrigation and Agricultural Mechanization Engineers (Tashkent, Uzbekistan) was the partner university at the Conference.

The main partner organisations in preparing and organising the Conference in Krasnoyarsk as well as of the previous scientific events [1-15] were Siberian Federal University and Krasnoyarsk Scientific Centre of the Siberian Branch of the Russian Academy of Sciences [16-18].

The program of the Conference consisted of keynote and plenary talks and numerous presentations (figure 1). Key speakers represented countries such as Kazakhstan, Uzbekistan and Russia.

Figure 1. Keynote and plenary talks.

The Program Committee and Editorial Board included prominent professors and scientists from the Russian Academy of Sciences, Liaoning Normal University (China), Ruhr University in Bochum (Germany), Siberian Federal University (Russia), Tashkent University of Information Technologies (Uzbekistan), SDE II Microsoft (USA),Tashkent Institute of Irrigation and Agricultural Mechanization Engineers (Uzbekistan), Voronezh State Technical University (Russia), Baku State University (Azerbaijan), University of Maribor (Slovenia) as well as specialists from Russian and foreign leading industrial enterprises (figure 2).

IOP Conf. Series: Materials Science and Engineering 1155 (2021) 011001 doi:10.1088/1757-899X/1155/1/011001

Yury V. Gulyaev - Academician of the Russian Academy of Science, Doctor of Physics and Mathematics, Professor, President of the International and Russian Union of Scientific and Engineering Associations, Moscow, Russia

Igor V. Kovalev – Doctor of Technical Sciences, Professor, President of Krasnoyarsk Regional Union of Scientific and Engineering Associations, Director of Krasnoyarsk Science and Technology City Hall, Krasnoyarsk, Russia

Sergey Drukarenko - Candidate of Science, Vice-President, First Secretary of Russian Union of the International and Russian Union of Scientific and Engineering Associations, Moscow, Russia

Uktam P Umurzakov – Dr. Sc., Professor, Rector of the Tashkent Institute of Irrigation and Agricultural Mechanization Engineers, Tashkent, Uzbekistan

Figure 2. International Programme Committee Chairs.

Members of the International Programme Committee were as follows:

- Alena A. Stupina - Doctor of Technical Sciences, Professor of the Siberian Federal University, Krasnoyarsk, Russia
- Bahadir S. Mirzaev - First Vice-Rector of the Tashkent Institute of Irrigation and Agricultural Mechanization Engineers, Doctor of Technical Sciences, Professor, Tashkent, Uzbekistan
- Takhir Z Sultanov - Vice-Rector for Research and Innovation of the Tashkent Institute of Irrigation and Agricultural Mechanization Engineers, Doctor of Technical Sciences, Professor, Tashkent, Uzbekistan
- Kadyrzhan A Shavazov - Dean of the Faculty of Agricultural Mechanization of the Tashkent Institute of Irrigation and Agricultural Mechanization Engineers, Candidate of Technical Sciences, Associate Professor, Tashkent, Uzbekistan
- Komil D Astanakulov - Head of the Department of Agricultural Machinery, Tashkent Institute of Irrigation and Agricultural Mechanization Engineers, Doctor of Technical Sciences, Senior Researcher, Tashkent, Uzbekistan
- Maja Fošner - PhD, Deputy Dean of the Faculty of Logistics, University of Maribor, Slovenia
- Alexander Ivanov - PhD, Researcher at the Federal Polytechnic School of Lausanne (EPFL), Lausanne, Switzerland
- Zhang Fengan – Associate Director of International Exchange Office of Liaoning Normal University, China
- Dr Gundula Herwig – Head of Staff Unit for International and European Affairs, Ministry of Science, Research and Culture, Potsdam, Germany
- Aleksey M. Zyuzin - Chairman of the Board of Directors of the House of Science and Technology of Russian Union of Scientific and Engineering Associations, Honored Inventor of the Republic of Mordovia, Russia.

IOP Conf. Series: Materials Science and Engineering 1155 (2021) 011001 doi:10.1088/1757-899X/1155/1/011001

The Editorial Board consisted of the following members:

The international scope of the Conference was confirmed by the participation of representatives from Kazakhstan (S. Utebaev Atyrau University of Oil and Gas, Atyrau; M. Kozybayev North-Kazakhstan State University, Petropavlovsk; Karaganda State Technical University), China (Central South University, Changsha, Hunan; China Aviation Industry General Aircraft Zhejiang Institute Co., Ltd), Uzbekistan (Tashkent State Technical University; Namangan Institute of engineering and technology), Kyrgyzstan (Central-Asian Institute for Applied Geosciences, Bishkek).

The participants from Russia represented more than 100 universities, scientific institutes and organisations, industrial enterprises:

- All-Russia Research Institute of Hydraulic Engineering and Land Reclamation of A N Kostyakov, Moscow
- Anton Chekhov Taganrog State Institute (branch) Rostov State University of Economics, Taganrog
- Bashkir State University, Ufa
- Bauman Moscow State Technical University
- Belgorod State Technological University named after V.G. Shukhov
- Blagonravov Institute of Mechanical Engineering, Russian Academy of Sciences, Moscow
- Boreskov Institute of Catalysis SB RAS
- Branch of Ufa State Petroleum Technological University in Salavat, Salavat, the Republic of Bashkortostan
- Buryat State Agricultural Academy named after V.R. Filippov, Ulan-Ude
- Center for Physical and Technical Problems of Energy of the North of the Federal State Budgetary Institution of Science of the Kola Scientific Center of the Russian Academy of Sciences, Apatity
- Chechen State University
- Chelyabinsk State University
- Department "Welding Production Technology" of the Ural Federal University named after B. N. Yeltsin, Yekaterinburg
- Department of Chemical and Food Production Machines and Apparatuses (Orenburg State University), Orenburg
- Department of Petroleum Engineering Technology, Ufa State Petroleum Technological University

- Derzhavin Tambov State University
- Don State Technical University, Rostov-on-Don
- Faculty of Biology and Chemistry Chechen State University
- Far Eastern State Technical Fisheries University, Vladivostok
- Federal Research Centre the Southern Scientific Centre of the Russian Academy of Sciences, Rostov-on-Don
- Federal State Budgetary Educational Institution of Higher Education "Murmansk Arctic State University"
- GS Nanotech, Gusev
- Institute of Informatics and Telecommunications, Reshetnev Siberian State University of Science and Technology, Krasnoyarsk
- LLC TEHOSNASTKA-RTD, RF
- Novosibirsk State Technical University
- Institute of Metal Physics named after A. N. Mikheev, Ural Branch of the Russian Academy of Sciences, Yekaterinburg
- Institute of Physics and Technology, Petrozavodsk State University
- Institute of Precision Mechanics and Control, Russian Academy of Sciences, Saratov
- Institute of service and business (branch) of Don state technical University in Shakhty, Rostov Region
- Irkutsk State Agricultural University named after A.A. Ezhevsky
- Irkutsk State National Research Technical University
- Irkutsk State University
- ITMO University, St. Petersburg
- Ivanovo State University
- JSC "Academician M F Reshetnev Information satellite systems",
- Kabardino-Balkarian state agricultural university named after V.M. Kokov, Nalchik
- Kabardino-Balkarian State University
- Kazan Federal University
- Kazan National Research Technological University
- Kemerovo State University
- Krasnoyarsk State Agrarian University
- K-SkAI LLC, Petrozavodsk
- Kuban State Agrarian University named after I.T.Trubilin, Krasnodar
- Kuban State Technological University, Krasnodar
- Kuban State University, Krasnodar
- Lipetsk State Technical University
- Lobachevsky State University of Nizhny Novgorod
- Mendeleev University of Chemical Technology of Russia, Moscow
- Military Academy of Logistics, St Petersburg
- Military Scientific Educational Center of Military-Air Forces "N.E. Zhukovsky and Ju.A. Gagarin Military-Air Academy", Voronezh
- MIREA—Russian Technological University, Moscow
- Moscow Power Engineering Institute (MPEI)
- Moscow State University of Technology "STANKIN"
- Nizhny Novgorod State Agricultural Academy
- Nizhny Novgorod State Technical University R.E. Alekseev
- North-Caucasian State Academy, Cherkessk
- North-Caucasus Federal University
- Novosibirsk State University
- Orel State University named after I.S. Turgenev
- Penza State University
- Perm Institute of the FPS of Russia
- Perm Military Institute of National Guard Troops of the Russian Federation
- Perm National Research Polytechnic University
- Perm State Agro-Technological University named after Academician D N Pryanishnikov
- Petrozavodsk State University
- Platov South-Russian State Polytechnic University, Novocherkassk
- Research and Design Institute of Environmental Problems LLC, Orenburg
- Research and production enterprise «Antares», Saratov
- Reshetnev Siberian state university of science and technologies, Krasnoyarsk
- Russian State Agrarian University-Moscow Agricultural Academy named after K.A. Timiryazev
- Saint-Petersburg State University
- Saint-Petersburg University for Aerospace Instrumentation
- Samara National Research University
- Samara Scientific Center of the Russian Academy of Sciences
- Samara State Technical University
- Siberian Federal Scientific Centre of Agrotechnology, Russian Academy of Science
- Siberian Federal University

- Siberian Fire and Rescue Academy of the Ministry of Emergency Situations of Russia, Zheleznogorsk
- Southern Federal University
- Southern Federal University, Rostov-on-Don
- Stavropol State Agrarian University
- T.F. Gorbachev Kuzbass State Technical University, Kemerovo
- Tambov State Technical University
- Thermal Electrical Systems LLC
- Togliatti State University
- Ufa State Aviation Technical University
- Ufa State Petroleum Technological University
- Ulyanovsk Research Institute of Aviation Technology and Production Organization
- Ulyanovsk State Technical University
- University of the Ministry of Internal Affairs of St. Peterburg
- Ural engineering center, Chelyabinsk
- Ural Federal University named after the first President of Russia B.N. Yeltsin, Ekaterinburg
- Volga Region State University of Service
- Volgograd State Agricultural University
- Volgograd State Technical University
- Voronezh Institute of High Technologies

The event has offered a platform for bringing together students, postdocs, innovative academics and industrial experts to exchange their ideas and contribute new engineering approaches to research mechanical and automation processes in various technological, production and information systems. Great interest was aroused by the section about Cybernetics, Computation and Simulation for Organization of Mechanical Engineering Production.

The Conference provided the premier interdisciplinary and multidisciplinary forum for researchers, practitioners and educators to present and discuss the most recent innovations, trends, concerns, practical challenges encountered and the solutions adopted in the fields of materials sciences, mechanical engineering, aerospace technologies, metallurgy, chemical, ecological and power engineering, cybernetics, data protection in automation systems, etc. The participants who could not come to the Conference due to different reasons including the problem of global cases of COVID-19 prepared video reports and presented on-site presentations which are available on the web-site of the Conference. A wide range of fundamental and applied problems in various production and technological systems has undergone consideration during the video conference.

The Conference also was broadcast in Zoom which allowed participants all over Russia as well as from abroad to participate and make reports as well as to ask and answer questions.

The programme of the Conference included the following 5 sections:

- Material Science and Aerospace Technology
- Mechanical and Automation Engineering for Industry 4.0
- Metallurgy, Chemical, Ecological and Power Engineering
- Processing, Engineering and Applications
- Information Technologies, Reliability and Data Protection in Automation Systems
- Cybernetics, Computation and Simulation for Organization of Mechanical Engineering Production

All participants were invited to present their papers in this Volume and all submitted manuscripts went through the independent peer review process. We are very grateful to all reviewers from Russia, China, USA, Slovenia, Germany, Azerbaijan and Uzbekistan for their time and highly professional comments. We deeply believe that their reviews gave opportunity to improve the scientific quality of the presented papers which may be useful for academic, scientific and industrial partners.

MIP-III-2021 is a highly demanded event among the scientists and specialists with a competitive submission process. More than 500 reports were submitted to the Organizing Committee of the Conference MIP-III-2021. 105 reports were selected for inclusion into this Volume after the review process. All the papers were subjected to rigorous peer-review by conference committee members and international reviewers. The papers were selected based on their quality, novelty, scientific significance and relevance to the Conference directions. The proceedings present to the readers the recent advances in the field of material science, mechanical and automation engineering for industry 4.0, data protection in automation systems, cybernetics, information technologies, chemical, ecological and power engineering. These papers reflect modern engineering approaches in all Conference directions, broaden the researches of the previous Conferences [3, 10].

We chose the IOP Conference Series: Materials Science and Engineering to provide all contributors with the opportunity to publish their papers in an international, peer-reviewed journal. This is understood and appreciated by all the participants of our Conference, and therefore this Volume provides an excellent overview of the main topics of our Conference. The Proceedings of the I Conference MIP: Engineering-2019 were published in IOP Conference Series in June 2019 [3] and the proceedings of the II Conference MIP: Engineering-2020 were published in IOP Conference Series in May 2020 [10].

Acknowledgements

The Conference was sponsored by JSC "Academician M. F. Reshetnev Information satellite systems", Zheleznogorsk, Russia; International Union of Scientific and Engineering Associations and Russian Academy of Engineering, Moscow, Russia (figure 3). The Organizing Committee is grateful to the sponsor organizations for sponsor support, financial contribution and cooperation.

JSC ACADEMICIAN M.F. RESHETNEV
Information Satellite Systems

Zheleznogorsk, Russia
http://www.iss-reshetnev.com/

The International Union of
Scientific and Engineering Public Associations

Moscow, Russia
http://www.rusea.info/

Russian Academy of Engineering

Moscow, Russia
http://www.info-rae.ru/

Figure 3. Conference Sponsor Organizations.

MIP Conference Committee highly appreciate the contribution of Professor Oleg Kravets (Voronezh State Technical University, Russia) for cooperation in organization of MIP: Engineering-III-2021 and preparation of the papers for this Volume. We are grateful to scientists and professors of Siberian Federal University, Saint Petersburg State University of Aerospace Instrumentation, Institute of Computational Modeling of Siberian Branch of Russian Academy of Sciences, National Research University Moscow Aviation Institute, South Ural State University (National Research University), Siberian State University named after M.F. Reshetnev, IrGUPS, ITMO University, SKFU, Novosibirsk State University for active participation in the Conference.

On behalf of the Conference Committee and organizers, we would like to thank all the authors who contributed to this Volume as well as to the reviewers, speakers and all the conference participants for their support to MIP-III-2021.

We express gratitude to IOP Publishing for an opportunity to publish the Proceedings of the Conference to provide open access and to make them available for worldwide recognition.

References

[1] 2018 *IOP Conf. Ser.: Mater. Sci. Eng.* **450** 011001
[2] Kovalev I V, *et al.* 2019 *IOP Conf. Ser.: Earth and Environ. Sci.* **315** 011001
[3] Kovalev I V, *et al.* 2019 *IOP Conf. Ser.: Mater. Sci. Eng.* **537** 011001
[4] Kovalev I V, *et al.* 2019 *J. Phys.: Conf. Ser.* **1399** 011001
[5] Kovalev I V and Voroshilova A A 2020 *IOP Conf. Ser.: Mater. Sci. Eng.* **734** 011001
[6] Antokhina Yu A, *et al.* 2020 *J. Phys.: Conf. Ser.* **1515** 011001
[7] Kovalev I V, Pyzhikova N V and Voroshilova A A 2020 *IOP Conf. Ser.: Earth and Environ. Sci.* **421** 011001
[8] Kovalev I V *et al* 2020 *IOP Conf. Ser.: Earth Environ. Sci.* **548** 011001
[9] Kovalev I V *et al* 2020 *IOP Conf. Ser.: Mater. Sci. Eng.* **919** 011001
[10] Kovalev I V and Voroshilova A A 2020 *IOP Conf. Ser.: Mater. Sci. Eng.* **862** 011001
[11] Kovalev I V *et al* 2020 *J. Phys.: Conf. Ser.* **1679** 011001
[12] Kovalev I V and Voroshilova A A 2020 *J. Phys.: Conf. Ser.* **1691** 011001
[13] Kovalev Igor V *et al* 2021 *IOP Conf. Ser.: Mater. Sci. Eng.* **1047** 011001
[14] Kovalev I V *et al* 2021 *IOP Conf. Ser.: Earth Environ. Sci.* **677** 011001
[15] Antokhina Yu A, *et al.* 2021 *J. Phys.: Conf. Ser.* **1889** 011001
[16] Testoedov N A, *et al.* 2018 *IOP Conf. Ser.: Mater. Sci. Eng.* **450** 022004
[17] Testoedov N A, *et al.* 2018 *IOP Conf. Ser.: Mater. Sci. Eng.* **450** 022002
[18] Testoedov N A, *et al.* 2018 *IOP Conf. Ser.: Mater. Sci. Eng.* **450** 022008

Peer review declaration

III International Conference on Modernization, Innovations and Progress (MIP-III-2021), 29-30 April 2021, Krasnoyarsk, Russian Federation

All papers published in this volume of *IOP Conference Series: Materials Science and Engineering* have been peer reviewed through processes administered by the Editors. Reviews were conducted by expert referees to the professional and scientific standards expected of a proceedings journal published by IOP Publishing.

- **Type of peer review: Double-blind**

All papers came through the basic review which included an initial technical criteria check (paper field, structure of submission, adherence to the submission instructions, English language usage and a check for the similarity rate). Any papers out of the scope or containing plagiarism, including self-plagiarism, were rejected.

We used a double-blind system for peer review; both reviewers' and authors' identities remained anonymous. The submitted papers were reviewed by at least two experts: one editorial staff member as well as at least one external reviewer. The third/fourth reviewer is involved in case two reviewers have disputes about the content of the papers or the authors do not agree with the review result. The review process took from 7 to 15 days as a rule. The reviews were conducted to the professional and scientific standards.

The decision to accept or reject the paper was based on the suggestions of reviewers. Acceptance/rejecting notifications were sent to the corresponding author(s).

- **Conference submission management system:**

Conference submission is carried out through the conference management system which was developed by the Organizer. It easy to use, and has features to make it suitable for simultaneous conference calls.

More than 16,500 registration forms have been already processed by the system which showed it stability and flexibility.

The conference management system is integrated with the conference website of the Organizer which helps the Organizer to cope with the complexity of the submission, reviewing, refereeing and accepting processes. The current version supports:

1. management of the access of reviewers and editors to papers;
2. automatic registration for conferences and paper submission;
3. list of the current, forthcoming and held conferences with all necessary information;
4. list of all conference materials with online presentations and/or video reports of participants, list of published proceedings with direct links to publications and journals;
5. sending emails to programme committee members, reviewers and authors;
6. preparation and publishing of conference programmes;

7. all authors submitting the papers are assumed to accept the terms of Publication Licence;
8. all requirements for papers, templates and sample papers are accessible for the authors.

- **Number of submissions received:** 502
- **Number of submissions sent for review:** 502
- **Number of submissions accepted:** 105
- **Acceptance Rate (Number of Submissions Accepted / Number of Submissions Received X 100):** 21
- **Average number of reviews per paper:** 2
- **Total number of reviewers involved:** 45
- **Any additional info on review process:**

The Checklist for the reviewers included the following criteria: relevance to the scope of the conference topics, the scientific quality, assessing the errors, organisation of the paper, the title, the references, the level of the English language, the quality of the figures and tables, etc.

The reviewers could give the following recommendation: to publish the paper as it is, to publish it after mandatory/optional minor/major revision, to reject the paper.

- **Contact person for queries:**

Anna Voroshilova, Krasnoyarsk Science and Technology City Hall, 61, Uritskogo Street, Krasnoyarsk, 660049, Russia, E-mail: krasnio@bk.ru. Tel + 7 913 538 6990

To the question of the dependence of the change in the speed of ultrasound and the natural frequency of vibration of atoms in monatomic planes during a polymorphic transition in iron

Stepan A Yurchenko

LLC "TEHOSNASTKA-RTD", 2 Novaya Str., Petropavlovka, Krasnoyarsk, 662925, Russian Federation

E-mail: stepan1974.yur@yandex.ru

Abstract. Iron is one of the most used metals in industry and construction. Machines and mechanisms made of iron operate over a wide range of temperatures and voltages. Checking their performance during operation is carried out using non-destructive testing. One of the methods of non-destructive testing is the method of molecular acoustics, which links the internal structure and physical and mechanical characteristics with the values of the longitudinal and transverse ultrasound velocity. The change in the speed of sound upon deformation of the material is characterized by a change in the number of defects in the crystal lattice or by the rotation of mesoscopic volumes. However, a change in the speed of ultrasound also occurs during polymorphic transformation, which is associated with a change in the geometric dimensions of the cell of the crystal lattice and can be both under the influence of temperature and under pressure. Therefore, when analyzing a material by the method of molecular acoustics, it is necessary to understand what exactly happened in the material - a polymorphic transition or an accumulation of defects. This is possible only with an accurate determination of two parameters - the natural frequency of atomic vibration, which changes during the polymorphic transition, and a change in the ultrasonic wave velocity due to a change in the interatomic interaction of atomic shells. Therefore, the main goal of this work is to derive a theoretical equation relating the change in the natural vibration frequency of atoms during the polymorphic transition of iron from αFe to γFe, with a change in the specific area per atom in the atomic plane.

Until now, it has been generally accepted that a change in the ultrasound velocity in a material is associated with dislocations [1-3] or with a turn of mesoscopic volumes [4,5] occurring during deformations. However, neither the theory of defects, nor the theory of mesoscopic volumes can explain, from the point of view of physical processes, the change in the natural vibration frequency of the atom and the speed of ultrasound in the material during the polymorphic transition associated with a change in the structure of the cell of the crystal lattice. In [6], the author provides a mathematical formula linking the natural frequency of vibration of iron atoms during polymorphic rearrangement from αFe to γFe, with a change in the coordination number K (the number of neighboring atoms) and a change in the volume of primitive cells from which the "skeletons" of unit cells are built. a centered cubic cell and a face-centered cubic cell of the iron crystal lattice.

However, the author did not offer a physical mechanism or law to which this mathematical formula corresponds. Thus, the main goal of this work is to find the relationship between the change in the parameters of primitive cells of the BCC and HCC crystal lattices of iron with the physical mechanism (law), which is responsible for the change in the natural vibration frequency of atoms and the speed of ultrasonic waves.

To analyze the mechanism of the polymorphic transformation of αFe (BCC crystal lattice) into γFe (HCC crystal lattice), which can occur both under the action of temperature and pressure [7], let us define the concept of constructing the crystal lattice of a material. The founder of the theory of the structure of materials was O. Brave. In his work [8], he proposed to consider the crystal lattice as an ordered arrangement of atoms in space and introduced the concept of syngony and types of lattices.

However, according to Academician E.S. Fedorov [9], this theory has drawbacks: "Almost until very recently, no attempts were made to correct or continue this construction, and it towered above others in its complete integrity. However, at its very foundation there were weak points that required alteration. " At present, in the course of experiments, one-atomic two-dimensional materials have been discovered (obtained) [10], the existence of which was not taken into account in the theory of O. Brave. Although when determining the structure of primitive cells in [11, 12] it is indicated that atomic planes can be distinguished in the crystal lattice. Thus, today it can be assumed that atomic planes (or monoatomic two-dimensional materials) are the building material for building the smallest structural material - a primitive cell. The "skeleton" of the unit cell of the BCC and HCC crystal lattice of iron is built from primitive cells (see figure 1 and figure 2).

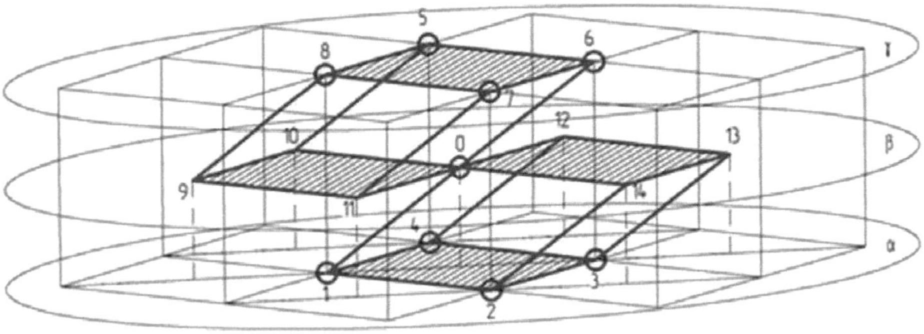

Figure1. The skeleton of the "elementary" 1-2-3-4-5-6-7-8 BCC cell of the crystal lattice is formed from two primitive cells formed from three parallel atomic planes α, β and γ.

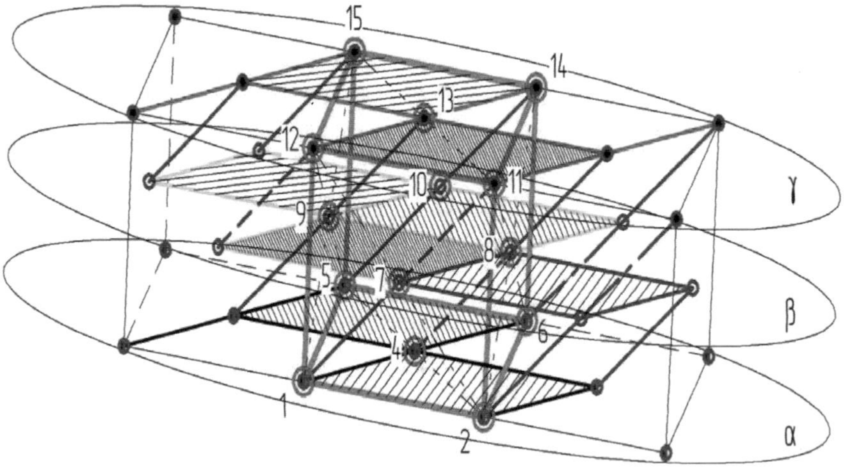

Figure 2. The skeleton of the "unit" cell of the HCC crystal lattice is formed of four primitive cells formed by three parallel atomic planes α, β and γ.

The mechanism of transmission of ultrasonic waves, in the model of interatomic interaction by collectivization of electrons located in outer orbits, by deformation of shells, was proposed by Academician Ya.I. Frenkel in [13]. In [14], it was proposed to refine the model of collectivization of electrons by pulling atoms together with a magnetic force loop, due to which it was possible to obtain a universal equation for determining the longitudinal, transverse and ultrasound velocity in the rod:

$$\vartheta_i = l_i \cdot \omega \cdot \frac{1}{n} \quad (м/с) \tag{1}$$

where: l_i – the length of the interatomic bond (the edge of the primitive cell) expressed in terms of the crystal lattice parameter; $\bar{a}$ – the length of the interatomic bond in the atomic plane; ω – vibration frequency of atoms in the volume of a primitive cell, n – the number of interatomic interactions involved in the transmission of ultrasound in different planes of the selected material cell.

It should be noted that equation (1) reflects the properties of the material associated with the structure of a new cell of two adjacent parallel atomic planes in the form of the parameter n. For a primitive BCC cell (1-2-3-4-0-12-13-14 in Fig. 1) and a primitive HCC cell (figure 3), n = 1 crystal lattices. In this case, the transformation of the BCC primitive cell can occur during deformation during the rotation of mesoscopic volumes, and due to a pure shift of atoms in atomic planes leading to the HCC structure of the primitive cell in accordance with figure 3.

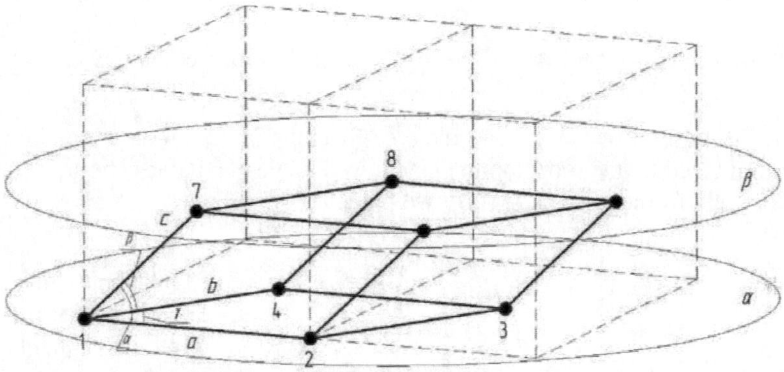

Figure 3. Primitive cells of an fcc lattice formed from atoms 1-2-3-4-5-6-7-8 located in two parallel planes α and β.

It should be noted that Eq. (1) makes it possible to calculate the ultrasound velocities for the "cores" of the unit cells of the BCC and HCC crystal lattices, taking into account the fact that the coefficient n = 2 due to the fact that atoms transfer the energy of an ultrasonic wave from the plane α to the plane γ only through interatomic interaction with atoms of the β plane. As a consequence, on the basis of this equation (1) in [14], theoretical values of the longitudinal, transverse and speed of sound in a rod were obtained for iron, with a BCC crystal lattice, and copper, with an HCC crystal lattice, which with a high accuracy of 95-100 % coincided with the experimental values of the longitudinal, transverse and ultrasound velocity in the rod for these materials determined on the basis of the molecular acoustics method.

This confirmed the idea of Academician E S Fedorov [9], that it is the choice of symmetry conditions (construction of a primitive cell) that is an important point for constructing an interconnected theory of considering all cumulative processes occurring in crystal lattices, which was presented in [15].

Considering the rearrangement of primitive cells from αFe to γFe as a system of arrangement of atoms in atomic planes, see figure 1 and figure 3, which form the volume of a primitive cell by interatomic interaction with each other, it can be noted that atoms in planes α and β form quadrangles. In the α and β planes for the BCC primitive crystal cell, the atoms form a square, and in the α and β planes for the HCC primitive cell, the atoms form a parallelogram. Therefore, it can be argued that under the influence of external forces during a polymorphic transition, the arrangement of atoms in atomic planes changes, which leads to a change in the geometry of the primitive volume of a crystal lattice cell and, accordingly, to a change in the physical properties of the material - the vibration frequency of atoms, the ultrasound velocity and Young's modulus.

It is known that the volume of a primitive cell of the crystal lattice accounts for the mass of one atom [11, 12], which is formed by taking into account all parts of the atoms cut out in space by the configuration of the cell. However, the cell mass can be represented as a set of atomic masses located on the cell area in atomic planes (see figure 4).

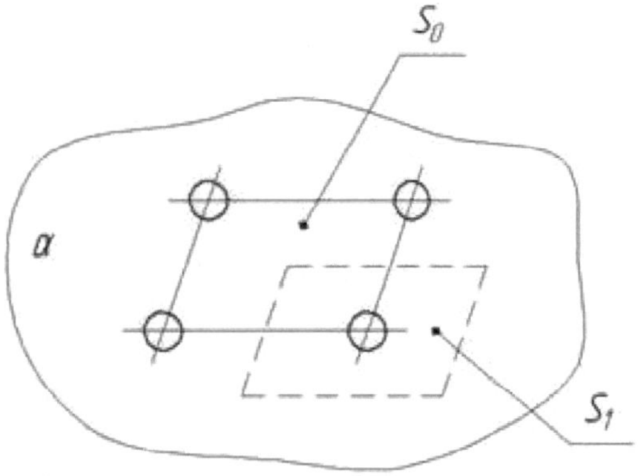

Figure 4. Area of a primitive cell in the atomic plane S_0. The area of the primitive cell selected by the Winger-Seitz method is S_1.

The area S_0 is the mass of one atom, and the area S_1 is the mass of one atom. However, since the mass of an atom is divided by the monatomic plane α into two equal halves (figure 5), the volume of a primitive cell is actually formed from atoms of two planes (figure 6).

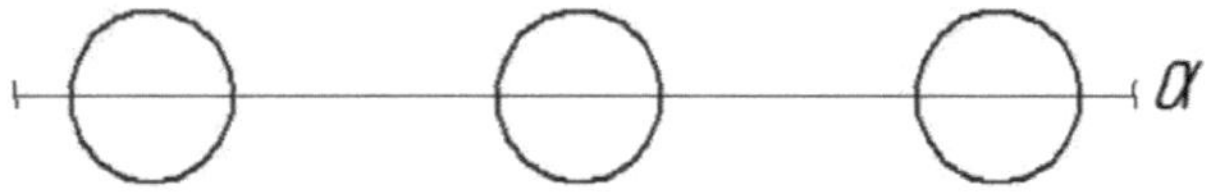

Figure 5. the mass of an atom is divided by the monatomic plane α into two equal halves.

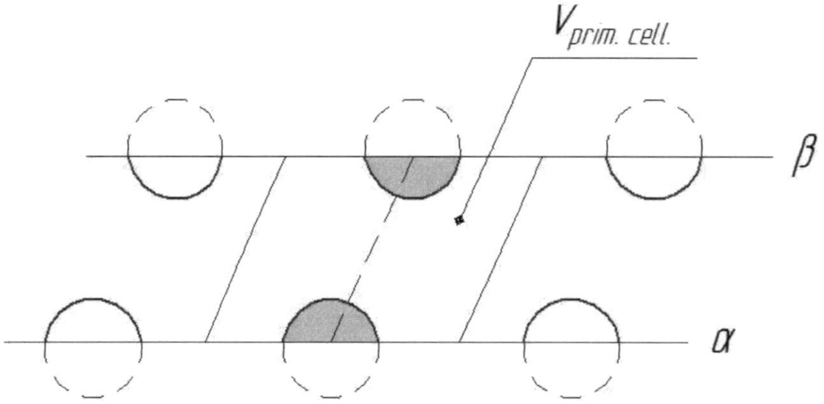

Figure 6. the volume of a primitive cell is actually formed from atoms of two planes.

Considering the above, consider the mechanism of the process described by the following equation, which was presented in [6]:

IOP Conf. Series: Materials Science and Engineering 1155 (2021) 012001 doi:10.1088/1757-899X/1155/1/012001

$$\frac{V_{0BCC}}{K_{BCC} \cdot \omega_{BCC}} = \frac{V_{0HCC}}{K_{HCC} \cdot \omega_{HCC}} \quad (m^3 * s) \tag{2}$$

This equation determines the dependence of the natural frequency of vibration of iron atoms during the polymorphic transition from αFe to γFe with a change in the volume of the primitive cell and a change in the coordination number K, which determines the number of neighboring atoms. It is known that the bulk density of iron - ρ_{Fe} changes during the polymorphic transition, but the atomic mass has a constant value m_{at} = Const. As a consequence, we transform equation (2) so that the value of the atomic mass is present:

$$\frac{V_{0BCC} \cdot \rho_{BCC}}{K_{BCC} \cdot \omega_{BCC} \cdot \rho_{BCC}} = \frac{V_{0HCC}}{K_{HCC} \cdot \omega_{HCC}} \cdot \frac{\rho_{HCC}}{\rho_{HCC}} \tag{3}$$

$$\frac{m_{at}}{K_{BCC} \cdot \omega_{BCC} \cdot \rho_{BCC}} = \frac{m_{at}}{K_{HCC} \cdot \omega_{HCC} \cdot \rho_{HCC}} \tag{4}$$

Let us carry out a further transformation of equation (4) - we reduce in the left and right sides the equality of the mass of the atom and after an additional transformation we obtain the equation:

$$K_{BCC} \cdot \omega_{BCC} \cdot \rho_{BCC} = K_{HCC} \cdot \omega_{HCC} \cdot \rho_{HCC} \tag{5}$$

However, given that the natural frequency of vibration of atoms can be represented in terms of time in accordance with the formula:

$$\omega = \frac{1}{t} \quad (1/s) \tag{6}$$

then equation (5) can be transformed into the following equation:

$$K_{BCC} \cdot \frac{\rho_{BCC}}{t_{BCC}} = K_{HCC} \cdot \frac{\rho_{HCC}}{t_{HCC}} \tag{7}$$

or into the equation:

$$K_{BCC} \cdot \frac{\rho_{BCC}}{t_{BCC}} - K_{HCC} \cdot \frac{\rho_{HCC}}{t_{HCC}} = 0 \tag{8}$$

The physical meaning of equation (5) when transformed into equation (7) can be expressed as follows: the mass of an atom oscillating about an equilibrium position, with a certain frequency ω, for a certain time interval *t* in the volume V_0 of a primitive cell, creates a bulk density ρ_v. Thus, during the polymorphic transition, due to the transformation of the initial primitive volume V_{BCC} of the bcc cell into the volume V_{HCC} of the primitive cell of the HCC crystal lattice, which is accompanied by a change in the bulk density, and taking into account the condition of continuity of the medium reflected by equation (8), a change in the oscillation frequency ω should occur in the material atom.

Moreover, the change in the coordination number K indicates that neighboring atoms also participate in the formation of a constant volumetric density of the primitive cell of the crystal lattice (that is, the process occurs simultaneously in the entire deformed volume, which corresponds to the Van der Waals principle).

Due to the fact that it was shown above that the volume of a primitive cell can also be formed from two atoms located in two adjacent parallel atomic planes, the process of formation of the surface density due to the vibration of an atom, and its change, should be similar to that shown above the process of forming bulk density. Therefore, equation (5) reflecting the change in bulk density, for the surface density, is represented by a similar expression:

$$U_{BCC} \cdot \omega_{BCC} \cdot \rho_{S\ BCC} = U_{HCC} \cdot \omega_{HCC} \cdot \rho_{S\ HCC} \tag{9}$$

And since the surface density ρs changes during the polymorphic transformation of the cell in the atomic plane, due to the change in the cell area, see figure 7, then equation (9) can be represented through the constant value of the atomic mass m_{at} = Const as:

$$\frac{m_{at} \cdot U_{BCC}}{S_{BCC} \cdot t_{BCC}} = \frac{m_{at} \cdot U_{HCC}}{S_{HCC} \cdot t_{HCC}} \tag{10}$$

Equation (10) is an analogue of equation (9). However, it represents the surface density in terms of the ratio of the mass of the atom to the surface area of the cell:

$$\rho_{Si} = \frac{m_{at}}{S_i} \tag{11}$$

where: Si is the cell area in the atomic plane (figure 4).

Or, taking into account the dependence between the natural frequencies of vibrations of atoms, in the atomic planes of the BCC and HCC lattices shown in figure 7.

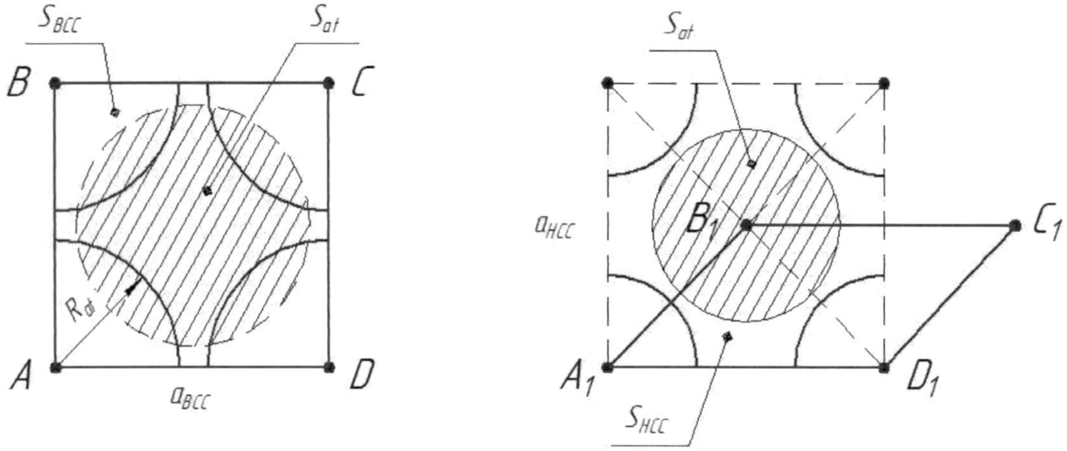

Figure 7. Conversion of area from square ABCD (for BCC cell) to parallelogram $A_1B_1C_1D_1$ (for HCC cell) at polymorphic transition.

Knowing the parameters of the crystal lattices of iron (for the BCC crystal lattice $\bar{a}_{BCC}$=2.86*10^{-10} m, and for the HCC crystal lattice $\bar{a}_{HCC}$ = 3.64*10^{-10} m and the radius of the iron atom R_{at} = 1.26*10^{-10} m), you can calculate the area of the cell S_{BCC} and S_{HCC}, as well as the area of the Sat atom itself.

$$S_{BCC} = a_{BCC}^2 = 8{,}1796 \cdot 10^{-20} м^2 \tag{12}$$

$$S_{HCC} = a_{HCC}^2 \cdot \frac{1}{2} = 6.625 \cdot 10^{-20} м^2 \tag{13}$$

$$S_{at} = a_{at}^2 \cdot \frac{1}{2} = 4{,}985 \cdot 10^{-20} м^2 \tag{14}$$

In equation (13), the coefficient (1/2) arises because there are two atoms per HCC cell area (see figure 7).

Coefficient U - defines compactness, which is defined as the ratio of the area of an atom to the area of a cell in the atomic plane:

IOP Conf. Series: Materials Science and Engineering 1155 (2021) 012001 doi:10.1088/1757-899X/1155/1/012001

$$U = \frac{S_{at}}{S_i} \tag{15}$$

$$U_{BCC} = 4{,}985*10^{-20}\text{м}^2 / 8{,}1796*10^{-20}\ \text{м}^2 = 0{,}61 \tag{16}$$

$$U_{HCC} = 4{,}985*10^{-20}\text{м}^2 / 6{,}625*10^{-20}\ \text{м}^2 = 0{,}747 \tag{17}$$

Having determined the values of the cell areas SBCC and SHCC, as well as the compactness coefficients U_{BCC} and U_{HCC}, we substitute their values into equation (10), while the value of the natural vibration frequency of the atom for the bcc crystal lattice is assumed to be equal to $\omega_{BCC} = 1.98*10^{13}$ Hz, which was established in [14].

Thus, the natural frequency of vibrations of atoms in the HCC atomic plane will have a value:

$$\omega_{HCC} = \frac{U_{BCC}\omega_{BCC}S_{HCC}}{S_{BCC}\cdot U_{HCC}} = 1.98\cdot10^{13}\,Hz\cdot0.61\cdot6.625\cdot10^{-20}\text{м}^2/(8.1796\cdot10^{-20}\ \text{м}^2*0.747) = \tag{18}$$

$$= 1.299\cdot10^{13}\text{Гц} \approx 1.3\cdot10^{13}\text{Hz}$$

In equation (18), the theoretical value of the natural frequency of vibrations of atoms in the atomic plane of the HCC crystal lattice is obtained.

To check this value, it is necessary to calculate the value of the longitudinal ultrasound velocity based on equation (1) and compare it with the experimental value of the longitudinal ultrasound velocity. Substitute in equation (1) the obtained frequency value ωHCC = $1.3*10^{13}$Hz

$$\vartheta_{HCC} = a_{HCC}\cdot\omega_{HCC}\cdot\frac{1}{n} = 3{,}64\cdot10^{-10}\,м\cdot1{,}3\cdot10^{13}\,Гц\cdot\frac{1}{1} = 4732\,\frac{м}{с} \tag{19}$$

Equation (19) gives the theoretical value of the longitudinal ultrasound velocity for the atomic plane of the HCC cell. Let's check this value based on the equation which was obtained in [16] to determine the longitudinal ultrasound velocity for st.9KhF containing 96% iron [17], depending on the heating temperature based on experiments:

$$c_i = \sqrt{\frac{E\left(1 - \frac{a_t}{4\cdot10^{-2}}T\right)(1-\mu)}{\rho(1+\mu)(1-2\mu)}} \tag{20}$$

Where: E – modulus of normal elasticity, MPa; μ – Poisson's ratio; ρ – material density, kg/m³; $\alpha_т$ – coefficient of linear expansion,10^{-6} °C;

Let us calculate the value of the longitudinal ultrasound velocity for iron at T = 912 °C, since the polymorphic transition from αFe to γFe occurs at a temperature of 911 °C, substituting the values given in [16] according to equation (20): E = $2*10^{11}$ Pa, $\alpha_т = 12*10^{-6}$ C◦, μ = 0.28 and ρ = 7850 kg/m³.

$$c_i = \sqrt{\frac{2\cdot10^{11}\,Па\cdot(1-0{,}2736)\cdot(1-0{,}28)}{7850\,\frac{м}{с^2}\cdot(1+0{,}28)\cdot(1-2\cdot0{,}28)}} = \sqrt{0{,}0002366\cdot10^6\,\frac{м^2}{с^2}} = 4864\,\frac{м}{с} \tag{21}$$

The difference between the theoretical value of the longitudinal ultrasound velocity calculated by equation (1) from the value of the longitudinal ultrasound velocity obtained by equation (20) is:

$$\psi = \frac{4864-4732}{4732}\cdot100\% \approx 2\% \tag{22}$$

The coincidence of the values of longitudinal velocities with an accuracy of ψ <5% is a good result. Thus, we can assume that the theoretical value of the frequency of vibrations of atoms in the atomic plane of the HCC crystal lattice ωHCC = $1.3*10^{13}$Hz obtained in equation (18) coincides with high accuracy with the true value of vibrations of atoms in the HCC crystal lattice of iron.

Based on a theoretical analysis of the polymorphic transition from αFe to γFe, it was found that a change in the cell area in the atomic plane leads to a change in the natural vibration frequency of atoms, which in turn leads to a change in the value of the longitudinal velocity of ultrasound propagation in these atomic planes. Thus, the established dependence, in [6], of the change in the frequency of atomic vibration and the ultrasound velocity in the volumes of primitive cells during the polymorphic transition from the BCC cell to the HCC cell of the crystal lattice, could be explained by the change in the area per atom in the atomic plane, which is associated with the physical the process of formation of the surface density of a material as a process of oscillation of the mass of an atom on the area of a cell.

In fact, in the work, it was possible to show that the surface density forming the bulk density is formed by the vibration of an atom having a constant mass value, and also to establish a general relationship between the surface densities during the transformation (polymorphic transition) of cells in atomic planes:

$$\frac{m_{at}}{\partial S}\cdot\frac{1}{\partial t}\cdot U = Const \tag{23}$$

Taking into account that equation (23) can be transformed to the following expression:

$$\frac{Const}{\partial S\cdot\partial t}\cdot U = Const \tag{24}$$

which, in turn, taking into account equation (15) takes the value:

$$\frac{\partial\omega}{\partial S}\cdot\frac{S_{at}}{\partial S} = Const \tag{25}$$

However, since the area of an atom can be taken as a constant value S_{at} = Const, then equation (25) takes its final form:

$$\frac{\partial\omega}{\partial S}\cdot\frac{Const}{\partial S} = \frac{\partial\omega}{\partial S^2} = Const \tag{26}$$

Equation (26) will correspond to the nomogram of the dependence of the area "swept out" by an atom in the atomic plane on the frequency of vibration of the atom shown in figure 8.

Based on the above analysis, it was possible to obtain equation (23), which, when transformed into equation (26), not only reflects the totality of processes occurring during the polymorphic transition, but also the totality of processes occurring in the elastic deformation region, which indicates that In the work, a new mechanism was found that combines these two processes into a single physical process, shown in figure 7 in the form of a nomogram. Where in each of the quarters of the Cartesian coordinates there is a hyperbola reflecting a certain constant dependence.

Point A_0 and its mapping A_0^1 correspond to the initial area S_0 per atom in the atomic plane. The mapping of point A_0^1 to the ω axis is the point $\omega(1)$ which corresponds to the initial value of the natural frequency of vibration of the atom in iron.

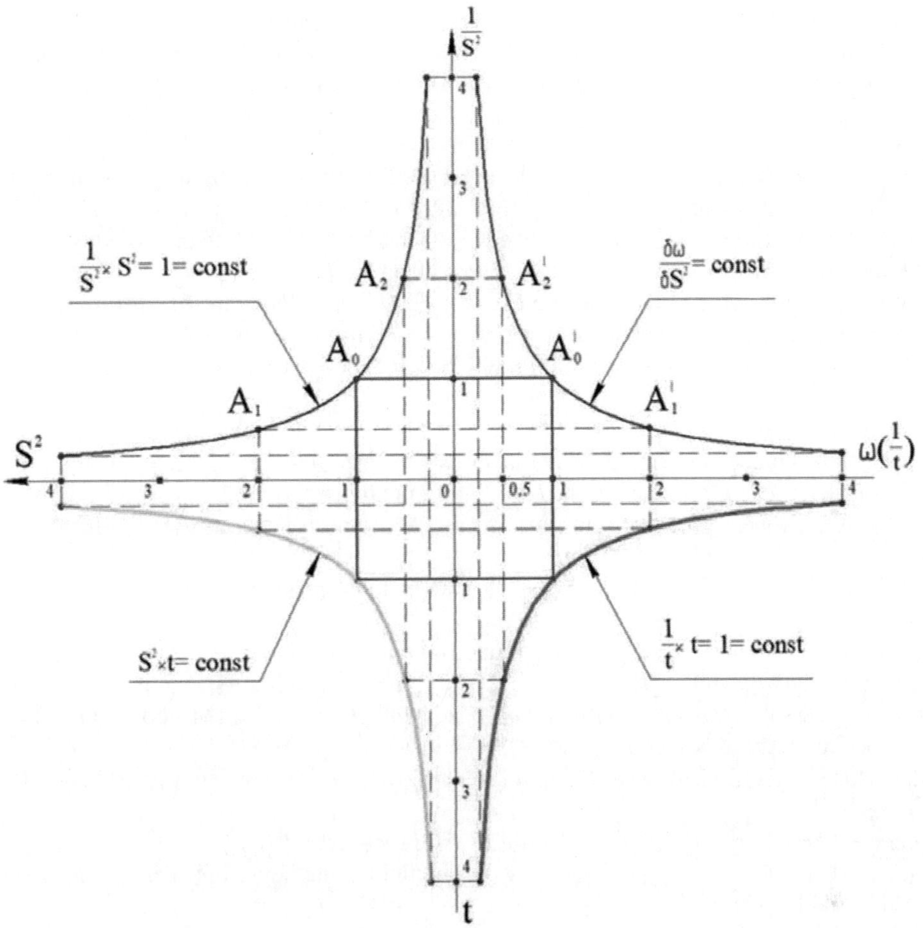

Figure 8. Nomogram of the dependence of the area "swept" by an atom in the atomic plane on the vibration frequency of the atom.

Points A_2 and A_2^1 correspond to the process of reducing the area per atom in the atomic plane, i.e. polymorphic transition of the primitive iron cell from αFe to γFe. In this case, the point A_2^1 on the ω axis will correspond to the point $\omega(0.5)$, which is less than the initial value of the natural vibration frequency of the atom, which was considered above. Point A1 and its display A11 are of interest. This point corresponds to the process of increasing the area of the cell per atom in the atomic plane. The mapping of point A_1^1 to the ω axis will correspond to the point $\omega(2)$, which is higher than the initial value of the natural vibration frequency of the atom. Thus, an increase in the area of the cell per atom leads to an increase in the value of the natural vibration frequency of the atom, which, accordingly, should be reflected in the form of an increase in the value of the speed of ultrasound passing through the iron. Analysis of the set of these conditions indicates that these changes can correspond to the process of elastic deformation of the bar.

Joule's work [18] confirms that elastic deformation leads to an increase in the volume of a primitive cell and, accordingly, the entire volume of the test material, iron. In it, Joule points out that in the course of thermal studies of the region of elastic deformation of iron, he found that the temperature of the material falls from the initial value. He explained this process by the adiabatic expansion of the entire volume of the test material, which confirms the increase in the volume of the primitive cell and, accordingly, the increase in the area of the cell in the atomic plane. An increase in the natural frequency of atomic vibrations in the region of elastic deformation was established in [19], in which the authors point out: "In our experiments, an increase in frequency is always observed during elastic deformation". The last confirmation is the work [20], in which it was found that in low-carbon steel in the region of elastic deformation, an increase in the value of the ultrasound velocity is observed.

References

[1] Semukhin BS, Zuev LB and Bushmeleva KI 2000 Ultrasound velocity in low-carbon steel deformed at the lower yield point *Applied mechanics and technical physics* **41(3)** 197-201
[2] Zuev L B, Semukhin B S and Bushmeleva K I 2000 Change in ultrasound velocity during plastic deformation Al *Journal of technical physics* **70(1)** 52-7
[3] Zuev L B, Semukhin B S and Bushmeleva K I 1999 Dependence of the ultrasound velocity on the acting stress in the plastic flow of polycrystals *Journal of technical physics* **69(12)**
[4] Grinyaev Yu V and Chertova N V 2000 Field theory of defects *Physical mesomechanics* **3(5)** 19-32
[5] Panin V E and Grinyaev Yu V 2003 Physical mesomechanics - a new paradigm at the junction of physics and mechanics of a deformable body *Physical mesomechanics* **6(4)** 9-36
[6] Yurchenko S A 2020 Influence of the coordination number and volume of a primitive cell on the vibration frequency of an iron atom during a polymorphic transition *Journal of Physics: Conference Series* **1679**
[7] Ogibailov M P and Kiyko I A 1966 *Essays on high-parameter mechanics* (From Moscow University)
[8] Bravais A 1849 Note sur les polyèdres simétriques de la géometrie *Journal de mathématiques pures et appliquées* **19**
[9] Fedorov E C 1949 *Symmetry and structure of crystals. Basic work.* (Moscow: Publishing House of the Academy of Sciences of the USSR)
[10] Katsnelson M I 2012 *Graphene: Carbon in Two Dimensions* (New York: Cambridge University Press)
[11] Ashcroft H and Merlin H 1979 *Solid State Physics* (Moscow: Mir)
[12] Vintaykin B E 2008 *Solid State Physics* (Moscow: Publishing house of the Bauman Moscow State Technical University)
[13] Frenkel Ya I 1972 *Introduction to the theory of metals* (L .: Publishing house "Nauka") p 424
[14] Yurchenko S A 2018 Refinement of the diatomic model of the interatomic interactions of Frenkel for the improvment of methods of non-destructive quality control *Periodico tche quimica* **15 (1)**
[15] Yurchenko S A 2021 *IOP Conf. Ser.: Mater. Sci. Eng.* **1100** 012043
[16] Alabyshev A P 2021 *Calculation of the propagation velocity of ultrasound velocity depending on the temperature of the material* Retrived from: scienct-bsea.bgita.ru/2014/mashin.
[17] Anuryev V I 2001 *Handbook of a mechanical engineer* (Moscow: Mechanical engineering)
[18] Joule J P *Proc.Rog.Soc* **8** (London) 564-9; *Phil.Mag.* **15(4)** 538-44
[19] Lebedev A B, Kustov S B and Kardashev B K 1992 On internal friction and Young's modulus defect in the process of deformation of crystals *Physics of Solids* **34(9)**
[20] Semukhin B S, Zuev L B and Bushmeleva K I 2000 Ultrasound velocity in low-carbon steel deformed at the lower yield point Applied *Mechanics and Technical Physics* **41(3)** 197-201

Comparison of characteristics of St3 in a free and annealed state

O A Nasibullina* and M Yu Pechenkina

Department of Petroleum Engineering Technology, Ufa State Petroleum Technological University, 1, Kosmonavtov str., Ufa, 450062, Russia

*E-mail: ksu33@bk.ru

Abstract. For the manufacture of field pipelines, low-carbon steel grade St3 is often used. It welds well, almost never hardens, and therefore is very convenient for working in structural elements. Studies of samples from this steel grade were carried out for the initial and heat-treated state by the following methods: study of mechanical properties (tensile test, hardness test); microstructural analysis. At the beginning of research, annealing was carried out in the work to obtain an equilibrium structure. According to the results of the experiment, it was noted that after annealing, the steel structure acquired low strength and hardness with high plasticity.
After annealing, tensile tests were carried out on steel samples. Tensile tests of the metal were carried out in order to determine the main indicators of mechanical properties and to compare these properties in the initial and heat-treated state. Hardness measurements were carried out according to the Rockwell method. The microstructure of steel samples was studied using an optical microscope "Metam RV-21-1", which made it possible to study the microstructure at a magnification of 200 times. It is shown that the structure of steel is ferritic-pearlite. In the initial state, there is a segregation banding and segregation streamer. The structure of the steel under study in the initial state is fine-grained. After annealing, it becomes more homogeneous, and the deformation texture disappears.

1. Introduction

For the manufacture of field pipelines, low-carbon St3 is often used [1-3]. It welds well, almost never hardens, and therefore is very convenient for working in structural elements.

The problem of corrosion of oilfield equipment has been particularly relevant in recent years [2, 4-6]. This is due to the fact that the application of methods for intensifying the development of oil fields is accompanied by a continuous increase in the water cut of well production and an increase in its corrosiveness. Under such operating conditions, there is a significant decrease in the service life of oilfield equipment, an increase in the number of failures due to corrosion [7-9], and product loss. In these conditions, the losses from the elimination of ruptures and the premature replacement of pipelines are also very significant. In addition, the outflows of oil products and wastewater causes irreparable damage to the environment [10-14], and reduces the efficiency and pace of development of oil fields.

Field pipelines are capital engineering structures designed for a long service life and designed for the uninterrupted transportation of natural gas, oil, water and their mixtures from their production sites to integrated treatment facilities and further to the points of tie-in into the main pipeline or for supply to another mode of transport - railway, river, sea.

The technical equipment of pipeline transport and the pace of pipeline construction are inextricably linked with the development of the gas, oil and oil refining industries [15-18]. Pipeline transportation of gas, oil and oil products is the most economical. In this regard, both in Russia and abroad, there has been a tendency for the outstripping growth of pipeline transport in comparison with its other types.

2. Experimental procedures

2.1. Materials
The object of the study was samples from steel St3. According to its chemical composition (table 1), the steel under study can be deciphered as follows: about 97% contains iron, and the amount of carbon is 0.14-0.22%. The strength and hardness of steel primarily depends on the carbon content. However, carbon reduces plasticity and weldability, therefore, in building steels, which, according to the conditions of their work in structures, must be sufficiently plastic, it is allowed a little - no more than 0.22%. This also ensures good weldability of these steels.

Table 1. The chemical composition of steel.

C, %	Si, %	Mn, %	S, %	P, %	Cr, %	Ni, %	Cu, %	N, %	As, %
0.14-0.15	0.15-0.30	0.40-0.65	<0.050	<0.040	0.03	<0.30	<0.30	<0.010	0.08

The purpose of steel St3 is sheets for electric-welded pipes operating at temperatures up to 300°C and pressures up to 1.6 N/mm^2. Parts of boilers and pipelines made of sheet up to 12 mm thick, and forged parts designed for operation at temperatures up to 200°C and pressures up to 1.6 N/mm^2. Table 2 shows the temperatures of the critical points of the studied steel.

Table 2. Temperature of critical points, °C

Ac_1	Ac_3	Ar_1	Ar_3
735	850	680	835

Investigations of samples from steel grade St3 were carried out for the initial and heat-treated state by the following methods:

1) study of mechanical properties:
- tensile test;
- hardness test;

2) microstructural analysis.

2.2. Experiment Techniques
At the beginning of research, annealing was carried out in the work to obtain an equilibrium structure. Annealing of the II type of steel products was carried out at a temperature of 880°C, followed by slow cooling in a furnace. According to the results of the experiment, it was noted that after annealing, the structure of St3 steel acquired low strength and hardness with high plasticity.

After annealing, tensile tests were carried out on samples of steel St3. Tensile tests (GOST 1497-84) make it possible to determine the strength and ductility characteristics of metals under static uniaxial loading. The testing machines are equipped with lever-type (or indicator) devices for recording the tensile diagram (change in the length of the sample depending on the applied stress).

The determination of the yield point of the metal is carried out at a constant rate of relative deformation of 1 mm/min. Tensile and universal machines that comply with GOST 7855-74 are used as testing machines.

Hardness measurements were carried out according to the Rockwell method (GOST 9013-59). The essence of the method consists in indenting an indenter with a hardened steel ball with a diameter of 1.5875 mm into the sample, under the action of two successively applied loads - preliminary P_0 and main P_1, which is added to the preliminary one, so that the total load $P = P_0 + P_1$. After holding for several seconds, the main load is removed and the residual depth of penetration of the indenter is measured, which is still under the action of the preload.

Microanalysis is an indispensable tool for assessing structural changes in the size, shape and orientation of grains, individual structural and phase components. The microstructure of the St3 samples was studied using an optical microscope (GOST 10243-75) "Metam RV-21-1", which made it possible to study the microstructure at a magnification of 200 times. The structure before and after heat treatment was studied in the surface and deeper layers of the metal.

3. The results of studies and their discussion

Tensile tests of the metal were carried out to determine the main indicators of mechanical properties and to compare these properties in the initial and heat-treated state.

The values of the following characteristics were determined: ultimate strength (σ_u), yield point (σ_y), relative elongation (δ), relative contraction (ψ).

The measurement results are shown in Table 3.

Table 3. Results of tensile tests of samples

	σu, MPa	σy , MPa	δ, %	Ψ, %
The initial state	564.1	344.7	19.4	60
Annealing	461.7	298.2	24.2	60

The tensile curves of steel samples are shown in Figure 1.

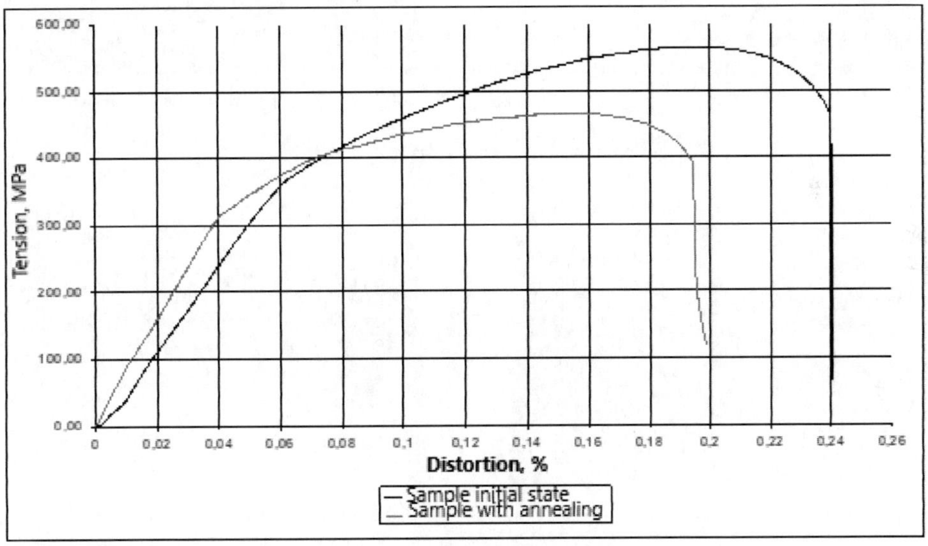

Figure 1. Tension curves of St3 specimens in the initial and heat-treated state.

The results of the tests carried out showed that the samples without heat treatment have higher strength properties than the heat-treated samples.

Determination of hardness at its moderate values on products of small thicknesses, as well as when using a steel ball as an indenter, the number of hardness HRB is determined on the B scale, i.e. at P_0 = 98 H, P_1 = 882 N, P = 980 N. Rockwell hardness number for the B scale is recorded as HRB. Table 4 shows the results of measuring the hardness of steel samples.

IOP Conf. Series: Materials Science and Engineering 1155 (2021) 012002 doi:10.1088/1757-899X/1155/1/012002

Table 4. Results of hardness measurement

Measure number	The initial state		Annealing	
	HRB	HB	HRB	HB
1	88	170	83	152
2	88	170	84	156
3	89	174	84	156
4	88	170	84	156
average	88.25	171	83.75	155

Looking at the data presented, we can draw a conclusion. That after annealing, the St3 structure acquired low hardness and strength, but high plasticity.

Microstructural analysis of the steel was also performed. The microstructure is shown in Figure 2-3.

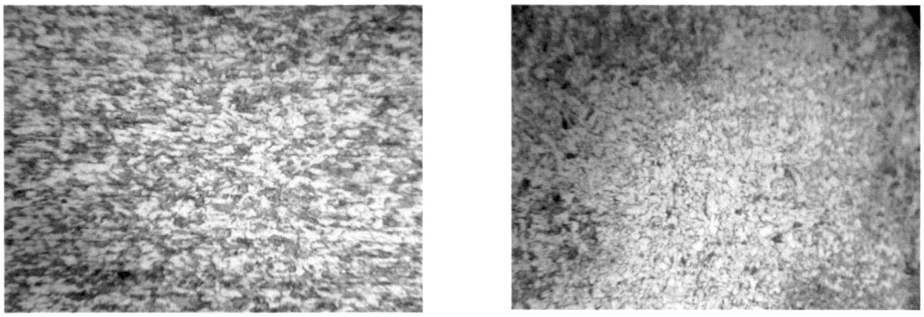

ferrite-pearlite structure decarburized portions of ferrite

Figure 2. Structure of St3 in the initial state

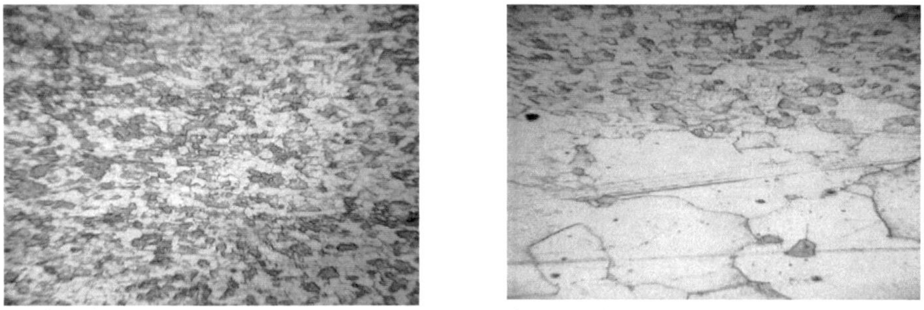

equilibrium structure decarburized portions of ferrite

Figure 3. Structure of St3 after annealing

The structure of the steel is ferrite-pearlite. In the initial state, there is a light segregation banding. The structure of St3 in the initial state is fine-grained. After annealing, it becomes more homogeneous, and the deformation texture disappears. The decarburized areas, which in the initial state are visible as light areas of ferrite, increase the grain size upon annealing to form large ferrite grains.

4. Conclusions

As a result of the studies (tensile and hardness tests), the main mechanical properties of St3 steel in the initial and heat-treated state were determined. The ultimate strength in the initial state was 564.1 MPa, the hardness was 171 HB, in the heat-treated state, 461.7 MPa and 155 HB, respectively.

As a result of conducting annealing of the II kind at a temperature of 880°C, the grain size increased in comparison with the initial state.

As a result of the conducted microstructural analysis, it was revealed that the structure of the steel is ferritic-pearlite. In the initial state, there is a light segregation banding and segregation streamer. The structure of St3 in the initial state is fine-grained. After annealing, it becomes more homogeneous, and the deformation texture disappears. The decarburized areas, which in the initial state are visible as light areas of ferrite, increase the grain size upon annealing to form large ferrite grains.

Acknowledgments
The authors are grateful to Alexei Gabdullovich Gareev, Professor of the Ufa State Petroleum Technological University (Russian Federation) for his advice and assistance in conducting experiments.

References

[1] Nasibullina O A and Rizvanov R G 2020 *Materials Science Forum* **992** 695-9
[2] Zaripov M Z, Fairushin A M and Karetnikov D V 2019 *Materials Science Forum* **946** 883-8
[3] Nasibullina O A and Gareev A G 2019 *Materials Science Forum* **946** 20-4
[4] Rizvanov R G, Mulikov D Sh, Karetnikov D V, Fairushin A M and Tokarev A S 2018 *IOP Conf. Ser.: Mater. Sci. Eng.* **317** 012077
[5] Yakhin A V, Karetnikov D V, Cherepashkin S E and Rizvanov R G 2019 *Chemical and Petroleum Engineering* **55(7-8)** 681-6
[6] Nasibullina O A and Tyusenkov A S 2020 *IOP Conf. Ser.: Mater. Sci. Eng.* **862** 022007
[7] Karetnikov D V, Rizvanov R G, Fairushin A M and Kolokhov K S 2013 *Welding International* **27** (7) 557-60
[8] Nasibullina O A and Rizvanov R G 2020 *Materials Science Forum* **989** 28-2
[9] Yakhin A V, Karetnikov D V, Rizvanov R G, Abakacheva E M and Gareev A G 2019 *Chemical and Petroleum Engineering* **54(11-12)** 801-5
[10] Vakulenko M V and Zharinova N V 2020 *IOP Conf. Ser.: Mater. Sci. Eng.* **862** 062022
[11] Kuzeev I R, Ibragimov I G, Bayazitov M I, Davydov S N and Khairudinov I R 1985 *Chem. and Technology of Fuels and Oils* **22(3)** 111-3
[12] Zharinova N V and Vakulenko M V 2020 *IOP Conf. Ser.: Mater. Sci. Eng.* **862** 032014
[13] Tyusenkov A S and Nasibullina O A 2019 *IOP Conf. Ser.: Mater. Sci. Eng.* **687** 066021
[14] Gaysin E Sh, Frolov Y A and Nasibullina O A 2020 *IOP Conf. Ser.: Earth and Environ. Sci.* **459** 032055
[15] Rizvanov R G, Mulikov D Sh, Karetnikov D V, Cherepashkin S E and Shirgazina R F 2017 *Nanotechnologies in Construction* **9(4)** 97-15
[16] Abdulganiev M R, Khasanov R N and Gafarova V A 2019 *Materials Science Forum* **945** 569-73
[17] Krioni N K, Mingazheva A A, Kononova A Yu, Mingazhev A D and Gafarova V A 2017 *Solid State Phenomena* **265** 215-20
[18] Skuridin N N, Latypova D R, Pechenkina M Yu, Bugay D E and Ryabukhina V N 2018 *Neftyanoe Khoz.: Oil Ind.* **5** 84-6

Protective structures on carbon steel in aqueous-salt solutions containing catholyte

M Yu Pechenkina* and O A Nasibullina

Department of Petroleum Engineering Technology, Ufa State Petroleum Technological University, 1, Kosmonavtov str., Ufa, 450062, Russia

*E-mail: map-66@mail.ru

Abstract. The process fluid of oil fields, with a huge amount of dissolved gases, mechanical impurities, salt ions, mineral acids and bases present in them, has a destructive effect on the internal surface of oil collection equipment. The simultaneous presence of carbon dioxide and hydrogen sulfide pose the greatest danger, since they are catalysts for the cathode reaction of hydrogen depolarization of low-alloy steel, and lead to the formation of soluble corrosion products - iron hydrogencarbonate. One common way to prevent electrochemical corrosion of oil equipment is to neutralize acidic components of process media. Sodium hydroxide is often used as a neutralizer, however, the content of sodium salts contaminates the obtained oil products and catalysts and leads to disruption of the technological process of its further processing. Accordingly, there is provided a process for the non-reactive neutralization of acidic components by introducing a solution of a hydroxide-saturated catholyte. In the scientific and technical literature, the problem of forming protective films on the surface of carbon steel by introducing catholyte into the process liquid remains insufficiently studied. In scientific research work, a method of forming surface structures by selecting the concentration of a catholyte solution has been developed. The thickness of the obtained protective structures obtained in the model of produced water formed on the surface of low-alloy steel A516-55 in solutions of sodium hydroxide and catholyte was calculated. The study revealed the dependence of the corrosion rate of steel on the concentration of sodium hydroxide solution and catholyte solution at various pH values.

1. Introduction

It was established that most of the damages of oil field pipelines are associated with two main factors: high corrosion activity of working media and relatively low corrosion resistance of steels used for the manufacture of pipelines. The most vulnerable equipment at oil fields are oil collection pipelines [1-3]. The use of corrosion inhibitors, which are effective in many water-salt solutions, often does not yield the expected result in oil collection pipelines due to the need to pump large volumes of them into the water-oil emulsion, control the residual content of reagents in the prepared oil and the high cost of inhibitors [4-5]. Therefore, the development of non-reactive methods for preparing technological media is a very promising area of research in the oil and gas industry. Carbon low-alloy steel A516-55 [6-8] is often used to manufacture oil collection pipelines.

The presence of acidic components in the formation water leads to the development of local corrosion processes of oil collection pipelines, as the most vulnerable equipment in oil and gas fields [9-11]. In previous studies, I showed the undisputed result of introducing a 30% concentration catholyte solution with pH = 10 into the produced water model, as a result of which the corrosion rate of samples in the presence of catholyte decreased by almost 5 times (from 0.235 to 0.0485 mm/year), which amounted to 80% of the protective effect [12-14]. This efficiency is commensurate with the use of a corrosion inhibitor [15-17]. However, the catholyte obtained by modifying water-salt solutions from tap water completely excluded the use of chemical reagents. To create it, a specially designed unit was used to modify the flow of liquid. As a result of the electro activation of the water-salt solution with the redistribution of ions, atoms and molecules, an alkaline solution in the near cathode space was obtained - catholyte, and in the near anode area a solution with an acidic medium - anolyte. Change of pH value of created solutions in the unit is performed by control of flow rate and incoming current strength [18-19].

2. Experimental procedures

2.1. Materials

Studies were carried out on samples of low-alloy steel A 516-55 widely used for the construction of oil and gas facilities. To create a model of produced water No. 3, which is as close as possible to the composition of the corrosive medium, sodium hydrogen carbonate ($NaHCO_3$) - 138 mg/l, calcium chloride ($CaCl_2$) - 333 mg/l and sodium sulfate (Na_2SO_4) - 213 mg/l were introduced into distilled water

Thus, the introduction of catholyte into the aqueous-saline solution characteristic of oil fields contributes to a significant reduction in the corrosion rate of low alloy steel. Thus, at pH = 10, the introduction of 30% catholyte solution into the produced water model leads to a decrease in the corrosion rate of low alloy steel A516-55 almost 5 times, which corresponds to a protective effect of 80%. This is comparable to the protective efficacy of corrosion inhibitors. Note that the catholyte was obtained by modifying the aqueous salt solutions from tap water, which excluded the use of chemical reagents.

Surface structures on steel were A 516-55 formed by adding to the corrosive solution 30% of the total volume in one case of sodium hydroxide solution (at pH = 10 and at pH = 12), and in the other, a catholyte solution at the same pH = 10 and pH = 12. The change in the microstructure of the steel surface was then evaluated using a METAM RV-22 metallographic microscope. However, there was no change in microstructure after corrosion tests, and this is due to the protective effect of adsorbed surface structures.

Figure 1 shows photos of steel A 516-55 samples after exposure.

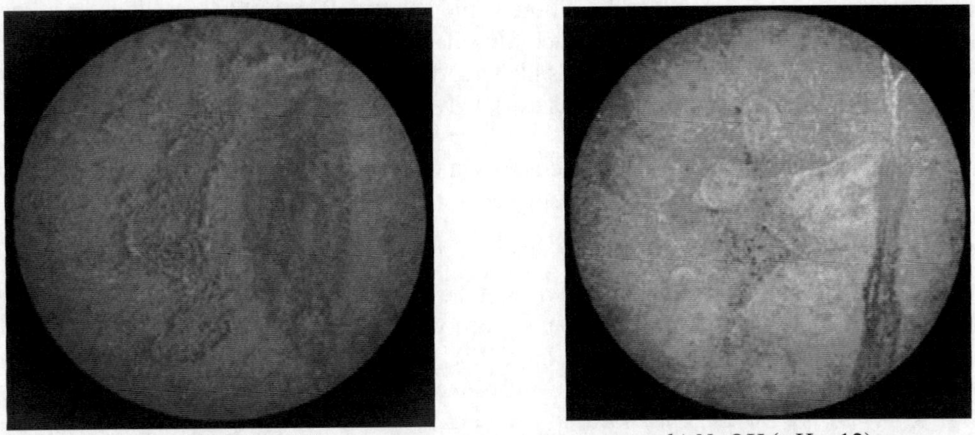

a) *NaOH* (*pH* = 10) b) *NaOH* (*pH* = 12)

Figure 1. Photographs of the surface of steel samples A 516-55 after exposure in the produced water model No. 3 with the addition of 30% of the total volume of sodium hydroxide solution to the corrosive solution.

By means of a three-dimensional microscope of MBS-10 carried out the analysis of the created superficial structures on steel A 516-55. In the presented photographs of the surface structures of the steel samples Figure 1, it can be observed that the addition of 30% sodium hydroxide solution to the corrosive solution results in the formation of non-uniform and non-uniform coatings compared to the samples shown in Figure 2. Most likely, this is caused by the differentiation of the influence of hydroxyl ions on adsorption processes on the surface of steel A 516-55. Such films have a lower barrier effect, whereby the corrosion rate of the samples after testing was higher.

However, after the addition of catholyte with $pH = 10$, continuous protective films are formed Figure 2, a).

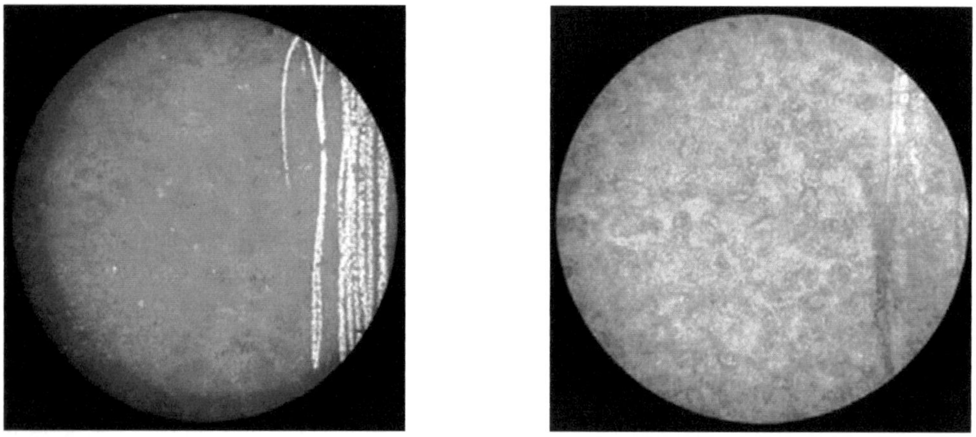

a) catholyte ($pH = 10$) b) catholyte ($pH = 12$)

Figure 2. Photographs of the surface of steel samples A 516-55 after exposure in the produced water model No. 3 with the addition of 30% catholyte solution of the total volume of the solution to the corrosive solution.

As a result of inspection of the appearance of the formed surface structures, it can be noted that after adding a solution of sodium hydroxide to the corrosive solution, thicker and looser iron hydroxide films are formed. This is probably caused by the differentiation of the effect of hydroxyl ions on adsorption processes on the surface of steel A 516-55. Such films have a lower protective effect, and therefore the corrosion rate of the samples after the test was higher. After the addition of catholyte, on the contrary, continuous protective films are formed. This is especially clearly seen in Figure 2 a), where a catholyte with pH = 10 was added as a neutralizer.

Microscopic studies have shown that in the presence of catholyte, a thinner (plastic) film is formed with high corrosion resistance and protective effect.

2.2. Experiment Techniques

The next stage of a research was studying thickness of the created superficial structures on samples from steel A 516-55. After corrosion testing in the produced water model with addition of 30% of the total volume of sodium hydroxide solution (pH 10 and 12) and catholyte solution (pH 10 and 12) to the corrosive solution, the surface of the samples was scanned using a confocal laser scanning microscope "LSM-5-Exciter" (Zeiss, Switzerland). Prior to scanning with a polyvinyl chloride spatula, a portion of the metal surface was cleaned from the film and a three-dimensional scan was carried out at the site capturing the oxide film and the stripping site. Four samples were 3D - Figure 3.

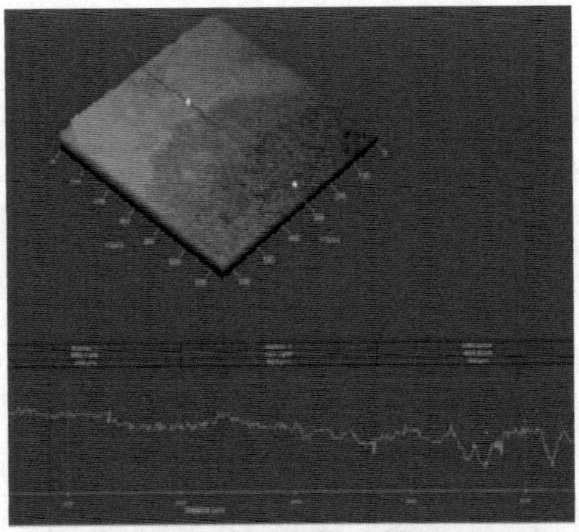

a) *NaOH* (*pH* = *10*)

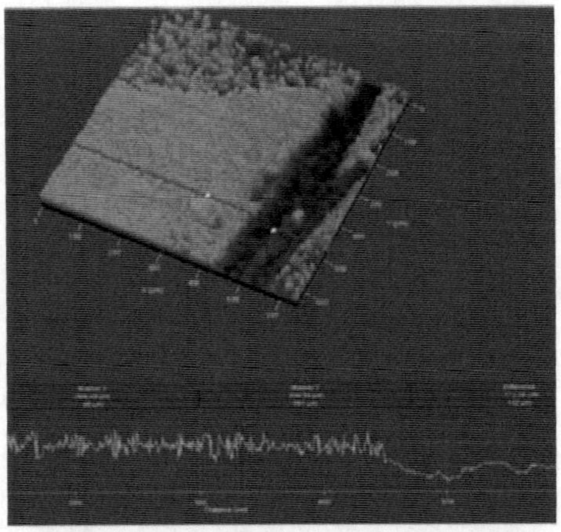

b) *NaOH* (*pH* = *12*)

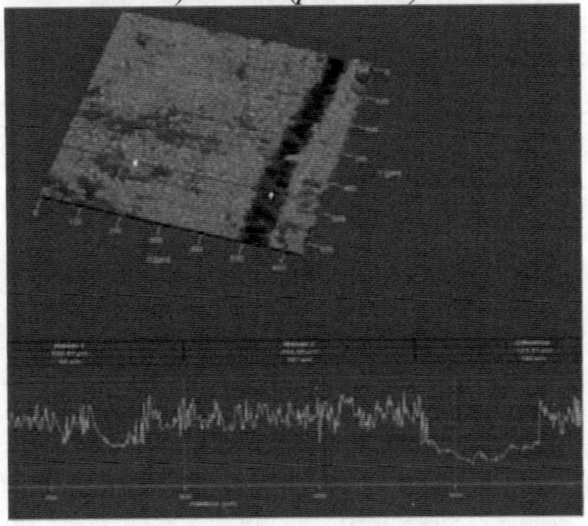

c) catholyte (*pH* = *10*)

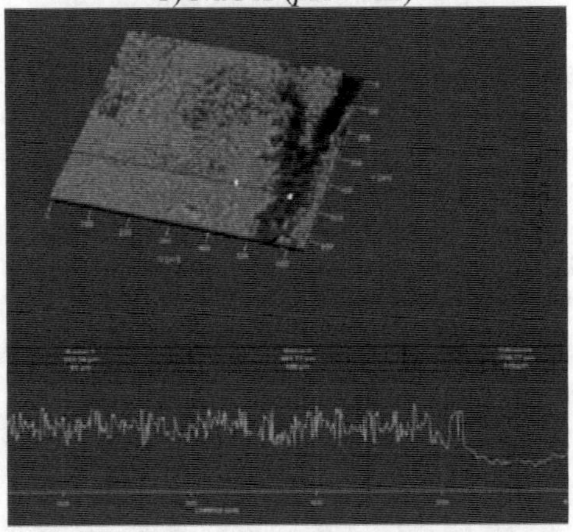

d) catholyte (*pH* = *12*)

Figure 3. Results of 3D scanning of steel A516-55 samples on a confocal microscope after exposure in a model of produced water with addition of NaOH and catholyte solutions.

X axis is the length of the profile section in the direction of the X axis. Z axis is the height of the estimated profile at any position of the X axis. White points in the figure show the points located on the surface of the hydroxide film and at the place of metal grinding. At the top of the figure are two white dots. The point on the light portion of the pattern corresponds to Marker 2 (the point on the surface of the hydroxide film). Point on the dark portion of the pattern corresponds to Marker 1 (the point on the surface of the metal). At the bottom of each pattern is shown the plane of the imaginary cross-section of the sample portion in which the spatula of polyvinyl chloride was cut, where the blue vertical line (marker 1) indicates the average height of the formed oxide film, and on the right the red vertical line (marker 2) indicates the average height of the metal surface.

X axis is the length of the profile section in the direction of the X axis. Z axis is the height of the estimated profile at any position of the X axis. White points in the figure show the points located on the surface of the hydroxide film and at the place of metal grinding. At the top of the figure are two white dots. The point on the light portion of the pattern corresponds to Marker 2 (the point on the surface of the hydroxide film). Point on the dark portion of the pattern corresponds to Marker 1 (the point on the surface of the metal). At the bottom of each pattern is shown the plane of the imaginary cross-section of the sample portion in which the spatula of polyvinyl chloride was cut, where the blue vertical line (marker 1) indicates the average height of the formed oxide film, and on the right the red vertical line (marker 2) indicates the average height of the metal surface.

The measurement results are shown in Table 1.

Table 1. Results of 3D scanning of the steel A516-55.

Reagent and value, *pH*	Marker 1, μm	Marker 2, μm	Amplitude, μm
NaOH, *pH*=10	436	603	167
catholyte, *pH*=10	62	197	135
NaOH, *pH*=12	49	191	142
catholyte, *pH*=12	80	198	118

From the data obtained in Table 1, it can be concluded that the formed films are thick because they exceed a thickness of 5000Å. In addition, it is noticeable that the films formed by the addition of catholyte are thinner than with the addition of sodium hydroxide, which suggests their greater ductility and, accordingly, strength. Thus, the height difference of the peaks and troughs between the metal and the surface with the formed surface structures can be calculated. Comparative analysis of the protective properties of the surface structures showed that the oxide film, when added with catholyte, is thinner by about 30 microns than with the addition of chemical reagent NaOH, therefore, it is stronger and has higher corrosion resistance.

3. The results of studies and their discussion

Studies were conducted to study the properties of the catholyte solution and its positive effect on the corrosion rate of A 516-55 steel used in water-salt solution. Gravimetric tests gave the chance to compare steel A 516-55 corrosion rates in water-salt solution to addition chemical reagent NaOH and a catholyte. The corrosion rate of the metal samples with the addition of catholyte was significantly lower (about 1.5 times) than with the addition of sodium hydroxide.

The results of the analysis of the thickness of the A 516-55 surface layers formed on steel showed that the hydroxide film formed on the metal in a corrosive medium with the addition of catholyte is about 30 microns thinner than when used instead of sodium hydroxide catholyte, and also has a higher continuity. This determines its effectiveness in counteracting corrosion of low-alloy carbon steel in water-salt solutions, which mimic the composition and properties of technological liquids of oil and gas facilities.

4. Conclusions

The article deals with the issue of combating internal corrosion of oil pipelines. The creation of scientific management parameters of the processes responsible for the kinetics of corrosion in pipes, as well as artificial intervention in the mechanisms of the corrosion process will lead to their prevention, a significant decrease damage and increase resource oil pipelines.

The article presents a comparative analysis of the protective properties of surface structures. Studies of the continuity and thickness of surface structures formed in solutions of sodium hydroxide and catholyte are carried out. The authors suggested using a 30% solution of catholyte with a pH of 10 as a neutralization of acidic components, which reduces the corrosion rate by 5 times.

The authors established the formation of a thinner film, which has a high protective effect when the catholyte is injected into the process medium, than when adding a solution of sodium hydroxide.

Thus, the use of catholyte as a neutralizer of water-salt solutions during pipeline operation will reduce the number of gusts and, accordingly, reduce the cost of their elimination, increase the reliability and extend the service life of pipelines, as well as ensure the environmental safety of oil and gas facilities

References

[1] Skuridin N N , Latypova D R, Pechenkina M Yu, Bugay D E and RyabukhinaV N 2018 *Neftyanoe Khoz.: Oil Ind.* **5** 84-6

[2] Nasibullina O A and Gareev A G 2019 *Materials Science Forum* **946** 20-4
[3] Zaripov M Z, Fairushin A M and Karetnikov D V 2019 *Materials Science Forum* **946** 883-8
[4] Kuzeev I R, Ibragimov I G, Bayazitov M I, Davydov S N and Khairudinov I R 1985 *Chem. and Technology of Fuels and Oils* **22(3)** 111-3
[5] Zharinova N V and Vakulenko M V 2020 *IOP Conf. Ser.: Mater. Sci. Eng.* **862** 032014
[6] Gaysin E Sh, FrolovY A and NasibullinaO A 2020 *IOP Conf. Ser.: Earth and Environ. Sci.* **459** 032055
[7] Karetnikov D V, Rizvanov R G, Fairushin A M and Kolokhov K S 2013 *Welding International* **27**(7) 557-60
[8] Tyusenkov A S and Nasibullina O A 2019 *IOP Conf. Ser.: Mater. Sci. Eng.* **687** 066021
[9] Vakulenko M V and Zharinova N V 2020 *IOP Conf. Ser.: Mater. Sci. Eng.* **862** 062022
[10] Abdulganiev M R, Khasanov R N and Gafarova V A 2019 *Materials Science Forum* **945** 569-73
[11] Nasibullina O A and Rizvanov R G 2020 *Materials Science Forum* **989** 28-2
[12] Krioni N K, Mingazheva A A, Kononova A Yu, Mingazhev A D and Gafarova V A 2017 *Solid State Phenomena* **265** 215-20
[13] Nasibullina O A and Rizvanov R G 2020 *Materials Science Forum* **992** 695-9
[14] Rizvanov R G, Mulikov D Sh, Karetnikov D V, Fairushin A M and Tokarev A S 2018 *IOP Conf. Ser.: Mater. Sci. Eng.* **317** 012077
[15] Yakhin A V, Karetnikov D V, Cherepashkin S E and Rizvanov R G 2019 *Chemical and Petroleum Engineering* **55(7-8)** 681-6
[16] Nasibullina O A and Tyusenkov A S 2020 *IOP Conf. Ser.: Mater. Sci. Eng.* **862** 022007
[17] Yakhin A V, Karetnikov D V, Rizvanov R G, Abakacheva E M and Gareev A G 2019 *Chemical and Petroleum Engineering* **54(11-12)** 801-5
[18] Kuzeev I R, Ibragimov I G, Bayazitov M I, Davydov S N and Khairudinov I R 1985 *Chem. and Technology of Fuels and Oils* **22(3)** 111-3
[19] Tokarev A S, Karetnikov D V, Rizvanov R G, Fayrushin A M andZaripov M Z 2020 *Materials Science Forum* **991** 70-6

IOP Conf. Series: Materials Science and Engineering 1155 (2021) 012004 doi:10.1088/1757-899X/1155/1/012004

Features of melting high-chromium flux-cored wire in pulsed arc welding

Yuri Davydov, Yuri Korobov and Alexander Davydov

URFU – Ural Federal University, Mira St., 19, Ekaterinburg, Russian Federation

E-mail: yu.s.davydov@urfu.ru, yu.s.korobov@urfu.ru, a.iu.davydov@urfu.ru

Abstract. The results of studies of the process of pulse-arc surfacing with flux-cored wire are presented. It is shown that with pulsed arc control, it is possible to increase the main parameters of the process to the values characteristic of a solid wire. The difference between the melting character of flux-cored and solid wire is shown.

1. Introduction

The use of flux-cored wires for welding and surfacing has a number of advantages compared to solid-section wires. However, there is also a certain list of problems that make it difficult to use them and, sometimes, lead to refusal to use them [1-6].

Flux-cored wires are characterized by uneven properties in cross-section, the charge (wire core) has a much lower electrical and thermal conductivity than the shell. The active arc spot is localized on the shell, and the wire core is heated by heat transfer from the shell, which changes the shape of the arc, the nature of thermal and physico-chemical processes at the end of the electrode, and also leads to different conditions of arc burning, formation and transfer of droplets than with a solid wire [7, 8, 9].

This leads to a decrease in the stability of arc burning, an increase in metal spattering, and a deterioration in the formation of the seam. At the same time, the typical welding modes used in modern welding machines often do not give an optimal result. This is especially true for wires that have an "unconventional" composition of the charge: with a small number of slag-forming components and, conversely, with a large number of metal components for a high degree of alloying of the deposited metal [10, 11].

The use of the pulse-arc process to regulate the stability of the arc often does not give a tangible result, which leads to the rejection of it in enterprises.

For welding alloy steels, wires with a high chromium content (12-27%) are used, which creates additional difficulties in developing the technology, which are associated with a decrease in the stability and uniformity of the melting of the flux-cored wire.

The aim of the study is to analyze the influence of the pulse-arc process on the characteristics of the melting process of high-chromium flux-cored wires.

2. Materials and methodics

As the object of the study, a wire with a diameter of 1.2 mm was selected, which allows to obtain a deposited metal of the 50Cr18 type, additionally alloyed with nitrogen. A mixture of the K-18 type (82% Ar+18% CO_2) was used as a protective gas. Studies of the characteristics of the welding process, the choice of its parameters and the methods used are described in detail in the works [10, 11].
Surfacing performed on the Storm-Lorch S5 SpeedPulse semi-automatic welding machine.

To study the melting process, the formation of droplets and the nature of the transfer of the electrode material, several variants of the technological process were used: the standard mode with continuous arc burning in a process with natural short circuits; SpeedPulse-the imposition of a modified pulse during the drop discharge period.

The nature of droplet formation was assessed visually by observing the arc, as well as indirectly by assessing the nature of splashing.

The spatter was evaluated by weighing the plates before and after surfacing and comparing them with the estimated weight of the spent wire. The quality of the roller formation was determined according to the developed method in points (from 0 to 5), evaluating the uniformity of the seam size, the presence of defects such as pores, slags, shells, the smoothness of the surface, the uniformity of weld surface with bead ripples (if any).

The arc current and voltage were oscillographed using the AWR224 welding parameter logger (ZAO Laboratory of Electronics, St. Petersburg).

The parameter ranges are shown in the table. 1. For each mode, five surfacings was made on plates with a thickness of 12 mm made of low-carbon steel.

Table 1. Ranges of welding parameter values.

Welding Parameters	Standart	SpeedPulse
Current, A	120-200	100-200
Wire Feed Rate, m/min	5-9	7-12
Voltage, V	20-28	26-32
Electrode extention, mm	20-25	27-32

To determine the characteristics, an experiment was implemented in accordance with the orthogonal plan of the second order. In accordance with the plan of the experiment, surfacing was performed at 9 combinations of the main parameters of the mode (at the central point of the plan - 6 parallel experiments to determine the reproducibility indicators). After mathematical processing, regression equations are obtained that connect the determined characteristics with the parameters of the mode.

3. Results and discussion

In Figure 1 the obtained calculated dependences of splashing and seam formation on the arc voltage at different values of the wire feed rate (over 32V - extrapolation) are shown.

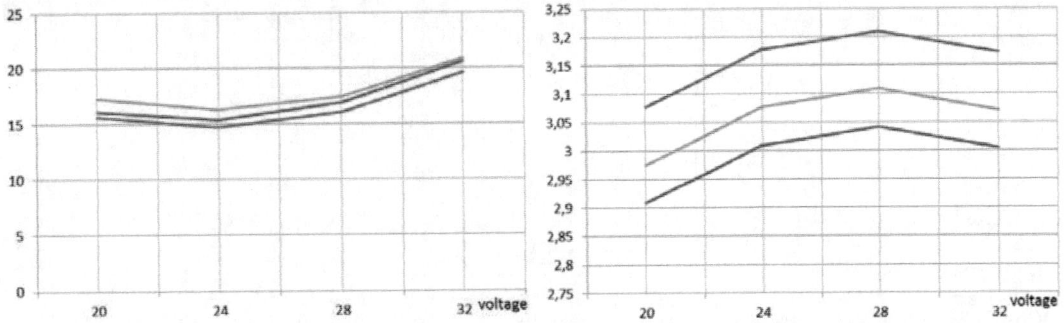

a) b)

Figure 1. The effect of arc voltage and the wire feed rate on the loss of electrode material Ψ (%) (a) and seam formation (points) (b) during surfacing in standard modes.

It can be seen that over the entire range of the studied currents, an increase in voltage has an almost opposite effect on splashing and seam formation: thus, it is almost impossible to distinguish a voltage range characterized by both low losses of the electrode material and good weld formation.

Figure 2 shows the stringer beads obtained in standard modes. The beads are narrow (up to 10 mm wide), have swells and turns (deep undercuts) at the edges. Visible pores that come to the surface, and slag inclusions. There are many large splashes up to 4 mm in size. The level of metal losses reaches 15-18%, the formation of the seam is on average below satisfactory. The measured value of the surfacing coefficient is 16-20 g/A*h.

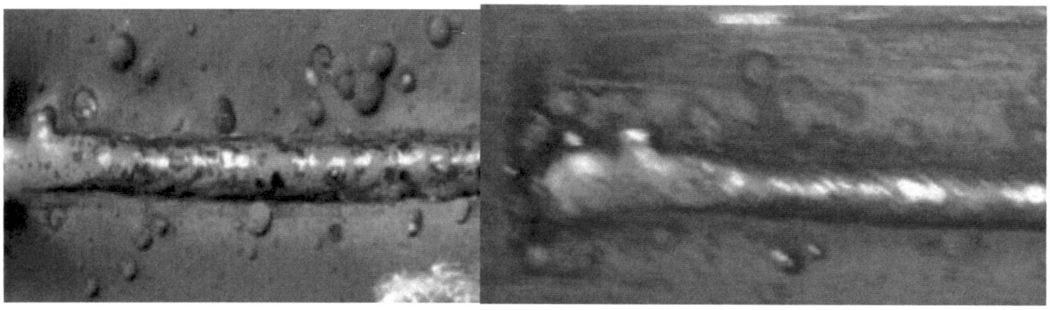

Figure 2. Appearance of the deposited beads. Standard modes.

When switching to the SpeedPulse modes, there are positive changes in comparison with previous experiments: the width of the obtained beads is from 12 to 16 mm; the formation of the seam is noticeably improved; the beads are uniform, without surges; the surface is smooth, there are no pores and slag inclusions; there are no large drops and slag particles fused in; there is a small amount of small splashes, up to 0.5 mm in size.

Figure 3 shows the calculated dependence of the loss coefficient on the arc voltage at different values of the wire feed rate. It can be seen that when the voltage increases from about 32 to 34 V, there is no noticeable deterioration in the spray characteristics, and the loss coefficient itself is 12-17%. It is interesting that the very nature of the change in the loss value is different: in standard modes, when the voltage increases, the losses increase, and in the pulse mode, on the contrary, they decrease. The measured value of the surfacing coefficient is 27-30 g/A*h.

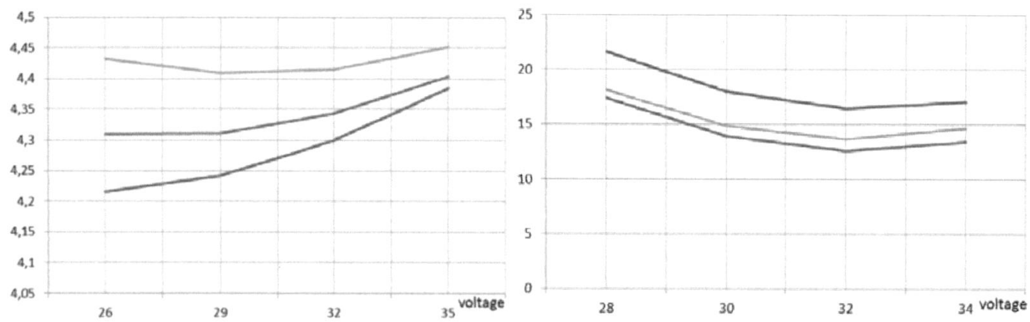

Figure 3. The effect of arc voltage and the wire feed rate on the loss of electrode material Ψ (%) (a) and seam formation (points) (b) during surfacing in SpeedPulse modes.

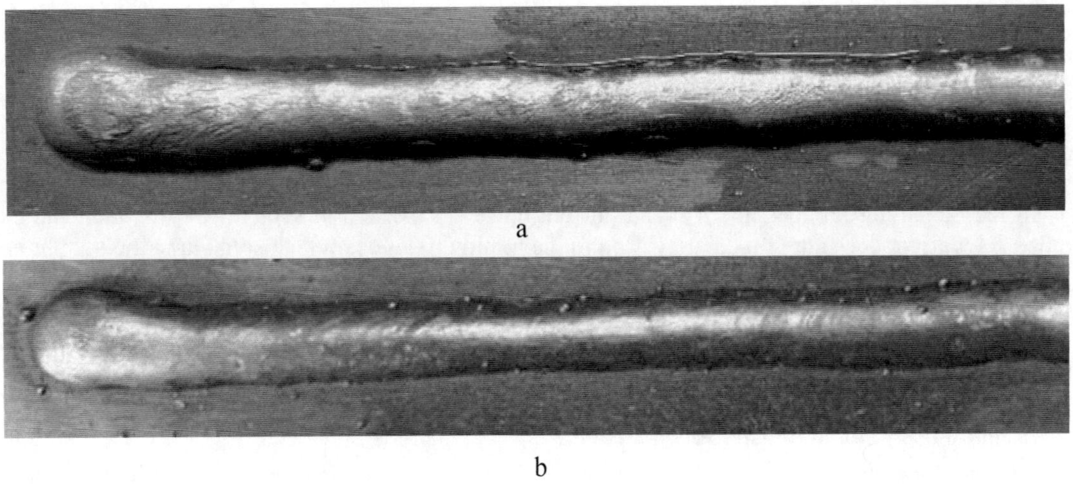

Figure 4. Appearance of the deposited bead. SpeedPulse mode: a) voltage 32 V, b) voltage 28 V.

The analysis of the melting process indicators shows the following:

- at close current values, the surfacing performance in pulse modes is 1.60 -1.80 times higher than in pulse-free modes;
- the value of the surfacing coefficient when using the pulse mode is higher by 30-35%;
- losses (burnuig, spattering) are reduced by 1.5-2 times in comparison with the pulse-free mode, while ensuring better weld formation.

These differences in the characteristics of the welding process can be explained by the different nature of the melting of the electrode, the dynamics of the formation and transfer of droplets. A special feature of the powder wire is that it consists of a metal shell and a core (charge), while the heat and, most importantly, the electrical conductivity of the charge is noticeably lower than that of the shel [9]. This leads to the fact that the charge melts only due to the heat transfer from the shell, on which, in fact, the arc burns. On the basis of these thermophysical features, possible melting schemes of the charge and shell are proposed. Figure 5 shows the two melting mechanisms of the charge and shell. In the implementation of option (a), the shell, melting, falls into the molten charge in small droplets, forming a metal-slag drop, which grows to a relatively large size and then, mainly under the influence of gravity, separates from the electrode.

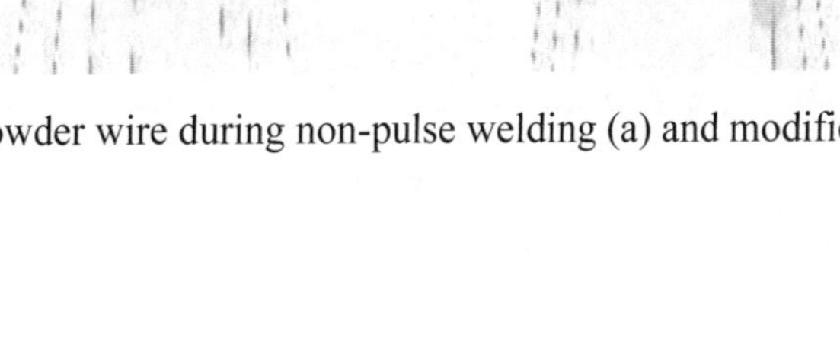

Figure 5. Melting of powder wire during non-pulse welding (a) and modified pulse welding (b).

In option (b), small droplets from the shell are separately moved to the bath in the fine-drop transfer mode, and a drop of charge melted due to heat transfer from the shell grows until it breaks away from the electrode in the large-drop transfer mode.

It is obvious that option (b) in the non-pulse mode can be implemented only at large current values corresponding to the small-drop transfer. For cored wires, due to the smaller cross-section of the metal shell than for solid wires, this current can be relatively small: 150-200A.

In the SpeedPulse modes, the welding current pulse is modified in such a way that after the main pulse, the current does not immediately drop to the minimum value, but for some time the current is set just in the range of fine-drop transfer. This facilitates the transfer of a certain amount of shell metal in the form of small droplets without mixing them with the molten charge, which ultimately leads to a more stable flow of the process as a whole. Large drops of molten charge are separated, probably, under the influence of gravity with a frequency of 0.4-0.5 Hz, even at a low average current value-from 100-120A [11].

In non-pulse welding, mode (b) is not achieved, since there is no current lifetime corresponding to the fine-drop transfer.

According to the standard mode waveform (Fig.6) it can be seen that the process is a series of chaotic short circuits, which explains the high spatter.

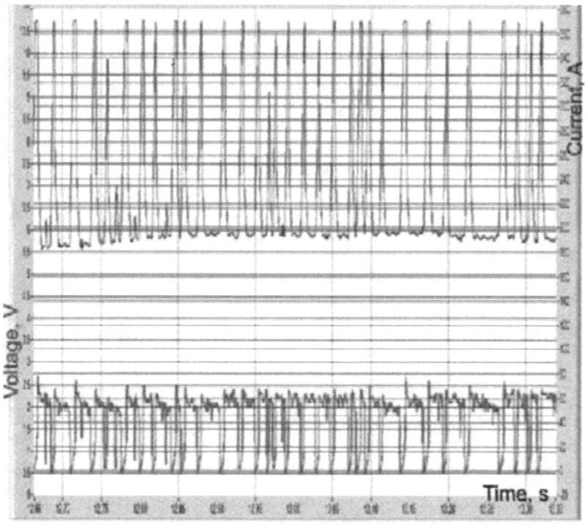

Figure 6. Waveform of the welding process with natural short circuits (Standart mode) ($V \times 10^{-1}$).

The SpeedPulse mode waveform (Fig. 7) shows relatively rare periodic short circuits, apparently corresponding to the transfer of larger droplets. A characteristic step on the back edge of the pulse (Fig. 7b) is used to activate an additional stage of small-drop transfer after the main drop is detached.

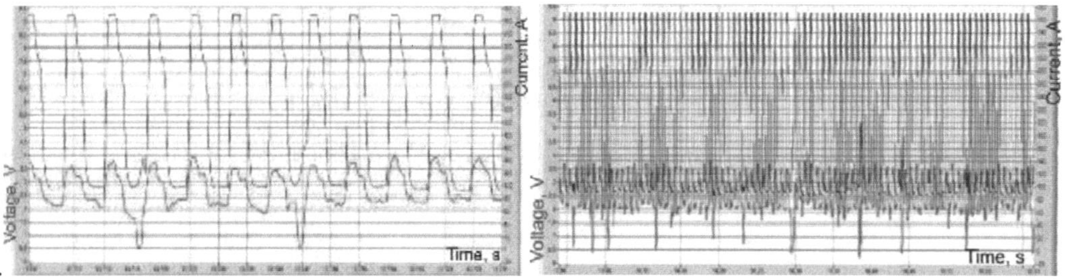

a b

Figure 7. Waveform of the pulse-arc welding process (SpeedPulse mode): a - the flow of the process as a whole; b – the increased time scale.

The lower frequency of the appearance of large droplets probably reflects the different mechanisms of melting of the shell and charge during welding with cored wire.

The formation of a series of small droplets without a short circuit is caused by the melting of the metal shell due to the heat of the active spot and the arc column. The periodic appearance of large droplets corresponds to the gradual melting of the wire charge due to heat transfer from the molten part of the metal shell.

4. Conclusions

In the pulse mode, when using a modified pulse, there are two types of transfer simultaneously: small-drop transfer of the shell metal and large-drop transfer of the charge material. The formation of droplets of different sizes is due to the difference in the thermophysical properties of the charge and the shell of the powder wire.

In this case, the transfer is ordered, which leads to increased arc stability and formation of the seam.

The range of stable modes, in which better seam formation and reduced splashing is achieved, is located in the area of higher arc voltages (32V or more) than when using standard modes and (or) solid wire.

Increasing arc stability reduces metal losses on burning and spattering up to 1.7 times, the surfacing coefficient and productivity increase up to 1.5 times, compared with welding with natural short circuits.

References

[1] Rosert R and Karasev M V 2012 *Powder wires - trends, development and their application in industry* (SPb: Publishing house-Polytechnic. un.) pp 220-230

[2] Shlepakov V N 2014 Physico-metallurgical and welding-technological properties of gas-shielded wire rosettes for welding structural steels *Automatic welding* **6-7** 56-9

[3] Pokhodnya I K *et al.* 1985 Parameters of melting and use of powder wires of various designs *Welding* **8** 33-4

[4] Minnick W H 2009 *Flux Cored Arc Welding Handbook* (The Goodheart Willcox Company Inc.)

[5] Potapievsky A G *et al.* 2012 *Welding of steels in protective gases with a consumable electrode. Technique and technology of the future* (Tomsk: TPU)

[6] Dos Santos E. B. F. *et al.* 2017 High frequency pulsed gas metal arc welding (GMAW-P): The metal beam process *Manufacturing Letters* **11** 1-4

[7] Davydov Yu S *et al.* 2017 Estimation of the parameters of pulse-arc welding with a high-chromium powder wire *Welding production* **6** 14-20

[8] Asnis A E, Gutman L M, Poqladiy V R, *et al.* 1982 *Welding in a mixture of active gases* (Kiev: Naukova Dumka)

[9] Gubin B A *et al.* 1992 Determination of the coefficient of heat transfer from slag to electrode wire for electroslag welding of steel *Automatic welding* **6** 26-9

[10] Davydov Yu S, Korobov Yu S and Davydov A Yu 2019 The influence of the electrode extension on the welding-technological characteristics of pulsed arc welding using cored wire *Materials Science Forum* **946** 895-900

[11] Davydov Yu S and Davydov A Yu 2019 Comparison of melting parameters of high-chromium flux-cored wire in pulsed arc welding and with natural short circuits *Welding. Renovation. Tribotechnics: proceedings of the IX Ural scientific-practical conference* (Nizhny Tagil: NTI (branch) Urfu) pp 40-6

IOP Conf. Series: Materials Science and Engineering 1155 (2021) 012005 doi:10.1088/1757-899X/1155/1/012005

Formation of a nanostructured diamond-like phase from C_{48} clusters: Ab initio calculations

V A Greshnyakov[*] and E A Belenkov

Physics Department, Chelyabinsk State University, Chelyabinsk 454001, Russia

[*]E-mail: greshnyakov@csu.ru

Abstract. A new fullerene-like cluster C_{48} and a nanostructured cubic diamond-like phase formed from an ordered condensate of these clusters were theoretically investigated. Calculations of structures, properties and stability were performed using the density functional theory method. It was found that the new cluster should be stable under normal conditions. As a result of the hydrostatic compression of the C_{48} clusters, a high-strength semiconducting diamond-like phase, called CA7, can be obtained at 26.5 GPa. The CA7 phase can be unambiguously identified experimentally using a calculated powder X-ray diffraction pattern.

1. Introduction

Nanostructures consisting of carbon atoms are used in the manufacture of lithium-ion batteries, hydrogen storage systems, supercapacitors, catalysts, sensors, and composite materials [1]. Such carbon nanostructures as graphene layers, nanotubes, and fullerenes are also used for the experimental production of superhard diamond-like materials at high pressures [2]. However, there are a number of hypothetical nanostructures of three-coordinated carbon atoms predicted in [3], which are more promising precursors for the formation of new nanostructured diamond-like phases [2]. These precursors include fullerene-like clusters, on the surface of which there are quadrangular or octagonal faces, but there are no pentagonal faces [3-5]. The most promising fullerene-like cluster is C_{48}, the shape of which is similar to that for the synthesized $B_{24}N_{24}$ nanostructure [6]. Therefore, this study is devoted to *ab initio* calculations of the structure and thermal stability of the new fullerene-like C_{48} cluster, as well as to the study of the crystal structure, properties, and the synthesis process of a nanostructured diamond-like phase formed from these clusters.

2. Methods

Calculations of the crystal structures and electronic properties of diamond, fullerene-like C_{48} clusters and nanostructured diamond-like phase were carried out by the density functional theory (DFT) method in the generalized gradient approximation (GGA) [7] using the Quantum ESPRESSO software package [8]. The norm-preserving Troullier-Martins pseudopotentials [9] and $8 \times 8 \times 8$ *k*-point grids were used for the calculations. The cutoff energy (E_{cutoff}) in the plane wave basis was 800 eV. The study of the thermal stability of the C_{48} cluster was carried out using the DFT-GGA and molecular dynamics methods.

The bulk modulus of the diamond-like phase was calculated using the modified Kumar-Sharma equation of state [10]. The Vickers hardness was calculated according to the method proposed by Gao et al. [11]. A theoretical study of the structural transformation of the C_{48} cluster condensate into the nanostructured diamond-like phase was carried out under the hydrostatic compression of the initial condensate according to the method described in [12, 13]. Also, X-ray diffraction patterns of carbon compounds were calculated for the Cu-$K_{\alpha 1}$ characteristic radiation according to the standard procedure [14].

3. Results

The optimized structure of the new C_{48} cluster, which has the shape of a truncated cuboctahedron and belongs to the *m*3*m* point symmetry group, is shown in figure 1. The surface of this cluster consists of 4-, 6-, and 8-sided faces and differs from the C_{48} fullerene, on the surface of which there are only 5- and 6-sided faces. The DFT-GGA calculations showed that all atoms on the surface of the fullerene-like C_{48} cluster are in equivalent crystallographic positions. Three different lengths of interatomic bonds (0.1397, 0.1478, and 0.1488 nm) and three angles between these bonds (90.0, 120.0, and 135.0°) correspond to each atomic position. The diameter and the Wells ring parameter (*Rng* [16]) of this cluster are 0.6710 nm and $4^1 6^1 8^1$, respectively.

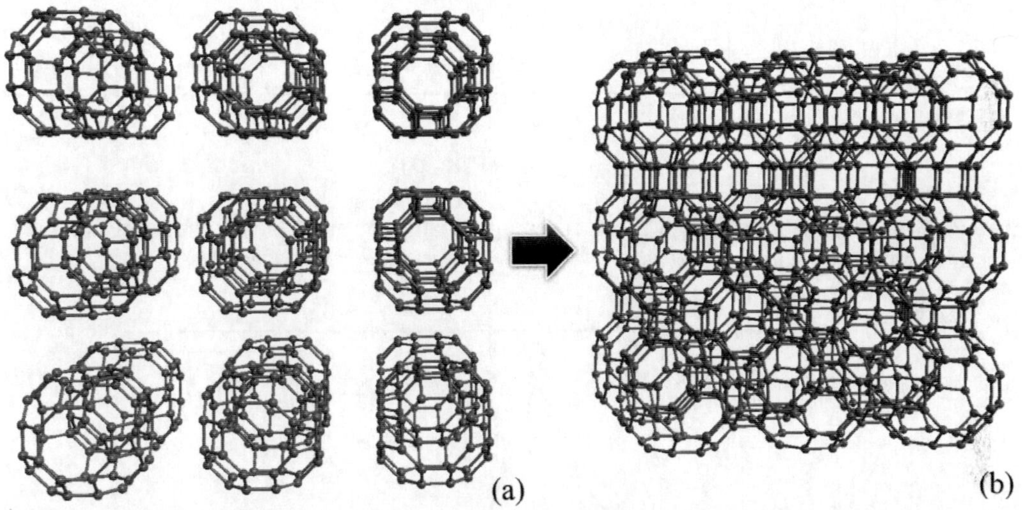

Figure 1. Model transformation of the structure of the fullerene-like C_{48} cluster condensate (a) into the structure of the diamond-like CA7 compound (b).

The cohesion energy (E_{coh}) of the C_{48} cluster was calculated as the difference between the total energies of isolated atoms and atoms in bound states. The calculated E_{coh} value is 7.00 eV/atom and is 10.9 % less than the corresponding value for 3C diamond, which was calculated by the DFT-GGA method in [15, 16]. In addition, the cohesion energy of the C_{48} cluster is lower than those for almost all diamond-like T- and S-phases (with the exception of SA2 and SA4); therefore, the molecular dynamics study of the stability of this cluster was carried out at room temperature. The isothermal annealing was simulated for 8.5 ps with a time step of 1 fs. The graph of the change in the total energy (E_{total}) versus the time of heat treatment is shown in figure 2. As a result of the annealing, the fullerene-like C_{48} cluster turned out to be stable at normal temperature. However, a slight deformation of the structure and a significant rotation of the C_{48} cluster with respect to the equilibrium position at the center of a simple cubic unit cell are observed during the annealing process.

An analysis of the C_{48} cluster structure showed that a porous material with a diamond-like structure of the CA7 phase described in [2] can be obtained on its basis (figure 1b). To obtain the CA7 phase, a condensate from the C_{48} clusters is required having a simple cubic (sc) lattice belongs to the *Pm*3*m* space symmetry group (figure 1a). Figure 3 contains graphs of the dependences of the difference total energy ΔE_{total} (relative to the total energy of the C_{48} condensate at zero pressure) on the atomic volume V_{at} under bulk uniform deformations. It was found that the formation of the diamond-like CA7 phase is possible upon hydrostatic compression of the sc C_{48} condensate, when the pressure reaches 26.5 GPa. The formation pressure value of the CA7 phase is much less than the corresponding values for the formation of diamond-like compounds from graphite [12, 13, 17-21]. During the direct phase transition, it is necessary to overcome the energy barrier of 0.22 eV/atom.

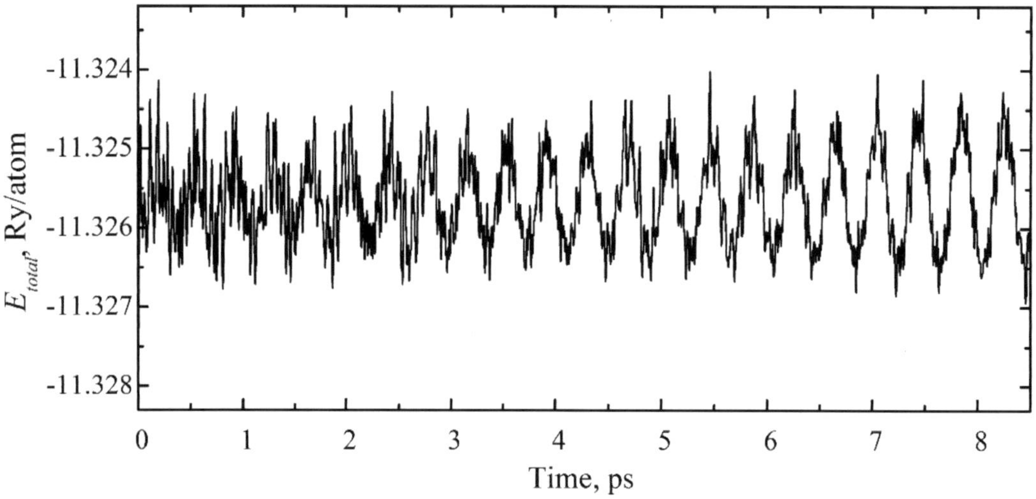

Figure 2. Calculated dependence of the total energy change on the time of annealing for the C_{48} cluster at room temperature.

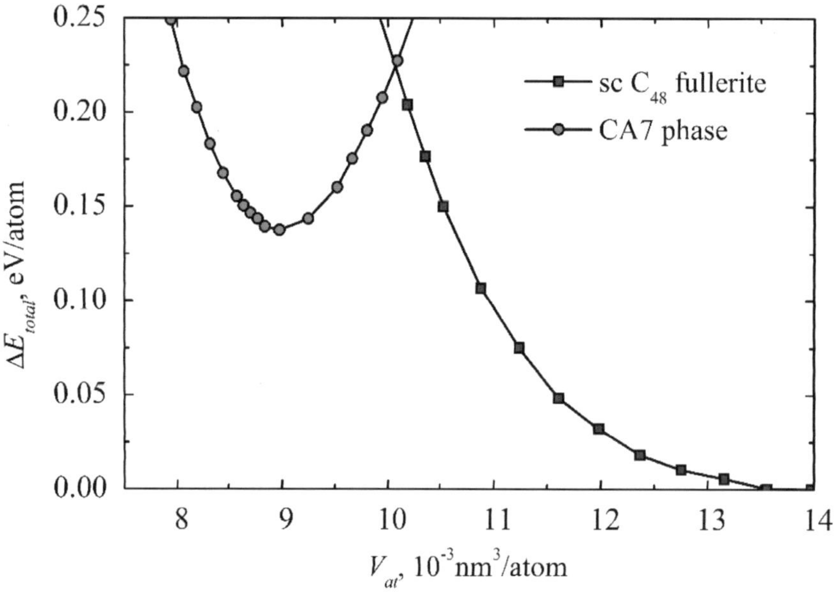

Figure 3. Dependences of the difference total energy on the atomic volume, characterizing the structural transformation "sc C_{48} condensate → CA7 phase".

At the next stage of the work, the mechanical and electronic properties of the new nanostructured phase from the polymerized C_{48} clusters were investigated. The bulk modulus (B_0) was defined in the range of relative volume change from 0 to 5.8 %. The B_0 value was 270 GPa, which is 39 % less than the corresponding value for diamond. The calculated Vickers hardness for the CA7 phase is 66.4 GPa. This value is only 26.2 % less than the calculated value for cubic diamond. The band structure of this phase, calculated by the DFT-GGA method, is shown in figure 4. The electron energies were calculated over six intervals (ΓN, NP, PN, NH, HP, and PΓ) between four high symmetry points (Γ, N, P, and H) in the Brillouin zone. The minimum direct band gap of the diamond-like CA7 phase, calculated as the difference between the electron energy of the conduction band bottom and the electron energy of the valence band top, is observed at the Γ point (i.e., at the center of the Brillouin zone) and is equal to 3.75 eV. This value is less than those for cubic diamond by 36.6 %. The electronic density of states (DOS) of the diamond-like CA7 phase was also calculated (figure 4). Using the calculated DOS-pattern, the minimum band gap value of the nanostructured CA7 phase was determined to be 3.34 eV. Therefore, this cubic diamond-like phase should be a wide-gap semiconductor with an indirect band gap.

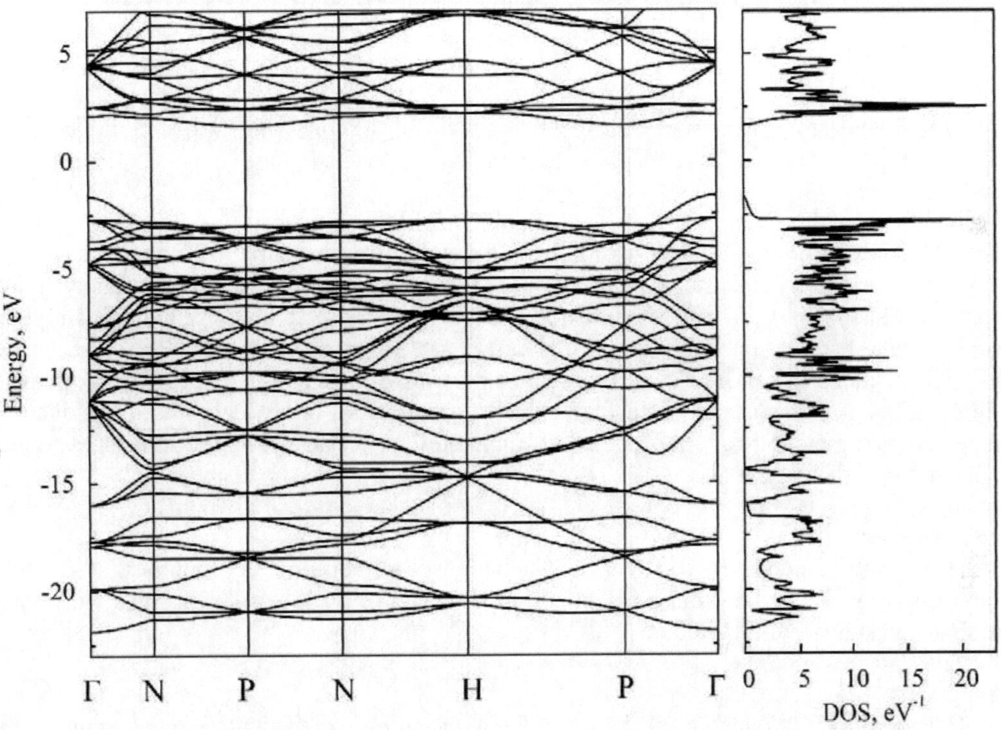

Figure 4. The band structure and the density of electronic states of the CA7 phase, calculated using the DFT-GGA method (zero on the energy scale corresponds to the Fermi energy).

For the experimental identification of the studied nanostructured CA7 phase in synthesized carbon materials, the calculation of its powder X-ray diffraction pattern was performed. Figure 5 shows the calculated X-ray diffraction patterns of cubic diamond, initial condensate from the fullerene-like C_{48} clusters, and diamond-like CA7 phase. It was found that the X-ray diffraction pattern of the CA7 phase contains only three most intense diffraction maxima corresponding to the following interplanar distances: 0.3083, 0.3776, and 0.5340 nm. The positions of these diffraction maxima differ significantly from the positions of the diffraction maxima of cubic diamond and the initial condensate from the C_{48} clusters; therefore, the new nanostructured CA7 phase can be unambiguously identified in experimentally obtained carbon materials.

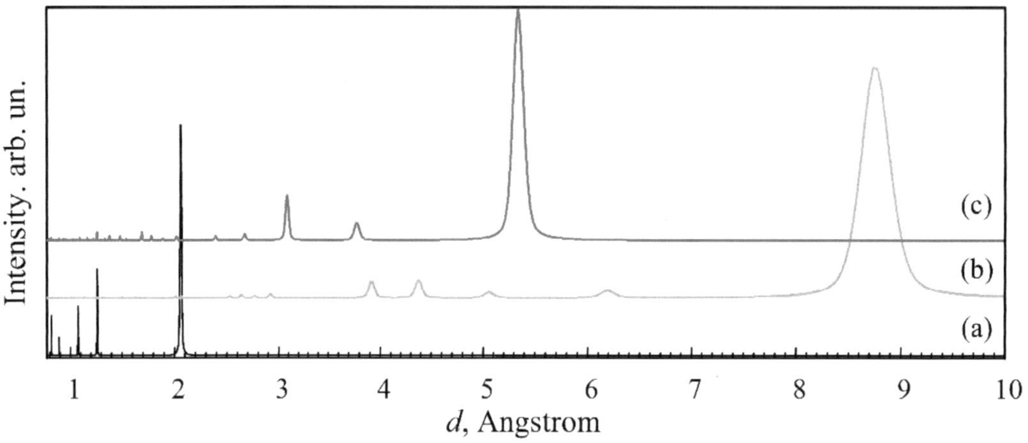

Figure 5. Calculated powder X-ray diffraction patterns of the following carbon compounds: (a) cubic diamond; (b) C_{48} cluster condensate; (c) nanostructured CA7 phase.

4. Conclusions

An *ab initio* study of a new fullerene-like C_{48} cluster and a nanostructured diamond-like CA7 phase based on polymerized C_{48} fullerite was carried out in this work. The DFT calculations showed that the C_{48} cluster, consisting of 4-, 6-, and 8-membered structural units, should be stable at room temperature. It was also found that the transformation of the condensate from these fullerene-like clusters into the porous cubic diamond-like phase CA7 can occur at a pressure of 27 GPa. This diamond-like phase must have high mechanical characteristics and can exhibit the properties of a wide-gap semiconductor. The calculated X-ray diffraction pattern of the nanostructured CA7 phase can be used for its experimental identification in carbon materials.

Acknowledgments

The research was funded by RFBR and Chelyabinsk Region, project number 20-43-740015. Greshnyakov V.A. thanks the Young Scientist Support Fund of the Chelyabinsk State University for partial financial support of this study.

References

[1] Morandi V and Ottaviano L 2017 *GraphITA: Selected Papers from the Workshop on Synthesis, Characterization and Technological Exploitation of Graphene and 2D Materials Beyond Graphene. Carbon Nanostructures* (Cham: Springer International Publishing AG) 221 p
[2] Belenkov E A and Greshnyakov V A 2016 *Phys. Solid State* **58** 2145
[3] Belenkov E A and Greshnyakov V A 2013 *Phys. Solid State* **55** 1754
[4] Krylova K A, Baimova Y A, Dmitriev S V and Mulyukov R R 2016 *Phys. Solid State* **58** 394
[5] Kang S, Xiang Z, Mu H and Cai Y 2020 *Phys. Lett. A* **384** 126035
[6] Pokropivny V V, Smolyar A S, Ovsiannikova L I, Pokropivny A V, Kuts V A, Lyashenko V I and Nesterenko Y V 2013 *Phys. Solid State* **55** 878
[7] Perdew J P, Burke K and Ernzerhof M 1996 *Phys. Rev. Lett.* **77** 3865
[8] Giannozzi P *et al.* 2017 *J. Phys.: Condens. Matter.* **29** 465901
[9] Troullier N and Martins J L 1991 *Phys. Rev. B* **43** 1993
[10] Greshnyakov V A and Belenkov E A 2014 *Russ. Phys. J.* **57** 731
[11] Gao F, He J, Wu E, Liu S, Yu D, Li D, Zhang S and Tian Y 2003 *Phys. Rev. Lett.* **91** 015502
[12] Belenkov E A and Greshnyakov V A 2018 *Phys. Solid State* **60** 1294
[13] Greshnyakov V A, Belenkov E A and Brzhezinskaya M M 2019 *Phys. Status Solidi B* **256** 1800575

[14] Guinier A 1964 *Theorie et technique de la radiocristallographie. 3rd edition* (Paris: A. Dunod) 740 p
[15] Belenkov E A and Greshnyakov V A 2015 *J. Mater. Sci.* **50** 7627
[16] Belenkov E A and Greshnyakov V A 2015 *Phys. Solid State* **57** 1253
[17] Baimova J A and Rysaeva L K 2018 *J. Struct. Chem.* **59** 884
[18] Rysaeva L K, Baimova J A, Lisovenko D S, Gorodtsov V A and Dmitriev S V 2019 *Phys. Status Solidi B* **256** 1800049
[19] Baimova J A, Galiakhmetova L K and Mulyukov R R 2021 *Comput. Mater. Sci.* **192** 110301
[20] Fan D, Lu S, Golov A A, Kabanov A A and Hu X 2018 *J. Chem. Phys.* **149** 114702
[21] Zhou Y, Chen X, Liu S-Li and Gan Li-H 2021 *Solid State Commun.* **323** 114095

Technological assurance of the quality of deformed steel products in mechanical engineering

I A Shveyov and E I Shveyova[1]

Kazan Federal University, Kazan, Russia

[1]E-mail: Shveeva_Katya222@mail.ru

Abstract. The article shows the influence of fibrous structure and non-metallic inclusions on the mechanical properties of steel products. The macrostructure of forging the crankshaft was investigated and the places of exit of the elimination zones to the surface of the car part were established. Based on this, there are proposed ways to improve the quality of deformable steel products in the machine building industry.

One of the main disadvantages of heavily loaded vehicle parts such as gears, shafts and other parts is the dispersion of their strength and durability. The duration of the work of the same type of parts made of the same steel grade, processed and hardened in the same production conditions differs several times. The reasons for the significant dispersion in the durability of parts are different, namely, they can be a consequence of the presence of metallurgical defects in the alloy, design and technological features and operating conditions of the vehicle. Therefore, solving the problems of stabilizing the strength and durability of various parts at the maximum level can be achieved on the basis of an integrated engineering approach, linking indicators and criteria of metallurgical, design, technological and operational types into a single system. The reasons that cause destabilization of the properties of finished products were studied. A change in metallurgical defects of steel during the technological redistribution of metal in machine-building production was revealed and the possibility of improving the properties of parts by forming a favorable macro- and micro-structure in the cross section of a metal product is shown [1].

At the engineering design stage of the part, the main emphasis on the part of the workers is on the selection and initial state of the stock of material that is used to manufacture the metal product. The physical and mechanical properties of the articles depend to a large extent on the manufacturing scheme of the parts, since they decisively affect the process of forming the metal structure.

High level of properties of parts is achieved in case of application of plastic deformation method in process cycle. This operation has a positive effect on the macro- and microstructure of steel. The crystalline dendritic structure of the cast steel workpiece undergoes significant changes during plastic deformation. Properly assigned and performed modes of plastic deformation contribute to elimination of porosity, infusion of microdefects, crushing and rational orientation of non-metallic inclusions and carbides, grain grinding and formation of optimal arrangement of fibrous structure and deformation texture [2]. The complex effect of the process parameters of hot plastic deformation processing (temperature, speed and degree of deformation, etc.) has a significant effect on the shape of the structural components, and as a result, increases all the indicators of the mechanical properties of the metal.

The direction of the fibers, or the macrostructure of the metal products, has a significant effect on the strength of the steel parts, since the mechanical properties of steel during tests along or across the fibers differ sharply. In the example of a steel sample 45 in the longitudinal direction, the strength and ductility of the steel are higher than in the transverse direction (normalization 860°C, tempering of 640 °C).

To make a strong part by forging or hot forging from rolled stock, it is necessary to use high strength of rolled stock in the longitudinal direction. In the finished part, the fibers should be axially directed and enveloped so that there are no break points or cut fibers during subsequent machining. The incorrect direction of the fiber along the main working sections of the part causes a decrease in mechanical strength and premature destruction of the finished products. During the engineering phase, it is also necessary to consider the possibility of radial arrangement of fibers, which, due to several circumstances, can inevitably be obtained on the part.

The central zone of rolled stock deserves special attention, since it concentrates non-metallic inclusions, there is porosity and other disadvantages that adversely affect the properties of the metal. This fact causes lower mechanical properties of the central zone with respect to the peripheral zone. As practice shows, when such a zone is shifted to the most loaded sections of the part, premature destruction of the product during operation is observed [3].

An illustrative example is the tooth breakage of a gear in the base of which there is a liquor square displaced from the central zone during plastic deformation of the metal in the forging manufacturing step (figure 1).

a b

Figure 1. Macrostructure of section of parts with location of liquation square at tooth base (a) and in central part (b) of gear.

In addition, the release of liquor zones to the surface can cause "spotty" hardness and the formation of quench cracks on forgings and parts during heat treatment (figure 2) and after grinding.

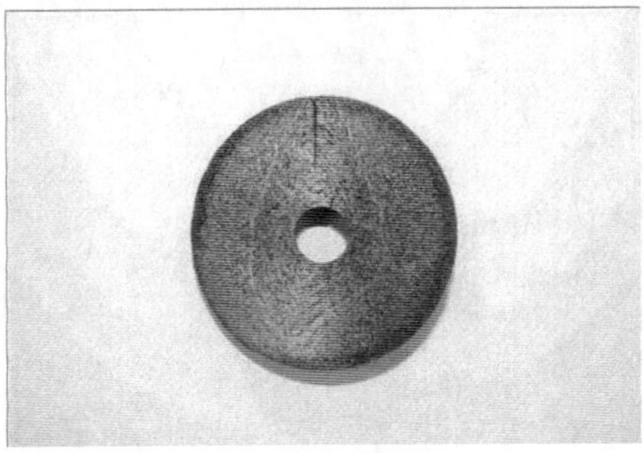

Figure 2. The liquefaction square in the section of the part macro profile.

IOP Conf. Series: Materials Science and Engineering 1155 (2021) 012006 doi:10.1088/1757-899X/1155/1/012006

As a result of experimental studies, on a specially prepared workpiece subjected to hot bulk stamping on KGShP, they were clearly convinced about the possibility of internal metal layers reaching the forging surface (figure 3).

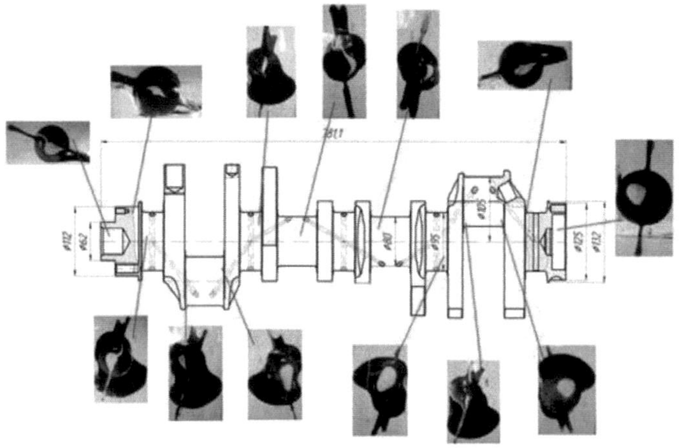

Figure 3. Macrostructure in cross section of various parts of crankshaft.

The metal blank was a square of 150 × 150 mm, in which the metal was removed from the central zone by drilling and a rod of Ø50 mm from another steel brand was installed in this place, which differed in degree of etchability from the metal of the original blank. In this way, a contaminated rolled stock zone was simulated with, and these are the liquor square, central porosity and other imperfections that reduce the mechanical properties of steel [4]. After heating up to 1230 ± 10°S of the combined billet, volumetric stamping was carried out. Samples were taken from various sections of the obtained forging and templets were made that corresponded to the cross-sections or longitudinal sections of the part. As a result of hot macrotreating of templets in a 50% solution of hydrochloric acid, the location of heterogeneous materials was revealed. In separate cross sections, the exit to the surface of the central zone is not visible [5].

However, in most cases, such a zone comes out or is close to the surface, including it is present in the zone of fillets and oil paths. Metal of rod on surface of forging is clearly manifested on templets of longitudinal section. This is shown in figure 4.

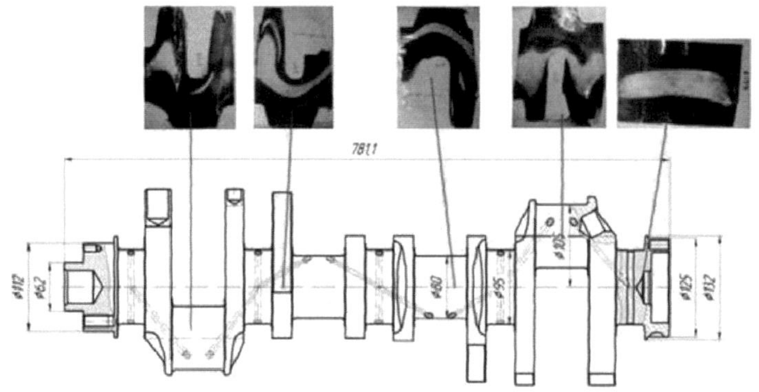

Figure 4. Macrostructure in longitudinal section of various parts of crankshaft.

To a large extent, the viscosity of steel depends on the content of sulfur: the temperature threshold of cold water in steels with a sulfur content of 0.005% and 0.03% differs by 30-40°C.

In any steel in some quantities there are elements that are gases - oxygen, nitrogen and hydrogen. Gases, even if contained in hundredths and thousandths of a percent, have a significant effect on the properties of the metal. Penetration impurities (nitrogen and oxygen), concentrating in grain-boundary volumes and forming emissions of nitrides and oxides, along the grain boundaries, increase the cold flow threshold and reduce resistance to brittle destruction. Oxides are stress concentrators and significantly reduce, especially in case of their accumulation, the endurance limit and can serve as a center of destruction of the part in operation.

These arguments show a significant decrease in the structural strength of parts that have non-metallic inclusions on the surface and especially if they are present in areas of loaded sections (fillets, holes, splines) [6].

Premature failure of metal products is often the result of brittle destruction of the material. The size of the grain, the amount, size, shape and nature of the distribution of non-metallic inclusions, central porosity, liquation manifestations of various kinds, etc., have a great influence. Many of these factors are inherited by the metal of the finished products from the cast state of the alloy and often have a negative effect. This is evidenced by the revealed center of destruction of the part, the beginning of which was a non-metallic inclusion, located even in the depth of the metal (figure 5).

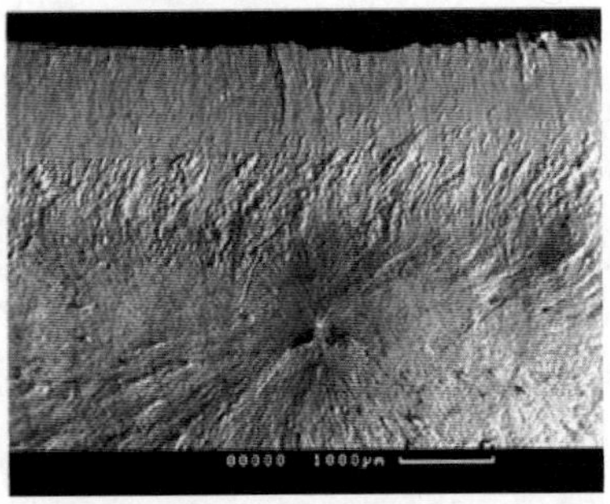

Figure 5. Non-metallic inclusion- the center of part destruction.

Thus, when developing the technology and designing the die tooling, it is necessary to achieve the correct metal flow during plastic deformation and create a rational fibrous structure, as well as prevent the central zone from leaving the area of the most loaded zones and to the surface of the part.

References

[1] Tikhonov A K 2008 *Metal. of mech. engineering.* **3** 38
[2] Nikitin V I, Nikitin K V 2005 *Heredity in cast alloys (Moscow)* vol 2 (Moscow: Engineering) p 476
[3] Neuchev A L 1996 *Kuzn.-stampp.* **11** 21-2
[4] Kurchatkin V V, Telnov N F 2000 *Reliability and repair of cars (Moscow)* vol 5 (Moscow: Academia) p 344
[5] Sorokin V M 1996 *Improvement of quality of a surface and durability of details of cars the shock and pulse and combined processing (N. Novgorod)* vol 1 (N. Novgorod: Nauka) p 408
[6] Tushinskiy L N 2000 *Theory and technology of metal alloys hardening (Novosibirsk)* vol 1 (Novosibirsk: Science) p 306

IOP Conf. Series: Materials Science and Engineering 1155 (2021) 012007 doi:10.1088/1757-899X/1155/1/012007

Analytical study of the non-stationary temperature field of a thermally thin plate

I I Latypov[1,3], L A Bigaeva[1], G S Mukhametshina[1], N A Shaikhutdinova[2] and A Y Gilev[1]

[1] Bashkir State University, 32, Validy str., Ufa, Russia
[2] Ufa State Petroleum Technical University, 1, Kosmonavtov str., Ufa, Russia

[3] E-mail: LatypovII@rambler.ru

Abstract. The report analytically investigates the unsteady temperature field of a plate experiencing intense thermal action. The problem is reduced to solving a singularly perturbed boundary value problem of the nonstationary thermal conductivity equation with nonlinear boundary conditions on movable borders. The approximate solution of which is obtained in the form of an asymptotic decomposition of the solution in the sense of Poincaré in degrees of small parameters, depending on the proximity of the point under consideration to the borders.

1. Introduction

Modern development of equipment and technology is impossible without the use of high-intensity energy sources. These processes are fleeting and local in nature. Solid-state lasers are one of the common sources of energy. These laser systems are used in technological operations related to thermal effects on materials, and are also used for research and testing of various materials. In this case, both the impact materials and the installation themselves or its elements experience strong thermal loads [1-3,6]. Therefore, when studying the temperature fields arising in the material, determining the thermal mode of operation of energy sources, it is necessary to take into account thermal and optical distortions arising in the material or in structural elements.

The object of the study is a plate, which is subjected to intense thermal action in the form of a flat front, in the cross section of which is approximated by a Gaussian function. At the same time, heat exchange with the environment is carried out in the mode of protective heating, taking into account the radiation component. Thermal and optical distortions are taken into account by introducing into the mathematical model the terms described by the heat flux and the temperature dependence of the absorption capacity of the material. Thermoelastic deformations of a material are observed in the form of a change in its profile when the surface of the material is irradiated with laser radiation at intensities insufficient for melting and evaporation of the metal. In this case, the body experiences significant thermal stress, which is strongly manifested in the `` *intermediate " layer*, i.e. in the layer between the surface and the inner regions of a given body, which greatly affects the performance and quality of mirrors of high-intensity lasers [2].

2. The problem Statement

The report poses the problem of determining the non-stationary temperature field of a plate irradiated by a powerful heat source in a short period of time, the thickness of which is much smaller than the other dimensions. The above assumptions admit a one-dimensional coordinate formulation of the problem. Thus, the problem in the one-dimensional approximation is reduced to a nonstationary boundary value problem for the heat equation

$$\frac{\partial U(\xi,\tau)}{\partial \tau} = a^2 \cdot \frac{\partial^2 U(\xi,\tau)}{\partial \xi^2} + K(\tau) \cdot \frac{\partial U(\xi,\tau)}{\partial \xi} + A[U(\xi,\tau)],\ (\xi,\tau) \in \Omega,$$

$$U(\xi,\tau) = U_0(\xi), \qquad \tau \to 0,$$

$$\lambda \frac{\partial U(\xi,\tau)}{\partial \xi} = -q_1(\tau) + \sigma_1 \left[U^4(\xi,\tau) - U_1^4\right] + A_v I(\tau),\ \ \xi = 0,$$

$$\lambda \frac{\partial U(\xi,\tau)}{\partial \xi} = q_2(\tau) - \sigma_2 \left[U^4(\xi,\tau) - U_2^4\right],\ \ \xi = h,$$

$$\Omega = \left\{(\xi,\tau):\ 0 < \xi < h,\ 0 < \tau \le t_{00}\right\},$$

$$I(\tau) = I_0 \cdot \exp\left\{-\frac{(\tau - t_0)^2}{t_{10}^2}\right\};\ \sigma_i = c_i \cdot \sigma;\ c_i, t_{00}, t_0, t_1, t_{10}, A_v, I_0 - const,\ \ i = 1,2.$$

Where $U(\xi,\tau)$ is the temperature, the required function; $U_0(\xi)$ - initial temperature distribution; h - plate thickness; U_1, U_2, $q_1(\tau)$, $q_2(\tau)$ - the temperature of the medium and the heat flux on the corresponding edges of the plate; λ, $a^2 = \lambda / c_v\rho$ - coefficients of thermal conductivity and thermal diffusivity; σ, c_i - Stefan-Boltzmann constant and grayness coefficients of plate edges; $I(\tau)$ is a function that approximates the cross section of a plane heat front. The response of a material to a thermal effect is described by two terms in the differential equation (1). Term $K(\tau) \cdot \frac{\partial U(\xi,\tau)}{\partial \xi}$ takes into account the contribution of thermooptical distortions, where the coefficient determines the dynamics of distortions and has a very specific physical meaning [1, 12]. The second term characterizes the actual absorption capacity of the material by a dependence of the type $A[U(\xi,\tau)] = A_0 + A_1 \cdot U(\xi,\tau)$, $A_i = const$, $i = 0,1$ [1,3,4].

After changing variables, $\mu = \xi + \int_0^\tau K(z)dz$, $\eta = \tau$, introducing dimensionless variables, $\mu = \bar{x} \cdot x$, $\eta = \bar{t} \cdot t$ and notations, $U(\bar{x} \cdot x - \bar{x} \cdot \psi(\bar{t} \cdot t), \bar{t} \cdot t) = T(x,t)$, $U_0(\bar{x} \cdot x) = T_0(x)$, we obtain a singularly perturbed boundary value problem of the heat equation with a small parameter $Fo \ll 1$, which is a consequence of the transience of the process (small time scale $\bar{t}$) and a small thickness of the plate (smallness of the spatial scale $\bar{x}$), which has the following form

$$\frac{\partial T(x,t)}{\partial t} = Fo \cdot \frac{\partial^2 T(x,t)}{\partial x^2} + A[T(x,t)], \tag{1}$$

$$T(x,t) = T_0(x), \quad t \to 0, \tag{2}$$

$$\frac{\partial T(x,t)}{\partial x} = -\varphi_1(t) + \gamma_1 \cdot \left[T^4(x,t) - U_1^4\right] + \bar{A}_v I(t), \quad x = \psi_1(t), \tag{3}$$

$$\frac{\partial T(x,t)}{\partial x} = \varphi_2(t) + \gamma_2 \cdot \left[T^4(x,t) - U_2^4\right], \quad x = \psi_2(t), \tag{4}$$

$$\left(x,t\right)\in\Omega'=\left\{\left(x,t\right):\psi_1(t)<x<\psi_2(t),\ 0<t\le\bar{t}_{00}\right\},\tag{5}$$

where $Fo=\dfrac{a^2\bar{t}}{\bar{x}^2}$, $0<Fo<<1$; $\bar{t}_{00}=\dfrac{t_{00}}{\bar{t}}$, $H=\dfrac{h}{\bar{x}}$, $\varphi_i(t)=\dfrac{\bar{x}}{\lambda}\cdot q_i(\bar{t}\cdot t)$, $\bar{A}_9=\dfrac{\bar{x}}{\lambda}A_9$,

$\gamma_i=(-1)^{i+1}\dfrac{\bar{x}}{\lambda}\sigma_i$, $q_i(\bar{t}\cdot t)=q_{i0}=const$, $\varphi_i(t)=\varphi_{i0}=const$, $i=1,2$; $\psi(t)=\dfrac{1}{\bar{x}}\int\limits_0^{\bar{t}\cdot t}K(z)dz$, $\psi_1(t)=\psi(t)$,

$\psi_2(t)=\psi(t)+H$; $I(t)=I_0\exp\left\{-\dfrac{\left(t-\bar{t}_0\right)^2}{\bar{t}_{10}^2}\right\}$.

It is assumed that the irradiation time and thermophysical characteristics do not change significantly.

Thus, the original boundary value problem has been reduced to a singularly perturbed boundary value problem for the heat equation with nonlinear boundary conditions on moving boundaries.

For the asymptotic study of this problem, we use the method proposed by G.A. Nesenenko [7], which he called the "*geometric-optical*" asymptotic method. The main provisions of this method are reflected in the works [7-12].

Using the above scheme, the Green's function and the solution of the boundary value problem are obtained in the form of asymptotic expansions in powers of small parameters in the sense of Poincaré.

It should be noted that the form of the asymptotic expansion of both the Green's function and the solution of the boundary value problem depends on the "*proximity*" of the point under consideration to the boundary of the domain. In accordance with the introduced concept of "*proximity*" of a point to the boundary, the area is divided into "*boundary*", "*intermediate*" and "*remote*" from the boundaries of the zone, each of which uses its own asymptotic scale [9-12].

In each of these selected zones, the Laplace method (a special case of the saddle point method) is used for the asymptotic analysis of the corresponding integrals. Moreover, in the area of points remote from the boundaries, the classical version of the Laplace method is used, where the reference integral has the form: $I=\int\limits_0^\infty x^\alpha\exp\left\{-\lambda x^2\right\}dx,\ \lambda>0$.

Inthe other zones, the Laplace method is used with some modification, which consists in the fact that an integral of the form is chosen as the reference integral:

$$J=\int\limits_0^\infty x^{\alpha-1}\exp\left\{-\gamma\cdot x-\frac{\beta}{x}\right\}dx,\ \gamma,\beta>0.$$

This follows from the fact that integrals of the form are analyzed in these zones, where the value at which the smallest value is reached (at) and the condition of "proximity" of the point to the boundary is satisfied, merges with the end of the integration interval (that is), and therefore the classical Laplace method with a reference integral of the form cannot be applied. Therefore, in [7], it was proposed to use an integral of the form as a reference integral in the asymptotic analysis of integrals of the form.

3. Solution of the boundary value problem

Let us assume that the boundary changes according to a linear law, which is quite possible due to the physical meaning of the coefficient, i.e. then

$$\psi(\bar{t}\cdot t)=\bar{k}_0\frac{\bar{t}}{\bar{x}}\cdot t=k_0\cdot t=\psi_0(t),\ k_0=\bar{k}_0\frac{\bar{t}}{\bar{x}},\quad\begin{cases}\psi_1(t)=\psi_0(t),\\ \psi_2(t)=\psi_0(t)+H.\end{cases}$$

According to [3,7,10], we write down the integral representation of the solution to the problem (1)-(5)

$$T(x,t)=\int_{\psi_1(0)}^{\psi_2(0)} T_0(y)G(x,t;y,0)dy+Fo\cdot\varphi_{10}\int_0^t G(x,t;\psi_1(s),s)ds$$

$$-Fo\cdot\bar{A}_\upsilon I_0\int_0^t \exp\left\{-\frac{(s-\bar{t}_0)^2}{\bar{t}_{10}^2}\right\}G(x,t;\psi_1(s),s)ds$$

$$-Fo\cdot\gamma_1\int_0^t \nu_1(s)G(x,t;\psi_1(s),s)ds+Fo\cdot\varphi_{20}\int_0^t G(x,t;\psi_2(s),s)ds$$

$$+Fo\gamma_2\int_0^t \nu_2(s)G(x,t;\psi_2(s),s)ds+\int_0^t\int_{\psi_1(s)}^{\psi_2(s)} A[T(y,s)]G(x,t;y,s)dyds$$

$$=T_L(x,t)+T_I(x,t)+T_{NL1}(x,t)+T_{NL2}(x,t)+T_A(x,t)=T^*(x,t)+T_A(x,t)\,. \quad (6)$$

Where $G(x,t;y,s)$ -Green's function of the corresponding homogeneous boundary value problem

$$\frac{\partial G(x,t;y,s)}{\partial t}=Fo\cdot\frac{\partial^2 G(x,t;y,s)}{\partial x^2},$$

$$G(x,t;y,s)=\delta(x,y),\quad t\to s,$$

$$\frac{\partial G(x,t;y,s)}{\partial x}=0,\quad x=\psi_i(t),\ i=1,2,$$

$$(x,t)\in\Omega'=\{(x,t):\psi_1(t)<x<\psi_2(t),\ 0<t\le\bar{t}_{00}\},$$

the solution of which can be found explicitly [9,12,14]:

$$G(x,t;y,s)=\frac{1}{2\sqrt{\pi Fo(t-s)}}\exp\left\{-\frac{[x-y]^2}{4Fo(t-s)}\right\}$$

$$+\sum_{m=0}^{\infty}\left[\frac{1}{2\sqrt{\pi Fo(t-s)}}\left[\exp\left\{-\frac{[x+y-2k_0s+2mH]^2}{4Fo(t-s)}+\frac{k_0[y-k_0s+mH]}{Fo}\right\}\right.\right.$$

$$+\exp\left\{-\frac{[2k_0s-x-y+2(m+1)H]^2}{4Fo(t-s)}-\frac{k_0[k_0s-y+(m+1)H]}{Fo}\right\}$$

$$\left.+\exp\left\{-\frac{[x-y+2(m+1)H]^2}{4Fo(t-s)}+\frac{k_0(m+1)H}{Fo}\right\}+\exp\left\{-\frac{[y-x+2(m+1)H]^2}{4Fo(t-s)}-\frac{k_0(m+1)H}{Fo}\right\}\right]$$

$$+\frac{k_0}{2Fo}\left[-\left\{-\frac{k_0[y-k_0s+mH]}{Fo}\right\}erfc\left[\frac{x+y-2k_0s+2mH}{2\sqrt{Fo(t-s)}}\right]\right.$$

$$\left.\left.+\exp\left\{-\frac{k_0[k_0s-y+(m+1)H]}{Fo}\right\}erfc\left[\frac{2k_0s-x-y+2(m+1)H}{2\sqrt{Fo(t-s)}}\right]\right]\right].$$

To simplify calculations, assume that, then you can get the following expressions

$$T_L(x,t)=T_0+\frac{Fo}{2k_0}\cdot\sum_{i=1}^{2}\varphi_{i0}\cdot A_i(x,t), \tag{7}$$

where

$$A_i(x,t)=\mathrm{e}rfc\left[\frac{x-(i-1)H}{2\sqrt{Fot}}\right]+\exp\left\{(-1)^i\frac{k_0\left[x-k_0t-(i-1)H\right]}{Fo}\right\}\times erfc\left[\frac{x-2k_0t-(i-1)H}{Fo}\right]$$
$$+\sum_{m=0}^{\infty}\left[\exp\left\{-\frac{k_0\left[x-k_0t+(m+i-1)H\right]}{Fo}\right\}erfc\left[\frac{x-2k_0t+(2m+3-i)H}{2\sqrt{Fot}}\right]\right.$$
$$-\exp\left\{\frac{k_0\left[k_0t-x+(m+i)H\right]}{Fo}\right\}erfc\left[\frac{2k_0t-x+(2m+i+1)H}{2\sqrt{Fot}}\right]$$
$$+(-1)^i\exp\left\{(-1)^{i-1}\frac{k_0(m+1)H}{Fo}\right\}erfc\left[\frac{(-1)^{i-1}x+(2m+i+1)H}{2\sqrt{Fot}}\right]$$
$$-\frac{k_0\left[(-1)^{i-1}x+(2m+i-1)H\right]}{Fo}\exp\left\{(-1)^{i-1}\frac{k_0mH}{Fo}\right\}erfc\left[\frac{(-1)^{i-1}x+(2m+i-1)H}{2\sqrt{Fot}}\right]$$
$$+(-1)^{i-1}\left[1+\frac{k_0\left[x+(-1)^i(2m+3-i)H\right]}{Fo}\right]\exp\left\{\frac{(-1)^ik_0(m+1)H}{Fo}\right\}erfc\left[\frac{(-1)^ix+(2m+3-i)H}{2\sqrt{Fot}}\right]$$
$$+2k_0\sqrt{\frac{t}{\pi Fo}}\left\{\exp\left\{(-1)^{i-1}\frac{k_0mH}{Fo}-\frac{\left[(-1)^{i-1}x+(2m+i-1)H\right]^2}{4Fot}\right\}\right.$$
$$\left.\left.+(-1)^{i-1}\exp\left\{(-1)^i\frac{k_0mH}{Fo}-\frac{\left[(-1)^ix+(2m+3-i)H\right]^2}{4Fot}\right\}\right\}\right],\ erfc[z]=\frac{2}{\sqrt{\pi}}\int_z^{\infty}e^{-u^2}du\,.$$

$$T_I(x,t)=-Fo\cdot\bar{A}_{\upsilon}I_0\int_0^t\exp\left\{-\frac{\left(s-\bar{t}_0\right)^2}{\bar{t}_{10}^2}\right\}G\left(x,t;\psi_1(s),s\right)ds\,.$$

Introducing notations

$$G\left(x,t;\psi_1(s),s\right)=\frac{1}{2\sqrt{\pi Fo(t-s)}}\exp\left\{-\frac{\left[x-\psi_1(s)\right]^2}{4Fo(t-s)}\right\}+\sum_{m=0}^{\infty}G_m\left(x,t;\psi_1(s),s\right),$$

$$T_I(x,t)=-Fo\cdot\bar{A}_{\vartheta}I_0\left[J_I^0+\sum_{m=0}^{\infty}J_{I,m}\right]=T_I^0(x,t)+\sum_{m=0}^{\infty}T_{I,m}\,,$$

let us carry out an asymptotic analysis of each of the integrals.
Then for the first integral we have

$$J_I^0(x,t)=\frac{1}{2\sqrt{\pi Fo}}\int_0^t\frac{1}{\sqrt{(t-s)}}\exp\left\{-\frac{\left(s-\bar{t}_0\right)^2}{\bar{t}_{10}^2}\right\}\exp\left\{-\frac{\left[x-k_0s\right]^2}{4Fo(t-s)}\right\}ds\,,$$

$$J_I^0(x,t)=\frac{1}{\sqrt{\pi Fo}}\exp\left\{-\frac{\left(t-\bar{t}_0\right)^2}{\bar{t}_{10}^2}-\frac{k_0\left[x-k_0t\right]}{2Fo}\right\}\sum_{n=0}^{\infty}C_{n,0}\left(x,t\right).$$

$$C_{0,0}=\sqrt{\pi\cdot Fo}\frac{2\bar{t}_{10}}{A_0}\left[-\exp\left\{\frac{x-k_0t}{2Fo}\cdot\frac{A_0}{\bar{t}_{10}}\right\}erfc\left[A\sqrt{t}+\frac{x-k_0t}{2\sqrt{Fo\cdot t}}\right]\right.$$

$$\left.+\exp\left\{-\frac{x-k_0t}{2Fo}\cdot\frac{A_0}{\bar{t}_{10}}\right\}erfc\left[-A\sqrt{t}+\frac{x-k_0t}{2\sqrt{Fo\cdot t}}\right]\right],\ A_0^2=k_0^2\bar{t}_{10}^2-4\cdot Fo\left(t-\bar{t}_0\right),\ A^2=\frac{A_0^2}{4Fo\cdot\bar{t}_{10}^2},$$

$$C_{1,0}=6\sqrt{\pi\cdot Fo}\frac{2\bar{t}_{10}}{A_0^3}\left[\left(\frac{\bar{t}_{10}^2}{A_0^2}Fo^2-\frac{\bar{t}_{10}}{A_0}Fo^2\left[\frac{x-k_0t}{2Fo}\right]+\frac{1}{3}Fo^2\left[\frac{x-k_0t}{2Fo}\right]^2\right)erfc\left[A\sqrt{t}+\frac{x-k_0t}{2\sqrt{Fo\cdot t}}\right]\right.$$

$$-\left(\frac{\bar{t}_{10}^2}{A_0^2}Fo^2+\frac{\bar{t}_{10}}{A_0}Fo^2\left[\frac{x-k_0t}{2Fo}\right]+\frac{1}{3}Fo^2\left[\frac{x-k_0t}{2Fo}\right]^2\right)erfc\left[-A\sqrt{t}+\frac{x-k_0t}{2\sqrt{Fo\cdot t}}\right]\Bigg]$$

$$+\exp\left\{-\frac{x-k_0t}{2Fo}\cdot\frac{A_0}{\bar{t}_{10}}\right\}erfc\left[-A\sqrt{t}+\frac{x-k_0t}{2\sqrt{Fo\cdot t}}\right]\Bigg].$$

The other integrals are defined in a similar way.

To find the nonlinear component in expression (6), we introduce the notation $T_{LI}(x,t)=T_L(x,t)+T_I(x,t)$, $T_{NL}(x,t)=T_{NL1}(x,t)+T_{NL2}(x,t)$, then $T^*(x,t)=T_{LI}(x,t)+T_{NL}(x,t)$. We find the unknown functions $\nu_1(t)$, $\nu_2(t)$ by solving the system of integral equations written out from the boundary conditions (3) - (4)

$$\nu_i(t)=\gamma_i\left[F_i\left(\nu_1(t),\nu_2(t)\right)-\bar{u}_i\right],\quad i=1,2,$$

where
$$F_j(\nu_1(t),\nu_2(t))=\left[T_{LI}^{(j)}(t)+Fo\sum_{i=1}^{2}(-)^i\gamma_i\cdot K_i^{(j)}\left(\nu_i(t)\right)\right]^4,\quad j=1,2;$$

$$K_i^{(1)}\left(\nu_i(t)\right)=\int_0^t\nu_i(s)\cdot G\left(\psi_1(t),t;\psi_i(s),s\right)ds,\ T_{LI}^{(j)}(t)=T_{LI}(x,t)\big|_{x=\psi_j(t)},$$

$$K_i^{(2)}\left(\nu_i(t)\right)=\int_0^t\nu_i(s)\cdot G\left(\psi_2(t),t;\psi_i(s),s\right)ds,\ \bar{u}_i=(-1)^{i+1}\frac{\varphi_i}{\gamma_i}+U_i^4,\ i=1,2.$$

We seek the solution of this system of equations by the method of successive approximations according to the scheme $\nu_i^{(n+1)}(t)=\gamma_i\left[F_i\left(\nu_1^{(n)}(t),\nu_2^{(n)}(t)\right)-\bar{u}_i\right],\ i=1,2;\ n=0,1,2,\ldots$.

Assuming $\nu_1^{(0)}(t)=0,\ \nu_2^{(0)}(t)=0$, for the first approximation we have $\nu_i^{(1)}(t)=\gamma_i\left[F_i(0,0)-\bar{u}_i\right]$, $i=1,2$. After carrying out an asymptotic analysis of the corresponding integrals for each approximation, we obtain asymptotic expansions in powers of the small parameter Fo

$$\nu_i(t)=\sum_{j=0}^{N}c_j^{(i)}(t)\cdot Fo^j+O\left(Fo^{N+1}\right),\ i=1,2,\qquad (8)$$

where the coefficients are calculated explicitly.

Substituting the obtained expansions into expression (6), we carry out an asymptotic analysis of the obtained integrals by the Laplace method [7-9], taking into account the conditions of "*proximity*" of the point (x,t) under consideration to the boundaries of the domain Ω'.

$$T^*(x,t)=T_{LI}(x,t)-Fo\cdot\gamma_1\int_0^t\left[\sum_{j=0}^{N}c_j^{(1)}(s)\cdot Fo^j+O\left(Fo^{N+1}\right)\right]G\left(x,t;\psi_1(s),s\right)ds$$
$$+Fo\cdot\gamma_2\int_0^t\left[\sum_{j=0}^{N}c_j^{(2)}(s)\cdot Fo^j+O\left(Fo^{N+1}\right)\right]G\left(x,t;\psi_2(s),s\right)ds\,.$$

The solution to the required boundary value problem is determined by the method of successive approximations according to the scheme

$$T^{(m+1)}(x,t)=T^*(x,t)+\int_0^t\int_{\varphi_1(s)}^{\varphi_2(s)}A\left(T^{(m)}(y,s)\right)G\left(x,t;y,s\right)dyds,\quad T^{(0)}=T^*(x,t),\ m=0,1,2\cdots. \quad (9)$$

After carrying out an asymptotic analysis of the integrals in expressions (9), using the Laplace method with reference integrals [7-9,15-16], we obtain an asymptotic expansion of the solution in scales of the form $\left\{Fo^j\right\},\left\{\left(\left|x-\psi_i(t)\right|/Fo\right)^p\right\},\left\{\left(\left|x-\psi_i(t)\right|/Fo\right)^p Fo^j\right\}$.

Then the following is true.

Statement. The asymptotics of the solution to the boundary value problem (7) - (11) has the following form:

1) In the "*boundary layer*" of the boundary $x=k_0t$, where the condition is met $x-k_0t=O\left(Fo^p\right)$, $p>1$,

$$T_{pg0}(x,t)=\sum_{i=0}^{N}\sum_{j=0}^{\bar N}d_{ij}^{(0)}(x,t)\left(\frac{x-k_0t}{Fo}\right)^j Fo^i+O\left(\left(\frac{x-k_0t}{Fo}\right)^{\bar N+1}Fo^{N+1}\right),\quad Fo\to 0\,.$$

2) In the "*intermediate layer*" of the boundary $x=k_0t$, where $x-k_0t=O\left(Fo\right)$, $Fo\to 0$,

$$T_{pr0}(x,t)=\sum_{i=0}^{M}d_i^{(1)}(x,t)\cdot Fo^i+O\left(Fo^{M+1}\right).$$

3) In the "*area remote from the borders* $x=k_0t$, $x=k_0t+H$ *points*", (if $p<1$), $x-k_0t=O\left(Fo^p\right)$, $k_0t+H-x=O\left(Fo^p\right)$, $Fo\to 0$, $\quad T_{ud}(x,t)\approx T_0+T_A(x,t)$.

4) In the "*intermediate layer*" of the boundary $x=k_0t+H$, where $k_0t+H-x=O\left(Fo\right)$, $Fo\to 0$,

$$T_{prH}(x,t)=\sum_{i=0}^{M}d_i^{(2)}(x,t)\cdot Fo^i+O\left(Fo^{M+1}\right).$$

5) In the "*boundary layer" of the border* $x=k_0t+H$, $k_0t+H-x=O\left(Fo^p\right)$, $p>1$, $Fo\to 0$

$$T_{pgH}(x,t)=\sum_{i=0}^{N}\sum_{j=0}^{\bar N}d_{ij}^{(3)}(x,t)\left(\frac{k_0t+H-x}{Fo}\right)^j Fo^i+O\left(\left(\frac{k_0t+H-x}{Fo}\right)^{\bar N+1}Fo^{N+1}\right).$$

An approximate solution is obtained in the form of asymptotic decompositions in the sense of Poincaré, the coefficients of which are calculated explicitly [8,12-14]. The resulting decompositions satisfy the initial and boundary conditions of the problem posed.

Remark 1. In practice, they are often limited to the first few coefficients of the asymptotic scale $\left(\left(\left|x-\psi_i(t)\right|/Fo\right)^j Fo^i\right)$, $i,j=0,1,\ldots,N$; $i+j\le N$. For example, when simulating a high-current sliding discharge of a capacitor over the surface of a dielectric (fluoroplastic-4), where,

$a^2 = 1{,}03 \cdot 10^{-7}\,\mathrm{m}^2 / cek$, $\bar{t} = 5 \cdot 10^{-9} cek$, $\bar{x} = 10^{-6}\,\mathrm{m}$, parameter Fo takes on a value $5{,}15 \cdot 10^{-4}$, to obtain an approximate solution, it is enough to take $N = 2$.

Remark 2. The approximate solution $T(x,t)$,obtained in the form of an asymptotic expansion in the sense of Poincaré allows differentiation and integration operations, therefore, we can find expressions for the heat flux $T_x(x,t)$ and the rate of restructuring of the temperature field $T_t(x,t)$.

Remark 3. The obtained analytical expressions for the approximate solution of a singularly perturbed boundary value problem allow us to perform a numerical-analytical analysis of the influence of material parameters (plate thickness, thermophysical properties, reflectivity and absorption capacity of the plate material, characteristics of the radiation source) on the dynamics of the temperature field in the plate.

Remark 4. The asymptotic method proposed and substantiated by G A Nesenenko [7,11], is applicable to find the asymptotics of solutions of both one-dimensional and multidimensional singularly perturbed linear and nonlinear boundary value problems of unsteady heat conduction in a domain with a complex (sufficiently smooth) movable geometry, with nonlinear boundary conditions.

4. Conclusion

An approximate solution of the posed singularly perturbed boundary value problem is obtained in the form of asymptotic decompositions in the sense of Poincaré, and the decomposition coefficients are calculated explicitly, which allows a parametric analysis of the problem posed: to reveal the influence of the irradiation regimes, the radiation component and the initial temperature, the absorption capacity of the material on the temperature distribution thermally thin plate.

References

[1] Mezenov A V, Soames L N and Stepanov A I 1986 *Thermooptics of solid-state lasers* (Leningrad: Mashinostroenie)

[2] Alpatiev A N, Danilov A A, Nikolsky N Yu, Prokhorov A N, Tsvetkov V B and Shcherbakov I A 1990 Features of thermal and generation modes of optically dense active media *Proceedings of the Institute of General Physics* **26** 107-7

[3] Prokhorov A M, Konov V I, Ursu I and Mikheilesku I N 1988 *Interaction of laser radiation withmetals* (Moscow: Nauka)

[4] Uglov A A, Smurov I Yu, *et al.* 1991 *Modeling of thermophysical processes of pulsed laser action on metals* (Moscow: Nauka)

[5] Tikhonov A N 1938 *On Volterra functional equations and their applications to some problems of mathematical physics* Moscow State University Ser. A **1(8)** 1-24

[6] Arakelian, S, Kucherik A, *et al.* 2016 Laser-induced synthesis of nanostructured metal–carbon clusters and complexes *Optical and Quantum Electronics* **48(11)** 505

[7] Nesenenko G A 1972 *On the asymptotics of the Green's function of a heat-conduction equation withsmall parameter* Mat. Sb. **87(2)** 204-11

[8] Nesenenko G A and Latypov I I 1992 *The asymptotics of the temperature field of a singularly perturbed boundary-value problem for the heat-conduction equation with nonlinear boundary conditions on the moving boundary* In: Partial Differential Equations. General Theory and Applications (St. Petersburg: Obrazovanie, Len. Gos. Ped. Univ.) 106-16.

[9] Kravchenko V F, Nesenenko G A and Latypov I I 1999 An application of integral equations to asingularly perturbed nonstationary boundary value problem for the heat equation in a domainwith moving boundaries *Differential Equations* **35(9)** 1184-8

[10] Kartashov E M 2001 Analytical methods for solving boundary value problems of nonstationary heat conduction in a region with moving boundaries *Engineering Physics Journal* **74(2)** 171

[11] Nesenenko G A, Pustovoit V I and Kravchenko V F 2006 *Asimptotika Poincaré resheniy zadac neregulyarnogo teplo- imassoperenosa* (Moscow: Fizmatlit) p 419

[12] Latypov I I, Shikirov R A and Ulitin N V 2014 Approximate solution of the problem of findingthe temperature in the active elements of solid-state lasers *Bulletin of Kazan Technological*

University **17(5)** 80-6

[13] Latypov I I 2017 Approximate solution to a singular perturbed boundary value problem ofthermal shielding *J. Phys.: Conf. Ser.* **918** 012005

[14] Latypov I I, Bigaeva L A, Chudinov V V, Gilev A Y and Gaisin F R 2019 Material evaporation with ultrashort laser exposure *IOP Conf. Ser.: Mater. Sci. Eng.* **537** 022068

[15] Whittaker E T and Watson J N 1963 *Course of modern analysis* (Moscow: Nauka) **2**

[16] Gradshtein I S and Ryzhik I M 1963 *Tables of Integrals, Sums, Series and Products* (Moscow: GIFML)

On the issue of determination of the mechanism of ultrasonic velocity change in the process of uniaxial deformation of the rod

S A Yurchenko

LLC TEHOSNASTKA-RTD, 2 Novaya Str., Petropavlovk, 662925, Russia

E-mail: stepan1974.yur@yandex.ru

Abstract. The molecular acoustics method is one of the types of non-destructive testing of materials during the operation of machines and mechanisms. Assessment of the performance of this or that material for further operation is carried out on the basis of changes in the ultrasonic velocity passing through the volume of the material. In this case, this assessment is carried out on the basis of a comparison of experimental data with data on the change in the ultrasonic velocity (UV) in a rod made of the same material subjected to uniaxial deformation. Thus, this method correlates the value of the ultrasonic velocity in the material, after a certain operating time as part of a machine or mechanism, with one or another area of deformation of the rod. They try to explain the changes in the ultrasonic velocity on the basis of the displacement of defects or the rotation of mesoscopic volumes in the material. However, the authors describing in their works the process, use expressions "most likely", "probably", which indicates that their explanations are an assumption, and not a strictly established mechanism for changing the ultrasonic velocity. This work shows that in the process of deformation, together with a change of the ultrasonic velocity, both the natural frequency of atomic vibration and the volume of the primitive cell change. The author proposes to estimate the change in the ultrasonic velocity during deformation on the basis of changes in the geometric parameters of the primitive cell in which ultrasound is transmitted along the edges of the primitive cells through the interaction of electron shells of atoms.

Until now, iron (Fe) and alloys based on it remain the most used materials in industry and construction. The performance control of parts made of Fe and its alloys is carried out using non-destructive testing methods. One of these methods is molecular acoustics. "The main task of molecular acoustics is to study the structure and physical properties of gases, liquids and solids by acoustic methods" [1]. Therefore, on the basis of molecular acoustics, researchers are testing mathematical models in which they try to establish the relationship between the acoustic properties of a material, depending on its crystalline structure, and the changes occurring with it during deformation. To date, a large amount of experimental data has been accumulated on the change in the ultrasonic velocity in various materials during deformation. The main source for assessing the ongoing changes in the volume of the material is the analysis of experimental data on the change in the ultrasonic velocity obtained in the process of uniaxial tension of a rod made of the material under study. In the course of the experiments, the correspondence between the values of the ultrasonic wave velocity for different regions

IOP Conf. Series: Materials Science and Engineering 1155 (2021) 012008 doi:10.1088/1757-899X/1155/1/012008

of deformation is established. Figure 1 shows an experimental graph of the change of UV for low-carbon steel 09G2S deformed at the lower yield point obtained in [2]

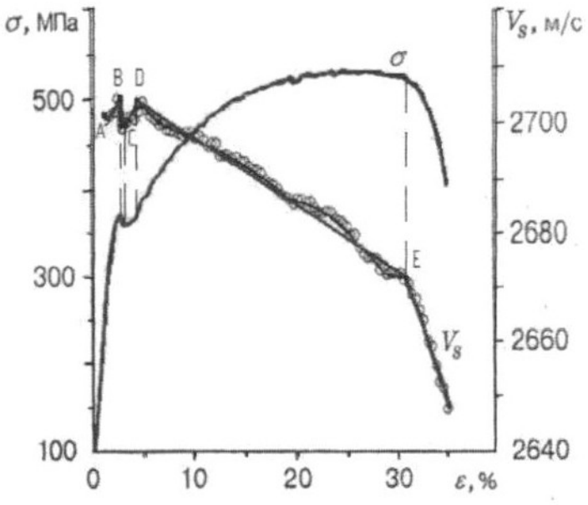

Figure 1. Experimental graph of the UV change during deformation of 09G2S steel at the lowest yield point.

Figure 2 and figure 3 show the experimental graphs of the UV changes for high-alloy steel 40X13 under active loading (strain rate 6.67 * 10^{-5} s^{-1}) of the samples in the state of delivery (figure 2) and after thermal annealing (high tempering $T^0C = 600$) (figure 3) obtained in work [3].

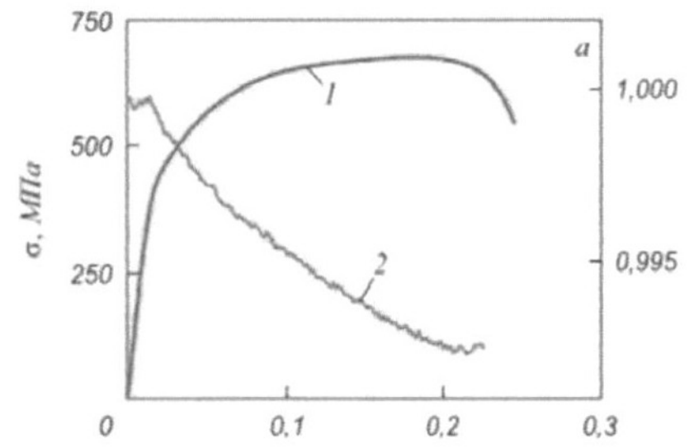

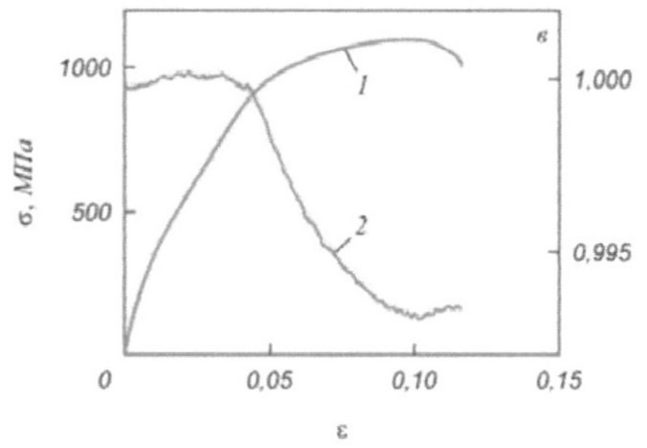

Figure 2. Experimental graph of ultrasonic speed in 40X13 steel under active loading as delivered.

Figure 3. Experimental graph of ultrasonic speed in 40X13 steel under active loading after heat treatment.

The difference between the graph of low-carbon steel and the graphs for high-alloy steel is the presence of a small plastic deformation area and a hardening area. This is due to the conditions of deformation, or rather to the rate of deformation.

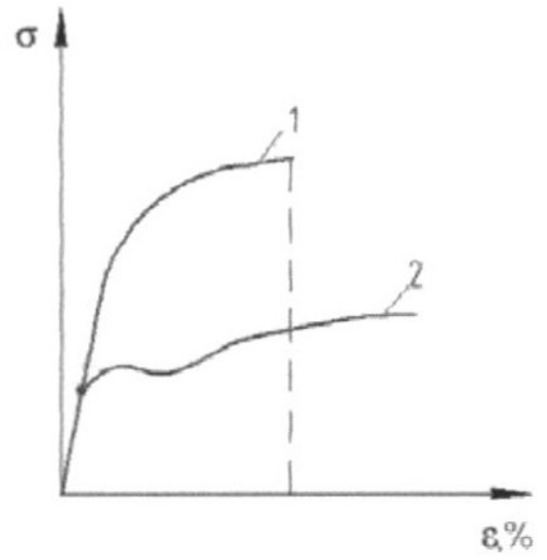

Figure 4. Stresses for steel depending on the deformation rate.

"When a body is loaded above the elastic limit, the magnitude of plastic deformation is determined by the loading rate: the higher the loading rate, the shorter the time of plastic deformation and the greater the stress at which the transition from elastic to plastic deformation occurs" [4]. By changing the loading conditions of low-carbon steel, increasing the strain rate, it is possible to obtain identical stress profiles similar to high-alloy steel. This is confirmed by the theoretical studies of RI Nadeeva [5] (see figure 4) and the experiments of Meijong [6].

A characteristic feature of the graphs shown in figure 1-3 is the behavior of the ultrasonic velocity in the area of plastic deformation, which, according to the authors of [3], is characterized by the fact that "... at the parabolic stage, the velocity decreases in proportion to the total deformation." The authors of [3] offer an explanation of this dependence on the basis of dislocations: "The change in the ultrasonic velocity during deformation is determined by the behavior of mobile dislocations directly involved in the sliding process." A similar explanation is given by the authors of [2]: "... the reason for the decrease in the ultrasonic velocity, apparently, is the accumulation of defects such as dislocations and their ensembles as the total deformation increases." However, the authors of [2, 3] do not show the model of ultrasound transmission in the material, and also do not pay attention to the experimentally established fact that in the course of uniaxial tension of the rod, the natural vibration frequency of atoms changes both in the region of elastic deformation and in areas of plastic deformation. The authors of [7] clearly indicate: "The resonant frequency rises in the elastic and falls in the plastic area". If we take into account the model of Academician Ya.I. Frenkel [8] about the transmission of ultrasound by deformation of the electron shells of atoms, then an increase in the natural frequency ν_{at} of atomic vibrations in the elastic region of deformation will lead to an increase in the propagation velocity of ultrasound in the material. Decrease vat of the natural frequency of atomic vibration, in the plastic region, leads to a decrease in the propagation velocity of ultrasound in the material. That is, on the basis of the model of ultrasound transmission through the deformation of atomic shells, it is possible to explain the process of changing the UV during deformation, for this it is necessary to establish the reasons for the change in the natural frequency of atomic vibration.

In this regard, it is proposed to carry out a theoretical analysis of the change in the natural frequency of vibration of atoms ν_{at} and the UV of the wave υ under uniaxial tension on the basis of a set of known models - the model of the transmission of ultrasonic velocity by deformation of atomic shells, the model of the structure of primitive cells of the theory of O. Brave [9] and the model stress state as "... forced displacement of an atom from a position of stable equilibrium" [10]. It is proposed to focus on the analysis of changes in the parameters of the primitive cell of the crystal lattice in the process of deformation, since "... the process of deformation and destruction of solids cannot be understood without a specific analysis of the real structure of the body and its changes during loading" [11].

Let us carry out a theoretical analysis of the process of changing the UV during uniaxial deformation of a low-alloy steel rod. To do this, we represent the rod as a set of sequentially located one after another "backbones" of unit cells of the Body-Centered Cubic (BCC) crystal lattice, see figure 5.

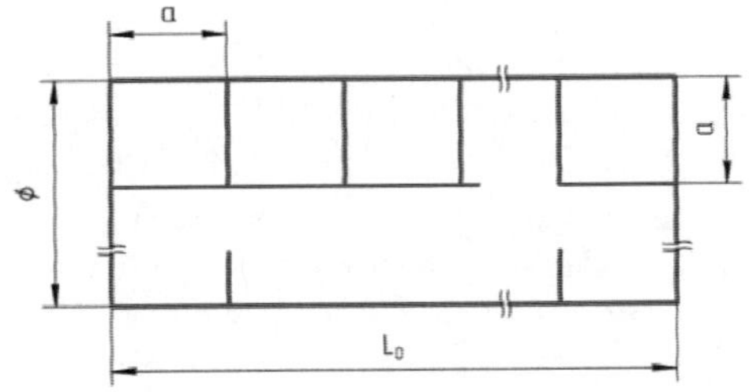

Figure 5. A rod as a sequence of N "backbones" of elementary cells located one after another with cell parameter a.

In this case, the length of rod L_0 can be represented by the following expression

$$L_0 = N * a \tag{1}$$

where a is the length (parameter) of the "backbone" of the unit cell of the BCC crystal lattice of iron. N - the number of "backbones" sequentially located one after another.

The area of elastic deformation of the rod, due to small relative deformations reaching 1-2%, can be represented as a sequential shift of atoms, in atomic planes from which a primitive cell is built [12], in the direction of the applied external force F_{vn} see figure 6

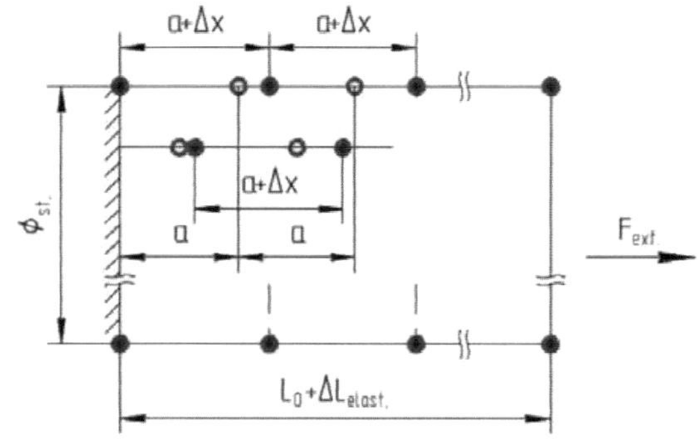

Figure 6. Deformation of the rod in the region of elastic deformation as a shift of atoms in atomic planes.

Then the new length of the rod, in the region of elastic deformation, will be expressed as

$$L_{elast} = L_0 + \Delta L_{elast} = (a + \Delta x) * N \tag{2}$$

and the relative elongation ΔL_{elast} can be expressed from equation (2) by the following equality:

$$\Delta L_{elast} = \Delta x * N \tag{3}$$

To determine the change in the volumes of primitive cells in the area of elastic deformation, we will use the thermal research method. Let us assume that the process of stretching a rod in the region of elastic deformation occurs in such a short time interval τ (c), that, in fact, this process can be attributed to an adiabatic process. Then, based on the adiabatic equation:

$$T * V^{(k-1)} = Const \tag{4}$$

where k>0

it can be found out that with adiabatic expansion of the body (increasing the volume of the rod V_{st}), the body cools, and with adiabatic compression, the body heats up. Joule's researches [13] on uniaxial tension of the elastic rod made of iron found that the rod in the elastic deformation region cools a little. Nadai [14] directly points out that during adiabatic volumetric expansion of a solid body, heat is absorbed. Thus, on the basis of [13, 14], it can be argued that the volume of the rod V_{st} increases. Taking into account the Van der Waals principle, an increase in the volume of the rod V_{st} can be carried out only by increasing all volumes V_{0Cell} "backbones" of unit cells of the BCC crystal lattice.

Let us assume that the increase in the volume of the "backbone" of the unit cell V_{0Cell}, in the elastic region of deformation, is carried out only by increasing the parameter of the interatomic distance a_{BBC} for some distance Δx in atomic planes along the vector of the applied external force F_{ext} while maintaining the geometric dimensions along the Y axis, as well as in the perpendicular plane (along the Z-axis). Figure 7 shows the deformation scheme of the "backbone" of a unit cell in the XY plane.

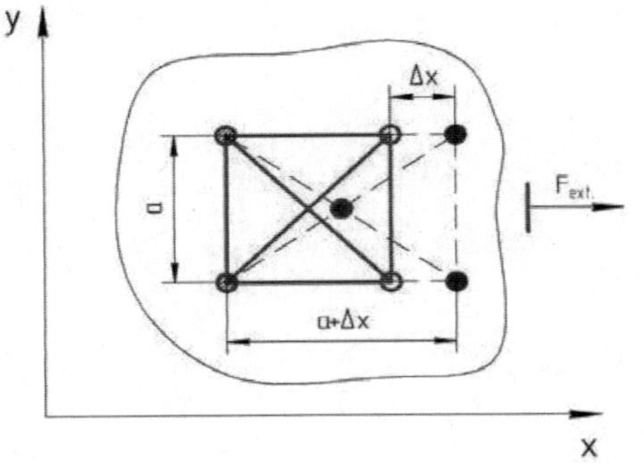

Figure 7. Deformation of the "backbone" of the unit cell in the atomic plane.

An increase in the length of the interatomic interaction (edge of the cell base) in the atomic plane leads not only to a change in the geometry and volume of the primitive cell V_{0Cell}, but also to a change in the density of iron.

Currently, reference books indicate the values of the material density as the ratio of the mass of the Mat $_{at}$ to the volume of the primitive cell V_{0Cell}. So the theoretical value of the density of iron is calculated based on the volume of the primitive cell of the BBC crystal lattice V_{0BBC}

$$p_{BBC}{}^{T} = \frac{M_{at}}{V_{0BBC}} = 9{,}3*10^{-26}кг/11{,}69*10^{-30}м^3 \approx 7955кг/м^3 \tag{5}$$

The theoretical value of density, 7955 kg / m3, differs from the experimental value of density ρ_{exper} = 7870 kg / m3 by only 1%. However, the density of the iron atom

$$p_{at} = \frac{M_{at}}{V_{at}} = \frac{M_{at}}{\frac{4}{3}*\pi*R_{at}^{3}} = 9{,}3*10^{-26}кг/8{,}375*10^{-30}м^3 \approx 11100кг/м^3 \tag{6}$$

Thus, the true density of the physical element - Fe differs from the density of the crystal lattice of iron. However, if we take into account that the atom oscillates around the equilibrium position (crystal lattice node) with equal frequencies ν_x and ν_y along the X and Y axes, see figure 8, then for a certain interval (oscillation period) T_0 see figure 9, " the swept-out "volume V_{zam}, is formed see figure. 10. It is larger than the volume of the atom and only on the basis of the given movement of the atom, it is possible to refer the mass of the atom to the larger, swept-out volume.

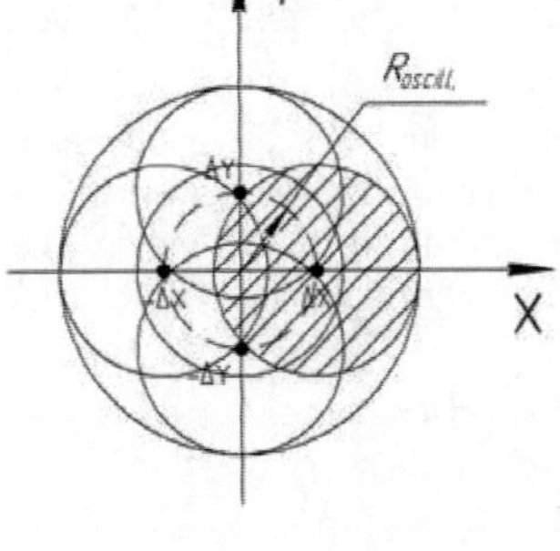

Figure 8. Oscillation of an atom in an atomic plane near the equilibrium position.

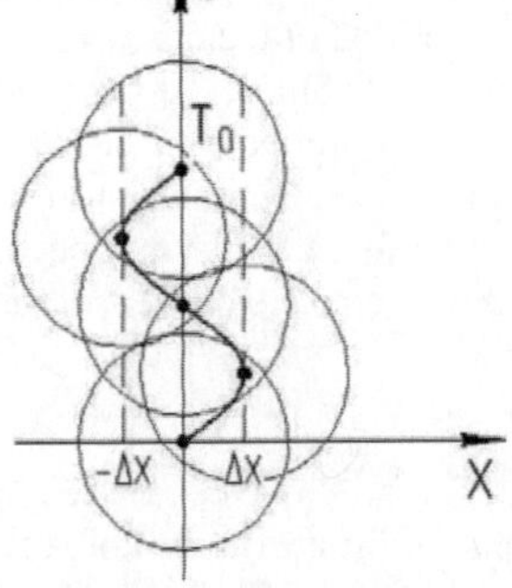

Figure 9. Oscillation of an atom for one period T_0.

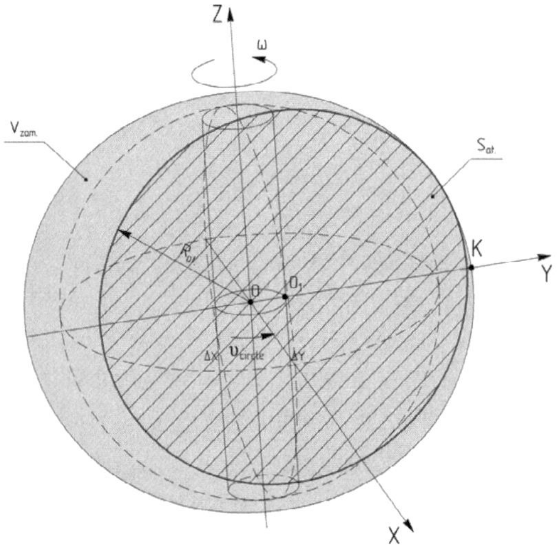

Figure 10. Formation of a "swept-out" volume by V_{zam} atom, in the space of the crystal lattice, in one revolution of the center of mass of the O_1 atom near the crystal lattice node O.

However, then it is necessary to call the density of the given "swept-out" volume V_{zam} for the period T_0, the "time-distributed" density ρ_t.

To determine the changes occurring with the "distributed" density ρt in the process of elastic deformation, we use the fundamental law of conservation of mass

$$m_{at} = Const \tag{7}$$

where the mass of an atom remains unchanged in any primitive cell - deformed, not deformed during a polymorphic transition for any time interval t_i at any frequency of atomic vibration ν_{at} in any volume of a primitive cell V_i.

Based on the foregoing, about the conservation of the mass of an atom in any state of the material, we transform equation (7) as

$$\frac{m_{at}}{V_i * T_i} = \frac{m_{at}}{V_i} * \nu_i = \rho_t * \nu_i = Const \tag{8}$$

where V_i is the volume of the cell

ν_i is the natural frequency of vibration of the atom

T_i is the period of vibration of the atom

ρ_t is "distributed in time" volume density

Suppose that equation (8) reflects the change in the "distributed" density of the material with a change in the primitive volume of the cell, as well as the change in the natural frequency of vibrations of the atom in the process of deformation. Let us write down the condition for the coherence of the initial parameters of the material with the parameters of the material in the region of the elastic deformation

$$\frac{m_{at}}{V_0} * \nu_0 = \frac{m_{at}}{V_1} * \nu_1 = Const \tag{9}$$

where V_0 is the volume of the cell in the unloaded state

ν_0 is the natural frequency of vibration of the atom

V_1 - volume of the cell in the area of elastic deformation

ν_1 is the natural frequency of vibration of the atom in the region of elastic deformation.

We transform equation (9) by reducing the mass of the atom

$$\frac{1}{V_0} * \nu_{0at} = \frac{1}{V_i} * \nu_{iat} = Const\left(\frac{1}{M^3 * c}\right) \tag{10}$$

Based on equation (10), it is possible to theoretically estimate the change in the natural frequency of vibration of the volume of the atom in the process of the elastic deformation. With an increase in the volume of a primitive cell, in the region of the elastic deformation, up to a new value $V_i > V_0$, the natural oscillation frequency of atom $\nu_{iat} > \nu_{0at}$ should increase in order to preserve condition (10). Work [7] is the confirmation of the theoretical conclusions on the growth of the natural frequency of atomic vibration. The authors point out: "In our experiments, an increase in the frequency was always observed during the elastic deformation." Growth of the natural frequency of vibration of atoms ν_{iat} can be explained by the change in the trajectory of the center of mass of the atom from a circle to an ellipse. It is known that when the amplitudes of two mutually perpendicular oscillations are equal, with equal frequencies, the point moves in a circle with a certain radius R_0 (see figure 8). With two mutually perpendicular oscillations of equal frequency but with different amplitudes and phase displacement, the movement of the point occurs along the eleps with the semiaxes A and B, which lie in the directions corresponding to the oscillations, see figure 11.

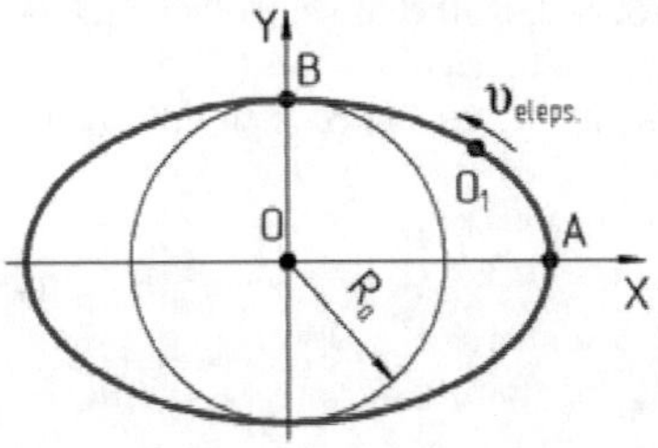

Figure 11. Changing the trajectory of the center of mass of an atom from a circle to an ellipse in the area of the elastic deformation.

Thus, when the interatomic bond is deformed in the direction of the X axis under the action of an external force F_{ext}, the trajectory of the center of the atom тO_1 changes to an ellipse. Taking into account that the period of oscillation of the center of atom T_0 should not change, then the speed of the center of the atom along the ellipse υ_{el} increases in comparison with the same speed of the center of the atom circumferentially. This is due to the increase in the length of the perimeter of the ellipse in comparison with the perimeter of the circle. We find the speed of movement along the ellipse by formula

$$\vartheta_{el} = \frac{2 * \pi * \sqrt{\frac{B^2 + A^2}{2}}}{T_0} \tag{11}$$

The length of semiaxis A is assumed to be equal to half of the length of the crystal lattice parameter in the elastic deformation region, taking into account the increase in the lattice parameter by 1%

$$A = \frac{a * 1{,}01}{2} = \frac{2 * R_0 * 1{,}01}{2} = 1{,}01 * R_0 \tag{12}$$

Knowing the turnover rate υ_{el} of the center of atom point O_1, it is possible to determine, with some assumption, the new natural frequency of atom ν_{1at} oscillation at point B relative to the node of the crystal lattice point O, assuming that point B is located on a circle with radius R_0

$$\nu_1 = \frac{\vartheta_э}{2 * \pi * R_0} = \frac{\sqrt{\frac{B^2 + A^2}{2}}}{R_0 * T_0} = \frac{\sqrt{\frac{R_0^2 + (1{,}01 * R_0)^2}{2}}}{R_0 * T_0} = \frac{R_0 * \sqrt{1{,}01}}{R_0 * T_0} = \frac{1{,}005}{T_0} \tag{13}$$

Thus, in the region of the elastic deformation, the frequency of natural oscillations of the atom increases, which is explained by a change in the movement trajectory of the center of the atom relative to the node of the crystal lattice and the preservation of the turnover time (period T_0) of the atom around the node of the crystal lattice.

An increase in the natural oscillation frequency of atom ν_{at} and an increase in the length of the crystal lattice parameter (the edge of the primitive cell) should lead to an increase in the velocity of the ultrasonic wave in the material. The value of the ultrasonic velocity in the elastic deformation region can be calculated by equation

$$\vartheta_{def} = l_i \cdot \nu_i \cdot \frac{1}{n} \ (\text{м/с}) \tag{14}$$

where l_i is the length of the interatomic bond (the edge of the primitive cell) expressed in terms of the crystal lattice parameter - the length of the interatomic bond in the atomic plane.

vi is the oscillation frequency of atoms in the volume of the primitive cell.

n - the number of interatomic bonds used for the transfer of ultrasonic energy in the plane and between adjacent parallel atomic planes forming the primitive and "backbone" of the elementary cell of the crystal lattice of the material.

Equation (14) was obtained in work [15] and allows only on the basis of the physical parameters inherent in the atom and the primitive cell of the crystal lattice to calculate the longitudinal, transverse and ultrasonic velocity in the rod. On the basis of this equation, in work [12], it was possible to refine the spatial design of primitive cells of BCC and Face-Centered Cubic (FCC) crystal lattices and introduce an additional requirement to the existing three requirements for choosing a primitive cell, on the passage of an ultrasonic wave along the interatomic bonds actually existing in space - the edges of the cell.

All changes - natural frequency, UV, volume of the primitive cell can be associated with the condition of continuity of the medium

$$\frac{m_{at}}{V_i} * \nu_i = Const = \rho_i * \nu_i \ \left(\frac{1}{м^3 * c}\right) \tag{15}$$

where ρ_i is the "time-distributed" density.

V_i is the volume of the primitive cell at some point in time

ν_i is the vibration frequency of atoms in the volume of the primitive cell at the same time which remains until the destruction of the material. A joint analysis of equations (9), (10) and (15) can be carried out on the basis of the nomogram, see figure 12

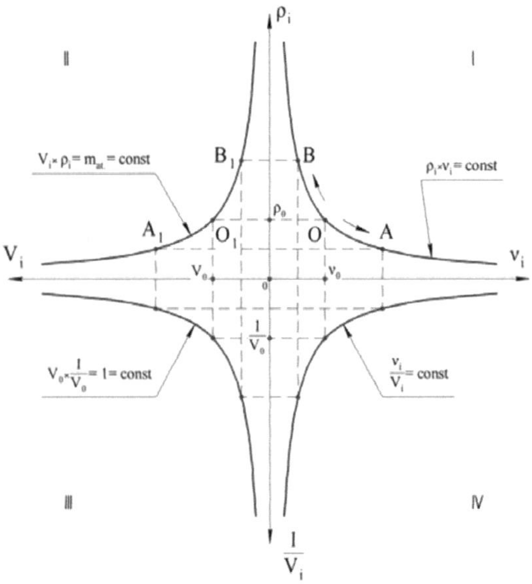

Figure 12. Nomogram of changes in the distributed density, primitive volume, and natural vibration frequency of atoms under loading.

In each of the quadrants of the Cartesian coordinate system, a constant hyperbolic function is presented. Point O_1 in the second quadrant corresponds to the initial conditions in the material before loading. An increase in the volume of the primitive cell (point A_1) under the action of external forces in the area of the elastic deformation corresponds in the first quadrant (point A) to an increase in the natural oscillation frequency of the atom, which corresponds to [7], and a decrease in the values of the "distributed" density.

Then, taking into account the fact that in the region of the plastic deformation [7], the natural oscillation frequency of the atom decreases, one can identify point B in the first quadrant as a region of the plastic deformation. As a consequence, point B_1 (display of point B) in the second quadrant indicates a decrease in the primitive volume of the cell in comparison with the initial parameter. A decrease in the primitive cell volume V_{0Cell} is possible with the all-round compression of the body or with the rearrangement of atoms in atomic planes under the uniaxial deformation, see figure 13.

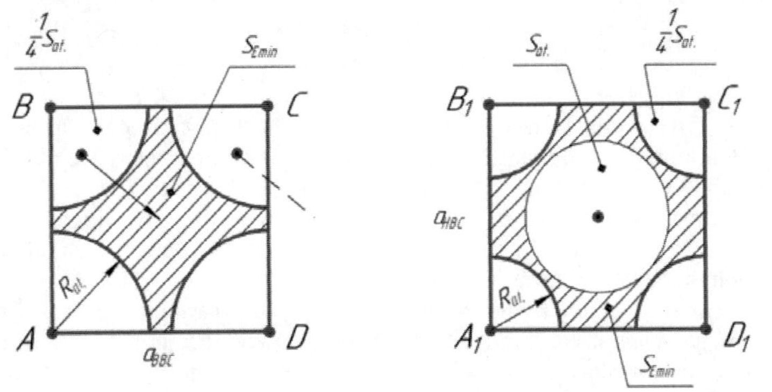

Figure 13. Rearrangement of atoms in the atomic plane into a region with low potential energy during deformation.

The process of the rearrangement of atoms, in atomic planes, corresponds to the mechanism of occupying space with a minimum potential energy [10], ie, the movement of atoms into the S_{Emin} region, see figure 13. This changes the parameter of the primitive cell. In accordance with work [15], the diagonal of the resulting crystal lattice must meet the requirement

$$a * \sqrt{2} \geq 4 * R_{at} \tag{16}$$

which results in value $a \geq 3{,}564 * 10^{-10} м$. The closest actually existing value of the "backbone" parameter of the elementary crystal lattice of iron, satisfying condition (16), is the parameter of the FCC crystal lattice. Figure 14 shows the "backbones" of BCC and FCC cells consisting of primitive cells given from work [12].

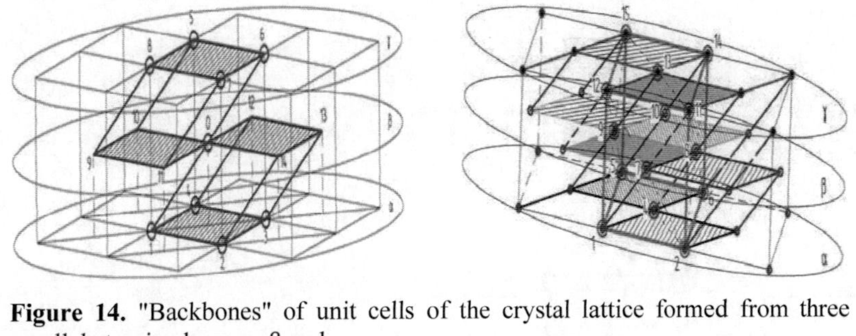

Figure 14. "Backbones" of unit cells of the crystal lattice formed from three parallel atomic planes α, β and γ.

The existence of FCC crystal lattice, that is, the transformation of primitive cells from αFe в γFe, is confirmed by [17, 18]. "At pressures above 105 kg / cm2, a polymorphic transition was found in iron by measuring the compressibility at the shock front."» [18]. A theoretical analysis of the change in the natural frequency of atomic vibration vat during the transformation of primitive cells from αFe to γFe was carried out in [16], in which the dependence of vat on the coordination number K was established.

$$\frac{V_{0BBC}}{K_{BBC} \cdot \nu_{BBC}} = \frac{V_{0HBC}}{K_{HBC} \cdot \nu_{HBC}} \tag{17}$$

where $V_{0\ BBC}$ is the volume of the primitive cell of the BCCcrystal lattice

K_{BBC} - coordination number of the BCC cell

ν_{BBC} is the oscillation frequency of atoms in a primitive BCC cell
$V_{0\,HBC}$ - the volume of the primitive cell of the FCCcrystal lattice
K_{HBC} - coordination number of the FCC cell
ν_{HBC} is the oscillation frequency of atoms in a primitive FCC cell.

It was found from equation (17) that the oscillation frequency of the atom in a primitive cell of the FCC crystal lattice decreases in comparison with the natural oscillation frequency of atoms in a primitive cell of the BCC crystal lattice

$$\nu_{HBC} \approx \frac{2}{3} * \nu_{BBC} \tag{18}$$

Thus, in the process of the theoretical analysis, it was found that the reasons for the change in the natural vibration frequency of atom ν_{at} are the change in the position of the atom in the space of the atomic plane both in the elastic region of deformation and in the plastic region of deformation. A change in the position of an atom in the atomic plane changes the spatial structure of the primitive cell, which leads not only to a change in the volume of the primitive cell, but also to a change in coordination number K characterizing the spatial configuration of the arrangement of atoms.

To clarify the characteristics (graph) of the change in the UV in the plastic region of deformation, let us consider, from a theoretical point of view, the process of polymorphic transformation of cells from αFe to γFe for a certain time interval Δτ during uniaxial deformation of the rod in the plastic region. Let the rod, in the region of elastic deformation, consist of N successively deformed BCC cells and at time τ_0 the first polymorphic transformation (plastic deformation begins) of the vertical plane occurs, see figure 15.

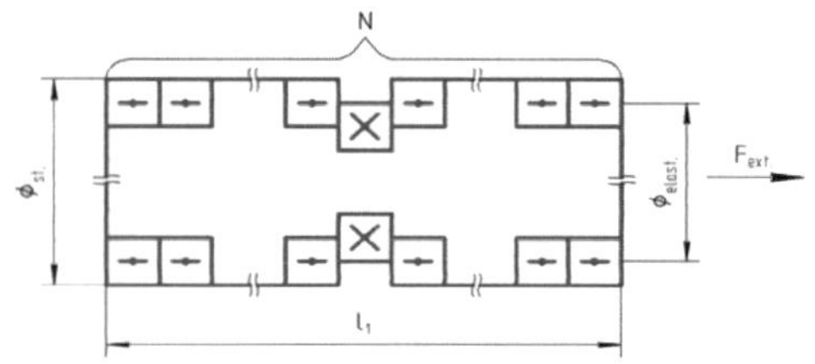

Figure 15. Timepoint τ_0.

For a certain period of time $\Delta\tau = \tau_1 - \tau_0$, a rearrangement of P vertical planes will occur, see figure 16.

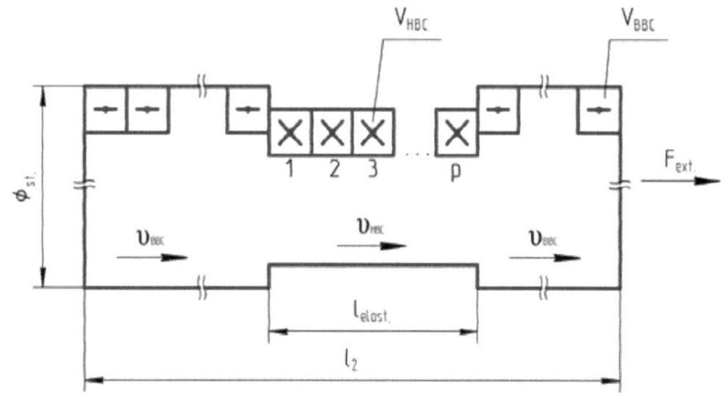

Figure 16. Time moment τ_1.

Most likely, the rearrangement does not occur sequentially, as shown in figure 16, but on defects of the crystal lattice, where the potential energy is minimal; however, to simplify the theoretical analysis, we will consider the rearrangement of cells occurring one after another, sequentially.

The UV of the wave in the rod in the region of the elastic deformation can be presented as

$$\vartheta_{st(def)} = \frac{L_{st}}{\sum \tau} = \frac{N * a^{1}_{BBC}}{N * \tau^{1}_{BBC}} \tag{19}$$

where L_{st} is the length of the bar in the area of the elastic deformation

τ^1_{BBC} is the time of passage of an ultrasonic wave of one length of the edge of the primitive cell in the region of the elastic deformation

a^1_{BBC} is the length of the edge of the primitive cell in the area of the elastic deformation.

The length of the rod at the polymorphic transition from αFe to γFe can be expressed by the following sum

$$L^1_{st(plast)} = (N-P)*a^1_{BBC} + P*a_{HBC} \tag{20}$$

where N is the initial number of cells of the FCC crystal lattices forming the length of the rod in the region of the elastic deformation.

P is the number of cells of fcc crystal lattices formed in the rod during polymorphic rearrangement in the region of the plastic deformation.

a^1_{BBC} is the length of the edge of the BCC crystal lattice in the elastic deformation region.

N a_{HBC} is the length of the edge of the FCC crystal lattice

Then the total time of passage of the ultrasound by the wave of the material through different structural units, different crystal lattices, can be expressed as

$$\sum\tau^1 = (N-P)*\tau^1_{BBC} + P*\tau_{HBC} \tag{21}$$

τ_{BBC} is transit time of 1 cell of the BCC crystal lattice

τ_{HBC} is the transit time of 1 FCC cell of the crystal lattice.

Accordingly, the UV of the wave in the rod in the region of plastic deformation will be expressed as

$$\vartheta_{st(plast)} = \frac{L^1_{st}}{\sum\tau^1} = \frac{(N-P)*a^1_{BBC} + P*a_{HBC}}{(N-P)*\tau^1_{BBC} + P*\tau_{HBC}} \tag{22}$$

Let us estimate the change of the UV according to equation (22) at two extreme points P = 0 and P = N.

$$\vartheta_{st(P=0)} = \frac{(N-0)*a^1_{BBC} + 0*a_{HBC}}{(N-0)*\tau^1_{BBC} + 0*\tau_{HBC}} = \frac{a^1_{BBC}}{\tau^1_{BBC}} = \vartheta^1_{BBC} \tag{23}$$

The UV of the wave in the rod will correspond to the speed of sound in the deformed bcc crystal lattice ($\upsilon_{Elast.def.BBC}$). At P = N, the ultrasonic velocity will be equal to the UV in the FCC crystal lattice.

$$\vartheta_{st(P=N)} = \frac{0*a^1_{BBC} + N*a_{HBC}}{0*\tau^1_{BBC} + N*\tau_{HBC}} = \frac{a_{HBC}}{\tau_{HBC}} = \vartheta_{HBC} \tag{24}$$

Based on the two extreme points, taking into account the uniform gradual transition of primitive cells from αFe to γFe,, we construct a theoretical graph of the change of the ultrasound in the rod, see figure 17.

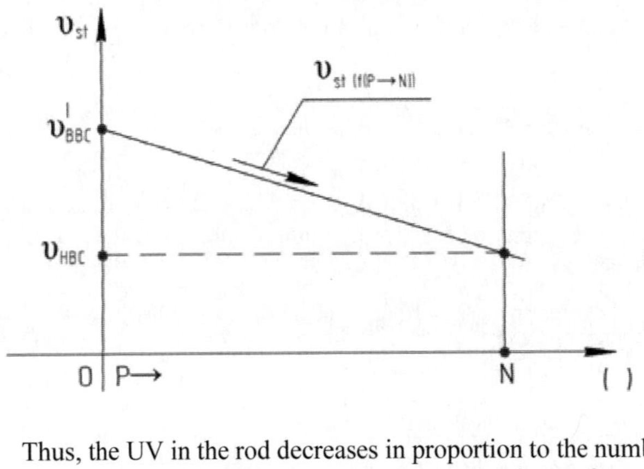

Figure 17. Theoretical change of the UV during polymorphic rearrangement of primitive cells from αFe в γFe.

Thus, the UV in the rod decreases in proportion to the number of changing crystal lattices of cells from BCC to FCC. Accordingly, knowing the mechanisms of speed change in the area of elastic and plastic deformation, it is possible to construct a theoretical graph of the changeof the UV in the rod, see figure 18

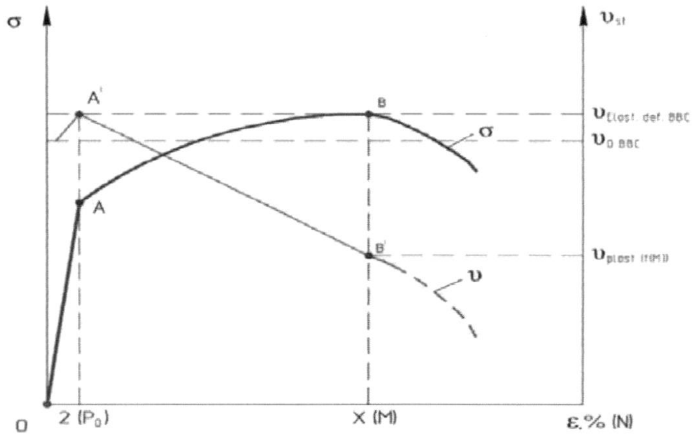

Figure 18. Theoretical graph of the change in the ultrasonic velocity in the rod during active loading.

The region of elastic deformation OA is an increaseof the UV from value υ_0 to the velocity value in the elastic region $\upsilon_{Elast.defBBC}$. The area of the plastic deformation AB is the drop of the UV. Point A1 corresponds to P = 0, and point B1 corresponds to some intermediate value P = M of polymer transitions of crystal lattices from αFe to γFe. In the region of plastic deformation (parabolic stage of stress), the UV decreases in proportion to the relative elongation, i.e. proportional to the growth in the number of changing crystal lattices BCC-FCC (P-M) to a certain value of the UV $\upsilon_{plast\ (f(M))}$.

Thus, the theoretical graph of the change of the UV in the rod made of iron, with the uniaxial deformation, has the same profile with the graphs shown in figure 2 and figure 3. This can be explained by the fact that the substitution of iron atoms for chromium atoms in the crystal lattice of 40X13steel does not change the type of the crystal lattice, since both materials have the same BCC crystal lattice, with practically the same value of the crystal lattice parameter $a_{Cr} = 2{,}855 * 10^{-10} M$, $a_{Fe} = 2{,}866 * 10^{-10} M$. The radius of the chromium atom $R_{Cr} = 1.3 * 10^{-10}$m is approximately equal to the radius of the iron atom $R_{Fe} = 1.26 * 10^{-10}$m (the difference is 3%). The valence of chromium and iron is the same. Also, the values of the ultrasonic velocity in the rod almost coincide $\upsilon_{Cr} = 5140$ m / s and $\upsilon_{Fe} = 5170$ m / s. Therefore, the deformation of the crystal lattice in the elastic deformation region and the polymorphic transformation from αFe to γFe with chromium atoms in the plastic region should be identical, which is confirmed by the coincidence of the theoretical graph of the ultrasonic velocity change with the experimental ones.

A theoretical analysis of the change of the wave ultrasonic velocity, based on the model of ultrasound transmission by deformation of the shells of atoms, along the edges of primitive cells of the crystal lattice of iron and their changes, during uniaxial deformation of the rod, shows that all changes of the UV are associated with the change in the natural oscillation frequency of atom ν_{at} which changes during the deformation of the interatomic bond, which is associated with the change in the trajectory of the center of mass of the atom near the node of the crystal lattice. In turn, the change in the motion of the center of mass of the atom is associated with the condition for the formation of "distributed" density ρ_t of the material, that is, the condition of continuity of the medium, which is expressed by the equation for the region of elastic deformation

$$\rho_0 * v_0 = \rho_1 * v_1 = Const = \frac{\rho_{it}}{\tau_i} \qquad (25)$$

Equation (17), which reflects the process of changing the "distributed" density during the polymorphic transformation of primitive cells (plastic flow regions), can be transformed into a similar equation, if the following mathematical transformations are carried out

$$\frac{\rho_{BBC} * V_{0BBC}}{\rho_{BBC} * K_{BBC} * v_{BBC}} = \frac{\rho_{HBC} * V_{0HBC}}{\rho_{HBC} * K_{HBC} * v_{HBC}} \qquad (26)$$

$$\frac{M_{at}}{\rho_{BBC} * K_{BBC} * v_{BBC}} = \frac{M_{at}}{\rho_{HBC} * K_{HBC} * v_{HBC}} \qquad (27)$$

with a reduction in the mass of the atom in both psrts of the equality, it is possible to obtain the final equality of the distributed densities for primitive cells of the BCC and FCC crystal lattices

$$\rho_{BBC} * v_{BBC} * K_{BBC} = \rho_{HBC} * v_{HBC} * K_{HBC} = Const \qquad (28)$$

Equation (28) differs from equation (25) only by taking into account the coordination number K. The coordination number K indicates that the formation of the "distributed" density of the BCC primitive cell ρt_{BBC} and the FCC primitive cell ρt_{HBC} involves the near neighboring atoms. Figure 19 shows a diagram of the formation of a "distributed" density of a primitive cell of a material by neighboring atoms.

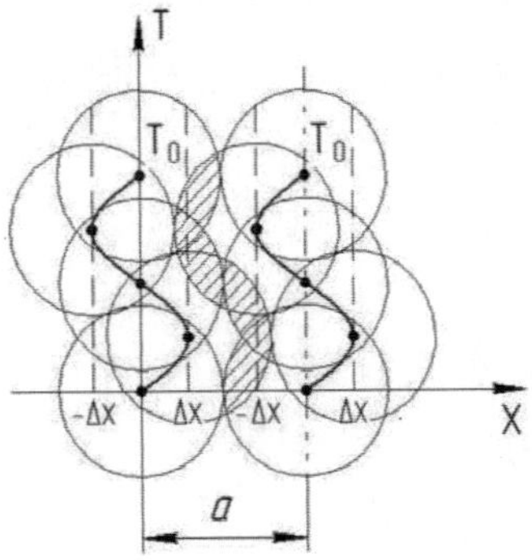

Figure 19. Overlapping of the same volume by neighboring atoms during oscillation period T_0.

by overlapping the same volume of the primitive cell for one period of oscillation T_o but at different points in time of this period.

Equation (25), reflecting the change in the "distributed" density in the elastic deformation region, should be supplemented with the coordination number K_{BBC}, which reflects the participation of neighboring atoms in the formation of the volume of the primitive cell of the BCC crystal lattice

$$K_{BBC} * \rho_0 * v_0 = K_{BBC} * \rho_1 * v_1 = Const \qquad (29)$$

Thus, all changes in the ultrasonic wave velocity during uniaxial deformation of the rod are explained on the basis of the continuity condition of the medium and are reflected in the nomogram in figure 12. Defects in the crystal lattice (such as vacancies) do not transfer the energy of the ultrasound wave. However, vacancies are regions with low potential energy and it is easier for atoms to displace precisely to this region upon deformation under the action ofthe external force; therefore, the polymorphic transition is most likely to begin precisely on the voids (vacancies) of the crystal lattice. It should be noted that the theoretical graph of the change of the UV shown in figure 18 will depend on the deformation rate of the rod. In this case it coincides with the active loading of the steel (the deformation rate is 6.67 * 10^{-5} $_s{}^{-1}$). The process of the material deformation as a displacement of atoms in atomic planes, depending on the loading rate, will be considered in a separate work.

References

[1] Mikhailov I G, Soloviev V A and Syrnikov Yu L 1964 *"Fundamentals of molecular actiics"* (Moscow: Publishing house "SCIENCE")
[2] Semukhin B S, Zuev L B and Bushmeleva K I 2000 "Ultrasound Velocity in Low-Carbon Steel Deformed at the Lower Yield Strength" *Applied Mechanics and Technical Physics* **41(3)** 197-201
[3] Barannikova S A, Bochkareva A V, Shlyakhova G V and Zuev L B 2016 "Change in the speed of ultrasound during plastic deformation of high-chromium steel" *From Izvestiya of higher educational institutions Ferrous metallurgy* **59(8)**
[4] Ionov V I and Ogibalov P M 1953 *"Stresses in bodies under impulsive loading"* (Moscow: "High School")
[5] Nadeeva R I 1953 "On the determination of the dynamic relationship between stresses and deformations" *Vestnik MGU* **10**
[6] Manjoyne M 1976 *Complex stress state and destruction* **3** (Moscow: Mir)
[7] Lebedev A B, Kustov S B, Kardashev B K 1992 *"On internal friction and the Young's modulus defect in the process of deformation of crystals"* (Zh-l: Physics of a solid) **34(9)**
[8] Frenkel Ya I 1972 *Introduction to the theory of metals* (L .: Publishing house "Science") p 424

[9] Bravais A 1849 Note sur les polyèdres simétriques de la géometrie *"Journal de mathématiques pures et appliquées"*
[10] Gubkin S I 1947 "Theory of metal forming by pressure" *State scientific and technical publishing house of literature on ferrous and non-ferrous metallurgy* (Moscow)
[11] Indenbom V L and Orlov A N 1962 "Physical theory of plasticity and strength" Zh-l *"Advances in physical sciences"* **LXXVI(3)** 557-90
[12] Yurchenko S A 2020 Construction of primitive cells of the cubic syngony on the basis of the molecular acoustics method *IOP Conf. Ser.: Mater. Sci. Eng.* **1100** 012043
[13] Joule J P *Proc.Rog.Soc.* **8** 564-9; *Phil.Mag.* **15(4)** 538-44
[14] Nadai A Plasticity and Fracture of Solids **2**
[15] Yurchenko S A 2018 Refinement of the diatomic model of the interatomic interactions of Frenkel for the improvment of methods of non-destructive quality control *PERIODICO TCHE QUIMICA* **15(1)**
[16] Yurchenko S A 2020 «Influence of the coordination number and volume of a primitive cell on the vibration frequency of an iron atom during a polymorphic transition» *Journal of Physics: Conference Series* **1679**
[17] Ogibailov M P and Kiiko I A 1966 *"Essays on the mechanics of high parameters"* From Moscow University
[18] Bancroft D, Peterson E and Minshall S 1956 *Journ Appl.Phys.* **27(3)** 1956

IOP Conf. Series: Materials Science and Engineering 1155 (2021) 012009 doi:10.1088/1757-899X/1155/1/012009

Calculation of the main parameters of the press extruder

G G Klasner[1], **V F Kremyansky and A M Spiridonov**

Kuban State Agrarian University named after I. T. Trubilin, Kalinina st. 13, Krasnodar, 350044, Russian Federation

[1]E-mail: egor.klasner.91@mail.ru

Abstract. In this work the issues of extrusion of leguminous crops are considered, information on modeling of process processes of press-extruders is given, analysis of the modern market of press-extruders is carried out and patent search of existing inventions in the field of extradited production is performed. A press extruder for processing leguminous crops has been developed and detailed information about the developed press extruder is presented. The ratio of the extruder power to the number of revolutions per minute was calculated.

1. Introduction

One of the main factors of rapid growth and health of farm animals is their nutrition. Progress in animal husbandry also depends on progress in the production of high-protein feed using environmentally friendly technologies.

Cost-effectiveness of production processes is the key to a successful economy. The use of special preparation of raw materials for further processing in specialized machines allows not only to achieve high quality of the final product. It also allows you to spend less energy and resources on production. Press extruders, which are designed on the basis of energy-intensive technologies that take into account not only the physical and mechanical, but also the rheological properties of high-protein feed, are very well suited for these purposes.

It is also worth noting that the efficiency of press extruders for the preparation of high-protein feed directly depends on their competent design.

2. Materials and methods

Description of the extrusion process. Extrusion is a common and effective method of heat treatment of feed components, the main task of which is deep gelatinization of starch, while the starch macromolecules are decomposed to form various dextrins and sugars, because of which the digestibility of feed is significantly increased, and the assimilation of nutrients occurs with less energy costs.

Extrusion is ideal for producing high-protein feed, as the process increases the nutritional value of the processed material by an order of magnitude. In addition, extrusion makes it possible to use by-products of feed production, which increases the yield of finished products and contributes to more resource-intensive agricultural production. The extrudate is almost completely decontaminated and meets the strict environmental requirements typical of modern feed production.

The design of the press extruders for the preparation of high-protein feed (table 1) determines the technological process. At the moment, the use of press extruders almost completely negates the negative factors that occur during the heat treatment of feed: the dry extrusion process takes less than 30 seconds, while the raw material is sterilized and decontaminated (pathogens, fungi, mold are completely destroyed); increases in volume, increases the digestibility of feed. It is tomogenized (the processes of grinding and mixing of raw materials in the extruder barrel continue, the product becomes completely homogeneous); it is stabilized (the action of enzymes that cause the product to burn out, such as lipase and lipoxygenase, is neutralized, anti-nutrients are inactivated.

Table 1. technical characteristics of extruders.

Full name	Capacity, kg/h
ETR extruders (LLC "Stimul", Kirov)	
Co brand	
ETR	Up to 100
ETR	Up to 200
ETR	Up to 350
Etrs	Up to 500
Etrs	Up to 600
ETR 700/55 Turbo	Up to 800
KFSO brand	
ETR	Up to 150
ETR	Up to 300
Etrs	Up to 400
Etrs	Up to 500
ETR 700/55 Turbo	Up to 700
PE series extruders (JSC "Zhasco", Volgograd)	
1100, 1100U, 1100S	-
900, 900U, 900S	-
750, 750U, 750S	-
370U	-
300U	-

Modeling of technological processes of press extruders.

A diagram describes a mathematical model of extruders for the preparation of high-protein feed (figure 1). The Diagram provides a visual representation of the dynamic state of the system, and what is especially important – it does not affect the differentiation of the internal time characteristic of the system.

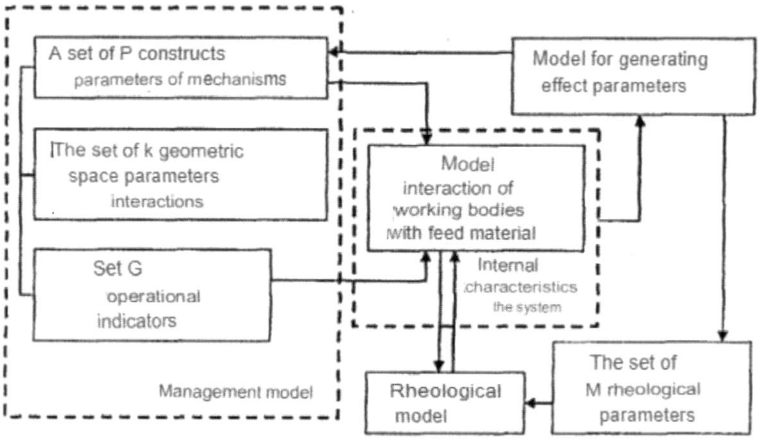

Figure 1. structure of the mathematical model of the feed preparation process using a press extruder.

Modeling of the processes of press extruders for the preparation of high-protein feed is possible using the pressing scheme, which is shown in figure 2.

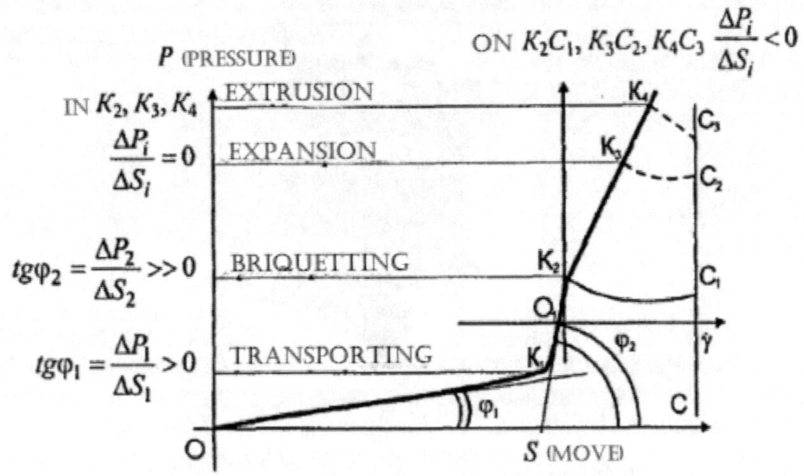

Figure 2. pressing Scheme for the extruder.

The basis of process modeling is the structure-forming approach, which characterizes the system of interaction between external and internal factors. This makes it possible to find optimal values for the effect parameters that demonstrate a symbiosis of output results.

Scientific studies of the working process of technical means for the preparation of high-protein feed, conducted by A. A. Artyushin, I. Z. Barfakov, V. G. Gopka, B. I. Vagin, G. M. Kukta, L. M. Kutsyn, S. M. Dotsenko, V. Yu.Frolov, A.V. Burmaga and others, have become decisive in the development and improvement of existing feed preparation equipment. The lack of universal equipment for the preparation of feed in the conditions of medium and small livestock farms was noted.

Modern market of press extruders.

Figure 3. KMZ-2U Press extruder.

Figure 4. Press-extruder of ETR «Stimul».

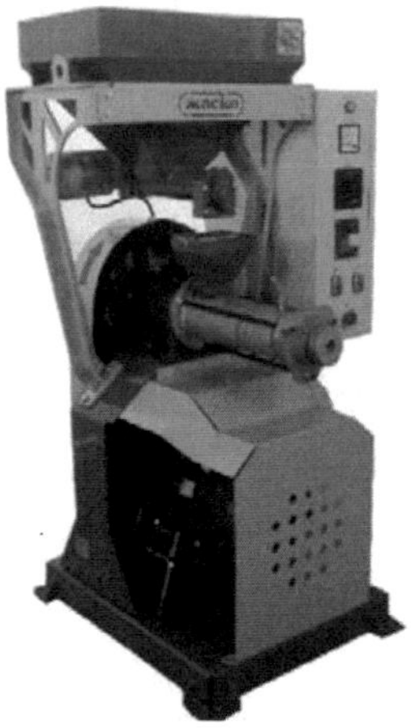

Figure 5. Zhasco PE-110 Press extruder.

Designed for the production of high-performance extruded feed from grain and straw. Universal feed extruders ETR of the KFSO brand are designed for the production of prebiotic feed from grain, grain mixture, including rye, with the possibility of adding 30-50% hay, straw, oats, bran, husk, meat and bone meal, cake and meal. Complete with extruders of the KFSO brand is an annular die, a strand separator, a hopper with a stirrer. By changing the geometry of the augers and the auger part itself: productivity increases by up to 50%; energy is saved by about 25-30%; the screw part itself works longer; the spare parts for the extruder are cheaper.

Designed for the preparation of complete feed: from grain crops (wheat, barley, corn, etc.); of the various mixtures of grains, including from a mixture of grains and straw (reeds); waste from the processing of meat, poultry and fish in mixture with vegetable filling; from the humid stale grains, having the smell of ammonia. The PE-100 extruder is equipped with: a screw feed of raw materials with a frequency controller; a magnetic catcher; a one-piece centered shaft; vibration supports.

Technical characteristics: capacity 80-130 kg / h; installed power 11.37 kW; extrusion process temperature 115-170 °C; overall dimensions 850x850x1500 mm; weight 250 kg.

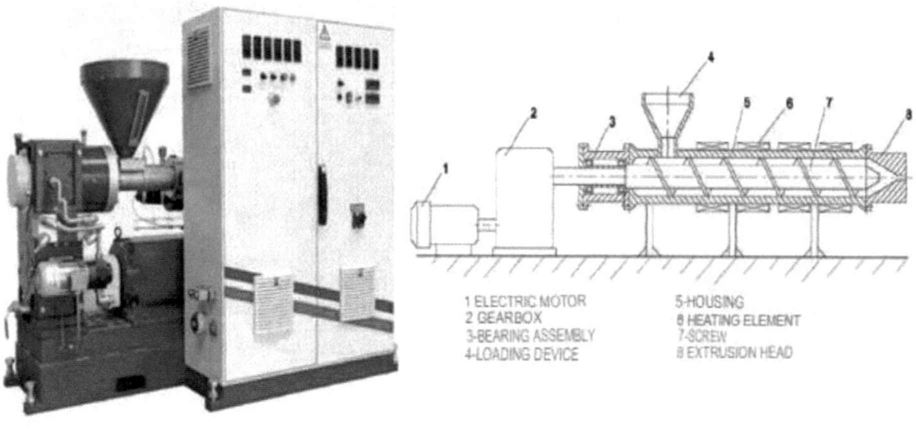

Figure 6. press extruder of ENCE GmbH: a – press extruder, b-diagram of the press extruder device.

IOP Conf. Series: Materials Science and Engineering 1155 (2021) 012009 doi:10.1088/1757-899X/1155/1/012009

The working part of machines of this type includes a set of screws, between which retaining washers are sequentially located. Augers and washers are fixed on the stud by means of a bolt having a conical head. On the inner surface of the casing of the working part, longitudinal channels are made so that the raw material does not rotate as it moves. At the outlet, the casing is equipped with a cone nut with a hole. Manipulation of the gap between the cones of the bolt and nut at the outlet, or the diameter of the outlet hole, allows you to adjust the temperature of the processed material.

In the food industry, extruders equipped with full-gearing screws that perform rotational movements in one direction are widely used. In such a design, the tops of one auger go into the depressions of the other.

Twin-screw self-cleaning extrusion machines are characterized by faster start-up of the screw and work at increased speeds. In such units, the pressure rises less often, because the product does not accumulate. In single-screw units, because material can accumulate in the coils, pressure rises occur more often. The result is an uneven output of products from the extruder.

The extruder is widely used for processing soybeans, barley, peas, wheat, corn, as well as for various mixtures of legumes and cereals. The principle of operation of the extruder consists in squeezing the processed product into special spinners arranged in the extruder barrel, while the processed mixture is subjected to high temperature (100-150 degrees) and high pressure (over 50 atmospheres).

Because of the extrusion process, the grain, due to the large pressure drop at the outlet of the extruder, increases many times in volume and increases the energy value of the product.

Technical characteristics: productivity, with a feedstock density of 0.65 kg / l is 250-400 kg / h; installed power 55 kW; extrusion process temperature 110-135 °C; overall dimensions 1630x1500x1500 mm; weight 1065 kg; specific energy consumption per 1 kg of product 0.085-0.22 kW.

3. Discussions and results

Let's turn to the patent search for press extruder designs. A device for treatment of feed containing the boot is equipped with a hollow housing in which is the auger having a stepped shaft and set it on helical coils with different inner diameter, is wound around the respective level of the shaft and the coil ends facing unloading the neck, mounted on a shaft with the possibility of reciprocating it in the axial direction (and.with. No. 1123626, A23N 17/00, 1981).

The disadvantages of the existing device are the high-energy intensity of the process, the heterogeneity of particles in the material composition of the resulting compound feed.

The essence of the invention is explained by drawings, where: in figure 7 General view of the press extruder is shown, in figure 8 fragment of the inner side of the housing on which the longitudinal ribs are located is shown.

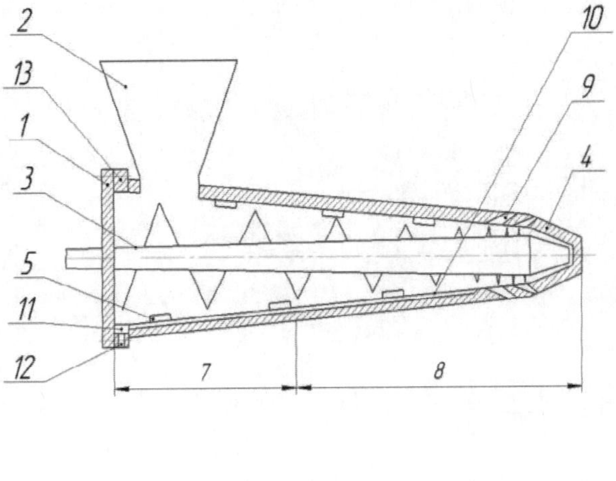

Figure 7. General view.

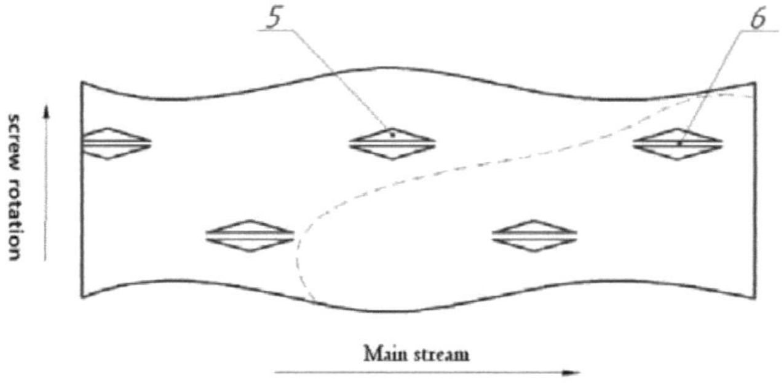

Figure 8. Feed hopper.

The press extruder works as follows. When the feed hopper (figure 8) is filled with the initial grain mixture, the press extruder drive is switched on (not shown in the drawings). When the screw 3 rotates, the mixture is captured by its turns and moves from the feed zone 7 to the plasticization zone 8 and then to the outlet of the matrix 4. As the mixture moves in the press extruder housing, due to the friction forces caused by the decreasing winding pitch of the screw 3 and the diameter of the housing 1, pressure is created in the extrudate, the longitudinal ribs 5 with holes 6 additionally deform the processed material. As a result, the finished product exits through the cone-shaped openings 9, and the squeezed liquid flows freely along the groove 10 into the receiver in the form of a niche 11, and then through the outlet 12 is removed from the housing cavity. The pressure value is regulated by changing the size of the outlet opening of the cone nozzle 4.

Cone-shaped press-extruder, the technical result of which depends on the strength and reliability of the screw installed in the cavity of the housing, made with longitudinal ribs on the inside and having areas for feeding and plasticizing the grain mixture. The screw is made with a decreasing diameter in the direction of its rotation. When working at the location points of the ribs made in the form of paired diamond-shaped elements, a torque occurs on the screw shaft:

Total torque:

$$\sum_{i=1}^{u} T_i \tag{1}$$

The shaft is calculated for strength:

$$\sum_{i=1}^{u} T_i = W_p[\tau] \tag{2}$$

where W_p is the polar moment of resistance.

For a solid shaft:

$$W_p = \frac{\pi d^3}{16} \tag{3}$$

$$d = \sqrt[3]{\frac{16\sum T_i}{\pi[\tau]}} \tag{4}$$

For a hollow shaft, where $d = DD\frac{d}{D}$:

$$D = \sqrt[3]{\frac{16\sum T_i}{\pi[\tau](hd)}} \tag{5}$$

$$T_i = F_i R_i \tag{6}$$

where R_i is the distance of the rib relative to the shaft axis
F_i – Circumferential force.
It is often necessary to determine the shaft diameter from the power *N* and the number of revolutions per minute *n*.

$$\tau_{max} = \frac{\sum T_{imax}}{W_p} \leq [\tau] \tag{7}$$

Where $\sum T_{imax} = 7162\frac{N}{n}$ Nm, if *N* in KW:
$\sum T_{imax} = 9550\frac{N}{n}$ Nm;
$\frac{\pi d^3}{16}[\tau] = \sum T_{max} = 7162\frac{N}{n};$

$$d = \sqrt[3]{\frac{7162 * 16 * T_{max}}{\pi[\tau]n}} = \sqrt[3]{\frac{36000}{[\tau]}} * \sqrt[3]{\frac{N}{n}} \tag{8}$$

It follows from the formula that for a given power *N*, an increase in the number of revolutions *n* leads to a decrease in the shaft diameter.
Taking $[\tau] = 20.0$ MPa (reference data), we get the value:

$$d = 12\sqrt[3]{\frac{N}{n}} \tag{9}$$

To reduce the deflection, we take instead of the specified $[\tau] = 20.0$ MPa $[\tau] = 12.0$ MPa, therefore:

$$d = 14{,}4\sqrt[3]{\frac{N}{n}} \tag{10}$$

This formula is used in practice.
In addition, in order to avoid shaft vibrations, it is required that the twist angle does not exceed ${}^{1}/_{4}{}^{\circ}$ per 1 linear meter.
In this case, the calculation is based on the permissible deformation:

$$\varphi_i^{\circ} = \frac{180^{\circ}}{\pi} * \frac{T_{max} * 100}{GY_p} = {}^{1}/_{4}{}^{\circ}$$

Substituting the values $Y_p = \frac{\pi d^4}{32}$, Tmax $T_{max} = 7162\frac{N}{n}$ and $G = 8 * 104^4\ MPa$, we get:

$$d = 12\sqrt[4]{\frac{N}{n}} \tag{11}$$

It is necessary to calculate the shaft so that it meets the strength condition and the torsional stiffness condition. In order to meet the strength condition for the shaft, it is necessary to calculate using the formula $d = 14{,}4\sqrt[3]{\frac{N}{n}}$, and according to the condition of permissible deformation, the value $d = 12\sqrt[4]{\frac{N}{n\ is\ required}}$.
We can find the value $\frac{of\ N}{n}$at which both conditions are met:

$$14{,}4\left(\frac{N}{n}\right)^{\frac{1}{3}} = 12\left(\frac{N}{n}\right)^{\frac{1}{4}}$$

Or

$$\frac{12}{14{,}4} = \left(\frac{N}{n}\right)^{\frac{1}{12}}$$

Where from

$$\frac{N}{n} \approx 0{,}11$$

Thus, the ratio of the power N of the press extruder to the number of revolutions per minute n should be 0.11.

4. Conclusion

The conducted studies show, that the developed press extruder for processing leguminous crops is able to perform the required operations at a high level, and does not have the disadvantages found in similar press extruders. The calculation of one of the main parameters of the press extruder-the ratio of the press extruder power to the number of revolutions per minute – is made.

The state-of-the-art press extruder market, with significant growth over the past decade, is far from optimal. The feed extrusion process is currently a very promising technology for producing high-protein feed for farm animals. In addition, machine-building companies focus their attention on the development of press extruders for agricultural production.

Extrusion is the most advanced method of feed processing due to the environmental friendliness of the process and an increase in the concentration of useful substances and vitamins compared to the starting material.

References

[1] Dotsenko S M 2012 Problems of the Soya Processing for Food *Food Manufacturing Industry.* **7** 18-21

[2] Dotsenko S M Skripko O V and Tilba V A 2014 The Use of Soya in Technology of the Protein - Vitamin and Protein - Carbohydrate Products *Storage and Processing of the Agricultural Raw Materials.* **4** 45-9

[3] Gupta P C 2014 Evaluation of Antifertility Potential of Ficus bengalensis (Linn.) In Male Albino Mice *International Journal of Pharmacy Research & Technology* **4(2)** 5-9

[4] Slozhenkina M I, Kuznetsova E A and Starodubova Yu V, Chepelenko M.N. 2015 The feasibility of using vegetable protein texture in the production of ham products *News of the Lower Volga Agro-University Complex: Science and higher professional education.* **1(37)** 161-4

[5] Frolov V Ju, Sysoev D P and Klasner G.G, 2015. Optimization of Parameters of the Chopper of the Soaked Soya Grain *Farm Machinery Operator.* **3** 24-25

[6] Kovaleva O A 2014 Meat By-Products as Raw Material for New Types of Canned Functional Foods *Proceeding of "Technical Sciences-From Theory to practice" XXXVIII Intern. Scientific-Practical Conference, Novosibirsk: Sibak,* **9**

[7] Klasner G G, Taras V S and Baranov V P 2020 Determination of safe conditions for moving ears to the working surface of the plant for sorting ears of seed corn *IOP Conference Series: Earth and Environmental Science* **421(2)** 022014

[8] Klasner G G, Tarasov V S and Baranov V P 2020 The study of the moment of resistance of the grinder of legumes *IOP Conference Series: Earth and Environmental Science* **421(2)** 022013

[9] Frolov Ju V, Klasner G G and Kremyansky, V F Justification of design-mode parameters of the chopper soaked soybean grain *IOP Conference Series: Earth and Environmental Science* **315(6)** 062014

Metric transformation of the surface of a tight punch for forming sheet shells of double curvature

V A Mikheev[1], F V Grechnikov[2], Roberto de Alvarenga[1] and V V Bodrov[3]

[1] Samara National Research University, 34, Moskovskoye shosse, Samara, 443086, Russia
[2] Samara Scientific Center of the Russian Academy of Sciences, 3A, Studenchesky pereulok, Samara, 443001, Russia
[3] Ural engineering center, 6, ul. Rozhdestvenskogo, Chelyabinsk, 454007, Russia

E-mail: vamicheev@rambler.ru

Abstract. For the first time, the problem of parametric reduction of the circumscribing surface of the shell of double curvature to the main axes and to the planes of symmetry defined by Gaussian coordinates is considered in detail. The solution of this problem is essential for ensuring the conditions of symmetrical shaping of the double-curvature circumflex shells by tight fit. The tight punch is presented in the form of a technological electronic layout, containing metric information of the surface of double curvature and storing information of the geometric alignment of the skin that extends to the outer circumferential surface of the unit. In this case, the circumferential surface of the shell of double curvature remains invariant with respect to the coordinate transformation.

1. Introduction

The aerodynamic shape of the aircraft is described by two basic concepts – it is the outline and contour of the surface. The outline refers to the surface of the airframe, streamlined by the air flow in flight, and the contour – the line of intersection of the outline with the coordinate planes of the aircraft XYZ (figure 1).

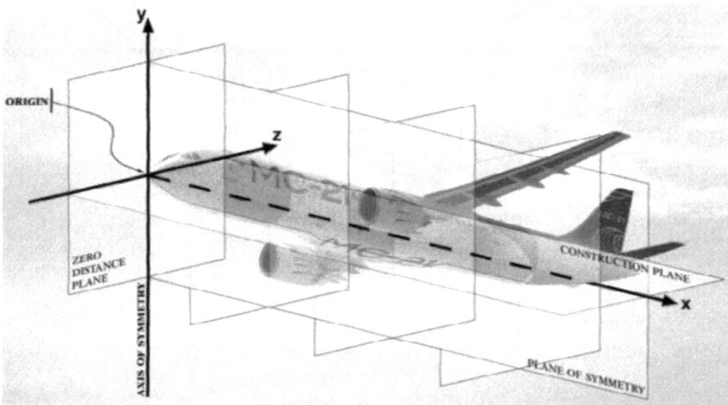

Figure 1. Diagram of the coordinate planes.

The outline of the aircraft is formed by the external skins. This is an external structural shell element that serves to give the outline of the airframe a certain (for example, streamlined) shape and is involved in the perception of the aerodynamic load. The outer skins used in aircraft are made of a sufficiently strong and rigid sheet material; they can resist normal and tangential forces and work for all types of loads. Currently, when designing and calculating the shells, the skin is taken into account as a load-bearing power element of the aircraft structure.

Most of the skins have the external shapes of the outline-forming surfaces corresponding to the theoretical contours of the aircraft, set by the aerodynamic model of the airframe surface [1]. The fuselage of the aircraft is designed entirely as a large circumscribing object by means of contours of parallel cross-sections passing through a series of points of the longitudinal X-axis, as in figure 1. After determining the cross-sections, you can also build longitudinal curves that connect them into a single three-dimensional shape. This is a way to define curves in a space in which a Cartesian coordinate system is introduced and consists in projecting these curves onto the plane of symmetry (x, y) and onto the construction plane (x, z) of the plane. Traditionally, these contour curves were drawn in full size at points using mechanical splines. The process is called the plaz method of constructing surfaces [2].

The specificity of the design and the complexity of the external forms of the aircraft surfaces does not allow us to coordinate the geometric parameters of the airframe assembly units, skins, technological equipment and other mating parts and to link them in the aggregate-assembly production with the help of traditional machine-building drawings. The only room, the size of which allowed to do this, is a structural plaz, which was drawn in projections on the plane (x, y) and (x, z) at a scale of 1:1 contours of units, assemblies and individual skins. The constructive plaz is used for making templates. Templates as information carriers are copies of contours that transfer the specified shapes and sizes to the sheaths and their volumetric equipment.

A feature of modern aggregate and assembly production is the increasing complexity of contours and increasing requirements for their quality, including the accuracy of reproducing shapes and sizes [3]. In turn, the reproduction of the shapes and sizes of aggregates, assemblies and sheaths is characterized by the specificity of the methods and means used. There are three groups of size matching methods:

- methods based on the use of only rigid media of shapes and sizes (templates);
- methods based, on the one hand, on the use of analytical descriptions of surfaces associated with aerodynamic contours, and on the other hand, on the use of structural plazas and rigid carriers of shapes and sizes (templates) for linking structural elements that are not related to the aerodynamic contours of the product;
- methods based on the plazeless production of the circumference equipment on the basis of analytical (numerical) setting of all geometric parameters of the skin and other structural elements, components and assemblies of the product, including the use of the holographic principle of linking information.

For the construction of surfaces began to use a computers, because serial testing required, as a rule, the use of more accurate analytical and graphical methods of linking. The geometric linking of structural elements remained the same complex multi-factor task, and it was performed first on the basis of the plaz-template method using only rigid carriers of shapes and sizes, and then using the calculation-plaz method using analytical methods used in drawing plaz lines using drawing machines [4].

At present, automated systems have appeared based on the development of the apparatus for mathematical modeling of objects and production processes using modern computer technology and CNC equipment. These are CAD/CAM systems, on the basis of which new methods have been developed that allow describing the vast majority of surface variants based on the parametric representation of curves and surfaces. New methods use analytical curves as systems of lines lying on the projected surface, but not the surface itself, which was a breakthrough in the design and construction of surfaces of complex shapes [5].

Parametric representation of curves and surfaces has advantages. Since we have to calculate tangents, normals, curvatures, etc., we need a parametrization for which the differentiation operation is easily performed. Parametric equations have certain advantages. For example, to obtain a graphic image of a curve or surface, the coordinates of their points corresponding to certain parameter values are calculated sequentially. The parametric method of defining curves frees you from being bound to any particular coordinate system. In this case, the shape of the curve or surface does not depend on the choice of the coordinate system. Therefore, the development of the parametric method can be considered natural, but without a computer, its use is impossible.

With the advent of computers, more universal methods were developed, moving away from traditional graphical methods. Modern operators, who have replaced the plaz surface specialists, should use various systems for determining surfaces. Systems of this type were made fully automatic, so that the only input information was the coordinates of the points through which the projected surface should pass. The principle of such systems is based on the method of two-dimensional interpolation. Thus, the spread of analytical methods for setting contours, automation of calculations and recording programs for CNC equipment contributed to the development of an independent method for the formation of shapes and sizes of mating structural elements, i.e., the method of plazeless linking of structural elements of an aircraft airframe.

The plazeless linking of the dimensions is carried out by means of an aerodynamic model of the water-forming surface obtained by calculation. The formation of interconnected working contours of technological equipment is ensured by its precise production on CNC machines. When using a plazeless coupling, the accuracy of the coupling of the mated structural elements is directly dependent on the accuracy of their manufacture. Since the external contours of the aircraft are set analytically, the use of plazeless linking means extends to the elements of the technological equipment, including the tight punches associated with the theoretical contours of the products. However, a complete three-dimensional description of the product is the basis for a plazeless production preparation [3].

Over the past decade, information technology has come a long way in its development. The emergence of «heavy» integrated CAD/CAM/CAE systems allows aviation engineers to perform an electronic description of the product with an accuracy of 1-3 microns, using the principle of three-dimensional parameterization [7]. The essence of parameterization is that the part is performed in a computer assembly as an object not with specifically specified dimensions, but with interrelated variables, when one of them changes, the entire part is rebuilt. The use of the parameterized model principle allows you to modify a huge number of assemblies of the production object within a few hours and is limited only by the capabilities of computer technology.

Such capabilities are inherent in the electronic description of the product – an electronic layout (EL) and the process of modeling the aircraft begins, as with traditional methods, with the creation of a geometric model of the surface, which was used to build cross-section contours when splitting plazas and developing programs for manufacturing outline equipment on CNC equipment. Geometric models of the surface were presented both in the form of a rigid immutable model and in a parametric form. Depending on the functions performed, the description of the surface was presented in the form of a frame model, or a surface model, or in the form of a solid.

We can assure you that the production technologies of aviation equipment, even with the use of modern digital technologies, remain the same, despite the fact that the number and complexity of manufacturing plazs and outline equipment during the spatial linking procedure on the EL has sharply decreased. Tight punches can be made not according to control casts from the surface standard (the basic standard), but directly on CNC machines without drawing the plaz and without making templates. For individual sections of the surface of the tight punch, its contours are equidistant with respect to the external theoretical outline and represent a real breakdown of the lines of frames, stringers and other power elements of the airframe.

However, due to the need to solve new geometric problems related to the shaping of a flat sheet by fitting it into a shell part, we need a different position in space of the surface of the tight punch, the shape of which will correspond to the first and second fundamental quadratic forms that define the surface up to its position in space. The meaning of the first and second quadratic forms is that they allow us to construct a surface under certain conditions [6]. Then, as the coordinate lines on the surface, it is better to take a network of lines of curvature.

In the context of differential geometry, these lines of curvature are contours of normal surface sections that have a geodesic origin according to the Gauss metric [7]. They are consistent with the internal geometry of the surface, the metric properties of which remain unchanged when it bends. This is the great importance of metric concepts, and some surface properties related to the structure of a differential quadratic form for a curve element on a surface can be investigated independently of space. Such properties of surfaces are called internal, and geometry based on the study of differential quadratic forms is called internal geometry.

Then, following Gauss, we define a surface in three-dimensional Euclidean space in a parametric way by setting three equations $x^i = x^i(q^1,q^2)$, where i = 1,2,3. The parameters q^1, q^2 define a grid of curves on the surface, called curvilinear, or Gaussian coordinates. This has become the main tool for measuring surface curvature. Such tools made it possible to determine two types of surface curvature, known today as the Gaussian curvature K and the mean curvature H. Therefore, the geometry of the shell does not coincide with the geometry of the skin for the assembly of the aircraft unit, since these geometries were tied to different categories of curves and surfaces: contour-contour-forming and Gaussian-parametric. Switching from one category to another and back again is not difficult when using modern digital technologies.

There is still no regulatory and technical documentation (RTD) in production, which determines the information content and status of EL in the technological documentation. Therefore, the production units still work with drawings, plazas and templates, and the EL practically «hang» and do not participate in the pre-production processes. Even when developing control programs for the manufacture of parts, their geometry is verified on the basis of plaz information, including tight punches. The main problem is that in order to ensure interchangeability, the volumetric contour-forming equipment, including the tight punches, are directly connected to the theoretical contours of the aerodynamic contours of the aircraft units.

However, there are already methods that allow you to describe any contours on a computer and build any parametric lines, including the conjugate orthogonal grid of surface curvatures. Therefore, we propose to create a technological electronic layout (TEL) that contains the geometric information that is necessary to solve a specific technological problem of obtaining the details of the cladding by shaping the skin. This primarily concerns the tight punch, the geometry of which will differ from the original electronic layout of the EL. Then the design of the technological process will be carried out using application systems, and the initial information will be transmitted in the form of TEL and drawings and supported by the information environment of pre-production.

2. Geometric elaboration of a tight punch

The process of forming the sheet material with a tight fit remains the main one in the production of shell parts of skins. The created metal forming equipment (tight presses) is unique, specialized and must be program-controlled. Much attention is paid to the problems of automating this process. As it was found, the surfaces of the tight punches are directly related to the theoretical contours of the aerodynamic contours of the aircraft units. This leads to the fact that the longitudinal line of curvature of the surface of the tight punch does not coincide with the direction of the load action, or rather with the direction of the forming contour of the shell. This circumstance violates the symmetry of the tightness forming, which leads to a violation of monotony and to the manifestation of instability of plastic deformation in the form of folds and rupture of the sheet billet [8].

The proportionality of the tensile loads and the monotonicity of the deformation of the sheet billet in the process of forming the skin under the conditions of a flat stress state will be preserved when its main directions coincide with the directions of the lines of curvature. They are consistent with the internal geometry of the surface, the metric properties of which remain unchanged when it bends. The study of internal geometry can be performed without considering its surface in space, including in the spatial rectangular coordinate system formed by the coordinate planes of the aircraft. Figure 2 shows the original tight punch, on the surface of which the shell part of the skin is located according to the contours of the outline.

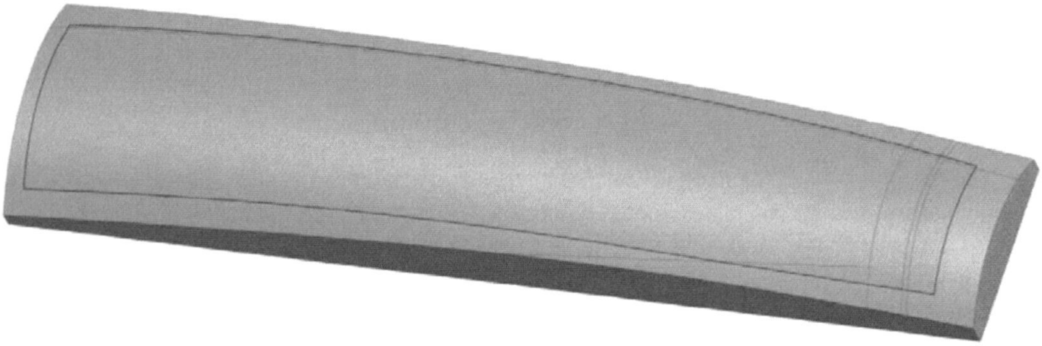

Figure 2. The original tight punch, on the surface of which the shell part of the skin is located according to the contours of the outline.

Figure 3 shows a converted tight punch, on the surface of which the shell part of the skin is also located according to the contours of the outline, but the upper ridge of the surface points is occupied by the forming contour of the longitudinal curvature of this shell. This is clearly seen when combining two tight punches on the outer surface (figure 4). The outer surface is one, because it is invariant with respect to its representation in different coordinate systems. In the spatial rectangular coordinate system formed by the coordinate planes of the aircraft, a tight punch is shown, the surface of which corresponds to the contours of the outline. In a curved coordinate system formed by two lines of curvature on a surface with a pole point as the origin, a tight punch is shown after our geometric study.

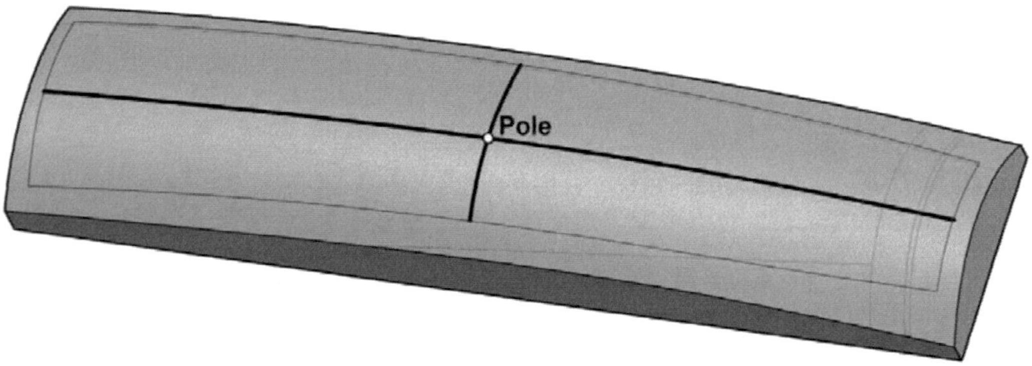

Figure 3. A converted tight punch, on the surface of which the shell part of the skin is also located, where the upper ridge of the surface points occupied the forming contour of the longitudinal curvature of this shell.

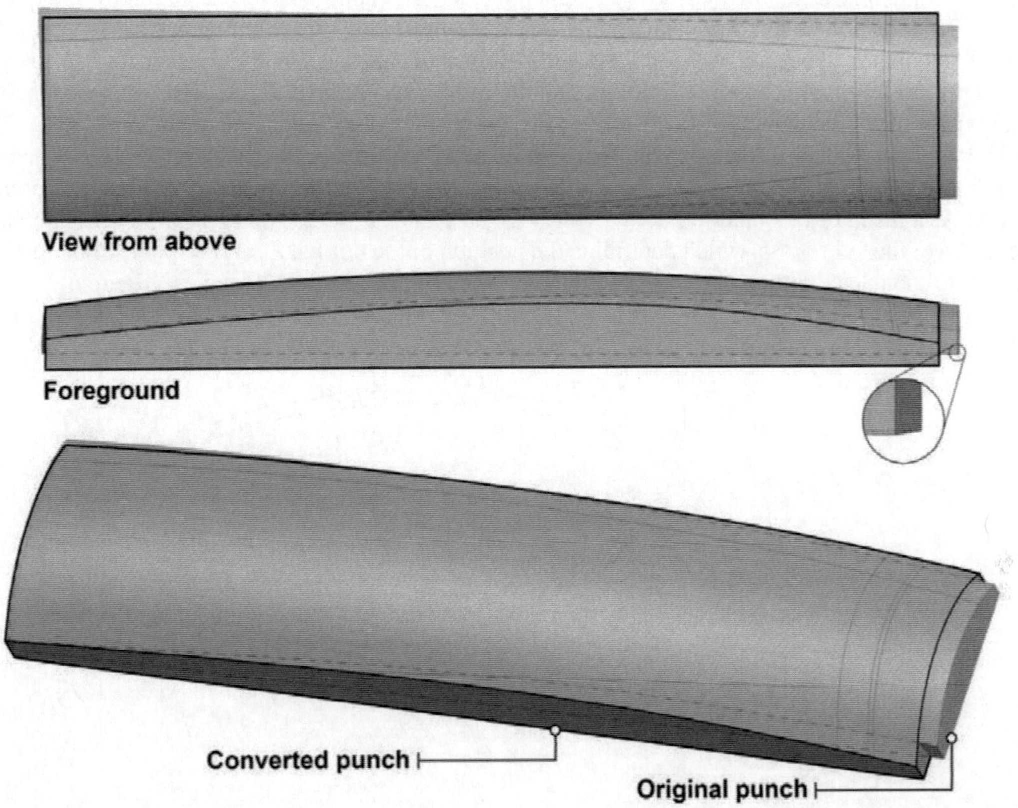

Figure 4. The combination of the original and converted tight punches on the outer surface, on which the shell part of the skin is located.

Figure 5 shows the results of the geometric study with the indication of the values of the geometric parameters in normal cross-sections, the upper points of which along the forming contour correspond to the coordinates of the points of descent of the sheet blank from the tight punch at any stage of the step-by-step wrapping.

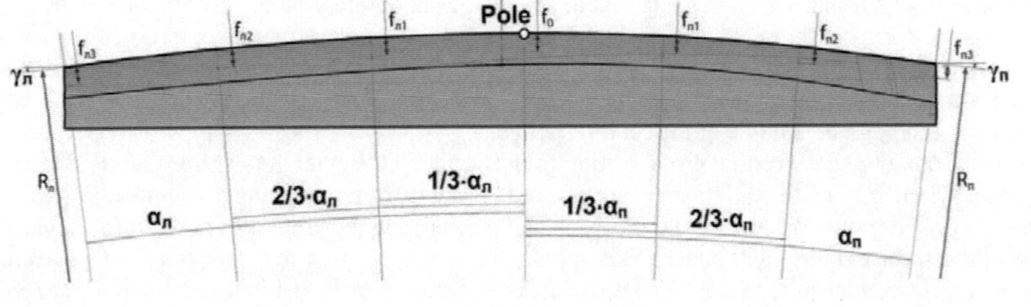

Figure 5. The results of the geometric study with the indication of the geometric parameters f and R in the normal cross sections of the tight punch.

The procedure for the geometric study of a tight punch is associated with the transition from the outer circumferential surfaces to the surface of double curvature, tied to the grid of the main lines of curvature. Means of parametric determination of surfaces in terms of Gaussian coordinates with the help of Gaussian parameters *u* and *v* were introduced, and the installation of the tight punch should be made so that the position of the shaping contour of the surface is aligned with the first vertical plane F_1. The second vertical plane F_2, orthogonal to the first, contains the position of the contour of the central cross-section that intersects the shaping contour at the "pole" (point *O*) of the surface of the shell of double curvature. As a result of this binding, both the orthogonal planes F_1 and F_2 pass into the vertical planes of symmetry of the tight press, which determine the position of the normal Z_0 in the "pole" of the surface of the shell of double curvature and the directions of the lines of curvature X_0 and Y_0 (figure 6).

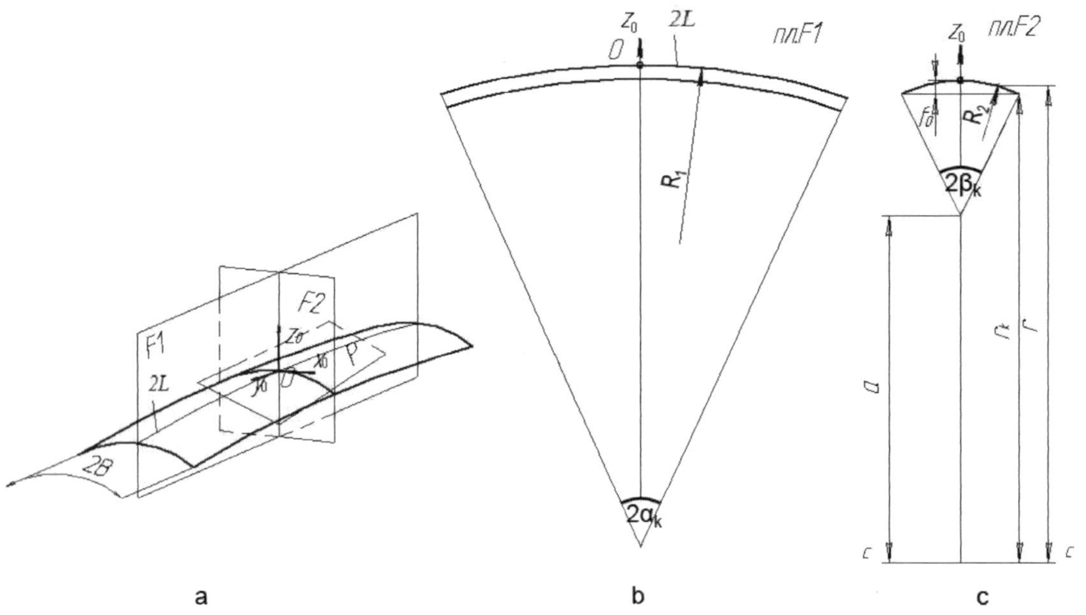

Figure 6. a) the surface of the shell of double curvature; b) the forming longitudinal contour (2L); c) the central transverse contour (f_0).

The installation of the base of the tight punch on the press table should ensure a constant position of the forming contour in the vertical plane of the press symmetry. This is the first time we have made such a transition. As a result, we have gained great advantages in automating the design and control of the processes of forming skin-covering shells of any surface, as well as complex geometric shapes. It became possible to distinguish on such a surface the position of the forming contour of the tight punch, which will be combined with the longitudinal vertical plane of the press, including the pole of the surface and passing through the middle of the central sponge.

For the contour of the central cross-section of the surface, it is enough to simply set the location of the axial symmetry line for the classical surface of the second order touching at the pole, as part of the toroidal surface (figure 7). As a result, it is possible to determine the properties of the local shape at the pole of the surface of the tight punch. This approach can also be used for other shapes, by selecting a classical second-order surface that touches it at the pole, which greatly facilitates the analytical approach to modeling, which adequately reflects the real process of shaping the skin.

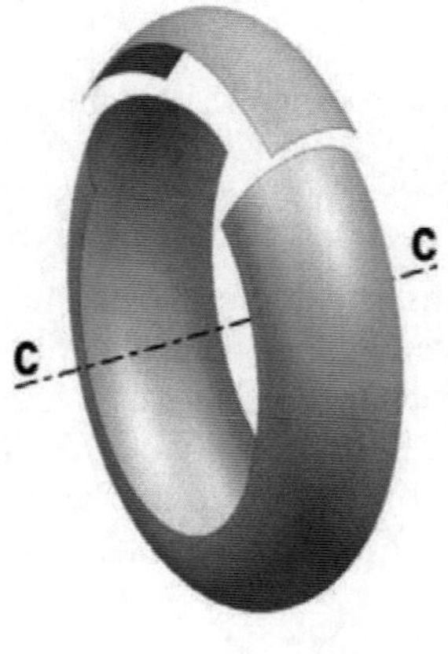

Figure 7. Position of the axis of axial symmetry c-c for the surface of the shell of double curvature.

For the central cross-section, you can set a cylindrical coordinate system: r, β and the «imaginary» axis of axial symmetry *c-c* of this contour. Then we will refer to the main geometric parameters of the double curvature shell (see figure 6):

- the radii of curvature of the contours of the surface sections by the vertical planes of symmetry F_1 and F_2 at the point O: R_1 (longitudinal) and R_2 (transverse);
- length of the contour of the longitudinal section of the surface $2L$ by the vertical plane of symmetry F_1 (forming contour);
- the length of the contour of the edge section of the surface $2l_к$ with a plane parallel to F_1;
- deflection of the contour of the central cross-section f_o in the vertical plane of symmetry of the surface F_2;
- longitudinal $2\alpha_к$ and transverse $2\beta_к$ angles of the contours of the cross-sections of the surface by the vertical planes of symmetry F_1 and F_2 of the tight punch;
- the value of a, equal to $(R_1 - R_2)$ in figure 4
- shell width $2B$.

The contours of the cross sections of the shell surface with the vertical planes of symmetry F_1 and F_2 are plane curves that intersect at the point O with the radii R_1 and R_2. The values that are the inverse of the radii:

$$k_1 = \frac{1}{R_1}; k_2 = \frac{1}{R_2} \tag{1}$$

are the normal curvatures of these lines on the surface that intersect at the point O and have the properties of extremality: one of them is the maximum, and the other is the minimum. In this case, the product of the curvature values determines the Gaussian curvature K at the point O. The directions of the tangents to these lines of the shell surface at the point O determine the main directions x_o and y_o, which lie simultaneously in the tangent plane and in the planes of symmetry F_1 and F_2.

Then the geometric shape of the shell of the double curvature will be locally characterized at the point O by the value of the Gaussian curvature. The Gaussian curvature as a metric parameter characterizes the deformation of the shell of a given geometric shape in the «pole». The Gaussian curvature of a surface is considered an invariant of the curvature tensor of plane lines passing through the point O and retains its value when bending. This deformation is called isometric bending, in which the lengths of the lines drawn on the surface of the shell and the angles between them are preserved. Only the geometric shape of the double-curvature shell changes, but its surface remains isometrically similar to the surface before bending.

In the coordinate system of the main directions x_o and y_o lying in the tangent plane, the surface of the shell in the vicinity of the point O can be represented as a function:

$$z_0 = \frac{1}{2} \cdot (k_1 \cdot x_0^2 + k_2 \cdot y_0^2) \quad (2)$$

The explicit form of specifying the surface in this form is used to find the characteristic metric properties of the shell. Just as some circle is close to the point of the curve in infinite proximity, so some classical quadratic surface of the second order will be close to the surface of the shell in the vicinity of the point O. It can include, in particular, a part of the external toroidal surface.

The considered method of parametric representation of the surface in terms of coordinates with the help of Gaussian parameters u and v, will allow to ensure the stress-strain state of the workpiece during the fitting in the main axes of stresses and deformations, but under the condition that the direction of the fitting is aligned with the vertical plane of symmetry F_1 of the surface of the fitting punch. To do this, it is sufficient to ensure that the plane in which the central clamping sponge of the tight press lies is perpendicular to the tangent at the point of convergence of the forming contour of the tight punch and the constant position of this contour in the vertical plane of symmetry F_1.

3. Geometric provision of symmetrical fit

Modeling of the processes of forming the skin is based on the geometric model of the shell, shown in figure 5. It should provide the calculation at any stage of the step loading when the sheet billet is stretched, the coordinates of the point of its descent from the punch, the angle of inclination at the point of descent, the length of the blank laid on the punch and the trajectory of the point near the press clamp relative to the forming contour of the punch, etc. The step process in the control programs is based on the increments of either the angle of descent of the workpiece from the punch, or the length of the workpiece laid on the punch. At the same time, the technological features of the process and the design models of shaping are such that the fitting can only be performed along strictly convex punches in the longitudinal direction.

Essentially, the surface of the shell is given by a grid of plane curves lying on the secant normal planes. A set of such lines is a discrete frame of the surface. The planes of symmetry F_1 and F_2, shown in figure 6, are used as the main planes on which the cross-section contours are projected. The gaps between the frame lines are filled graphically by projections using two-dimensional interpolation methods. Thus, mathematical modeling of the geometric shape of the shell is reduced to obtaining an approximating curve for a given discrete set of points that meets the requirements of continuity and smoothness.

In addition to the requirements for the approximating curve, it is necessary to take into account two features: first, the source data may contain gross errors (when measuring coordinates, operator errors when entering coordinate values into computer memory, etc.); second, the source data inevitably contains sampling errors. Since classical methods of curve approximation, such as Lagrange polynomial interpolation, root-mean-square interpolation, cubic equation interpolation in the form of Beziers or *B*-splines, may not always meet the above requirements, it becomes necessary to develop a special apparatus for geometric modeling of the class of problems under consideration.

In computer-aided design, the development of methods for determining the smoothness of surfaces is of great interest. Currently, the best mathematical methods for determining the smoothness of a surface use orthogonal meshes of minimum and maximum curvature or Gaussian curvature K. For the parametric representation of surfaces, the Gaussian parametric coordinates u and v are used and the differential geometric parameters are determined at the intersection point of two lines defining the neighborhood of the synthesized surface. A surface is defined by sets of flat lines written in parametric form:

$$\begin{aligned} x &= x(t); \\ y &= y(t); \\ z &= z(t), \end{aligned} \quad (3)$$

where t – the parameter, when changing from -1 to 1, the point runs along the line from beginning to end, have continuous derivatives of the 2nd order (figure 8).

The coordinates of the nodes in the network are calculated directly from (3) for known values of the parameter t, and the derivatives from (3) in the form of tangents:

$$\dot{r}_u = (\dot{x}_t, \dot{y}_t, \dot{z}_t) \text{ and } \dot{r}_v = (\dot{x}_t, \dot{y}_t, \dot{z}_t) \quad (4)$$

The vector m is the unit vector of the normal to the surface and is calculated from the values of the tangents known from (4) in the form:

$$m = \frac{\dot{r}_u \times \dot{r}_v}{||\dot{r}_u \times \dot{r}_v||}, \quad (5)$$

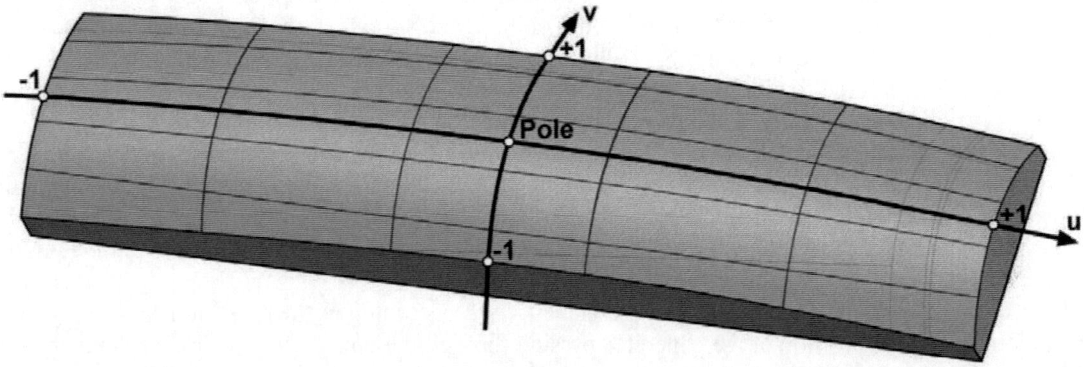

Figure 8. Parametric representation of the surface using the Gaussian parameters u and *v*.

where the notation of the form $||\dot{r}_u \times \dot{r}_v||$ is the norm of the vector.

Having the nodal parameters (4) and (5), it is already possible to build efficient algorithms for reducing the surfaces of double curvature to the lines of curvature. In general, it is necessary to have a surface equation of two variables:

$$x = x(u, v); \\ y = y(u, v); \\ z = z(u, v), \quad (6)$$

in the neighborhood of each grid node and find the second derivatives for a more accurate description of the surface geometry based on the application of mixed derivatives:

$$\ddot{r}_{\mathrm{uv}} = (\ddot{x}_{\mathrm{uv}}, \ddot{y}_{\mathrm{uv}}, \ddot{z}_{\mathrm{uv}}) \quad (7)$$

However, since these equations do not exist, we can estimate (7) by the values of the first and second derivatives of a pair of lines intersecting at a given node. The basis for this estimate is the assumption that in the vicinity of the node, the lines of the discrete frame differ slightly from the corresponding normal sections of the surface to be constructed.

Take any of the tangents to the line *u* or *v*, for example $\dot{r}_u$, then we can write:

$$\dot{r}_u \cdot m = 0 \quad (8)$$

Differentiating this ratio by the length of the arc S of the selected curve, we get:

$$\ddot{r}_u \cdot m + \dot{r}_u \cdot \dot{m} = 0 \quad (9)$$

Given that $\ddot{r}_u = N_u$ is the normal to the curve at a given point and, by substituting in the coordinate directions, we get:

$$\dot{m}_u = \frac{-N_u \cdot m}{\dot{r}_u}; \dot{m}_v = \frac{-N_v \cdot m}{\dot{r}_v}; \tag{10}$$

From the expression of the coefficients of the second differential quadratic Gauss form, we write:

$$\ddot{r}_{\text{uv}} \cdot m = \frac{-\dot{r}_u \cdot \dot{m}_u + \dot{r}_v \cdot \dot{m}_v}{2} \tag{11}$$

Substituting (10) in (11), multiplying on the right by m and taking into account that $m^2=1$, we get:

$$\ddot{r}_{\text{uv}} = \frac{1}{2} \cdot \left[N_u \cdot \frac{\dot{r}_u \cdot \dot{r}_v}{(\dot{r}_u)^2} + N_v \cdot \frac{\dot{r}_u \cdot \dot{r}_v}{(\dot{r}_v)^2} \right] \tag{12}$$

It can be seen from (12) that the mixed derivative is a weighted average of the normals to the coordinate lines at their intersection. The weight coefficients are complexes of tangent vectors. Many orthogonal lines can be drawn through any nodal point of the surface, only two of them are the main curvatures that satisfy the condition $\ddot{r}_{\text{uv}} = 0$. In practice, when constructing a discrete framework, one tends to define one of the families of surface lines. As a rule, this is a set of cross-sections. The second family is defined kinematically as the family of equidistant lines.

It is known that the kinematic method of forming a water-forming surface consists in determining a one-parameter family of curves forming the desired surface. If we are dealing with a family of straight generators, for example, one-line surfaces, then one of the curvatures, in our case k_2 is 0, then the Gaussian curvature K at all points is also zero and the surface is unfolding, i.e. it can be unfolded into a plane.

For the surface of the shell of a biconvex shape (K>0), according to the condition $K = |k_1 \cdot k_2| =$ max, you can set the position of the generator passing through the point O and the radii of curvature at this point R_{1o} and R_{2o} in the main directions of the surface x_0 and y_0.

Here, the generating, for example, v–curve moves along some guiding u-curve, and the surface is given by the equation:

$$r(u, v) = r(v) \cdot [T(u)], \tag{13}$$

where $r(v)$ – parametric equation of the generatrix ($0 \leq v \leq 1$);

$T(u)$ – means the transformation of the generatrix $r(v)$ and determines its shape in the process of moving and changing along the guide.

In our case, the shape of the generatrix is determined by the magnitude and direction of the deflection of the cross-section of the shell surface (see figure 6). If we take the length of the arc or the angle of the arc measured from the center point as the parameter t, then we have:

$$|\dot{r}(\alpha)| = 1; |\ddot{r}(\alpha)| = k_1;$$
$$|\dot{r}(\beta)| = 1; |\ddot{r}(\beta)| = k_2; \tag{14}$$

This consideration of the surface does not contradict, but, on the contrary, facilitates the search for the directions of the main axes x_0 and y_0. It was noted above that the combination of curvatures of any node point of a discrete frame:

$$K = k_{\min} \cdot k_{\max} \tag{15}$$

characterizes the local shape of the surface, and at point O the shape of the contiguous quadratic surface as a whole. Thus, we get a smooth convex surface. In the reasoning, we used a vector parametric representation that is compatible with the use of spatial transformations of three-dimensional coordinates.

Thus, in order to study the circumferential surfaces of a tight punch, taking into account the properties of the classical quadratic shell in contact, the geometric data during processing must go through the following stages: construction of a discrete frame; setting geometric parameters in the cell nodes; transformation of the discrete frame into a model of a continuous surface; representation of the surface in the main axes; setting the curvature contours with second-order curves.

4. Summary

To determine the directions of the main axes x_0 and y_0 at the point O, we considered the surface of the shell as a whole, when it is defined as a one-parameter family of curves of transverse contours located at their upper point along the shaping contour in normal planes. Moreover, in the vicinity of the point O, the surface of the shell is close to the quadratic surface (2), which belongs to the surface of the toroid type.

For the practical implementation of the tightness scheme under conditions of symmetry, it is necessary, first of all, to solve the problem associated with the representation of the shell surface in the main axes. The solution of the problem is achieved by using a combination of the following techniques when forming a biconvex shell.

Based on the 3D model of the shell surface, the vertex (point O), the main axes x_0 and y_0, and the coordinates of the points of the forming contour and the central cross-section located in the central planes of symmetry F_1 and F_2 of the tight press are determined.

The symmetry of the working surface can be created by «building up» the surface on the computer due to «smooth crosslinking» in the direction of the x_0 and y_0 axes. Then the forming contour will be located in the plane of symmetry F1 and will be aligned with one of the lines of curvature passing through the point O.

Such a setting of the forming contour of the tight punch provides, at any stage of step loading, the calculation of the coordinates of the vanishing points of the workpiece from the punch, the angle of inclination at the vanishing points, the length of the blank laid on the punch and the trajectory of the point near the press clamp relative to the forming contour.

The step-by-step process should be constructed for the surface reduced to the main axes and planes in the control programs for forming tight shells under the conditions of symmetry and their combination with the main axes of anisotropy of the sheet billet and the coordinate planes of the tight press.

References

[1] Kuzmin V F 2002 *Ensuring the requirements for the aerodynamic contours of the aircraft in aviation production* (Moscow: Mashinostroenie) p 272

[2] Ivanov Yu L, Kuzmin V F, Mar'in B N, Volkov I V, Makarov K A, Murav'ev V I, Odinokov V I, Pekarsh A I and Shport V I 1999 *Modern technological processes of airframe assembly* (Moscow: Mashinostroenie) p 304

[3] Pekarsh A I, Tarasov Yu M and Krivov G A 2006 *Modern technologies of aggregate and assembly production of aircraft* (Moscow: Agraf-press) p 304

[4] Weisburg V A, Medvedev B A and Bakumsky A N 1985 *Automation of aviation production preparation processes based on computers and CNC equipment* (Moscow: Mashinostroenie) p 216

[5] Fox A and Pratt M 1982 *Computational geometry. Application in design and manufacturing* (Moscow: Mir) p 304

[6] Filin A P 1975 *Elements of shell theory* (Leningrad: Stroyizdat) p 384

[7] Mikheev V A, Dement'yev S G, Samokhvalov V P, Savin D V and Surudin S V 2013 Isometric conditions for the formation of a tight shell of double curvature with minimal thickness difference *Izvestia of Samara Scientific Center of the Russian Academy of Sciences* **15(6-1)** 161-6

[8] Mikheev V A and Surudin S V 2017 Fundamentals of the calculation of the process of forming tight thin shells of double curvature *Izvestia of Samara Scientific Center of the Russian Academy of Sciences* **19(1-3)** 555-62

Simulation of heat transfer in a flow of over-critical nitrogen and oxygen in a horizontal circular tube

S V Borodkin[1,3], I L Bataronov[2], A V Ivanov[1] and V I Ryazhskikh[2]

[1] Military Scientific Educational Center of Military-Air Forces “N.E. Zhukovsky and Ju.A. Gagarin Military-Air Academy”, Voronezh, Russian Federation
[2] Voronezh State Technical University, 394026, Voronezh, Moskovsky prospekt, 14, Russian Federation

[3] E-mail: borodkinstanislav@yandex.ru

Abstract. In the framework of the k-ε turbulence model with enhanced wall treatment, heat transfer in a flow of over-critical nitrogen and oxygen in a horizontal circular tube with heat boundary conditions of the third kind in the pressure range of 20÷40 MPa, mass flow rates of 0.03÷0.12 kg/s, and heat transfer coefficients to the environment of 500÷2000 W/(m^{2}K) is simulated. The distributions of the bulk temperature, calculated as the thermodynamic temperature of the average mass enthalpy of the supercritical fluid flow, are obtained. Analysis of the heat transfer coefficient of the flow showed that at pressures less than three times the critical one, a reduced heat transfer mode appears, so the near-critical region is not recommended for use in oxygen and nitrogen gasifiers. In the framework of the multiplicative model for the Nusselt number, insignificant simplices of thermophysical properties are established and correlations for oxygen and nitrogen in the over-critical region are constructed, which agree with the result of the computational experiment with an average relative accuracy of 1÷1.5%. On the basis of the one-dimensional heat transfer equation, the adequacy of the obtained correlations is shown. It is established that when using the bulk temperature based on the mass average enthalpy, a separate account of the initial heat transfer section is not required.

1. Introduction

High-pressure nitrogen and oxygen are widely used in the power and life support systems of aerospace aircraft. The production of these materials from cryogenic liquid is carried out by gasification using on-board or ground installations [1]. Optimization of the gasification process in the supercritical fluid state (SCF) requires the use of the heat transfer coefficient between the SCF and the heating system [2–5]. For ordinary liquids and gases, universal dependences for calculating this coefficient are obtained [3, 5], while for SCF there are a large number of theoretical and experimental studies [6-8], which give very heterogeneous results for the Nusselt number, depending on the type of substance and experimental conditions. At the same time, data on nitrogen is very limited, and on oxygen is practically absent. In the near-critical region, which is characterized by a sharp change in the thermophysical properties of SCF with temperature, processes occur that lead to reduced heat transfer [5]. Therefore, for gasification, an over-critical region of SCF states is usually recommended, in which the change in properties is fairly smooth. In this case, sufficiently reliable results on the local heat transfer coefficient can be obtained by calculating the turbulent heat and mass transfer in the framework of approved models [9–11].

2. Problem statement

To calculate the conjugate heat and mass transfer in a turbulent flow, k-ε turbulence models [10, 11], based on the Reynolds-averaged transport equations (RANS), are currently most widely used, including for the SCF [12–15].

To simplify the computational problem, we will neglect the buoyancy effects and the swirling of the flow in the heat exchanger. These effects can be accounted for by the corresponding correction factors [2, 3]. Then the problem becomes axially symmetric. The use of an axially symmetric formulation makes it possible to significantly reduce the numerical dimension of the problem, which is important for performing calculations in a long, thin tube.

As shown by experimental observations and theoretical calculations [12-15], the pressure drop in the SCF-flow in the evaporator is less than 0.1% of the operating pressure, so the influence of the pressure change in the SCF flow on the thermophysical properties can also be ignored in the calculation.

The standard k-ε model with enhanced wall treatment, which has higher accuracy [12–15], is adopted for the simulation.

The numerical computations were performed using ANSYS CFD FLUENT software.

The region of solution of the problem was the longitudinal section of the circular tube of radius $R = 5$ mm and length $L = 4$ m. To exclude the influence of boundary effects at the inlet and outlet of the tube, buffer regions with a length of 20 mm were added at the ends with the condition of thermal insulation of the tube surface (figure 1).

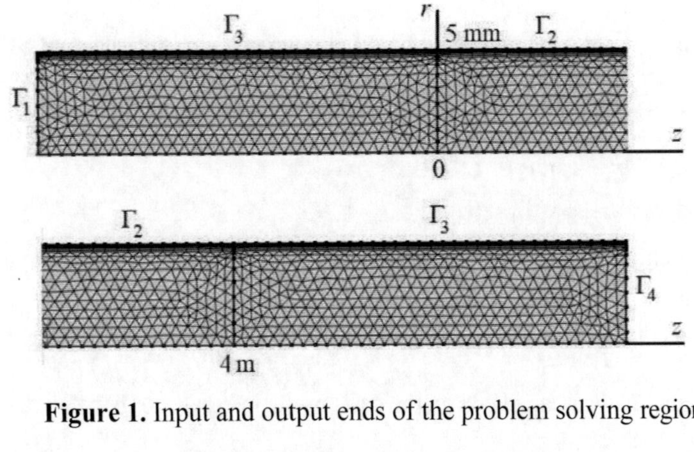

Figure 1. Input and output ends of the problem solving region.

In accordance with the selected geometry, the following boundary conditions were set on the lines Γ_i (figure 1).

1) On the line Γ_1 of the input flow of SCF: uniform temperature distribution $T(r) = T_{\text{in}}$ and self-similar distribution of the velocity of the turbulent flow. Using this self-similar profile reduces the size of the input buffer zone required to relax the input distributions of variables.

2) On line Γ_4 of the SCF output flow: no conductive heat transfer and uniform pressure distribution $p(r) = P_{\text{out}}$. The exact conditions at the output from the heat exchange zone are not known in advance, so usually a buffer zone is introduced without heat exchange with a coolant of such length that at the exit from it the flow can be considered stable, that is, such that the formulated conditions are fulfilled. Then, at the actual boundary of the output from the heat transfer zone, natural boundary conditions will automatically be obtained. The length of the required buffer zone is determined by the selection in the computational experiment.

3) On the line Γ_2 of heat exchange with the environment, the Newton-Richman condition with a given heat transfer coefficient α_0 is used:

$$\lambda \frac{\partial T}{\partial r}(R,z) = \alpha_0 \cdot \left(T_{\text{ex}} - T_{\text{w}}(z)\right). \tag{1}$$

Here λ is the thermal conductivity coefficient of SCF, T_{ex} is the temperature of the environment, T_{w} is the temperature of the flow of SCF on the wall Γ_2. Usually, the theoretical calculation of convective heat transfer is made for one of two modes: constant heat flow through the wall and constant wall temperature [5]. However, in a real technical device, none of these conditions are fulfilled, and the physical conditions of heat exchange are closer to the general condition of heat balance (1). For practical purposes, it is sufficient to study a problem with a constant heat transfer coefficient α_0, which, thereby, becomes an additional variable parameter of the model.

4) On the lines Γ_3 of the buffer zones, a thermal isolation condition is applied to ensure the stabilization of the input and output SCF flows.

The functions of the temperature dependence of the thermophysical parameters were used in the form of specially constructed approximations [16].

For the correct solution of the problem near the wall, a boundary mesh was used, consisting of 8 layers of thin rectangular elements with a decreasing thickness, formed in automatic mode (figure 1). The rest of the solution area was covered with triangular elements with a maximum size of 0.4 mm.

The constructed grid contained about 200,000 finite elements, the number of degrees of freedom is about 820000, the time to solve one problem on a 6-core processor with a frequency of 3.6 GHz was 25 minutes.

3. Computational experiment for calculating heat transfer in a one-dimensional SCF-oxygen flow

The computational experiment was carried out for the following parameter values covering the operating modes of the gasifiers:

- Input temperature $T_{\text{in}} = 90$ K.
- Output pressure $P_{\text{out}} = 20 \div 40$ MPa.
- Environment temperature $T_{\text{ex}} = 333$ K.
- Mass flow rate SCF $G = 0{,}03 \div 0{,}12$ kg/s.
- Heat transfer coefficient $\alpha_0 = 500 \div 2000$ W/(m^{2}s).

The bulk temperature T_{h} was calculated on the basis of the integral energy balance equation:

$$\Delta h(z) = \frac{2\pi R \alpha_0}{G} \int_0^z \left(T_{\text{ex}} - T_{\text{w}}(z')\right) dz', \tag{2}$$

where $\Delta h(z)$ is the change in the average mass enthalpy of the SCF flow from the input section to the section with the z-coordinate. Based on the thermodynamic dependence of the change in the specific enthalpy on the temperature at constant pressure, an approximating formula $T - T_{\text{in}} = \Theta(\Delta h)$ was constructed, using which the bulk temperature was calculated:

$$T_{\text{h}}(z) = T_{\text{in}} + \Theta\left(\Delta h(z)\right). \tag{3}$$

The calculation is also performed by calculating the increment of the average mass specific enthalpy in a given flow section with respect to the value at the input to the heat exchange zone according to the formula:

$$\Delta h(z) = \frac{2\pi}{G} \int_0^R h(r,z) v_z(r,z) \rho(r,z) r dr - h_0. \tag{4}$$

Here $h(r,z)$ is the local value of the specific enthalpy of SCF, $v_z(r,z)$ is the z-component of the local flow rate of SCF, $\rho(r,z)$ is the local value of the density of SCF. Then the temperature value T_h was calculated using the formula (3).

Within the limits of the error of the computational experiment and the error of the physical model, both calculated values of the bulk temperature coincided with each other up to the value of the viscous dissipation, which was insignificant in this experiment. This also justifies the neglect of viscous dissipation in calculating the bulk temperature, commonly used in SCF-flow simulation [12–15].

Using the equations (2) and (3), from the numerical solution of the problem for a given set of values of the parameters G, α_0, P_{out}, the distributions of the bulk temperature and the wall temperature along the flow length, selectively presented in figure 2, were calculated. For each set of parameters, the temperatures are represented by curves of the same type: the lower curve is the bulk temperature, the upper curve is the wall temperature.

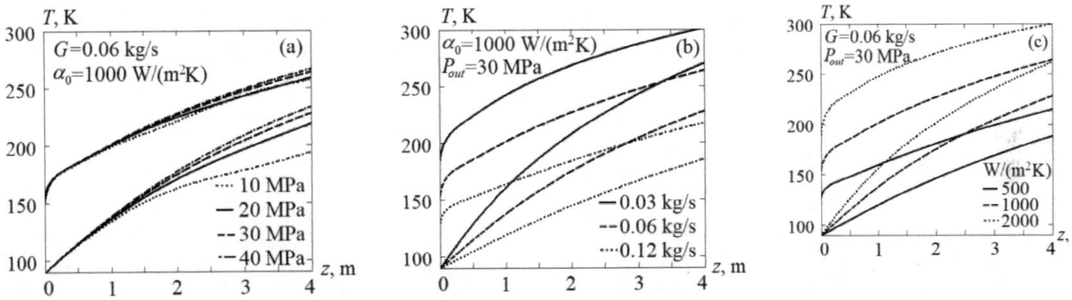

Figure 2. Dependence of the distribution of the bulk temperature (lower curves) and the wall temperature (upper curves) on the pressure (a) and mass flow rate (b) of SCF-oxygen and the heat transfer coefficient to the environment (c).

For comparison, a calculation was also made with a pressure of $P_{out} = 10$ MPa from the near-critical region, the result of which is shown in figure 2(a) by a dotted line. As can be seen, in this area there is a significant decrease in the intensity of heat transfer and an increase in the temperature factor.

On the distribution of the wall temperature, a characteristic initial section of thermal stabilization with a length of about 30 cm is observed. However, this section is not visually detected on the distribution of the bulk temperature. Therefore, the further analysis was carried out only for data from the heat-stabilized zone.

As can be seen from figure 2, the change in SCF pressure slightly affects the difference between the bulk temperature and the wall temperature, as well as the character of the change in these temperatures along the tube. However, an increase in the heat transfer coefficient leads to a significant change in the temperature profile and an increase in the output temperature. In turn, the change in the mass flow rate of SCF has a similar, but inverse, effect on the temperature profile.

4. Correlation of the heat transfer coefficient of SCF-oxygen

To calculate the local heat transfer coefficient of SCF, the heat flow density through the wall was used, which is found from the equation (1):

$$q(z) = \alpha_0 \left(T_{ex} - T_w(z)\right). \tag{5}$$

The local heat transfer coefficient in SCF, corresponding to the temperature T_h, was determined from the condition of the heat flow balance on the wall:

$$\alpha(z) = \frac{q(z)}{T_w(z) - T_h(z)} = \alpha_0 \frac{T_{ex} - T_w(z)}{T_w(z) - T_h(z)}. \tag{6}$$

Note that here the heat transfer coefficient is determined relative to the bulk temperature (3), so, generally speaking, it may not coincide with the values found from other correlations.

The local Nusselt number was determined on the basis of equation (6) by the formula

$$\mathrm{Nu} = \frac{2R\alpha}{\lambda(T_\mathrm{h})}. \tag{7}$$

This value was also determined for the thermal conductivity coefficient taken at the bulk temperature.

As a result, for each combination of parameters, a table $(\mathrm{Nu}, T_\mathrm{h}, T_\mathrm{w}, P, \alpha_0)$ is found for the dependence of the Nusselt number on the flow temperatures of the SCF and the wall, the pressure of the SCF, and the environment heat transfer coefficient. Some results of processing the computational experiment according to equations (6), (7) are presented in figure 3. A comparison is also made with the calculation of heat transfer in the near-critical region (figure 3(a), curve 10 MPa). As can be seen from the comparison, the exit to the near-critical region leads to a significant irregular decrease in the heat transfer coefficient. Therefore, the operating mode of the pacificators should be selected in such a way that the working area of the evaporator remains in the over-critical pressure region ($P \geq 3P_\mathrm{c}$, where P_c is the critical pressure).

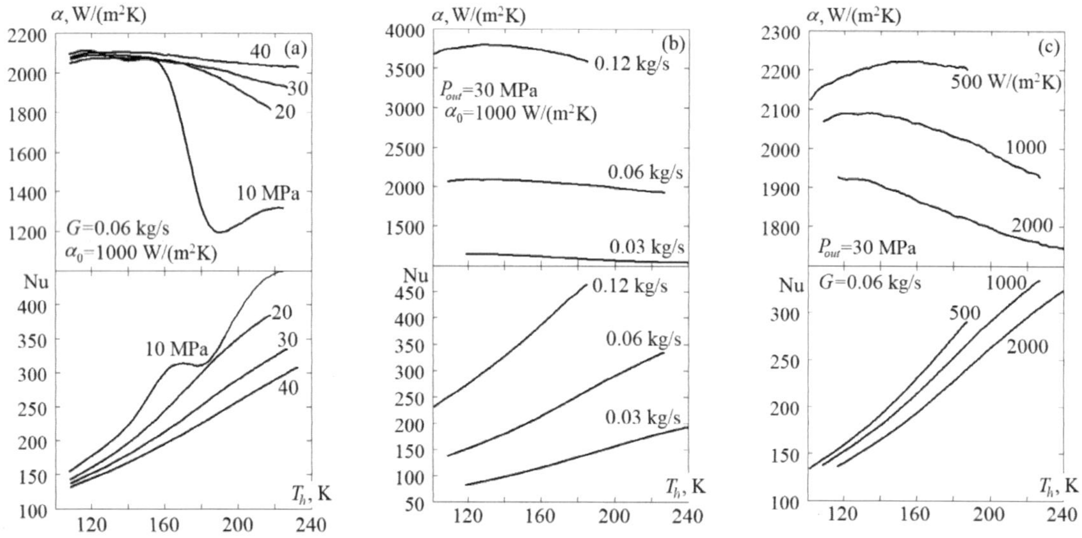

Figure 3. Temperature dependence of the heat transfer coefficient and the Nusselt number in SCF-oxygen at different pressure (a), mass flow rate (b), and heat transfer coefficient to the environment (c).

To obtain the correlation, a multiplicative model was used, in which all commonly used factors were included:

$$\mathrm{Nu}(T,P) = a_0 \,\mathrm{Pr}^{a_1}\,\mathrm{Re}^{a_2}\left(\frac{\mu}{\mu_\mathrm{w}}\right)^{a_3}\left(\frac{\lambda}{\lambda_\mathrm{w}}\right)^{a_4}\left(\frac{c_\mathrm{P}}{c_\mathrm{Pw}}\right)^{a_5}\left(\frac{\rho}{\rho_\mathrm{w}}\right)^{a_6}\left(\frac{\bar{c}_\mathrm{P}}{c_\mathrm{P}}\right)^{a_7}\left(\frac{2R\alpha_0}{\lambda}\right)^{a_8}\overline{\mathrm{Pr}}^{a_9}, \tag{8}$$

where $\bar{c}_\mathrm{P} = \dfrac{h_\mathrm{w} - h}{T_\mathrm{w} - T_\mathrm{h}}$ is the average integral heat capacity of SCF, $\mathrm{Pr} = \dfrac{\mu c_\mathrm{P}}{\lambda}$ is the Prandtl number in SCF, $\overline{\mathrm{Pr}} = \dfrac{\mu \bar{c}_\mathrm{P}}{\lambda}$ is the average heat capacity of the Prandtl number, often used [5, 6, 8] in constructing the correlations for SCF, c_P, ρ, μ are the specific heat capacity, density, and dynamic viscosity of SCF, which are functions of pressure and temperature [16]. The "w" index means the calculation of the value at the wall temperature T_w.

The coefficients a_i were determined by regression analysis of the linearized model obtained by logarithm of equality (8). The analysis showed that the coefficients a_1 and a_9 are strongly correlated, with the coefficient a_1 having less significance in the model. Also, the coefficients a_3, a_4 and a_8 were essentially insignificant. As a result of filtering out these coefficients, the model was reduced to the form

$$\mathrm{Nu}(T,P) = 0.0292\,\overline{\mathrm{Pr}}^{0.542}\,\mathrm{Re}^{0.773}\left(\frac{\rho_w}{\rho}\right)^{0{,}464}\left(\frac{\overline{c_P}}{c_P}\right)^{0.663}\left(\frac{c_P}{c_{Pw}}\right)^{0.357}. \quad (9)$$

In this model, all the coefficients showed a high level of significance, and the relative error in determining the natural (not logarithmic) values of the Nusselt number was: the maximum value is 3%, the average value is 1%.

5. Computational experiment and correlation for the heat transfer coefficient of SCF-nitrogen

The second series of calculations was performed for SCF-nitrogen with the same set of parameters as for oxygen. The results of processing the computational experiment according to equations (2)–(7) showed that the effect of pressure, mass flow rate of SCF and heat transfer coefficient for SCF-nitrogen qualitatively coincides with similar results obtained for SCF-oxygen.

To approximate the dependence of the Nusselt number in nitrogen, the model (8) was also used, but the coefficients a_i were considered linear functions of pressure. In the framework of the regression analysis, the same insignificant coefficients were established as in the case of oxygen, and an additional insignificant coefficient a_5 was established. In addition, the coefficient a_7 found a significant dependence on pressure. The resulting correlation for nitrogen has the form:

$$\mathrm{Nu}(T,P) = 0.03862\,\overline{\mathrm{Pr}}^{0.500}\,\mathrm{Re}^{0.745}\left(\frac{\rho_w}{\rho}\right)^{0{,}442}\left(\frac{\overline{c_P}}{c_P}\right)^{0.223-1.392\frac{P}{P_c}}, \quad (10)$$

where P_c is the critical nitrogen pressure (3.4 MPa). For equation (10), the relative error in determining the natural values of the Nusselt number was the maximum value of 5%, the average value of 1.5%.

6. Discussion

In the high-temperature limit, the last three factors in the left side of equations (9), (10) tend to one, and $\overline{\mathrm{Pr}} \to \mathrm{Pr}$, as a result, equations (9), (10) show good agreement with the classical Dittus-Boelter correlation, which is important for calculations in a wide temperature range. This also confirms the validity of equations (9), (10). The difference in the numerical coefficients is due to the fact that for the Reynolds numbers under consideration, the Dittus-Boelter correlation itself is inaccurate [5], and our correlations are also constructed for a range of high pressures.

The criterion equations (9), (10) do not explicitly include the factor of the environment heat transfer coefficient, its influence is automatically taken into account through the self-consistent formation of the wall temperature in the process of heat transfer. This means that a self-similar mode of heat transfer from SCF is formed in the considered over-critical region, which allows using universal dependences of the form (9, 10) for a wide range of pressures and mass flow rates.

To verify equations (9), (10), the bulk temperature distribution was calculated in the framework of a one-dimensional model of heat transfer in the flow [4]

$$Gc_P\frac{dT_h}{dz} = \pi\frac{T_{ex}-T_h}{R_t(T_h,T_w)};\quad R_t(T_h,T_w) = \frac{1}{2R\alpha_0} + \frac{1}{2R\alpha(T_h,T_w)}. \quad (11)$$

Here R_t is the thermal resistance between the SCF and the environment. To determine the wall temperature, we use the ratio following from the equation (6):

$$T_w = \frac{\alpha_0 T_{ex} + \alpha(T_h, T_w) T_h}{\alpha_0 + \alpha(T_h, T_w)}. \quad (12)$$

Equation (12) forms a rapidly convergent scheme of simple iterations for determining the wall temperature at a given bulk temperature. The result of solving the system (11), (12) with equations (7), (9) for the heat transfer coefficient in SCF-oxygen is shown in figure 4.

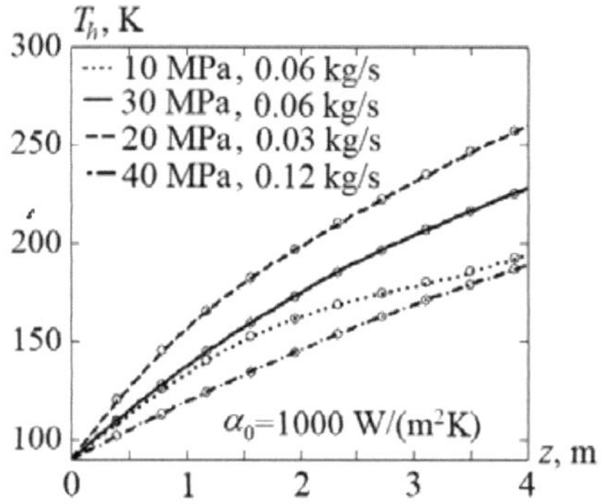

Figure 4. Solution of the one-dimensional equation (11) (lines) in comparison with the result of a computational experiment (circles).

Similar results are obtained for equation (10).

Figure 4 shows that the correlations (9) and (10) allow us to obtain an adequate distribution of the volume temperature of the flow of supercritical nitrogen and oxygen in the gasifier evaporator. It should be noted that equations (9) and (10) are constructed without taking into account the initial heat transfer section. However, the solution obtained with these equations coincides with the result of a computational experiment, in which this section is automatically taken into account. Hence, it can be concluded that when using the bulk temperature, determined by the average mass enthalpy, in the heat transfer equation (11), a separate account of the initial heat transfer section is not required.

Figure 4 also shows an example of calculating the bulk temperature distribution at a pressure of 10 MPa, which is located in the near-critical region and does not fall into the pressure range for which equations (9) and (10) are constructed. Nevertheless, even in this case, a good agreement is obtained with the result of the computational experiment. However, this fact needs to be experimentally verified [5].

Correlations (9), (10) were used to analyze and optimize the operation of the gasification unit SGU-7KM-U and showed results that are in good agreement with experimental observations.

References

[1] Dobrolensky Y P 1989 *Aviation equipment* (Moscow: Military Publishing House)
[2] Kutateladze S S 1991 *Handbook of Heat Transfer and Hydrodynamic Resistance* (Moscow: Energoatomizdat)
[3] Rohsenow W M, Hartnett J P and Young Cho Y I 1997 *Handbook of Heat Transfer* (New–York: McGraw–Hill)
[4] Avduevskiy V S, *et al.* 1992 *Fundamentals of Heat Transfer in Aviation and Rocket-Space Technology* (Moscow: Mashinostroenie)
[5] Petukhov B S, Genin . G and Kovalev S A 1974 *Heat transfer in nuclear power plants* (Moscow: Atomizdat)
[6] Petrov N E and Popov V N 1988 Heat transfer and hydraulic resistance with turbulent flow in a tube of water under supercritical parameters of state *Thermal Engineering* **35** 557-80

[7] Pioro I L and Duffey R B 2005 Experimental heat transfer in supercritical water flowing inside channels (survey) *Nuclear Engineering and Design* **235** 2407-30
[8] Huang D, Wu Z, Sunden B and Li W 2016 A brief review on convection heat transfer of fluids at supercritical pressures in tubes and the recent progress *Applied Energy* **162** 494-505
[9] Kays W M and Crawford M E 1993 *Convective heat and mass transfer* (New York: McGraw-Hill)
[10] Pope S B 2000 *Turbulent flows* (Cambridge, UK: Cambridge University Press)
[11] *ANSYS Fluent User's Guide* 2014 (Canonsburg, PA: ANSYS Inc.)
[12] Yang J, Oka Y, Ishiwatari Y, Liu J and Yoo J 2007 Numerical investigation of heat transfer in upwind flows of supercritical water in circular tubes and tight fuel rod bundles *Nuclear Engineering and Design* **237(4)** 420-30
[13] Du Z, Lin W and Gu A 2010 Numerical investigation of cooling heat transfer to supercritical CO2 in a horizontal circular tube *The Journal of Supercritical Fluids* **55** 116-21
[14] Cheng H, Ju Y and Fu Y 2018 Thermal performance calculation with heat transfer correlations and numerical simulation analysis for typical LNG open rack vaporizer *Applied Thermal Engineering* **149** 1069-79
[15] Cheng H, Ju Y and Fu Y 2019 Experimental and simulation investigation on heat transfer characteristics of supercritical nitrogen in a new rib tube of open rack vaporizer *International Journal of Refrigeration* **111** 103-12
[16] Borodkin S V, Bataronov I L and Ivanov A V Approximation of thermophysical parameters of supercritical nitrogen and oxygen 2021 *Proc. XXII Int. Sem. "Physical and Mathematical Modeling of Systems"* (Voronezh: VSTU) p 60-9

IOP Conf. Series: Materials Science and Engineering 1155 (2021) 012012 doi:10.1088/1757-899X/1155/1/012012

Calculation of the optimal power indicators for the engine of a small spacecraft

A Podolskiy [1], V Volynets[1], F Knyazhev[2], E Begendzhov[3] and A Guk [4]

[1] Bauman Moscow State Technical University, Hoisting and transport systems Department, 2nd Baumanskaya st., 5/1, Moscow, 105005, Russian Federation
[2] Moscow Power Engineering Institute (MPEI), Department of Electronics and Nanoelectronics, Krasnokazarmennaya st., 14, Moscow, 111250, Russian Federation
[3] Bauman Moscow State Technical University, Department of Ecology and Industrial Safety, 2nd Baumanskaya st., 5/1, Moscow, 105005, Russian Federation
[4] Bauman Moscow State Technical University, Electronic Technologies in Mechanical Engineering Department, ul. Baumanskaya 2-ya, 5/1, Moscow, 105005, Russian Federation

E-mail: ms.dokk@yandex.ru

Abstract. The use of small space satellites is perspectively for the development of sciences, development of new technologies, conducting small engineering tests, and tracking global earth processes. In this article the main dependencies between various parameters of a small space satellite, which operates on outboard air, have been determined. It make possible to carry out its most optimal design. An assessment of the possibility of extrapolation in the calculations of a small number of thermalized particles to volumes close to the real number is carried out. The influence of changing the parameters of the air intake of the installation on the power of its engine and the friction force acting on the installation has been investigated. As a conclusion restrictions on engine parameters are given. They were reduced to a dependence on the diameter of the air intake.

1. Introduction

Small spacecraft are used for testing and demonstrating technologies: systems and components of rocket and space technology, conducting research in the field of solar and space physics, planetology, astronomy and astrophysics, earth sciences, and space biology. They carry out the study of the boundaries of the earth's atmosphere, daily and global observations of earth processes that change during the day, the study of the physicochemical properties of the surfaces of planets or their atmospheres, the study of the survival and adaptation of organisms to outer space (to microgravity and to high levels of radiation).

The propulsion system for small spacecraft, which operates on the outboard air, collects fractions of the residual atmosphere and uses them as a working fluid for the electric rocket engine to compensate for the braking force acting on the spacecraft. The system theoretically allows the satellite to stay in orbit for an unlimited amount of time without a reserve of working fluid.

IOP Conf. Series: Materials Science and Engineering 1155 (2021) 012012 doi:10.1088/1757-899X/1155/1/012012

Calculations for a propulsion system were made in the article. System is capable to maintain the orbit of a small spacecraft in a circular orbit of the Earth with a height of 200 km, with a lifetime of at least 5 years. The main task of the device is to conduct radar reconnaissance.

1.1. Assumptions and baseline data

A typical installation scheme for outboard air consists of an air intake, a collector, and an electric rocket engine. Accepted assumptions:

- Incoming fractions enter the air intake without hindrance. Then we can consider a problem in which a given flow rate m falls into the volume of the collector N. This flow rate at a known altitude will depend only on the inlet diameter of the air intake.
- The flow of incoming fractions does not interact with the sidewalls of the initial path.
- The reverse flow and the flow of fractions to the engine are thermalized.
- There are no collisions of fractions with each other.

The return current from the air intake is represented by three components:

- Thermalized flow, driven by a pressure gradient.
- Fractions reflected from the walls of the air intake.
- Fractions that have reversed their direction as a result of collisions with other fractions.

Obviously, the greater the degree of pressure increase in the air intake, the greater the contribution of the first and third components.

The return current from the air intake is determined by the pressure difference created in it. In the calculations, we will neglect the effect of the incoming particle flow, which reduces the magnitude of the reverse current. In this case, we can use the usual equations of vacuum technology to estimate the flow.

Consider the initial design data, which were taken from the NRLMSIS-E model and reduced to the daily average:

Table 1. Initial data for the calculation.

Title	Numerical value
Inlet fractions temperature	835K
Temperature of fractions in the collector	300K
Average molar mass	20.56 g / mol
Density of fractions in orbit	$2.68*10^{-10}$ kg / m^3
Pressure	$9*10^{-5}$ Pa

At given pressures $p = 10^{-2}..10^{-5}$ Pa, a free-molecular flow regime is realized.

2. Main part

2.1. Air intake length

The efficiency of a mesh air intake is defined as the ratio of the number of fractions entering it to the number of fractions leaving the engine. To preserve the validity of the assumption of the motion of incoming fractions without collisions, the length of the air inlet L must be less than the free path of the fractions. In this case, the design pressure will be taken as the average between the inlet and the pressure in the manifold $p_{sr} = 0.001$ Pa, then the average free path will be $\lambda = 10$ m.

So, L should not exceed 10 m. Since with an increase in the length of the air inlet, its conductivity decreases and, accordingly, the reverse current, it is worth taking the maximum possible length of the air inlet.

To increase the efficiency of the air intake, we propose the use of a mesh throughout the entire length of the inlet line. The minimum grid step is set, on the one hand, by technological conditions, and on the other, by optical transparency and the absence of particle collisions with the wall. After several iterations, the author found the following parameters that satisfy all boundary parameters: the minimum grid spacing $a = 11$ *cm*, the length of the air intake $L = 3$ *m*. In this case, the average number of particle collisions with the wall is also two. The mean free path along the inlet path will be 10 m.

2.2. Selecting the diameter of the air intake

An important parameter of the air intake, which determines the operation of the entire system, is its diameter. There are two restrictions on its size. The minimum air intake diameter is set by the air mass flow rate, at which a helicon discharge can be ignited at an acceptable level of input power. The maximum diameter is limited by the electrical power, which is allocated for the needs of the electric rocket engine. The larger the diameter of the air intake, the greater the force of aerodynamic drag and, accordingly, the greater the required engine thrust and power consumption.

On the other hand, it is important to choose the diameter of the discharge chamber corresponding to the diameter of the inlet. As shown earlier, the larger this diameter, the smaller the diameter of the air inlet can be. It is also important that the cross-section of the discharge chamber is significantly less than the diameter of the air intake. The larger the size of the discharge chamber, the more fractions fly through the system.

According to the author's analysis, the maximum diameter at which a helicon discharge can be ignited at an acceptable power level (up to 1 kW at a flow rate of no more than 1.5 mg / s) was 100 mm.

2.3. Numerical modeling of a collector

A collector is provided for thermalization of fractions entering the system. In this collector fractions repeatedly collide with the walls and lose their energy. However, it is not possible to estimate its size analytically. For this purpose, as well as to check a number of assumptions of the model, numerical simulations were carried out.

It is necessary to ensure that the numerical model corresponds to real conditions. This correspondence is determined by three key parameters: the concentration of fractions, the interaction of fractions with the wall, and the interaction of fractions with each other.

Within the framework of numerical calculations, a limitation is imposed on the concentration of fractions, since a large number of fractions requires huge volumes of computer RAM. In our case, for 32 GB of RAM, the upper limit for the number of modeled fractions became $N = 5*10^5$, a similar number of fractions was used by other authors. At the same time, the concentration of fractions in the studied orbit is $N = 8*10^{15}$. On the other hand, the calculation of the flight path of a large number of fractions is associated with time constraints. So, it took 4 days to calculate 1 ms for 10.000 fractions, which makes it impossible for this investigation to obtain a large amount of data.

In this regard, the first modeling task was to estimate the error in modeling a much smaller number of fractions.

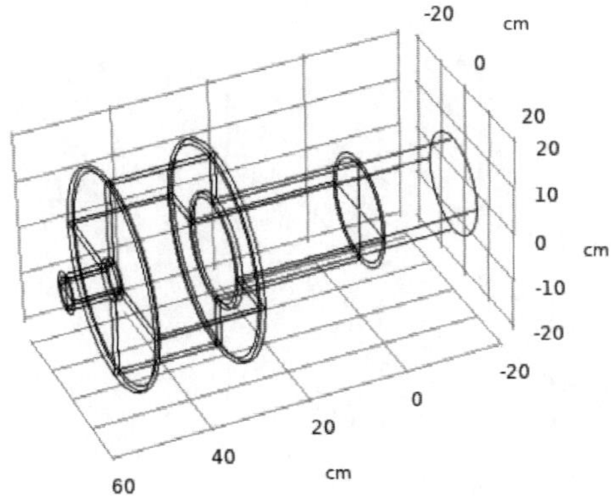

Figure 1. Calculation model type.

At the initial moment of time, the entire volume of the object under study contained 1000 randomly located fractions, they moved in the same direction as the axis of the air intake at a speed of 7.8 km / s.

At the same time a "cylinder" appears in front of the investigated object, containing the investigated number of fractions. All the fractions that appeared at a given speed moves from the cylinder to the air intake. The cylinder is replenished with fractions at that moment, when the last particle leaves its limits. This allows you to simulate a continuous stream of fractions.

In this model, the fractions do not interact with each other, since their free path lies in the range from 1 m to 380 m, which is much larger than the characteristic dimensions of the air intake. This means that the fractions mainly interact with the walls of the air intake.

2.4. *Influence of the number of fractions on the model*

To assess the effect of particle concentration on the calculation results, a calculation was performed with a different number of incident fractions: 500, 750, 1000, 2000, 5000, 10000.

In this problem, the estimated parameter is the number of fractions entering the hole on the back of the collector per unit time. This hole acts as a motor. Below (figure 2) is a graph of the dependence of the number of fractions in time for different numbers of incident fractions.

The initial burst of particle flood is due to the exist of initial fractions in the air intake. It can be seen that the particle flood reaches the "shelf" in about 1 μs.

After analyzing the obtained result, we can see that the flood grows linearly with an increase in the number of fractions. This allows us to conclude that in the absence of a grid, an increase or decrease in the number of fractions does not cause additional effects. This means that the inlet volume and the meshless inlet in general can be modeled using fewer fractions and extrapolated to a larger number.

2.5. *Particle scattering*

All equations of the vacuum technique are obtained with the condition of diffuse collision of fractions with the wall. The validity of this condition has been confirmed by numerous experiments in the case of a thermalized flow.

However, in our case, the flow is thermalized only in the collector. Due to the high speed, a significant number of fractions can also mirror reflect from the wall of the air intake or from the walls of the mesh. Roughness also determines the magnitude of this effect. The smoother the surface, the more fractions are mirrored.

Since the influence of this effect is difficult to evaluate analytically, a numerical experiment was carried out. In this experiment was estimated the effect of the fraction of diffusely scattered fractions from specularly reflected ones. This ratio was taken to be 0 (diffuse scattering), 0.1, 0.5, 0.9, 1 (specular reflection).

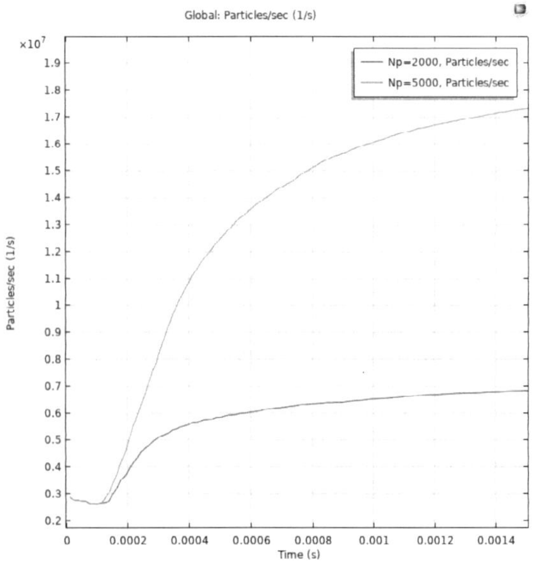

Figure 2. Particle flow into the engine with different numbers of incident fractions

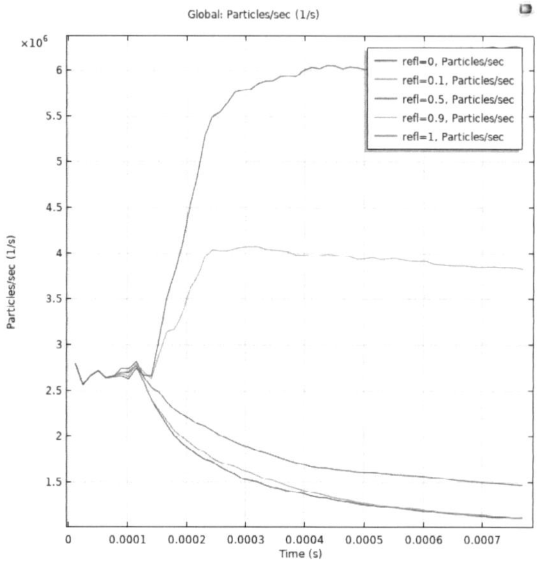

Figure 3. Particle flood into the engine at different values of the reflection coefficient

Figure 3 shows the change in the particle flood at different values of the coefficient. It can be seen that in the range of 0.5-1 its effect is insignificant, but in the range from 0 to 0.5 there is a sharp jump.

It can be concluded that the effect of specular reflection of fractions insignificantly affects the particle flood, even with a significant number of specularly reflected fractions. For all future calculations the specular reflectance is assumed to be 0.8.

2.6. Optimal geometry

An important task solved by a numerical experiment is to find the maximum compression ratio in the reservoir.

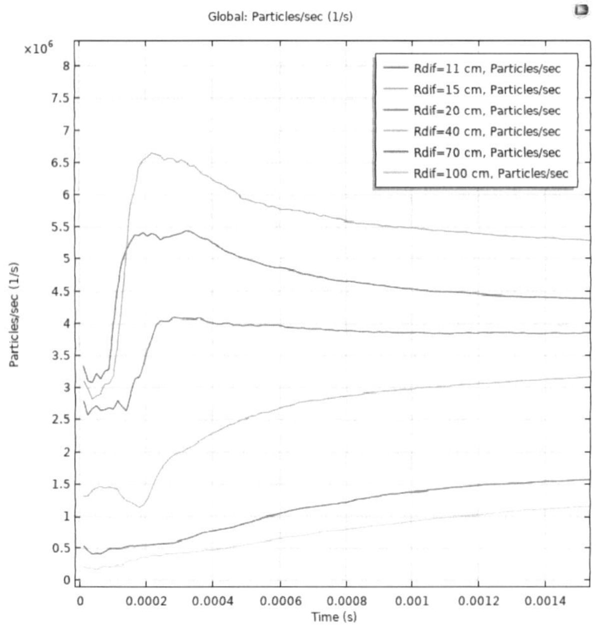

Figure 4. Particle flow into the engine with different collector diameters.

At the maximum compression ratio the flow of fractions into the engine will also be maximum, since other parameters of the inlet channel do not change.

In the developed model was investigated a case with a constant reservoir length and variable diameter. The semidiameters were investigated: 11 cm (equal to the pipe diameter), 15 cm, 20 cm, 40 cm, 70 cm, 100 cm.

It can be seen from the figure 4 that there is an optimal ratio of the collector diameter to the air intake diameter of 1.2.

2.7. Dependence of the characteristics of the installation on the parameters of the air intake

After calculating a remote control operating only on outboard air, it is necessary to assess the dependence of characteristics on the parameters of the installed confuser (air intake).

The coefficient of the collection efficiency of molecules (capture coefficient) can be defined as the ratio of fractions emitted through the nozzle of the propulsion system to the number of fractions that have passed through the confuser:

$$\eta_c = \frac{N_T}{N_i} \tag{1}$$

From the number of fractions it is possible to count the specific mass flow rate. Consequently the mass of fractions entering the confuser per unit time is:

$$\dot{m}_i = \rho \cdot V \cdot S_1 \tag{2}$$

where S_1 is the area of the inlet section of the confuser. So we can get:

$$\eta_c = \frac{\dot{m}_T}{\rho \cdot V \cdot S_1} \tag{3}$$

Ehe engine thrust $F_T = \dot{m}_T \cdot g \cdot I$, so a relation between specific impulse and particle capture coefficient:

$$\eta_c \cdot g \cdot I = \frac{1}{2} \cdot V \cdot C_d \tag{4}$$

The capture ratio η_c is determined by the geometric parameters of the air intake, which were described earlier.

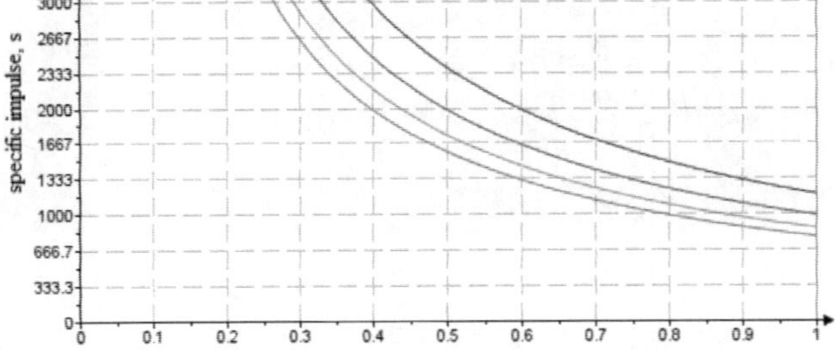

Figure 5. Dependence of the unit specific impulse on the efficiency of the air intake at different values of the shape factor.

From the above, it is possible to obtain a dependence that makes it possible to assess the possibility of using various types of engines in outboard air systems to compensate for the friction force acting on the spacecraft.

In this work it was also shown that the ratio of thrust to friction force is constant. From this it can be concluded, that if any engine is capable of providing the required specific impulse on an oxygen-nitrogen mixture, then it is possible to find an orbit in which the thrust force for it will be equal to the friction force - there will be an orbit in which it will be able to compensate the frictional force.

In this calculation, however, the increase in the midsection due to the collector was not taken into account. Previously, the optimal ratio of the collector diameter to the diameter of the inlet path was obtained.

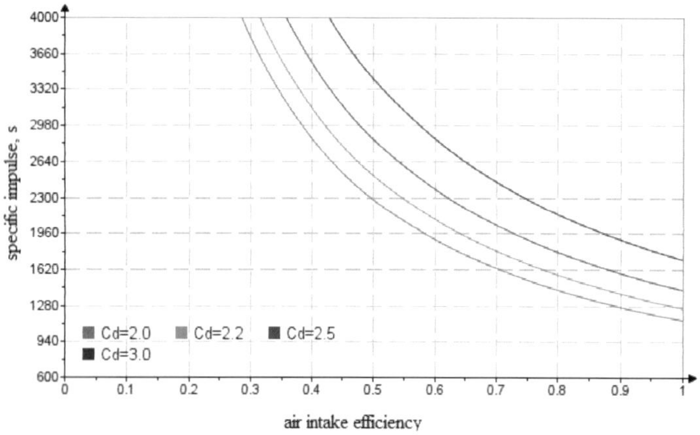

Figure 6. Dependence of the required specific impulse of the installation on the efficiency of the air intake.

2.8. Calculation of aerodynamic drag and engine power

In the simplest case, the friction force acting on the spacecraft can be calculated as:

$$F_d = \frac{1}{2}\rho \cdot V^2 \cdot S \cdot C_d \tag{5}$$

where C_d is the shape factor, depends on the spacecraft shape and flight altitude.

For heights of 200-300 km, it lies within 2.1-2.7, S is the spacecraft cross-sectional area, V is the orbital velocity, ρ is the density of the atmosphere.

In this case, the assumption of the continuity of the medium is used. However, the friction force calculated by this method is in good agreement with the existing data on satellite flights at low altitudes.

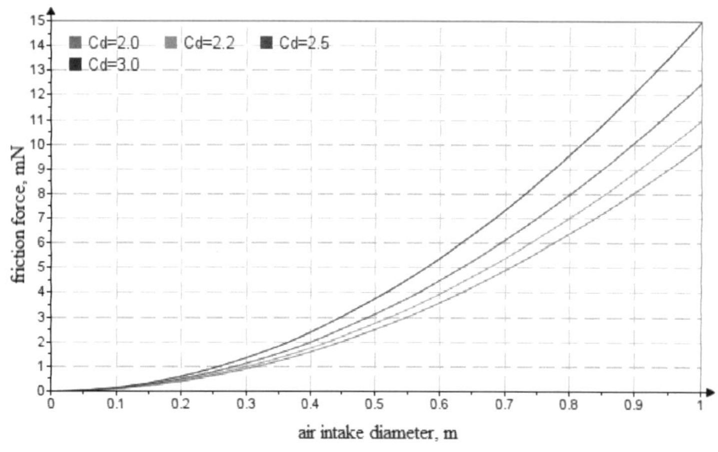

Figure 7. Calculation model type.

More complex models, in which the friction force is calculated when considering the fractions bombarding the spacecraft, gives an underestimated result. The results of calculating the friction force are shown in figure 7.

Knowing the price of engine thrust and taking into account the midsection of the collector, you can go to the required power:

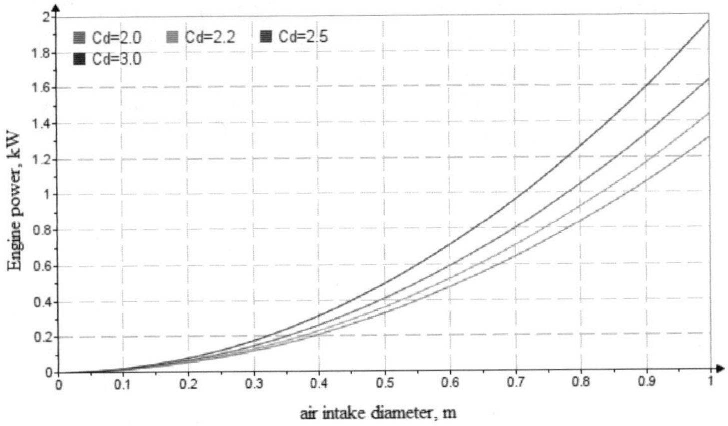

Figure 8. Calculation model type.

3. Conclusion

From the analysis, restrictions on the parameters of the engine follow, which were reduced to a dependence on the diameter of the air intake:

- Required specific impulse: as the diameter of the air inlet increases, the efficiency of collecting fractions decreases. Accordingly, the larger the diameter of the air inlet, the greater the particle velocity must be to compensate for the friction force.
- Flow limitation: to ensure a minimum operating flow of 0.1 mg / s, the air intake diameter must be at least 0.28 m.
- Power limitation: with an increase in the midsection, the aerodynamic drag of the spacecraft increases, and the required thrust to maintain the orbit correspondingly increases. Knowing the energy price of thrust, you can get the electric power of the engine.

References

[1] Longmier B W *et al.* 2011 VX-200 magnetoplasma thruster performance results exceeding fifty-percent thruster efficiency *J. Propul. Power* **27(4)** 915-20

[2] Stenzel R L 1976 Whistler wave propagation in a large magnetoplasma *The Physics of Fluids* **19(6)** 857-64

[3] Boswell R W *et al.* 1997 Helicons—the early years *IEEE Transactions on Plasma Science* **25(6)** 1229-44

[4] Hole M J and Simpson S W 1997 Performance of a vacuum arc centrifuge with a nonuniform magnetic field *Physics of Plasmas* **4(10)** 3493-500

[5] Barbashov N N and Chetvernin M Y 2021 Methods for improving accuracy in measuring deviations from roundness and cylindricity *IOP Conference Series: Materials Science and Engineering* **1047** 012032

[6] Chen F F and Boswell R W 1997 Helicons—the past decade *IEEE Transactions on Plasma Science* **25(6)** 1245-57

[7] Takahashi K *et al.* 2010 Characterization of the temperature of free electrons diffusing from a magnetically expandig current-free double layer plasma *Journal of Physics D: Applied* Physics

43(16) 162001-1-4
[8] *World Wind Energy Association report for 2019* Retrieved from: https://wwindea.org/
[9] Masson-Delmotte V *et al.* 2018 *Summary for Policymakers. In: Global Warming of 1.5°C. An IPCC Special Report on the impacts of global warming of 1.5°C above pre-industrial levels and related global greenhouse gas emission pathways, in the context of strengthening the global response to the threat of climate change, sustainable development, and efforts to eradicate poverty* (Geneva, Switzerland: World Meteorological Organization) p 32
[10] Pan X and Kraines S 2001 Environmental Input-Output Models for Life-Cycle Analysis *Environmental and Resource Economics* **20** 61-72
[11] Barbashov N N and Abdullina L R 2020 Selection of effective criteria for determining the volume of measurements *Journal of Physics*: *Conf. Ser.* **1515** 052029

IOP Conf. Series: Materials Science and Engineering 1155 (2021) 012013 doi:10.1088/1757-899X/1155/1/012013

Research and optimization of the hydraulic performance of bench lines for testing of low thrust spacecraft engines

V P Nazarov, A A Zuev, D A Zhuikov and M I Tolstopyatov

Department of Aircraft Engines, Reshetnev Siberian State University of Science and Technologies, 31 Krasnoyarsky Rabochy Av., Krasnoyarsk, 660037, Russian Federation

E-mail: nazarov@sibsau.ru

Abstract. Low thrust engines perform important functional tasks in control systems of motion, orientation and stabilization of spacecraft on various paths and space orbits. They are subject to high-reliability and long operating time in pulsed modes with a large number of starts. Bench tests of the engines are a special part of the program development of low thrust engines with high reliability and efficiency. The article analyses wave processes in hydraulic lines of test benches, which prevent obtaining reliable parameters and performances of engines during tests. Authors show that the exclusion of oscillation processes in test bench systems is the main condition for the organization and planning of tests. Similarity conditions for bench hydraulic lines of spacecraft propulsion systems are developed. A methodology for estimating the frequency responses of the bench lines during pulsed test modes is presented. Mathematical functions of geometric parameters of the bench hydraulic lines on the pulse responses of low thrust engines are obtained, allowing optimization of the bench propellant feeding system to avoid resonance oscillation processes. Research results provide reliable estimates of the parameters and reliability of low thrust engines.

1. Introduction

Low thrust engines (LTE) are designed for spacecraft motion control on space paths, spacecraft stabilization, orientation and correction in space. Design features of the LTE defined by their special mass and dimensions characteristics, specific processes of mixing and combustion of propellant components, absence of regenerative cooling of the engine chamber. They may operate in continuous and pulse modes and the design of LTE provides for multiple firings with varying durations (from several hundredths of a second to tens of seconds). The total number of cycles of LTE can amount to as many as one million starts, with the active life of a spacecraft up to 15 years.

A specific feature of LTE is the non-stationarity of its operating mode. When the engine thrust is increased at build-up time and reduced at cut-off time, the mixing, combustion and exhaust processes take place under off-nominal conditions, engine efficiency is significantly reduced. The specific impulse in such modes is lower than in continuous modes: $J_{non-stat\ mode} < J_{cont\ mode}$; its reduction can reach up to 50% [1].

The energy parameters and thermal condition of the LTE during its operation are determined by factors whose influence cannot often be determined by theoretical calculations. In these cases, only experimental studies make it possible to make a final choice of the optimum rocket propellant mixing schemes and parameters, design of units and the engine configuration as a whole. It should be noted that nowadays there is no acceptable theory of experimental testing of complex technical systems to optimize the design parameters and other characteristics of rocket-space technology products [2].

2. Research task

In the process of designing experimental (prototype) development of LTE for various applications, much attention is paid to bench testing methodology, technical equipment of benches simulating the impact of physical conditions in space and the use of diagnostic methods and equipment for various physical investigations and measurements.

The requirements for firing tests benches of LTE are specific, the main of which are: to achieve the required high-altitude environment; to compliance with the laws of variation of engine inlet pressures and combustion chamber pressures, ensuring that the temperature of the propellant components is within the specified limits (both negative and positive).

The dynamic processes occurring in the propellant lines depend on many factors. It is known [3, 4, 5] that the type of dynamic processes in lines has a significant influence on engine parameters, which is especially relevant for LTE, as operation in pulse modes, causing dynamic processes, is one of the most typical features of operation of such engines. It is therefore necessary to ensure that the dynamic processes occurring in the test bench lines are consistent with those in the propellant lines of propulsion system in order to reliably determine the performance of the engine during the test run.

A simple solution to this problem only by the identical design of the propellant feeding system for LTE on the bench and in the propulsion, system does not appear to be possible for a number of reasons:

- the need for a high degree of safety in bench testing, which requires sufficient distance between the supply tanks and the vacuum chamber;
- the possibility of implementing various test programs on the test bench, which is possible due to the installation of additional fittings (valves, throttling devices and measuring instruments) in the propellant line;
- design features of the test bench.

Popov and other authors considered the task of velocity propagation when viscous fluid velocity increases in a circular pipe under effect of suddenly appeared constant differential pressure [6, 7, 8]. Solutions to the Navier-Stokes equation are obtained for an incompressible fluid assuming that the velocity distribution is determined beyond the section where the flow stabilises along the length of the pipe. As the friction, effects propagate towards the pipe axis, the unevenness of the velocity distribution increases and the velocity profile asymptotically approaches parabolic.

In order to realize the condition of similarity of hydrodynamic processes it is necessary to ensure similarity of boundary and initial conditions, on which solutions of equations depend. Initial and boundary conditions are determined by the values of pressure p and flow velocity c at the initial moment of time and the patterns to be satisfied at the end sections of the tube during the non-stationary process.

An analysis of the equations of fluid flow in a circular pipe show that hydrodynamic similarity of propellant lines in the propulsion system and on the test bench is achieved by compliance of the numbers in the equations: St = idem, Eu = idem, Fr = idem and Re = idem, as well as equality of Mach number $\mathrm{M} = C/a$ (where a is reduced speed of sound in system "line - liquid"), dimensionless wave resistance $a = p\ c\ a\ /\ p^{*}$ and relative friction losses $\Delta p\ /\ p^{*}$. It is also necessary to ensure that the hydraulic lines and boundary conditions on which the solutions to the equation depend are geometrically similar.

Therefore, the similarity conditions are criteria that should be the same for the test bench and the propulsion system at a specified propellant components flow rate $\dot{m}$.

The task of oscillations of a viscous incompressible fluid is considered in a number of papers [6, 8, 9]. Solutions to this problem show that the law of influence on the fluid flow significantly affects the distributions in non-stationary flows. Therefore, when modelling non-stationary hydrodynamic processes, it is very important to ensure the identity of the type and mode of the processes, in addition to compliance with the similarity criteria. For the purpose of mathematical models, fluid flow is described by equations in cross sectional averaged flow but time-dependent variables. By means of this hydraulic method, a one-dimensional mathematical description of fluid flow is obtained. In which quasi-stationary coefficients are used, having at each moment in time the values obtained by steady-state motion of the medium at velocities equal to the instantaneous velocities. It is noted in [9] that the use of quasi-stationary coefficients does not lead to significant errors in calculations of non-stationary hydrodynamic processes in the case where the distribution of velocities over the flow section over time is not significant.

Two types of equation solutions are used when analysing the wave performance of lines: for forced oscillations of the system, which give the frequency response, and for free oscillations, which define the transient motion of the system.

A special feature in the creation of test bench lines for feeding propellants into the engine under pulsed mode conditions with a frequency f is the matter of the natural frequency of oscillation of the fluid in the line f_o. The natural (free) oscillations in a system arise after the termination of the action of external disturbances. The free oscillations, and first of all the frequencies of these oscillations, are of interest for explanations of the peculiarities of the dynamical performances of the system.

It is obvious that at $f_o \approx f$, processes will be initiated in the test bench lines, which will significantly affect the performance of the tested liquid propulsion system, which will inevitably lead to unreliability of the engine parameter estimates based on the test results.

3. Optimise the performance of the bench's hydraulic lines

The pulsed modes of LTE generate turbulent and non-stationary flow processes of the propellant components in the lines. The features of non-stationary flows in lines are discussed in [7, 8]. The dynamic operations occurring in the propellant lines depend on many factors, determined by the properties of the propellants, the design of the lines and the sequence of the LTE [10]. The nature of non-stationary processes is particularly influenced by the inertial properties of the fluid at small cross-sections and long line lengths [6, 11].

Optimisation of firing test processes for LTE requires solving the task of ensuring identity of the propellant supply system for LTE including compliance of inertial, wave and hydraulic performance of feeding lines.

Therefore, the basic criteria for optimising the performance of the test bench lines to obtain reliable information on the functional capability of LTE based on the results of fire tests should be formulated as follows:

- the propellant flows in the bench and spacecraft lines should be similar;
- the oscillating processes resulting from the pulsed operation of the engine to be tested does not have to introduce additional errors in the evaluation of the performance of the engine.

Similarity conditions of hydrodynamic processes occurring during non-stationary flow of an incompressible fluid could be described by means of the equations of flow and continuity of the fluid [12].

The proximate use of differential equations leads to considerable and usually insurmountable difficulties in solving specific tasks. The complexity of the problem of calculating turbulent currents is caused by the fact that turbulent currents are non-stationary, with changes in magnitudes over time having the character of a multitude of pulsations of different frequencies and amplitudes. The paper [9] states that the coefficients of the equations could be expressed using known numbers:

$$\mathrm{St} = \frac{L}{GT} - \text{Strouhal}\,;\ \mathrm{Eu} = \frac{\rho^*}{\rho C^2} - \text{Euler};\ \mathrm{Re} = \frac{CL}{\upsilon} - \text{Reynolds};\ \mathrm{Fr} = \frac{C^2}{gL} - \text{Froude}$$

where p and υ are density and kinematic viscosity of the fluid.

Equation of dynamics of the test bench line supplying the propellant component to the engine is written in the form:

$$j \cdot \frac{d\dot{m}}{dt} + \xi \cdot \frac{\rho c^2}{2} = p_{in} - p_{out} + p_m, \tag{1}$$

where $j = l/F$ is inertial loss coefficient; l is the length; F is the cross-sectional area of the line section; $\dot{m}$ is the rate of fluid flow; p_{in} and p_{out} are inlet and outlet pressures of the line section; p_m is the pressure of the mass load of the liquid column on the area F; ξ is the hydraulic resistance coefficient.

The hydraulic resistance coefficients for steady and non-stationary flow of the fluid have different values. The steeper the front of the liquid flow rate is rising at time and the higher the frequency of oscillation of the flow parameters, the greater the difference of these values. However, the hypothesis of equality of friction coefficients for steady and non-stationary modes is used in the study of dynamic modes in hydraulic systems with low-viscosity fluids [13, 14].

The practice of mathematical modelling [15, 16] shows that most tasks of low-frequency dynamics (up to ≈ 20 Hz), including those related to investigation of fluid motion in bench hydraulic lines, can be described with a sufficient degree of accuracy by simple dependences if we consider a system with lumped parameters. In this case the length of the line subdivision section l_{max} could be an order of magnitude smaller than the wavelength of the highest process frequency f_{max} taken into account in the calculations. Taking the speed of propagation of longitudinal waves equal to the speed of sound in a liquid $v_{long} = c$, and considering that the wavelength $\lambda_{long} = c/f = c \cdot T$, we can write:

$$l_{max} \leq \frac{2\pi a}{\omega_{max} n} = \frac{a}{f_{max} n}, \tag{2}$$

where n is the safety factor; $n \leq 6 \dots 12$.

The paper [8] notes that the hydraulic resistance of the line causes only a slight reduction of the resonance frequencies.

The resonance frequencies have the values below:

$$f_p = \frac{1+2n}{4\pi d} \cdot \left(\frac{E_{liq}}{\rho_{liq}}\right)^{1/2}, \tag{3}$$

where $n = 0,1,2 \dots$; $c = (\frac{E_{liq}}{\rho_{liq}})^{1/2}$ is the velocity of propagation of disturbances in the line (speed of sound); l is the length of line.

In order to avoid the phenomenon of resonance causing negative (in terms of influence on the test results) non-stationary processes, the natural frequency of the lines must be significantly different from the frequency of forced response excited by the pulse operation of LTE being tested. The following frequency dependence f_o of the maximum frequency of forced response $f_{engin\,max}$ is assumed, taking into account the safety factor $f_o \geq f_{eng\,max}$.

Using the safety factor $n = 10$, we obtain $f_{o\,min} = 10 \cdot f_{eng\,max}$, and then we write formula (3) like this:

$$l_{bench\,max} = \frac{0.1a_\infty}{\pi f_{eng\,max} (2 + \frac{2D}{\delta})^{1/2} \cdot \frac{E_{liq}}{E_{line}}},$$

where $l_{bench\,max}$ is the maximum length of the bench line.

As the value of a_∞ is significantly greater than β_t, i.e. given the inequality $a_\infty \gg \beta_t (T - T_0)$, , we can write:

$$l_{bench\,max} = \frac{0.1a}{\pi f_{eng\,max} (2+\frac{2D}{\delta})^{1/2} \cdot \frac{E_{liq}}{E_{line}}}. \tag{4}$$

In the paper [13] dependence of optimum length of the line l_{line} and wavelength λ_{long} of the highest frequency of the process of forced longitudinal oscillations $l_{line} = 0.1 \cdot \lambda_{long}$ is given. Using the dependence $\lambda_{min\ long} = a/f_{max}$, where a is the speed of sound of the liquid, write the equation for the design of bench lines:

$$l_{bench} \leq \frac{0.1a}{f_{eng\ max}} \text{ or } l_{1bench\ max} = \frac{0.1a}{f_{eng\ max}}. \quad (5)$$

Thus, a restriction is defined: the length of the test bench line could be such that its natural frequency is at least 10 times the maximum frequency of LTE pulse mode during testing [17]. Considering the two conditions for the maximum length of the bench line, it should be noted that the permissible value will be the one that satisfies the condition:

$$l_{allowable\ bench} = min(l_{bench\ max}, l_{1bench\ max}). \quad (6)$$

Based on the mathematical equations and conditions obtained, it is possible to calculate design values for the lengths of the bench lines for the conducted tests of LTE, ensuring reliable values of the test results.

4. Conclusion

On the basis of the conducted research and based on the results of scientific works of other authors, possible methods of conducting tests of LTE with simulation of operating conditions of engines in the propulsion systems of spacecrafts are presented.

The analysis of features of pulse modes of operation of LTE as part of spacecraft defining the requirements to the design of special test equipment of hydraulic systems of the bench, the mechanisms of influence of these modes on the reliability of the results of bench tests at design development of the engine were investigated.

A method of estimating the frequency responses of the hydraulic lines of a test bench for LTE during pulse modes is presented.

References

[1] Nazarov V P, Zhuravlev V Y, Kraev M V and Kraeva E M 2015 *Spacecraft Rocket Engines* (Krasnoyarsk: Siberian Aerospace University) p 200

[2] Yatsunenko V G, Nazarov V P and Kolomencev A I 2016 *Bench Tests of Rocket Engines* (Krasnoyarsk: Siberian Aerospace University and Moscow Aviation Institute) p 248

[3] Kudryavtsev V M and Vasiliev A P 1993 *Fundamentals of Theory and Calculation of Liquid Propellant Rocket Engines* (Moscow: Higher School) p 364

[4] Moshkin E K 1970 *Non-Stationary Modes of Operation of the Liquid Propellant Rocket Engine* (Moscow: Mashinostroenie Publishers) p 336

[5] Polukhin D A, Oreshenko V M and Morozov V A 1987 *Testing of Pneumatic and Hydraulic for Propulsion Systems* (Moscow: Mashinostroenie Publishers) p 186

[6] Popov D N 1969 Hydraulic resistance of pipelines during a transient turbulent flow of the fluid *J. Proc of High. Educ. Inst.* **9** 89-93

[7] Popov D N 1987 *Dynamics and Regulation of Hydraulic and Pneumatic Systems* (Moscow: Mashinostroenie Publishers) p 201

[8] Popov D N 1970 Features of non-stationary flows in pipes *J. Proc of High. Educ. Inst.* **7** 78-92

[9] Popov D N, Panaiotti S S and Ryabinin M V 2002 *Hydraulic* (Moscow: Bauman Moscow State Technical University) p 462

[10] Glickman B F 1979 *Non-Stationary Flows in Hydraulic and Pneumatic Circuits* (Moscow: Mashinostroenie Publishers) p 125

[11] Annenkova G K and Glickman B F 1973 Effect of viscosity on forced responses of a compressible fluid in a cylindrical tube *J. Proc. of the USSR Acad. of Sci. Energy and Transport* **4** 138-145

[12] Haustov A I 1998 Modelling viscous fluid flow in dimensional pipes *J Rep. of the Acad. Of Sci.* **6** 1-4

[13] Navruzov K 1986 *Hydrodynamics of Pulsating Flows in Pipelines* (Tashkent: Academy of Sciences) p 205
[14] Faizulayev D F, Navruzov K and Fattayev F N 1981 Pulsating flow of a viscous incompressible fluid in a circular branch pipe *J. Rep. of the Uzbek SSR Acad. of Sci.* **10** 20-2
[15] Belyaev E V and Chvanov V K 1999 *Mathematical Modelling of Liquid Propellant Rocket Engines* (Moscow: Moscow Aviation Institute) p 324
[16] Belyaev E V, Belik N P and Uvarov E I 1979 *Spacecraft Jet Control Systems* (Moscow: Mashinostroenie Publishers) p 232
[17] Kraev M V, Krushenko G G, Kaychuk L N and Yatsunenko V G 2008 Development of the Basic Systems of a Firing Test Bench for Low Thrust Liquid Propellant Rocket Engines *J. Inst. of Computational Modelling, Siberian Branch of the Russian Acad. of Sci.* **1** (*Preprint* icm.krasn.ru/rprojects_refs.php?year=2008&kind=8&lang=rus)

Improvement of the processes of forming tight shells of double curvature

V A Markovtsev[1], E G Karpukhin[2], V A Mikheev[3], Roberto de Alvarenga[3] and A Yu Kosolapova[3]

[1] PC «Ulyanovsk Research Institute of Aviation Technology and Production Organization», 34, str. Vracha Mihailova, Ulyanovsk, 432057, Russia
[2] Ulyanovsk State Technical University, 32, Severny Venetz str., Ulyanovsk, 432027, Russia
[3] Samara National Research University, 34, Moskovskoye shosse, Samara, 443086, Russia

E-mail: vamicheev@rambler.ru

Abstract. The article analyzes the process of shaping double curvature shells made of aluminum sheet material used for covering the outer contour of an aircraft. The main rejection signs of casings made by the method of shaping by wrapping are considered. The main problems arising during the shaping of double-curvature shells by wrapping are presented, and the ways of their solution are considered and proposed.

1. Introduction

The modern development of air transport depends mainly on the creation of technology aimed at the development of methods, methods and means of obtaining shells of double curvature of complex spatial forms, the availability of production conditions and modern equipment with program control, the use of new materials from aluminum alloys, despite the great popularity composite materials on the same aluminum base. Most often, these are thermally or precipitation hardened aluminum alloys with a complex phase composition Al - Mg - Cu, including an aluminum-lithium alloy of the Al - Mg - Li system.

At the same time, the technological process is organized to achieve the goal: to improve the quality of the shells, by ensuring a more uniform distribution of the thickness of the workpiece after shaping by the wrap and to reduce the likelihood of its waviness and folds formation. The technological process (TP) is characterized by parameters that are divided into the following groups: material properties; friction conditions; dimensions of the sheet blank; heat treatment modes; tightening punch parameters; kinematic conditions of the shaping process by the skin. TP management is built on the basis of the analysis of the specified information, which is compared with the management objectives and a program is formed based on the results of the comparison.

However, first you need to carefully study the TP, realize the achievement of a certain goal, what exactly should result from the implementation of the TP, present each operation or action, the conditions for their implementation, and finally, describe in detail the sequence of actions. This sequence of actions leading to the achievement of a specific goal can be called a control algorithm. When compiling an algorithm, complex actions are broken down into simpler actions, up to elementary movements. Such detailing allows the most complex action to be presented as a set of simpler actions, and the implementation of the simplest actions can be entrusted to program control, this will be a direct path to the automation of the technological process. In the general case, the technological route of wrapping is presented in the form of stages (figure 1):

Stage 1. The sheet blank is installed in the clamps of the press and fixed in them.

Stage 2. The sheet blank is pre-stretched in order to bring the material beyond the yield point.

Stage 3. The sheet blank is wrapped around the surface of the covering punch, forming the top of the double curvature shell. This is ensured by both the movement of the clamping devices and the lifting of the press table. In this case, the area of contact between the workpiece and the covering punch expands. This stage continues until the sheet blank comes into contact with the surface of the tightening punch [1].

Stage 4. The sheet blank is calibrated until it completely adheres to the surface of the covering punch by applying additional tensile deformations necessary to stabilize the geometric shape of the skin.

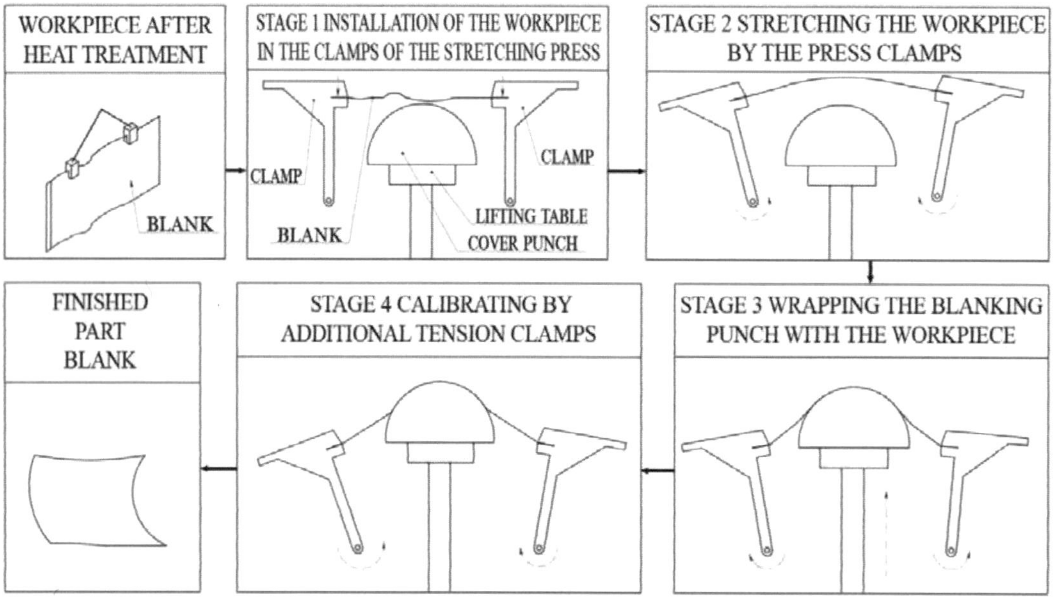

Figure 1. General case of the shaping process.

At the first stage, the aluminum sheet blank comes either in a «soft», pre-annealed state, or in a freshly hardened «viscous» state. Research by scientists from VIAM showed that some aluminum alloys after quenching for some time (up to several hours) have increased plastic properties, after which aging of the material occurs, as a result of which the plasticity decreases, but the strength properties increase [2]. This is the reason for the use of sheets in a freshly hardened state. For sheet blanks, the stage of rapid cooling is performed in water or 40% polyalkylene glycol solution [3].

During quenching, wavy warping occurs - the flat shape of the sheet blank deviates greatly. Therefore, it is very difficult to tuck it into the jaw of the press clamp. A clamping device [4] has been developed, which makes it possible to mechanize the laborious process of sheet threading without additional movements of the clamping devices and to reduce the time required for processing. The warping of the rest of the billet is eliminated at the second stage by straightening the sheet billet by stretching in the clamps of the stretching press. Shells of double curvature, complex in shape, are obtained by shaping by wrapping in several transitions. Each of the transitions of the skin, as a rule, includes all four stages.

Thus, before the first transition, the sheet blank is usually annealed or freshly hardened. Then, between the transitions, annealing is performed, and before the last one - quenching [5]. The accuracy of the tension calibration step depends both on the size of the leash and distortions of the shell contour during quenching, and on the elastic springback after shaping by the tightness, due to the difference in the stress-strain state of the material in different zones of the sheet blank. Therefore, the calibration operation does not always ensure the stability of the shape with deviations within the tolerances, and the details of the skin after calibration sometimes require manual finishing [6].

The main blanking and stamping equipment used for shaping are hydraulic shaping presses. Today, aircraft manufacturing enterprises have manual and programmed stretch presses. Here are the stretch presses installed and operating at JSC «Aviastar-SP». The equipment with manual control includes stretching and stretching presses RO-3M and RO-630-11. The equipment with programmed control (CNC) includes longitudinal stretching press FEL; transverse stretch stretch presses FEKD and FET.

The work [7] presents ideas on the modernization of domestic equipment for shaping with a close-fitting with manual control. To transfer it to the program control mode, programmable logic controllers (PLC) and personal computers (PC) are used. SMART 2 controllers were used as PLCs. To combine PLCs and PCs, a standard high-speed interface network PROFIBUS was used. An interface network based on high-speed buses makes it possible to integrate wrapping equipment, PLCs and computing facilities into a single system. The auto-control system is configured for a specific type of press and the wrapping process is performed by placing the corresponding control program in the PLC memory. The developed program control system allowed to achieve the level of automation comparable to the French presses of the ACB company.

At the same time, obtaining skin parts on presses with programmed control allows an unlimited number of times to reproduce a debugged control program for shaping by displacements, and invariant to a large proportion of disturbing factors. But in order to carry out shaping by wrapping on CNC presses, it is necessary to control all process parameters, practically excluding any chances, in conditions of instability of the properties of the initial sheet materials and the conditions of the deformation process of each individual sheet. In this regard, on the covering equipment, the control system also provides for manual control. However, when using the control system on existing strapping equipment, the following problems can hardly be solved:

- inhomogeneity of plastic deformations in the resulting shell, including thinning of the sheet blank, due to the complexity of the geometric shape, external friction forces and a great tendency to thinning of the aluminum alloy sheet;
- the unpredictability of the behavior of the sheet blank after preliminary heat treatment, carried out immediately before the shaping by the wrapping;
- the need to perform manual finishing in some places where the workpiece does not fit to the tightening punch.

In modern conditions, it is already unacceptable to use manual control of the shaping processes by wrapping and with manual finishing operations. This is due to the fact that high requirements are imposed on the aircraft skin in terms of the quality of the aerodynamic contour and in terms of strength and resource, which are difficult to ensure without solving the above problems.

2. Analysis of defects and the choice of ways to improve the shaping process with a wrap

For the wrap shaping process, the efficiency and feasibility of improvement can be assessed in the following ways:

- comparison of the costs for the quality of the cladding before improvement and after achieving the accuracy of the geometric shape of the surface of double-curvature shells (by comparing its waviness) [8];

- comparison of production costs before improvement and after.

The first assessment method affects the performance of the aircraft. The second method of estimation is determined by the production time with the minimization of the number of personnel involved in the production, as well as the cost of blanks and production costs. Both methods are united by the fact that they directly depend on the rejection signs that arise both after and during the shaping process. The main rejection signs arising in the process of shaping sheet blanks by covering are:

1. Folding. Corrugations. Waviness.

Folds are an irreparable sign of rejection. The appearance of folds is preceded by corrugations. Corrugations arise as a result of local loss of stability of the edge of the workpiece from compressive forces during bending along the tightening punch or during compression of the section of the workpiece located (in front of the gap) between adjacent press jaws. The main method for eliminating corrugations is additional stretching of the workpiece with simultaneous manual finishing with a rubber mallet. As a rule, the corrugations cannot be completely smoothed out, thereby leaving waviness in the skin. Waviness is a valid rejection characteristic within certain values. The limiting value of the relative height of waviness for the wing skins, empennage, engine nacelles, pylons and the front part of the aircraft fuselage was about 0.02 mm [9].

2. Breaks in the sheet blank. Localization of plastic deformation and uneven distribution of the thickness of the skin part.

The break of the workpiece is preceded by local thinning. Localization occurs as a result of excessive stretching of the workpiece material and high frictional forces. The reason for excessive stretching of the workpiece can be both the elimination of the corrugation that occurs accidentally, and an incorrectly selected kinematic mode of shaping. The influence of friction on the localization of the thinning of the workpiece during shaping by a tight-fitting is manifested in the slowing down of the plastic flow of the material of the workpiece, during its stretching, at the points of contact with the tightening punch when entering the section between its end face and the press clamps. As a result, the friction-free edge of the sheet blank undergoes excessive tensile deformations, leading to the localization of thinning and rupture of the blank. If there is no rupture, there is an uneven distribution of the thickness of the skin. This rejection feature is acceptable, but not required for modern aircraft.

The bypass-forming surfaces of the skins, associated with the spatial shape of the shells of double curvature and the parts of the aircraft frame in contact with them, have become more complicated, the requirements for their compatibility during assembly have increased due to the difference in thickness of the skin walls. It is required to minimize the thickness difference to the recommended limits, but this requires a new technology to increase the degree of deformation during stretching, which is achieved practically in one transition without localizing the thinning deformation and equalizing their values in different areas of the sheet blank.

3. Notches and scratches, as well as their filling.

This feature occurs as a result of friction of the aluminum sheet on the surface of the covering punch, especially if it is cast from secondary aluminum. It is successfully eliminated by using antifriction coatings and special technological lubricants [10]. If they appear after shaping, it must be eliminated by cleaning out the defective places, with further filling. The use of a putty for finishing defective areas is permissible only on the inside of the skins due to the high requirements for the outer contours of the aircraft. At the same time, there is a permissible proportion of filled defects from the total surface area of the skin, which is also set by the aircraft designers [9].

Therefore, when writing a technical task for the development in the field of shaping by wrapping skin parts, it is necessary to unequivocally know the limiting minimum of parameters that must be taken into account. Among these parameters, two groups can be distinguished:

1. Parameters affecting the performance of skin parts:

- the amount of waviness of the skin, taken as a percentage, affecting the aerodynamic drag of the aircraft as a whole;
- the proportion of filled defects, which impairs the uniform strength of the skin parts. Therefore, companies operating an aircraft with filled defects need to periodically check and repair these elements. This defect is no longer acceptable for modern aircraft.

2. Parameters affecting the cost of aircraft production:

- uneven distribution of the thickness of the skin, which ultimately affects its cost. As a result, with its high value, the designer is forced to lay the increased thickness of the sheet blank;
- the proportion of notches and scratches from the total surface of the workpiece obtained after shaping. The more marks and scratches, the higher the labor intensity of their cleaning and filling, which increases the labor intensity of the plating production;
- the laboriousness of the shaping of the skins, depending on the duration of manufacture, the amount of highly paid manual labor for the manual removal of corrugations, as well as the number of personnel involved in the process.

Obviously, both groups of the above parameters are completely dependent on rejection signs. Therefore, the elimination of rejection signs is one of the main tasks of research and engineering work dedicated to improving the shaping process with a wrap. To eliminate the aforementioned rejection signs, various methods have been developed to intensify the shaping process with a wrap.

3. Intensification of the shaping process with a wrap

The intensification of sheet stamping processes is usually understood as a set of measures or actions that can improve the quality of a part, reduce the labor intensity of its manufacture, and reduce the number of equipment and tooling used [11].

Distinguish between direct and indirect intensification of the shaping processes by the covering. With direct intensification, the intensifying factor affecting the sheet blank is presented in an explicit form. It can be heating, additional force, high-speed, electromagnetic and other effects. Indirect methods of intensification are outwardly invisible, but they significantly change the traditional process of stretching and are usually associated with special preparation of the initial workpiece: optimization of its dimensions, a given distribution of mechanical properties, obtaining the desired anisotropy [12].

Changes to the die tooling are required. First of all, this concerns the tightening punch, the geometry of which must differ from the original design version, which is directly related to the theoretical contours of the aerodynamic contours of the aircraft units. The transition to a grid of curvatures of the surface of the covering punch gives great advantages in design automation and control of the processes of shaping by covering shells of any surface, as well as complex geometric shapes. It became possible to select on such a surface the position of the shaping contour of the tightening punch, which will be aligned with the longitudinal vertical plane of the press, including the surface pole and passing through the middle of the central jaw [13, 14]. However, this requires some external structural change of the tightening punch.

Based on the physical principle of action, we distinguish groups of intensification methods:

- changing the friction conditions on the contact surface of the workpiece and the covering punch;
- external design changes to the tightening punch;
- changing the kinematic scheme of loading sheet blanks;
- a change in the nature of anisotropy due to the presence of certain preferential crystallographic orientations in the sheet material;
- change in the kinematic scheme of loading sheet blanks.

By changing the friction conditions on the contact surface of the workpiece and the covering punch, it is meant a decrease in friction between the workpiece and the covering punch [15, 16], or by local increase in friction in the central part of the workpiece, practically blocking the movement of the metal in this zone under the action of a container with an elastic medium [17]. Reducing friction at the edges of the workpiece has a positive effect on the cladding, maintaining the locations of the greatest deformations in the central region of the sheet workpiece with a minimum probability of sheet breakage in one of its free areas between the edge of the punch and the press clamps and providing a more uniform distribution of the workpiece thickness.

Uniform distribution of thickness is ensured due to the zonal plastic flow of the metal during deformation. First, the central part of the blank is stretched, and then the outer parts. In this case, in order to avoid excessive deformation of the central part, it is possible to use a local increase in friction due to a profiled elastic medium having a high coefficient of friction with aluminum alloys. Also, this method, due to the profiled elastic medium, makes it possible to deform the concave parts of the skins having alternating curvature.

By changing the kinematic loading scheme, it is meant that additional forces are applied to the sheet blank, both in the longitudinal direction and in the transverse direction. Also, this group includes methods based on manipulating the angles of coverage of the workpiece of a tight-fitting punch, as in the case of a stepped covering [18] or, as in the case of covering over two tight-fitting punches [19]. In the transverse direction, forces are applied through the use of a clamp. As a rule, the clamp is resorted to in the case of the formation of the concave parts of the shells with alternating curvature. In the case of longitudinal-transverse tightening, there is a differentiated application of forces to the side edges of a wide workpiece by means of additional stretching devices. This makes it possible to smooth out the corrugations arising in the process of shaping the shell [20] or by sequential stretching of the shell, preliminary stretching of the flat billet to a full angle, unloading and unbending of the shell to a small angle of wrap, followed by repeated wrapping of the shell in the isometric state [21].

There are two main types of wrap angle manipulation:

- pre-bent at a small angle of coverage, the sheet blank is tightened along the tightening punch, stepwise, with a certain step, increasing the angle of coverage to the full value [18] and with longitudinal-transverse covering, stretching the central flat part of the workpiece when tightening trough-shaped parts [20];
- the sheet blank is preliminarily tightened along the tightening punch to the full angle of coverage, then unloaded and unbent to a small angle of wrap, followed by re-tightening along the second tightening punch [19] or with subsequent re-tightening of the shell in the isometric state to the full angle of coverage along the same tight punch [21].

Both types of manipulation of the wrap angles allow to obtain shells with a more uniform distribution of thickness, both in the longitudinal and in the transverse directions.

The plasticity of the material of the sheet blank can be increased by treating it with a pulsed electric current [22] or by heating during the forming process [23]. These heating methods make it possible to obtain titanium alloy shells with large wrap angles. This is achieved by further increasing the degree of ultimate deformation in one pass. By changing the resistance to deformation of individual sections of the workpiece is meant intensification by changing the values of the required forces, for the shaping of certain individual areas of the workpiece in the process of shaping by the wrap. In this case, workpieces of various widths, local heating, as well as linings made of various materials [13, 24] can be used, which increase the thickness of the workpiece in places where the tensile deformations are maximum.

The developments given in [21] apply the conditions of symmetrical tightening due to external structural changes of the tightening punch. They provide for the correspondence of the texture axes of symmetry of the properties of the sheet blank and the directions of the lines of curvatures of the shell of double curvature intersecting at the geometric pole of the surface of the drawing punch, as well as the location of the vertical plane of symmetry of the drawing press. This plane will pass through the geometric pole of the surface of the covering punch and the middle of the central clamp of the sheet blank.

As a result, not only the mechanical and deformation properties of the sheet material are taken into account, but also their directionality. The directionality determines the magnitude and nature of the anisotropy of the properties of sheet metal, which will ensure stable control of the processes of shaping by wrapping shells of double curvature with a uniform change in thickness in different areas of the sheet blank without folds and ruptures. To do this, it is necessary to perform a geometric study of the covering punch in another operating system using metric transformations.

These methods make it possible to provide a more uniform distribution of the thickness of the workpiece by:

- an additional increase in the degree of local deformation, in the area where the workpiece had not been stretched before;
- reducing the degree of local deformation in the area where the workpiece is excessively stretched;
- the use of one or another kinematic scheme, which is accompanied by the oncoming movement of the boundaries of the deformation zone while maintaining the location of the greatest deformations in the central region of the sheet blank.

Obviously, most of the above methods for intensifying the skin are aimed at solving the problem of uneven distribution of skin thickness. As can be seen from the above, most of the methods solve the problem of uneven thickness distribution. At the same time, only one method aimed at eliminating corrugations with the help of additional stretching devices partially solves the problem of smoothing corrugations. However, as practice has shown, the shaping of a sheet blank under conditions of symmetric wrapping significantly reduces the likelihood of corrugations and folds [13, 14]. At the same time, additional stretching devices are designed for thin skin [20] and additional allowances for the width of the workpiece are required.

In our opinion, the main way to solve the above problems is to develop a system for adaptive control of the process of shaping sheet parts by wrapping. The adaptation consists in automatic real-time correction of the original control program, depending on the actual mechanical properties and the required deformations of the workpiece in the characteristic places of the skin, taking into account the geometric shape.

4. Adaptive wrap forming control

When creating conditions for symmetric wrapping due to external structural changes to the wrapping punch, great advantages can be obtained in design automation and control of the shaping processes of wrapping shells of any surface, as well as complex geometric shapes. Now it is possible to select on such a surface the position of the shaping contour of the tightening punch, which will be aligned with the longitudinal vertical plane of the press, including the surface pole and passing through the middle of the central jaw. For the contour of the central section of the surface, it is quite easy to establish the location of the axis of axial symmetry for a classical second-order surface contacting at a pole. Therefore, the properties of the local shape at the pole of the surface of the tight-fitting punch are easily determined, which greatly facilitates the analytical approach to modeling, which adequately reflects the real process of shaping by the tight-fitting.

However, the implementation of stable control of the shaping processes by wrapping shells, close in geometric shape of the shell, with a uniform change in thickness in different areas of the sheet blank without folds and rupture, is possible only on stretch presses with programmed control. In addition, it was provided for the correspondence of the texture axes of symmetry of the properties of the sheet blank and the directions of the lines of curvatures of the shell of double curvature, intersecting at the pole of the surface of the covering punch.

Stable control of the shaping processes by the wrap is based on the initial position of the deformation zone in the central region of the sheet blank. The shaping of the sheet blank according to one or another kinematic loading scheme is accompanied by the counter movement of the boundaries of the deformation zone while maintaining the location of the greatest deformations in the central region of the sheet blank. This allows you to increase the degree of shaping, practically in one transition with a minimum probability of sheet breaking in one of its free sections between the edge of the punch and the press clamps.

At the same time, the use of computer and CNC facilities for the independent manufacture of a covering punch for the implementation of symmetric covering conditions in practice has increased the efficiency of using modern digital technologies. For this, it is required to develop a toolkit for converting the coordinates of the clamping elements into the parameters of the working bodies of the press and vice versa - the parameters into coordinates. It is necessary to have a kinematic model of the covering press, which will allow observing the position of the working bodies of the press in each frame of the control program. Now it is possible to perform the necessary adjustment of the parameters of the control program directly in the Unigraphics software environment based on the results of modeling a particular wrapping method using the PAM-Stamp or ANSYS/LS-DYNA software package. Therefore, the development of an automated system for the production of bypass-forming elements of aircraft became possible using modern simulation tools, calculating the tensile and thinning deformations of the sheet at individual characteristic points of the double curvature shell surface.

Accordingly, when designing a system for adaptive control of shaping by wrapping a sheet blank, it is necessary to determine the following:

- Rejection signs by which the control program will be optimized.
- A method for measuring the values of physical processes that affect the occurrence of rejection signs.
- Techniques for reducing and eliminating the likelihood of rejection signs.
- The method of constructing a mathematical model, on the basis of which the efforts and displacements of the working bodies of the covering equipment will be recalculated.

Ideally, the amount of waviness and uneven thickness distribution should be minimized, and wrinkling and tearing should be completely eliminated. In order to minimize these values, it is necessary to track the deformations of the workpiece using the digital image correlation (DIC) method, together with the movements and forces on the working bodies of the pressing equipment using sensors. In this case, the recorded deformations of the workpiece and the data from the sensors must be compared with the data available in the system software.

The model underlying the digital twin is calculated by the finite element method, and has the ability to almost instantly automatically update and correct in the case of not only deformation of the workpiece, but also displacements and efforts of the working bodies of the pressing equipment. When updating the finite element model on the digital twin, the control program is automatically recalculated, which is then transmitted to the sensors of the working bodies of the pressing equipment. Such a scheme of adaptive control of the shaping process by a wrap is shown in figure 2.

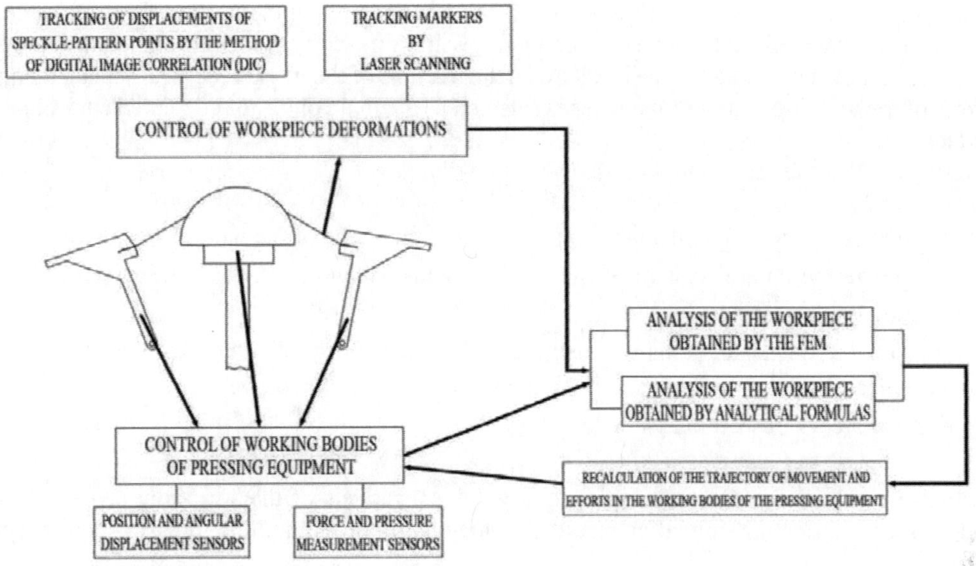

Figure 2. Ideal system of adaptive control of pressing equipment for shaping by wrapping double curvature skins.

Problems of the proposed system of adaptive control of wrapping equipment for shaping wrapping of double curvature shells:

1. FEM implies a long calculation time. In order to reduce the computation time, the digital twin will need to work on a supercomputer with a processor with 54,000 cores, each of which has a frequency of at least 3 Hz. The cost of such a supercomputer is tens of millions of rubles.

2. The speckle pattern applied with a water-based paint will need to be removed very carefully as it is applied to the outer surface of the skin overlooking the theoretical contour of the aircraft. The second disadvantage of the speckle pattern is the complexity of the application when covering with annealing between operations. The problem is that before annealing, it is necessary to remove the speckle pattern, which increases the duration of the plating manufacturing process by at least 20 minutes (it is necessary to wait until the wash swells the speckle pattern for further removal). After annealing, it is necessary to apply a speckle pattern and carry out a polymerization reaction using drying installations (for example, infrared lamps), which also adds 10 minutes to the labor intensity of the process. Plus, for the application of the speckle pattern, a special area must be allocated, equipped with exhaust ventilation and personnel qualified to carry out painting work. The cost of a Digital Image Correlation (DIC) system is significantly lower than that of a supercomputer and comparable to the cost of a stretch press equipment.

As a result of the above, one of the ways to solve the problem is a number of optimal pre-calculated finite element models, according to which the wrapping process is performed without preliminary and inter-transition heat treatment. Due to the lack of heat treatment, the parameters of the wrap forming process become more predictable. Thus, the measured sheet deformations using cameras in the digital image correlation (DIC) system will be within a certain range of scatter. Having a range of spread of sheet deformations during stretching at a certain stage of shaping, it will be possible to calculate a number of finite element models in advance within 1-2 days on a computer that is fairly average in terms of processor characteristics (for example, having 6 cores, 3 Hz each).

The process itself will look like this: the workpiece is inserted into the clamps of the press, stretched with clamps, and a tightening punch. In the process of deformation, the digital image correlation (DIC) system captures the process. The process itself is recorded by high-speed cameras, which is

IOP Conf. Series: Materials Science and Engineering 1155 (2021) 012014 doi:10.1088/1757-899X/1155/1/012014

automatically calculated in advance and compared with the available finite element models. Also, using feedback sensors from the working bodies of the pressing equipment, the data of forces and displacements are transmitted to a personal computer with finite element models and control programs. The closest finite element model to the measured deformations of the workpiece and displacements with the efforts of the working bodies of the press is selected as the main one and the related control program is activated.

It is believed that the use of the proposed scheme will allow:

- exclude or minimize manual labor on fine-tuning the shape of the workpiece;
- modernize the control system of the motion of the actuators (clamps, lifting table, portal) of existing tying presses;
- improve the quality of skins by ensuring a more uniform distribution of the thickness of the workpiece after deformation and reducing the value of its waviness;
- exclude breaks of workpieces.

Further research is aimed at developing finite element models of the wrapping processes, with a method for recalculating the simulation results into the control program of the pressing equipment.

5. Summary

The focus is already on the research carried out by the finite element method in modeling the processes of shaping a sheet blank by a tight-fitting punch placed on the table of a virtual stretch-drawing press. The use of computer and CNC facilities for the independent manufacture of a covering punch to implement the conditions of symmetric covering in practice will increase the efficiency of using modern digital technologies.

It is necessary to create a complex model and a computational algorithm by the processes of shaping by wrapping, accumulating many years of experience in operating domestic stretch presses, theoretical and experimental developments, patents that allow calculating the tensile and thinning deformations of the sheet at individual characteristic points of the double curvature shell surface.

However, the achievements of modern information technologies, raising the level of technologies based on the digital twin - the digital representation of the object, have led to the need to use the accumulated production and technological experience, the available developments in the field of storing and operating with information and applying in various fields. This made it possible to determine the main directions of improving the processes of shaping by wrapping shells of double curvature.

References

[1] PI 1.4.1854-88 1989 Technology of skin shaping by the method of longitudinal stretching on stretching and stretching presses of the RO type (NIAT)

[2] Kishkina S I and Fridlander I N 1982 *Aviation Materials* (Moscow: ONTI)

[3] Belyanin P N 1979 *Production of Wide-Body Aircraft in the USA* (Moscow: Mashinostroenie)

[4] Mikheev V A *et al.* 1982 A.C. No. 927364. Clamping device publ.

[5] Tishchenko I I 2007 *Improvement of technologies for covering large-sized shells on the basis of finite element modeling of shaping processes*: Ph.D. thesis: 05.03.05 (Voronezh: VGTU)

[6] Borisov V G 2004 *Processes for the Manufacture of Thin-Walled Aircraft Parts by Methods of Plastic Shaping (*Kazan: Kazan Publishing House. State Tech. University)

[7] Mikheev V A, Malyshev B S *et al.* 1999 Development of auto-control systems for covering equipment based on PC, PLC SMART2 and PROFIBUS network *Izvestia of Samara Scientific Center of the Russian Academy of Sciences* **2**

[8] TR 1.4.1974-89 1991 Formation of plating parts of complex shape of the difshaping (NIAT)

[9] PI 1.4.1854-88 Technology of skin shaping by the method of longitudinal stretching on stretching and stretching presses of the RO type (NIAT) 1989

[10] Karpukhin E G and Koshkina A O 2019 Evaluation of the antifriction properties of coverings of tight-fitting punches for the possibility of using them in the shaping of coverings with tight-fitting on the FET and FEL coverings *XXIV Tupolev readings (School of young scientists) texts of reports of the participants of the International youth scientific conference 6*
[11] Ershov V I, Popov O V, Chumadin A S *et al.* 1999 *Sheet Stamping: Calculation of Technological Parameters* (Moscow: MAI Publishing House)
[12] Grechnikov F V 1998 *Deformation of Anisotropic Materials* (Moscow: Mechanical engineering)
[13] Mikheev V A, Dementyev S G, Samokhvalov V P, Savin D V and Surudin S V 2013 Isometric conditions for the formation of a tight shell of double curvature with minimal thickness difference *Izvestia of Samara Scientific Center of the Russian Academy of Sciences* **15(6-1)** 161-6
[14] Mikheev V A and Surudin S V 2017 Fundamentals of the calculation of the process of forming tight thin shells of double curvature *Izvestia of Samara Scientific Center of the Russian Academy of Sciences* **19(1-3)** pp 555-562
[15] Maksimenkov V I, Molod M V, Lyakhova N E and Reshetnikov E V Pat. No. 2459680 RF MKI C2. Method of covering sheet materials publ. 27.08.2012, No. 24.
[16] Groshikov A I and Malafeev V A 1976 *Procurement and Stamping Works in Aircraft Construction* (Moscow: Mechanical Engineering)
[17] Mikheev V A and Shchurovsky D V Pat. No. 2275266 RF MKI C2. Method of forming parts of double curvature publ. 27.04.2006, No. 12.
[18] Chistyakov V P and Chistyakov E V A.C. 707647 USSR, MKl B 21 D 11/20 Method of shaping of parts of double curvature No. 2459522/25-27 appl. 03.04.1977 publ. 04.30.1979
[19] Chistyakov V P, Mikheev V A and Chistyakov E V A.C. 659238 USSR, MKI B 21 D 11/20 Method of forming parts of double curvature No. 2459522/25-27 appl. 03.04.1977 publ. 04.30.1979
[20] Sosov N V, Komarova L G and Sosov N N Pat. No. 41 652 RF MKI U2. Press for covering sheet parts with double curvature publ. 10.11.2004.
[21] Mikheev V A, Savin D V and Surudin S V Pat. No. 2573859 RF MKI C2. Method of shaping thin-sheet parts of shells of double curvature of biconvex form by covering publ. 27.01.2016
[22] Popov O V, Tannenberg D Yu, Vlasenkov S V, Maryin B N and Makarova E A Pat. No. 2057607 RF MKI C2. Method of covering sheet blanks publ. 10.04.1996
[23] Ershov V I, Glazkov V I and Kashirin M F 1990 *Improvement of the Forming Operations of Sheet Stamping* (Moscow: Mechanical Engineering)
[24] Kazakov I A, Kotov O E and Abramov P A Patent No. 2272690 RF MKI C1. Method of covering sheet blanks publ. 27.03.2006, No. 9.

Artificial intelligence for imaging data analysis in materials science: microscopy and behind

A G Okunev[1,2], M Yu Mashukov[1], N N Sankova[1,2] A V Nartova[1,2] and A V Matveev[1]

[1] Novosibirsk State University, Pirogova str., 1, Novosibirsk, 630090, Russia
[2] Boreskov Institute of Catalysis SB RAS, pr. Ac. Lavrentieva, 5, Novosibirsk, 630090, Russia

E-mail: matveev.nsu@mail.ru

Abstract. In various fields of microscopy, a fairly common task is to count and determine the size of the studied objects. Sometimes this is a one-time tedious task that is done by lab personnel without the programming skills. In this paper, we present an approach that allows researchers to perform automated processing of research imaging data using our services and even to train deep neural networks on their own in no code mode.

1. Introduction
Advances in imaging technologies have had a significant impact on all areas of science and technology, providing the opportunity to study the objects with high resolution in space and time. Advances in the physical foundations of microscopy have made it possible to create devices that can be used to produce images at the level of individual atoms. Often, when studying physical, chemical, and biological objects, it is important to determine their statistical parameters (quantity, size, and area). At the same time, the operator needs to process thousands of objects to obtain statistically reliable results. To solve these problems, various methods and technologies based on computer vision, machine learning, neural networks, etc. are being actively developed, primarily for biology purposes [1-3].

We have developed two services based on the use of deep machine learning:

- an online service ParticlesNN for recognizing nanoparticles in images of scanning tunnelling microscopy (STM) and transmission electron microscopy (TEM).
- cloud service DLgram, using Telegram messenger as an interface, which can be used for training a neural network and recognizing any type of object.

2. The online service ParticlesNN
We have developed the online service ParticlesNN [4], http://particlesnn.nsu.ru, which is based on the Cascade Mask-RCNN neural network.

2.1. Scanning tunnelling microscopy
Training was performed on a dataset containing 23 STM images with 5157 nanoparticles. As a result, the trained neural network recognized nanoparticles in the verification set with 0.93 precision. The

accuracies for mean particle size calculated from predicted contours compared with ground truth were in the range of 0.87–0.99, figure 1 [5, 6].

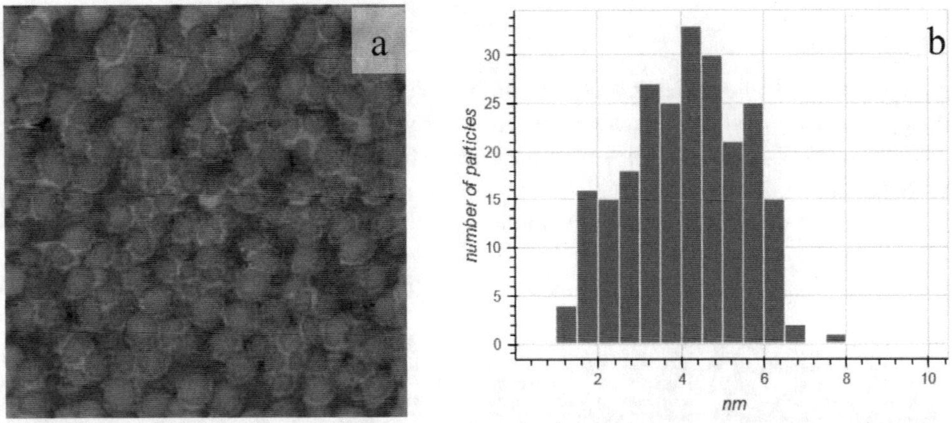

Figure 1. Example of nanoparticles recognition on a STM image (a). Boundary particles are highlighted in pink, which can be excluded when calculating the average particle size to get a more accurate result. A histogram of the size distribution of recognized nanoparticles is shown on (b). STM image of Pt nanoparticles deposited on highly-oriented pyrographite was obtained using an SPM 100 VT (RHK Technology, USA).

2.2. Transmission electron microscopy
Training was performed on a dataset containing 21 TEM images with 1114 nanoparticles. As a result, the trained neural network recognized nanoparticles in the verification set with 0.71 precision. The accuracies for mean particle size calculated from predicted contours compared with ground truth were in the range of 0.73–0.84, figure 2.

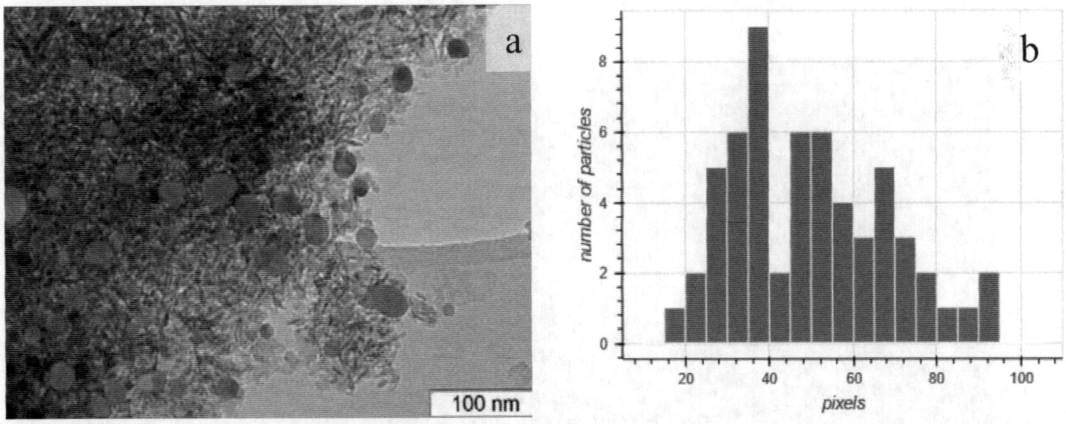

Figure 2. Transmission electron microscopy image of Pt nanoparticles on alumina catalyst: 223 particles were recognized by the network (a). A histogram of the size distribution of recognized nanoparticles is shown on (b). TEM image was obtained using a JEM-2010, Jeol Co., Japan, with the lattice resolution of 0.14 nm with accelerating voltage of 200 kV.

The online service ParticlesNN processes the results obtained by determining the size of the contours and their statistical parameters – the degree of coverage, concentration, average size, and distribution of objects by size.

The Particles NN online service differs from other software products (CellProfiler [1], Ilastik [2], ImageJ/Fiji [3]) in the following features:

- it is possible to process images that contain significant noise and artifacts without pre-processing.
- a user can adjust automatically defined contours in order to refine the recognition results.
- the option of simultaneous statistical processing of several images is implemented.
- the processing results are displayed in the form of a histogram and tables, in which information on all identified objects is available – their coordinates, dimensions.
- the refinement of particle contours is implemented using the fitting procedure with a Gaussian 2D distribution.
- the functions of correct accounting of objects on the image border are implemented.

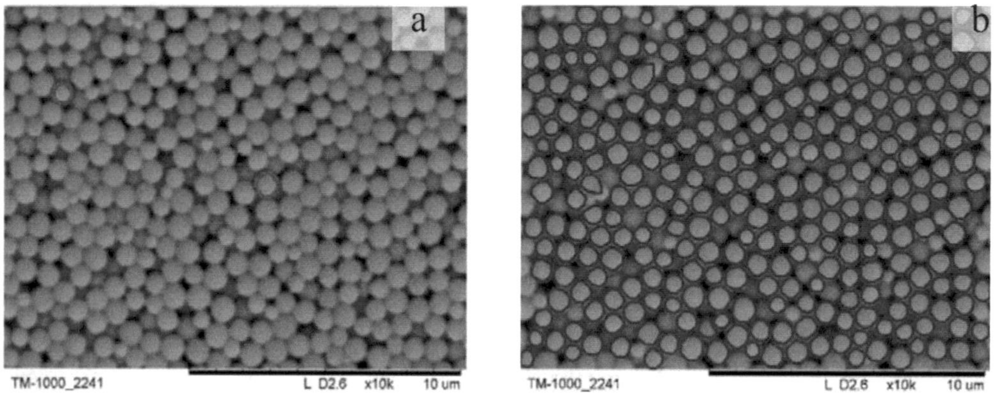

Figure 3. Scanning electron microscopy (SEM) images obtained using a Hitachi TM-1000 microscope of polystyrene microspheres: a training image with 2 labeled particles (a); 281 particles recognized by network, 35 are not identified (b).

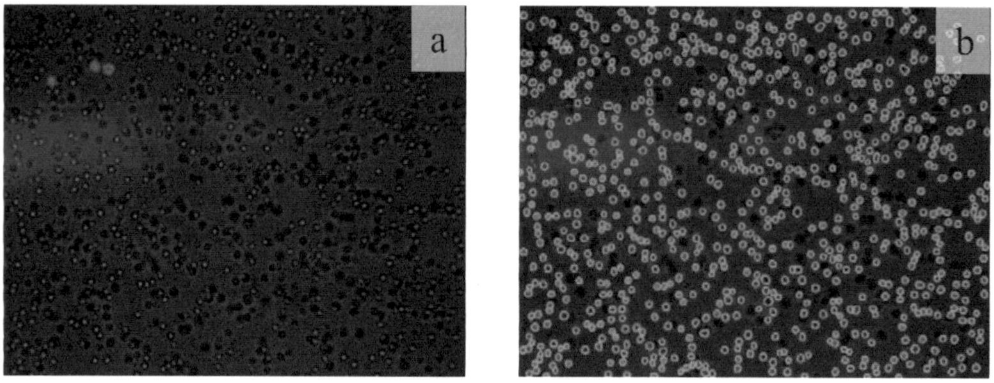

Figure 4. Optical microscopy images obtained using an optical microscope Biomed-5 with Levenhuk M1000 Plus camera at a magnification of x100 of polystyrene microspheres: a training image with the 3 labelled particles (a); 807 particles recognized by network, 104 are not identified (b).

3. The cloud service DLgram

We implemented a revolutionary subshot learning approach that allows researchers to train a deep neural network using a handful of the instances of objects of interest. The trained network able to effectively recognize the rest of the object instances on the same image as well as correctly segment similar objects on other images, as shown in figures 3-4. Training and inference can be implemented in no code mode and perform automated processing of scientific data using a cloud service in Telegram messenger group referred as t.me/nanoparticles_nsk.

The realized cloud service is especially user-friendly for the following reasons:

- There is no need to install special software.
- it is possible to train a neural network on any type of objects.
- only a few objects should be labelled for training.
- the trained neural network can be used for recognizing identical objects on other similar images.
- a user can adjust automatically defined contours in order to refine the recognition results.
- the statistical processing of the obtained data is implemented.

4. Conclusion

We have introduced two services for automatic image data processing based on the use of deep neural networks:

- online service ParticlesNN with a pre-trained neural network for recognizing nanoparticles in scanning tunneling and transmission electron microscopy images.
- the cloud service DLgram based on the Telegram channel that allows a user to train the network on its own and to use it in future work.

The obtained data can be used to determine their statistical parameters (quantity, size, and area).

Acknowledgments

This work was supported by Ministry of Science and Higher Education of the Russian Federation within the 5-100 Excellence Programme, as well as within the governmental order for Boreskov Institute of Catalysis SB RAS (project AAAA-A21-121011390011-4).

References

[1] McQuin C, Goodman A, Chernyshev V, Kamentsky L, Cimini B, Karhohs K, Doan M, Ding L, Rafelski S, Thirstrup D, Wiegraebe W, Singh S, Becker B, Caicedo J and Carpenter A 2018 CellProfiler 3.0: Next-generation image processing for biology *PLOS Biology* **16** 1-17

[2] Berg S, Kutra D, Kroeger T, Straehle C, Kausler B, Haubold C, Schiegg M, Ales J, Beier T, Rudy M, Eren K, Cervantes J, Xu B, Beuttenmueller F, Wolny A, Zhang C, Koethe U, Hamprecht F and Kreshuk A 2019 ilastik: interactive machine learning for (bio)image analysis *Nat Methods* **16** 1226-32

[3] Schindelin J, Arganda-Carreras I, Frise E, Kaynig V, Longair M, Pietzsch T, Preibisch S, Rueden C, Saalfeld S, Schmid B, Tinevez J-Y, White D J, Hartenstein V, Eliceiri K, Tomancak P and Cardona A 2010 Fiji: An open-source platform for biological-image analysis *Nat Methods* **9** 676-82

[4] Particles N N http://particlesnn.nsu.ru

[5] Okunev A G, Nartova A V and Matveev A V 2019 Recognition of nanoparticles on scanning probe microscopy images using computer vision and deep machine learning *International Multi-Conference on Engineering, Computer and Information Sciences (SIBIRCON)* Novosibirsk Russia 0940-3

[6] Okunev A G, Mashukov M Y, Nartova A V and Matveev A V 2020 Nanoparticle recognition on scanning probe microscopy images using computer vision and deep learning *Nanomaterials* **10** 1-16

IOP Conf. Series: Materials Science and Engineering 1155 (2021) 012016 doi:10.1088/1757-899X/1155/1/012016

Effect of tungsten nanoparticles on interaction of Sn-Cu-Co metallic matrices with diamond

A V Ozolin, E G Sokolov and D A Golius

Kuban State Technological University, 2, Moskovskaya St., Krasnodar, 350072, Russia

E-mail: ozolinml@yandex.ru

Abstract. The effect of tungsten nanoparticles on interaction of sintered Sn-Cu-Co metallic matrices with diamond has been studied. Nanoparticles were obtained by grinding standard tungsten powder in a planetary-centrifugal mill. Commercially pure tin, copper, and cobalt powders were mixed with synthetic diamonds and the milled tungsten. Samples were shaped by static pressing in a mold and sintered in vacuum at 820°C. The sintered samples were subjected to static compression test, and hardness of the metallic matrices was measured. The structure of fractures and distribution of the elements over them were explored by the SEM and energy dispersive X-ray microanalysis methods. It has been determined that introducing tungsten nanoparticles into the composite material contributes to dispersion hardening of the metallic matrix and higher strength of its adhesion to diamond.

1. Introduction

Diamond abrasive tools with metallic matrices are widely used in construction and mechanical engineering industries. One of the principal technologies for obtaining diamond abrasive tools is powder metallurgy. Using it, diamond cutting and grinding wheels, drill bits, and other tools are manufactured. Operational properties of such tools depend on physical and mechanic properties of the metallic matrices and the strength of diamond-to-matrix adhesion.

An efficient method for improving physical and mechanic properties of composite materials with metallic matrices is to introduce nanosized particles into them [1-6]. For this purpose, particles of metals, oxides, carbon materials, etc. are used [2, 3]. The effect of nanoparticles on the structure and properties of composite materials is associated with several causes. The presence of distributed finely dispersed particles in the material leads to hardening under the Orowan mechanism [2]. In sintering of some materials, finely dispersed high-melting particles play the part of crystallization centers, which ensures the formation of a fine-grained structure contributing to improvement of mechanical properties [6]. According to works of RAS academician I. F. Obraztsov's school and some other researchers [5, 8, 9], nanosized particles present in the composite matrix have an essential effect on properties of the interfacial layer. This phenomenon allows scaling up physical and mechanic properties of composite materials massively.

In a number of works, it is demonstrated that introducing nanosized additives allows significantly improving operational properties of diamond metallic composite materials [6, 10-15] used for abrasive tools. The positive effect described in these works is achieved by enhancing mechanic properties of the metallic binder of the composite material mostly under the Orowan mechanism.

IOP Conf. Series: Materials Science and Engineering 1155 (2021) 012016 doi:10.1088/1757-899X/1155/1/012016

The contemporary studies show that nanoparticles in diamond-containing composite materials influence interfacial interaction. In particular, in work [14] it has been found that nanosized particles of tungsten carbide influence the interaction of cobalt and iron micro-powders with diamond grains: they separate these particles and thus block catalytic graphitization of diamond. In work [15], it is demonstrated that modification of the Next100 metallic matrix with tungsten carbide nanoparticles provides stronger fixation of boron nitride crystals in the binder.

At present, there are almost no information in literature about the effect the size of finely dispersed high-melting particles has on adhesion activity of the liquid phase to diamond in sintering of composite materials. The effect of finely dispersed high-melting particles on the formation of structure of metallic binders in liquid phase sintering and, in particular, on the structure and properties of interfacial layers, is studied little, too. However, findings of the above research works give grounds for assuming that the liquid phase containing finely dispersed particles features a higher adhesion activity to diamond. They also make possible the hypothesis about better strength properties of the resulting diamond-containing composite material at the expense of modified properties of interfacial layers in its structure.

The objective of this work is to study the effect of tungsten nanoparticles on the interaction of the Sn-Cu-Co metallic matrices with diamond.

2. Materials and methods

For preparation of the samples, commercially pure metal powders were mixed at the following proportion, % (wt.): 20 of Sn, 43 of Cu, 30 of Co, and 7 of W. The authors used tungsten of grade W16,5 (TU 48-19-417-8) which they ground in a planetary-centrifugal mill. The technology and ball milling modes were used as described in work [16]. A part of the samples were prepared with ground tungsten (particle size of 25 nm–0,8 μm), the other part – with non-ground tungsten (particle size of 19–24 μm). Into the resulting mixtures, the SDB1065 diamonds (grain fineness – 50/60 mesh) were added at the quantity of 25% of volume of the metal powders. The samples of 10 mm diameter and 25 mm height were compacted by static pressing in a mold and sintered in vacuum at 820°C for 20 min exposure time. The sintered samples were then subjected to static compression test. Hardness of the matrices was measured by the Rockwell method, scale "B". The fracture structure of the samples, as well as distribution of elements in them were studied by methods of scanning electron microscopy and energy dispersive X-ray microanalysis using the EVO HD 15 (Zeiss) scanning electron microscope.

3. Results and discussion

With the said proportion of the components, the sintered matrices consist of the following phases: cobalt and tungsten particles, solid solution of tin in copper (Cu), and intermetallic compound Cu_3Sn [17]. Sintering of the matrices was accompanied by the formation of a considerable amount of the liquid phase which was crystallized as solid solution (Cu) and intermetallic compound Cu_3Sn during the subsequent cooling down [18]. Particles of cobalt and tungsten remained in the solid state during sintering. After sintering, tungsten nanoparticles were distributed within copper and tin phases mainly, with a certain part of them being inside cobalt particles.

The study of the external surface of the sintered samples has shown that the binders containing ground tungsten wet and cover the diamond grains better (figure 1). Nanoparticles of tungsten distributed in the liquid phase feature a high surface energy, which ensures more intensive chemical interaction of the liquid phase with the surface of diamond.

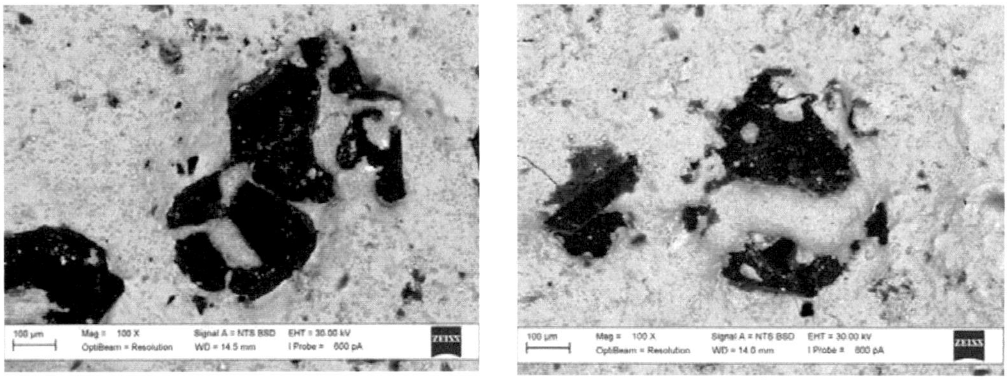

a b

Figure 1. Surfaces of sintered composites: (a) containing tungsten particles sized 1–8 μm; (b) containing tungsten particles sized 25 nm–0,8 μm.

The hardness of the matrix containing non-ground tungsten amounts to 96–98 HRB. The binder with ground tungsten features a higher hardness of 103–105 HRB, which is apparently explained by dispersion hardening according to the Orowan mechanism. Under static compression, the samples were deteriorated by brittle fracture, spalling along inclined and longitudinal planes. So, compressive strength of the samples containing non-ground tungsten is 206 MPa, that of the samples with ground tungsten – 243 MPa.

Figure 2 shows fractures of diamond-containing composite materials sintered with non-ground and ground tungsten, respectively. In both cases, there are particles of the metallic matrix on the surface of diamonds, which gives evidence about strong adhesion of the matrix to diamond. Destruction took place not only along the diamond-matrix interface, but also along the matrix itself.

a b

Figure 2. Elements distribution over fractures of composites containing tungsten particles sized (a) 19–24 μm and (b) 25 nm–0,8 μm.

In figure 2a, the residues of matrix substance on diamond grains are copper and tin phases (colored green) and individual cobalt particles (colored red). In figure 2b, alongside the matrix substance residues of similar composition, one can also observe finely dispersed tungsten particles sintered to the surface of diamond. It can be seen that the size of cobalt particles is different in these materials: in the material with non-ground tungsten, it is 9–15 μm, and in the one with ground tungsten – 3–10 μm. In sintering, tungsten nanoparticles prevent recrystallization of cobalt through the liquid phase and growth of cobalt particles [19]. Owing to dispersion and uniform distribution of adhesion active tungsten and cobalt particles, a large area of their contact with the surface of diamond is achieved.

4. Conclusion
Thus, introducing tungsten nanoparticles into the composite material contributes to dispersion hardening of the metallic matrix and higher strength of its adhesion to diamond.

Acknowledgments
The authors express their gratitude to the Grants Council of the President of the Russian Federation for financial support of the study (application SP-5863.2021.1).

References
[1] Vityaz P A (ed.) 2011 *Modification of materials and coatings with nanosized diamond-containing additives* (Minsk: Belaruskaya Navuka) (In Russ.)
[2] Casati R and Vedani M 2014 *Metals* **4** 65-83
[3] Ma K, Lavernia E J and Schoenung J M 2017 *Rev. Adv. Mater. Sci.* **48** 91-104
[4] Turan M E, Sun Y and Akgul Y 2018 *J. Alloys Compd.* **740** 1149-58
[5] Kostikov V I, Agureev L E and Eremeeva Zh V 2015 *Russ. J. Non-Ferr. Met.* **56**, 325-8
[6] Sudnik L V, Vityaz P A and Ilyushchenko A F 2012 *Diamond-containing abrasive nanocomposites* (Minsk: Belaruskaya Navuka) (In Russ.)
[7] Vityaz' P A, Zhornik V I, Kovaleva S A and Kukareko V A 2016 *Russ. J. Non-Ferr. Met.* **57** 135-40
[8] Obraztsov I F *et al.* 2004 *Mekhanika Kompositsionnykh Materialov i Konstruktsii* **10(4)** 596-612 (In Russ.)
[9] Lurie S, Volkov-Bogorodskiy D, Solyaev Y, Rizahanov R and Agureev L 2016 *Comp. Mat. Sci.* **116** 62-73
[10] Polushin N I, Kudinov A V, Zhuravlev V V, Stepareva N N and Maslov A L 2013 *Russ. J. Non-Ferr. Met.* **54** 412-6
[11] Andreev V A, Gureev A I, Sevast'yanov P I, Loginov V I, Levashov E A, Loginov P A, Rupasov S I and Kurbatkina V V 2014 *Russ. J. Non-Ferr. Met.* **55** 652-6
[12] Sun Y, Wu H, Li M, Meng Q, Gao K, Lü X and Liu B 2016 *Materials* **9(5)** 343
[13] Loginov P, Mishnaevsky Jr L, Levashov E and Petrzhik M. 2015 *Mater. Design* **88** 310-9
[14] Sidorenko D A, Zaitsev A A, Kirichenko A N, Levashov E A, Kurbatkina V V, *et al.* 2013 *Diamond Relat. Mater.* **38** 59-62
[15] Loginov P A, Sidorenko D A, Levashov E A and Andreev V A 2018 *Russ. J. Non-Ferr. Met.* **59** 341-51
[16] Ozolin A V, Sokolov E G and Golius D A 2020 *IOP Conf. Ser.: Mater. Sci. Eng.* **862** 022057
[17] Sokolov E G, Ozolin A V and Arefieva S A 2020 *Mater. Sci. Forum* **992** 511-6
[18] Sokolov E G and Ozolin A V 2018 *Mater. Today Proc.* **5** 26038-41
[19] Sokolov E G, Ozolin A V and Svistun L I 2019 *JP J. Heat Mass Transf.* **16(2)** 297-305

IOP Conf. Series: Materials Science and Engineering 1155 (2021) 012017 doi:10.1088/1757-899X/1155/1/012017

Torsion of a thin-walled structure of a waveguide with rectangular cross-section

I V Kudryavtsev, M V Brungardt, E A Goncharova, O A Li and S I Troshin

Siberian Federal University, 79 Svobodny Pr., Krasnoyarsk, 660041, Russia

E-mail: ikudryavcev@sfu-kras.ru

Abstract. The paper has considered the problem of determining the stress state of thin-walled waveguide structures during torsion. An analytical solution using the Bredt's formula has been considered and its comparison with a numerical solution using the finite element method has been performed. The numerical solution has been carried out using a created volume model, and the load has been modeled as a set of concentrated forces distributed over the entire cross-section of the waveguide. It has allowed us to reduce the local effects in the form of complex stress state and increase the accuracy of the results. Comparison of the results has shown that the stresses calculated by the finite element method are always higher than those obtained by the Bredt's formula and it is necessary to take into account their maximum values. The deformation evaluation also has shown that the FEM calculations give larger twist angles than those obtained by the Bredt's formula for all the considered section sizes.

1. Introduction

Calculation of the action of almost any load for thin-walled beams is a very complicated task [1-15]. Contrary to the theory of beams, a thin-walled cross-section cannot be considered as absolutely rigid and unchangeable; therefor its deformation must be taken into account. This requires the application of a more complex theory of plates and shells that leads to system of differential equations for the two unknown values. Due to the complexity of differential equations, it is not always possible to obtain the accurate analytical solutions.

The development of computers, numerical methods and universal calculation programs based on them made it possible to automate the calculations of thin-walled sections and in many respects pushed analytical calculation methods into the background. Among the numerical methods, the most widespread is the finite element method due to its versatility, ease of use and implementation in many ready-made software products.

However, numerical methods have their own peculiarities and limitations, which require modification of the original design scheme to obtain a correct solution. For example, from the accuracy point of view it is preferable to use volume models and volume finite elements [16-20]. But this approach does not allow us to specify any torques, since volume finite elements do not have rotational degrees of freedom, they have only translational ones. As a result, the calculator needs to convert the torque load into an equivalent distribution of concentrated forces. The accuracy of such a conversion will greatly affect the results obtained, especially in the local area of application of this equivalent load, which makes it difficult to calculate when there is a torque load.

This paper considers the problem of torsion of a thin-walled beam with rectangular cross-section as applied to waveguide structures. The example of analytical and numerical solutions with a comparison of the results is given. In the numerical solution, a rather complicated method is proposed for decomposing the torque into an equivalent set of concentrated forces uniformly distributed over the cross-section of the waveguide.

2. Analytical solution according to R. Bredt's formula

We consider the problem of pure torsion of an extended beam, which is rigidly fixed on one side and loaded with a torque *T* on the other side (figure 1).

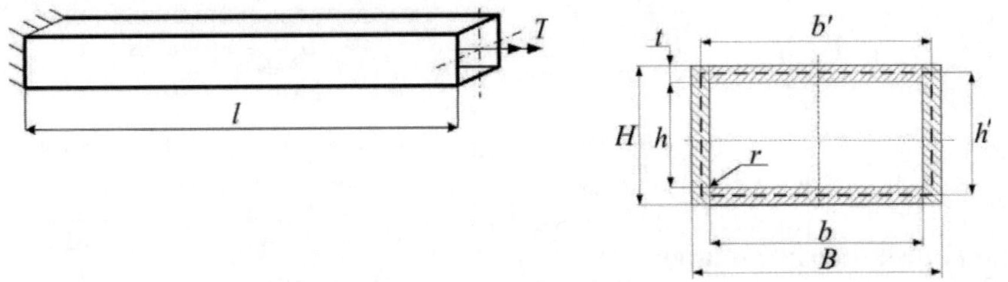

Figure 1. Torsion of the waveguide.

In 1896, R. Bredt proposed a simple solution to this torsion problem for thin-walled beams with almost any cross-section shape [5]. Using the analogy with the theory of beams, to determine the stresses and strains during torsion, it is necessary to know the polar moment of inertia and the polar moment of the cross-section during torsion. With regard to the rectangular thin-walled cross-section of the waveguide, the expressions for the polar moment of inertia and the moment of resistance are:

$$J_\rho = \frac{4\omega^2}{\oint \frac{ds}{t}}, \quad W_\rho = 2\omega t \; . \tag{1}$$

where ω is the cross-sectional area enclosed within the median line of the thin-walled cross-section; $\oint \frac{ds}{t}$ is the integral over the closed contour of the median line of the cross-section.

In our case:

$$\omega = b'h'; \quad \oint \frac{ds}{t} = \frac{2}{t}(b'+h') \; . \tag{2}$$

The substitution of the expressions (2) in the formulas (1) gives:

$$J_\rho = 2t\frac{(b'h')^2}{b'+h'}, \tag{3}$$

$$W_\rho = 2t \cdot b'h'. \tag{4}$$

Then the shear stresses and strains during torsion are determined by equations equivalent to the dependences of beam theory:

$$\tau_{cp} = \frac{T}{W_\rho}, \tag{5}$$

$$\varphi = \frac{T \cdot l}{G \cdot J_\rho}, \quad (6)$$

where G is the shear modulus.

Unlike the beam theory, formula (5) determines not the maximum, but the average stresses in the thin-walled section of the waveguide. This follows from the assumptions made by Bredt when deriving these simple analytical relations. The maximum stresses are supposed to act in the inner corners of a rectangular cross-section and depend on the radii of rounding of these corners r (figure 1, b). The values of these maximum shear stresses can be determined through the stress concentration factor according to the formula [X]:

$$\tau_{\max} = \tau_{cp} \cdot \alpha_K = \tau_{cp} \cdot \left(1 + \frac{t}{2r+t} - t \cdot \frac{b'+h'}{b'h'}\right), \quad (7)$$

Analytical dependencies (5-7) are valid for sufficiently long beams, i.e. when the ratio of length to maximum cross-section size is not less than 5...6.

3. Preparation of a numerical solution

The numerical solution of the torsion problem by the method is carried out for a three-dimensional model of a thin-walled waveguide, which will allow us to obtain a more accurate solution. This approach is complicated by the need to decompose the concentrated torque into a set of concentrated forces. This follows from the implementation of the finite element method for solid bodies that do not have rotational degrees of freedom. Therefore, we transform the torque into an equivalent load in the form of distributed forces F_i (figure 2).

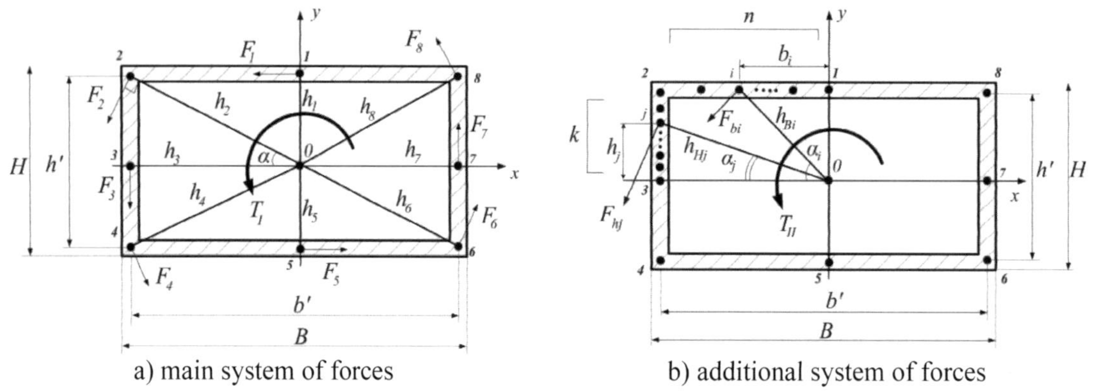

a) main system of forces b) additional system of forces

Figure 2. The system of the forces equivalent to the torque.

The value of the i-th force is calculated from the condition that the moment of this force is constant and equal to

$$T_i = F_i \cdot h_i = \frac{T}{N} = const, \quad (8)$$

where h_i is the moment arm for F_i

N is the specified total number of forces into which the moment T is expanded

A feature of the cross-sectional geometry of the waveguide is that H <> B, as a result of which the distribution of forces along the walls will be uneven. In the case of a thin-walled cross-section with a small wall thickness the points of application of forces are naturally chosen at points on its middle surface. Then the forces Fi, which simulate the torsion, will be distributed along the entire perimeter of the middle surface of the waveguide cross-section. Let's represent the external torque T as consisting of two components:

$$T = T_I + T_{II}\ ., \tag{9}$$

where T_I is the the component of the total torque from the concentrated forces F_i, located at 8 characteristic points of a rectangular cross-section (figure 2, a) and forming the main system of forces; T_{II} is the component of the total torque from concentrated forces F_i located in the intervals between 8 characteristic points of the cross-section and forming an additional system of forces.

3.1 Basic system of forces

The T_I component in equation (2) is expressed as

$$T_I == 8 \cdot T_i = \sum_1^8 F_i h_i = 4F_y \cdot \frac{2B'}{\cos\alpha} + 2F_B \cdot H + 2F_H \cdot B\ , \tag{10}$$

where $F_y = F_2 = F_4 = F_6 = F_8 = T_i \cdot \frac{\cos\alpha}{2B'}$ is the forces at the corners of the cross-section;

$F_B = F_1 = F_3 = \frac{T_i}{2H}$ is the forces on the lower and upper sides of the cross-section;

$F_H = F_3 = F_7 = \frac{T_i}{2B}$ is the forces on the lateral sides of the cross-section;

$\alpha = arctg\left(\frac{H}{B}\right)$ is the cross-section diagonal angle.

3.2 Additional system of forces

Using only the main system of forces F_i during translation, located only at 8 characteristic points of a rectangular cross-section (figure 2, a) will lead to a highly complicated loading of the cross-section and the incorrect resulting solution. Consequently, it is necessary to create an additional system of forces in the intervals between the forces of the main system, while, due to the rectangular waveguide cross-section, the amount of additional forces on different walls will be different.

Figure 2, b shows, for example, additional forces in one quadrant of the cross-section. Obviously, the appearance of one force in one quadrant causes the appearance of a similar force in all other quadrants of the cross-section. Thus, the total number of forces simulating the torque will be equal to:

$$N = 8 + 4 \cdot (n + k) \tag{11}$$

and the moment of each force F_i will be equal to:

$$T_i = \frac{T}{N} = \frac{T}{4(2 + n + k)}\ . \tag{12}$$

To determine the amount of additional forces, it is necessary to determine the moment arms to the points of their location. For the upper wall of the section (figure 1, b), the position of this moment arm is determined by the geometric sizes:

$$b_i = \frac{B \cdot i}{2(n+1)}\ ; \quad \alpha_i = arctg\left(\frac{H \cdot (n+1)}{B \cdot i}\right); \quad h_i = \frac{1}{2}\sqrt{H^2 + \left(\frac{B \cdot i}{n+1}\right)^2} \tag{13}$$

Knowing the moment arm and the torque value, the corresponding value of the force F_i we define as:

$$F_{bi} = \frac{2 \cdot M_i}{\sqrt{H^2 + \left(\frac{B \cdot i}{n+1} \right)^2}} . \tag{14}$$

Similarly, the geometric dimensions and the value of the forces for the vertical sides can be found:

$$h_j = \frac{H \cdot j}{2(k+1)} ; \quad \alpha_j = arctg\left(\frac{H \cdot i}{B \cdot (k+1)} \right) ; \quad F_{hi} = \frac{2 \cdot M_i}{\sqrt{B^2 + \left(\frac{H \cdot j}{k+1} \right)^2}} . \tag{15}$$

Using the obtained dependencies, a program was developed that, for the given loading conditions, creates a macro for the Ansys software. Let us perform calculations on analytical dependences based on the Bredt's formula, obtain a numerical solution with the expansion of the torque into forces, and then compare the results.

4. Results and discussion

A cantilever waveguide was calculated with the following sizes: l=1 m, B = 0.0374 mm, H = 0.0174 mm, t = 0.0012 mm, r = 0. Material: E=7E + 10 Pa, G = 2.665E + 10, Poisson's ratio 0.3. Load: T = 1 N·m, which was represented by a set of forces (figure 3).

The numerical calculation was carried out by the finite element method in the Ansys software using 208,793 finite elements Solid98. The graphical results are shown in the figure 4, and the numerical data are summarized in the table 1.

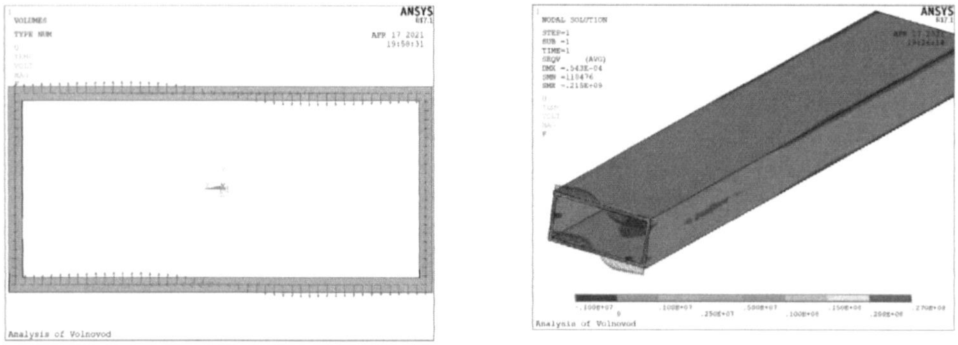

a) set of forces over the section b) stress state of the waveguide

Figure 3. Results of numerical calculation of the waveguide.

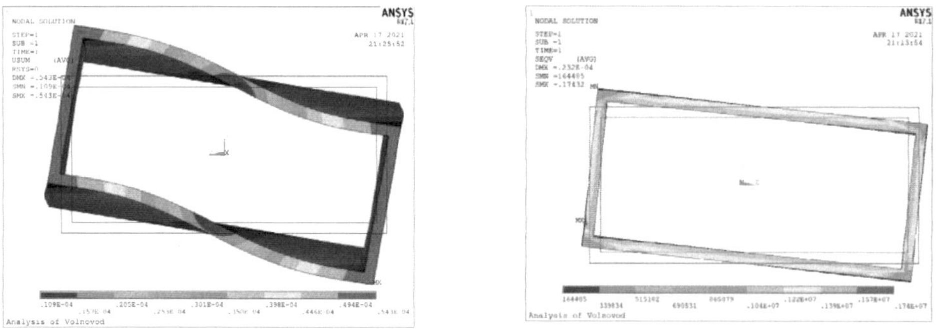

a) deformation b) stress intensity

Figure 4. Results of the numerical calculation in the cross-section of the waveguide.

Table 1. Calculation results.

	Average shear stress, MPa	Maximum shear stress, MPa	Angle of twist, radian
Bredt's solution	0.71	1.34	0.0024
FEM solution	0.78	1.35	0.0025
Deviation, %	9.14	0.383	5.34

The results of calculating the stresses in the waveguide show that the Saint-Venant principle is fulfilled and the edge effect from the application of the load is persist at a distance of one section width (figure 3, b), despite the model used in the form of a distribution of a large number of concentrated forces.

Comparison of the numerical values of the results in table 1 shows that the volume model predicts a little higher stresses and strains than the Bredt's formula does. Such a discrepancy can be explained by the influence of the cross-section deformations on the solution. The change in the values of shear stresses along the wall thickness is also noticeable.

The conducted studies of the dependence of the obtained calculation results on the number of points N showed that with an increase in their number from N=18 to N=656, the obtained stress-deformed data changes by only 3-5%. The variation in the number of finite elements from 8,000 to 46,690 showed that the results of calculating also changed insignificantly, by 3-4%. Thus, there is no significant refinement of the results due to a change in the FE mesh size or the number N of forces modeling the torque. The practice of applying the developed technique shows that the number of areas and the corresponding distributed forces in the cross-section must correspond to the number of finite elements lying in the plane of the cross-section.

5. Conclusion

In this work, analytical and numerical calculations of an extended waveguide of rectangular cross-section for the action of a torque were carried out. The calculation results showed a fairly good convergence of the results of analytical and numerical calculations. Thus, the results obtained make it possible to recommend the Bredt's formula for calculating stresses and strains in waveguides with rectangular cross-section.

Acknowledgements

The research was funded by RFBR, Krasnoyarsk Territory and Krasnoyarsk Regional Fund of Science, project number 19-41-240005.

References

[1] Grant C 1995 *Proc.of the Institution of Mechanical Engineers* (University of Newcastle upon Tyne) 133-40
[2] Rbodes J and Walker A C 1987 *Developments in thin-walled structures* (Applied Science Publishers, London)
[3] Murray N W 1984 *Introduction to the theory of thin walled structures* (Clarendon Press, Oxford)
[4] Morassi A 1995. *J Elasticity* **39** 213-27
[5] Bredt R 1896 Kritische Bemerkungen zur Drehungse astizitat *Zeitschri des Vereines deutscher ngenieure* **40** 813-917
[6] Schmidrathner C 2019 *Acta Mech.* **230** 4035-47
[7] Weigang C and Wierzbicki T 2000 *Thin-Walled Structures* **36** 181-96
[8] Gere J M 2004 *Mechanics of materials* (Cengage Learning, Toronto, Canada)
[9] Hibbeler R C 2016 *Statics and Dynamics* (Pearson Education, NJ, USA)
[10] Timoshenko S 1940 *Strength of materials. Part I. Elementary theory and problems* (D. Van

Nostrad Company, NY, USA)
[11] Ugural A C 2010 *Stresses in beams, plates and shells* (CRC Press, Taylor & Francis Group)
[12] Vlasov V Z 1961 *Thin-walled Elastic Beam* (English translation. Published for NSF and Department of Commerce by the Israel Program of Scientific Translations, Jerusalem)
[13] Meek J L and Lin W J 1990 J *Struct Eng-ASCE.* **116** 1473-89
[14] Carrera E 2011 *Beam Structures. Classical and Advanced Theories* (John Wiley & Sons, Ltd)
[15] Djugash, A C 1994 *J. Construct. Steel Research* **31** 289-304
[16] Hamming R W 1987 *Numerical Methods for Scientists and Engineers* (Dover Publications)
[17] Chapra and Canale R 2014 *Numerical Methods for Engineers* (McGraw-Hill Education)
[18] Burden R L 2015 *Numerical Analysis* (Cengage Learning)
[19] Shrikhande M 2008 *FEM and computational structural dynamics* (PHI Learning Private Limited, New Delhi)
[20] Moaveni S 2015 *Finite element analysis* (Pearson, USA)

Effect of titanium and chromium diffusion coatings on corrosion resistance of sintered iron

E G Sokolov

Kuban State Technological University, 2, Moskovskaya St., Krasnodar, 350072, Russia

E-mail: e_sokolov.07@mail.ru

Abstract. The authors have studied the effect of titanium and chromium diffusion coatings on corrosion resistance of sintered iron in solutions of acids and salts. The samples of 85% density were obtained from powdered iron by static pressing in press dies and sintering in vacuum. The diffusion coatings were applied in molten lead with titanium and chromium additives at the temperature of 1100°C and 2 hours exposure time. Microstructure of the samples was then explored by the optical metallography method. The content of elements in the coatings was measured using energy dispersive X-ray microanalysis. Studies of corrosion resistance of the samples in solutions of acids and salts were conducted by the gravimetric method. Corrosion resistance was also evaluated by time to the first corrosion foci emerging and the area of corrosion damage. Diffusion titanium and chromium coatings have ensured higher corrosion resistance of sintered iron in all the aggressive media studied. Chromizing is the most efficient for protecting sintered iron in solutions of sulphuric and nitric acids, as well as in media containing acetic acid and hydrogen peroxide. Meanwhile, titanizing is the most efficient for protecting sintered iron in sea water.

1. Introduction

Corrosion resistance is one of the major characteristics of machine parts; knowing it allows using the materials correctly and selecting reliable and cost-efficient coatings for them. As a rule, sintered materials feature lower corrosion resistance than the cast and rolled stocks [1]. Corrosion resistance of sintered materials decreases as the residual porosity increases. The presence of pores and porous channels leads to a larger active surface of the material, which triggers the development of corrosion processes on the external and internal surfaces simultaneously. The studies conducted [2-4] demonstrate that it is pores that are the point where corrosion emerges in sintered stainless steels in various aggressive media. The pores cause local crevice corrosion which spreads along the particle separation boundaries, which ultimately results in partial destruction of the material. According to the work [5], the speed of corrosion is by a factor of 100 higher in sintered steels than that in the massive ones.

For maintaining operability of sintered items in corrosive media, it is expedient to use protective coatings which ensure sealing open porosity of the sintered stock and feature a high strength of adhesion to the base, thus eliminating the coatings peeling from the substrate.

Such properties are found in diffusion titanium, chromium, nickel, and copper-nickel coatings applied in the medium of liquid-metal solutions [6, 7]. This method of coating application relies on the phenomenon of isothermal mass transfer of solid metals through the readily fusible "transport" melt [8, 9]. The coating technology consists in the following: isothermal holding of items in the medium of low-melting metal is conducted, and the metal or alloy to be making up diffusion coating with the material of the items is introduced into the said medium in the form of powder, pieces, or in the compact state. The low-melting metal acting as the transport medium has to dissolve the diffusion additives while not penetrating into the bulk of the item metal. According to the iron-lead diagram [10, 11], solubility of lead in solid iron at 1530°C amounts to 2.7×10^{-4} % at. Meanwhile, solubility of iron in liquid lead is insignificant: it is 0.08–0.21 % at within the temperature range of 1302–1495°C [11]. So, extremely low mutual solubility of the components eliminates penetration of lead into the surface of sintered iron and prevents liquid lead from dissolving the surface of the parts. This allows using lead as the "transport" melt for applying diffusion coatings on iron.

The studies conducted have shown that this method allows obtaining uniform titanium and chromium coatings on sintered iron which seal off its open porosity [6, 12, 13].

With regard to this, the objective of this paper was to study the effect of titanium and chromium diffusion coatings on corrosion resistance of sintered iron in solutions of acids and salts.

2. Materials and methods

The samples have been manufactured of powdered iron PZhV2.160.26. The blanks in the form of bushings having 16 mm OD, 8 mm ID, and 14 mm high were obtained by static pressing in press dies. The samples were then sintered in vacuum at the temperature of 1100°C. Density of the sintered samples was 85%. Part of the samples were subjected to diffusion titanizing and chromizing. The diffusion coatings were applied in molten lead with titanium and chromium additives at the temperature of 1100°C and 2 hours exposure time. Microstructure of the coated samples was investigated using the MMR-4 metallographic microscope. For making microstructure apparent, the samples were etched with 3% solution of nitric acid in the ethyl alcohol. The content of elements in the coatings was studied by the method of energy dispersive X-ray microanalysis (EDXMA) using SEM JEOL JSM-7500F.

The studies of corrosion resistance of the diffusion coated samples in solutions of acids and salts have been conducted by the gravimetric method. The samples were weighed before and after exposure to aggressive media using the Adventurer AR2140 analytical balance. Before weighing, the samples were carefully cleaned of the corrosion products according to the technique [14] and rinsed with ethyl alcohol. According to the test results, the intensity of corrosion was evaluated as the loss of mass per surface area unit within a certain time.

Corrosion resistance of the coatings was also assessed according to the technique [14], by the time to emergence of the first corrosion foci and by the area of corrosion damage.

Corrosion resistance of the coatings obtained was tested in the following media:

- 3% aqueous solution of NaCl;
- 20% CH_3COOH +20% H_2O_2 + 60% H_2O;
- 10% aqueous solution of H_2SO_4;
- 0.5% aqueous solution of HNO_3.

3. Results and discussion

3.1. Chemical composition and structure of the coatings

Microstructure of the samples with titanium and chromium coatings is shown in figure 1. On the surface of the titanium coating, there is a layer of about 10 μm thickness consisting of the TiFe intermetallic compound with the titanium content of 48 wt % [11]. Under this layer, there is a solid solution of titanium in α-iron with the 2.2–3 wt % content of titanium. At the temperature of 1100°C, diffusion of titanium in γ-iron leads to the γ→α transformation. This results in the diffusion layer forming which consists of α-phase columnar grains aligned with the diffusion flow direction. It can be seen from figure 1 that the coating is differentiated from its base by the line of phase recrystallization. The total thickness of the titanium coating amounts to 120 μm.

The chromium coating represents a solid solution of chromium in α-iron. The content of chromium is 38 wt % on the surface. With distance from the surface, it goes down to making up 12 wt % at the line of phase recrystallization. The average thickness of the chromium coating is 105 μm, but on individual areas, chromium has penetrated into the base somewhat deeper owing to diffusion along defects of the sintered iron structure.

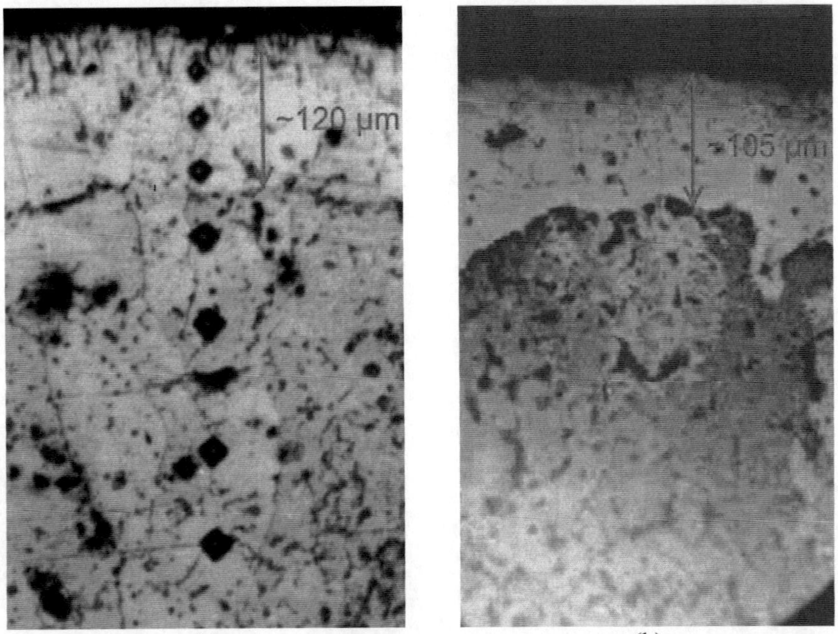

(a) (b)

Figure 1. Microstructure of diffusion coatings on sintered iron, magnified at ×200: (a) titanium coating; (b) chromium coating.

3.2. Corrosion resistance of the coatings. Test results

Figure 2 shows the results of corrosion tests conducted on the 15 hours basis for 85% density iron. It can be seen that chromium coatings feature the highest resistance to the solution containing 20% of CH_3COOH + 20% of H_2O_2 + 60% of H_2O. As for the titanized parts, corrosion foci of around 100–300 μm diameter and approximately the same depth have been found on them after exposure for 15 hours. Obviously, their emergence must be associated with open pores being present in titanium coating. The quantity of corrosion foci amounts to 2–3 per cm^2. Meanwhile, no such damage occurs if the test duration is up to 1.5 hours.

In the 3% aqueous solution of NaCl imitating sea water, the most efficient protection of sintered iron is ensured by titanium coatings. However, this is only achieved with the high quality of coating having no discontinuities. In the opposite case, the process of electrochemical corrosion develops at the points of discontinuities, and corrosion pits form on the surface of the items. On materials of various densities, resistance of the titanium coatings to 3% aqueous solution of NaCl does not differ significantly.

Diffusion chromizing improves corrosion resistance of sintered iron in 10% aqueous solution of H_2SO_4 and 0.5% solution of HNO_3. It should be noted that chromizing in liquid metal solutions ensures a higher corrosion resistance as compared to solid phase chromizing. In particular, diffusion coating on powdered iron obtained by the solid phase method can only protect the items in 0.5% solution of H_2SO_4 and 0.5 % solution of HNO_3 during time-limited work (less than 1 hour) [8]. Meanwhile, with chromium coatings obtained from liquid metal solutions, no noticeable damage is observed in case of over 6 hours of operation.

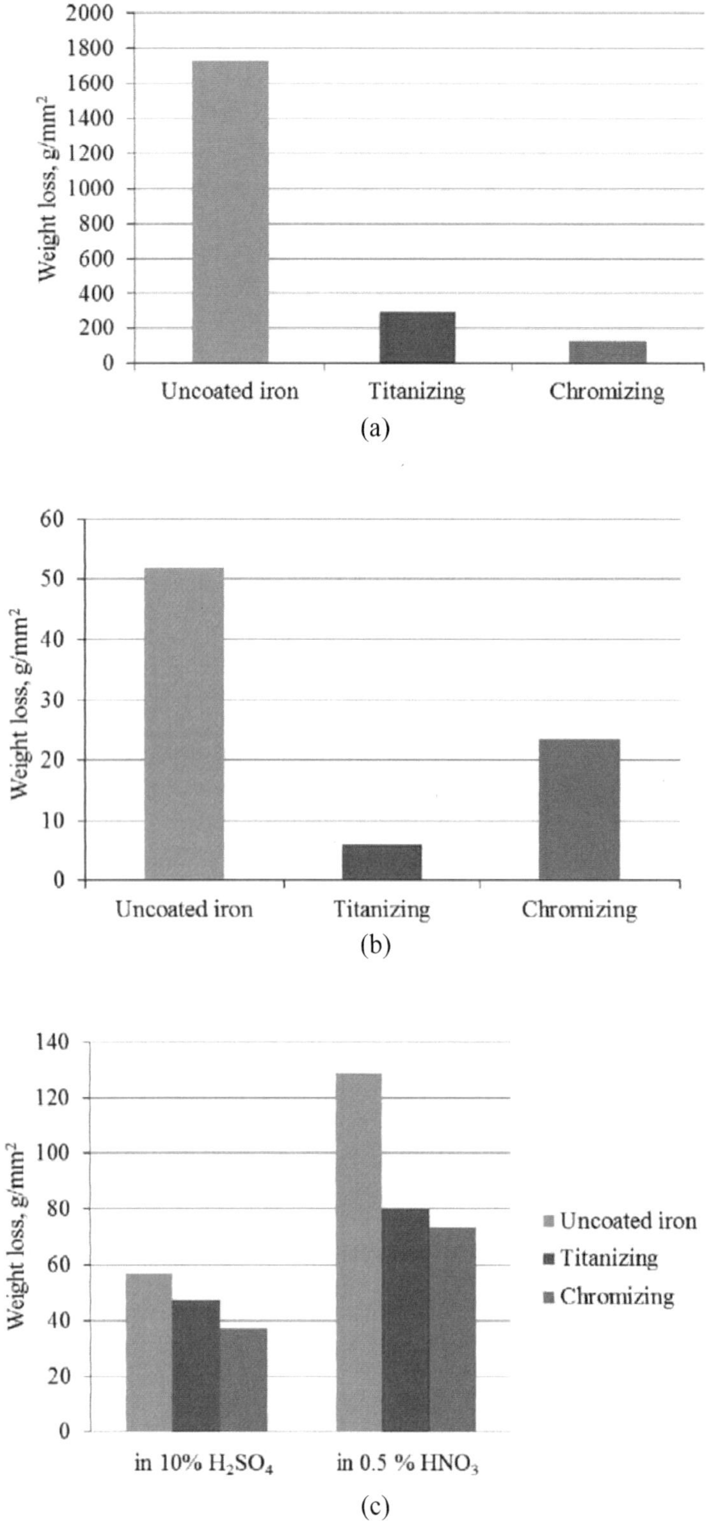

Figure 2. Effect of titanizing and chromizing on the weight loss of sintered iron. (a) in solution containing 20% of CH_3COOH + 20% of H_2O_2 + 60% of H_2O; (b) in 3% aqueous solution of NaCl; (c) in 10% aqueous solution of H_2SO_4 and 0.5% solution of HNO_3.

4. Conclusion

Diffusion titanium and chromium coatings have ensured higher corrosion resistance for sintered iron in all the aggressive media under study. Chromizing is the most efficient for protecting sintered iron in solutions of sulphuric and nitric acids, as well as in media containing acetic acid and hydrogen peroxide. Meanwhile, the most efficient coating for protecting sintered iron in sea water is titanizing.

References

[1] Bartenev V V and Grigor'ev V P 1993 *Powder Metal. Met. Ceram.* **32** 260-2
[2] Otero E, Pardo A, Sáenz E and Utrilla M V 1997 *Can. Met. Quart.* **36(1)** 65-72
[3] Meesak T and Thedsuwan C 2018 *Mater. Today: Proc.* **5(3)** 9560-8
[4] Moral C, Bautista A and Velasco F 2009 *Corros. Sci.* **51(8)** 1651-7
[5] Otero E, Pardo A, Utrilla M V, Perez F G and Merino C 1997 *Corros. Sci.* **39(3)** 453-63
[6] Yurchik S M and Artem'ev V P 2002 *Met. Sci. Heat Treat.* **44** 456-7
[7] Bobylyov E E 2019 *J. Phys.: Conf. Ser.* **1399** 044055
[8] Sokolov E G, Ozolin A V and Svistun L I 2019 *JP J. Heat Mass Transfer* **16** 297-305
[9] Artem'ev V P and Chaevskii M I 1986 Diffusion titanizing in a medium of liquid-metal melt *Adhesion of Melts and Soldering of Materials* **16** (Kiev: Naukova Dumka) p 82-5 (In Russ.)
[10] Okamoto H, Schlesinger M E and Mueller E M (eds.) 2016 *ASM Handbook: Alloy Phase Diagrams* (Materials Park, Ohio: ASM International)
[11] Lyakishev N P (ed.) 1996 *Diagrams of Binary Metallic Systems* (Moscow: Mashinostroyeniye) (In Russ.)
[12] Artemyev V P, Sokolov E G and Yurchik S M Pat. of the Russian Federation 2174059 IPC B22F 3/24; C23C 10/22 / No. 2000100894/02
[13] Artemyev V P and Sokolov E G 2018 *Tekhnologhiya Metallov* **4** 9-12 (In Russ.)
[14] Fokin M N and Zhigalova K A 1986 *Methods for corrosion testing of metals* (Moscow: Metallurghiya) (In Russ.)
[15] Dubinin G N 1964 *Diffusion chromizing of alloys* (Moscow: Mashinostroyeniye) (In Russ.)

IOP Conf. Series: Materials Science and Engineering 1155 (2021) 012019 doi:10.1088/1757-899X/1155/1/012019

Effect of microalloying on the stability of the endurance characteristics of hot-deformed powder steels

V Y Dorofeyev[1,4], A N Sviridova[1], Y M Berezhnoy[1], E N Bessarabov[1], S V Sviridova[2], K S Kochkarova[3] and I D Derlugyan[1]

[1]Platov South-Russian State Polytechnic University (NPI), 132 Prosveshcheniya str, Novocherkassk, Rostov oblast', 346428 Russia
[2]Derzhavin Tambov State University, 33 Internatsional'naya str., Tambov, 392036, Russia
[3]North-Caucasian State Academy, 36 Stavropol'skayastr, Cherkessk, KChR, 369000, Russia

[4]E-mail: dvyu56.56@mail.ru

Abstract. The stability of the characteristics of the contact and low-cycle fatigue life of steels obtained from iron powders with different contents of alloying and microalloying elements, as well as impurities, has been studied. The carbon content in powder steels was varied. It has been established that doping Na or Ca microadditives contributes to an increase in the stability and absolute value of the parameters of the contact and low-cycle fatigue life of hot-deformed powder steels due to the activation of cohesive interaction on interparticle surfaces and carbon diffusion, which contributes to the formation of a homogeneous structure. Doping microadditives of Al contributes to an increase in the contact and low-cycle fatigue life of the powder steels in comparison with the samples-witnesses. However, the formation of zones of structural heterogeneity in steels containing more than 0.4 wt. % C, causes a decrease in the stability of these characteristics.

1. Introduction

The stability of the mechanical properties of materials largely makes it possible their use in manufacturing parts for various purposes. The importance of this factor increases in the case of manufacturing products operating at low temperatures and exposed to dynamic, fatigue, static and wear loads. Brittle materials (ceramics, sitalls, glass, cast iron, etc.) are characterized by low stability of mechanical properties, which is associated with the heterogeneity and defectiveness of the structure caused by the instability of technological processes of their preparation [1].

The cause for the instability of the compact steel's properties can be the presence of internal stresses and areas of metastable austenite. In this regard, heat treatment is an effective method for increasing the stability indicators of such steels. As an additional heat treatment of high-strength strip steel, it is recommended to use artificial aging, which provides a significant decrease in the value of internal stresses and promotes the precipitation of dispersed carbides and carbonitrides that stabilize dislocations [2].

The influence of technological modes of deformation processing, test conditions, as well as alloying on the mechanical properties of metastable austenitic steels and indicators of their stability is often ambiguous. In particular, if the crack resistance decreases with the introduction of alloying elements, then the indicators of its stability increase. On the contrary, an increase in the degree of deformation leads to an increase in crack resistance and a decrease in its stability [3]. An objective criterion for the stability of austenite and the mechanical properties of steels of this class is the coefficient characterizing the degree of decomposition of austenite.

Powder steels are characterized by low stability of mechanical properties, which is associated with the presence of a large number of impurities, chemical and structural inhomogeneities, discontinuities and defects of various origins (pores, microcracks, delamination, etc.). Indicative in this respect is the diagram, which shows the change in the scatter of the values of the nickel content in powders of the Distaloy type at different time stages of their production, given in [4]. This diagram clearly illustrates the fact that the development of technologies for obtaining Distaloy powders was aimed at reducing the spread of nickel content values. This made it possible to significantly reduce the instability of the values of the mechanical properties of the materials obtained, as well as the dimensions and characteristics of the shape and the relative position of the surfaces of products.

The instability of the sizes and properties of powder materials and products is caused by the segregation of the components of the initial mixtures during their preparation [5]. Each technological operation of obtaining a powder product has a hereditary effect both on the absolute values of the characteristics of mechanical properties and dimensional accuracy of the products obtained, and on the indicators of their stability [6].

The cause for the instability of the sizes and properties of powder materials and products based on them may be the unsatisfactory quality of interparticle jointing. With regard to powder materials obtained by pressing-sintering technology, this leads to uneven localization of the setting bridges, which depends on the direction of the acting forces during cold pressing and causes anisotropy of shrinkage during sintering [7, 8].

The instability of the properties of hot-deformed powder steels (HDPS) is often associated with the unsatisfactory quality of interparticle jointing, which is formed at the stage of repressing of the heated porous workpiece in the die matrix. Earlier it was shown that the introduction of microadditives of alloying elements provides an increase in the absolute value and stability of the values of the mechanical properties of HDPS by improving the quality of interparticle jointing [9]. The stability of the mechanical properties of steels based on unalloyed iron powders was studied in relation to the characteristics of impact toughness and crack resistance. Determination of the impact strength implies the possibility of using smooth samples and does not cause difficulties. On the contrary, the assessment of crack resistance is associated with the need to apply a stress concentrator, which is not always possible. In particular, when testing samples with a hardened surface layer, the application of a notch will lead to an incorrect estimate of the value of crack resistance.

Various types of fatigue tests are widely used to assess the performance of materials under cyclic loading [10]. However, the currently available information on the fatigue characteristics of powder steels is not systematized, despite the large amount of experimental data obtained. This does not allow using it in the well-known software package nCode based on Ansys in order to predict the resource of powder materials and products. As applied to powder materials obtained by hot forging of porous blanks, this problem is of particular importance due to the low reproducibility of properties.

In addition, the study of the behavior of a hot-deformed powder material under the influence of cyclic loads is of scientific importance. This is due to the fact that a significant part of the research carried out to date has been devoted to the study of the dependence of the fatigue characteristics of powder steels on density. The upper limit of the studied values of the density of powder steels (ρ), as a rule, was limited to the values $(7.2 - 7.4) \cdot 10^3$ kg / m^3. The number of studies devoted to the study of the endurance characteristics of powder steels in a high-density state is significantly less. This indicates the feasibility of further research on HDPS endurance. The availability of relevant information will provide the possibility of plotting the dependences of the fatigue strength of powder steels on the density in a wider range of its values. In this case, the obtained values of HDPS density and properties can be (with certain adjustments caused by the presence of impurities and structural features) a standard. In particular, in the German-language literature P. Beiss proposed a generalized formula of M. Yu. Bal'shin:

$$P = P_0(\frac{\rho}{\rho_0})^m,$$

where: P – material property with porosity; P_0–property of a compact material; ρ – density after sintering; ρ_0 – density of the compact material; m–exponential multiplier.

The problem lies in determining the values of P_0 and ρ_0. The well-known technique for ρ_0 determining from the volume fraction of constituent components does not take into account changes in density due to diffusion, phase and structural transformations. The method for ρ_0 determining, which provides for the melting of powder molding, also looks not quite correct. The liquation of the melt components, the formation of slag on the surface of the melt, the occurrence of crystallization, phase and structural transformations during its cooling may have a significant effect on the ρ_0 values and the properties of such a reference material.

This work is a continuation of research of the HDPS structure and properties with microadditions of alloying elements [11]. Its purpose is to study the possibility of increasing the stability of the HDPS endurance characteristics by introducing microadditions of alloying elements.

2. Research methods

For fatigue tests, samples were made on the basis of PZhV2.160.26 iron powder manufactured by the Sulinsky metallurgical plant, as well as AstaloyCrM chromium-molybdenum iron powder and ABC100.30 unalloyed iron powder manufactured by HöganäsAB (Sweden). The chemical composition of iron powders is given in research [11]. The technological scheme for obtaining samples is shown in figure 1.

The carbon content in powder steels (C_{Cm}) was varied. Carbon was introduced in the form of pencil graphite powder GK-1 GOST 4404-78. Sodium was introduced in the form of bicarbonate $NaHCO_3$ GOST 2156-76, calcium in the form of carbonate $CaCO_3$ GOST 4530-76. During microalloying with aluminum powder of crushed ferroaluminium FA-50 TU U 27.3 - 13533123 - 001 - 20 fraction (-0.063) mm was used. The content of microadditives corresponded to the optimal values established earlier: $C_{Na} = 0.2$ wt. %; $C_{Ca} = 0.3$ wt. %; $C_{Al} = 0.4$ wt. %.

To determine the characteristics of endurance of powder steels and carry out structural analysis, prismatic samples with a size of 5 × 10 × 55 mm, as well as cylindrical samples with a size of ø 26 × 6 mm, were obtained. To compensate for the cooling of the surface layers of the workpiece, hot forging was carried out in a mold, the heating temperature of which was $T_D = 600$ °C. Hot-forged samples were carburized to compensate for decarburization, followed by heat treatment. To reduce the risk of the delayed destruction, the release duration was increased to the usually recommended values and was 3 hours.

The method for determining the characteristics of contact (N_{90}) and low-cycle fatigue life (N_{lcf}) corresponded to that described earlier [11, 12]. The stability of mechanical properties was assessed by the value of the coefficient of variation [13]

$$V = \frac{S}{\bar{X}},$$

where $\overline{X} = \frac{1}{n}\sum n X_i$– arithmetic mean value; $n = 10$ – number of experiments; X_i -absolute value of a mechanical property in the experiment; $S = \sqrt{D}$ –standard deviation; $D = \overline{(X_\iota - \bar{X})^2}$ – dispersion.

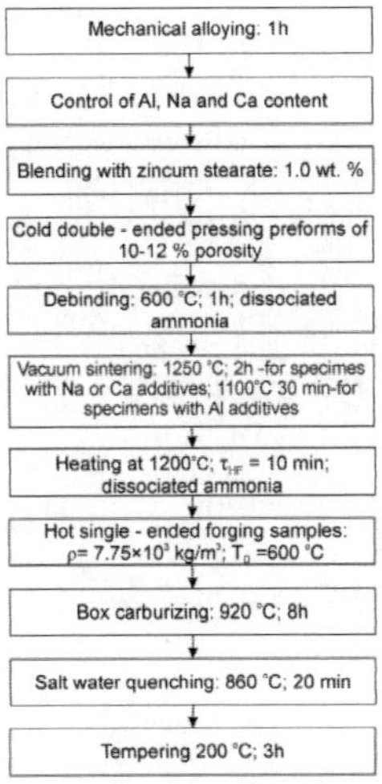

Figure 1. The flow diagram of sample production.

Metallographic analysis was carried out using an AltamiMET-1M optical microscope (Altami LLC, Russia) on etched (3% Nital) and non-etched thin sections. The fractures of the samples were studied using a Quanta 200 i 3D scanning microscope-microanalyzer.

3. Results and discussion

Figure 2 shows the dependences of the endurance characteristics of microalloyed powder steels and their stability on the carbon content. The N_{90}, N_{lcf} (C_{Cm}) dependences for steels with microadditives of calcium were published earlier. The indicated dependences are non-monotonic. An increase in the C_{Cm} values to 0.8 wt. % contributes to an increase in the characteristics of endurance, which is associated with an increase in the amount of martensite in the structure of steels. With a further increase in the C_{Cm} value, the N_{90} and N_{lcf} values decrease due to the precipitation of excess cementite.

The largest values of N_{90} are observed on samples of chromium-molybdenum steel of eutectoid composition with a microaddition of *Na* (figure 2, a; curve 6). The samples based on undoped iron powder ABC100.30 showed slightly lower N_{90} values (curve 5). The lowest contact endurance was demonstrated by reference samples based on PZhV2.160.26 powder without microadditions of alloying elements (curve 8). Microalloying made it possible to increase the contact endurance of the samples obtained on the basis of this powder, despite the relatively high number of impurities contained in it (compare curves 8 and 4).

In contrast to contact fatigue, in the process of testing for low-cycle fatigue life, the best results were demonstrated by samples based on unalloyed ABC100.30 iron powder with a microaddition of *Na* (figure 2, b; curve 5). The lowest values of N_{lcf} are observed on the reference samples. Similarly, to contact endurance, microalloying made it possible to increase the N_{lcf} values (figure 2, b; compare curves 4 and 8).

When microalloying *Ca* and *Al*, the highest indices of contact and low-cycle fatigue life are observed for samples obtained from AstaloyCrM powder (figure 2, e, f; in figure 2, c, d, the data for N_{90} and N_{lcf} are not shown). The introduction of microadditives *Ca* and *Al*, as well as microadditives of *Na*, makes it possible to increase the values of N_{90} and N_{lcf} in comparison with the reference samples. The mechanism of the positive effect of microadditives on the characteristics of contact and low-cycle fatigue life is associated with the activation of cohesive interaction on interparticle surfaces [11]. The introduction of *Na* or *Ca* microadditives contributes to the refinement of grain boundaries and particles from impurities. In addition, the presence of *Na* or *Ca* microadditives makes it possible to activate the diffusion of carbon in powder steels [13]. This contributes to the formation of a more homogeneous structure, which is a significant factor in terms of ensuring the stability of the indicators of mechanical properties [9]. Aluminum, being localized on interparticle surfaces in the composition of a solid solution in iron, provides an increase in free energy and activation of recrystallization processes that control the formation of cohesive bonds.

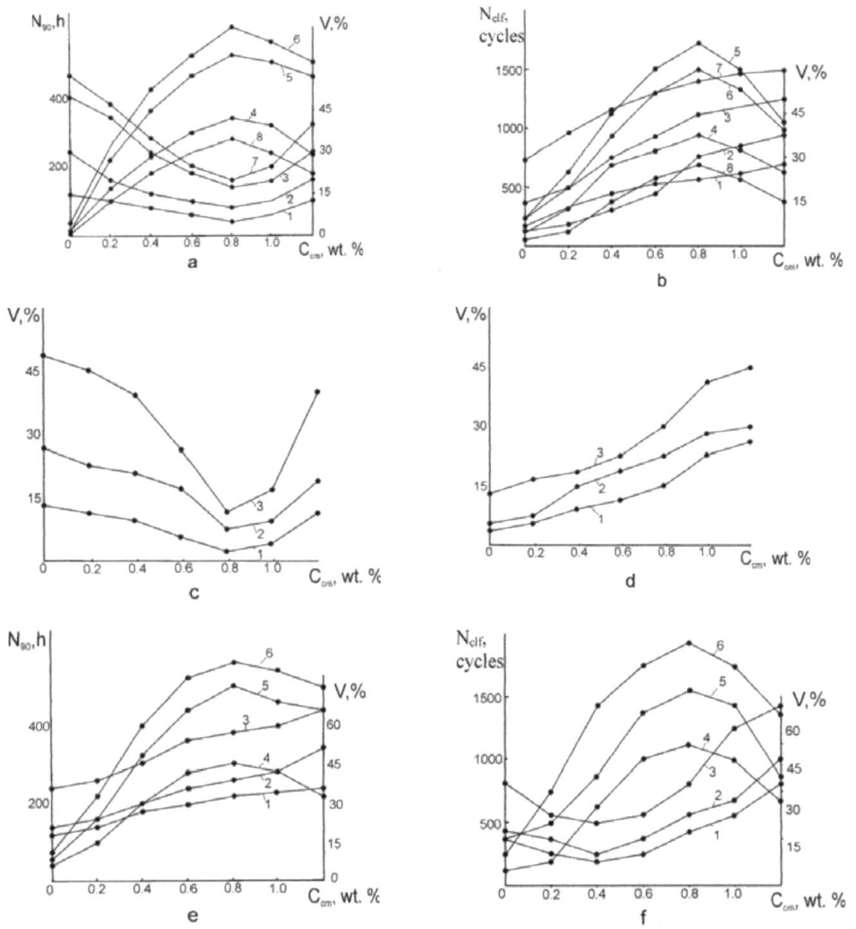

Figure 2. The effect of carbon content on rolling contact (a, c, e) and low-cecle (b, d, f) fatigue life of hot-deformed powder steels and their stability.
Base powder: ABC100.30 (1, 5); AstaloyCrM (2, 6); PZhV2.160.26 (3, 4, 7, 8).C_{Na} = *0,2 wt. %* (a, b);C_{Ca} = *0,3 wt. %*(c, d);C_{Al} = *0,4 wt. %*(e, f).*V* (1 – 3, 7); N_{90}, N_{lcf} (4 – 6, 8). 7, 8 – samples-witnesses without microadditives of alloying elements

The effect of carbon and microalloying additives included in the composition of powder steels on the stability indicators of the characteristics of contact and low-cycle fatigue life described above is somewhat different. For steels with microadditives of *Na* or *Ca*, the dependences of the stability indicators of the values of contact fatigue on the carbon content have a non-monotonic character (figure 2, a, c; curves 1 - 3). With an increase in C_{Cm} to 0.8 wt. % the *V* values decrease, reaching a minimum. At $C_{Cm} < 0.8$ *wt. %* the structure of steels is martensite with troostite patches. Despite the presence of a surface cemented layer with a homogeneous martensitic structure and individual inclusions of cementite, deformation of the wear tracks occurs in the process of contact loading of such samples.

The inhomogeneity of the distribution of troostite and martensite sections in the sublayer of the samples predetermines the corresponding inhomogeneity of the deformation of the wear tracks. In this regard, for low-carbon steels, the *V* values are at a rather high level. The development of pitting cracks is characterized by branching with the formation of a large number of fracture products. At C_{Cm} = *0.8wt. %*, the relief of the fracture surfaces decreases, which determines the presence of a corresponding minimum in the V (C_{Cm}) dependences. Microalloying made it possible to increase the stability of the parameters of the contact fatigue of steels based on PZhV2.160.26 powder with a high content of impurities in comparison with the reference images (figure 2, a; compare curves 3 and 7).

The dependences of the stability indicators of the low-cycle fatigue life values of steels with microadditives of *Na* or *Ca* on the carbon content are monotonic (figure 2, b, d; curves 1 - 3). As C_{Cm} increases, the stability of the low-cycle fatigue life characteristics decreases (the *V* values increase). In this case, in the process of testing low-carbon steels, a fatigue crack originating in the surface layer near non-metallic inclusions, during its development, forms a fracture zone characterized by the presence of areas of shear transcrystalline and ductile fracture.

The presence of a viscous component in the fracture contributes to an increase in the stability of the N_{lcf} values. The result obtained is consistent with the known concepts of the dependence of the stability of strength indicators on the absolute value of the latter: as strength increases, its stability decreases [14].

The effect of the *Al* microadditions on the stability indicators of the endurance characteristics of powder steels is ambiguous. In contrast to steels with microadditives of *Na* or *Ca*, the stability of the values of contact fatigue decreases monotonically with an increase in the carbon content: the values of V increase in the entire studied range of the C_{Cm} values (figure 2, e; curves 1 - 3). This is due to the fact that aluminum and carbon during mixing and subsequent cold pressing of the blanks are localized near the interparticle boundaries. In the process of sintering and hot forging, aluminum, being a graphitizing element, causes the formation of zones of structural inhomogeneity near the former interparticle boundaries depleted in carbon. The distribution of these zones in the structure of the samples does not have an ordered character, which predetermines the corresponding instability of the contact endurance indices, which increases with an increase in C_{Cm}.

The impurity content is an additional factor that determines the level of stability of properties. Therefore, in all cases, their most stable indicators were obtained on samples made of ABC100.30 powder with a low content of impurities (curves 1). The least stability of properties was demonstrated by samples of PZhV2.160.26 powder with a high content of impurities (curves 3).

The dependences of the stability parameters of the low-cycle fatigue life characteristics on the carbon content are non-monotonic (figure 2, f; curves 1-3). As C_{Cm} increases in the range of 0-0.4 wt. % values of *V* fall, and with a further increase in C_{Cm} the *V* values increase. This is due to the fact that the N_{lcf} characteristic is more determined by the quality of interparticle jointing compared to the contact endurance. Therefore, at C_{Cm} = 0.4 wt. % the factors ensuring the activation of the processes of cohesive interaction dominant.

The mechanism of the positive effect of *Al* has been described above. The action of small additions of carbon can also have a positive effect in connection with the processes of reduction of oxides on interparticle surfaces [9]. The combined effect of these two factors (microadditives of *Al* at $C_{Cm} \leq 0.4$ wt.%) provided an increase in the stability of the N_{lcf} values. Obviously, in the process of testing for contact endurance, the action of this mechanism was leveled by the action of another mechanism described above and associated with the "pushing" of the solid surface cemented layer by balls of the test stand into the body of the softer sublayer and the core of the sample, as a result of which a large number of wear products were formed.

With a further increase in C_{Cm} above 0.4 wt. %, the stability of the N_{lcf} values decreases, which is associated with an increase in the structural inhomogeneity near the interparticle boundaries.

4. Conclusion

1. The introduction of *Na* or *Ca* microadditives contributes to an increase in the stability and absolute value of the indicators of the contact and low-cycle fatigue life of HDPS due to the activation of cohesive interaction on interparticle surfaces and diffusion of carbon, which contributes to the formation of a homogeneous structure.

2. The presence of a viscous component in the fracture of samples of microalloyed low-carbon powder steels obtained as a result of testing for low-cycle fatigue provides a higher stability of the durability characteristics in comparison with high-carbon steels destroyed by the mechanism of transcrystalline cleavage. When testing for contact fatigue, a higher stability of the durability indicators is observed on high-carbon steels, which is associated with the specifics of the stress-strain state of the surface layers of the samples under contact loading.

3. The introduction of *Al* microadditives contributes to an increase in the contact and low-cycle fatigue life of HDPS in comparison with the reference samples. However, the formation of zones of structural heterogeneity in steels containing above 0.4 wt. % *C*, causes a decrease in the stability of these characteristics.

4. The optimal combination of the absolute value and stability of the contact and low-cycle fatigue life indicators is demonstrated by samples of eutectoid composition steels with *Na* or *Ca* microadditives, obtained on the basis of AstaloyCrM atomized chromium-molybdenum powder.

Acknowledgment

The study was carried out with the financial support of the Russian Foundation for Basic Research, Grant No. 19-08-00107 A. The authors would like to thank the Höganäs Eastern Europe company for the provided iron powders manufactured by HöganäsAB.

References

[1] Vazhentsev Y G and Sedokov L M 1975 Stability of the mechanical properties of sitalls *Bulletin of the Tomsk Polytechnic Institute* **263** 76-80

[2] Lutsenko A N, Arabey A B, Nemtinov A A, Tsvetkov D S, Popova S D, Korchagin A M, Tikhonov S M and Glebov A G 2010 Stability of the Mechanical Properties of High-Strength Steel Strip over Time *Steel in Translation* **40(5)** 393-7

[3] Bhandarkar D, Zackay V F and Parker E R 1972 Stability and Mechanical Properties of Some Metastable Austenitic Steels *Metallurgical Transactions* **3** 2619-31

[4] Lindskog P *Proc. European Congress and Exhibition on Powder Metallurgy Euro PM2013 (Gothenburg)* vol 2 (European Powder Metallurgy Association) 231-43

[5] Lindskog P 1975 Dusting and Segregation in the Handling of Powder Mixes *Powder Metallurgy***19(2)** 79-81

[6] Himmelstein P and Müller N 1973 Equations Describing Dimensional and Density Variations in Sintered Bodies: Application for Predicting Optimum Production Practice*Powder Metallurgy* **16(32)** 459-68

[7] Molinari A, Menapace C, Torresani E, Cristofolini I and Larsson M 2013 Working hypothesis for origin of anisotropic sintering shrinkage caused by prior uniaxial cold compaction *Powder Metallurgy* **56(3)** 189-95

[8] Zavaliangos A, Missiaen J M, and Bouvard D 2006 Anisotropy in Shrinkage during Sintering *Science of Sintering* **38** 13-25

[9] Dorofeyev V Y, Kochkarova K S 2014 *Proc. 11-th Int. Conf. on New Materials and Technologies: Powder Metallurgy, Composite Materials, Protective Coatings and Welding (Minsk)* (Minsk: Belaruskaya Navuka) 78-82

[10] Schneider M, Maassen R, Wawoczny D and Radis C 2021 Fatigue data for PM steels: Yow the world's largest PM part producer is facing the challenges of data generation and organization *Powder Metallurgy Review* **10(1)** 57-81

[11] Dorofeyev V Y, Sviridova A N, Berezhnoy Y M, Bessarabov E N, Kochkarova K S and

Tamadaev V G 2019 *IOP Conf. Ser.: Mater. Sci. Eng.* **537** 022046

[12] Dorofeyev V Y, Sviridova A N, Berezhnoy Y M, Bessarabov E N, Kochkarova K S, Pustovoit V N and Sviridova S V 2020 *IOP Conf. Ser.: Mater. Sci. Eng.* **862** 022045

[13] D'yachkova L N, Kerzhentseva L F and Markova L V 2004 *Iron-based powder materials* (Minsk: Tonpik Press) p 228

[14] Kotrechko S A and Meshkov Y Y 2011 The Role of Strength and Plasticity in the Brittleness of Metals *Questions of Atomic Science and Technology Series: Physics of Radiation Damage and Radiation Materials Science* **98(4)** 63-5

Estimation of the thermal gap when soldering waveguide elements

I V Kudryavtsev, M V Brungardt, E A Goncharova, O A Li and A V Kolotov

Siberian Federal University, 79 Svobodny Pr., Krasnoyarsk, 660041, Russia

E-mail: ikudryavcev@sfu-kras.ru

Abstract. This paper proposes a method for determining the required gap between the elements of the waveguide path during its assembly by soldering. The need for a thermal gap is defined by difference in mass and dimensional parameters of the elements to be connected, due to which a significant no uniformity of heating and thermal deformations occur. The no uniformity of thermal deformations leads to jamming of one element in another, therefore, the heating time and the size of the thermal gap are need to be determined. The diameter of the solder wire is determined according to the size of the thermal gap and the required fillet of the soldered seam. The performed numerical calculations of the stress state of the obtained soldered seam showed the absence of stress concentration. Mechanical tests have confirmed the high strength of soldered seam.

1. Introduction

Soldering of thin-walled waveguide elements is a crucial stage in their manufacture [1-6]. The soldering of the waveguide path is carried out sequentially from one connection of the elements to another. Soldering of aluminum elements of the waveguide structure is carried out by using special microwave equipment with high-frequency currents. With such heating, heat is distributed mainly along two heating circuits, which are shown in figure 1.

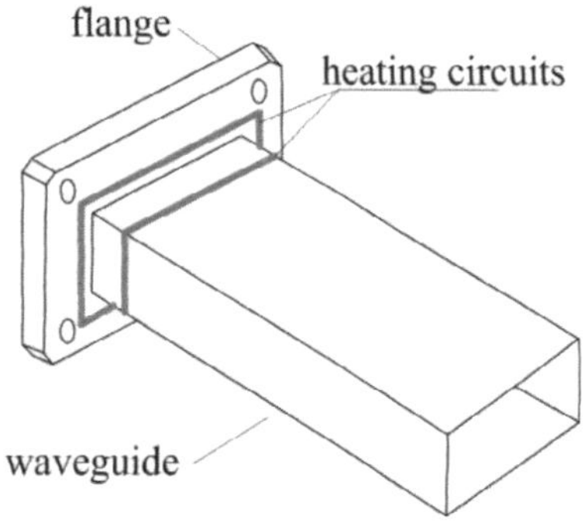

Figure 1. Heating circuits.

The flange and the sleeve are more massive than the thin-walled waveguide. This leads to the fact that when heated, the thin-walled waveguide will heat up and expand faster than the flange. Since some part of the thin-walled waveguide is located inside the recess in the flange, it is possible for the thin-walled element to jam with its subsequent unwanted deformation.

In this paper, a method is considered for calculating temperature deformations during heating of waveguides of various cross-sectional sizes in the process of soldering. Based on the data obtained, the thermal gap was justified and the volume of the solder to obtain a high-quality soldered seam was determined. The strength of the soldered seam is confirmed by calculations and mechanical tests.

2. Statement of the problem

The simplest and most effective solution to the problem of no uniform deformation is a thermal gap between the walls of the brazed elements. A thermal calculation of the joint is necessary to choose the value of the gap.

The solution to the problem consists of the following stages:

1) estimation of the temperature fields of the flange and thin-walled waveguide during heating at selected points in time;

2) estimation of the stress-strain states of a separate flange and a thin-walled waveguide at selected points in time;

3) creation of diagrams of changes in the position of the points under study on the flange and thin-walled waveguide during heating.

4) determination of the required volume of solder and calculation of the required diameter of the welding wire.

The heating of the elements during soldering is carried out according to figure 2. The temperatures of both heating circuits (figure 1) change synchronously.

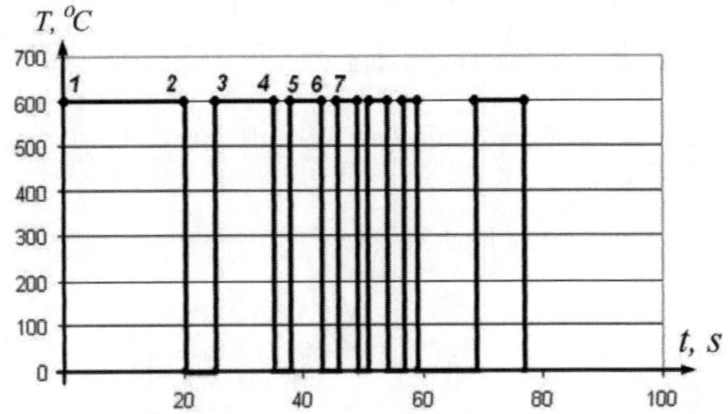

Figure 2. Diagram of heating load.

Figure 2 shows the heating mode: periods of heating alternate with periods of relaxation, i.e. stopping heating, in order to obtain a more uniform temperature field.

The calculation of the elements will be carried out separately, since our goal is to exclude their influence on each other during heating. Having calculated the thermal deformations of the elements, and superimposing them on top of each other, we will find the value of the required gap. This gap will exclude the influence of the elements on each other during heating.

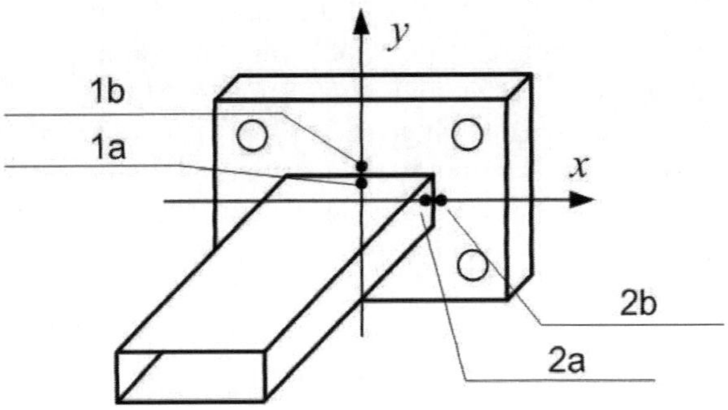

Figure 3. Location of control points.

Further, the calculation of the thermo-deformed state of the joint elements during the heating process was carried out and the displacements of control points 1 and 2 were determined (figure 3).

The indices "a" and "b" in the designation of points indicate that the points belong to the thin-walled waveguide and flange, respectively. Having applied the heat load along the heating circuits according to the heating cycle, we determine the position of the control points in time. By converting the absolute displacements of the control points into their displacements relative to each other, we obtain a diagram of the change in the gaps between the flange and the waveguide during the soldering process.

3. Calculation of the gap during the soldering process

The calculation is carried out using Ansys and Nastran software [7-15]. Figure 4 shows a computational finite element model of a thin-walled waveguide and a flange with an applied thermal load.

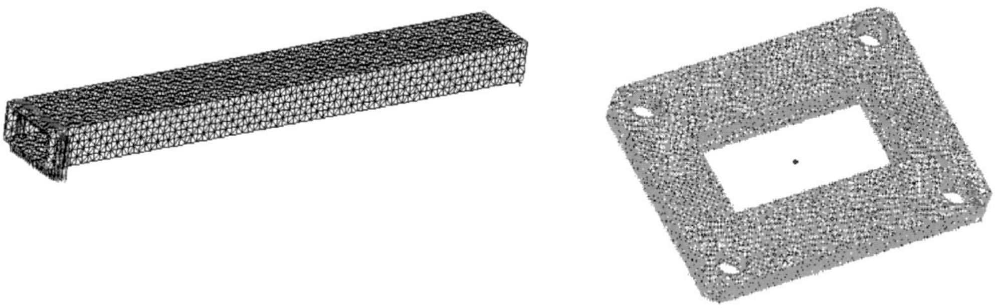

Figure 4. Finite element model of the waveguide and flange.

Figure 5 shows a typical result of calculating the temperature field of a thin-walled waveguide and a flange.

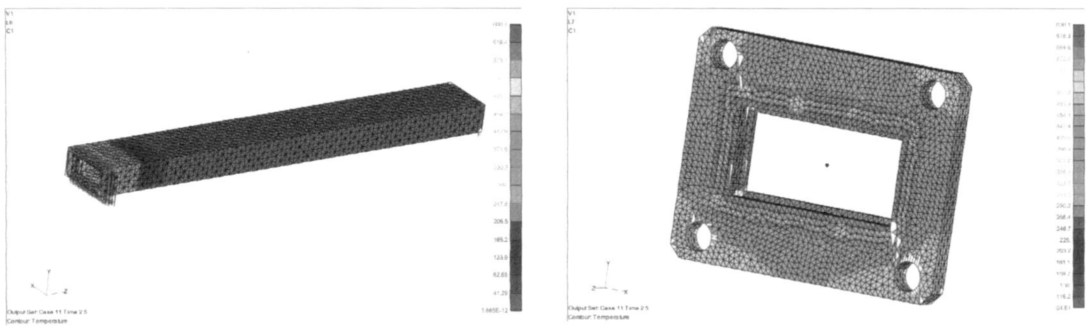

Figure 5. Temperature field of the waveguide and flange.

A noticeable no uniformity of heating and temperature values of these elements occur due to the difference in their geometry. Temperature fields for all standard sizes of the waveguide have a similar appearance, differing only in temperature values. We are interested in a very small part of a thin-walled waveguide located near the waveguide heating circuit. The calculation confirms that the heating of a thin-walled waveguide occurs much faster than the heating of a flange and this can lead to its jamming.

We determined the thermal displacements of the control points in the joint and plotted diagrams of the change in the values of the gaps between the flange and the thin-walled waveguide for all the investigated standard sizes of the section. Figure 6 shows a typical diagram for a 15x35 waveguide.

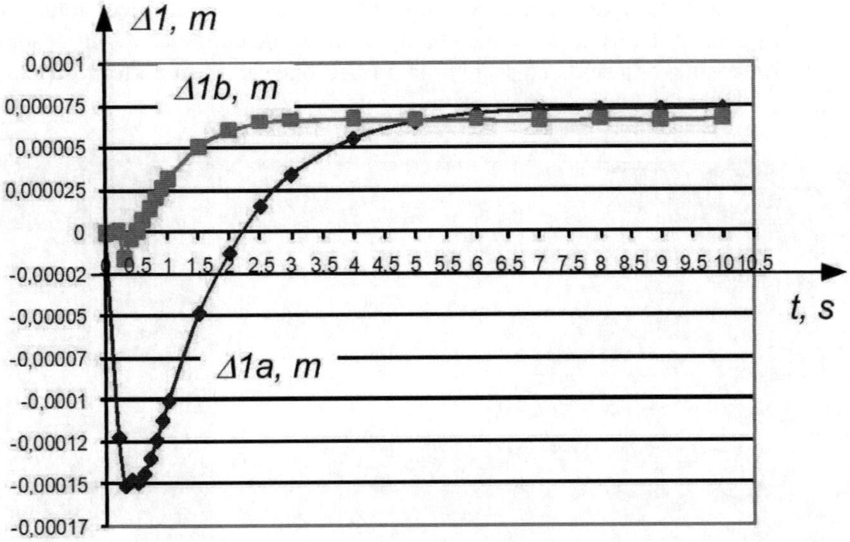

Figure 6. Diagram of the gap changing.

By analyzing the diagrams of changes in the gap, we can determine the moment of jamming and its duration, as well as the size of the required gap. The diagram of the gap change is convenient as it allows us to find the values of the greatest mutual penetration of the control points from the maximum deviation from the horizontal axis to the negative area. The found values represent the required thermal gap between the elements; this gap prevents the mutual influence of elements during heating.

The value of the thermal gap, determined for different standard sizes of the section, varied from 0.11 mm to 0.15 mm. Knowing the perimeter p, depth h and the size of the gap s, it is possible to determine the volume of the solder V needed to fill the gap (figure 7). This value must be increased by the size of the fillet. Let us consider the example of a fillet in the form of a right-angled triangle with the length of the legs 1mm.

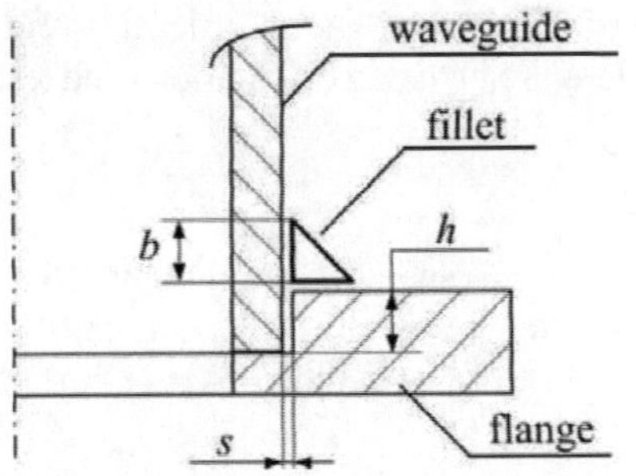

Figure 7. Geometry of a gap.

The required volume of solder is:

$$V = p \cdot (h \cdot s + \frac{b*b}{2}). \tag{1}$$

For solder in the form of a round wire, the required diameter is:

$$d = 2\sqrt{\frac{V}{\pi p}} \,.. \tag{2}$$

The calculation results were taken into account when developing the technology of soldering waveguide elements and this let us improve the quality of the soldered joints. Figure 8 shows the result of calculating the stress state of a soldered seam with a fillet; one can see a uniform stress state without areas with stress concentrations.

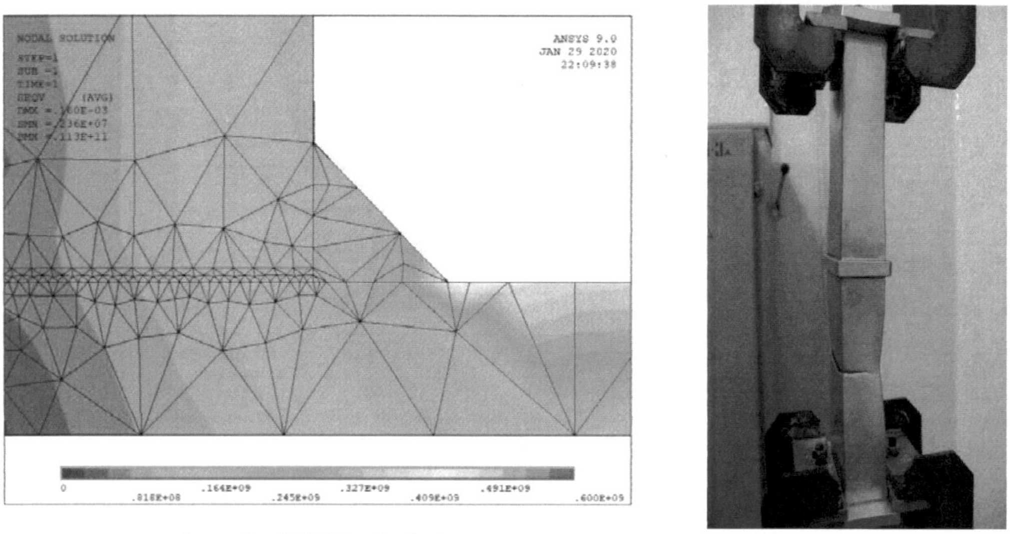

a) result of FEM calculation b) mechanical test

Figure 8. Strength of the soldered joint.

Mechanical tests of the brazed structure of the waveguide confirmed that failure occur outside the zone of the soldered seam.

4. Conclusion

In this work, the study of thermal deformations of the elements of the waveguide is carried out. The proposed approach to the study of the thermoelastic state of the assembly in the process of soldering let us substantiate the thermal mode of heating and the size of the thermal gap in order to avoid jamming of the brazed elements. Studying the stress state of the soldered seam using calculations and tests showed its high strength.

Acknowledgements

The reported study was funded by Russian Foundation for Basic Research, Government of Krasnoyarsk Territory, Krasnoyarsk Regional Fund of Science, to the research project no. 20-48-242914: "Development of methods of calculation of a thermoelastic state of waveguide system of the spacecrafts of communication at operation".

References

[1] Collin R E 2001 *Foundations for microwave engineering* (John Wiley & Sons, Ltd)
[2] Chang K 2005 *Encyclopedia of Rf and Microwave Engineering* (John Wiley & Sons, Ltd)
[3] Orfanidis S J 2008 *Electromagnetic Waves and Antennas* (Rutgers University)
[4] Davidson D B 2011 *Computational electromagnetics for RF and microwave engineering* (Cambridge University Press)
[5] Helszajn J 2000 *Ridge waveguides and passive microwave components* (The Institution of Engineering and Technology, London, UK)
[6] Marcuvitz N 1951 *Waveguide handbook* (Dover Publications, NY, USA)
[7] Hamming R W 1987 *Numerical Methods for Scientists and Engineers* (Dover Publications, NY,

USA)
[8] Chapra S and Canale R 2014 *Numerical Methods for Engineers* (McGraw-Hill Education)
[9] Burden R L 2015 *Numerical Analysis* (Cengage Learning)
[10] Cifuentes A O 1989 *Using MSC/NASTRAN: Statics and Dynamics* (Springer, NY USA)
[11] Dechaumphai P 2021 *Finite Element Analysis with PATRAN / MSC NASTRAN* (Alpha Science)
[12] Tickoo S 2016 *NX Nastran 9.0 for Designers* (CADCIM Technologies)
[13] Madenci E 2006 *The finite element method and applications in engineering using Ansys* (Pearson Education, NJ, USA)
[14] Stolarski T 2018 *Engineering analysis with Ansys Software* (Butterworth-Heinemann)
[15] Xiaolin C 2015 *Finite Element Modeling and Simulation with ANSYS Workbench* (CRC Press)

Differential thermal analysis of the results of isothermal discrete scanning of sheet aluminum material

V A Mikheev[1], D V Agafonova[1], G P Doroshko[2] and M A Krasheninnikov[2]

[1] Samara National Research University, 34, Moskovskoye shosse, Samara, Russia, 443086

[2] Samara State Technical University, 244, Molodogvardeyskaya street, Samara, Russia, 443100

E-mail: vamicheev@rambler.ru, gen_dor@mail.ru

Abstract. The article presents an application of temperature and thermal analysis of discrete scanning data of heat perception of alloy 1163 to take into account thermal effects during plastic deformation.

1. Introduction

Sheathing parts made of aluminum alloy sheets are widely used in rocket and aircraft construction. The relatively low specific strength and corrosion resistance attracts designers, but the high cost and problems with achieving the necessary technological characteristics of shaping sheet material by wrapping in the manufacture of shells of complex configuration, limits the use of sheets of aluminum alloys with improved chemical composition due to microalloying with rare earth metals.

The characteristics of the shaping of the sheet material by the wrap in the manufacture of shells of complex configuration are determined by the changes in the material composition of the material, the level of hardening during plastic deformation and the anisotropy of the properties obtained during the tensile test of the samples, depending on the crystallographic texture of the sheet material. When the sheets are rolled and the shell parts are formed by the wrapping, heat is generated from them as a result of the dissipation of a part of the energy of mechanical action on the workpiece in these processes, performed at room temperature. This contributes to plastic deformation of the sheet blanks, which until now has not been considered a possible effect.

It is necessary to know the dynamics of heat perception of sheet material; its registration requires a higher measurement accuracy. It is possible to implement a variant of such a study when studying by methods of differential thermal analysis - DTA, with processing by methods of discrete analysis of data obtained by isothermal discrete scanning - IDS. There is already experience in rolling lithium alloys, when the thermal effects of deformation were taken into account [1].

The aim of the research is to show how the thermal effects during plastic deformation can be taken into account by the methods of temperature analysis. The object of the study was a sheet material of an aluminum alloy of grade 1163, which belongs to the class of duralumin with a very low content of impurities.

2. Experimental technique

Typical, as in heat treatment, the use of differential thermal analysis involves a continuous change in external temperatures. If it is combined with a gap in each act of iso-temperature discrete scanning, then the obtained data in the complex represent the changes occurring in the alloys in the form of distribution functions.

The possibilities of the method of isothermal discrete scanning are in expanding the application of temperature analysis for the study of structures [2]. Due to special heat treatment, it is possible to successively change not only the mechanism of dissipation of a part of the energy of mechanical action on the workpiece, but also to influence changes in the internal boundaries of the material composition of the alloy with control over the change in the sample volume during heating [3].

The change in the crystallographic texture of the material is accompanied by heat release. The properties of the subsequent resistance to corrosion cracking, intergranular and exfoliating corrosion depend on the cooling rate during quenching, technological background and sheet thickness. This is what is observed with DTA, while both processes and the movement of heat fluxes can be controlled with one device.

The choice of sheet material for research is the 1163 RDMV alloy of the Al-Cu-Mg system of regulated thickness and annealed is considered preferable as one of the main structural metallic materials in aircraft construction. Compared to other alloys of this system, it has increased strength, endurance, and higher fracture toughness, due to which it is actively used in products requiring a long service life in elements of aircraft structures. Alloy 1163 has a high level of fracture toughness and endurance characteristics; it is used in a naturally aged state (mode T) and after thermomechanical treatment (mode T7) [4]. But its working temperatures are only up to 80 ° C, which is insufficient, and since sheets of alloys D16-T, D16ch.-T and 1163-T have a reduced corrosion resistance, the problem of their heat perception requires a solution.

Samples were taken from the delivery lot of OJSC "Kamensk-Uralsky Metallurgical Plant". Plated sheets of regulated thickness - soft, cladding. The mechanical characteristics of this alloy are given in GOST 21631-76, TU 1-801-84-83. For heating, 10 samples were made in the form of plates 1.8 mm thick, 250 * 20 mm in size. Pieces were cut into squares 5 * 5 mm, marked, subjected to discrete heating, scanned (continuous heating) and sent to storage in a dry desiccator. The thermograms were taken on a DTA-500 of the Trunin-Meshalkin scheme, a furnace designed by V.P. Egunov. The required modes of isothermal scanning were achieved by multiple holding of the samples separately.

In a laboratory oven in the thermostat mode at a given temperature, one sample of 1163 alloy of the same size with a set heating step was placed one by one. After holding, the samples were quickly removed from the oven chamber. The oven temperature is then raised to the next value. Heating cycles are repeated until the entire diagram is plotted, keeping constant: sample size, weight and holding time.

After measurements, each sample was subjected to isothermal holding for 1 minute in a DTA-500 at temperatures of 25, 171, 343, and 514 ° C [3], cooled in air was held for the second time by continuous heating in the range of 100-550 ° C, followed by cooling in the furnace chamber to 100 ° C, then withdrawn for testing and characterization.

Preliminary processing of the results included the construction of a diagram of the change in the rate of heating and cooling of the furnace chamber, changes in the ΔthermoEMF (difference in the electromotive force of a thermal nature) of heating and cooling. The sections of the curves are presented in the form of volumetric histograms and section surfaces [4].

3. Processing of the obtained results of alloy 1163

Figure 1 shows the change in the heating rate of the samples during scanning. The vertical axis is the heating rate, the horizontal axis is the continuous heating temperature. The curves represent the temperature of isothermal annealing: 171, 25, 343 and 514 °C. A sharp rise and a saddle in the region of 150 ° C reflects the start of heating, then the rate of temperature rise increases in small steps, and short displacements of internal thermal processes occur in the sample. The lines show all the stages of what happens to the sample. Differences in the position of the lines relative to each other are associated with the processes of absorption and release of heat due to differences in isothermal holding. For specimens heated at a higher temperature (343, 514 ° C), there is also a higher start at startup (more by 5 ° C) and a higher heating rate and a continuing increase in the heating rate. From 425 ° C, the steps in all graphs increase in size and, accordingly, their frequency decreases, the curve of the 514 ° C sample differs from all the others, starting to separate at 343 °C.

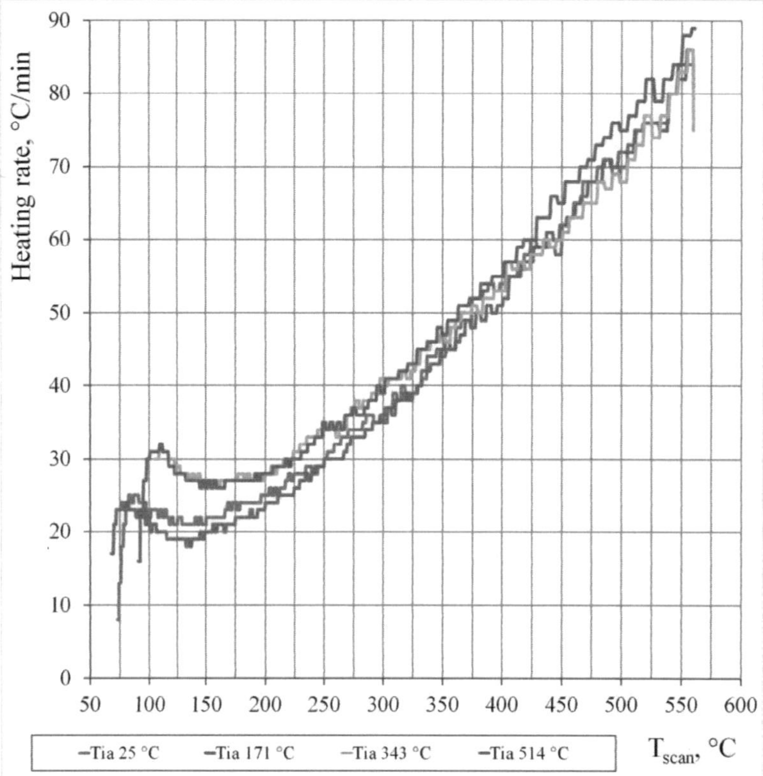

Figure 1. Diagram of the change in the heating rate of samples of alloy 1163; Tia - isothermal annealing temperature, Tscan - continuous heating temperature.

Of particular note is the running of lines with a change in relative position. The most important factor in T^MA (Temperature Analysis) analysis is a sign of a change in states. Despite the similar general nature of the lines, when viewed in a section diagram, they diverge, showing additional inflections - they are the desired ones. The temperature of preliminary holding (Tia) affects, the curves of 171, 25, 514, 343 °C are not arranged in order. In another section, there may be a different sequence. In order to see this, a special technique for representing images has been developed [5]. This is especially required when analyzing response lines during cooling.

Figure 2 shows the change in the cooling rate of the samples, which according to the indicated curves begins in the field of positive values - heating continues even when the furnace is turned off, which is associated with the inertia of the heaters. The heating rate decreases rapidly, turning into cooling - a field of negative values, thermal analysis recommends keeping in mind the boundary temperature of 514.5 ° C. In contrast to thermal processes during heating, during cooling, there is no noticeable difference in the nature and quantitative parameters of the curves. It is the modes of cooling in thermography that are recommended for the diagnosis of chemical processes [6]. Here, the same sequence of thermal analysis - overlapping and overlapping does not mean data coincidence.

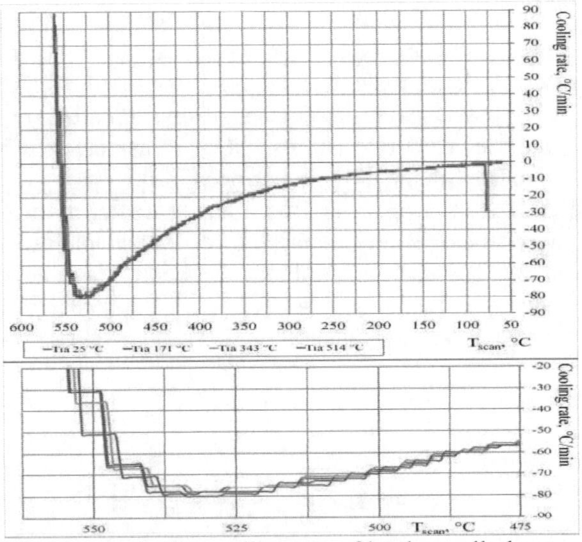

Figure 2. Diagram of the change in the cooling rate of isothermally kept samples of alloy 1163; Tia is the isothermal annealing temperature, Tscan is the continuous cooling temperature.

Figure 3 shows the effect of the temperature of isothermal holding on the nature of the reactions of the samples under continuous thermal exposure. Stationary temperatures of 171.5 °C and 514.5 °C delimit heat perception in the sample, representing it in two separate states, up to 171.5 °C and after 514.5 °C. Stationary temperatures of 171.5 °C and 514.5 °C delimit heat perception in the sample, representing it in two separate states, up to 171.5 °C and after 514.5 °C . The temperature for 343 °C is considered as characterizing the force changes in the middle of the interval 250-343 °C of line 514, and two domes 250 °C and 450 °C, minima 171.5 °C and 514 °C. This layout shows the relationship with the redistribution of graininess in the volume.

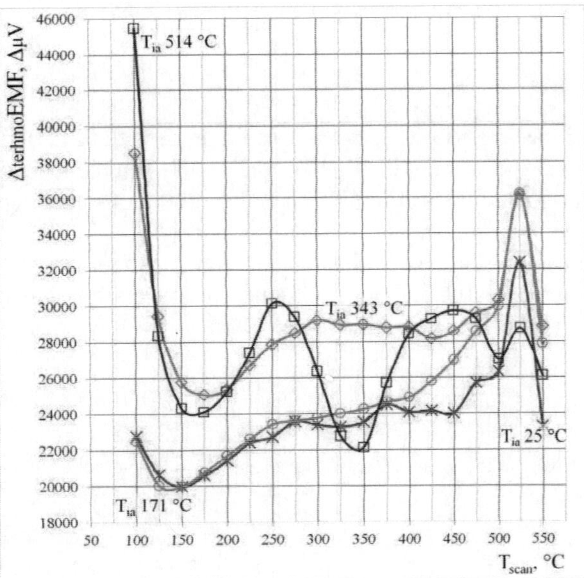

Figure 3. Fragment of heating 1163, the effect of isothermal exposure on the difference in electrical potential during differential thermal scanning - TTS; Tia - isothermal heating temperature, Tscan - continuous cooling temperature.

The nature of the lines below 171.5 °C shows the thermal dynamics of the entire sample. Above 514.5 °C, fragmented melting already begins, with holding for one minute there is not enough energy to cover the entire volume of the sample during this time. At 343 °C, small anomalies and orthogonality of the transition, processes at the contact of crystals are noticeable. According to the temperature analysis, the processes occurring at 514.5 °C are characterized by the transformation of the stationarity of the regime in the volume of the substance [7, 8].

On figure 4 it can be seen how the emphasis shifts in the character of the curve near the stationary temperature of 514.5 °C, which is especially noticeable for the sample processed at 514 °C. At the same time, if for treatments 25, 171 and 343 °C before 514.5 °C there appears a step with a sharp increase in Δthermo EMF, then after treatment 514 °C this is a saddle before lifting. The heat transfer area is expanded - this is a sign of structural changes.

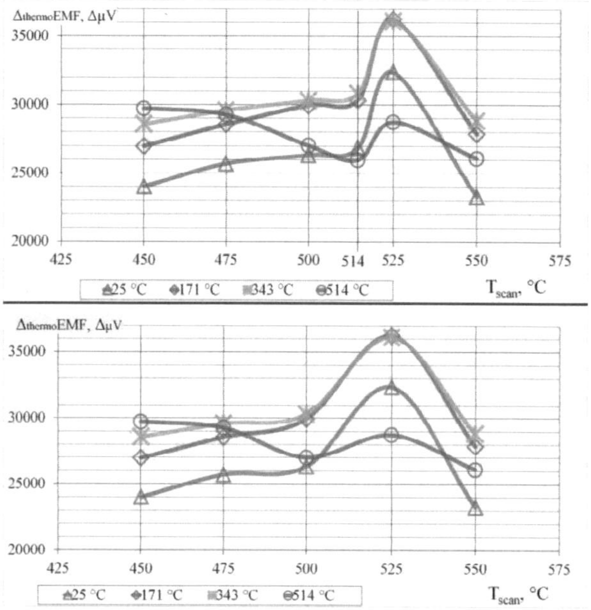

Figure 4. Comparison of the 450-550 °C section from Figure 3 with the 514.5 °C point (half-interval of stationary temperature) and without it. 25, 171, 343 and 514 in the legend - designations of the temperature of isothermal heating; Tscan is the continuous cooling temperature.

Figure 5 shows the cross-sections of the heating diagrams of the samples by the temperatures of continuous heating. The curves of 100 and 525 ° C stand out especially - they can be used to determine the boundary of the contact surface.

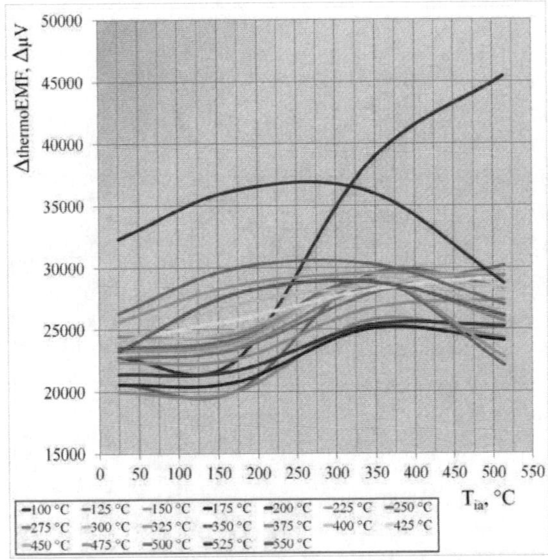

Figure 5. Diagram of the influence of the isothermal holding temperature on the difference in the electrical potential of the sample during heating; curves - temperature of continuous heating. The horizontal axis of temperatures of isothermal heating (Tia) axis.

Figure 6 shows an analogue of the 3D model combining figure 3 and figure 4. Peaks at 343 and 514 ° C at the beginning of heating are clearly shown, and peaks at the end of heating at 25, 171 and 343 ° C. Curve 343 ° C has a high peak at the beginning of heating and at the end, while it does not have a "pit" in the middle of the range. A more convenient version of a histogram for viewing data in a three-dimensional graph is a surface (Figure 6). The advantage of such histograms is the ability to accurately determine the parameters of the secondary scan.

Figure 6. Fragment of heating, histogram of the effect of the temperature of continuous heating on the difference in the electrical potential of the sample during isothermal holding; Tia is the isothermal heating temperature, Tscan is the continuous cooling temperature.

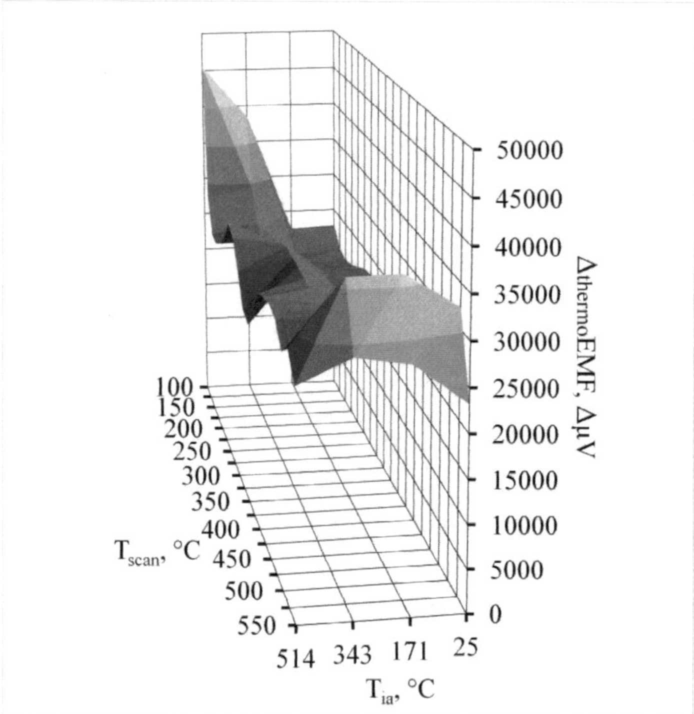

Figure 7. Fragment of heating, histogram-surface of the influence of the temperature of continuous heating on the difference in the electrical potential of the sample during isothermal holding; Tia is the isothermal heating temperature, Tscan is the continuous cooling temperature.

In figure 7 that an increase in the iso-holding temperature increases ΔthermoEMF, being especially pronounced at 100 and 550 ° C (high peaks), reaching a general maximum at 343 ° C and decreasing at 514 ° C, except for the highest peak at 100 ° C. The diagram in figure 7 according to TMA very clearly reflects the general tendency of transformation to the stationarity temperature.

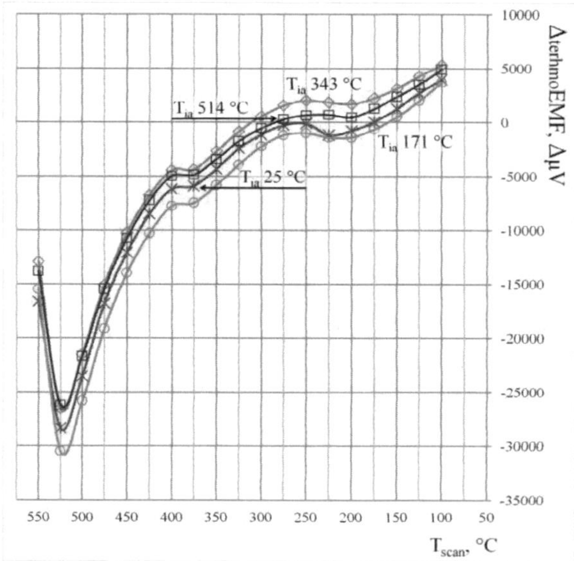

Figure 8. Influence of isothermal exposure on the difference in the electrical potential of the sample during thermal scanning for 1163; fragment - cooling; Tscan is the continuous cooling temperature.

IOP Conf. Series: Materials Science and Engineering 1155 (2021) 012021 doi:10.1088/1757-899X/1155/1/012021

On figure 8, the cooling for all curves has a uniform character, steps are visible at 400-450 and 250-275 ° C. The peculiarity of cooling is that after 125 ° C the electric potential of the samples grows and is restored by 40-45% of the initial one at the beginning of cooling. And these lines clearly show the nature of the accuracy of the work of the IDS method when optimizing the modes - the lines strictly follow their own trajectories, the states are characterized as steady.

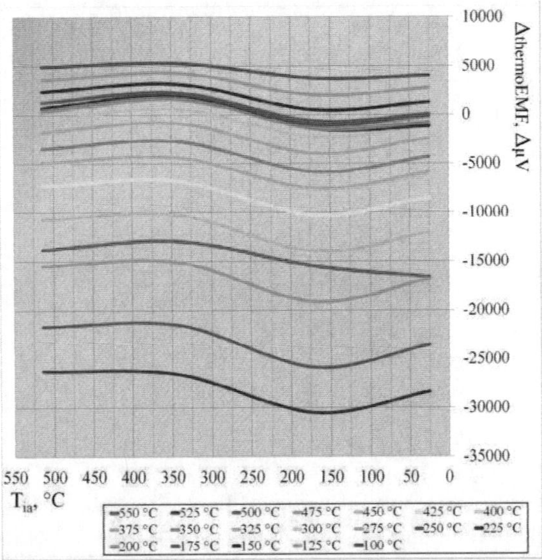

Figure 9. Diagram of the effect of the temperature of isothermal holding on the difference in the electrical potential of the sample during cooling; curves - temperature of continuous heating. The horizontal axis of temperatures of isothermal heating (Tia) axis.

Figure 9 shows the cross sections of the cooling curves of the samples. The curves are smooth and homogeneous. The lowest levels on the Δ-axis are thermoEMF at temperatures of 125 and 150 ° C. The 550 ° C curve has the highest ΔthermoEMF. The highest density of curves is in the region of zero difference in electric potentials.

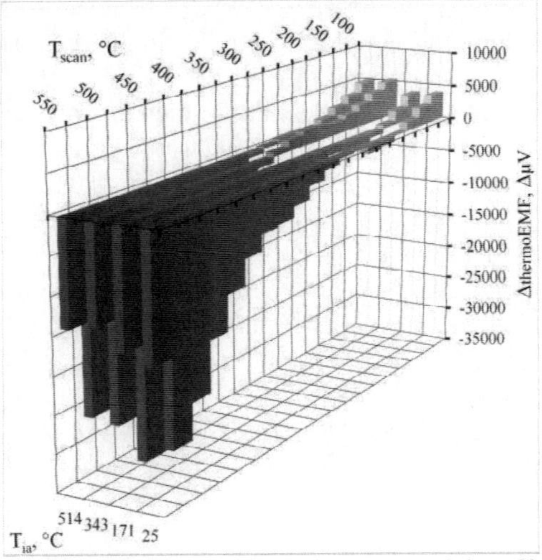

Figure 10. Fragment of cooling in the form of a histogram of the effect of the temperature of continuous heating on the difference in the electrical potential of the sample during isothermal holding; Tia is the isothermal heating temperature, Tscan is the continuous cooling temperature.

Upon cooling, the electrical potential of the samples passes into the field of negative values with a tendency to recover in the field of positive values upon further cooling after 125 ° C.

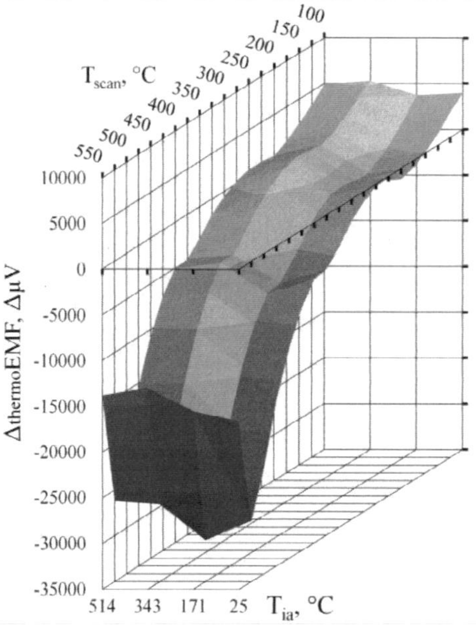

Figure 11. Histogram-surface of the influence of the temperature of continuous heating on the difference in the electrical potential of the sample during isothermal holding for 1163; fragment - cooling. Tia is the isothermal heating temperature, Tscan is the continuous cooling temperature.

Figure 11 shows the cooling of samples previously discretely fired at 25, 171, 343 and 514 ° C and then continuously heated to 550 ° C. The nature of the cooling is uniform, the processing temperature has almost no effect on the internal processes during the cooling of the sample. Two steps and a saddle are clearly visible, after which the electropotential of the sample tends to recover upon cooling to less than 125 ° C. Key temperature 171.5 ° C.

4.Summary

The processes of heat perception in the samples during heating and cooling have characteristic effects at 171, 343, 514 °C.

A sharp increase in the heating rate at the beginning, then a decrease, the formation of a “saddle” and then a linear increase in the rate shows that the heat perception of the sample refers only to the outer part of its volume, has a layer that has established itself in temperature, and then continues along the classical gradient. Such data can be scaled by reference to the dimensions of the sample. This also means the parameters of plastic deformation.

The difference in thermal processes in the dynamics of heating and cooling is associated with the redistribution of grains by size. Upon cooling, there is no noticeable difference in the nature of structural changes, the heating is completed, and the heat perception is leveled, in fact, the material has become more homogeneous.

The DTA data showed that the greatest dynamics of changes in the heat perception of alloy 1163 occurs during processing at 514 ° C.

References

[1] Agafonova D V 2021 *The pragmatism of applied temperature analysis to determine the rolling* parameters *under thermal exposure* (IOP Conference Series: Materials Science and Engineering vol 1) ed Mikheev V A, Doroshko G P pp 1047(1), 012062

IOP Conf. Series: Materials Science and Engineering 1155 (2021) 012021 doi:10.1088/1757-899X/1155/1/012021

[2] Krasheninnikov M A *Investigation of the effect of alumina additives on the sintering of low-melting clay by the method of section DTA diagrams // Urban Planning and Architecture. - 2018. - T. 8. - No. 2. - S. 51-55. / Vestnik.2018.02.8*

[3] Doroshko G P 2007 *Introduction to the temperature analysis of the properties of materials: monograph* (Samara: Samara State University of Architecture and Civil Engineering Samara) p 396

[4] *Micheev V A 2015 New method for producing metal alloys with interphase boundaries formed Col. Papers of 4th International Scientific and technical conference. Melallophysics of materials and deformation processes (Samara: Samara State University) pp 70-74*

[5] Garkushin I K 2005 *Equilibriums diagrams in systems with metavanadates of some alkali metals* (Moscow: Machine-building industry) ed Gubanova T V, Petrov A S, Anipchencko B. V p 118

[6] Holba P 2013 *Heat transfer and phase transition at DTA experiments. Chapter 5 in book Thermal analysis of Micro-, nana- and non-crystalline materials* (Springer, Berlin) ed Šesták J, Sedmidubský D pp 99-134

[7] Zvereva I A 2013 *Development of methods of thermal analysis to study nanostructured layered systems. Col. Papers of XIV International conference on thermal analysis and calorimetry in Russia* (RTAC -2013). SPb) pp 16-19

[8] Pavlechko E V 2013 *Thermodynamic properties and polimophism of low-temperature ion liquids.Col. Papers of XIV International conference on thermal analysis and calorimetry in Russia* pp 20-24

Influence of full-scale climatic tests on the properties of aluminum alloys in zones of river and marine water mixture

N I Bulysheva[1,3], E A Varchenko[2], A V Nazarenko[1], G Yu Glushchenko[1], A A Goncharov[2] and A V Gladkikh[2]

[1]Federal Research Centre the Southern Scientific Centre of the Russian Academy of Sciences, 344006, st. Chehova, 41, Rostov-on-Don, Russian Federation
[2]G.V. Akimov Gelendzhik Climate Test Center of All-Russian Scientific Research Institute of Aviation Materials, 353466, st. Pochtovaya, 20, Gelendzhik, Krasnodar region, Russian Federation

[3]E-mail: bulysheva@ssc-ras.ru

Abstract. Within nine months, comprehensive studies were conducted on the effect of full-scale climatic tests on the properties of samples of aluminum alloy AMg6 when fully submerged in the estuary region of the Don River (the basin of the Sea of Azov) at two stations. The features of the formation of micro- and macro-fouling communities on experimental plates at one, three, six and nine months after the start of the experiment are shown. It is shown that microfouling communities on the experimental plates begin to form within a month of exposure. Diatoms were the background-forming species during the entire practical period of the study. Single macrozoobenthos organisms begin to appear on the samples after three months and only at second station. After six months, macro-fouling organisms are fixed on plates at both stations. As the exposure period increases, the number and biomass of macro-foulers increases. The corrosion resistance assessment of the samples revealed the absence of intercrystalline corrosion during the entire test period in water, at both stations, as well as pitting corrosion during the first six months of exposure.

1. Introduction

Corrosion processes are irreversible, and therefore, they must be identified at early stages, quantified corrosion damage, and predicted the danger of development in the absence of measures for anticorrosion protection. Economic calculations show that preventing damage, accidents and disasters is much cheaper than eliminating them [1]. Attention to monitoring of the state of structures and predicting the corrosion resistance of materials is paid all over the world. In the aquatic environment, fouling, corrosion, aging and biodeterioration of materials and structures can hinder their operation, up to its complete impossibility [2]. Therefore, the search for solutions to improve the corrosion resistance of metals is of scientific and practical interest.

On apply scale in the national economy, one of the most commonly used metals are aluminum and its alloys, which are easily deformed and are highly processability. From the kinetics of corrosion processes prospective, aluminum alloys are a short-circuited system of multi-electrode elements [3]. Experts pay close attention to the study of the corrosion resistance of the AMg6 alloy, which is in demand in ship and rail car building, aerospace and construction industry. There are some research data of identification of the biological and corrosion resistance of this alloy during both full-scale and accelerated tests in the marine environment [4–6]. Studies of the corrosion resistance of aluminum alloys have not previously been carried out in estuaries, therefore the purpose of this experiment was to study the combined effect of climatic and biological factors on experimental plates made of AMg6 aluminum alloy in zones of river and marine water mixture.

2. Material and methods

December 19, 2018 the stands with experimental plates of 15×15 cm in area, 2 mm thick made of AMg6 aluminum alloy were installed at a depth of 2 meters on the SSC RAS testing ranges (Coastal scientific expeditionary base "Kagal`nik" (Station 1, St. 1) and hydrometeorological post "Donskoy" (Station 2, St. 2)) at the mouth area of the River Don in zones of river and marine water mixture (figure 1). Uninstalling of plates was partial carried out after one, three, six and nine month exposure in water respectively. Each plate was photographed from both sides, placed in a separate container with water, within 10-15 minutes was delivered to the laboratory at the coastal base for visual inspection and sampling of the outgrowth. The selection and treatment of sessile microbiota was carried out in accordance with standard hydrobiological methods [7].

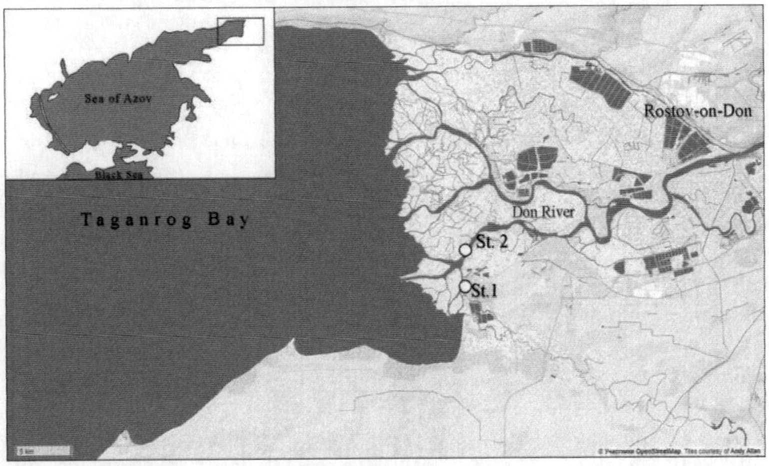

Figure 1. The map-diagram of the area of research

The study of the surface microstructure and determination of the local chemical composition of corrosion products was carried out in 2 stages. At first stage of the study (before the experiments even started), the quality of the experimental plates was monitored for the content of third-party components. The measurement results showed that the used aluminum alloy AMg6 meet state standards and do not contain impurities. The second stage was elementary analysis of experimental plates, which was carried out on a scanning electron microscope Carl Zeiss EVO 40 (EHT = 20 kV, I = 1 nA) with an X-ray attachment for energy dispersive microanalysis INCA at a magnification of x400. Calibration of the current value was carried out on a reference sample of cobalt. Spectrum accumulation time 50 seconds.

Determination of hardness was carried out according to ASTM E2546-15 by the Vickers method. Hardness was determined with a Vickers diamond pyramid, at a force of 100 N and a loading rate of 0.5 mm/s.

3. Results and discussion

The hydrochemical conditions of the water area where the experiment was carried out were determined by the interaction of river and sea factors in the formation of the hydrological-hydrochemical regime. The ionic composition of water at Station 1 is predominantly of the sulphate class of the calcium group. The exceptions are cases of increased underground feeding of the Don

Delta branches and the inflow of the Kagal`nik River in conditions of stable stratification in winter. At this time, the water of this area belongs to the sulfate class of the sodium group [8]. Station 2 recorded water of the bicarbonate class of the calcium group. In addition, the surface water layer at both stations was well aerated. It is known that parts of ships and hydraulic structures moving in aerated seawater, in addition to being directly corroded by seawater, are also subject to erosion and cavitation. [5]. Apart from hydrological and hydrochemical conditions, the sessile microbiota influence to the immersed in natural reservoirs materials.

Thru the experiment, it was revealed that the colonization of the samples begins with the formation of micro-fouling communities represented mainly by diatoms. The number of species recorded in microfouling is varied from 23 to 36 at St. 1 and from 18 to 32 at St. 2 (table 1). After a month of exposure, the most common representatives of the species were Bacillaryophyta. Biomass and abundance of microphytoperiphyton on the AMg6 plates were 0,2 μg/cm^2 and 67 cell/cm^2 respectively at St. 1, number of species – 25; while 0,2 μg/cm^2 and 53 cell/cm^2 respectively at St. 2.

Table 1. Number of species of microphytoperiphyton on experimental plates at stations.

Station	1 month	3 months	6 months	9 months
Kaga'lnik (St. 1)	25	27	23	36
Donskoy (St. 2)	18	23	26	32

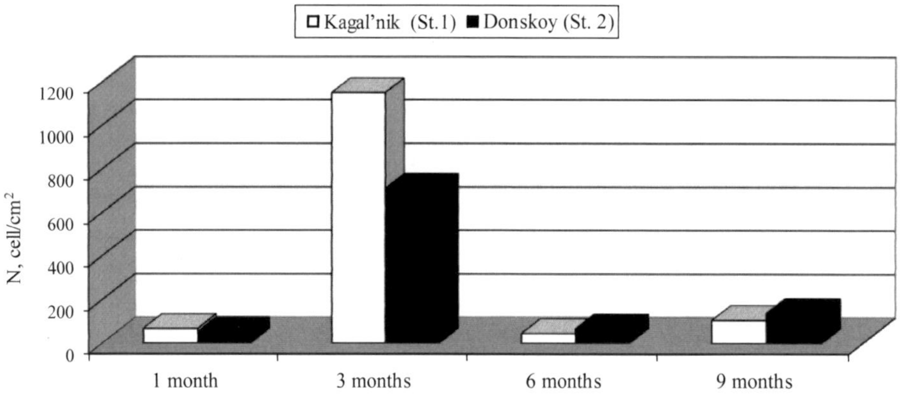

Figure 2. Dynamics of total density of microphytoperiphyton on experimental plates after 1, 3, 6 and 9 months of exposition.

Three months later, diatoms continued to dominate on the AMg6 plates. Their quantitative indicators, such as frequency of occurrence, number of species (27 at St. 1 and 23 at St. 2), abundance (1150 cells/cm^2 at St. 1 and 712 cells/cm^2 at St. 2) and biomass (2.2 μg/cm^2 at St. 1 and 1.6 μg/cm^2 at St. 2) have significantly increased.

After six months, at both stations a sharp decrease in the abundance to 44 and 66 cell/cm^2 and biomass of 0.1 and 0.2 μg/cm^2 was noted on the plates, while the dominance was retained by diatoms. The species composition of phytoperiphyton has changed. There was a relative decrease in the abundance of attached diatoms, which is probably caused by the action of the intense pressure of phytophages. The number of species on the plates has decreased. The species list was supplemented by planktonic algae from the Chlorophyta type.

After nine months of exposure of the plates an increase in the species diversity of microalgae from five types was noted in the development of phytoperiphyton (up to 36 at St. 1 and up to 32 spices at St. 2). Biomass values decreased to 0,0001 μg/cm^2 at St. 1 while to 0,0003 μg/cm^2 at St. 2. At the same time, the values of the number increased to 106 cell/cm^2 at St. 1 while to 139 cell/cm^2 at St. 2. The basis of the biomass was composed of a few large-celled diatoms, the abundance of small-celled representatives of the Chlorophyta and Cyanobacteria types.

A macro-fouling community begins to form on the experimental plates after 3 months of exposure. Single individuals of *Tubifex tubifex* (O. F. Müller, 1774) noted at plates exhibited at St. 2 (table 2, figure 3). Their number reached 89 ind/m^2, 0,04 g/m^2. Macroinvertebrates appeared at St. 1 after six months of exposure. Molluscs *Dreissena polymorpha* (Pallas, 1771) and *Fredericella sultana* (Blumenbach 1779) formed the basis of abundance and biomass. Its total number was 1019 ind/m^2, while total biomass was 97,4 g/m^2. After six months of exposure at St. 2, in addition to *D. polymorpha* and *F. sultana* the crustaceans of *Corophium volutator* (Pallas, 1766), which form specific corofiid silts in the Don Delta, were a mass species. Its total number was reached 8417 ind/m^2 while total biomass was 156,4 g/m^2.

Table 2. Number of species of macrozoobenthos on experimental plates at stations.

Station	1 month	3 months	6 months	9 months
Kaga'lnik (St. 1)	-	-	4	4
Donskoy (St. 2)	-	1	6	4

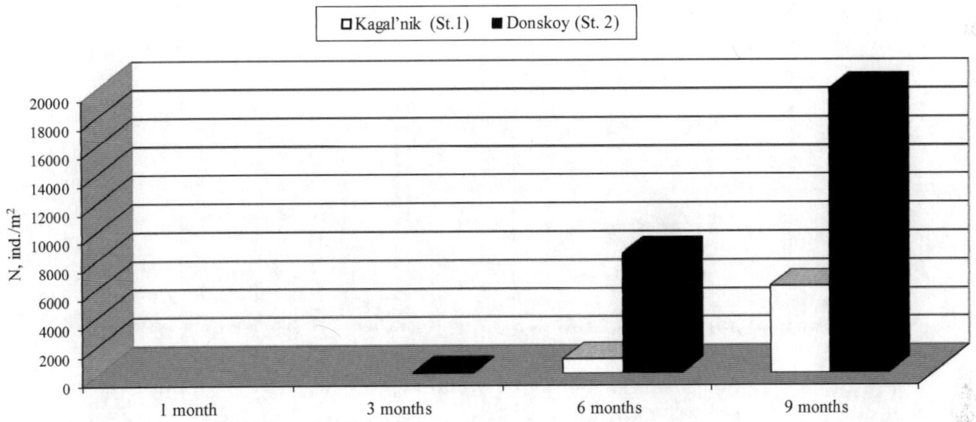

Figure 3. Dynamics of total density of macrozoobenthos on experimental plates after 1, 3, 6 and 9 months of exposition.

After nine months, four taxa of macroinvertebrates were recorded on the plates. *D. polymorpha* formed the basis of fouling communities at both St. 1 and St. 2. The total number and biomass were 6116 ind/m^2 and 161,6 g/m^2 respectively at St. 1, while 19888 ind/m^2 and 312,88 g/m^2 respectively at St. 2.

It should be noted that the population at St. 2 was 3-7 times higher than at St. 1 (figure 3). A similar picture was observed for biomass.

The assessment of the effect of a combination of abiotic and biotic factors on corrosion processes occurring in a natural reservoir was carried out on the basis of the weight loss of metal plates (corrosion rate), metallographic analysis, and the loss of mechanical properties of the aluminum alloy during tensile tests.

Measurements of the elemental composition of the cleaned areas of the sample surface, performed using energy-dispersive X-ray microanalysis, were of a local nature and were repeated many times. As mentioned earlier, according to the data of the studies carried out before the start of the experiment, it was revealed that the AMg6 sample meets state standards and does not contain impurities (figure 4a). After three months, in addition to the main elements Al and Mg, such elements as O, Si, Mn are noted in the spectrum (figure 4b). After six months Si, O, C, Ca, Na, S, Cl, Fe were identified (figure 4c). On the samples that were in water for nine months, C, Ca, K, O, Cl, Na, Si, S are found (figure 4d). Silicon, calcium and carbon come from organisms of the fouling community, the rest of the elements are included in the ionic composition of water in the surveyed water area.

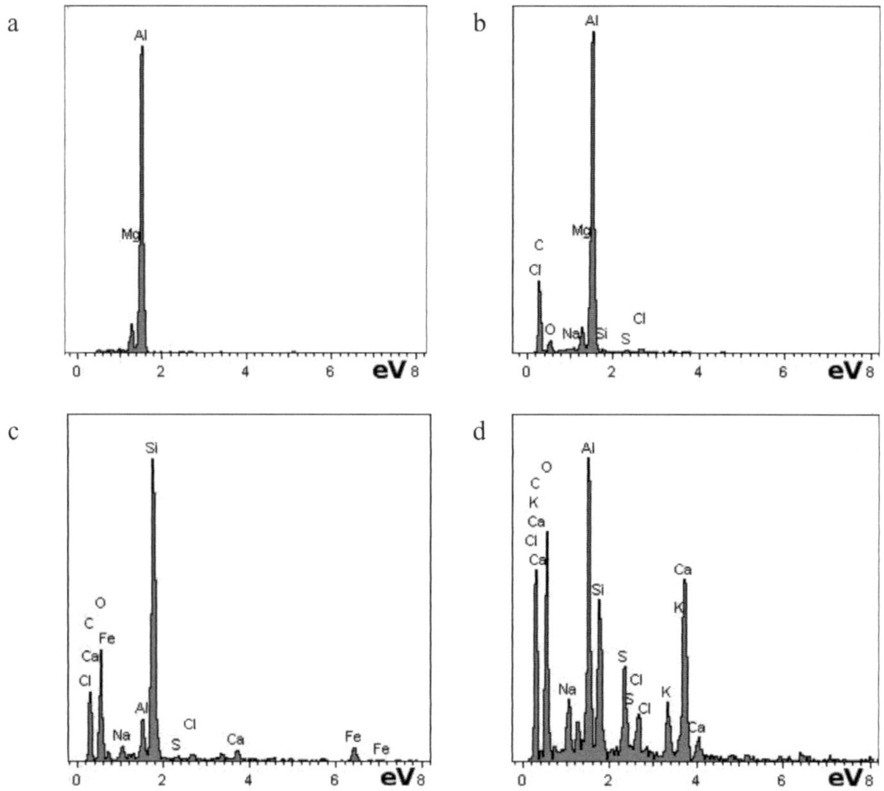

Figure 4. Analysis of the elemental composition of the local area of the AMg6 sample: a – before the start of the experiment; b – after 3 months of exposition; c – after 6 months of exposition; d – after 9 months of exposition.

Metallographic studies were carried out with an assessment of the depth of intergranular and pitting corrosion, during which it was noted that after three months of exposure to water on the AMg6 alloy samples, no corrosion damage and, accordingly, intergranular corrosion were not detected after tests at both stations. After six months of the experiment, pitting corrosion was detected on the AMg6 alloy samples only after exposure at St. 2, its depth was 0.02 mm, intergranular corrosion was not detected after testing at both stations. Nine months after the start of the research, it was revealed that the largest area of corrosion damage (40 %) for the AMg6 alloy was obtained at St. 2. But the area of corrosion damage does not exceed 10 % of the surface at St. 1. According to the results of metallographic studies, it was found that after nine months of exposure in water at both stations, the depth of pitting corrosion of the samples is 0.07 (St. 1) and 0.022 mm (St. 2) respectively. Also there is no intergranular corrosion.

Table 3. The results of measuring the Vickers hardness of AMg6 samples in the initial state and after 3, 6, 9 months of exposition by stations (HV).

Exposure conditions	3 months	6 months	9 months
Initial data	91.00	91.00	91.00
Station 1	89.50	88.50	88.00
Station 2	89.25	88.25	88.00

The change in the hardness index after three months of the experiment for the AMg6 samples was 1 % (within the error of the device), while after six and nine months it was 3 %.

As a result, for the AMg6 alloy samples, the absence of intergranular corrosion was revealed throughout the entire testing period in water, at both stations, as well as pitting corrosion during the first six months of exposure. The development of corrosion damage on the samples is uniform. According to the results of determining the composition of corrosion products, all tested samples are dominated by surface oxidation products, elements that make up the salts of the aqueous medium of exposure sites (calcium, potassium, sodium, magnesium), as well as elements of organic origin: spongy particles containing a large amount of silicon, oxygen and calcium and carbon-based fibrous elements (microalgae).

Acknowledgements
This work was conducted with partial support of the Russian Foundation for Basic Research (No 18-29-05078 мк).

References

[1] Varchenko E A and Kurs M G 2017 Full-scale testing of metallic materials in seawater: key approaches to assessing corrosion and biodeterioration resistance *Electr. sci. j. Proceedings of VIAM* **11** 12

[2] Karpov V A, Kovalchuk Yu L, Poltarukha O P and Il`in I N 2007 *A comprehensive approach to protection against marine fouling and corrosion* (Moscow: T-vo nauchnyh znanij KMK) p 156

[3] Karimova S A, Gilikov V P, Mikhailov A A, Chesnokov D V, Igonin T N and Karpov V A 2012 Full-scale accelerated testing of aluminum alloys under the influence of the marine atmosphere *Corrosion: Materials, Protection* **10** 1–3.

[4] Varchenko E A and Kurs M G 2018 Crevice corrosion of aluminum alloys and stainless steels in seawater *Electr. sci. j. Proceedings of VIAM* **7**(67) 96-105

[5] Varchenko E A and Kirichok P F 2020 Investigation of biological and corrosion resistance of aluminum alloy samples after field tests in Gelendzhik Bay. Part 2 *Electr. sci. j. Proceedings of VIAM* **9**(91) 105-114

[6] Plotnikov A D, Korneeva E Y, Alekhova T A and Zagustina N A 2017 Investigation of the influence of biocorrosive lesions on the mechanical characteristics of samples made of aluminum alloys AMg6 and 1570S as applied to the operating conditions of the Russian segment of the ISS *Space Engineering and Technology* **4**(19) 36-45

[7] Guidance on methods of hydrobiological analysis of surface water and bottom sediments 1983 (Leningrad, Russia: Gidrometeoizdat) p 239

[8] Matishov G G, Bulysheva N I, Kleschenkov A V, Gluschenko G Yu, Varchenko E V, Goncharov A A and Gladkikh A V 2020 Estuaries of the Sea of Azov as unique polygons for research of biocorrosion. experience in setting up an experiment *Science in the South of Russia* **16**(2) 63–75

IOP Conf. Series: Materials Science and Engineering 1155 (2021) 012023 doi:10.1088/1757-899X/1155/1/012023

Forecasting and operational assessment of the quality of repaired units

A K Apazhev, Y A Shekikhachev, L M Hazhmetov and L Z Shekikhacheva

Kabardino-Balkarian state agricultural university named after V.M. Kokov, Nalchik, Russia

E-mail: shek-fmep@mail.ru

Abstract. Evaluation alone is not enough to improve product quality. It is necessary to have not only the estimated indicator itself, but also to declare the factors that influence the formation of quality and to investigate this influence. Only then does it become possible to carry out quality management. To study the quality of repaired products, we use the economic and statistical method, which is as follows. Product quality is considered in two aspects: production and consumer. Manufacturing quality is a set of product properties (physical, chemical, physical and mechanical, etc.) obtained by it in the process of manufacturing or repair. The consumer quality of products manifests itself in the process of consumption (operation) and is determined by the cumulative impact of all characteristics of production quality and conditions of consumption. With the help of the correlation level, two main tasks can be solved: operational quality assessment at the stage of production (repair), based on predicting the consumer quality of the projection according to the known values of its production quality indicators; management of the production quality of products in accordance with a given level of its consumer quality. The most characteristic indicators of the production quality of the repaired units are the errors in the dimensions of the shape of parts, the relative position of their surfaces, the size of the washed out links of the dimensional chains of mechanisms, the physical and mechanical properties of materials, etc. At the repair enterprise, by means of appropriate measurements, the values of the indicators of the production quality of batches of products are determined. Then, in operation, these products are monitored in order to determine their consumer quality. The consumer quality of repaired products can be expressed by a number of indicators, but the most expedient is the use of such an indicator as winding to the first failure caused by aging processes (wear, fatigue failure, etc.). Considering the above, research aimed at predicting and quickly assessing the quality of repaired units is an urgent problem.

1. Introduction

The unit is considered as a complex product consisting of a number of subsystems. The study of the relationship between production and consumer quality is carried out separately for each subsystem [1-6].

As a result of research, we obtain information about the values of the indicators of the production quality of subsystems and the values of the operating time of these subsystems to failure. Based on this information, mathematical models are built that reflect the correlation between the production and consumer quality of individual subsystems. After that, a general mathematical model of the system as a whole is constructed. Research and assessment of the quality of repaired products are considered on the example of an engine [7-13].

2. Research results

The production quality of engines is finally formed during the assembly process (running-in of units is not considered). The input of the assembly process is the state of the parts, which is characterized by a set of primary indicators x'. The production quality of the assembled product is characterized by a set of complex indicators x'' that depend on the primary and quality of assembly operations. Complex indicators carry more information about the state of the object, but it is not always possible to measure them. In these cases, primary indicators are used to assess production quality.

In quality research, the engine is considered as a system consisting of a number of subsystems. Each subsystem is considered a simple product. Indicators of its production quality are determined, a connection with consumer quality is found, connections with other subsystems are analyzed. Figure 1 shows a diagram of the study of the quality of a complex product.

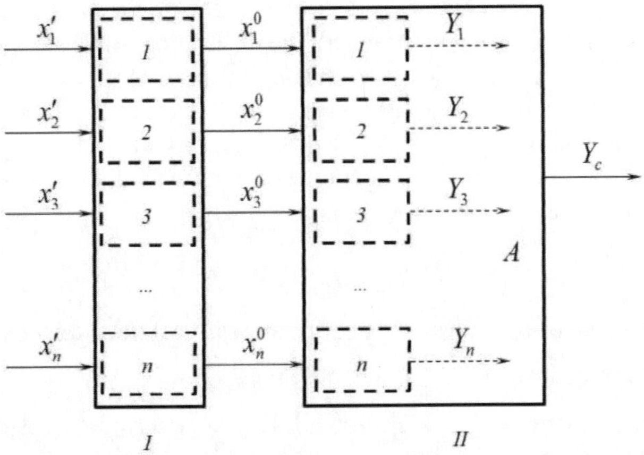

Figure 1. Scheme of quality research: x'_i, x^0_i – the aggregate, respectively, of the primary and complex indicators of the production quality of the *i-th* subsystem ($i = 1, 2,..., n$); Y_c – a set of indicators of the consumer quality of the system; Y_c – a set of indicators of consumer quality of the *i-th* subsystem; A – operator forming the general indicator; I – product assembly; II – product operation.

The main indicator that objectively characterizes the consumer quality of engine repair is the value of its operating time to failure of one of the subsystems caused by aging processes.

Often, an economic assessment of the quality of such complex products as an engine cannot take into account a number of important properties; therefore, additional indicators are required, on which appropriate restrictions are imposed.

For engines, such indicators are power, stability of operation at low speeds, oil pressure in the line, noise, vibration, the content of harmful impurities in exhaust gases, the fatigue strength of parts, etc. is considered suitable and its quality is not evaluated.

The operating time of the system to failure under the conditions of a specific implementation is determined by the operating time of the "weakest" subsystem. Having received an equation for this subsystem that connects its production and consumer quality, a mathematical model is built for the entire system. To solve this problem, the Monte Carlo method and methods of planning experiments are used.

The general method for studying the quality of engines is as follows:

1. The nomenclature of the main and auxiliary indicators characterizing the consumer quality of the system is determined.

2. Within the system, separate subsystems are distinguished and their mutual connections, transmission channels and the nature of signals are considered.
3. Indicators of the production quality of each system are determined.
4. Correlation equations are built connecting the consumer and production quality of the subsystems.
5. Indicators of consumer quality of the system as a whole are determined.
6. A mathematical model is constructed that reflects the dependence of the indicators of the consumer quality of the system on the indicators of the production quality of the subsystems.
7. In accordance with the existing requirements, restrictions are imposed, but the indicators of the consumer quality of the system and the permissible values of the indicators of production quality are determined.

The working conditions of the subsystem, and, consequently, the terms of their service are determined by both internal and external factors. Internal factors are indicators of the production quality of the subsystem itself, external factors are a manifestation of the influence of the environment. A group of factors characterizing the quality of repairs is distinguished from the total set of factors for the study.

Considering the engine as a complex product, the following subsystems can be distinguished according to their functional characteristics: power supply; gas distribution; ignition; cylinder-piston group; connecting rod and piston group; crankshaft - main bearings; clutch; lubricants; cooling. For each subsystem, a nomenclature of indicators is determined that characterize, mainly, their production quality; It is going to the baseline information about the production and the consumer as this subsystem, and then constructed a mathematical model:

$$\bar{y}_i = f(x_{i1}, x_{i2}, \ldots, x_{in}), \quad (1)$$

where: $\bar{y}_i$ – the average value of the consumer quality indicator of the *i-th* subsystem; $x_{i1}, x_{i2}, \ldots, x_{in}$ – indicators of the production quality of the *i-th* subsystem.

After obtaining such mathematical models for all subsystems, the next stage is performed - the construction of a mathematical model of the system as a whole.

To determine the operating time of the entire product (L) by the known values of the operating time of its subsystems (L_i), the Monte Carlo method is used. The essence of this method lies in the use of random numbers earned in one way or another with a given distribution law.

The solution of the problem of constructing a mathematical model of the system as a whole is carried out in two stages on a computer. At the first stage, an algorithm is developed for determining the operating time of the system (L) for given urinations of indicators of the production quality of individual subsystems (x). At the second stage, by changing the initial data (x), the corresponding values of the operating time of the system (L) are determined, and on the basis of this information, the mathematical model of the system is already developed directly.

To perform the first stage, it is necessary to have mathematical models that reflect the relationship between the production and consumer quality of all product subsystems:

$$L_i = f(x_{i1}, x_{i2}, \ldots, x_{in}). \quad (2)$$

After that, you need to know the scheme of failure, i.e. diagram of the conditional connection of elements in the system.

The operating time of the unit before the first failure of any subsystem is taken as an indicator of the consumer quality of the engine. That is, we are dealing with a serial connection of elements. The operating time of each specific engine is determined by the operating time of its "weakest" subsystem, ie:

$$L = \min(L_1, L_2, \ldots, L_n), \quad (3)$$

Where: L – engine operating time to the first failure; $L_1, L_2, \ldots, L_n$ – the value of the operating time of its individual subsystems.

The block diagram of the algorithm for determining the engine operating time is shown in figure 2.

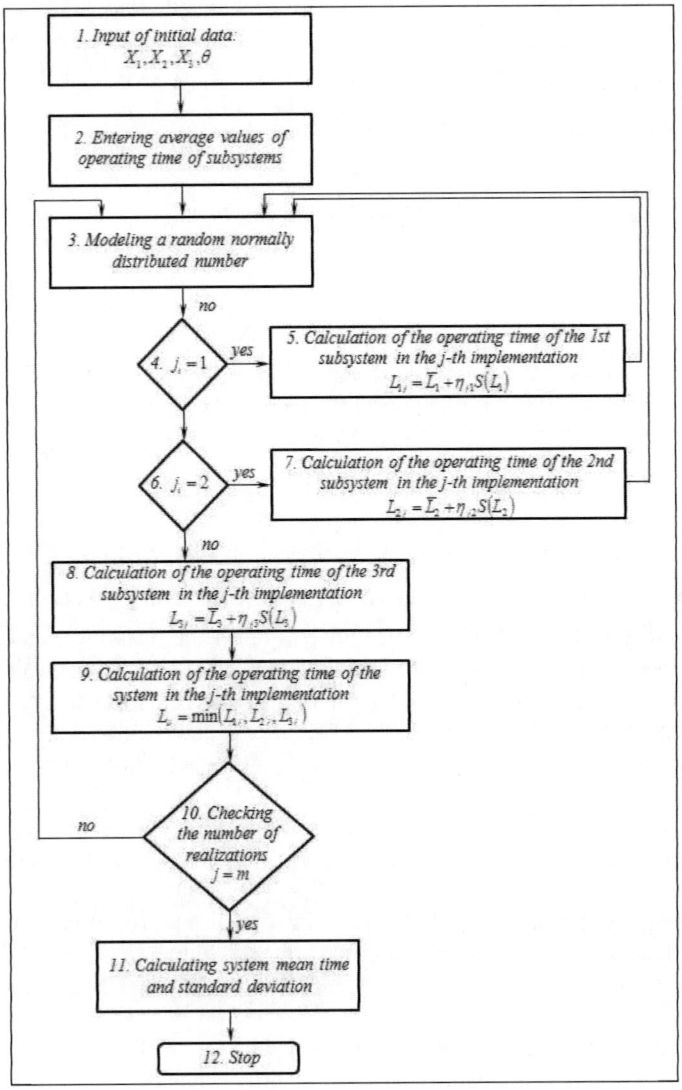

Figure 2. Block diagram of the algorithm for solving the problem of modeling a system by the Monte Carlo method.

At the second stage, for the operational assessment of the quality of the repaired engines, a nomogram was built (Figure 3), which allows predicting the average engine operating time using the known values of two production quality indicators.

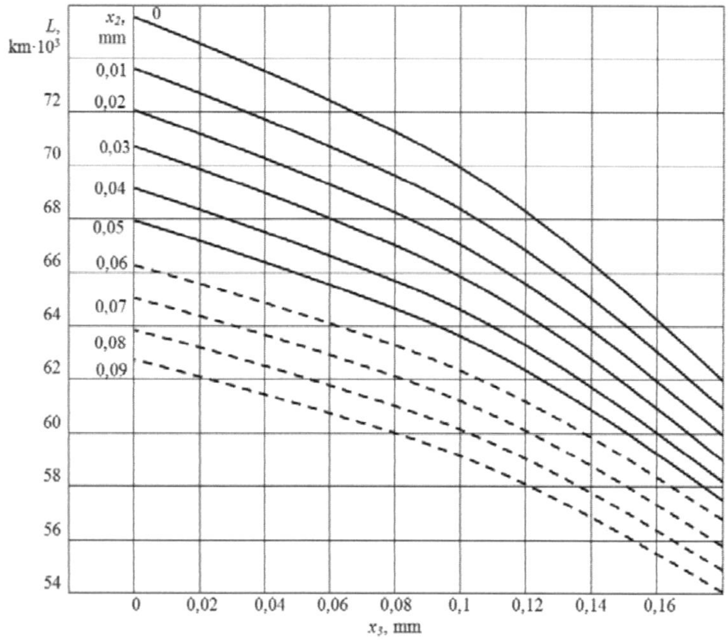

Figure 3. Nomogram for predicting engine operating time: x_2 - deviation of the diameter of the connecting rod journals from the nominal value, mm; x_3- crankshaft runout, mm.

To predict the engine operating time in accordance with the value of the indicator x_2, the required curve is selected on the graph and the value $\bar{L}$ is determined from the value x_3. An operational assessment of the quality of repaired engines is carried out on the basis of predicting their operating time. Having determined, based on the results of production control, the average values of the production quality of engines repaired at a given enterprise, it is possible to assess the quality of products by predicting the operating time.

3. Conclusion

Forecasting methods make it possible to determine the resource distribution of the elements of machines and assemblies depending on the quality of manufacture (repair), operating conditions and operating modes.

When implementing the above approach to assessing the service life of products, it is necessary to use the following characteristics: fatigue resistance of parts under high-cycle loading; loading of machine parts; wear resistance in specific operating conditions and probabilistic and statistical research methods.

Methods for predicting the durability of machine parts and the durability of units and assemblies in terms of durability of their elements can serve as a scientific basis for determining the suitability of repaired products.

References

[1] Galiev I G and Khusainov R K 2016 Results of determining the optimal values of overhaul times of tractors in agricultural production, taking into account the level of their operation *Bulletin of Kazan State Agrarian University* **11 2 (40)** 87-90

[2] Khusainov R K and Galiev I G 2013 Justification of the resource consumption of tractor units and systems, taking into account the differentiated approach when assigning technological operations for the planned period *Bulletin of Kazan State Agrarian University* **8 2 (28)** 73-6

[3] Galiev I G and Khusainov R K 2014 Substantiation of the choice of the option of repair actions taking into account the intensity of resource consumption of tractor units *Bulletin of Kazan State Agrarian University* **9 2 (32)** 68-71

[4] Galiev I G and Khusainov R K 2012 Determination of the weight of technological operations and

the level of resource consumption of tractor units and systems *Bulletin of Kazan State Agrarian University* **7 3 (25)** 74-7

[5] Galiev I G, Ziganshin B G, Abdrakhmanov R K and Khusainov R K 2017 Substantiation of the level of differentiation of agricultural work by tractors *Machinery and equipment for the village* **10** 28-31

[6] Khaliullin F Kh, Matyashin A V, Akhmetzyanov R R, Medvedev V M and Lushnov M A 2019 Prospects for using the bayes algorithm for assessing the technical condition of internal combustion engines *IOP Conference Series: Materials Science and Engineering electronic collection* 012016

[7] Yahin S, Gabdrafikov F, Khaliullin F, Khusainov R and Naficov I 2020 Improving the operational efficiency of tractors by ensuring their ability to perform work *BIO Web of Conferences. International Scientific-Practical Conference "Agriculture and Food Security: Technology, Innovation, Markets, Human Resources"* 00111

[8] Sinitsky S A, Medvedev V M, Lukmanov R R, Pikmullin G V and Makarova O I 2020 Investigation of the effect of air supply on the effective engine performance of a machine-tractor unit under unsteady load *BIO Web of Conferences. International Scientific-Practical Conference "Agriculture and Food Security: Technology, Innovation, Markets, Human Resources"* 00025

[9] Galiev I, Khafizov C., Adigamov N and Khusainov R 2018 Increase of efficiency of tractors use in agricultural production *Engineering for Rural Development. Proceedings* 373-7

[10] Apazhev A K, Shekikhachev Y A, Egozhev A M, Shekikhacheva L Z and Egozhev A A 2020 Improving the durability of machine parts connections *IOP Conference Series: Materials Science and Engineering* **862(3)** 032005 DOI: 10.1088/1757-899X/862/3/032005

[11] Apazhev A K, Shekikhachev Y A, Batyrov V I and Shekikhacheva L Z 2020 Influence of non-uniformity of fuel supply parameters on diesel engine performance *Journal of Physics: Conference Series (JPCS)* **1679** 042063 DOI: 10.1088/1742-6596/1679/4/042063

[12] Apazhev A K, Shekikhachev Y A and Batyrov V I 2020 Technological support for the accuracy of the assembly of mechanisms *Journal of Physics: Conference Series (JPCS)* **1679** 042062 DOI: 10.1088/1742-6596/1679/4/042062

[13] Shekikhachev Y A, Batyrov V I, Shekikhacheva L Z, Bolotokov A L and Gubzhokov H L 2020 Prediction of service life of auto-tractor engine parts *IOP Conference Series: Materials Science and Engineering* **862(3)** 032001 DOI: 10.1088/1757-899X/862/3/032001

Gear tooth action in a plane synthesis based on the criterion of constant curvature

G A Timofeev and A A Barkova

Bauman Moscow State Technical University, Theory of machines and Mechanisms Department, 2nd Baumanskaya st., 5/1, Moscow, 105005, Russian Federation

E-mail: timga@bmstu.ru

Abstract. The relevance of the problem of the gear tooth action in a plane synthesis based on the criterion of tooth profile wheel group curvature is justified. Based on a mathematical model of gear tooth action in a plane with a constant angular speed ratio, an algorithm for solving the problem with a given constant curvature is given. The obtained solutions allow us to perform an analysis of the general type of lantern gear in a plane.

1. Problem statement

Cogwheels with a constant curvature of the tooth profiles are mainly structurally performed by the pinion [1]. This makes it possible to increase the machinability, eliminate the relative slip at the contact point and, accordingly, reduce the wear of the profiles, the loss of power to friction and increase the efficiency by an order of magnitude [2-6].

2. Mathematical model and synthesis algorithm

A mathematical model of gear tooth action in a plane with a constant gear ratio can be represented as:

$$v_j = -a_j\omega_j sin\alpha + r(\dot{\alpha} - \omega_j) \tag{1}$$

$$\dot{r} + a_j\omega_j cos\alpha = 0 \tag{2}$$

$$k_j = (\dot{\alpha} - \omega_j)/v_j \tag{3}$$

where j=1, 2 - the index wheel, v_j - speed of the contact point along the tooth profile, a_j – centrode radius, ω_j – angular rate, α – angle of action, r - distance from the contact point to the pitch point, k_j - curvature of the tooth profile, a dot above a parameter means its time derivative t.

Let us consider the solution of the problem of gearing synthesis by the criterion of constant normal curvature $k_j = const$. We write (3) considering (1) in the form:

$$R_j = \frac{v_j}{\dot{\alpha} - \omega_j} = r - \frac{a_j\omega_j sin\alpha}{\dot{\alpha} - \omega_j}$$

from which

$$r = R_j + \frac{a_j \omega_j sin\alpha}{\dot{\alpha} - \omega_j} \quad (4)$$

where $R_j = 1/k$ – curvature radius and the second term is the line of action for the criteria $v_j = 0$. It is given by the formula:

$$\tilde{r}|_{vj=0} = -a_j sin\alpha \mp a_j \sqrt{sin^2\alpha + c_0}$$

where the parameter c_0 is:

$$c_0 = \frac{2\tilde{r}_0}{a_j}\left(\frac{\tilde{r}_0}{2a_j} + sin\alpha_0\right)$$

Thus,

$$r = R_j - a_j sin\alpha \mp a_j \sqrt{sin^2\alpha + c_0} \quad (5)$$

Here, the choice of the sign is determined by setting the initial conditions. Differentiating equation (5) in time and comparing with equation (2), we find that the function $\alpha(t)$ satisfies the equation:

$$\dot{\alpha} = \frac{\omega_j}{1 \mp \frac{sin\alpha}{\sqrt{sin^2\alpha + c_0}}} \quad (6)$$

Its solution has the form:

$$\varphi_1 = \alpha - \alpha_0 \mp \left[arcsin\frac{cos\alpha}{\sqrt{c_0 + 1}} - arcsin\frac{cos\alpha_0}{\sqrt{c_0 + 1}}\right] \quad (7)$$

Having the expressions (5) and (7), it can be written the formulas for the teeth profiles of the wheels in the form:

$$x_j = a_j cos\varphi_j + rsin(\alpha - \varphi_j)$$

$$y_j = -a_j sin\varphi_j - rcos(\alpha - \varphi_j)$$

and if the second link is a translationally moving racking, then

$$x_2' = rsin\alpha$$

$$y_2' = -a_j\varphi_j - rcos\alpha$$

The results obtained correspond to out-of-band lantern gear by introducing an additional geometric connection to the model.

$$(a_1 + e)sin\varphi_1 = -(r + \rho)cos\alpha \quad (8)$$

where ρ – pin radius, e - offset of the center of the pin relative to the pole.

Let's establish a connection between the constants c_0 and e. Comparing equations (7) and (8), we get:

$$e = a_1(\sqrt{c_0 + 1} - 1) \quad (9)$$

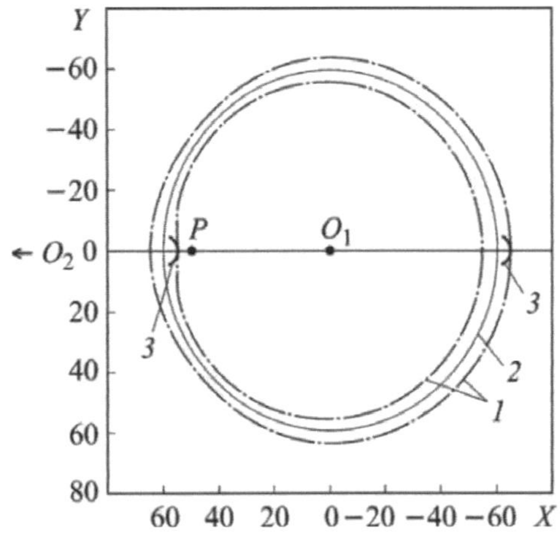

Figure 1. Lines of action for the case of k_1 = const (out-of-band pinion engagement, c_0> 0): 1 — line of action; 2 — line centers pin; 3 — profile of the first link (pinwheel); O_1 — axis rotation of the first link; O_2 - axis of rotation of the second link; P — pitch point.

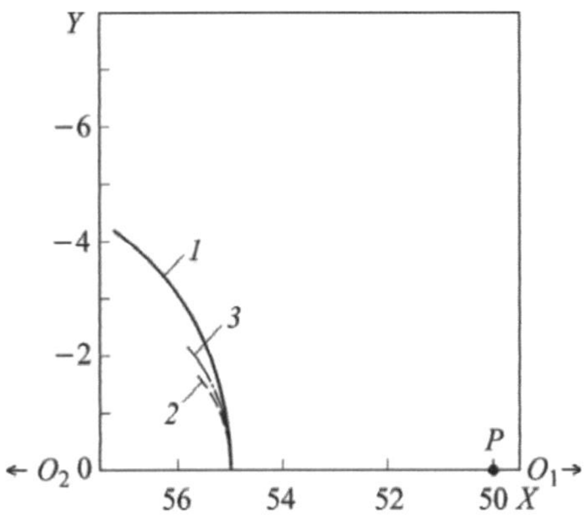

Figure 2. Link profiles for the case k_1 = const (out-of-pole pinion engagement, c_0 > 0): 1-profile of the first link (pinwheel); 2- profile of the second link; 3 - profile of the rail; O_1 - axis of rotation of the first link; O_2-axis of rotation of the second link; P- pitch point.

The lines of action for the constant curvature of the first link profile for out-of-band pinion engagement with a_1 = 50 mm and the gear ratio i_{12} = 2 are shown in figure 1. The two curves correspond to the gear sign "±" in equation (5). Line corresponding to the " + " sign, located inside the center line of the pin, and the line corresponding to the " - " sign is on the outside. In both cases, the profile of the pinion is the arc of the circle. In more detail, the profile of the pinion and the profiles of the second link are shown in figure 2.

Consider the case k_j = const, v_j = const. For the relative velocity, considering equations (5) and (6), it was obtained the following expression:

$$v_j = -a_j\omega_j sin\alpha + \left(R_j - a_j sin\alpha \mp a_j\sqrt{sin^2\alpha + c_0}\right)\left(\frac{\omega_j}{1 \mp \dfrac{sin\alpha}{\sqrt{sin^2\alpha + c_0}}} - \omega_j\right) =$$

$$= -a_j\omega_j sin\alpha + (R_j - a_j sin\alpha \mp a_j + \sqrt{sin^2\alpha + c_0})\frac{a_j\omega_j sin\alpha}{-a_j sin\alpha \mp a_j\sqrt{sin^2\alpha + c_0}} = \quad (10)$$

$$= \frac{R_j}{r - R_j} a_j\omega_j sin\alpha$$

Then, for v_j = const, there must be $r = R_j + Csin\alpha$. Substituting this in equation (10), we get the identity:

$$\left((C + a_j)^2 - a_j^2\right) sin^2\alpha - a_j^2 c_0 = 0$$

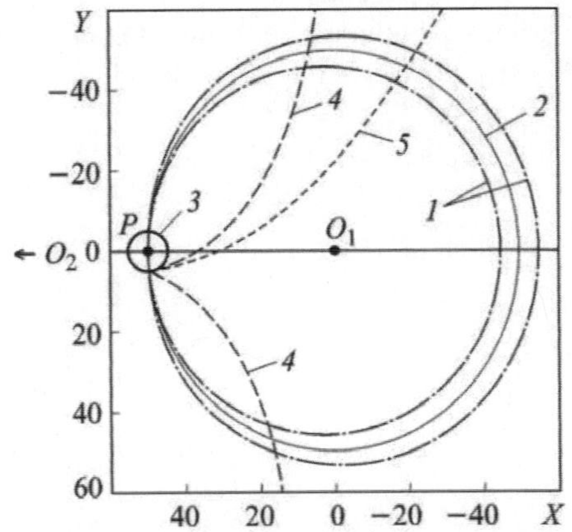

Figure 3. Lines of action and profile links for the case k_1= const (pole pinion engagement, c_0= 0):
1 — line of action; 2 — line centers pin; 3 — profile of the first link (pinwheel); 4 — profile of the second link; 5 — profile rail; O_1 - axis of rotation of the first link; O_2 - axis of rotation of the second link; P — pitch point.

It holds for any α only under the condition $C = -2a_j$, $c_0 = 0$. From equation (9) it also follows e = 0, i.e. the case of constant curvature and velocity corresponds to the pole pinion engagement. The engagement line has the form:

$$r = R_j - 2a_j sin\alpha \tag{11}$$

which is the polar equation of limacon of Pascal.

Line of action and profile links for constant curvature profile of the first link for the pin cycloid gear with a_1 = 50 mm and reduction ratio i_{12} = 2 is shown in figure 3. Line of action is limacon of Pascal, and the profile of the first line is a circle.

Consider the case of k_1 = const, k_2 = const. In this case, we obtain together with equations (1)–(3) a redundant system of equations, the solution of which leads to the imposition of an additional connection on the constant parameters of the engagement. We rewrite the relation (4) in the form:

$$\dot{\alpha} - \omega_j = \frac{a_j \omega_j sin\alpha}{r - R_j} \tag{12}$$

Hence, when R_1 = const, R_2 = const:

$$\omega_1 + \frac{a_1 \omega_1 sin\alpha}{r - R_1} = \omega_2 + \frac{a_2 \omega_2 sin\alpha}{r - R_2} \tag{13}$$

The formula for the engagement line in the case of k_1 = 0, k_2 = const is obtained from the equation (13) in the form:

$$r = R_2 + \frac{a_j \omega_j sin\alpha}{\omega_1 - \omega_2} \tag{14}$$

This, like equation (11), is the polar equation of Pascal snail. It follows from (12) that:

$$\dot{\alpha} = \omega_1 \tag{15}$$

Differentiating equation (14) with respect to time t and using equation (2), considering (15), it was obtained:

$$\frac{\omega_1 a_j \omega_j cos\alpha}{\omega_1 - \omega_2} = -a_j \omega_j cos\alpha$$

Hence $\frac{\omega_1}{\omega_1 - \omega_2} = -1$. This means that such transfer may be made in the form of a face with internal gear and gear ratio i_{12} = 1, 2[7-8]. The pinwheel is located on the second wheel, and the profiles of the teeth of the first wheel are straight, parallel, radial direct. In this case, the velocities have expressions, respectively $v_1 = -a_1\omega_1 sin\alpha$, $v_2 = R_2(\omega_1 - \omega_2) = const$.

If $k_1 = k_2$, then from equation (4) we have $\omega_1 = \omega_2$ and, consequently, $a_1 = a_2$, which corresponds to a coupling mechanism with zero center distance with arbitrary but congruent link elements. The same result corresponds to the solution for $k_1 = k_2 = const$, $k_1 = k_2 = \infty$ ($R_1 = R_2 = 0$).

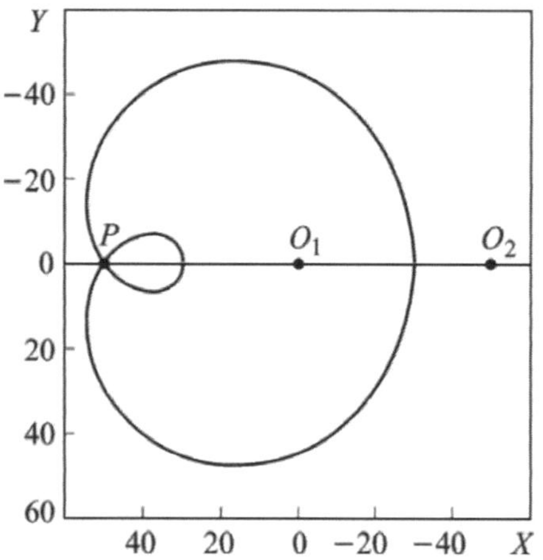

Figure 4. The line of action for the case $k_1 = 0$, $k_2 = const$:
O_1 — the axis of rotation of the first link; O_2 - the axis of rotation of the second link; P- pitch point.

Figure 4 shows the engagement line for the case $k_1 = 0$, $k_2 = const$.

3. Conclusion

The resulting formulas describe the entire set of pole and non-pole pinion gears in a plane with internal, external and rack-and-pinion gearing. As a result of solving the problem, it becomes possible to analyze them for the entire set of basic and derived performance criteria, as well as the ability to choose the optimal ratios of constant parameters — the center distance, gear ratio, the radius of the handguards and the radius of the installation of the handguards.

References

[1] Ivanov A S, Ermolaev M M and Chirkin A V 2020 Reliability functions for specific mechanical faults *Russian Engineering Research* **4** 275-82

[2] Osipkov A, Poshekhonov R, Arutyunyan G, Basov A and Safonov R 2017 Modeling of thermoelectric generator power characteristics for motorcycle-type engines *Journal of electronic materials* **10** 6195-203

[3] Shiriaev P, Shishov K, Osipkov A and Tishchenko L 2019 Development of the automotive thermoelectric generator electrical network *Journal of electronic materials* **4** 1998-2009

[4] Egorova O V, Samoilova M V, Timofeev G A and Evgrafov A N 2020 Force loading analysis of a cylindrical self-braking inverse gearing *International review of mechanical engineering* **2** 73-8

[5] Prokhorov V P and Chernyshev I N 1977 Selection of rational parameters of the end gear *News of higher education institutions. Mechanical engineering* **3**

[6] Prokhorov V P and Chernyshev I. N 1977 On the issue of synthesis of a variety of end pinion transmission *News of higher education institutions. Mechanical engineering* **12**

[7] Egorova O V, Panyukhin V V, Samoilova M V and Timofeev G A 2019 Involute self-braking inverse gears. Geometry of external and internal engagement *Mechanisms and machine science (book series)* **73** 1017-28

[8] Terekhova N Yu, Egorova O V, Brekalov V G and Spasskaya D D 2019 The design methodology of Bauman Moscow State Technical University *Mechanisms and machine science (book series)* **73** 843-51

Semi-automatic welding in the environment of protective gases of welded structures from hardening steels with regulation of thermal cycles

N V Zharinova* and M V Vakulenko

Ufa State Petroleum Technological University, Ufa, 450062, Russian Federation

*E-mail: zharinova_nata@mail.ru

Abstract. In order to increase the technological and operational strength of welded joints of type X12CrMo5 steels in the process of welding in a protective gas environment, it is possible to use vibration deformation treatment of a welded joint with a pearlite joint to form a metal structure of a welded joint with minimal sensitivity to the formation of cold cracks due to an increase in the rate of heating and cooling, reduction of residence duration and provision of conditions of high-temperature decay of austenitic metal structures in seam and near-seam zones above critical temperatures. The work carried out research on the process of semi-automatic welding in the medium of protective gases in the mixture of carbon dioxide and argon with pearlite and austenitic welding materials of welded structures from hardening steels of grades 12Kh2M1 and X12CrMo5 with regulation of thermal cycles (RTC) due to the use of vibration deformation treatment and accompanying forced cooling.

1. Introduction

For welded joints made by austenitic welding materials, welding in the environment of protective gases by multilayer welding with layer-by-layer cooling of butt joints is regulated. Control of the thermodeformation cycle of welding due to accompanying forced cooling ensures hardening of austenite seam metal on austenite and reduction of the width of hardened hard sections of the zone of thermal influence (ZTI) to a critical value, at which their presence practically does not adversely affect the bearing capacity of welded joints. Use of austenitic welding option is allowed for equipment operating in media that do not cause corrosion cracking [1-5].

This technology applies to the assembly and welding of structural elements of oil and gas chemical equipment from low-carbon chromomolybdenum hardening steels of grade 12Kh2M1 and X12CrMo5 according to GOST 20072-74 by semi-automatic welding in the environment of protective gases with uniform pearlite welding wires with subsequent thermal treatment and austenitic welding materials without heating and thermal treatment.

2. Experimental procedures

To increase the technological and operational strength of welded joints of steels of type X12CrMo5, welding of butt joints in a mixture of carbon dioxide and argon is provided. At the same time, in order to ensure the transition from large-drop to jet transfer of molten metal of welding wire, better formation of weld metal and less spraying, the best shape of the wire is regulated welding modes providing high current densities - 150 A/mm^2 or more. For welded joints made by homogeneous pearlite welding wires, it is possible to use vibration deformation treatment of the welded joint to form the structure of the welded joint metal with minimal sensitivity to cold crack formation due to ensuring the decay conditions of austenitic structures of the ZTI metal above critical temperatures and increasing the time between welding and subsequent thermal treatment according to the high tempering mode.

For thick-walled welded joints made by austenitic chromium-nickel wires with the addition of manganese, the use of concomitant forced cooling is regulated, which ensures the hardening of the austenitic seam on austenite and a decrease in the width of the hardened hard sections of the thermal influence zone to critical ones, in which the negative effect of these sections practically does not affect the bearing capacity of the welded joint. Due to exclusion of preliminary and concomitant heating during welding and possibility of mechanized welding at increased modes higher labor productivity is provided. Welding in the medium of CO_2 and in the mixture of protective gases (Ar + 20-25% CO_2) is carried out at direct current of reverse polarity from power sources with a rigid or falling characteristic [6-9].

Use of a welding wire of 2NM307 (GOST 2246-70) is allowed for the weld joints working at a temperature not above 450 °C. It is necessary to apply a high-nickel welding wire of Sv-05kh5n40g7m8t to structural elements which weld joints are exposed to heating to temperature above 450 °C. Preparation of welded parts edges for welding shall be performed mechanically in accordance with the requirements of GOST 5264-80 and GOST 16037-80, technical specifications, drawing, corresponding instructions. Thermal preparation of edges is allowed only in case of impossibility of mechanical treatment. Thermal cutting should be carried out with preliminary heating of cutting point up to 250-350 ° C and subsequent delayed cooling in heat insulation. After heat treatment, the edges shall be cleaned with abrasive tools to a depth of at least 2-3 mm, counting from the deepest depression of the crack surface, with subsequent inspection of the edges through a magnifying glass of 4-7 times increase. When cracks are detected, edges are subjected to additional treatment with abrasive tools until cracks are completely removed, followed by color flaw detection. Grind burrs throughout the machined contour.

In the process of welding with uniform pearlite electrodes, in order to increase the time between welding and thermal treatment in order to prevent cracking before the start of high-temperature thermal treatment, the welding process in the protective gas medium must be carried out using vibration (frequency 50Hz, amplitude up to 1 mm) treatment. Low-frequency vibration loading during welding operations ensures uniformity of residual stress removal in the welded joint area.

When welding the root layer, it is necessary to monitor the complete penetration of the edges. The surface of the root layer on the back side should be made with uniform reinforcement and smooth transition to the base metal. Before welding it is necessary to grind the tack points with a grinding wheel to pure metal with smooth transition to the main metal. When welding single-layer butt seams and the first layer of multi-layer seams, it is recommended to move the seam burner progressively without transverse oscillations. In case of accidental arc break, light it again, retreating 15-20 mm back from the crater, having previously cleaned the seam to pure metal; well boil the arc break places. In order to avoid defects in the root layer, it is necessary to ensure that the tack is completely melted.

Inclusive a welding wire of 2NM307 and wall up to 20 mm thick inclusive a welding wire of Sv-05kh5n40gm8t to make welding of joints wall up to 14 mm thick with layer-by-layer cooling. In this case, welding of each subsequent layer is recommended when the previous one cools to a temperature not higher than plus 100 ° C, and in case of welding at negative temperatures - to a temperature of plus 200-250 ° C [10-15].

In order to increase the technological strength of thick-walled welded joints, improve the structure of thermal influence zones and weld metal, it is recommended to weld the second and subsequent layers with accompanying cooling with water or water-air mixture (figure 1).

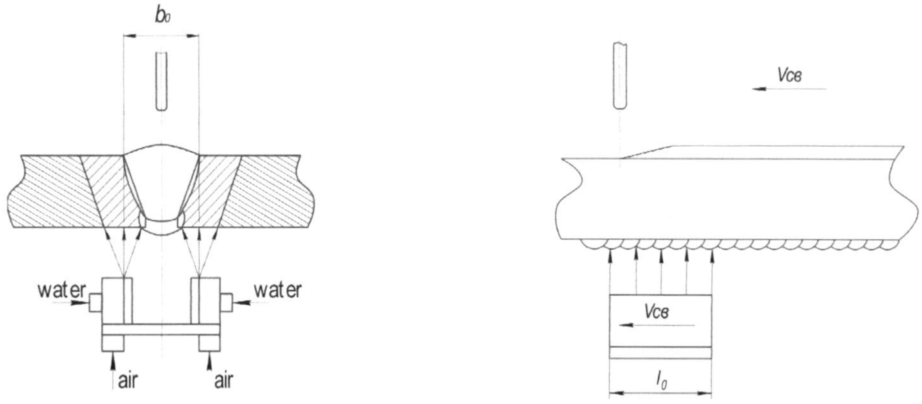

$v_{св}$ – welding speed; l_0, b_0 - cooling parameters

Figure 1. Schematic diagrams of accompanying cooling during welding.

Cooling device is installed directly under welding arc and moves during welding in direction of arc movement at welding speed. In order to more effectively affect the shear zones of the tucker, cooling is carried out by two liquid flows directed to the high-temperature regions of the zones of thermal influence adjacent to the fusing line.

3. The results of studies and their discussion

Effective length of cooling zone lo at cooling with water-air mixture is 8-10 mm. The width of the cooling zone bo is 10-20 mm. For multi-layer welding with filling of the preparation with layers for the entire width of the butt seam with a V-shaped bevel of edges at a preparation angle of 60 ° b_o, it is recommended to select from the graph shown in figure 2. The distance from the nozzle to the cooling surface is 35-40 mm. Cooling water flow rate as well as air pressure during welding shall be controlled by means of adjustment valves on the cooling device by liquid flow meter and pressure gauge. At accompanying cooling with water-air mixture water flow rate is 0.8-1.2 l/min, air pressure - 0.3-0.4 MPa.

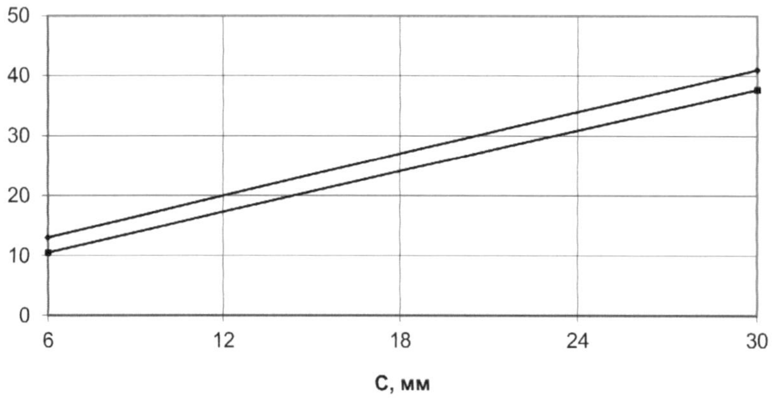

Figure 2. A graph to select the width of the cooling zone depending on the fill height of the "C" groove during semi-automatic welding CO_2 a multi-layer butt weld with rollers applied over the entire width of the groove.

The welding zone should be blown with protective gas before the welding arc is excited. The distance from the burner nozzle to the surface of the welded part shall not exceed 25 mm. The estimated flow rate of protective gas during wire welding with a diameter of 1.2 mm is 9-12 l/min; during wire welding with diameter of 1.6 mm - 14-18 l/min; at welding with wire with diameter of 2.0 mm - 16 - 20 l/min.

The order of layering during welding of horizontal joint of pipes is shown in figure 3.

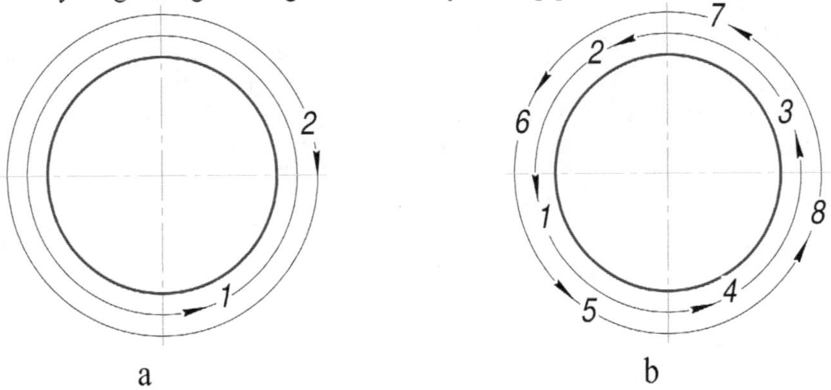

a -is the root layer; b - second and subsequent layers

Figure 3. How to apply layers when welding a horizontal pipe joint with a diameter up to 219 mm (a), diameter more than 219 mm (b).

After each pass of the multilayer seam, it must be externally inspected to detect cracks and other defects. If detected, the defects are eliminated before the next pass. To improve the quality of the weld joint, it is recommended to grind the previous layer to pure metal with a grinding wheel or a metal brush before applying the next layer. To provide an annealing effect on the metal structure of the thermal influence zone in order to eliminate the martensitic structure and increase operability, the order of seaming during multilayer welding must be performed with layer-by-layer cooling and compliance with the sequence of seaming. Weld layers on the metal edges must be made at maximum speed without oscillating the electrode wire. The recommended procedure for placing rollers when welding joints of vertically and horizontally located pipes is shown in figure 4 [16-18].

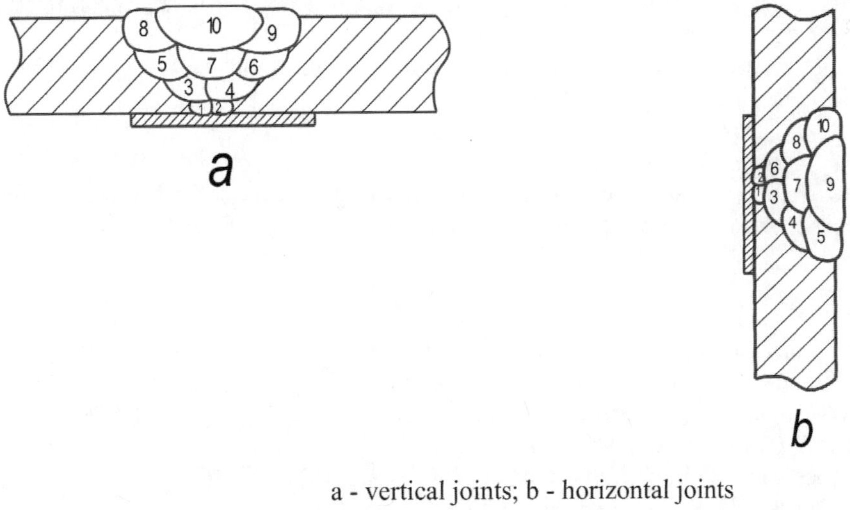

a - vertical joints; b - horizontal joints

Figure 4. Approximate Welding Roll Stacking Order.

When pearlite welding materials are used after welding works completion, structures are subject to mandatory heat treatment - high tempering, according to GOST 34347-2017. Tempering is carried out at a temperature of 730... 760°C. Holding time 2... 2.5 h (with wall thickness 10... 20 mm) and 4... 4.5 h (with thickness up to 30 mm). Next, cooling with the furnace to 500°C and final cooling in calm air. Heat treatment shall be carried out not later than 48 hours after completion of welding works. Do not perform any operations with the article before heat treatment. Heat treatment is not performed using chromium-nickel electrodes.

Quality control of welded joints is carried out operationally in accordance with the requirements of GOST 34347-2017, product specifications, drawings and relevant instructions.

Maximum permissible width of solid sections in ZTI h_t for welded joints from steels of type X12CrMo910 shall not exceed the value calculated according to the formula

$$h_{\mathrm{T}} \leq S\sqrt{\frac{1}{a^2}+\frac{\kappa_{\mathrm{B}}-1}{2a\kappa_{\mathrm{B}}^2(1-\kappa_{\mathrm{BT}})}}-\frac{1}{a},$$

where S – thickness of the structural element, mm;

$\kappa_{\mathrm{B}} = \sigma_{\mathrm{B}}^{\mathrm{T}} / \sigma_{\mathrm{B}}^{\mathrm{M}}$;

$\kappa_{\mathrm{BT}} = \sigma_{\mathrm{B}}^{\mathrm{M}} / \sigma_{\mathrm{T}}^{\mathrm{T}}$;

$\sigma_{\mathrm{B}}^{\mathrm{M}}$, $\sigma_{\mathrm{B}}^{\mathrm{T}}$ - respectively, the time resistance of the soft (main) and solid (zone of thermal influence) metals;

$\sigma_{\mathrm{T}}^{\mathrm{T}}$ - yield strength of solid metal;

a – correction factor ($a \approx 30$).

Removal of defective areas and their brewing are carried out in compliance with the same technological procedures that are provided for when welding products according to the relevant instructions. Correction of the same area is allowed not more than twice.

4. Conclusions

As a result of the work done, it has been shown that the control of the thermo deformation cycle of welding due to the accompanying forced cooling ensures the quenching of the austenite seam on austenite and the reduction of the width of the hardened hard sections of the thermal influence zone to critical ones, at which the negative effect of these sections practically does not affect the bearing capacity of the weld joint.

Due to exclusion of preliminary and concomitant heating during welding and possibility of mechanized welding at increased modes higher labor productivity is provided.

Acknowledgment

The authors are grateful for the advice and assistance in conducting experiments to Professor of the Ufa State Petroleum Technological University (Russian Federation) Andalis Garifovich Khalimov.

References

[1] Zaripov M Z, Fairushin A M and Karetnikov D V 2019 *Materials Science Forum* **946** 883-8
[2] Vakulenko M V and Zharinova N V 2020 *IOP Conf. Ser.: Mater. Sci. Eng.* **862** 062022
[3] Zharinova N V and Vakulenko M V 2020 *IOP Conf. Ser.: Mater. Sci. Eng.* **862** 032014
[4] Rizvanov R G, Mulikov D Sh, Karetnikov D V, Fairushin A M and Tokarev A S 2018 *IOP Conf. Ser.: Mater. Sci. Eng.* **317** 012077
[5] Yakhin A V, Karetnikov D V, Cherepashkin S E and Rizvanov R G 2019*Chemical and Petroleum Engineering* **55(7-8)** 681-6
[6] Nasibullina O A and Tyusenkov A S 2020 *IOP Conf. Ser.: Mater. Sci. Eng.* **862** 022007
[7] Karetnikov D V, Rizvanov R G, Fairushin A M and Kolokhov K S 2013 *Welding International* **27** (7)557-60
[8] Nasibullina O A and Rizvanov R G 2020 *Materials Science Forum* **989** 28-2
[9] Yakhin A V, Karetnikov D V, Rizvanov R G, Abakacheva E M and Gareev A G 2019 *Chemical and Petroleum Engineering* **54(11-12)** 801-5
[10] Nasibullina O A and Gareev A G 2019 *Materials Science Forum* **946** 20-4

[11] Kuzeev I R, Ibragimov I G, Bayazitov M I, Davydov S N and Khairudinov I R 1985 *Chem. and Technology of Fuels and Oils* **22(3)** 111-3
[12] Tyusenkov A S and Nasibullina O A 2019 *IOP Conf. Ser.: Mater. Sci. Eng.* **687** 066016
[13] Fatkullin M R, Fairushin A M andRizvanov R G 2020 *Materials Science Forum* **989** 747-52
[14] Nasibullina O A and Tyusenkov A S 2019 *IOP Conf. Ser.: Mater. Sci. Eng.* **537** 022023
[15] Gaysin E Sh, FrolovY A and NasibullinaO A 2020 *IOP Conf. Ser.: Earth and Environ. Sci.* **459** 032055
[16] Tokarev AS,Karetnikov DV,Rizvanov RG, FayrushinAMandZaripov MZ 2020 *Materials Science Forum* **991** 70-6
[17] Nasibullina O A and Rizvanov R G 2020 *Materials Science Forum* **992** 695-9
[18] Rizvanov R G, Mulikov D Sh, Karetnikov D V, Cherepashkin S E and Shirgazina R F 2017 *NanotechnologiesinConstruction* **9(4)** 97-15

IOP Conf. Series: Materials Science and Engineering 1155 (2021) 012026 doi:10.1088/1757-899X/1155/1/012026

The effect of machined wheel profile geometry on cutting force

N V Kanatnikov, A S Pashmentova and S Y Radchenko

Orel State University named after I.S. Turgenev, Komsomolskaya st., 95, Orel, 302026, Russia

E-mail: NKanatnikov@ya.ru

Abstract. The mechanisms based on standard geometry gears have reached their technological limits. To meet the needs of the market, it is necessary to meet new and new, often conflicting operational requirements. At the same time, it is necessary to reduce the mass of the node and increase the maximum torque, in addition, the noise level and increase the axial load must be reduced. To achieve the increasing requirements for the operational characteristics of mechanisms gears with specially designed tooth geometry are used. A tool for cutting such gears must be specially designed and manufactured for each new gear pair, given into account its design features. The introduction of a new tool in the technological processes of machining gear profiles leads to the fact that the surface quality decreases and tool wear increases. At present, the concept of this problem is scattered and composed of the experience of individual researchers. The work reveals the dependencies connecting physical phenomena accompanying the cutting process and the geometric parameters of the profile of the bevel gear. Based on the results of the project, it is possible to calculate the optimal profile parameters based on the technological parameters of the process.

1. Introduction

Mechanisms using transmission with gears of the standard geometry of the tooth profile have reached their technological limit. To achieve ever-increasing performance requirements, modern torque transmission mechanisms (gearboxes, reducing gears, differentials, etc.) use gears with specially designed tooth geometry. The main operational properties of a gear train are the perception of axial loads, the strength of the teeth, and the noise level. To control the strength of the teeth we change the angle of the tooth profile, increase the radius of rounding of the base of the tooth stem, and apply profile modifications. To reduce the noise level, we reduce the roughness and increase the contact area.

To achieve the ever-increasing performance requirements of torque transmission mechanisms in high-tech industries, gears with specially designed teeth geometry are used. The tool for cutting such gears must be specially designed and manufactured for each new gear pair, taking into account its design features. The introduction of a new tool in the proven technological processes of gear profile processing leads to a decrease in surface quality and increased tool wear. At present, the understanding of this problem is disjointed and consists of the experience of individual researchers [1, 2, 3].

To process the changed tooth profile, it is necessary to use a special cutting tool with a modified geometry of the cutting part. In practice, to use of a tool with a modified profile leads to a change in the predicted tool wear and machining accuracy. Currently, there are only scattered experimental data, systematic data and recommendations are not available in the reference and scientific and technical literature. It can be suggested that these effects are observed due to the fact that the physical conditions of cutting change, the geometry of the cutting wedge bar, the chip removal mechanism, the friction forces on the front and back surfaces of the tool are being changed. The result is an increase in the cutting force and deterioration of the heatsink. Thus, having the ability to predict the physical phenomena accompanying the cutting process, it is possible to determine the rational geometric parameters of the cutting part, as well as adjust the cutting modes to achieve the required values of tool life and accuracy of the processed product.

In this paper, we consider the influence of the geometry of the teeth of a bevel gear on the cutting force that occurs during milling with a tool head. A significant influence of the profile geometry on the considered physical characteristic of the process means that the geometry of the tooth profile affects the practical results of processing. Therefore, when designing new gear profiles, it is necessary to take into account the technological conditions.

2. Modelling

The basis for models that predict the physical processes accompanying cutting is the methods of analytical, empirical, numerical and hybrid modelling [4]. Analytical modelling allows us to predict the cutting force, friction force, stress, strain, temperature, etc. However, due to the fact that cutting is accompanied by many complex physical phenomena, the use of analytical modelling in some cases is difficult. To predict the cutting force, analytical models based on the theory of sliding lines are used [4].

Empirical models are characterized by high accuracy and ease of use, but their development requires an array of experimental data, and the scope of application is limited. Designing a new empirical model requires a lot of time and material costs. Currently, the most widely used empirical approach is "Mechanistic Force Modeling". The essence of the method is to determine the specific forces per unit length of the cutting blade and then sum them over all the cutting edges involved in cutting. This method can be used to calculate the forces for cutting processes with a tool with a non-standard profile, when several edges are involved in the chip removal process. L. Berglind and D. Plakhotnik developed a model for calculating the cutting forces that occur when processing parts of arbitrary configuration on five coordinate milling centres based on this approach [5].

The disadvantage of this approach is that if it is necessary to make any changes to the developed technological process (the use of new processing modes, the use of a new cutting tool, etc.), it is necessary to calibrate the model to clarify the correction coefficients. Calibration is carried out by conducting a set of experimental studies. After the model is calibrated, the forecast error will be less than 5% [6].

Since the late 70s, numerical methods have been used to simulate cutting. Numerical methods allow us to successfully model both small and large values of the cutting force, as well as solve problems in two-and three-dimensional space [1, 3, 7, 8]. Numerical models are easier and more convenient to study because of the ability to conduct computational experiments with them. The consistency and formality of numerical models allows us to identify the main factors that determine the properties of the object under study to research the response of the simulated physical system to changes in its parameters and initial conditions. The finite difference method and the finite element method have been used for the numerical solution of cutting thermal physics problems. Thanks to the development and increased availability of special software tools and computing power, it has only recently become possible to widely use numerical methods to solve cutting simulation problems.

K. D. Bouzakis [1] and others proposed a comprehensive method for modeling the process of gear milling with hobbing cutters. Calculations using the finite element method made it possible to predict temperatures, deformations, stresses, and a number of other physical variables that characterize the processes occurring in the cutting zone, and on their basis to predict the wear of the cutting tool.

F. Klocke [3] deal with the problem of the influence of the geometric parameters of the cutting tool, namely, the profile angle of the cutting bit and nose radius of the cutter tip on thermal and mechanical stresses in the tool and the chip in the processing of bevel gears. The cutting process was modelled using the finite element method. In these works, particular solutions for the processes of processing gears of a certain profile are obtained, which allow us to establish some dependencies that arise during the processing.

Existing models for predicting the physical parameters of the cutting process are not adapted for studying the processes of gear processing of bevel gears due to the following features: the design of the tool, the cutting part of which is designed specifically for the processed pair of gears; the kinematics of the cutting process: continuous change in the geometric parameters of the cut metal layer and cutting angles; the chips formed by the top and side cutting edges, converging on the front surface of the cutter, chip fragments overlap each other, causing additional loads in the tool.

To solve the problem of predicting the cutting force that occurs during the mechanical processing of bevel gears, a method of predictive modelling has been developed [9]. Input data for modelling: the geometry of the cutting edge of the tool, processing modes, the behaviour of the material during deformation and the conditions of contact between the workpiece and the tool, the kinematics of the cutting process. Accepted assumptions: the tool is absolutely homogeneous and solid, the tool surface has a uniform structure, the processed material has a geometrically and physically nonlinear structure, the deformation of the workpiece is described by the Johnson-Cook method. The calculation scheme is shown in figure 1.

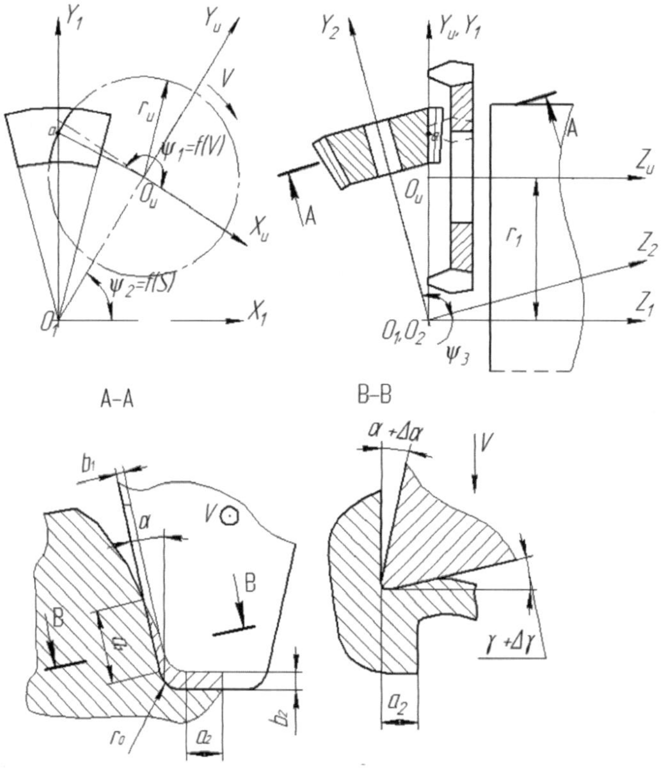

Figure 1. Calculation scheme for determining the parameters of the simulated process.

At the first stage, the parameters of the cut chip and the kinematic change in the geometry of the cutting part of the tool are calculated. To do this, an analytical method is used to calculate the parameters of the cut layer. At the second stage, three-dimensional modelling of the cutting tool and the workpiece is performed, considering the data obtained at the first stage. At the third stage, tests are conducted in a virtual environment. DEFORM 3D is used to simulate the processing.

The proposed approach allows us to calculate the axial components of the cutting force that occur during a single, most loaded cut. Next, we can change the system parameters under study and make repeated predictions to identify the influence of the parameters under study on the cutting force.

Verification of the adequacy of the simulation results was carried out by conducting experimental studies. Experimental work was carried out on the basis of a planning-machine (model GD320). For cutting, a special tool head was used, which allows modelling the cut under study within specified conditions [10]. To measure the axial components of the cutting force, a universal dynamometer UDM-600 was used. The dynamometer readings were recorded using a personal computer in the LabVIEW environment. The results of comparing the experimental and calculated values showed differences of no more than 12%. The proposed model can be considered adequate. Figure 2 shows a design scheme for a series of numerical experiments. Table 1 shows the initial data for carrying out numerical experiments.

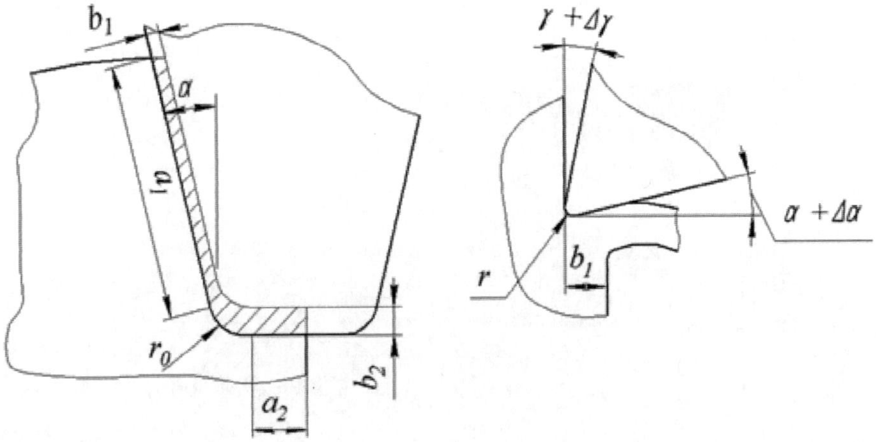

Figure 2. Parameters of the simulated process.

Table 1. Initial data for modeling.

Cutting process parameter	Value
Chip thickness b_1, mm	0.04
Chip thickness b_2, mm	0.06
Cutting edge length a_1, mm	2.9
Cutting edge length a_2, mm	0.4
Kinematic change of edge r, mm	0.01
Rake angle α_s, deg	5
Kinematic change of rake angle $\Delta\alpha_s$, deg	0.025
Clearance angle γ_s, deg	5
Kinematic change of Clearance angle $\Delta\gamma_s$, deg	0.015
Rake angle α_t, deg	10
Kinematic change of rake angle $\Delta\alpha_t$, deg	0.02
Clearance angle γ_t, deg	6
Kinematic change of Clearance angle $\Delta\gamma_t$, deg	0.012
Cutting speed V, m/min	100
Tool material	WC
Workpiece material	16MnCr5

3. Results

During the numerical experiment, options for changing the geometric parameters of the cutting part of the tool, depending on the geometry of the processed tooth profile, namely, the angle of the cutter profile and the radius of rounding of the cutter, were considered. The change in the angle of the cutter profile occurred in the range from 15 to 25 degrees. The change in the radius of rounding of the cutter occurred in the range from 0.1 to 0.2 mm. The selection of these ranges is based on practical recommendations. The values of the cutting forces are shown in the graphs, Figure 3. The purpose of this work was to establish the influence of the geometric parameters of the processed tooth on the cutting force, so figure 3 shows graphs of the change in the cutting force calculated as the root of the sum of the squares of the axial components of the cutting force. The expression for calculating the cutting force is represented by the dependence in equation 1.

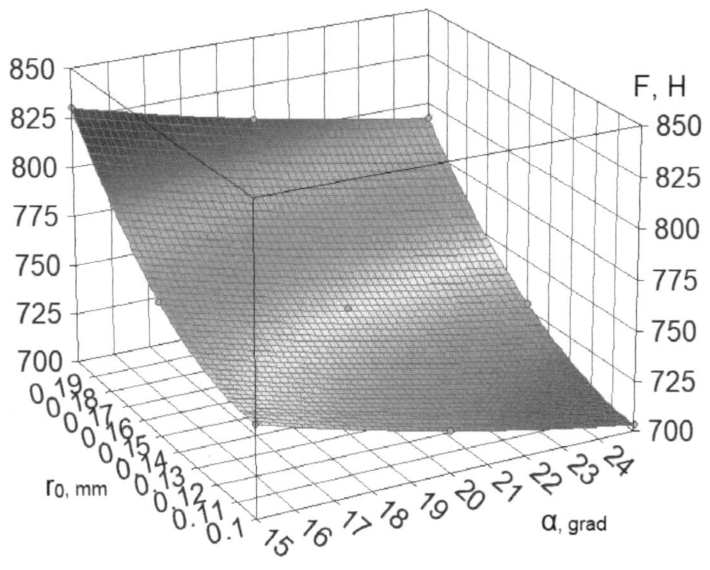

Figure 3. Graph of change in cutting force.

$$F = \frac{1047.5 - 7.4x + 2111.2y + 1065.9y^2}{1 + 6.5y - 17.3y^2}. \quad (1)$$

The relationship between the cutting force and the radius of rounding of the tip of the cutter can be explained by the fact that with increasing radius, the length of the cutting edge involved in cutting increases. Its significant influence can be attributed to the fact that the process of processing a tooth with a relatively small height (3 mm) was modeled, it can be assumed that with an increase in the height of the tooth, the influence of this parameter will decrease.

According to the study, when the profile angle decreases, the cutting force increases. This can be explained by the fact that when the angle decreases, the cutting conditions deteriorate, and chip removal is blocked.

Thus, the results of the simulation show that, depending on the change in the geometry of the processed wheel, the physical characteristics of the cutting process also change. In this case, changing the tooth geometry in the recommended range allows you to reduce the cutting force by more than 10 % under the same processing conditions. Of course, in order to establish the exact qualitative and quantitative characteristics of the influence of tooth geometry on the cutting process, it is necessary to conduct a set of additional studies, but already at this stage it can be argued that taking into account the technological conditions when designing new gear profiles will increase the processing efficiency.

The proof that the profile geometry of the bevel wheel teeth has a significant impact on the practical results of processing means that it is possible to determine the rational profile parameters at which it is possible to simultaneously achieve: the required quality parameters of the product, the durability of the cutting tool and the reduction of the main time. This will improve the processing efficiency of bevel gears, as well as provide practical recommendations for optimizing the profile according to technological parameters.

Acknowledgments
The reported study was funded by RFBR, project number 19-38-90126.

References

[1] Bouzakis K, Lili E, Michailidis N and Friderikos O 2008 Manufacturing of cylindrical gears by generating cutting processes: A critical synthesis of analysis methods *CIRP Annals-Manufacturing Technology* **57** 676–96
[2] Brecher C, Klocke F, Brumm M and Hardjosuwito A 2013 Simulation based model for tool life prediction in bevel gear cutting *Production Engineering* **7** 223–31
[3] Klocke F, Brumm M and Herzhoff S 2012 Influence of Gear Design on Tool Load in Bevel Gear Cutting *Procedia CIRP* **1** 66–71
[4] Arrazola P et al 2013 Recent advances in modelling of metal machining processes *CIRP Annals* **62** 695-718
[5] Berglind L, Plakhotnik D and Ozturk E 2017 Discrete cutting force model for 5-axis milling with arbitrary engagement and feed direction *Procedia CIRP* **58** 445-50
[6] Altintas Y and Merdol S 2007 Virtual High Performance Milling *CIRP Annals – Manufacturing Technology* **56** 81–5
[7] Fang N and Jawahir I 2002 An analytical predictive model and experimental validation for machining with grooved tools incorporating the effects of strains, strain-rates, and temperatures *CIRP Annals-Manufacturing Technology* **51** 83-6
[8] Jin X and Altintas Y 2012 Prediction of micro-milling forces with finite element method *Journal of Materials Processing Technology* **212** 542-52
[9] Kanatnikov N, Kanatnikova P and Pickalov D 2018 Thermal Processes Simulation during Processing Bevel Gears *Solid State Phenomena 284* 495–500
[10] Grigorev S et al 2020 Influence of Tool Design on the Cutting Force in the Machining of Noninvolute Gears *Russian Engineering Research* **40** 316–9

IOP Conf. Series: Materials Science and Engineering 1155 (2021) 012027 doi:10.1088/1757-899X/1155/1/012027

Classification of mechanical engineering products at the stages of their life cycle

B M Bazrov

Blagonravov Institute of Mechanical Engineering, Russian Academy of Sciences, Moscow, Russia

E-mail: modul_lab@mail.ru

Abstract. The article analyzes the classification of mechanical engineering products; it is shown that in the overwhelming majority of classifications, products are considered as objects of operation and there are practically no classifications of products as objects of production (except for the classification of parts) and objects of disposal, which negatively affects the efficiency of machine-building production. A methodological approach is proposed to construct a classification of products as objects of operation, items of production and disposal objects.

Product classification plays an important role in mechanical engineering at all stages of their life cycle: operation, production, and disposal.

Thus, the same set of products should have different classifications depending on the stage of their life cycle.

The classification of products, their elements, as objects of operation, is widely used for their unification, standardization, and creation of a database.

The classification of products and their elements as items of production allows, during the technological preparation of production, to develop standard technological processes, the element base of the means for technological support of products, to carry out their unification, etc.

The classification of products and their elements at the disposal stage can be used in a similar way.

Analysis of product classifications [1, 2] shows that in the overwhelming majority of cases, when classifying products, they are considered as objects of operation (except for the classification of parts) and there is practically no classification of products as items of production and disposal of objects.

The lack of product classifications such as items of production and disposal objects significantly reduce the efficiency of technological preparation for the production and products disposal.

This shows the need to build three product classifications of the same set of products at the stages of their life cycle: operation, production and disposal.

In connection with the abovementioned, it seems relevant to develop a methodological approach to building a classification of products at each stage of their life cycle.

Each stage of the product life cycle imposes its own requirements on the product, depending on the role it performs.

These requirements are represented by the corresponding characteristics of the product. They, when constructing a classification, should act as distinctive features.

The problem of the product classification, regardless of the stage of its life cycle, lies in their huge, continuously growing variety and a large number of characteristics of product designs. This problem, which appears during the operation phase, is further exacerbated during the production and disposal phases due to the wide variety of their technologies. In these conditions, it is necessary to abandon the promising hierarchical method of classification, using the facet method.

This paper proposes a solution to the problem of classifying products at different stages of the product life cycle using a hierarchical method.

Since the requirements for a product are different at different stages of its life cycle, then one should expect different classifications. These differences are manifested in a different composition of distinctive features, their sequence, as well as dividing the ranges of extreme values of characteristics included in different classifications of life cycle stages into a different number of ranges.

Let us consider the construction of the classification of products as objects of operation, items of production and disposal objects using the example of the classification of parts.

The construction of classifications of parts should begin with the determining the characteristics of the parts directly involved in the processes at each stage of the product life cycle and the subsequent construction of classifiers. The information on the characteristics of the part is given on the drawing. These characteristics include the geometry of the design of the part, dimensions, requirements for accuracy, roughness, and hardness of the material. The listed characteristics of the parts affect the processes of the stages of its life cycle in different ways.

At the stage of operation, the characteristics of the part are determined by the functions that are performed by the part in the product during its operation.

The part has two main functions:

- the role of the base part on which other parts are mounted; an example is the machine bed;
- direct participation in the work process performed by the part.

For example, a gear in a gearbox is designed to transmit torque. Each function performed by a part imposes corresponding requirements on it. The role of the base part imposes requirements on the geometry of the part structure, dimensions, accuracy, roughness, rigidity, and the part's participation in the working process additionally imposes requirements on strength, wear resistance, etc. The listed characteristics should be reflected in the form of distinctive features when constructing the classification of parts at the stage of operation ...

According to the abovementioned, the following characteristics are taken as distinguishing features of the classification of parts at the operation stage: the type of geometry of the part structure, dimensions, accuracy, surface roughness, material hardness, which largely determine such qualitative characteristics as rigidity, strength, wear resistance, etc. affects not only the quality of the basing, but also the rigidity, strength of the part; accuracy and roughness affect the quality of basing, fit when joining parts, workflow; hardness affects the strength, rigidity of the structure.

At the manufacturing stage, the characteristics of a part that affect the technology of its manufacture include the type of geometry of the structure, dimensions, accuracy, roughness, and machinability of the material. The geometry of the part structure influences the choice of machines that ensure the manufacture of all surfaces of the part; the characteristics of accuracy and roughness determine the choice of the level of accuracy of machine tools, and the dimensions determine the overall dimensions of the latter; the characteristics of the hardness of the material of the part and its machinability determine the required processing methods.

At the disposal stage, the disposal technology is influenced by the type of geometry of the part structure, determining the devices for its installation, the dimensions of the part, which affect the choice of technological equipment and the hardness of the material of the part, which affect the choice of the method of destruction of the part.

It should be noted that when classifying products, only the part was considered both as an object of operation and as an item of production.

Consider the structure of the technological classification of parts. The most common technological classification of parts is the classification given in the works [3, 4]. Its technological classifier of parts is a systematized set of tables of names of common features of parts. The structure of a complete technological classifier of parts, containing fourteen characters, has two sections: a constant of six characters and a variable of eight characters.

The constant section of the classifier reflects the dimensional characteristics of the part (three characters), material groups (two characters), and the technological method (one character) of the part.

The variable section of the technological classifier of parts takes into account the type of the primary blank (two characters), accuracy degree (two characters) and roughness, the degree of accuracy, the type of additional processing and weight (one character each).

The considered technological classification of parts has significant drawbacks.

From the listed characteristics, it is already clear that the technological classification of parts contains not only the characteristics of the part's structures, but also the technology for its production, which is a significant disadvantage, since the technology depends on many factors, including production conditions.

Another disadvantage is that with the help of the technological classification of parts, not only an approximate description of the design of the part is obtained, but also an ambiguous distribution of parts by classes.

Indeed, according to the unified system of design documentation (ESKD) [2], the classifier contains 100 classes, of which 51 classes are a reserve: classes 71-76 cover details of all industries. Their analysis shows that the distinctive features by which the parts are divided into classes do not always allow to do it unambiguously. For example, class 71 is rotation bodies such as wheels, disks, pulleys, etc.; class 76 is parts of technological equipment, tools, but they also can include parts like rotation bodies.

The main reason for the noted shortcomings of the technological classification of parts is that the part is considered as an object of operation, and not as a production item.

In connection with the abovementioned, it seems relevant to further develop the technological classification of parts.

Let us consider a methodological approach to creating a classification of parts at each stage of their life cycle.

As distinctive features at all stages of the product life cycle, we will take the following characteristics of the part: the type of geometry of the design of the part, dimensions, accuracy and hardness of the material.

The roughness of the surface of the part is not included in the listed characteristics, since in the overwhelming majority of cases it is associated with the accuracy of the part.

When creating the classification of parts at all stages of their life cycle, we will accept the distinctive features, their sequence, according to the above list of characteristics.

Under these conditions, the main difference in the classifications of parts at different stages of the life cycle will be the difference in dividing the ranges of the extreme values of each characteristic into a different number of ranges and their values.

It happens because of the different influence of characteristics on the processes of different stages of the product life cycle. For example, a change in the accuracy of a part by 2-3 accuracy degrees does not require significant changes in the manufacturing technology of parts, but these changes in accuracy during the operation stage may turn out to be unacceptable.

This is the reason why the range of the precision value of the part during an operation will be divided into more bands.

After determining the composition of the distinctive features and their sequence, it is required to divide the ranges of the extreme values of the distinctive features by the number of bands (groups) and their values. The greatest difficulties in dividing the range of characteristic values into bands arise when the characteristic is not described by a number. Geometry of the part structure is an example of such characteristic. For instance, a part can be a rotation object, a body, a lever, and so on.

If you try to solve this problem, taking into account each version of the geometry of the structure of the part, then there will be countless of them. Therefore, they must be grouped according to a certain similarity criterion.

In this case, the problem arises of determining the similarity criterion when dividing the types of geometry of the structure of parts into groups. But at the same time, at each stage of the life cycle of the part, due to the difference in their processes, the similarity criterion may be different.

At the operation stage, the determination of the similarity criterion should be based on the influence of the geometry of the structure on the operation process; in the production stage - the effect on the manufacturing technology, and during disposal, their influence on the recycling technology should be taken into account.

Another challenge is the selection of the dimensional characteristics of the part as a distinctive feature.

The dimensions of the part are divided into:

- the dimensions of the surface;
- coordinating dimensions that determine the relative position of surfaces;
- overall dimensions [5, 6].

There is a standard "Basic norms of interchangeability of A&A system. Adjustment fields and recommended allowances"[7], in which the dimensions are divided into bands, where the part is considered as an object of operation.

When dividing sizes into bands, the small size area is divided into a larger number of small size bands. As the size increases, the number of bands decreases and their values increase.

When determining the criterion for dividing the range of accuracy values into bands, the problem of taking into account the difference in accuracy rates for dimensions, relative position of surfaces and geometric shape of surfaces arises.

To solve this problem, we will use the average value of the accuracy degree.

When dividing the range of extreme values of accuracy of the part at the production stage, the size of the bands should be different, from high accuracy to the fifth accuracy degree. Bands should cover one accuracy degree at a time, starting from the fifth accuracy degree, bands should cover two or more degrees, since their achievement in the production does not require a significant change in technology.

Similarly, the classification of parts as of disposal objects should be built, where the geometry of the design of the part, the overall dimensions and the hardness of the material of the part are taken as distinctive features.

Additionally, the distinguishing features of the classification should take into account the characteristics of the process at the stages of the product life cycle [8].

As a conclusion we can state as follows:

- analysis of the classification of products showed that, in general, in the classifications, products act as objects of operation and there are practically no classifications of products as an item of production (except for the classification of parts) and classification of products as of disposal objects;
- analysis of technological classifications of parts showed their significant drawbacks, which reduce the effectiveness of their use;
- a methodological approach is proposed in building the classification of products at the stages of their life cycle, including determining the composition of distinctive features and dividing their values by the number of bands and their sizes;
- the creation of the classification at the stages of the life cycle is shown on the example of a part.

References

[1] Artobolevsky I I 1988 *Theory of mechanisms and machines* (Moscow: Nauka) p 640

[2] Classifier ESKD Classes 71, 72, 73, 74, 75 *Illustrated determinant of parts* 1988 (Moscow: Publishing house of standards) p 401

[3] *Technological classifier of mechanical engineering and instrument-making parts* 1988 (Moscow: Publishing house of standards) p 256
[4] *All-Russian technological classifier of mechanical engineering parts OKO21* 1995 (Moscow: Publishing house of standards) p 264
[5] Amirov F G 2012 Classification of parts by size, method of constructing a coordinate system of a part for geometric modeling *Mechanical engineering* **8** 32-5
[6] Volchkevich I L and Galiy V V 2015 On the problems of classifying parts at the early stages of designing technological complexes *Izvestia of the Tula State University Engineering Science* **10** 100-7
[7] GOST 25347-82 *Basic standards of ESDP interchangeability Tolerance fields and recommended landings* (approved and put into effect by the Decree of the USSR State Committee for Standards dated July 21, 1982 No. 2764) p 53
[8] Fetisov K V and Maksimov P V 2019 On the approach to the classification of parts for additive manufacturing *Aerospace engineering, high technologies and innovations* **2** 197-200

The layout method is an effective method for creating a product design

B M Bazrov and N A Rodionova

Blagonravov Institute of Mechanical Engineering, Russian Academy of Sciences, Moscow, Russia

E-mail: modul_lab@mail.ru

Abstract. The advantages of creating a product design by the layout method are shown. An example of using the layout method when creating modular machines is given. It is noted that the effectiveness of using the layout method largely depends on the quality of the element base; the characteristics of the element base are given. A methodological approach to the construction of element bases is proposed, including the division of the element base into two parts: the base of the main elements and the base of the auxiliary elements.

When creating products, the following methods of a diverse nature are used [1, 2, 3].

The method of succession is the gradual development of a design by introducing into it new or additional parts, units, assemblies to replace obsolete or those which don't meet modern requirements, in order to change the previous characteristics of the product. The method is based on improving the existing product design.

The standardization method is the creation of a structure based on the use of standard parts and units, elements with standard parameters. This allows to use already developed technical documentation.

The method of unification is the elimination of unnecessary diversity by reducing the list of permissible elements and solutions, bringing them to the same type, multiple use in the design of the same parts and units.

The basic unit method is the production of various products, united by the presence of the common, basic part. Typically, this element is the most complex part of future products. The development of the base part is carried out in such a way that, by attaching additional parts to it, it would be possible to create products with a changed appearance, the number of functions performed, and characteristics quite simply and quickly.

Modification method - alteration of a product in order to adapt it to new requirements, working conditions, technological process (method of manufacturing and assembly) without changing the most expensive and critical parts in it.

The inversion method is the creation of a new design based on changes in the functions, shapes or positions of parts of an existing product. For example, to replace the tension spring with a compression spring, to make the convex surface concave. The layout method is the design of the product structure by combining the elements that make up the element base.

Its application begins with establishing the type of the product with its layout diagram, and the elements corresponding to it. The layout of the product design is drawn up in the form of a preliminary assembly drawing of the product design. After revising the project, the final technical documentation for the product and its specification is drawn up.

This method of creating a product has found wide application in mechanical engineering. Almost the design process of any product includes an element of its layout. In cases where the proportion of the assembly process dominates the entire product design, the product can be considered to be designed by the layout method.

Layout is a complex and responsible design task that takes a lot of time. One has to analyse several possible options and choose the best. In this case, the layout is a stage that largely determines the quality of the product.

The construction of products by the layout method allows you to significantly increase the efficiency of product creation by reducing the cost of product design, design time, costs of manufacturing the product and testing the design of the product on the first copy, contributes to the automation of the design process, facilitates the repair of products, and expands the unification and standardization of elements.

It should be noted another significant advantage of creating a product by the layout method, which is called reversibility. It allows, after the failure of the product, to reuse elements that have retained their quality.

The layout method is based on the element base, the quality of which largely determines the effectiveness of its application.

Modular machines [4, 5, 6] are an example of creating products by the layout method.

Modular machines are those which are assembled from unified and partially special units and parts by combining them into a single assembly [1, 2, 3].

The layout of modular machines is carried out by forming the structure of modular machines using normalized units (figure 1).

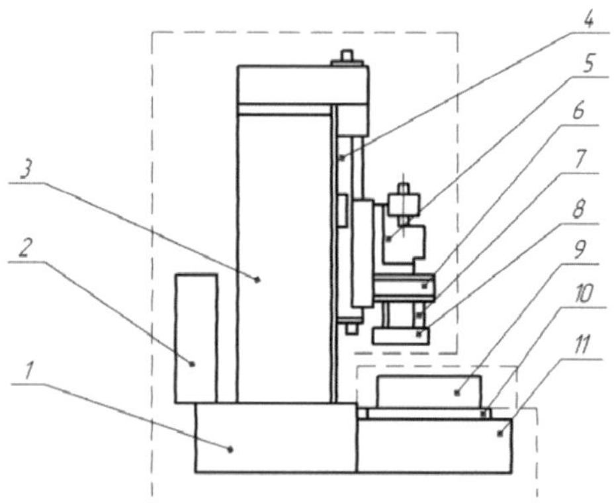

Figure 1. Layout of a single column modular machine with rotary table. 1 - side frame, 2 - electrical cabinet, 3 – rack, 4 - power table, 5 - power headstock, 6 - spindle box, 7 - instrumental, adjustment, 8 - bush plate, 9 –clamping device clamping device, 10 - rotary table, 11 - medium bed.

Such a design of modular machines implies the presence of the element base.

Elements of modular machines consist of four main groups of unified units: power units (power heads and tables), spindle units (power heads and spindle boxes), base units (bed, column, thrust squares), transport units (dividing tables).

Success in the creation of modular machines is determined by the presence of the listed elements in the base.

Thus, the creation of modular machines turns into their layout from normalized units. The technical documentation is known and is kept in the database.

When forming the element base of modular machines, several units of the same type (assemblies) of different sizes (called normalized or unified) are developed, which make it possible to design a machine that quite well corresponds to a given technological process of manufacturing a part.

Unified or normalized units of modular machines are developed before a particular machine is designed. These units can be used in machines of various designs.

Thus, the efficiency of building products by the layout method largely depends on the quality of the element base.

Taking into account the advantages of the layout method, it should be used whenever possible when creating all products. This requires the organization of the element base of products.

However, the unlimited variety of products for service purposes gives rise to an unlimited variety of product designs. It creates the problem of creating a single element base on the scale of mechanical engineering.

This is the reason why in future we will consider the problem of creating element bases at the enterprise level, when each enterprise forms its element base.

In this regard, an urgent task is to develop a methodological approach to the construction of element bases for products for various purposes.

The element base of products is a collection of units and parts required to create a product design. The element base is characterized by the range of elements, their quality, the level of element description, completeness and openness. The range of elements can be divided into two groups: the main elements directly involved into the performance of the product for its service purpose and auxiliary elements of general purpose, such as screws, nuts, bearings, pumps, and so on. It is the first group of elements that determines the difference between products for different purposes.

Another important characteristic of the element base is the quality of the elements, the indicators of which may be different depending on the purpose of the elements. At the same time, the quality element will be the element that corresponds to the modern achievements of science and technology. The completeness of the element base is characterized by the level of coverage of the range of product design elements. The completeness of the element base should cover, if possible, the entire range of structural elements of products for this purpose. Determining the completeness of the element base is also a difficult task, since there is inaccessibility of information about many elements.

Also, an important characteristic of the element base is the level of description of the designs of product elements. In some cases, the same elements can be presented in the form of sketches, and in another case - in the form of drawings with comments. The element base may include design options for the same elements.

The element base should be open, since its state changes over time due to obsolescence of elements that should be replaced with modern elements of a higher quality, as well as the introduction of elements of new structures. Therefore, should control over the state of the element base in order to introduce new elements on time.

Taking into account the abovementioned, we will consider the construction of element bases at enterprises.

The enterprise must organize two element bases: one containing the main elements and the other containing auxiliary elements.

The first element base may be different depending on the level of production of products. For example, the element base of vehicles at the level of the mechanical engineering sub-industry will include elements of aircrafts, ships, cars, motorcycles and other vehicles, and the element base of cars will include only elements that are part of the design of cars.

As the analysis of the elemental bases of various products shows, they mainly contain the main elements that directly ensure the performance of the products for their service purpose. For example, the element base of cars contains, first of all, bodies, cabins, chassis, engines, etc.

Element bases are formed according to the results of statistical studies of the frequency of use of the marked elements, which make it possible to establish the elements most often found in products and include them in the element base. Such formation of elemental bases significantly reduces their quality and efficiency of use.

In connection with the abovementioned, the following is proposed.

Firstly, all the elements of the product. should be included into both element bases.

Secondly, element bases should be formed before organizing the production.

Thirdly, all elements of the first and second groups should be included into the element bases.

Over time and as the design of the product develops, the composition of the element base should change by eliminating obsolete elements and including new elements.

Another significant drawback of element basis is the absence of auxiliary elements of the product design. For each type of product, the corresponding auxiliary elements are typical. The lack of an element base of auxiliary elements significantly reduces the efficiency of the layout method, since the designer has to spend time looking for them in other sources of information, and in the absence of them, offer his solutions.

So, the element base of any group of elements should contain a range of elements, as well as options for elements of the same name. Further, in the element base, different values of the characteristics of each element should be reflected, for example, dimensions, level of accuracy, and others. The construction of element bases should begin with the development of a classification system, including the classification of the names of the elements and the classification of each element. The presence of a classification of elements allows them to be encoded, which is necessary for creation an information retrieval system.

When organizing element bases, difficulties arise with the construction of the classification of elements, due to a wide variety of their designs.

In this regard, it is proposed to use the description of the product design in a modular representation [7, 8]. Then the functional technological module is taken as the main elements, and the functional serving modules, ensuring the functioning of the first modules, are taken as auxiliary elements.

For example, a screw-cutting lathe performs its service purpose with three functional technological modules: a chuck, a tailstock and a cutting tool holder.

Functional service modules include a gearbox, an electric motor, etc.

So, the conclusions are as follows:

- Application of the layout method increases the efficiency of product creation by reducing the cost of designing and production.
- The effectiveness of using the layout method largely depends on the quality of the element base.
- The element base should consist of two parts: the base of basic elements and the base of auxiliary elements.
- The element base is characterized by the range of elements, their quality, completeness and openness.

References

[1] Orlov P I 1988 *Fundamentals of Design* (Moscow: Mechanical Engineering) p 560
[2] Dietrich I 1981 *Design and construction* (Moscow: Mir) p 456
[3] Goldschmidt M G 2007 *Design Methodology* (Tomsk: TPU) p 173
[4] Feofanov A N and Manlyuk S Yu 2016 Development of a system for geometric modeling of the layouts of modular machine tools *Scientific visualization* **3(7)** 132-44
[5] Timofeev Yu V, Khitsian V D, Vaserman M S and Gromov V V 1985 *Modular machines of medium and small sizes* (Moscow: Mashinostroenie) p 248
[6] Feofanov A N and Potapov A V 2009 Development of a database of unified units of modular machines *Bulletin of the Moscow State Technical University Stankin* **3** 8-12
[7] Bazrov B M 2001 *Modular technology in mechanical engineering* (Moscow: Mechanical engineering) p 368
[8] Bazrov B M and Kheifets M L 2019 Product description with a structured set of modules *Proceedings of the X International scientific conference* (Kemerovo) pp 35-40

IOP Conf. Series: Materials Science and Engineering 1155 (2021) 012029 doi:10.1088/1757-899X/1155/1/012029

Review of previous Russian studies in the abrasive flow machining

V A Levko[1], O V Litovka, D I Savin and N S Terjaev

Reshetnev Siberian State University of Science and Technology, 31, Krasnoyarsky Rabochy Av., Krasnoyarsk, 660037, Russian Federation

[1]E-mail: levkosaa@mail.ru

Abstract. Abrasive flow machining (abrasive flow machining (AFM)) is increasingly used worldwide. It is used for finishing and removing surface stresses in parts, the finishing machining of which using traditional methods is difficult. By now there are a large number of publications on this topic. A special interest are the articles in which previous works on various aspects of this technology are systematized. Due to a number of reasons, there are no publications from the scientific periodicals of Russia. This article presents a review of the works of Russian researchers in modelling the machining process by abrasive flow. They are aimed at modelling the rheological characteristics of the medium and contact interactions in processing. Particular attention is paid to the machining of an abrasive flow with an aligning device, as well as parts from powder molybdenum. The purpose of this article is to acquaint the wide range of readers with the works of Russian studies in the treatment of the abrasive flow of high viscosity medium.

1. Introduction

The modern level of industrial production led to the widespread use of parts with a complex geometric shape. As a rule, high-alloyed steel, titanium alloys, cobalt, nickel, chromium and molybdenum are used as materials. The surface layer of such parts makes high requirements for the quality of the surface layer, but their finishing processing is difficult in traditional ways.

Therefore, as such a finishing operation, such an unconventional finish-treatment method is becoming increasingly used as treatment with an abrasive stream [1].

The basis of AFM underlies the flow in the processed part under the high pressure of the viscoelastic material filled with fine abrasive particles. The stream of abrasive material (medium) takes the shape of the channel being processed and has a compressive pressure on the surface being processed, providing abrasive contact [2]. To apply this technology in the finishing processing of specific parts, it is necessary to carry out a large amount of experimental studies. This is due to the fact that by the present time the generally accepted theoretical foundations of the machining process by the abrasive flow is not created. With abrasive contact, there is a number of physical phenomena that affect the quality and performance of AFM.

To date, a large amount of experimental and theoretical studies in this area has been carried out in the world. The systematization of such works is presented in the critical review of past studies and achievements in the process of finishing by abrasive flow [3]. This work caused great interest among readers. However, it does not consider publications under the subject matter in the Russian scientific periodical.

The first open publication in this area in the USSR appeared in 1980 [4]. Since then, in 40 years, a number of patents and scientific articles have appeared in Russia. However, almost all of these works are hard to reach for English-speaking readers. The main reason is that the practically all articles are published in Russian. The second reason is that in Russia the machining process with an abrasive flow is more known as abrasive-extrusion processing. This name is used as a keyword in most works, including translated into English. The purpose of this article is an addition to the review [3] analysis of Russian authors whose works were published in leading scientific journals of Russia.

2. Review of published works on modelling the machining by abrasive flow in Russia

Of the four types of machining by abrasive flow in Russia, bidirectional treatment with abrasive stream (Bidirectional AFM) received the greatest distribution. In the Soviet Union, 15 patents were recorded on the device for processing with an abrasive flow, 10 patents on methods for machining by an abrasive flow and 2 patents for the composition of the medium. As a polymer base of medium, synthetic silicone rubber SCT is used [5]. Environments based on this rubber belong to high viscosity environments.

The first publications were devoted to the development of recommendations for the processing of specific details. The basis of the recommendation was the resulting empirical dependencies that bind the pressure of the medium, the grit and the volume of the abrasive medium with the roughness of the treated surface Ra and the removal of the material [6], as well as the change in the stress state of the surface layer [7]. The empirical dependences of the AFM process made it possible to ensure the accuracy of the flow of fuel components through the channel channels [8], as well as the basis of the finishing technology of the profile holes [9]. Recommendations on the choice of equipment, the composition of the medium and processing parameters are proposed [10].

The critical analysis of the current state of this technology has shown that the absence of the theoretical foundations of the process built on the basis of generally accepted representations in the field of tribology and rheology does not allow to assess the boundaries of the applicability of the method from a scientific point of view [11].

2.1. Rheological properties of the work environment

To simulate the stream of the medium during machining by the abrasive flow, it is necessary to know the numerical values of the coefficients characterizing its rheological properties.

The technique of experimental determination of viscosity coefficients, elasticity and plasticity of the AFM medium [12] was proposed. It is based on the experimental determination of the numerical values of the maximum flow rate of the ω_{max} medium and the pressure drop ΔP on the section of the cylindrical channel L and diameter d. The device for the study of viscous and plastic properties of the medium is shown in figure 1. The following is described in the paper and a device and a device to determine the numerical values of the Young E module and the Poisson ratio μ of the high viscosity medium for machining by the abrasive flow [12].

The experimentally numerical values of the maximum flow rate of the medium and the calculated values of the effective viscosity of the medium are used to determine the values of the sliding velocity γ' of the medium rate. Further, using known formulas for the rheology of viscoelastic liquids, the values of tangent stresses τ and normal stresses σ, arising in the medium flow in the circular channel of the constant cross section were calculated [13].

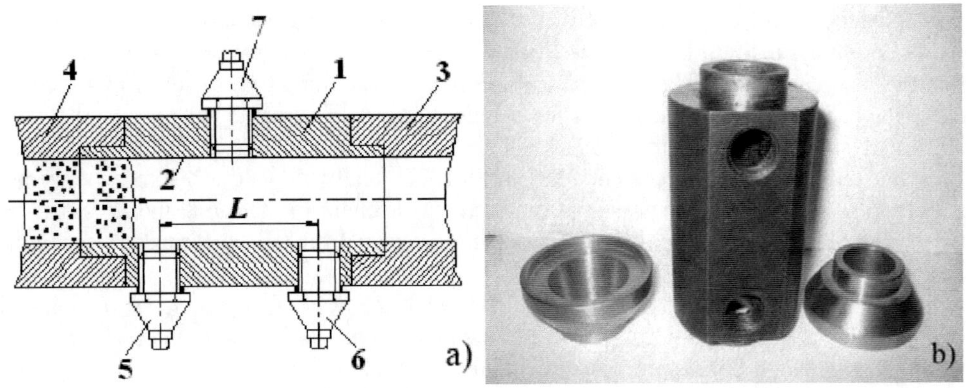

Figure 1. Device for the study of viscoelastic properties of the environment: a) scheme b) photo: 1 - body; 2 - channel with a diameter of 25 mm; 3 and 4 - UES-25 installation cylinders; 5 and 6 - pressure sensors; 7 - temperature sensor [12].

According to the [11] procedure for the shear flow of the medium filled with an abrasive value of 320 µm, at a pressure $P = 6$ MPa in the cylindrical channel $R = 0{,}0125$ *m* radius and length $L = 0{,}07$ *m* the maximum flow rate of 0.0204 m / s is determined, the value of an effective viscosity of 36900 *Па с* was obtained. Next, the rate of shear $\gamma' = 7{,}68$ c^{-1}, the value of tangent stresses τ = 283331 *Па*, the size of normal voltages $\sigma = 250100$ *Па,* the value of the components of the cutting forces $Fa = 2{,}28$ H and $Fr = 2{,}0$ H [13] were calculated.

To study the effect of the shape of the processable channel, visual studies of the abrasive flow in 28 channels were carried out. To perform them, a technique and a device has been developed, which allows you to fix the stream of medium through a window of organic glass with a thickness of 20 mm [14]. Before visual studies on the surface of the medium placed in the device, a rectangular mesh with a size of 15x15 mm and a depth of up to 2 mm, which was filled with a white electrocorundum F70, was applied by template. A white grid has a good contrast with the surface of the medium, which ensured the desired image contrast. During the studies, the degree of influence of the shape of the processed channel and local resistances on the features of the flow is established. The features of the flow are shown in the form of a series of photographs of the flow for each sample [14].

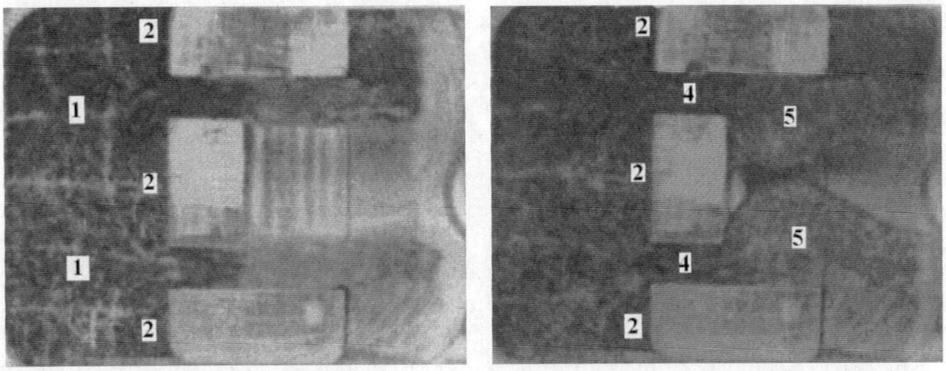

Figure 2. The flow process in the sample is the wall at the entrance [14].

So, for the flow in the channel (figure 2), the displacement of the stream of medium 4 is characterized from the central axis of the channel. At the same time, at the entrance to the channel, there are not two, but three zones of stagnation 2. Two at the entrance to the canal and one before the partition. Over the partition, the flow separation from the wall of the channel being processed is observed. Due to the effect of elastic recovery, there is a gradual filling of the entire volume of the channel under study, including for the partition. After filling this section of the channel, the medium in it occurs a stagnant zone. The main part of the flow slides along the border of this zone. Only the surface of the partition is processed, which is parallel to the main stream [14].

When analysing the movement of individual abrasive grains in the medium stream, it is established that they are moved along the current lines. With the steady flow, the distance between adjacent grains does not change. Thus, the assumption of an elastic chain formation in the stream is confirmed. It has been established that the grain in contact with the treated surface performs a rotational motion in the stream [14].

To simulate AFM channels of large length, a model of the flow of a viscoelastic medium filled with abrasive particles is proposed. Observed experimentally in the current line, the set of abrasive grains is presented as a chain consisting of sequentially connected identical elements. Each element is deformed regardless of the rest, and the continuity of the chain is ensured by the compound of the segments of the viscoelastic medium in the balls that simulate abrasive particles. Processing conditions are determined by the stress-strain state of the medium. The magnitude of the deformation depends on the flow rate gradient on the channel wall. It was revealed that with the steady flow, the ratio between normal and tangent stresses is stored along the entire length of the channel being processed [15].

To study the change in the flow pressure of the medium along the length of the channel being processed, a diagram is applied with a pressure registration with four sensors. The results of the study showed that the change in pressure along the length of the channel with the steady flow is insignificant and practically occurs simultaneously along the entire length of the channel [16].

In the study of the medium composition effect on the processing modes by the abrasive flow, it is established that at present, the medium is filled with an abrasive of one grain. In such media, during deformation, abrasive grains form linear chains moving only in their current lines along the entire length of the channel being processed. In this case, with the steady flow mode, the elastic voltage occurs between the current lines with a different gradient of the flow rate and on the surface of the channel being processed [17].

It is proposed to apply media filled with abrasive grains of several grain. They form a different spatial structure. In this case, larger grains play the role of centres around which more numerous grains of smaller sizes are grouped. In this case, abrasive balls occur in the medium, and the resulting abrasive linear chains have a smaller length. In this case, the number of abrasive chains increases. The dependence of the percentage in the medium of each abrasive grain is proposed depending on its value. Such environments were called polydisperse media [17].

2.2. *Contact interactions when machining abrasive flow*

One of the first in Russia was the experience of describing the contact interactions of abrasive grain with a treated surface based on the contact form of micro-grain microledge during microcutting. Contact options with a sharp cone, sphere, cone with a truncated vertex, a cone with a rounded vertex are considered. It is noted that the obtained formulas give only qualitative agreement with experimental data [18]. The creation of a model of motion of the abrasive grain at AFM will allow to calculate the intensity of the wear of the material at various combinations of all factors [19].

The description of contact processes at AFM from the point of view of tribology made it possible to attribute them to the processes of external friction of a rough and wavy body (billet) with a viscoelastic body (medium). At the same time, the abrasive grain itself in the stream of the medium is experiencing plastic contact. The microasperity of the treated surface at various levels of contact may experience elastic, elastoplastic, plastic deformations, or the microcutting process will be carried out [20]. For all types of contact, the formulas for calculating the actual area of contact of micron of abrasive grain and the treated surface are proposed. On their basis, the equations were proposed to determine the volume of the material being removed and the roughness of the treated surface [21].

When studying the energy processes in the cutting zone at AFM, the empirical dependence of the change in the temperature of the medium during treatment [22] is defined. The direct dependence of the thermal conductivity coefficients and the temperature of the medium is established on the degree of its filling with abrasive grains [23].

3. Overview of published works on the development of production processes based on abrasive flow machining

In Russia, the ability to process the abrasive flow of parts from aluminum alloys is confirmed. The empirical dependences of the scratch depth from the type of abrasive grain on the sample of aluminum alloy AMG-6 are determined. Recommendations are given on the choice of the composition of the medium and the mode of processing parts from aluminum alloys [24], including rectangular openings with a length of 10 mm and a cross section of 2 × 0.625 mm [25]. The possibility of using AFM is investigated to reduce surface stresses in the channels of parts after the electric erosive processing [26].

The technique of six stages is proposed, which allows optimizing the costs production of technological preparation [27]. An approach is proposed, in which for each stage of the life cycle of the AFM environment for AFM, a set of factors and response functions is determined. At the next stage of the life cycle, the response function of the previous stage becomes factors (source data). The life cycle includes five stages divided into thirteen steps. Based on the analysis of world experience for each stage, a set of factors is proposed to allow the characteristic of the processed part, the composition of the medium, processing modes, as well as contact interactions [28].

3.1. Machining with an abrasive flow of parts with a variable shape

Experimental machining of the abrasive flow of parts with a cylindrical and conical channel showed the unevenness of the removal of material and changes in roughness. Non-uniformity is caused by the restructuring of the flow profile. To ensure uniform processing of such channels, a device that align the channel cross-section by its length is proposed. In this case, there is a transition from the flow of medium through local resistance (cone) to the flow in the ring channel (figure 3) [29]. To confirm the experimental results and processing modelling, a device is made (figure 4) [30].

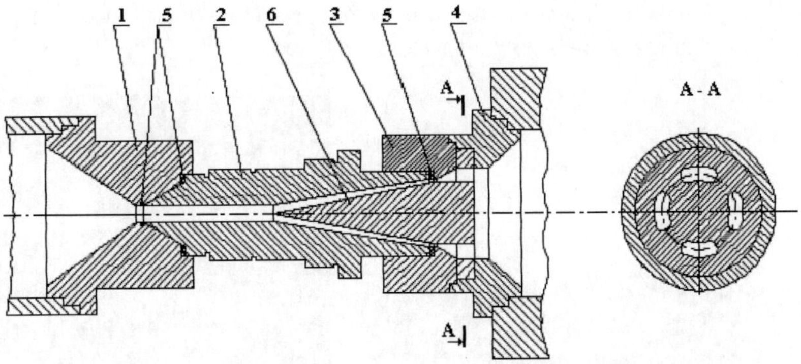

Figure 3. AFM scheme using an alignment device: 1 - upper adapter; 2 - detail; 3 - lower adapter; 4 - base; 5 - gaskets; 6 - an alignment device [29].

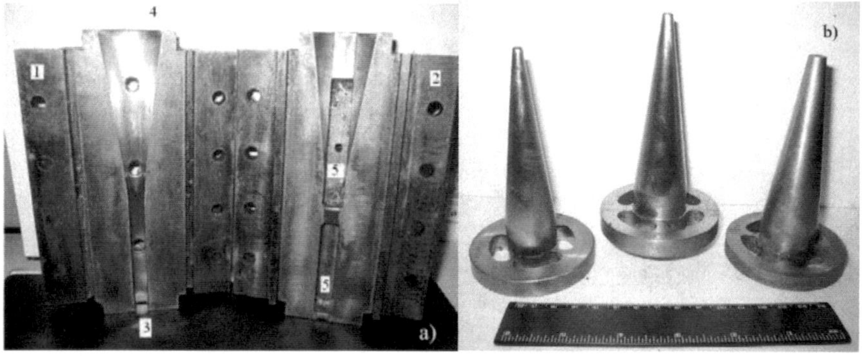

Figure 4. Device: a) 1 - cover; 2 - body; 3 - cylindrical part of the channel; 4 - conical part of the channel; 5 - grooves for removable samples; b) levelling devices that implement different cross-sectional area of the ring gap [30].

Figure 5. The original state of the sample surface before AFM: a) the general type of sample, b) an increase of 63 times [30].

Figure 5 shows a general view of the sample and the initial state of its surface before AFM. Figure 6 shows a general view of the sample and the roughness of its surface after 30 AFM cycles. The machining of the abrasive flow with the use of the levelling device made it possible to provide uniform roughness in the channels with a variable cross section.

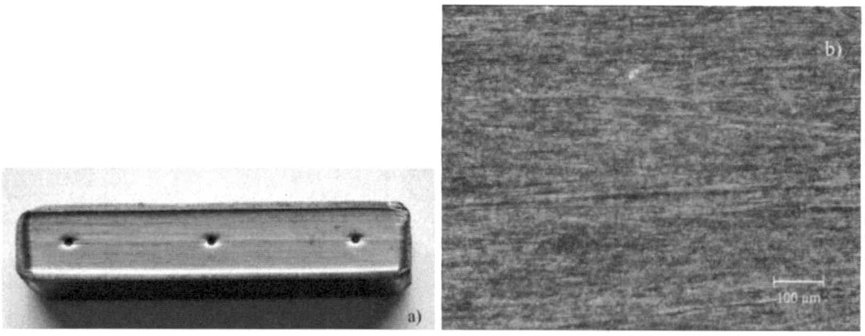

Figure 6. Status of the sample surface after 30 cycles of AFM: a) General view of the sample, b) an increase of 63 times [30].

3.2. *Machining by the abrasive flow of parts made of molybdenum alloys*

For finishing processing of billets from molybdenum powder alloys, the treatment powder with an abrasive flow of high-viscous medium is processed. To do this, the model for removing the material with a single abrasive grain is specified. In this model, the abrasive grain is a multi-season tool acting on the surface being processed, both by the actual contact area, and along the contour contact area.

Based on these features of the contact, the abrasive stream was treated in two stages of the billet from the powder molybdenum alloy. The initial state of the surface has a roughness of Ra = 1.6 μm [32].

Figure 7. The surface of the workpiece of molybdenum powder alloy after abrasive flow machining [32].

For processing at the first stage, a normal electrocorundum of graininess F40 and an extrusion pressure of 12 MPa is applied. It was possible to obtain the surface roughness Ra = 1.45 ... 1.5 μm. For AFM at the second stage, high-viscous medium is used, in which the grains of silicon carbide are used by the graininess of the F80. The magnitude of the extrusion pressure is 12 MPa. The roughness of the surface of the blank Ra = 1.3 ... 1.4 μm (figure 7) was obtained. Its texture is characterized by the absence of protrusions and wrappings, whose sizes would be much more average roughness. The possibility of the applicability of the Machining with an abrasive flow by high-viscous medium for finishing processing of billets from powder molybdenum [32] is experimentally confirmed.

4. Conclusions

The following conclusions can be drawn from the review presented in this article. In Russia, bidirectional abrasive flow machining received the greatest distribution. A number of devices, fixtures of the compositions of media and processing methods have been developed.

For successful AFM modelling in Russia, a number of studies on the rheology of the medium were conducted. Coefficients of effective viscosity, numerical values of the Young module and the Poisson rate for high viscosity medium are experimentally defined. Based on these data, the rate of shift, the value of tangent stresses and normal voltages of the high viscosity medium in the cylindrical channel is calculated.

Visual studies of the stream of medium in channels of various geometric shapes experimentally confirmed theoretical assumptions: the emergence of elastic (normal stresses) in the medium stream; the movement of abrasive particles in the stream of medium over the current lines to form an elastic (power) chain.

To simulate AFM channels of large length, a model of the flow of a viscoelastic medium filled with abrasive particles is proposed. It is proposed to use polydisperse media filled with abrasive grains of several grain. They are characterized by a more rigid structure.

For all types of contact, formulas are proposed to determine the actual contact area of micron of abrasive grain and the treated surface. On their basis, the equations are proposed to determine the volume of the material being removed and the roughness of the treated surface.

A new method of machining with an abrasive flow, which ensures uniformity of the removal of material and roughness in the conical channels is developed. To do this, an aligning device is applied, which provides the transition from the flow of medium through the local resistance (cone) to the flow in the annular slit.

The processing of the abrasive stream is successfully applied to the finishing of the part of the powder molybdenum.

References

[1] Rhoades L 1991 *Journal of Materials Processing Technology* 1991 **28(1-2)** 107-16
[2] McCarty R W 1970 *Method of honing by extruding* Pat. US No 3521412
[3] Petare A C and Jain N K 2018 *The International Journal of Advanced Manufacturing Technology* **97(1-4)** 741-82
[4] Popenko A I, Migunov V M, Shirkevich P V and Kovgan A I 1980 Pat. of the USSR No 738837
[5] Sysoev S K 1988 Pat. of the USSR No 1379063
[6] Sysoev A S, Sysoev S K and Lubnin M A 2002 *Mechanical Engineering Technology* **4** 28-32
[7] Sysoev A S, Sysoev S K and Lubnin M A 2003 Mechanical Engineering Technology 3 38
[8] Sysoev S K, Sysoev A S, Levko V A, Snetkov P A and Lubnin M A 2007 *Mechanical Engineering Technology* **6** 48-52
[9] Sysoev S K, Timchenko A I and Levko V A 1991 *Russian Engineering Research* **1** 65-67
[10] Sysoeva L P, Sysoev S K and Sysoev A S 2013 *Vestnik SibSAU* **2 (48)** 228-32
[11] Levko V A 2006 *Bulletin of the Tomsk Polytechnic University* **6 (309)** 125-29
[12] Snetkov P A, Levko V A, Pshenko E B and Lubnin M A 2009 *Vestnik SibSAU* **5(26)** 99-103
[13] Levko V A 2005 *Vestnik SibSAU* **4** 96-100
[14] Levko V A, Lubnin M A, Snetkov P A, Pshenko E B and Turilov D M 2009 *Vestnik SibSAU* **5(26)** 93-99
[15] Levko V A 2008 *I. Yakovlev Chuvash State Pedagogical University Bulletin* **5** 85-94
[16] Levko V A 2012 *Vestnik SibSAU* **4** 159-62
[17] Levko V A and Pshenko E B 2006 *Vestnik SibSAU* **4** 64-68
[18] Sysoev A S, Sysoev S K, Zverintseva L V and Alikin A S 2007 *Vestnik SibSAU* **2** 70-73
[19] Nazarov N G 2013 *Science & Education: Bauman Moscow State Technical University* **11** 257-68
[20] Levko V A 2008 *Metalloobrabotka* **3** 19-23
[21] Levko V A 2009 *Russian Aeronautics* **52(1)** 94-98
[22] Snetkov P A and Pshenko E B 2006 *Vestnik SibSAU* **6** 100-03
[23] Pshenko E B, Shestakov I Ya, Remizov I A and Veretnova T A 2019 *Siberian Journal of Science and Technology* **20(2)** 277-83
[24] Sysoeva L P 2015 *Bulletin of the South Ural State University Series: Mechanical engineering industry* **5(2)** 40-50
[25] Levko V A and Levko A A 2011 *Vestnik SibSAU* **4** 169-73
[26] Sysoeva L P and Sysoev A S 2013 *Russian Engineering Research* **33(1)** 51-52
[27] Levko V A, Lubnin M A, Sysoev S K, Sysoev A S and Sysoeva L P 2011 *Vestnik SibSAU* **4** 173-78
[28] Levko V A, Simakova A E and Savin D I 2017 *Siberian Journal of Science and Technology* **18(4)** 932-38
[29] Levko V A and Pshenko E B 2008 *Metalloobrabotka* **2** 7-10
[30] Levko V A, Snetkov P A and Pshenko E B 2011 *Vestnik SibSAU* **3** 140-45
[31] Levko V A and Pshenko E B 2012 Pat. of the Russian Federation No 2469832
[32] Levko V A, Savin D I and Litovka O V 2020 *Vestnik Irkutskogo gosudarstvennogo tehnicheskogo universiteta = Proceedings of Irkutsk State Technical University* **24(1)** 36-51

Model of analysis of the results of geometry scanning of turbine nozzles

E Yu Pechenina[1], V A Pechenin and M A Bolotov

Samara National Research University, 34, Moskovskoye shosse, Samara, 443086, Russia

[1]E-mail: ek-ko@list.ru

Abstract. The paper presents a mathematical model to perform computer calculation for open flow area of turbine nozzles. The model is used to work with objects in *.stl format when processing data after scanning. Using the developed model, calculations of the part quality parameters will not depend on the controller qualifications, and the proposed approach will not require unique adjustment, costly repair and storage of reference patterns for all sizes of nozzle devices. The calculation was performed in three sections, which improves accuracy while reducing labor intensity and cost of measurement. Increasing the number of sections practically does not affect the labor intensity of calculations, so there are no significant restrictions on performing virtual measurements in four or more sections. The model was implemented in the mathematical package MATLAB. The model was tested with the example of STL processing of the turbine nozzle blades sector.

1. Introduction

Turbine rotors and nozzles are particularly difficult to manufacture and assemble. Quality control of assemblies is also quite a time-consuming operation [1]. Currently, aircraft engine manufacturers are equipped with optical scanners to determine deviations of manufactured parts and assemblies from nominal models.

As a result of measuring parts and assemblies on optical scanners, faceted models in *.stl format appear, to which traditional CAD modeling tools cannot be applied, and the calculation of specific geometric parameters requires additional processing of the measured data. The paper presents a model for analyzing and calculating such a key turbine assembly parameter as the nozzle passage section area (PA).

The PA of the interblade channel is defined as the area of the surface area bounded by a contour, which consists of the curves of the blade feather cut faces forming the channel, and the transition radii from the feather to the flanges, which provide the smallest width cross section in the reamer. Ultimately, the assembly is interested in the irregularity of the PA of the channels in the entire nozzle apparatus, not in the absolute values of the PA of each channel. For this reason, PAs are changed in production by hand-held indicating devices as a deviation from the reference window, with individual parameters measured, allowing the calculation to be performed using an equation:

$$F_{\min} = K \cdot H \cdot (b_1 + b_2) / 2 , \quad (1)$$

being K – empirical correction coefficient, b_1 and b_2 – width of the measured interblade channel at distances R_1 and R_2 from the motor axis respectively, H – height of the interblade channel between b_1 and b_2.

Increasing the number of sections in equation (1) will increase the reliability of the calculations, but in manual measurement, it entails a significant increase in labor intensity. In the developed model, the PA calculation was performed in three sections and there are no significant restrictions on performing virtual measurements in four or more sections.

2. Model for calculating the cross-sectional area

The Stl model is a piecewise linear surface. It is characterized by the following parameters: $\mathbf{V}_{g\times3}$– matrix of coordinates of vertices of stl-model grid, $\mathbf{F}_{m\times3}$ – matrix of combinations of three vertices (rows contain ordinal numbers of vertices) that form facets of surfaces, $\mathbf{N}_{m\times3}$ – coordinate matrix of facet normals. The calculation model is designed for a pre-segmented model, that is, one in which there are areas of facet associations responsible for certain facets of the model. The stages of the model are shown in figure 1.

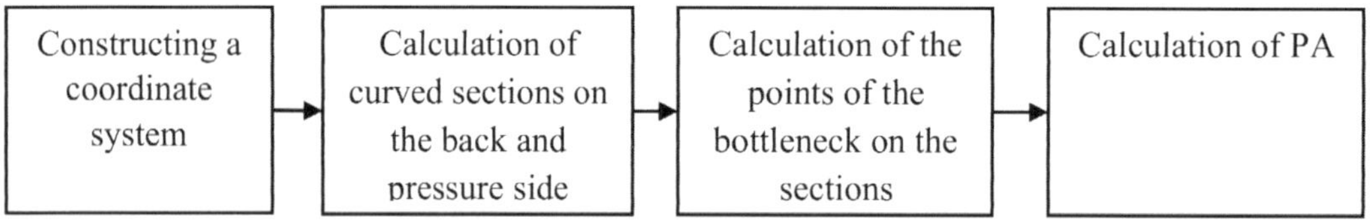

Figure 1. Stages of the mathematical model.

The coordinate system construction involves finding the motor axis center. To do this, the cylindrical surfaces (1 and 2 on figure 2) are inscribed with replacement "cylinder" elements using the method of least squares (MLS). The axis alignment is performed then by multiplying the STL vertex matrix by the rotation matrices.

$$\mathbf{V}_r = \mathbf{V} \cdot \mathbf{M}_{n\_y} \cdot \mathbf{M}_{n\_z} . \tag{2}$$

being $\mathbf{M}_{n\_y}$, $\mathbf{M}_{n\_z}$ – the rotation matrices around the axes *OY* and *OZ*. The rotation angles are calculated from the normal vector coordinates of cylindrical surface 2, figure 3.

Surface 1 (figure 3) shall lie "to the right" on the major axis, surface 2 "to the left", so the direction of the major axis is determined. The coordinate system center is at the midpoint $x_ц$ between the centers of the cylindrical surfaces. The aligned coordinates are moved by the coordinate value of this center according to the equation:

$$\vec{v}_{ni} = \vec{v}_{ri} - \vec{x}_c , \tag{3}$$

being $\vec{v}_{ni}, \vec{v}_{ri}$ – coordinate vectors *of the i-th* vertex of STL before and after the displacement. As a result, a matrix of aligned and displaced STL vertex coordinates is generated $\mathbf{V}_n$ for further calculations.

Figure 2. Base surfaces segmented on stl.

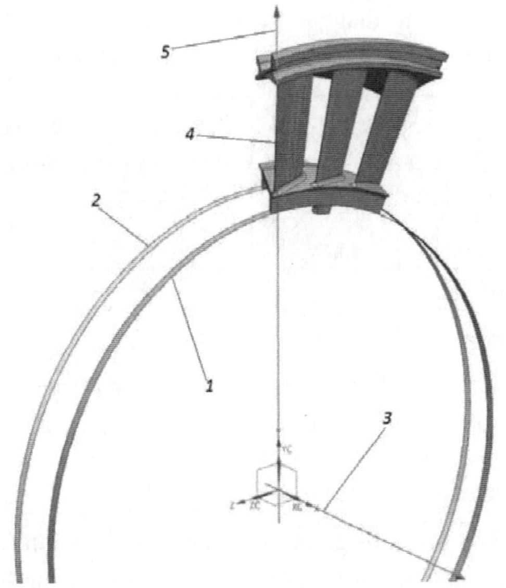

Figure 3. Creation of coordinate system 1, 2 – front and rear base surfaces, 3 – base axis, 4 – back surface; 5 – second major axis.

The second step is to calculate the points of intersection of the back and pressure side surfaces with the cylindrical surfaces. The axes of the cylindrical surfaces coincide with the principal axis of the coordinate system, the values of the radii are determined by measuring altitudes of the minimum section values between the surfaces of the back and pressure side. The more such measurements are performed, the more accurately it is possible to calculate the unevenness of the PA values of the inter-blade channels. In this work, the PA are calculated using three cross-sections (figure 4).

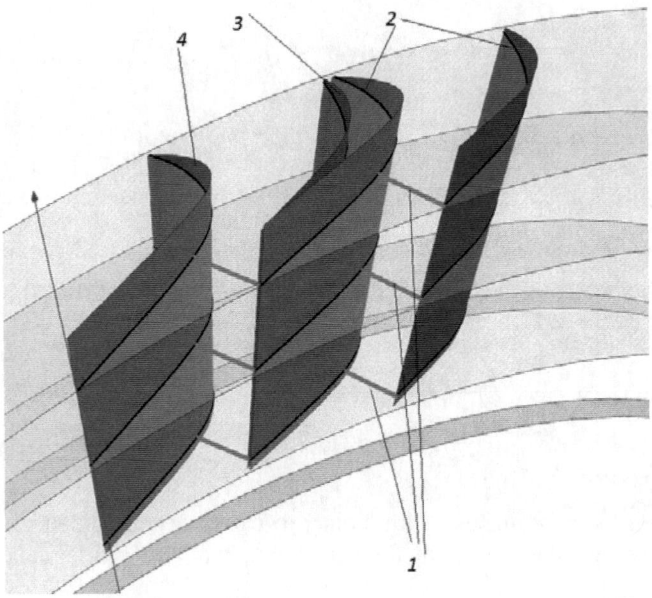

Figure 4. Calculation of the passage area for section 1 – widths of inter-blade channels at different altitudes; 2 – cross-sections of the pressure side and back surfaces; 3 – surface of the pressure side; 4 – surface of the back.

The block diagram of the algorithm for finding the point on the intersection line of the stl surface and the cylindrical surface is shown in Figure 4 (figure 5). The basic axis is OX.

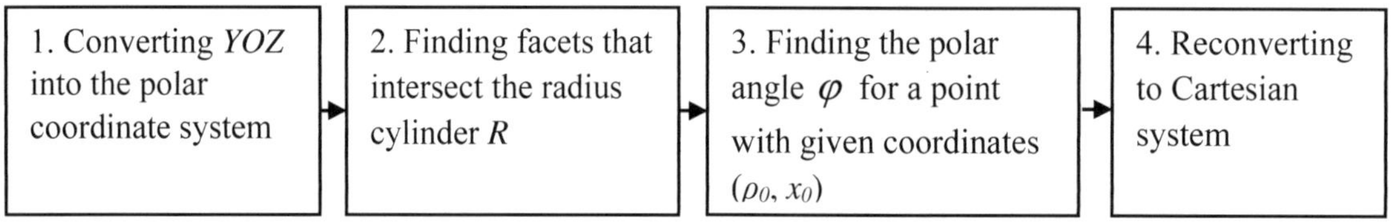

Figure 5. The algorithm for calculating the level line point on the surface.

The algorithm requires finding a point on the surface S_{stl} at the corresponding radius *R* from the coordinate system center with coordinate x_0 along the reference axis. In the first step, the coordinates of the two axes (except for the reference one) are converted to the polar coordinate system. Now, the vertices STL have the coordinates (x, φ, ρ).

The second step is to search for facets on the surface S_{stl}, on which there is a set of points with polar radius coordinate $R=\rho_0$, based on the condition that in the set:

$$\{(\rho_1 - \rho_0), (\rho_2 - \rho_0), (\rho_3 - \rho_0\}, \tag{4}$$

are both positive and negative numbers, being ρ_1, ρ_2, ρ_3 – the polar radii of the facet vertices. As a result, a significantly smaller area is selected from the surface S_{stl} $S_{stl \cap \rho_0}$ to find the level line point.

The third step is to find the polar angle for a point with given coordinates (ρ_0, x_0). The selection area for assigning the coordinate is x_0 chosen based on the minimum and maximum values of this coordinate for the vertices of the area $S_{stl \cap \rho_0}$. To check if the image point belongs to the facet, three values are calculated d_{ver} (in projection on the plane $XO\rho$):

$$\begin{cases} d_{ver1} = (x_1 - x_0)\cdot(\rho_2 - \rho_1) - (x_2 - x_1)\cdot(\rho_1 - \rho_0), \\ d_{ver2} = (x_2 - x_0)\cdot(\rho_3 - \rho_2) - (x_3 - x_2)\cdot(\rho_2 - \rho_0), \\ d_{ver3} = (x_3 - x_0)\cdot(\rho_1 - \rho_3) - (x_1 - x_3)\cdot(\rho_3 - \rho_0), \end{cases} \tag{5}$$

being $x_1, y_2, x_2, \rho_2, \rho_3, \rho_3$ – the coordinates of the vertices j of the – facet.

If the quantities d_{ver} in equation (5) are of the same sign, the point is inside the facet.

After determining the facet to which the point belongs, the missing polar angle coordinate of the point is searched for, based on the equation of the straight facet:

$$\varphi_0 = z_1 + \frac{n_1(x_1 - x_0) + n_3(\rho_1 - \rho_0)}{n_2}, \tag{6}$$

being $\{n_1, n_2, n_3\}$ – coordinates of the normal vector of the facet.

At the third stage of the PA calculation, the coordinates of the points to calculate the bottleneck between the sections are searched for. Consider the algorithm for calculating this bottleneck for a single section.

On the pressure side, the point P_{ps_PA} that has the largest coordinate along the axis of rotation is selected. Through the points of the back section, the equation of the interpolating normalized spline of degree 3 is given [2]. The coordinates of the spline point between the two set points P_0 and are P_1 determined from the equation:

$$P(u) = [1-3u^2+2u^3 \quad 3u^2-2u^3 \quad u-2u^2+u^3 \quad -u^2+u^3] \cdot \begin{bmatrix} P_0 \\ P_1 \\ P_0' \\ P_1' \end{bmatrix}, \tag{7}$$

here the vectors P_0', P_1' are vectors of derivatives at the points of P_0, P_1; ; u – dimensionless parameter lying in the range [0; 1].

All distances between the set points of the spline and the point P_{ps_PA} are calculated. Two spline segments in the vicinity of the set point closest to P_{ps_PA}. Calculation of the back point coordinates P_{bp_PA}, located at the section end of the smallest width, was carried out by the method of golden section [3]. The search criterion was to minimize the distance between the points P_{ps_PA} and P_{bp_PA}. The error of the algorithm ε was assumed to be 10^{-4}.

The final stage of calculating the PA is based on calculating the sum of the areas of the quadrangles formed by adjacent sections of the smallest width, the ends of which are connected by segments. Equation for calculating the area of the passage section with a set of section points $P_{ps\_PA\_i}$, $P_{bp\_PA\_i}$, $P_{ps\_PA\_i+1}$ and $P_{bp\_PA\_i+1}$ looks like:

$$S_{PA} = \sum_{1}^{m-1} d_1 \cdot d_2 / 2 \cdot \sin\beta, \tag{8}$$

being d_1 and d_2 – are the diagonals of each quadrilateral, distances $|P_{ps\_PA\_i}, P_{bp\_PA\_i+1}|$ and $|P_{ps\_PA\_i+1}, P_{bp\_PA\_i}|$; β_i – acute angle between the diagonals of the i-th quadrilateral; m – number of sections.

3. Segmentation of the scanned surface edges

Each point of a regular surface in three-dimensional Euclidean space is characterized by two extreme values of curvatures $k_{\min}$ and $k_{\max}$, called principal [2]. For a linear stl-surface, the calculation of the principal curvatures based on the derivatives is a nontrivial problem; various approaches have been developed to calculate them correctly [4,5,6], in particular, the results from [4] were used in the study. Based on the principal curvature tensor, the surfaces are segmented (divided into separate regions), in the current study, the algorithm was used from [7].

After model segmentation, recognition of the target faces (base, back and pressure side surfaces) was performed. Recognition can be performed manually, but this will increase the labor intensity of the calculations, so this stage has been automated by using computer vision techniques [8], in particular convolutional neural networks [9]. To train a neural network, a training sample is needed to prepare, and to assess the quality – a test sample. Various STL geometric parameters can be used as input data [10]. Taking into account the segmentation already performed, the task is simplified, and the projections of the model onto six planes (front, back, left, right, top and bottom view) parallel to the coordinate planes are used as input data to the neural networks. Stl object is placed as if in a cube the faces of which are parallel to the coordinate planes [11]. The Roberts algorithm was used to prepare the projections.

The output is also six projections of the parts, with the difference that the faces on them are marked. The target facets for recognition are the facets on which the assembly is mated and controlled. Accordingly, these facets are distinguished by different colors, the other facets of the body have a single and different color.

In order to perform recognition, we used U-net neural network [12], the output of which is a picture in "shades of gray". A view of one of the stl projections, the same projection marked for network training, and the output of the neural network is shown in figure 6.

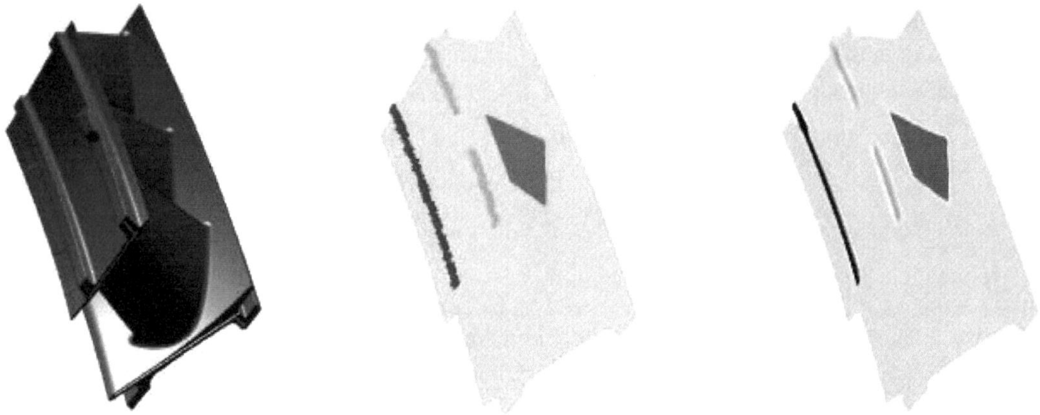

a b c

Figure 6. a) original stl projection; b) marked projection; c) recognition using U-net.

Segmentation and facet recognition is a necessary step to automate calculations of the PA parameter using the developed model.

4. The results of the experimental study

The developed model was implemented in the mathematical package MATLAB. For testing, a sector model of three nozzle blades was used. In the NX package, sections and PA were measured. The solid model was converted to STL with 411524 facets (figure 7). In order to perform the calculations, the STL surface was segmented into individual facets. The training sample for U-net was 1000 artificially generated cases, the test sample was 50 cases. Using the recognition results, the facets required for the calculations were selected (figure 8). The number of back facets was 24266, the pressure side – 28336.

Figure 7. STL sector model.

Figure 8. Segmented facets.

Measurement radii R_1, R_2 and R_3 are 205.5; 232.5 and 257.5 mm, respectively. The calculated area on the model was 701.234 mm^2. The boundaries of the segmented facets are not perfectly flat, since they represent many adjacent triangles in space. Therefore, it is important to select the necessary offset from the boundaries of the measurement sections, while at the same time, that it was not too big. Table 1 shows the calculation results of the cross sections for the smallest width and PA when changing the offsets from the pressure side boundary.

Table 1. Errors in changing the offsets from the surface boundaries.

Offset, %	Reference, mm		Calculated values, mm		Offset, mm	
	1 Flow area	2 Flow area	1 Flow area	2 Flow area	1 Flow area	2 Flow area
0	11.95	13.52	12.06	13.69	0.11	0.17
5			12.03	13.59	0.08	0.07
8			12.05	13.61	0.10	0.09
15			12.09	13.66	0.14	0.13
Offset, %	1 Flow area	2 Flow area	1 Flow area	2 Flow area	1 Flow area	2 Flow area
0	14.967	701.234	15.07	707.51	0.10	6.28
5			15.03	703.70	0.06	2.46
8			15.05	704.64	0.08	3.41
15			15.10	707.02	0.14	5.79

Optimal in terms of accuracy was the use of 5% offset from the boundaries of the surfaces. In addition, the calculation accuracy is also affected by the number of section points. Table 2 shows the calculation results of the sections for the smallest width and PA when changing the number of points on the splines (4).

Table 2. Errors in changing the number of spline points

Number of points	Reference, mm		Calculated values, mm		Offset, mm	
	1 Flow area	2 Flow area	1 Flow area	2 Flow area	1 Flow area	2 Flow area
10	11.95	13.52	12.04	13.59	0.09	0.07
20			12.03	13.59	0.08	0.07
30			12.03	13.59	0.08	0.07
50			12.03	13.59	0.08	0.07
Number of points	1 Flow area	2 Flow area	1 Flow area	2 Flow area	1 Flow area	2 Flow area
10	14.967	701.234	15.14	704.77	0.17	3.54
20			15.03	703.72	0.06	2.48
30			15.03	703.67	0.06	2.44
50			15.03	703.70	0.06	2.47

The deviation of the sections when using 30 points did not exceed 0.09 mm, the deviation of PA – 2.44 mm. Taking into account the measurement errors of scanners 0.05-0.1 mm and relatively low discreteness of the partition of the considered STL, the obtained calculation errors are acceptable.

The instrumental error in existing measurement technologies using indicators is from 1 mm^2, so that the computer calculation is comparable in accuracy with manual measurement, and in terms of performance is significantly higher.

5. Conclusion

The results allow us to automate and improve the accuracy of calculating the quality parameters of individual parts, as well as significantly reduce the labor intensity of control operations after assembling.

The considered model is one of the stages of nozzle assembly using the "virtual assembly" approach. The measured geometry can be used to perform a virtual assembly [13] that is to create a set of adequate digital mathematical models of the assembled product, taking into account the most essential properties and processes of this particular instance of any real product. Based on such a virtual product, assembly geometric and physical parameters can be determined prior to real assembly and testing. Thus, virtual assembly technology makes it possible to reduce the labor intensity of the process and to choose the optimal assembly option that provides the required quality indicators.

Acknowledgments
This paper was financially supported by the Ministry of Science and Higher Education of the Russian Federation as part of the Russian Federation Presidential Scholarship (number SP-262.2019.5). Experimental studies were performed on the equipment of the common use center of CAM-technologies (RFMEFI59314X0003).

References
[1] Zakharov O V, Balaev A F, Kochetkov A V 2017 *Procedia Engineering* **206** 1458-63
[2] Rogers D, Adams J 1990 Mathematical Elements for Computer Graphics (London: McGrawHill) p 604
[3] Powell M J D 1978 *ed. G.A. Watson, Lecture Notes in Mathematics* (Berlin: Springer Verlag) 144-57
[4] Cohen-Steiner D and Morvan JM 2003 *In Proceedings of the nineteenth annual symposium on Computational geometry* **8** 312-21
[5] Meyer M, Desbrun M, Schröder P and Barr 2003 In *Visualization and mathematics III* Berlin Heidelberg 35-57
[6] Taubin G 1995 In *Proceedings of IEEE International Conference on Computer Vision*. 902-07
[7] Lavoué, G Dupont F and Baskurt A 2005 *Comput. aided des.* **37** 975-87
[8] Zakani F R Arhid K Bouksim M Aboulfatah M and Gadi T 2018 Computer Optics **42** 312-19
[9] Rosebrock A Deep Learning for Computer Vision with Python 2017 (PyImageSearch.com) p. 332
[10] Qi C R Su H Mo K and Guibas L J 2017 In *Proceedings of the IEEE conference on computer vision and pattern recognition* 652-60
[11] Bolotov M A, Pechenin V A, Ruzanov N V, Kolchina E J 2019 *CEUR Workshop Proceedings* **2391** 342-49
[12] Ronneberger O Fischer P and Brox T 2015 October In *International Conference on Medical image computing and computer-assisted intervention* 234-41
[13] Rezchikov A F, Kochetkov A V and Zakharov O V. 2017 *MATEC Web Conf.* **129** 01054

Choosing a statistical method for predicting a quantitative indicator

R G Vlasov[1] and Yu S Korobov[1,2]

[1]Department "Welding Production Technology" of the Ural Federal University named after B. N. Yeltsin, Yekaterinburg, 620002, Russia
[2]Institute of Metal Physics named after A. N. Mikheev, Ural Branch of the Russian Academy of Sciences, Yekaterinburg, 620002, Russia

E-mail: novik904@yandex.ru

Abstract. The aim of the study is to find a machine learning method that allows achieving the maximum accuracy of the forecast when training on a specific data set. The article provides a comparative overview of strategies for choosing the optimal machine learning method. The process of training sample preparation is considered, the necessity of its modification is justified. The search for the optimal machine learning method was carried out in the statistical computing environment "R" using the package "CARET". The graphical results of training 96 algorithms that implement various statistical methods are presented, and the choice of the best one is justified on the basis of the proposed methodology.

1. Introduction

The main tool for data analysis is machine learning, and the number of methods used in modern machine learning is in the hundreds. In the process of problem solving of choosing the optimal machine learning method and the algorithm that implements it, we analyzed the strategies used by various authors to select the appropriate method.

- The most popular strategy is to analyze publications on the topic of interest and choose from several popular methods by empirically comparing their effectiveness in relation to the data under study [1-8].
- Selecting several of the most popular methods in a particular class and making an ensemble of them [9].
- The choice of the method depending on the type of data studied, in accordance with the set of some rules [10-12].
- The use of the most popular method without justifying the reasons for its choice [13-18].

From the point of view of the authors of the article, strategies 1, 3 and 4 cannot be considered satisfactory, since most of the machine learning methods (possibly more effective) are not even considered when implementing such a search.

2. Initial data

As the initial data for the mathematical model construction, data describing the process of testing technically complex products of several types in the assembly shop of a machine-building enterprise are used. Table 1 shows the dependence between the test duration of the product (response (y)) and the number of other products already in the test process at the start time of the of the product's test (vector of input parameters (x1…x10)). In total, there are 50 observations in the sample.

Table 1. Initial data for the simulation.

№	Vector of input parameters ($x_1 ... x_{10}$)										Response
p/p	x_1	x_2	x_3	x_4	x_5	x_6	x_7	x_8	x_9	x_{10}	(y)
1	2	1	0	2	1	1	0	0	2	0	40
2	0	0	0	0	2	0	0	1	2	3	54
3	1	0	0	0	2	0	0	1	2	3	47
4	3	0	0	0	2	0	0	1	2	3	52

The exploratory analysis of the training sample was carried out according to the protocol [19], which consists in a step-by-step verification of the statistical criteria of the original sample. It was found that some variables have near-zero variance (they have little variability-they contain many values of "0" and only a few values of "1"). Such variables can have a negative impact on the convergence of those statistical methods that calculate the covariance matrix in the process. In order to study the effect of zero variance dispersion on the convergence of various methods, three training samples were formed according to the following rules:

1. Sample "C" - the original set (Table 1. Initial data for modeling.).
2. Sample " A "(Table 2. Sample "A".): a vector consisting of a repeating sequence of numbers from 0.001 to 0.009 is added to each column;
3. Selection " B "(Table 3. Selection "B".): a vector consisting of a repeating sequence of numbers from 0.01 to 0.09 is added to each column.

Table 2. Sample «A».

№	Vector of input parameters ($x_1 ... x_{10}$)										Response
p/p	x_1	x_2	x_3	x_4	x_5	x_6	x_7	x_8	x_9	x_{10}	(y)
1	2.001	1.001	0.001	2.001	1.001	1.001	0.001	0.001	2.001	0.001	40
2	0.002	0.002	0.002	0.002	2.002	0.002	0.002	1.002	2.002	3.002	54
3	1.003	0.003	0.003	0.003	2.003	0.003	0.003	1.003	2.003	3.003	47
4	3.004	0.004	0.004	0.004	2.004	0.004	0.004	1.004	2.004	3.004	52

Table 3. Sample «B».

№	Vector of input parameters ($x_1 ... x_{10}$)									Response
p/p	x_1	x_2	x_3	x_4	x_5	x_6	x_7	x_8	x_9	x_{10}	(y)
1	2.01	1.01	0.01	2.01	1.01	1.01	0.01	0.01	2.01	0.01	40
2	0.02	0.02	0.02	0.02	2.02	0.02	0.02	1.02	2.02	3.02	54
3	1.03	0.03	0.03	0.03	2.03	0.03	0.03	1.03	2.03	3.03	47
4	3.04	0.04	0.04	0.04	2.04	0.04	0.04	1.04	2.04	3.04	52

3. Research method

The statistical computing language "R" and the software package "CARET" were used as a tool for constructing mathematical models [20]. The "CARET" package allows the automatically finding of the optimal values of hyperparameters by using cross-validation to test the quality of each model. The developers have provided the default values of hyperparameters for each algorithm, so using this package gives a significant gain in time, since it eliminates the need to configure a large number of hyperparameters of each model iteratively, in manual mode. At the time of writing this paper, the package included over 200 statistical learning algorithms from various packages. We have selected 98 algorithms that represent various machine learning methods and are most often used for solving regression problems.

4. Results and discussion

Using three training data sets, 98 algorithms were trained, and the corresponding mathematical models were constructed. The models were evaluated using the metrics R^2 (R-squared – coefficient of determination) and *MSE* (Mean square error – mean square error of the forecast). R^2 can be interpreted as a measure of the model's compliance with the data, and *MSE* as an offset-the deviation of the predicted values from the training sample values.

MSE calculated as:

$$MSE = \frac{1}{n}\sum_{i=1}^{n}(y_i - \hat{y}_i)^2 \tag{1}$$

where: $\hat{y}_i$ – – predicted value of the quantitative response;

y_i – quantitative response from the training sample;

n – number of observations.

R^2 can be calculated as:

$$R^2 = (r_{yx})^2 \tag{2}$$

where r_{yx} – the correlation coefficient, calculated as:

$$r_{yx} = \frac{\sum(x_i - \bar{x})(y_i - \bar{y})}{n\,\sigma_x\,\sigma_Y} \tag{3}$$

where σ_y and σ_x mean square deviations for y and x, respectively:

$$\sigma_y = \sqrt{\frac{\sum(y_i - \bar{y})^2}{n}};\sigma_x = \sqrt{\frac{\sum(x_i - \bar{x})^2}{n}} \tag{4}$$

It is important to note: in order to save resources, *MSE* (1) and R^2 (2) were calculated using only the training sample, without checking on the test sample.

Based on the results of training 96 models, diagrams in the *MSE* and R^2 axes are calculated, which are shown in Figures 1, 2 and 3.

The main goal of the study is to find a method that will allow us to obtain a model with minimal bias and most fully describe the dependence between x and y (match the data). That is: the model must have a maximum R^2 and a minimum *MSE*, and therefore, its coordinates will lie in the upper-left corner of the diagram. Also, the analysis of Fig. 1-3 allows us to draw conclusions about how stable the algorithms representing different machine learning methods work when using training samples with different forms of representation of input variables. Figure 1 shows a rectangular area where models with *MSE* = 0 are located. These models are obtained using algorithms known as "Nuclear Smoothing Methods". The value of *MSE* = 0 indicates the inadequate operation of these algorithms, confirms the assumption that variables with near-zero variance have a negative impact on the convergence of some methods, and confirms the need to modify the input variables. Most algorithms have built predictive models (oval area). Dashed lines show models with the minimum *MSE*, and an arrow marks the model with the maximum value of R^2.

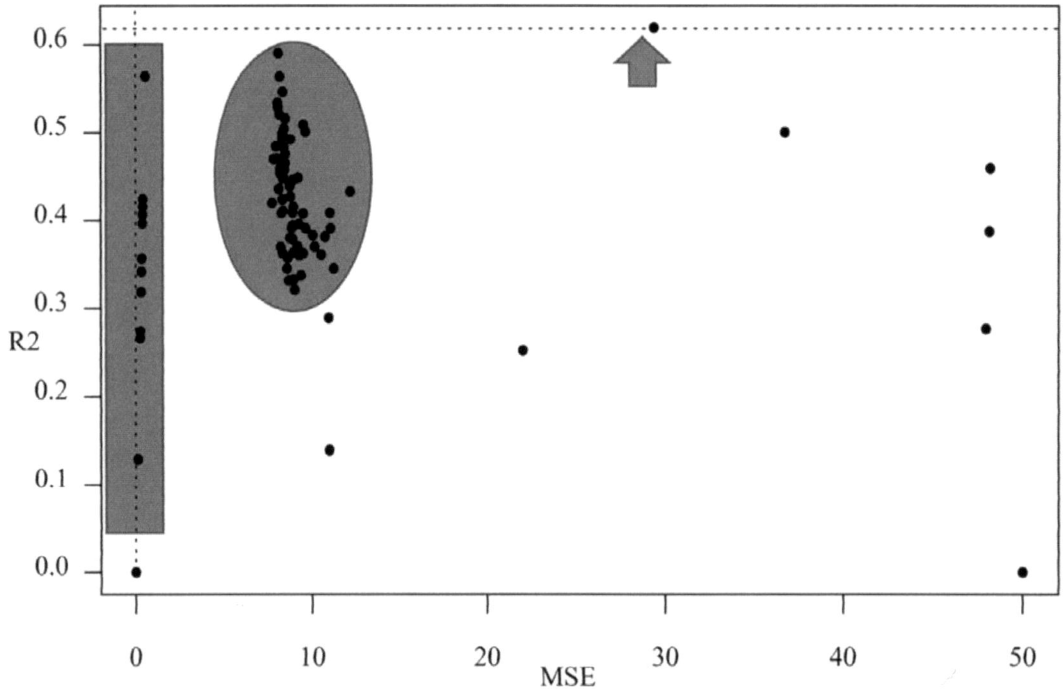

Figure 1. Results of training algorithms using the sample "C".

In figure 2: it is obvious that as a result of training using the "A" sample, most of the algorithms have built predictive models: there are no models with *MSE* = 0. The exception is two algorithms that have $R^2 = 0$, and these are again "Nuclear Smoothing Methods".

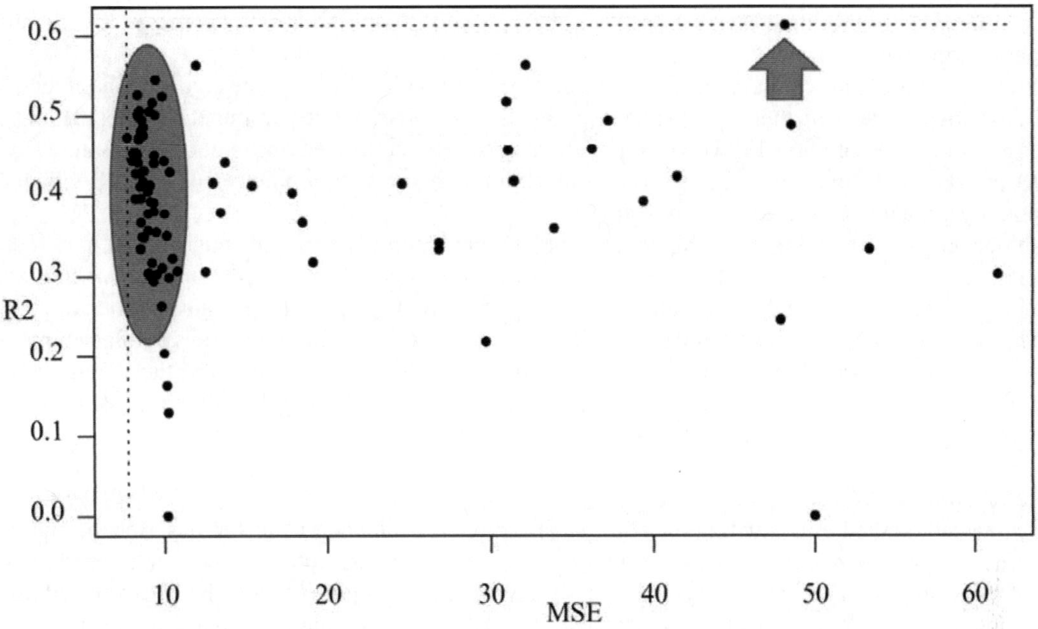

Figure 2. Results of training algorithms using the sample "A".

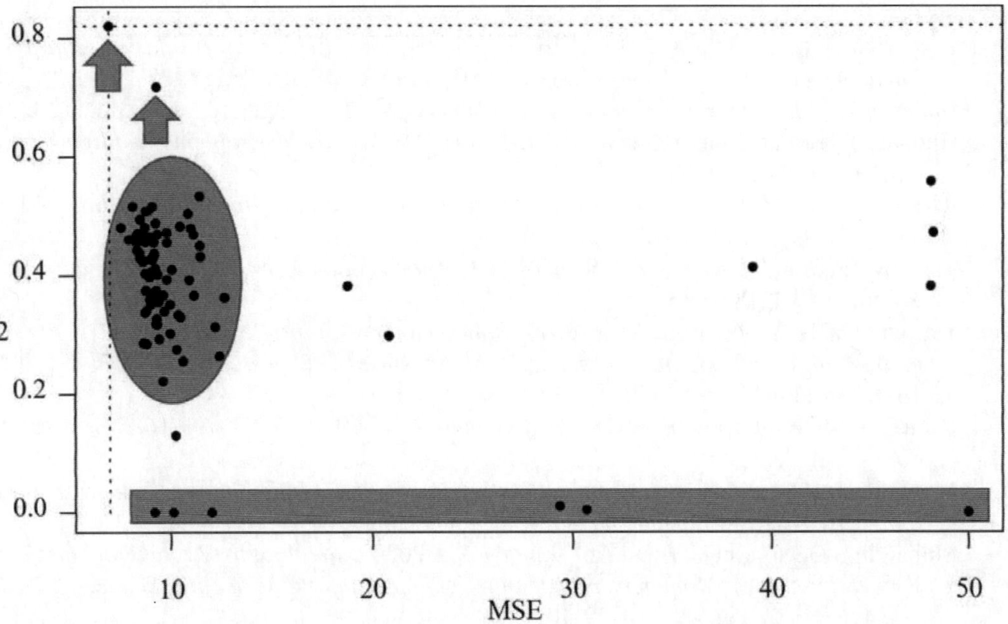

Figure 3. Results of algorithm training using sample "B".

There is a group of models with average accuracy (oval area), but in general, the results are worse than in figure 1, since there are much more models with a relatively large *MSE*. The model with the maximum value of R^2 is marked with an arrow.

Figure 3 clearly shows the winner-the model in the upper-left corner of the diagram, marked with an arrow. This model is obtained by a neural network-a multilayer perceptron with the topology 10 – 4 – 5

– 1 (10 neurons in the input layer, 4 and 5 in the first and second hidden layers, respectively, and 1 output neuron).

The coordinates of the best model $R^2 = 0.82$ and $MSE = 6.98$ indicate that the algorithm has identified and described quite well the dependence between the vector of input parameters (x1...x10) and the response Y. At the same time, it is worth paying attention to the model whose coordinates are slightly lower (marked with an arrow): this is a Bayesian neural network with topology 10 – 10 – 1, which can potentially get a model with similar indicators.

In our opinion, it will be preferable in the case of a more complex sample consisting of hundreds of observations and the presence of noise or outliers in the sample. The group of models with average accuracy values is somewhat more blurred than in figure 1 and figure 2. There is also a group of models that has the value $R^2 = 0$, which indicates a failure in the process of building models by algorithms. All these algorithms belong to different methods and you can determine the causes of the failure only by studying the principles of their operation in detail. At the stage of finding the optimal algorithm, these methods can simply be excluded from the list of tested ones.

5. Conclusion

The proposed method of evaluating the ability of machine learning algorithms to describe the analyzed data allows you to make an informed choice in the direction of the optimal algorithm, contributes to obtaining a more accurate model and significantly saves the time required to achieve the expected result. Also, the method allows you to assess the need for data pre-processing and the effectiveness of the processing itself. This distinguishes it from the strategies described above. The only significant drawback of the method is the resource intensity.

References

[1] Aizpurina J I and others 2017 IEEE *Conference on Electrical Insulation and Dielectric Phenomenon (CEIDP)*. **24(4)** 153-6. doi: 10.1109/CEIDP.2017.8257475

[2] Ballester P J 2020 *Drug discovery today: technologies*. **32** 81-7. doi:10.1016/j.ddtec.2020.09.001

[3] Guolang S, Mufan L and others 2020 *Hindawi: Mathematical Problems in Engineering* 1-13. doi:10.1155/2020/7842847

[4] Khaledian Y, Miller B and others 2020 *Applied Mathematical Modelling*. 81 401-18. doi:10.1016/j.apm.2019.12.01

[5] Tarasov O F and Turlakova S S 2018 *Industrial economics*. **3(83)** 57-75. doi: 10.15407/econindustry2018.03.057

[6] Staroverova K Y and Bure V M *2017 Bulletin of the Saint Petersburg University. Applied mathematics. Computer science. Management processes.* **13(1)** 51-60. doi: 10.21638/11701/spbu10.2017

[7] Dikij D I 2020 *Information and control systems*. 4 50-60. doi:10.31799/1684-8853-2020-4-50-60

[8] Mayfeld H J, Smith C and others. 2020 *Environmental Modelling and Software*. 131 1-10. doi:10.1016/j.envsoft.2020.104741

[9] Shibajkin S D, Nikulin V V and Abbakumov A A 2020 *Bulletin of the Astrakhan State Technical University. Series: Management, computer engineering and Computer Science*. **1** 29-40. doi: 10.24143/2072-9502-2020-1-29-40

[10] Doronina L S and Kaskinova D I 2017 *Proceedings of the Orenburg State Agrarian University*. **7(7)** 49-52

[11] Neminushhaya L A, Skotnikova T A and others. 2015 *Bulletin of the Technological University*. **18(2)** 377-82

[12] Narkevich A N and Vinogradov K A 2019 *Social aspects of public health* **65(4).** URL: http://vestnik.mednet.ru/content/view/1092/30/lang.ru. doi:10.21045/2071-5021-2019-65-4-9

[13] Ablaev M F and Salixova N M 2020 *Scientific Notes of the Kazan University. Series of Physical and Mathematical Sciences*. **162(3)** 41-58. doi: 10.26907/2541-7746.2020.3.241-258

[14] Xal`yasmaa A I 2020. *Bulletin of the Irkutsk State Technical University.* **24(5)** 1093–104 doi:10.21285/1814-3520-2020-5-1093-1104
[15] Bure N A, Grebennikova N L and Staroverova K 2018 *Bulletin of the Saint Petersburg University. Applied Mathematics.* **14(4)** 325-33
[16] Kadomcev M I and Razveeva I F 2020. *Engineering Bulletin of the Don.* **2 (62)** 12-8.
[17] Xajrullina A I, Madzhidov T I and others 2018. *Scientific Notes of the Kazan University.* Series of natural sciences. **160(2)** 200–13
[18] Sulimov P A 2016 *Bulletin of the Novosibirsk State University of Economics and Management.* **1** 285-306
[19] Alain F Zuur , Elena N Ieno and Chris S Elphick. 2010 *Methods in Ecology and Evolution.* **1(1)** 3–14. doi: 10. 1 1 1 1 /j .2041-2 10X.2009.00001 .x
[20] Site: The “caret” Package: https://topepo.github.io/caret/index.html

The motor brake of a compressor type

Viktor V Verbitsky and Vladimir M Pogosyan[1]

Kuban State Agrarian University named after I. T. Trubilin, 13, Kalinina st., Krasnodar, Krasnodar Territory, 350044, Russian Federation

[1]E-mail: pogosyn@gmail.com

Abstract. The accident rate in road transport remains very high and it is necessary to consider all the factors that affect this process to reduce it. In this regard, special attention should be paid to the process of long-term braking, the negative manifestations of which require the development of additional braking systems (retarding retarders) for vehicles operating in mountainous areas, primarily in the sphere of passenger transport. Transmission retarder brakes, which provide sufficient braking performance, have a number of disadvantages that hinder their use. The existing motor brakes-retarders show insufficient retardation and Kuban State Agrarian University conducted researches to improve their effectiveness. The compressor brake mode was experimentally studied after the theoretical analysis, in which an increased pressure was created in the inlet collector, and at the end of the compression stroke, the air from the cylinder was released through a special valve back into the system, which created the braking effect. The experiments have confirmed the possibility of a significant increase in the braking torque of the engine in the compressor brake mode, when both flaps are closed - the exhaust flap after the exhaust collector and the inlet collector before the carburetor, and compressed air is supplied to the inlet collector at different pressures. Then the braking torque increases in comparison with the engine braking more than 3 times.

1. Introduction

When driving a car on a lengthy descent or with repeated frequent braking, the brakes overheat in the friction zone, where various researchers record temperatures up to 300-370°C, and sometimes higher. In these conditions the overheating of brake pads, their charring, pitting and warping of the brake drum or brake disk occur [1].

When the brake pads are heated, released substances act as a lubricant and reduce the coefficient of friction by half or more. A decrease in deceleration during braking on a long descent with a slope of 3-4° and a length of 19 km was recorded from 5.6 m/s^2 to 2.6 m/s^2 in tests conducted by Russian Automobile and Automotive Research Institute.

At a high temperature of brake pads during operation, their wear increases dramatically – they have to be changed not after 70-90 thousand kilometers, but after 8-10 thousand kilometers. At the same time, the pad friction coefficient decreases and the braking distance of a car increases. Reducing the level of traffic safety decreases the labor productivity of a driver, as the average speed decreases [2].

Attempts to use the engine braking mode on a lengthy descent lead to an increase in the vacuum in the inlet collector and, as a result, to the suction of oil into the combustion chamber. In addition, the carbon monoxide content in the exhaust gases increases almost in 6 times in the forced idle mode of the engine when braking [3].

The exhaust brake is simple and cheap, since, in essence, it simply includes a flap after the exhaust collector. When braking, the ignition is switched off and the braking effect is created by the back pressure from the exhaust collector on the upward stroke of the piston. However, the braking efficiency of the exhaust brake is insufficient.

It is reasonable to install the effective retarding brakes, first of all, for passenger cars operating in mountainous conditions. It will allow us to slow down the car when there is no need for emergency braking. In addition, there is a backup brake system, which increases traffic safety.

Retarding brakes can be transmission and motor type. Transmission retarding brakes can be hydrodynamic and electrodynamic, the exhaust brake patented in the early twentieth century, mainly, is known as a motor brake [4].

The retarding brakes with radial blades of the Thompson Company are most often used as hydrodynamic retarding brakes. The brake is attached either to the frame or to the crankcase of the rear axle. The weight is 80-96 kg. The working fluid is a mineral oil that can be displaced by compressed air from the compensation cylinder into the brake housing. The absorbed power is controlled by the compressed air pressure, as it determines the fluid level in the brake housing.

There are other designs of hydrodynamic brakes. In this capacity, a torque converter can be used, in which the level of braking is regulated by the amount of oil in the housing. There was also developed the brake in which a gear mounted on the driveshaft was permanently engaged with two other gears, forming two oil pumps. When braking, the pressure valves of the pumps were locked, but this caused a strong increase in the oil temperature.

The most famous electrodynamic brake was developed by the French Company Telma. It consists of a stator mounted on a frame with induction coils that are powered by a battery and create a magnetic field. The brake rotor is connected to the driveshaft. The braking moment occurs as a result of the interaction of the magnetic fields of the stator and the eddy current fields of the rotor. The rotor has blades for cooling, since its disks can heat up to a cherry-red color during operation. The Telma brake weighs 100kg for a car weighing 6 tons.

Transmission retarding brakes provide deceleration up to 1.5 m/s^2, but have a number of disadvantages. The hydrodynamic brake is characterized by high cost, bulkiness and the need for additional maintenance. The electrodynamic brake, in addition to its high cost, bulkiness, and heavy weight, usually requires an additional battery pack. All these disadvantages prevent the widespread use of transmission retarding brakes.

2. Materials and methods

70 h/p carburetor engine at a nominal speed of 2800 rp/m and a maximum torque of 20 kg/m, it was installed on an electric brake stand, which during tests worked in the mode of an electric motor. To increase the number of revolutions of the crankshaft, the torque was transmitted through one of the lower stages of the gearbox. The clutch driven plate was removed and replaced with a strain gauge link.

Since the slip current collector operates reliably at a speed of no more than 2000 rpm, an end collector was used to transmit the pulse from the strain member at higher revolutions. The conductive line passed through the crankshaft, in which special channels were made by electric spark treatment and after the line was laid, special channels were filled with epoxy resin. The electrical impulse from the strain-link was removed by an end collector from the toe of the crankshaft and recorded on an oscillogram, where the rotation frequency was simultaneously recorded.

Braking performance tests were conducted with engine braking, exhaust brake, high efficiency exhaust brake, and compressor mode. In the exhaust brake mode, a damper was installed after the exhaust manifold and the braking torque was determined at different speeds of rotation of the crankshaft. In the high efficiency exhaust brake mode, two flaps were closed, one after the exhaust manifold and one in front of the carburetor.

Compressed air was supplied from a stationary compressor to the engine cylinders at different pressures, after which the braking efficiency was determined.

IOP Conf. Series: Materials Science and Engineering 1155 (2021) 012032 doi:10.1088/1757-899X/1155/1/012032

3. Discussions and results

The braking characteristics of the engine in the mode of the exhaust brake and the exhaust brake with increased efficiency at different speed modes are presented in Table 1.

Table 1. Braking characteristic of the engine in the mode of the exhaust brake with increased efficiency.

Mode	The number of engine revolutions per minute				
Braking mode	800	1200	1600	2000	2400
Engine braking - braking torque, Nm	40	48	54	60	72
Exhaust brake - braking torque, Nm	90	102	116	130	156
Exhaust brake with increased efficiency at p = 3 kg / cm2 - braking torque, Nm	104	116	128	132	162
Exhaust brake with increased efficiency at p = 4 kg / cm2 - braking torque, Nm	116	128	140	152	172
Exhaust brake with increased efficiency at p = 5 kg / cm2 - braking torque, Nm	128	138	152	168	178

Thus, the braking torque increases 2.5 - 3.2 times compared to engine braking in the mode of the exhaust brake with increased efficiency at a pressure of the supplied compressed air of 5 kg / cm2. The strength conditions of the engine parts are met in this case.

In order to improve the operating conditions of the engine and increase the braking efficiency, the compressor brake mode was tested, in which air was discharged from the cylinder at the end of the compression stroke through a special valve installed in the block head. The braking characteristic of the engine in the compressor mode was determined at a pressure of the supplied compressed air from 1 to 4 kg/cm^2.

The results of the compressor mode tests at different compressed air pressures are shown in Table 2.

Table 2. Braking characteristic of the engine in compressor mode at different pressures of the supplied compressed air.

Mode	The number of engine revolutions per minute			
Braking mode	800	1200	1600	2000
Compressor brake at p=1 kg/cm^2				

- braking torque, Nm	126	128	132	136
Compressor brake at p=2 kg/cm^2				
- braking torque, Nm	168	172	176	180
Compressor brake at p=3 kg/cm^2				
- braking torque, Nm	232	236	240	244
Compressor brake at p=4 kg/cm^2				
- braking torque, Nm	282	284	286	290

The condition of the strengthening of all engine parts during the transition to the compressor mode is met, since the operating pressure of the exhaust valve is significantly lower than the pressure of the end of combustion. The braking torque of the engine in this mode, when compressed air is supplied at a pressure of 196 kN/m^2, exceeds the braking torque in the normal braking mode of the engine by more than 3 times.

4. Conclusion

Transmission retarding brakes are not widely used in cars due to their high cost, significant weight and the need for additional maintenance.

Prolonged braking by the engine on lengthy descents leads to the flow of oil into the combustion chamber and additional carbon monoxide emissions into atmosphere, and in this case, the deceleration happened is small.

The compressor brake mode, which is characterized by its simple design and low cost, creating a high braking torque, can be recommended for vehicles operating in mountainous conditions.

References

[1] Kalimulin M R 2016 *Analysis of the process of forming the technical appearance of particularly light high-mobility wheeled vehicles for mountain operating conditions* (MSTU named after Bauman) **77** 48211
[2] Mazhitov B Zh 2011 *Methods of reducing the toxicity of diesel exhaust gases and the thermal stress of the braking system of cars during operation in mountain conditions* (Saint Petersburg)
[3] Knechtges H 2014 Mit Jeder Stufesteigt Lohnunternehmen *ATZ* **69** 6
[4] Kabanin P A 2012 Determination of the rational configuration of the brake system for a three-axle all-wheel drive agricultural vehicle *Technology of wheeled and tracked vehicles* **2** 43-64
[5] Stahn Charles R 1993 *Automobile construction and operation* (Toronto, London, Madrid, Paris: McGraw-Hill Ryerson Limited)
[6] Electronic braking system of cargo vehicles 2001 *Car engineering abroad* **9** 27-33
[7] Julian M M 1969 Frenos de disco *Ferrocarrilesytranvias* **33(380)** 119-26
[8] Khrostophorov E N 2006 Braking mechanism for vehicles *Tractors and agricultural machines* **9** 40-1
[9] Mamiti G I 2009 Braking mechanism of the drum type with self-installing pads *Bulletin of mechanical engineering* **11** 89-92

Special automatic control system for the accuracy of processing axisymmetric workpieces

O I Drachev[1], B M Gorshkov[2] and N S Samokhina[2]

[1]Togliatti State University, Togliatti, Russia
[2]Volga Region State University of Service, Togliatti, Russia

E-mail: kaf_ekis@tolgas.ru

Abstract. The article discusses a special automatic control system for the machining process as applied to metal turning lathe. In processing, generalized information about the precision parameters of the processing system is gathered that makes affordable to integrate four control loops into its system. The first control loop stabilizes the tool post body. The second control loop stabilizes the workpiece axis during cutting. The third control loop stabilizes the cutting edge and the fourth one stabilizes the tool-workpiece.

1. Introduction

The development is applied to machine-tool manufacture, especially to metal turning lathes, and can be used on automated metal turning lathes with adaptive systems.

There is a known method for automatic control of the processing of the workpiece that the position of the workpiece axis and the tip of the tool is automatically adjusted relative to an artificial gage length. To do this, measure the increment in the diameter of the workpiece to be machined and the position of the tool tip towards the artificial gage length and in the function of their measurement, a signal is generated to control the position of the tool tip and the axis of the workpiece on a predetermined artificial gage length [1]. However, the layout markings do not allow measuring the diameter and shape of the workpiece to be processed during the cutting. Furthermore, they do not control the dimensional depreciation and the mine hole wear, the elastic OX and OY axes' travels of the tool tip and the frequency of chip formation. Lack of information does not allow to control the cutting process and the position of the "tool-carriage" and "workpiece-carriage" forming subsystems [2-10].

2. The relevance and purpose

The article discusses the technological process of processing axisymmetric workpieces. Serious technological difficulties arise in manufacturing during machining, their main reasons are as follows: significant elastic compliance and low vibration resistance, as well as different flexibility of the technological system elements. The deteriorating effect of these factors in the in manufacturing of parts of the "shaft" type leads to a violation of the processing datum surfaces, errors in the shape and size of parts, limitation of cutting conditions and, ultimately, to the decrease in the operational reliability of finished parts.

An experience of the individual, preproduction and manufacturing production of the shafts analysis showed that traditional methods of processing low-rigid long-length axisymmetric workpieces are counterproductive.

Consequently, improving the machining during the axisymmetric parts manufacture is an important task that increases the production efficiency and the quality of engineering products.

Improving the accuracy of machining workpiece accuracy enchances the operational properties of the metal turning lathe and devices and leads to the machining allowances decrease. That, in turn, reduces the lab our intensity manufacturing parts and assembling machines, increasing their reliability.

The purpose of the work is increasing the accuracy, productivity and expand the technological capabilities of the metal turning lathe, using a special computer aided system machining accuracy.

According to the special computer aided system machining accuracy, the value of the dimensional depreciation, the tool back and surface where is the point of separation of the chips, the current diameter's value of the workpiece in the plane traversing through the tip of the tool are measured additionally separately and simultaneously. The elastic cross and cross-rail travels of the tool tip control the change the slicing frequency of the chip formation process and separately generates a control signal proportional to the increment in the diameter of the workpiece from the dimensional depreciation. In the function of changing the elastic mechanical system the position of the support-prism relative to the artificial gage length is controlled, maintaining the relative position of the tool tip and the axis of the workpiece constant. In this case, simultaneously, as the function of changing the elastic mechanical system of the tool tip during the cross and cross-rail travels the feed rate changes, keeping the position of the tool tip and the workpiece axis at a given level. When the surface depreciation is reached, the decrease in feed upon reaching of the surface depreciation of the critical state is limited.

3. The key discourse. Special ACS operations

The technology of machining using a special automatic control system for the turning process is as follows: the value of the dimensional depreciation, the tool back and surface where is the point of separation of the chips, the current diameter's value of the workpiece in the plane traversing through the tip of the tool are measured additionally separately and simultaneously; The elastic cross and cross-rail travels of the tool tip is measured and the change in the slicing frequency of the chip formation process is controlled. Separately, a control signal is generated in proportion to the increment in the diameter of the workpiece from dimensional depreciation and mechanical compliance of the elastic mechanical system as a function of increment in dimensional depreciation; The position of the self-centering rest (SCR) relative to the artificial gage length is controlled; At the same time, simultaneously, as a function of elastic displacements of the cross and cross-rail tool tip travels the amount of feed is changed and its decrease is limited when the surface depreciation is reached the critical state.

The figure 1 shows a functional flow diagram of a special automatic control system for the workpieces machining of the shaft type.

During processing the generalized information about the precision parameters of the technological system (TS) is collected that makes it possible to integrate four control loops into its system.

The first stabilization loop of the tool holder body includes the tool holder stabilization sensor 1, fixed on the tool holder body 2 structure, relative to the artificial gage length 3, differential-input amplifier 4, initial position sensor 5 of the electrohydraulic gear 6.

The second stabilization loop of the workpiece axis during the machining consist of the SCR 7, SCR initial position sensor 8, SCR indexing mechanism 9, SCR differential amplifier 10, SCR electrohydraulic gear 11, ultrasonic sensor 12 fixed on the tool, impulse generator 13, amplifier detector 14, modulator electrode 15, assembly test recording of the artificial gage length 16.

The third stabilization loop of the tip of the tool consist of, mentioned in the second stabilization loop, the ultrasonic sensor 12, impulse generator 13, amplifier detector 14, modulator electrode 15, dimensional depreciation unit 17, differential amplifier 18, electrohydraulic gear 19 of the tool, initial tool indexing mechanism 20, initial tool inspection sensor 21 (cutting depth indexing mechanism).

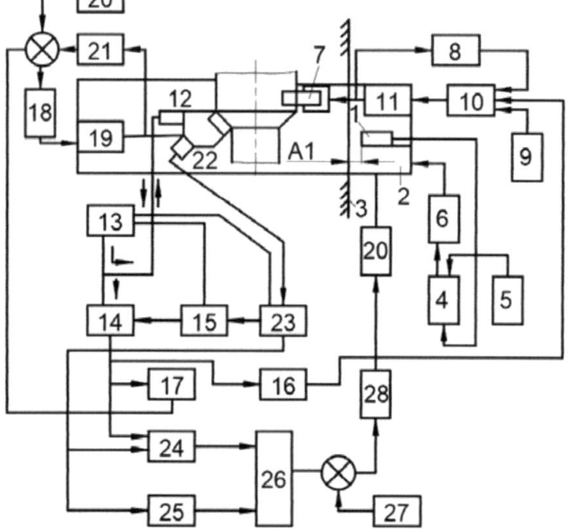

Figure 1. Functional flow diagram of a special automatic control system for the workpieces machining of the shaft type:
1 – tool holder stabilization sensor; 2 – tool holder body structure; 3 – artificial gage length; 4 – differential-input amplifier; 5 – initial position sensor; 6 – electrohydraulic gear; 7 – SCR; 8 – SCR initial position sensor; 9 – SCR indexing mechanism; 10 – SCR differential amplifier; 11 – SCR electrohydraulic gear; 12 – ultrasonic sensor; 13 – impulse generator; 14 – amplifier detector; 15 – modulator electrode; 16 – assembly test recording of the artificial gage length; 17 – dimensional depreciation unit; 18 – differential amplifier; 19 – electrohydraulic gear; 20 – initial tool indexing mechanism; 21 – initial tool inspection sensor; 22 – cutting force contour sensor; 23 – amplifier detector; 24 – block for generating a control signal as a function of elastic displacements of the cutter tip; 25 – dimensional depreciation unit on the surface; 26 – speed of the line feed indexing mechanism; 27 – speed control indexing mechanism; 28 – driver amplifier.

The fourth stabilization loop of the cutting force includes the sensor 22, amplifier detector 23, modulator electrode 15, control signal former in the function the tip of the tool elastic cross and cross-rail travels 24, dimensional depreciation unit on the surface 25, speed of the line feed control signal former 26, speed of the line feed indexing mechanism 26, driver amplifier 28.

During machining, the information on the precision behavior of the technological processing system (TPS) is read and summarized. The complex of the simultaneous four control loops operation shows the effect of simultaneous information directly from the tool zone about the current value of the machined workpiece diameter and sectional shape, its elastic travels, about the quality of the processed surface, about the dimensional depreciation and face wear of the tool, for example, the mine hole wear, the cross Y, X, Z axis travels of the tool tip and slicing of chip formation. Before the start of turning, the first stabilization loop of the tool holder housing position is switched on, while the contactless sensor 1 is rigidly fixed to the body structure 2relative to the artificial gage length3 (the latter is installed on the machine bed and can be replaceable and adjustable in the radial direction of the workpiece). In the case of the body structure 2 travel relative to the artificial gage length 3, the indexing mechanism I generates an electrical signal proportional to the amount of the tool holder travel, with account for to the differential-input amplifier 4, where this signal is compared with the indexing mechanism 5 signal, amplified and fed to the electrohydraulic gear 6. The latter travels the body structure 2 in the desired line, stabilized it relate to the artificial gage length 3 along the entire length of the workpiece.

The tool holder body structure stabilization 1 of along the entire path of its travel makes it possible to create an artificial reference and measurement base for the control loops of the tool tip and workpiece axis, eliminating all errors introduced by the guides and lead screws. In the second stabilization loop of the workpiece axis, before the process of cutting, the SCR 7 is set relative to the tool holder body structure, while the position of the control sensor 8 included in the feedback of the second control loop corresponds to the nominal diameter of the workpiece and parallelism to the axis of the working surface of the artificial gage length 3. The signal from the sensor 8 is compared with the indexing mechanism 9 signal and fed to the differential-input amplifier 10, the output of the latter is connected to the electrohydraulic gear, which controls the position of the SCR 7.

During cutting, the output signal from the sensor 12, proportional to the current diameter of the workpiece to be machined in a plane passing through the tip of the tool along the normal to the machined surface of the workpiece through the control unit where this signal is divided into a signal proportional to the increment in the diameter of the workpiece being machined from tool depreciation and the signal proportional to the value of the diameter increment due to the bending deflection of the workpiece. The latter feeds to the input of the differential amplifier 10, where a control signal is generated in proportion to the increase in the diameter of the workpiece being processed as a function of its resilient flexing in the cutting area and is fed to the electrohydraulic gear 11, while changing the position of the SCR 7 relative to the artificial gage length 3 so that the axis of the workpiece is parallel to the artificial gage length 3.

The sensor 12 emits flexing waves under the influence of the ultrasonic pulse generator 13 oscillations to the workpiece, receives the reflected flexing waves, converts the flexing oscillations into electrical ones and supplies them to the differential-input amplifier 14, then to the elastic travels registration unit 16 and then to the amplifier 10. The plate pulse indexing mechanism 15 ultrasonic generator 13 and amplifier detector 14 are coupled to the input.

The second stabilization loop work of the workpiece axis make it possible to compensate for the workpiece axis travel due to the action of cutting forces and to minimize vibrations arising during processing, as well as to increase the vibration resistance of the “workpiece-support” system.

The blocks related to the second control loop are also involved in the work of the third stabilization loop of the cutter tip (12, 13, 14 and 15). While the cutting the sensor 12 signals from the ultrasonic pulse generator 13 to the workpiece being processed, and the reflected from the workpiece signals pass through the sensor 12 and go to the amplifier detector 14, the output of which is coupled to the tool 17 depreciation unit that registers the tool dimensional depreciation and gives signal proportional to the amount of the dimensional depreciation. The output unit 17 signal is fed to the input of the amplifier detector 18 which generates a control signal proportional to the size of the tool dimensional depreciation, and it is fed to the input of the electrohydraulic gear 19. The latter corrects the tool tip position according to the set from the cutting depth indexing mechanism 20. The position of the tool tip is set from the tool holder body structure 2 and is controlled by the position sensor 21 that included in the feedback of the electrohydraulic gear. The operation of the second and third control loops allows maintaining a constant relative position of the tool and the axis of the workpiece.

The mine hole tool wear at the point of chip separation along the leading surface, the elastic cross and cross-rail travels and the change in the frequency of the chip formation process slicing are measured by the fourth control stabilizing cutting forces loop. At the same time in the function of changing the elastic cross and cross-rail travels of the tool tip tool (due to the increase in cutting forces from depreciation), the cross feed of the tool is controlled, maintaining the stable position of the tool tip and the axis of the workpiece at a given level and limiting the decrease in feed when the mine hole wear reaches a critical state and reaching the frequency of the chip friction in advance of its predetermined value. In this control loop, the deteching head 22 is mounted on the tool along the normal to the main cutting edge. Both in the second and in the third control loops, the measuring signal is generated by units 13 and 15 and also by a separate amplifier detector 23. The output of the latter is simultaneously connected to the unit 24 for generating a control signal as a function of the elastic Z-axis cross and X-axis cross-rail travels and to the control of the tool depreciation the mine hole wear unit 25 and the slicing frequency of the chip formation. The output signal from unit 26 is compared with the signal 27 the error signal is fed to the input of the driver amplifier 28 that controls the electric gear of the cross-rail feeds of the tool. While compensating for increased cutting forces by reducing feeds limits the cutting process in terms of productivity, then the mine hole wear and the slicing frequency of chip formation register these minimum feeds and stop the machining process.

The simultaneous four-controlled loop system operation make it possible to measure displacement in three directions, stabilizing the tool tip along the XYZ axes taking into account dimensional depreciation and the mine hole wear by controlling the position of the tool tip and also the control in the Y-axis direction is carried out by a separate servo drive as well as control of the cross feed in the X- and Y-axis directions. ACS makes it possible: To measure the diameter of the workpiece to be processed in the plane traversing through the tip of the tool along the normal and the workpiece surface; To measure the shape of the workpiece in the cross and cross-rail sections during processing; To roughness degree control of the machined surface and the surface in the cutting plane of the workpiece; To control the position of the workpiece axis, stabilizing it relative to the artificial gage length with a predetermined accuracy.

4. Results and conclusion

The developed technology for processing low-rigid shafts based on the automatic control technological system of the RV 104 lathe turning system, as well as the control systems and units for it, was implemented at OJSC "Azotremmash" in Togliatti during manufacturing of low-rigid shafts for pumping and compressor equipment of chemical units. Typical shafts produced had a length to diameter ratio of more than 40 and a centerline tolerance of 0.015 to 0.025 mm (runout 0.03 to 0.05 mm), which is less than 10 microns per linear meter of shaft length.

The existing technology, based on the factory TPD recommendations, provided for at least six turning passes using a standard travelling rest and three stabilizing operations carried out in the turning intervals and also the possibility of correction by bending. The diameter of the workpiece was 1.5-2 times larger than the diameter of the shaft so that it was possible to gradually "select" the turning operations for arising permanent technological deformations of the shaft axis bending and bring them to the specified straightness tolerance. According to the factory technology, the duration of the production cycle reached 57 hours, and the operations of the heat setting and flexible straightening took up to 80% of the cycle time. In addition, based on practically accumulated experience, to output one shaft with a given accuracy along the straightness of the axis, a deliberately increased batch of defective products was launched. This batch was launched as batch for shafts in which the length-diameter ratio 85 that reached a ratio of 11:1. New turning technology using an ACS at two, most three passes using a self-centering travelling rest, solves the problem of accuracy and productivity. At the same time, using the developed installation and calculation methods, a forecast of permanent technological deformations of the bending was carried out for each pass, the over dimension was distributed, the processing modes and tool parameters were appointed.

When the shafts were manufactured according to the factory technology, the change of permanent technological deformations of the bending was unstable and unpredictable that led to a large number of defective products. During the low-rigid shafts manufacturing, according the new technology, the maximum values of the permanent technological deformations of the bending were steadily decreasing. This led to the fact that additional bending corrections were excluded and also was ensured the specified accuracy of the shaft along the straightness of the axis.

The use of the proposed method on the RV 104 lathe when processing non-rigid workpieces makes it possible: To increase the processing accuracy by 6-8 times in the cross and cross-rail sections, depending on the cutting conditions and parameters of the control systems; To increase the processing productivity by 5-8 times by increasing the cutting conditions without reducing the requirements for processing accuracy; to reduce the production cost of shafts by 3.2 times due to the elimination of energy-consuming heat setting operations; To reduce the number of turning passes and defective products; To reduce material consumption by 1.8 times due to the use of workpieces with a diameter as close as possible to the diameter of the finished shaft; To reduce the energy-consuming processing by 82 times; To reduce the duration of the technological cycle by 2.4 times; To extend the operational life of the shafts in the units by 1.6 times due to the achieved accuracy and its preservation in the future during normal operation. This is due to the fact that the dynamic characteristics of the TS system are not decisive during processing. Artificially built-in control loops and their transient index determine the behavior of the TS.

The ACS of the mechanical shafts processing in which the length-diameter ratio 85 from structural and corrosion-resistant steels for pumping and compressor equipment of chemical units was designed and implemented at OJSC "Azotremmash", which made it possible to ensure the specified accuracy in the straightness of the shaft axis.

References

[1] Drachev O I 2005 *Manufacturing Techniques for Low-stiff Axisymmetric Parts* (Saint Petersburg: Politekhnika) p 289

[2] Drachev O I and Kravtsov A N 2012 *Automatic Control of the Process of Low-stiff Parts Turning* (Togliatti: CJSC ONIKS) p 250

[3] Starkov V.K. 1989 *Machining Stability and Quality Management in Automated Production* (Moscow: Mashinostroenie) p 296

[4] Svinin V M, Samsonov A, Savilov A V and Pyatykh A 2018 Self-oscillation suppression when turning non-rigid shafts using spring tools and the spring headstock center *Proc. Int. Conf. on Modern Trends in Manufacturing Technologies and Equipment* (Sevastopol: EDP Sciences) p 01059

[5] Palagnyuk G G 1995 Theoretical substantiation of the method of reducing the energy intensity of the cutting process and its multiparameter active control *Diagnostics and control in technical systems* 74 – 82

[6] Kraynev D, Bondarev A and Tikhonova Z 2020 Mathematical apparatus for predicting cutting tool life in turning process *Proc. Int. Conf. on Industrial Engineering* (Sochi: Springer International Publishing) pp 1107-14

[7] Tikhonova Z, Kraynev D and Frolov E 2020 Thermo-emf as method for testing properties of replaceable contact pairs *Proc. Int. Conf. on Industrial Engineering* (Sochi: Springer International Publishing) pp 1097-105

[8] Nesterenko P S, Tchigirinsky J L and Nesterenko E N 2020 Analysis of influence of strains of technological system elements on machining accuracy under turning of non-rigid shafts based between centers *Proc. Int. Conf. on Industrial Engineering* (Sochi: Springer International Publishing) pp 1193-201

[9] Zakovorotny V L and Gvindjiliya V E 2019 The dependence of tool wear and quality parameters of the surface being cut on dynamic characteristics *Obrabotka metallov (*tekhnologiya, oborudovanie, instrumenty) **21** 31–46

[10] Lapshin V, Zakovorotny V, Gvindjiliya V and Anh N D 2019 Tool wear due to deformation displacements during metal turning *XV Int. Scientific-Technical Conf. "Dynamic of Technical Systems"* (Rostov-on-Don: American Institute of Physics Inc) p 030002

IOP Conf. Series: Materials Science and Engineering 1155 (2021) 012034 doi:10.1088/1757-899X/1155/1/012034

Functional product map design in lifting and transportation machine building digitalization conditions

M A Prilutskaya, A D Murukina and O O Podoliak[1]

Ural Federal University named after the first President of Russia B.N. Yeltsin, Ekaterinburg, Russia

[1]E-mail: o.o.podoliak@urfu.ru

Abstract. In the Industry 4.0. mechanical engineering is undergoing major changes. Combining industry and digital technologies for the flexible and efficient production creating is a main idea of the Fourth Industrial Revolution. The article proposes a high-tech product functional map designing approach based on the author's functions classification, taking into account the specifics of products designed within the Industry 4.0 framework. Based on the warehouse floor transport market trends study, as well as the forklifts consumers survey results, a functional map of a product designed at the Russian lifting and transport engineering enterprise was proposed.

1. Introduction

The transition to Industry 4.0, as a means of the manufacturing industry competitiveness increasing, has been started in many countries all over the world. The Fourth Industrial Revolution aims to combine industry and digital technologies to create flexible and efficient goods production and high-quality services with the lowest cost [1]. This transition will make the market more personalized through the development and production of products individually for each customer.

Certainly, according to the Industry 4.0, not only mechanical engineering as a whole becomes a subject for a huge transformation, but also its subindustries are involved in this process (for example, the lifting and transport engineering). Consumers of this industry are increasingly in need of automated technology that can carry out logistics operations with the least human participation. So, automated storage and retrieval systems (AS/RS) of industrial, and then trading enterprises became a development trend since 1970s and continue their transformation until now [2]. A huge variety of equipment and its models are used for charging and discharging operations: trolleys, forklifts, trucks, reach trucks, etc. Consumer's requirements for equipment used in such warehouse complexes are the possibility of integration into warehouse system, the long-term battery life, complex operations automation functions (necessary to prevent the emergencies), etc. The increasingly complicated functionality of lifting and transportation equipment requires manufacturers of new flexible design methods, taking into account the customer's individual requirements.

This article proposes high-tech product functional map designing approach based on the author's functions classification. This classification takes into account the specifics of products designed within the Industry 4.0 framework. As an object of investigation, the electric forklift is considered.

2. Research method

One of the main approaches for forming the greatest products value with the lowest costs at the designing stage is value engineering [3]. Value engineering determines the product's functional structure, which is necessary for the formation of the product's consumer value [4]. This analysis involves functional map designing (including external and internal product functions), relationships and links between functions based on a functional approach.

Functional map designing is one of the most difficult tasks, because on the one hand, it should ensure the product competitiveness in terms of "price-quality" ratio, on the other hand, a rationally designed functional map will gives opportunities for the further product improvement with the lowest modernization cost.

To design a functional product map, it is necessary:

1. Product operating conditions have to be determined. For lifting and transportation equipment, this is the general organization characteristic (activity scope, cargo turnover, average weight and dimensions of the transported cargo, etc.); warehouse parameters; used equipment and its technical characteristics; operation simplicity evaluation of lifting equipment.

2. A list of the main technical characteristics have to be determined. These characteristics are important for consumers, according to this list consumers make a decisions about the brand and model of lifting equipment.

3. Technical characteristics of competitor products have to be analyzed.

4. The lifting equipment basic model characteristics have to be established as well as the optional parameters that increasing consumer properties and the product price.

Once all the necessary parameters for the designed product have been set and the function carriers have been defined - product design elements (parts, assemblies, aggregates, etc.), it is necessary to formulate functions that indirectly or directly will create the product's consumer value. All functions performed by the object and its components are formulated based on the definition that "function" is the content of the action or process for which the product is intended, as well as from the position of execution and providing them with the required consumer properties [5].

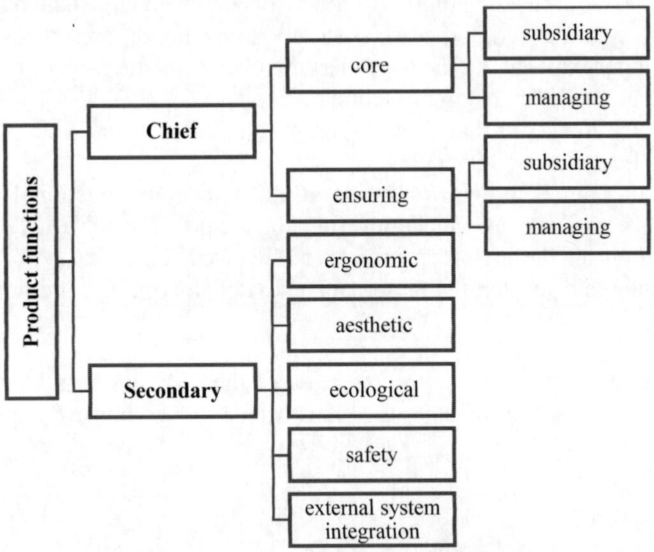

Figure 1. Product function classification.

After the product functions complete list is ready functions have to be divided into the main classification groups, which allow to describe the product and its component elements purpose and functioning, as well as their interaction with the external environment - the lifting equipment operator and the warehouse complex as a whole. The authors propose the following functions hierarchy (figure 1), which allows to describe a complex product and to realize the effective modernize it the functional or process areas.

All functions of a complex product can be divided into 2 groups, which in turn are also divided into their constituent subfunctions.

The chief function is a function that reflects the purpose of an object (the purpose of its creation). It answers the question what the object is doing or what the object is for [6].

Secondary function is a function that provides the consumer properties of an object together with the chief function. It answers the question how an object interacts with the external environment, performing the chief function.

All secondary functions can be divided into:

- ergonomic - functions that create convenient conditions for operator;
- aesthetic - functions that create a person's perception of an object (appearance);
- ecological - functions preventing environmental damage;
- safety - functions that create conditions for the object's (and its components) safe operation during its interaction with the external environment;
- external system integration - functions that allow to integrate the product into the warehouse complex information system.

The core functions are functions that contribute the chief function implementation through the work execution.

Ensuring functions - functions that allow to ensure the object serviceability, creating the necessary conditions for the chief function implementation.

Subsidiary functions - functions that allow to perform a top-level function. Functions that reflect actions and relationships within an object, due to the construction principle, execution peculiarities. Answer the question how (technologically, structurally) the main function is implemented [6].

Industry 4.0 focuses on production digitalization associated with large data base that need to be read, collected, analyzed, systematized, processed, stored, transmitted, presented and much more [7]. Consequently, the authors singled out the secondary functions category: " external system integration," as well as a separate group of "managing functions".

Managing functions - functions that allow to control the functions interaction both on the same level and on different levels (upper and lower).

Basic and supporting functions description will allow to detail the functional model up to the second level, description of subsidiary and managing functions- up to the third level. Therefore, proposed functions' classification for the high-tech product a will allow to design both the basic and optional product concept, taking into account both internal and external working conditions.

3. Results

The respondents' questionnaire survey (the structure of the companies participating in the survey is presented in figure 2) showed that recently there has been a to warehouse complexes automation level increasing tendency.

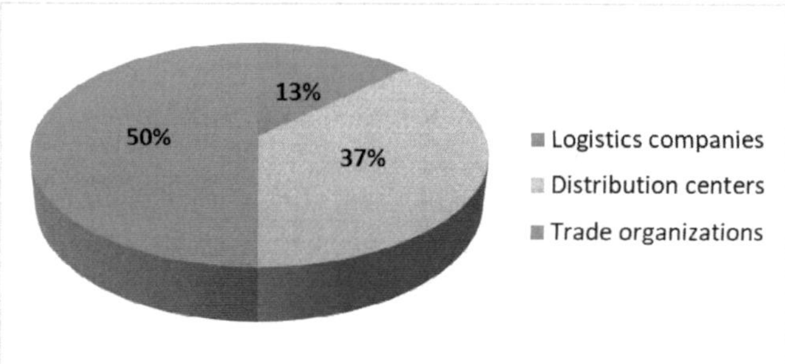

Figure 2. Target focus group structure.

The intra-warehouse floor-type transport market increasingly requires unmanned vehicles that are able to partially or completely replace the operator's work, that reduces costs, while increasing the speed and accuracy of order processing. Many foreign manufacturers offer automated analogues for the existing models of lifting equipment on the market, for example, Linde and Jungheinrich. So, 90% of the organizations participating in the survey have the warehouse information system installed. The examples of such systems are: Warehouse Management System, Logistic MANAGER 8, Axapta, 1C Warehouse Management and others. 50% of respondents have fully or partially integrated forklifts into this system. The result confirmed the necessity and relevance of applying the more complex hierarchy for the functions proposed by the authors.

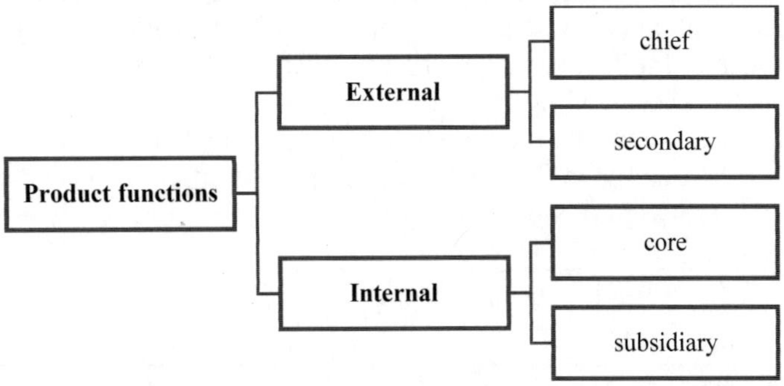

Figure 3. Functions' classification widely used in the value engineering.

In the existing function classification system (figure 3), which is usually used in value engineering and functional cost engineering [8], there are no functions responsible for the automation and management of the object. However, today the equipment is controlled not only by the operator and internal software, but also by an external warehouse system into which this object is integrated, that inevitably leads to a product functional map complication during its designing and requires functions' new classification.

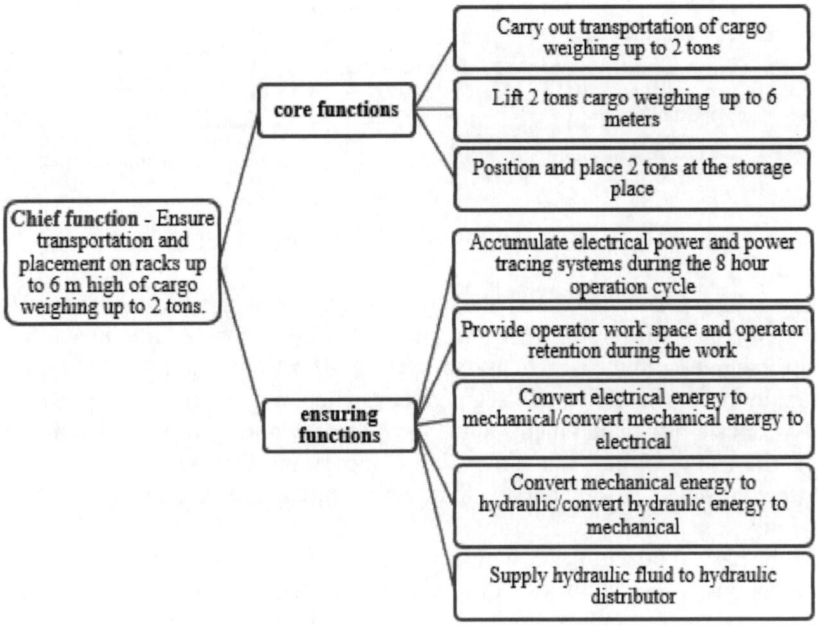

Figure 4. Electrical forklift main function diagram.

Based on the investigation the functional purpose of the product was determined - ensure transportation and placement on racks up to 6 m high of cargo weighing up to 2 tons. Based on the proposed classification of complex product functions, all electric forklift functions were divided into main classification groups. Figure 4 shows the diagram of the product's main function. The designed forklift has three main functions that facilitate the destination function implementation. The main functions of the upper level can also be divided into the lower level main functions, which, in turn, will consist of subsidiary and managing functions.

In addition to the trend of warehouse integration into a single information system for their remote control and management, a warehouse equipment consumers survey revealed another development trend - safety. More and more manufacturers are beginning to offer various solutions to increase the safety of work in the warehouses. As noted by many respondents, forklifts today are equipped with various sensors and systems that avoid collisions with racks and other obstacles, prevent the warehouse workers injury (for example, the RFID tag reading system on the clothes of employees). Therefore, ergonomics, safety and the forklifts integration into the warehouse complex information system became the main areas in the secondary functions map designing. Developed forklift secondary functions diagram is presented in figure 5.

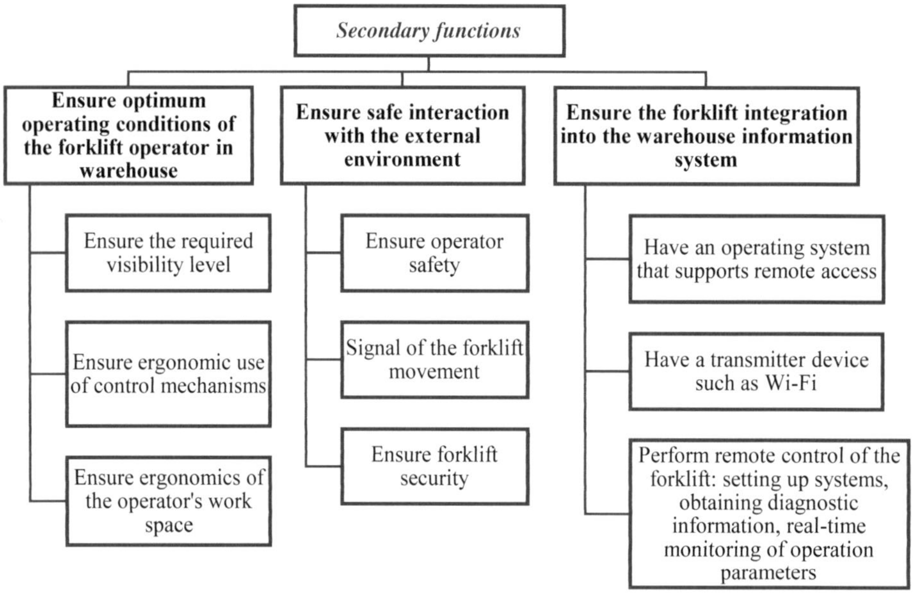

Figure 5. Electrical forklift secondary functions diagram.

Secondary functions associated with the object interaction with the external environment determine the product appearance, the convenience and safety of its operation, environmental friendliness, and the possibility of integration in the warehouse complex information system. The composition of functions is determined by the particular market segment consumers requirements and preferences and can vary significantly from simpler requirements to the most specific ones. Operating conditions (the warehouse parameters) will also determine the variety and composition of these functions. Therefore, secondary functions affect the object design, but do not affect the product destination and actions it must perform, and work quality.

Note that the basic model of the designed product can include both functions related to the main and secondary functions, which allows to design an individual functional map of the product for each customer. The personal product concept creation takes into account the individual consumer requirements and the product operation peculiarities that will help to increase the manufacturer's competitiveness in the market, as well as contribute the transition for Industry 4.0

4. Discussion

The proposed approach to the functional product map development can be used not only in the lifting and transport engineering, but also in other high-tech mechanical engineering industries. To describe the full composition and classification of the functional characteristics of the forklift, a survey of potential customers was conducted, which made it possible to determine:

- basic operating and maintenance conditions for forklifts;
- compile a list of the most significant technical characteristics with the greatest value for the consumers;
- categorize product parameters into "base parameter" and "optional parameter".

The author's functions classification developed for the complex products functional map designing allows to take into account the specifics of Industry 4.0, where new industries are created on the basis of digital technologies, the objects of which interact with each other on the basis of the large data base exchange. Multi-level management of such objects requires the corresponding functions of the products being designed, and therefore new approaches to such products development. It should be noted that the complication of products and their interactions does not allow creating a universal functions classification value engineering method, which could be unified for all products and industries. The proposed classification can be supplemented by other types of functions depending on the complexity of the designed product and its control structure during operation.

References

[1] Philbeck T and Davis N 2018 The fourth industrial revolution *Journal of International Affairs* **72(1)** 17-22

[2] Roodbergen K J and Vis I F A 2009 A survey of literature on automated storage and retrieval systems *European journal of operational research* **194(2)** 343-62

[3] Murukina A, Tipner L, Prilytskaya M and Kalinina N 2020 Marketing aspect in the 'Value Engineering' in the new products development of machine-building enterprises *IOP Conf. Ser: Mat. Sc. and Eng.* **966** 012097 doi: 10.1088/1757-899X/966/1/012097

[4] Leber M and Bastic M 2014 Value Analysis as an Integral Part of New Product Development *Procedia Engineering* **69** 90-8

[5] Fernandes M T 2015 Value Analysis Going into a further dimension Engineering *Technology & Applied Science Research* **5(2)** 781-9

[6] Karpunin M 1987 *The practice of functional-cost analysis in the electrical industry* (Moscow: Energoatomizdat) p 288

[7] Lasi H, Fettke P, Kemper H G, Feld T and Hoffmann M 2014 Industry 4.0 *Business & information systems engineering* **6(4)** 239-42

[8] Abdullah A and Adesta E 2015 Implementation of functional analysis using value analysis value engineering (VAVE) *ARPN J. Eng. Appl. Sc.* **10 (21)** 10072-6 doi: 10.10072-10076

IOP Conf. Series: Materials Science and Engineering 1155 (2021) 012035 doi:10.1088/1757-899X/1155/1/012035

Methods to improve vibroacoustic characteristics of mining machines

A I Podgorny, A V Kudrevatykh, N V Kudrevatykh, A S Ashcheulov and A S Ashcheulova

Federal State Budgetary Educational Institution of Higher Education "T.F. Gorbachev Kuzbass State Technical University", Vesennya St, 28, Kemerovo, Russia

E-mail ascheulovas@kuzstu.ru

Abstract. At machine-building enterprises, serious attention is paid to the issue of improving the products quality. Noise and vibration are one of the main indicators that characterize the quality, comfort and reliability of machines, so the reduction of noise and vibration level is one of the most important scientific and technical problems of national engineering. Along with the economic aspects, the reduction of machine noise contributes to the ecological state improvement, as well as the working and living conditions of workers. Noise and vibration are one of the most common adverse factors that worsen working conditions in production, construction and transportation.

Numerous medical studies have proven the harmful effects of prolonged high-intensity noise. The rhythm of heart activity changes, blood pressure increases, hearing deteriorates, and the process of fatigue accelerates under the influence of noise. Vibration disease and cochlear neuritis are the most common among occupational diseases and are about 30 % of the total number of occupational diseases. Noise is one of the factors that reduce labour productivity; in addition, it can cause accidents and injuries at work. It is established that the noise reduction from 5 to 10 dB can lead to the labour productivity increase by 5 %.

The necessity to increase labour productivity, achieve notable social and health effects in order to make work in transport, construction and production more attractive, makes the task to ensure comfortable working conditions, eliminate adverse factors that affect health such as noise and vibration the priority.

In addition, noise reduction in transport and industry is cost-effective. Estimates show that potential economic effect of noise reduction can reach 1-2 billion roubles per year. Noise reduction produced by the machine by 1 dB automatically increases its value by 1 % or more on the world market.

In this regard, it is natural that a lot of attention is paid to solving the problem of noise reduction all over the world. In many scientific works, the physical foundations of the radiation nature and sound propagation, the processes of creating noise by various mechanisms, methods for evaluating and measuring sound, methods and means of noise protection are described.

Much attention is paid to noise reduction in mining engineering. However, despite the efforts of scientists, designers, and machine builders, the problem of mining transport noise is currently very relevant. At least 30 to 50 % of produced and operated mining machines exceed the existing permissible noise.

Until recently, the attention of researchers and designers in the field of mining engineering was paid to ensuring normal acoustic environment at the operator's workplace. This was due to the traditionally higher noise requirements of the MM operator, the rationing of which is stricter in Russia than abroad. The cabin was used as a method of noise control, which is a reliable mean in terms of vibroacoustic qualities, but structurally complex with a high-cost control, but which effectively reduces noise in the workplace. At the same time, the problem of external noise, which has currently received increasing attention, is still unsolved.

Attempts to use "passive" methods of reducing external noise by noising over and shielding radiation sources for a number of economic and technological difficulties have not gone beyond the scope of researches and have not found wide application in mining engineering.

In addition, the structure analysis of various road construction machines allowed to conclude that the design and production of road construction equipment is carried out without taking into account its future acoustic parameters, i.e., the machine design does not initially include solutions that allow to obtain the specified acoustic characteristics of the created machine.

The methods used to combat the noise of already created machines, which do not provide differentiated approach to the study of the process of noise generation and radiation, necessitate the use of expensive and technically complex methods of noise reduction, involving complete isolation of radiation sources or the operator's workplace.

As practice shows, such methods of noise reduction are quite effective in terms of noise reduction, but are economically unprofitable, difficult to implement in mass production, and inconvenient to operate.

Therefore, the purpose of this study is to develop methods to predict and reduce the noise of the power units of mining machines at the design stage, as well as in the process of their experimental refinement by acoustic parameters.

The development, tuning and exploitation of 'low-noise' mining equipment should be based on clear ideas about the causes that lead to machines acoustic radiation, which should serve as a theoretical basis for specifying the requirements for machines design, as well as the principles of mining machines design (MMD) with specified vibroacoustic characteristics. Numerous studies in the field of machine noise reduction have established that in all cases, the solution of noise and vibration protection problem of models that came into production and are in operation is economically less feasible than at the design stage. At the same time, the task of developing or selecting a set of measures that provide the specified vibroacoustic parameters at the minimum (in the limit of the minimum possible) costs, taking into account the requirements of manufacturability, convenience and reliability in operation, should be already solved at the earliest stages of machine development. The introduction of methods to reduce the noise load even at the stage of production of mining machines will reduce the harmful effects of vibrations on both a machine operator and equipment reliability. The introduction of the same methods to already operating equipment is not always economically justified and technologically possible, in this regard, it is necessary to develop other methods to improve the machine vibration and noise characteristics. [1, 2, 3]

The complexity of the processes leading to acoustic radiation, a large number of radiators makes it difficult to obtain an objective picture of mining machines noise generation. The basis of such models is the following basic physical prerequisites: the noise occurs in the source because of the impact or other interaction of parts, during the flow and pulsation of gas flows in the systems and mechanisms of the power unit, during machine unit vibrations on the suspended span under the influence of various power factors. The sound field excited by the source is formed in various conjugate elements of the machine structure. The sound energy is partially reflected or absorbed [4, 5, 6]. The acoustic radiation occurs due to the transfer of the sound energy part from the point of disturbing force application, to the external surfaces of the machine that emit sound. It should be noted that for mining machines, it is the vibration of the external surfaces that is one of the most significant noise sources.

The practice of dealing with the machine noise has developed two interrelated methods of noise absorption. These are measures aimed at reducing noise in the excitation source and absorption of acoustic energy in the path of sound waves propagation. Noise reduction in the source is usually achieved by reducing the intensity of the power excitation, as well as reducing the sound conductivity and sound-emitting ability of the radiator. [7, 8, 9] Reducing the intensity of power actions in various sources is achieved in a number of cases due to difficulty to implement design measures aimed at reducing gaps in joints and units, reducing aerodynamic vortices and pulsations, balancing rotating parts, affecting the flow of the working process, etc. Reduction of sound conductivity and sound-emitting ability of the structures is achieved mainly by breaking the synchronism of vibrations of the radiating surfaces, reducing the radiation area, increasing the coefficient of inelastic losses in the parts materials.

Noise reduction in the way of its propagation is achieved by the use of sound insulation, sound absorption, vibration isolation, as well as a number of other means from the acoustics arsenal. [10, 11]

The practical possibilities of modern methods of machine noise reduction are quite wide. Individual measures, as well as their complexity, are able to provide any required reduction in machine noise. However, it is necessary to take into account not only the acoustic efficiency of the measures, but also the economic criteria that determine the ability to solve the problem of achieving the specified vibroacoustic parameters with acceptable material costs.

As practice has shown, this task is the most efficiently solved only if the machine is designed from the initial design stages, taking into account the necessity to obtain, along with the specified performance and efficiency characteristics, the specified vibroacoustic indicators.

At the same time, in relation to mining machines, the priority issue is the design or selection of a power generation system that meets technical and economic requirements of this machine type and provides the required vibration and acoustic characteristics. It should be noted that the possibilities of acoustic improvement of the internal combustion engine design used for mining machines are wider than in vehicle engines. A prerequisite for this is the possibility of a wider variation of the engine parameters in mining transport in order to obtain the required values of sound levels to the detriment of their weight and size characteristics. [12, 13, 14]

Considering this, in order to ensure the specified values of the effective power and vibration-acoustic parameters of an engine, the optimal combination of parameters must be based on the requirements for power generation system of a particular type of mining equipment, taking into account the specifics of their technological functioning. In this situation, it is possible to obtain the specified vibroacoustic parameters of an engine by reducing its rated speed (de-forcing by speed). The reduction in effective power can be compensated by the use of pressure charging.

In addition, for many mining machines, the requirements for weight and dimension are not among the primary ones. Here, the option of restoring the engine power to the required values by increasing the working volume is not excluded. Despite the weight increase, dimensions, and metal consumption, this method of noise reduction has a number of advantages compared to the expensive, structurally complex, as not sufficiently developed method of noising over and capsulating of power units.

One of the most significant reserves for reducing the noise level of mining machines power units is pressure charging. Thus, the use of pressure charging, according to experts, reduces the noise from the combustion process in diesels by 18 dBA and brings its levels to values lower than the levels of mechanical noise, which allows, all other things being equal, to reduce the sound power emitted by the engine by more than half.

The use of gas turbine pressure charging also has a positive effect on the noise levels generated by the exhaust system. The gas turbine in the exhaust system leads to the pulsation smoothing of the exhaust gas flow, which reduces the noise levels of the exhaust process by 10-12 dBA. However, as practice shows, the noise levels of non-muffled exhaust systems, even in the presence of turbocharging, are significant. Therefore, at present, it is accepted and considered reasonable to combat with their acoustic radiation by using autonomous anti-noise devices for this purpose.

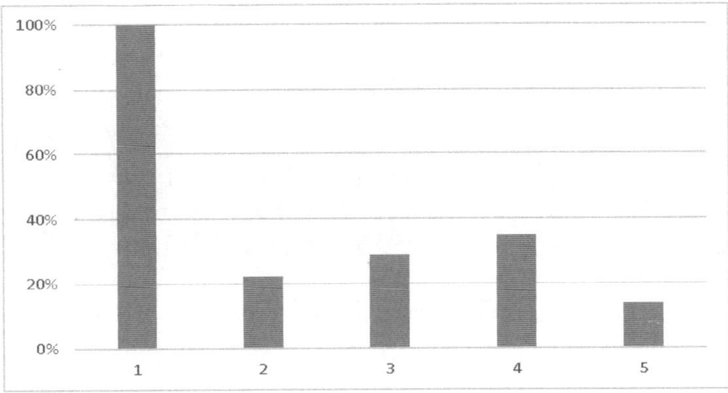

Figure 1. General balance of acoustic power of the motor grader: 1 - motor grader, 2 - transmission, 3 - engine, 4 - exhaust, 5 - intake

Under the existing standards for the noise of mining machines, the acoustic radiation of the exhaust system of the power unit does not significantly affect the external noise, if the acoustic efficiency of the silencer is at least 25 dBA. With the existing noise levels of engines used in mining engineering equal to 120-125 dBA, the noise levels of the exhaust process equipped with a silencer, not exceeding 95 dBA, practically exclude the exhaust system from among the noisiest sources of the machine.

Such muffling characteristics are provided with both reactive silencers and active silencers with sound-absorbing material. The effectiveness of active silencers, in which the sound energy is reduced due to its absorption by the sound-absorbing material, is quite high, but in practice, such silencers are used extremely rarely due to the rapid coking of the sound-absorbing material. In this case, reactive-active silencers are more acceptable.

As practice has shown, for the intake systems of mining machines power units, there is no need to use additional noise silencers, since the air purification systems of diesel engines, made in accordance with the standards requirements, are effective noise silencers.

Reducing the noise and vibration of machines in operation is technically more complicated. In this case, a significant reduction in noise and vibration levels at an operator's workplace is often achieved by using noise and vibration-absorbing coatings of the control cabin, but effective reduction of external noise requires the use of a set of measures, often difficult to implement without changing the machine design significantly.

In this situation, it is possible to solve the question of the possibility and rational strategy for improving the vibroacoustic characteristics of mining machines only on the basis of an analysis of the conditions for the sufficiency of certain influences on the structure in order to reduce the acoustic activity of the most intense sources that form its sound field.

Here, the necessary initial stage of the study of the machine acoustic qualities is the determination of its main contribution sources to the overall picture of radiation. The methodology of such studies should be based on a number of techniques that allow isolating the noise of individual systems and machine units. One of these techniques can be the method of sequential exclusion of sources, the essence of which is to install effective noise suppression structures on individual noise sources, as well as other measures that allow excluding the influence of this source to the overall noise of the machine. The specific content of the methodology is individual for each type of mining equipment.

Based on the assessment of the contribution of each of the sources and the determination of the most intense radiators, the question of appropriate means to influence them, allowing to exclude these sources from the number of the ‘loudest’ ones, is decided.

The practice of noise reduction in mining equipment, as well as in related areas of mechanical engineering, such as autotractor production, has developed a number of typical methods of influencing the most intense sources of machine noise, which, as a rule, are the exhaust system, as well as the internal combustion engine, as an independent source and as the exciter of the design of a vehicle itself. In some cases, they allow to achieve results that meet the requirements of legal documents at low costs and changes in the machine design.

In addition, the most successful design solutions found in the process of finishing measures to improve the acoustic qualities of the design of machines in operation can be used in solving problems of obtaining the specified vibroacoustic characteristics of newly designed machines.

References

[1] Kudrevatykh A, Ashcheulov A, Ashcheulova A, Karnadud O and Rattmann L 2019 Actual technical condition assessment of a motor-wheel gear of a dump truck belaz based on the operating oil parameters *E3S Web of Conferences IVth International Innovative Mining Symposium* **105** 03021

[2] Kudrevatykh A, Ashcheulov A, Ashcheulova A and Urazbaeva K 2020 Comparative analysis of gearboxes wear in excavators and mining trucks *E3S Web of Conferences International Innovative Mining Symposium* **174** 03017

[3] Markovnina A I, Mazunova L N, Makarov V S, Belyakov V V, Vakhidov U SH and Erasov I A 2019 Imitation modeling as a forecast of transport flows self-organization *IOP conference series: materials science and engineering* **632** 012001

[4] Klubnichkin V E, Dyakov A S, Klubnichkin E E, Zakharov A Y, Suchenina A S, Basmanov I V and Vakhidov U S 2019 Experimental evaluation of stability and controllability of domestic and foreign made utility terrain vehicles *Journal of physics: conference series* **1177** 012045

[5] Kudrevatykh A V, Ashcheulov A S and Ashcheulova A S 2019 Actual technical condition assessment of mine excavators' slewing gear based on the operating oil parameters *IOP Conf. Series: Materials Science and Engineering* **537** 032033

[6] Strizhak A, Vakhidov U, Lipin A, Dorofeev R, Sogin A V and Mazunova L S 2019 Modelling of vehicles with rotary-screw propulsion unit along water-flooded substructure *Journal of physics: conference series* **1177** 012039

[7] Vakhidov U, Molev Y, Slusarev A and Tuntyev I 2019 Load mode analysis of pipe-layers operating on swampy grounds *Journal of physics: conference series* **1177** 012037

[8] Manianin S, Molev Y, Vakhidov U, Makarov V, Belyakov V, Dorofeev R, Zonov I and Tyurikov P 2019 Experimental research of the effect of track chains pre-static tension on the mobility of a tracked vehicle *ICMT 2019 - 7th international conference on military technologies* 8870025

[9] Manyanin S E, Vaxidov U SH and Maslov K A 2020 Ways to improve the quality of products using hot isostatic pressing *Journal of advanced research in technical science* **22** 94–7

[10] Dadonov M, Kulpin A, Ostanin O and Sulei-menov E 2019 Distribution of static normal reactions to wheels of open-pit dump trucks depending on the longitudinal and cross sections of the open-pit road *E3S Web of Conferences. International Innovative Mining Symposium* **105** 03009

[11] Stenin D V, Kulpin A G, Kultayev E E, Kulpina E E and Borovtsov V A 2016 Influence of service conditions of quarry dump trucks on the thermal state large-size tires *Coal in the 21st Century: Mining, Processing and Safety* 116-19

[12] Kosolapov A and Krysin S 2017 The raising influence of information technologies on professional training in the sphere of automated driving when transporting mined rock *E3S Web of Conferences* **21** 03012

[13] Stenin D V, Stenina N A and Bakanov A A 2016 Evaluation of the open pit vehicles loading influence on the reliability of motor – wheel reducers *Coal in the 21st Century: Mining, Processing and Safety* (Kemerovo: ATLANTIS PRESS) pp 256-260

[14] Dadonov M, Kulpin A, Borovtsov V and Zhunusbekova A 2020 Effect of aerodynamic loads on redistribution of normal reactions of quarry dump trucks tires *E3S Web of Conferences Vth International Innovative Mining Symposium* **174** 03018

Traction mobile device for rotary tillage

M M Litvinova, I S Fedorchenko and N V Sukhenko

Reshetnev Siberian State University of Science and Technology, 31, Krasnoyarsky Rabochy Av., Krasnoyarsk, 660037, Russia

E-mail: marg32883@gmail.com

Abstract. The organization of the territory of a forest nursery is the division of its territory into parts that have different economic purposes in order to make the most efficient use of the nursery area and ensure maximum mechanization of work. The establishment of a new nursery begins with the establishment of its outer boundaries. The most optimal shape of the nursery area is square or rectangular. The area of the nursery is divided by a network of main and secondary roads into branches, fields and quarters. The width of the main roads - main roads, circumferential and located perpendicular to the length of the side of the fields - should be 6-8 m, and the secondary ones, located along the long side of the fields, should be 3-6 m. The branch of the nursery is located taking into account the soil, relief and hydrological conditions. Plots with the best structure and granulometric composition of fertile soils, flat relief, protected from adverse effects of winds are allotted for the sowing department. The auxiliary area of the nursery is 25% of the useful area for large nurseries, 30 - for medium nurseries, 40% - for small nurseries.

1. Introduction

For the most efficient use of machines and tools in large nurseries, the field length is recommended to be 300-500 m and more, and the width - 40-120 m, in medium and small nurseries, the field length is 50-200 m, width - 20-60 m.

The territory of the nursery is divided into parts with the help of a theodolite, after clearing it of shrubs, stumps, logging residues and planning the area.

The nursery is divided into fields and quarters with a network of roads. In small and medium-sized forest nurseries, the area of the quarter is equal to the area of crop rotation. To prevent soil washout and erosion on the slopes, the nursery fields are divided in such a way that the main soil cultivation is carried out across the slope. In areas damaged by wind erosion, crop rotation fields are located perpendicular to the prevailing direction of harmful winds or with a deviation from it by more than 30° [1].

In work [2], the authors proposed a new scheme for organizing a forest nursery with circular fields, shown in figure 1, as well as design solutions for mechanized tillage in them [3,4,5].

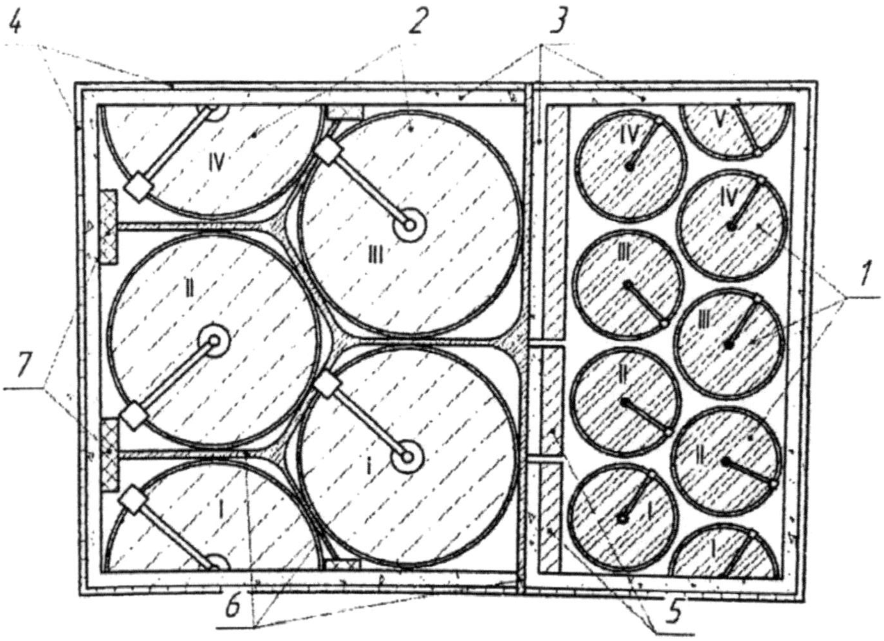

Figure 1. Scheme of the organization of a forest nursery with circular fields: 1 - sowing department; 2 - school department; 3 - forest belts; 4 - hedge; 5 - areas for digging planting material, composting and other economic purposes; 6 - roads and lanes with hydronates; 7 - technological sites; I-V - crop rotation fields.

According to the proposed scheme, the forest nursery consists of a sowing department, organized from round fields, equipped with traction devices I to V crop rotations, a school department, also organized from round fields, equipped with servo rotation devices. The sowing and school departments are separated by a forest belt and surrounded by hedges. The nursery is equipped with areas for digging in planting material, preparation of kospost, technological sites and a network of roads and strips with hydronates.

2. Materials and methods

With the aim of increasing the automation of technological processes for growing planting material in forest nurseries and partial replacement of machine-tractor traction with other types of drive, the author of [6] proposed methods for organizing circular forest nurseries.

Circular forest nursery with radius R_{max} consists of a central square with a radius R_c, consistently increases in diameter of ring-shaped crop areas with radius R_j^k, separated by tramlines of width a and consisting of rows of plants with a radius R_i^c.

To organize a circular field of the nursery, it is necessary to set the radius of the central site, it will be determined with the parameters a,b и h. Also determine the number of seeding lines with radii $R_{i,j}^{c,k}$ and the maximum radius of the nursery R_{max}.

Formula for the maximum radius of the first ring-shaped crop area:

$$R_{1\,max}^{k} = R_c + a + H \tag{1}$$

Where, a – the width of the technological track;

H – sowing field width.

The maximum radius of the second annular sown area is defined as:

$$R_{2\,max}^{k} = R_1^k + a + H = R_c + 2(a + H) \tag{2}$$

IOP Conf. Series: Materials Science and Engineering 1155 (2021) 012036 doi:10.1088/1757-899X/1155/1/012036

To grow planting material, it is also necessary to determine the radii of the seeding rows.

Radius of the first and subsequent sowing rows R_1^c a circular nursery is defined as:

$$R_1^c = R_c + a + b \tag{3}$$

Using the above formulas, it is possible to design a circular forest nursery of the required size with the required number of planted areas and the number of seeding rows.

The authors propose a new design solution [7] for performing mechanized technological operations during soil cultivation in forest nurseries when growing forestry crops.

The traction mobile device for circular tillage, figure 2, contains a central support with an arrow connected with the possibility of rotation relative to the support through a vertical shaft with a splined connection and a hinge, a working trolley, a chain drive sprocket for moving the working trolley connected by means of a bearing support to a vertical shaft. The vertical shaft is additionally equipped with a clutch capable of being connected to a sprocket and connected to a motor mounted on the boom by means of a vertical shaft drive, and configured to be disconnected from the vertical shaft drive. The central support is made in the form of a gearbox, the drive shaft, with the coupling of which is connected to the vertical shaft, and the driven shafts are connected to the drills.

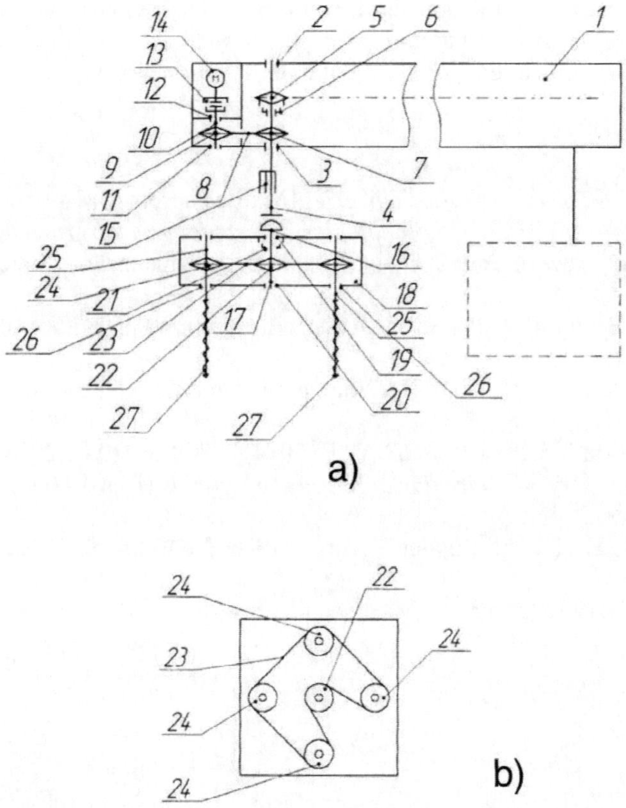

Figure 2. a) – diagram of a traction mobile device for circular tillage in side view; b) – diagram in a top view of the gearbox of a traction mobile device for circular tillage.

Traction mobile device for circular tillage works as follows. Before starting work, the traction mobile device is located in the center of the circular field to be processed. The operator closes the clutch 13, then opens the clutches 6 and 21. Turns on the engine 14. Torque is transmitted from the motor shaft 14 through the clutch 13 to the shaft 11. From the shaft 11 through the drive sprocket 9, chain drive 8 and the driven sprocket 7, the torque is transmitted to the vertical shaft 4. From the vertical shaft 4 through the splined connection 15 and the hinge 16, the torque is transmitted to the drive shaft 17 of the gearbox 18, where from the drive sprocket 22 through the chain transmission 23 the torque is transmitted to the driven sprockets 24 and the driven shafts 25 and drills 27 associated with them. When assuming the working position by the gearbox 18, i.e. after burying the driven shafts 25 with drills 27 into the soil to the required value, the operator turns off the engine 14, opens the clutch 13 on the shaft 10, and closes the clutches 6 on the vertical shaft 4 and 21 on the drive shaft 17 of the gearbox 18. Thus, the sprocket of the chain drive of the drive for moving the working carriage 5 becomes stationary relative to the gearbox housing 18. The mobility of the device is achieved due to the possibility of installing the device anywhere in the treated area without preliminary preparation of the base for fixing the central support. The use of the proposed invention improves the efficiency of processing a circular field [7].

3. Conclusion

In Russia, there are more than 1,450 nurseries with an area of over 15,730 hectares, the bulk of which are concentrated in the Altai Territory, the Republic of Tatarstan, Tver, Perm, and Vladimir regions. By the method of placing the territory and the nature of the technological process, circular and linear are distinguished. Circular nurseries have an area in the form of a circle, along the edges of which natural plantations are preserved. The proposed technical solution was specially developed for circular forest nurseries in order to replace wheeled and tracked tractors with environmentally friendly technologies. This installation works by means of a traction drive from an electric motor, which favorably affects the soil cover and reduces emissions into the atmosphere, and also provides an increase in productivity, mobility and efficiency due to the possibility of its installation at any point of the treated surface [8].

References

[1] Novoseltseva D I and Smirnov N A 1983 *Guide to forest nurseries* (M.: Lesn.prom-st) 140-56
[2] Nevzorov V N, Kholopov V N and Byrdin P V 2016 *Technologies and equipment for growing seedlings in nurseries with circular organization of fields* (Krasnoyarsk: Krasnoyarsk State Agrarian University) 15-42
[3] Kholopov V N , Nevzorov V N and Byrdin P V Pat. of the Russian Federation RU 2 608 810 C2 / No. 2015112417, 06.04.2015
[4] Byrdin P V , Nevzorov V N and Kholopov V N Pat. of the Russian Federation RU 2 591 035 C1 / No. 2015107764/13, 05.03.2015
[5] Byrdin P V Pat. of the Russian Federation RU 2014 100 768 A / № 2014100768/13, 10.01.2014
[6] Byrdin P V 2014 *Forest mechanized nursery with circular organization of fields* (Bratsk: BSU) 156-9
[7] Kholopov V N and Fedorchenko I S Pat. of the Russian Federation RU 2 646 050 C2 / No. 2016100313, 11.01.2016
[8] Litvinova M M and Fedorchenko I S 2019 Krasnoyarsk. Mech. engine. 244-8

Derivation of crack propagation velocity formula for investigation dynamic problems

I A Magomedov[1,3], Z A Dzhabrailov[1] and M V Abubakarov[2]

[1] Faculty of information technology, Chechen State University, 32 Sheripov Street, Grozny, 364024, Russia

[2] Faculty of Technology and Management in Education, Chechen State Pedagogical University, 62 Isaev Avenue, Grozny, 364068, Russia

[3] E-mail: ismwork@mail.ru

Abstract. The following thesis will illustrate derivations of crack propagation's formula for investigating brittle and elastic bodies. This work will look at the static and dynamic cases for crack initiation and propagation. Some approaches will be outlined in this work, which are the technics widely used in the field of mechanics for investigating the crack propagation and initiation related problems. At the end of the work crack propagation velocity formula will be derived.

1. Introduction

Currently, we live in the world of gadgets and software packages. Without a doubt, gadgets integrated our life thoroughly. The applications of such are everywhere. They can be found in the various activities of daily life routines, for example, mobile phones, vacuum cleaners, powerbanks, irons, TVs, etc. It is clear that every devicee/gadget is run by software. The complexity of software will depend on the basis of it use. One can assume and outline that iron will use simple software to do its functions. Whereas mobile phones will have more complex software embedded to perform complex tasks. Therefore, software plays a vital role in modern-day devices. Similarly, software packages also are greatly influencing our life and pushing the boundaries to the next level. Alike to gadgets they can be found everywhere. We use apps on our mobile phones, computers and etc. They are well used in the industry to perform some complex tasks that can not be done in real life. Also, they are utilized to fasten the analytical process and to minimize the cost. The following software packages can be seen in the industry as the standards, such as Solidworks, Inventor, Abaqus, and others. Each software package has its own unique technique when it comes to the analysis, but the basis is the same. The software packages use engineering formulas to analysis the problem. Hence the software packages are itself a combination of formulas (brought up to a certain degree of complexity) that are aimed to tackle dawn the given problem. As the work will look up on the fracture and particularly on crack propagation the formulas used in the analysis will be outlined. The crack propagation velocity formula will be derived by using different methods and formulas.

Fracture can be traced back to when human-made structures start occurring. And the problem getting worse each year as structures become more complex in the technological world. There is much to be learned, however understanding of the failure of material and the ability to prevent such failure is increased since World War II. Dramatic failure related to fracture happened during World War II with Liberty ships, these ships were fabricated during the war in a short time and sustained serious fractures as an outcome of shortage of time and money to build with consideration of standards, in this case mainly using the riveted method, but instead, large cylinders were welded together. This ominous event gave sufficient knowledge on how to avoid the cracks occurrence and their further spreading in the structure of ships. Therefore, transformed the way of perception and the attitude toward how the it have to be treated to avoid failures in general.

2. Different approaches to deal with the fracture

In this paragraph, different approaches will be outlined for both static and dynamic. To work crack propagation velocity formula the below presented approaches will be utilized. These approaches can be utilized in different scenarios. The problems (tasks) can be simple, complex, require some amendments such as idealizing the structure for the utilization of formulas and etc.

2.1 LEFM

Fracture mechanics (FM) is the part of mechanics which is mainly aimed at dealing with the formation and propagation of cracks in the different structures. LEFM is a tool of representing qualitative measure in the brittle material (BM), which resists the formation of erratic crack growth. Griffith was the first to propose the theory that the main contributors for the fracture occurrence are flows and existing cracks [1]. The tool can be defined utilizing Griffiths approach (Energy Balance). Similarly, it can be described by the Stress intensity Factor, which is used in this particular time. If one is about applying one of two before mentioned approaches, the correct mode should be selected for the valid outcome. The following three modes are chosen when dealing with a such problem: opening, sliding and tearing. Fracture in highly BMs occurs under mode I (opening), and brittle crack propagates with displacement surfaces on its planes in a manner to minimize the result of tensile stresses, which is shear loading. Earlier the the concept of FM was mostly valid to materials that obey Hooke's law [2]. Simply saying LEFM is a simple approach, yet sophisticated, theory of fracture, which capable of dealing exclusively with a sharp crack in elastic bodies [3]. Crack propagation comes in two forms: transgranular and intergranular. In the first case as the name suggests it goes through the material grains and the second one goes along the boundaries due to its weakness. So, the tool (LEFM) can be used if the following steps are taken into account, for instance, the presence of an ideal condition in which all materials should be elastic, except for a small area close the crack tip (SACCT). The area occurs to be in an inelastic state due to the high stress in the region. However, the plastic zone formed by the high stress near the crack tip is relatively small to the linear dimension of the structure and even the size of the crack, then inelastic region and its effect to the analysis can be neglected. Therefore, LEFM theory can be utilized to analyse the fundamental behaviour of any material, which tends to crack in the manner mentioned above.

2.2 Griffith Energy Balance

Griffiths applied the idea of the first law of thermodynamics to understand the crack formation, which states when a system changes from a non-equilibrium state to an equilibrium state, there is a net decrease in energy [1]. Around the 1920s he established study of fracture in the glass. The approach worked with the sharp cracked specimen with the existence of the idealized condition in which all materials (bodies) should be elastic and under static loading condition. Also, the following material should have to have brittle manner but for the small area close the crack tip. The idea of his work was focused on taking account of the global energy balance of the system rather than on crack tip stresses directly. As well to work out the energy's amount required to propagate the initial crack in a thermodynamically permanent way. He was aware of Inglis studies on stress concentration near the crack tip when he started his studies of fracture in the glass. However, Inglis work introduces mathematical complexity: stresses at the crack tip approaches infinity with a perfectly sharp crack and utilisation of such solution would predict near-zero strength in the material, which will lead to the rupture of bounds near the crack tip with a very small load applied. Stress at the crack tip of the stress plate was defined by Inglis as illustrated below in formula 1. Griffith illustrated the total energy of the system can be written as in formula 2. Where W is the system's total energy or (work done), dU_γ is the energy available for the formation of new crack surfaces and dU_E is the internal potential energy. Or it can be written as the following formula 3. Where the right side represents the fracture, resistance associated with $G_{critical}$, which is the crack growth resistance of the material R, and the left side shows the elastic strain release rate G. A crack will propagate with the energy release rate criterion $G \geq R$. Griffith was capable to illustrate that internal strain energy is as shown in formula 4 by utilising Inglis description for the stress at the tip of the crack. Where the potential energy of an uncracked plate is represented by U_{Eo} with the plate thickness B. $U_\gamma = 4aB\gamma$ as a crack requires a formation of two new surfaces, therefore factor 2 is required and where γ is the surface energy or energy required to create the new cracked surface. The surface energy value per unit area is constant, hence it is a linear function of a, but the stored strain energy released in the growth of crack is a function of a^2, and is, therefore, parabolic [4]. Using formula 3 it can be written as shown below in formula 5 by setting the derivative of total energy $U_E + U_\gamma = 0$. Finally, by solving it to get formula 6 for fracture stress. Where E is young's modulus.

$$\sigma_m = \sigma(1 + \frac{2a}{b}) \tag{1}$$

$$\frac{dW}{dA} = \frac{dU_\gamma}{dA} + \frac{dU_E}{dA} = 0 \tag{2}$$

$$-\frac{dU_\gamma}{dA} = \frac{dU_E}{dA} \tag{3}$$

$$U_E = U_{Eo} - \frac{\sigma^2 a^2 \pi B}{E} \tag{4}$$

$$\frac{dU_E}{dA} = -\frac{\sigma^2 a\pi}{E} \text{ or } \frac{dU_\gamma}{dA} = 2\gamma_s \tag{5}$$

$$\sigma_f = (\frac{2E\gamma}{\pi a})^{\frac{1}{2}} \tag{6}$$

2.3 Stress intensity factor (SIF)

Stress intensity factor (SIF) is a friendly user tool when dealing with the prediction of stress near the tip of the crack as a result of remote load. It also defines whether there will be propagation of the crack and measures the real forces, which are applied at the tip of the crack. There is three type of modes of loading and each of them produces $\frac{1}{\sqrt{r}}$ singularity at the tip of the crack. Hence the proportionality constants scaling factor k and function of θ depends on the mode and it is better to introduce stress intensity factor K, which equals to $k\sqrt{2\pi}$ or $K = k\sqrt{2\pi}$. Mode I loading is considered as loading stress is acting perpendicular to the crack plane and hence stress field ahead of the tip of the crack can be written as illustrated below formula 7. It can be observed that formula 7 is on the left side with the assumption of usage of an isotropic leaner elastic material and at the right side rearranged making the scaling factor K object [1]. The assumption is made for the non-linear behaviour of the material at the tip of the crack that the effect of the region on the solution can be negligible due to the small area compared to the crack length. Otherwise different methods are applicable such as: J-integral and crack opening displacement [4]. Once, $\theta = 0$, then shear stress equals zero and function $f = 1$ and the following formula 8 is assumed to be, which describes stress in the close vicinity of the crack tip. Formula 8 only valid close to the crack tip where the $\frac{1}{\sqrt{r}}$ singularity dictates the stress field as mentioned before [3]. Hence stresses near the tip of the crack grows in proportion to K.

$$\lim_{r\to 0} \sigma = \frac{K}{\sqrt{2\pi r}} f(\theta) \text{ or } K = \lim_{r\to 0} \sigma\sqrt{2\pi r}\, f,(\theta) \tag{7}$$

$$\sigma = \frac{K}{\sqrt{2\pi r}} \tag{8}$$

$$K = \sigma\sqrt{2\pi r} \tag{9}$$

2.4 J-Integral

Stress intensity factor was applied to analyse the problem with plastic zone near the crack tip, which is considered too small if to be compared with the size of the specimen (small scale yielding). However, in event of the presence of a plastic zone, but considered as large-scale yielding stress intensity factor is not applicable due to the providence of precise results. J-Integral comes as a significant tool when dealing with elastic-plastic FM. It can be defined analogous to the strain energy release when considered non-linear formula 10 or equal under liner elastic loading. When solving the problem, the J Integral is not dependent on the path around the tip of the crack [5] and it can be easily defined experimentally. J-integral can be defined as depicts formula 11.

$$J = \frac{dU_E}{dA} \tag{10}$$

$$\delta_c = \frac{4K_c^2}{\pi E\sigma_y} \tag{11}$$

*2.5 Crack opening displacement (*COD)

Wells was a pioneer of COD for the analysis of crack initiation with large-scale yielding. It was said that as the result of strain crack will extend with help of voids growth and coalescence to the tip of the crack. The approach measures the strain at the crack tip and the crack is estimated to extend at a critical δ_c, which is estimated to be as stated in formula 12 by Irwin. Where σ_y is yield stress and its can be assumed small scale yielding with

$$\frac{\sigma}{\sigma_y}. \tag{12}$$

COD can be related to (strain energy release) G by the following formula 13 which is also related to J-integral as it was mentioned before. However, one needs to consider that those relationships formula 14 valid only under linear elastic fracture mechanics.

$$G = \sigma_y\delta_c \tag{13}$$

$$G = J = \sigma_y\delta_c \tag{14}$$

2.6 Irwin

G. R. Irwin was an outstanding scientist in the fields of FM and he was the one who suggested a fundamentally new amendment of the Griffiths theory. Irwin suggested the concept of effective surface energy and changed the theory of Griffith in terms of singular stresses at the tip of the crack and showed that his approach is identical to Griffith's energy approach. Normal stress σ (σ_{yy}) can be defined as in formula 8, which is on the crack plane of linear elastic material. It can be said that when $\sigma_{ys} = \sigma_{yy}$ yielding occurs for plane stress conditions. Replacing the right-hand side with yield strength to before mentioned formula 8 therefore the one can solve for r gives a first-order estimation of the plastic zone size formula 15. It can be observed that the formula 15 has a stress intensity factor, which was introduced by Irwin to relate microstructure characteristics to a local mechanical state near the tip of the crack, which results in providing a very modest description of fracture processes as an uninterrupted process of steady (stable) crack growth to the inception of instability [6].

$$r = \frac{1}{2\pi}\left(\frac{K_I}{\sigma_{ys}}\right)^2 \tag{15}$$

2.7 Mott model

A lot of work has been done to examine BM under rapid loading, under which the material tends to shatter into small fragments. With the higher magnitude more pieces of the shattered brittle body occur. There is much more to be learned in the field of rate sensitivity of BM behaviour. Griffith was the one who introduced the energy balance formula 2 of the system of BM under static conditions describing crack propagation as mentioned before. Mott was the one who started analysing dynamic crack propagation using the analytical approach. Considering propagation of centrally located through a crack in an infinite plate. Similarly, as Griffith, Mott developed a model, which defines crack propagation with the dynamic energy release as illustrated in formula 16. Where F is the work done by external force and U represents the internal work. To enable the equation function as it was intended the following assumptions were suggested: constant speed for the traveling crack in the body under uniform time dependent stress, which is applied perpendicular to the plane of the crack, crack speed $\dot{a} = (v)$ is comparably small than shear wave speed and the displacement and stress fields are similar as in the electrostatic problem [7]. Using assumptions before mentioned Mott modified Griffiths formula 2 to formula 16 by including kinetic terms E_k, to reach a configuration of keeping the total energy constant. E_k for a cracked plate (assuming a unit thickness) was derived to be as depicted in formula 17 for a crack with repeat propagation and taking account of the stress field. Where k is constant dependent on crack speed and σ is stress perpendicular to the crack plane, and ρ is the mass per unit area of the plate. Substituting kinetic energy formula 17 and formula 4 to modified Griffiths energy balance formula 16 then one can get Mott's dynamic energy release rate formula 18 as illustrated below. Where γ is work of fracture or surface energy in ideally brittle material, but with $G(t)$ equal to constant γ at crack initiating stage assumed to be constant as well. The equation gets form as illustrated below. Now the right side can be differentiated with respect to a ignoring the kinetic terms and the one can have the following formula 20. The one can rearrange the equation making $a = a_o$ mane subject, which results in providing from fracture stress initial static crack length shown in formula 21 [4].

$$G(t) = \frac{dF}{dA} - \frac{dU_E}{dA} - \frac{dE_k}{dA} \tag{16}$$

$$E_k = \frac{1}{2}k^2a^2v^2\rho(\frac{\sigma}{E})^2 \tag{17}$$

$$G(t) = \frac{1}{2}\frac{d}{da}[\frac{\pi\sigma^2a^2}{E}] - \frac{1}{2}\frac{d}{da}[\frac{k}{2}\rho a^2v^2\left(\frac{\sigma}{E}\right)] = 2\gamma \tag{18}$$

$$2\gamma = \frac{1}{2}\frac{d}{da}[\frac{\pi\sigma^2a^2}{E}] - \frac{1}{2}\frac{d}{da}[\frac{k}{2}\rho a^2v^2\left(\frac{\sigma}{E}\right)] \tag{19}$$

$$2\gamma = \frac{\pi\sigma^2a}{E} \tag{20}$$

$$a_o = \frac{2E\gamma}{\pi\sigma^2} \tag{21}$$

2.8 Crack velocity

It is important in the dynamic investigation propagation of crack to be able to determine the resulting effect of the kinetic term to formula 21, which can be observed by the speed of the crack Equation. The equation of speed of the crack was first derived by Mott making wrong assumption that $\frac{dv}{dA} = 0$, but later Dulaney, Brace and Berry modified Mott's analysis and derived Equation 22 shown below. Where $c_o = \sqrt{\frac{E}{\rho}}$, which is the speed of sound for 1D wave propagation through the material. Wells and Roberts obtained an approximation for k and after few more assumptions, they illustrated that $\sqrt{\frac{2\pi}{k}} \approx 0.38$. When $a > a_o$ the crack speed reaches its value of limit and formula 22 becomes unstable, which is

IOP Conf. Series: Materials Science and Engineering 1155 (2021) 012037 doi:10.1088/1757-899X/1155/1/012037

$0.38c_o$ by Wells and Roberts. Freund was the one who done more in-depth numerical analysis and come up with the following relationship formula 23. Where c_r is Rayleigh (surface) wave speed and it has been observed performing experimental analysis that speed of the crack does not generally reach c_r and (maximum is $0.6c_o$) [4], which typically range from 0.2-0.4c_o. For the distinct event when $\gamma = 0$, a propagating crack is simply a disturbance on a free surface which have to move at the Raleigh wave velocity. In formula 22 and formula 23 the limiting velocity is independent of the fracture energy, hence the speed of the crack that can be reached should be equal to c_r for all γ [2]. A more detailed dynamic analysis of Freund assumed that crack length or crack speed does not rely on the fracture energy and the resistance of the material grows with the crack speed [8].

$$v = \sqrt{\frac{2\pi}{k}} c_o(1 - \frac{a_o}{a}) \tag{22}$$

$$v = c_r(1 - \frac{a_o}{a}) \tag{23}$$

3. Conclusion

To conclude, the following work was done to derive crack propagation velocity formula. Describing the methods used in the field of fracture mechanics some formulas were derived. Using the formulas and rearranging them in the right manner the crack propagation velocity formula was derived. The work illustrated that to derive the formula some conditions should be met, which are necessary to correctly utilize the method and derive the formula. It is obvious that to derive such a formula many researchers have contributed by performing various studies. With the right formulas and software packages run on powerful computing machines and yet it is not enough to analysis fracture. There is much more to be learned and improved in the field of mechanics.

References

[1] Wagner, Fleming and Petri (n.d.) *Static and dynamic crack propagation in brittle materials with XFEM*
[2] Anderson T 2005 *Fracture mechanics* (Boca Raton, FL: Taylor & Francis)
[3] Ksm.fsv.cvut.cz. *FRACTURE MECHANICS "WHAT IS FRACTURE MECHANICS"*. [online] Available at: http://ksm.fsv.cvut.cz/~sejnom/download
[4] Crump T, Ferté G and Mummery P 2015 Dynamic fracture effects on remote stress amplification in agr graphite bricks. SMiRT-23, 80
[5] Janssen M, Zuidema J and Wanhill R 2005 *Fracture Mechanics: Fundamentals and Applications*. (Taylor & Francis)
[6] Yarema S 1996 On the contribution of G R Irwin to fracture mechanics *Mater Sci* **31(5)** 617-23
[7] Watanabe M 2006 Higher Order Mode-I Elastodynamic Stress and Displacement Fields of Dynamic Crack Propagation in Brittle Materials *JSME Int. J.* **49(1)** 107-15
[8] Freund L B 1976 *Dynamic Crack Propagation* (The Mechanics of Fracture, American Society of Mechanical Engineers, New York) pp 105-34

Synthesis of the control loop for the coordinate of the horizontal plane of the quadcopter information-measuring and control system

Dinh Ba Phuong

MIREA - Russian Technological University, 78 Vernadsky Av., Moscow, 119454, Russia

E-mail: nhoht_ck49@yahoo.com

Abstract. The article discusses the translational motion of a quadcopter on a horizontal plane. Based on the analysis of this movement, a control loop for the horizontal plane coordinate of the information-measuring and control system of quadcopters was synthesized. The results obtained provide a stable flight of the quadcopter, a decrease in the influence of wind disturbances and monitoring of various objects when performing special tasks.

1. Introduction

Due to its compactness, high stability and maneuverability, quadcopters (QC) are currently widely used [1,2]. Control of the quadcopter is carried out by changing the rotation speed of one or more engines and, therefore, the corresponding torques and traction and lifting characteristics change [2]. Despite the advantages of quadcopters, control is a difficult problem. The main problems when controlling a quadcopter are: loss of balance of movement in case of damage to any engine; lack of high quality control loops; lack of appropriate sensors; relatively high power consumption and significant inertia in control.

To solve the above problems, an information-measuring and control system of a quadcopter (IMS QC) has been developed. This system is a set of technical means for processing measurement information and solving control problems [3]. Since the quadcopter has 6 degrees of freedom, including 3 for displacements (X, Y, Z) and 3 for orientation (R, I, P), to improve the IMS QC it is necessary to synthesize a control loop along the height Z, horizontally X with subordinate contour R (in Y it is similar to subordinate contour P).

In [3], the synthesis of the control loop for the Z-coordinate of the IMS QC is considered, and in this paper, the peculiarity of the synthesis of the control loop for the X coordinate in the horizontal plane with a subordinate contour along R of the information-measuring and control system of the quadcopter will be considered.

2. Research results

To synthesize the control loop of a quadcopter along the X axis, we will consider its motion only along the horizontal OXY plane. In this case, the quadcopter needs to rise to a certain height Z0, and then its movement along the X axis is carried out by changing the speeds of the two engines (ASD1 and ASD2) and the angles between the axes of these vectors and the local coordinate system. We assume that at the beginning of movement at altitude Z0 the quadcopter has zero velocity and the centre of mass coordinate is located at point O (0, 0, Z0) with orientation angles (0,0,0), and after a certain time dt the quadcopter is at point O1 (X1, 0, Z0) with angles (dr,0,0). Figure 1 shows a kinematic diagram of a quadcopter with ASD1 and ASD2 engines involved.

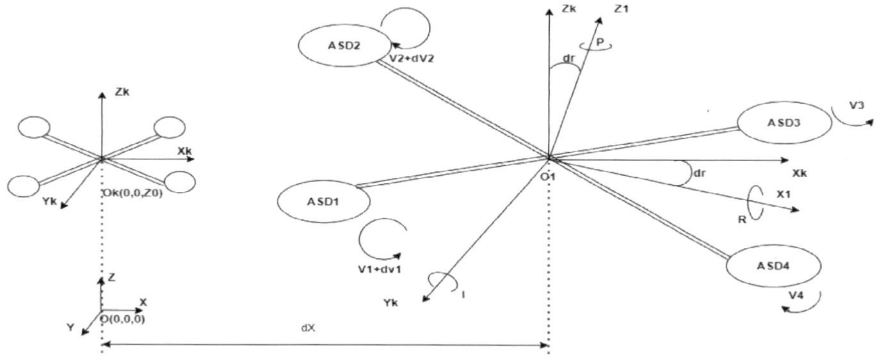

Figure 1. Kinematic diagram of a quadcopter with 2 engines involved.

Here OXYZ is the coordinate system of the centre of gravity of the spacecraft on the ground; OkXkYkZk – QC coordinate system at height Z0 at the initial time of movement along X; O1X1Y1Z1 – fixed coordinate system of the spacecraft is rigidly connected to the body; R, I, P – orientation angles; ASD1–ASD4 – speed-controlled electric drives; V1–V4 – rotation speeds respectively of electric motors ASD1 – ASD4; dv1, dv2 – additional speeds to ASD1 – ASD2; dr – is the deviation of the orientation angles when the spacecraft body is rotated relative to the position of the centre of mass; dX – distance between point O1 and point Ok.

Let us analyze the linearized differential equations of quadcopter motion along the X axis and get [3,4,5]:

$$\frac{d^2x}{dt^2} = Kmx \times I - \frac{dx}{dt} \times Kx \tag{1}$$

$$\frac{d^2r}{dt^2} = Km \times du \tag{2}$$

where Kmx is the coefficient of mutual influence of the quadcopter coordinates; I is the pitch parameter; Kx – coefficient of aerodynamic resistance along X; du is the difference in force rise between ASD1 and ASD3. Based on this equation, it will be possible to construct a structural diagram of the control loop of the IMS QC along the X coordinate with a subordinate contour R, which is shown in figure 2.

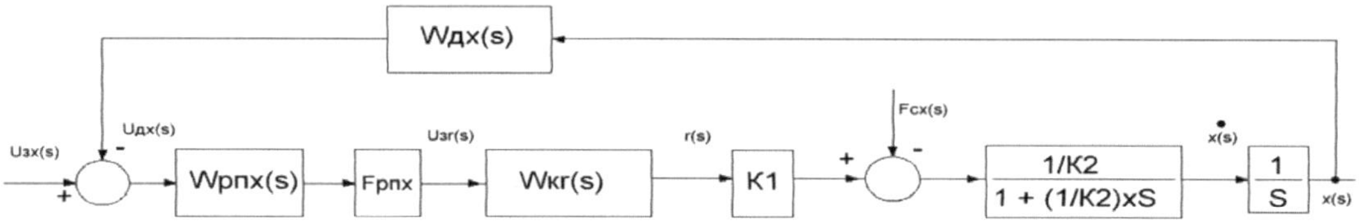

Figure 2. Block diagram of the control loop of the IMS QC along the X coordinate.

The following designations are adopted:

IOP Conf. Series: Materials Science and Engineering 1155 (2021) 012038 doi:10.1088/1757-899X/1155/1/012038

Wрпх (S) - transfer function of the position controller along the X coordinate; Fрпх - nonlinearity of the position controller; Wкr (S) transfer function of the slave control loop along R; K1, K2 - respectively, the coefficients of thrust, counteraction along the X coordinate; Wдх (S) - X position sensor transfer function; Uзх (S), Uдх (S) - Laplace images, respectively, of position setting signals and from position sensors along the X coordinate; Fcx (S) - Laplace image of the resistance force along the X coordinate. Let us synthesize the slave contour R with the transfer function Wкr (S). The block diagram of the control loop of the IMS QC along the R coordinate is shown in figure 3.

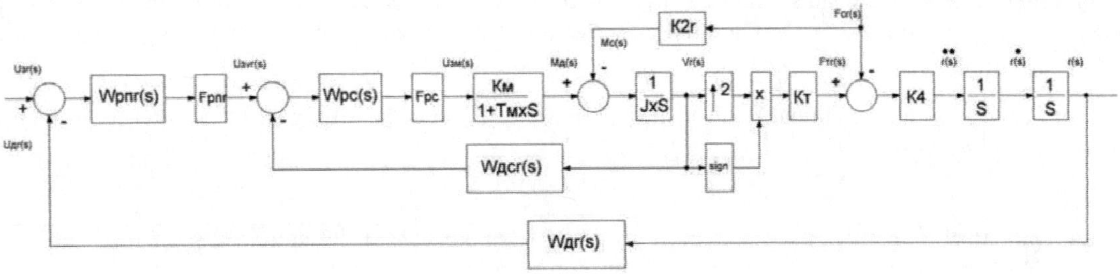

Figure 3. Block diagram of the control loop of the IMS QC along the coordinate R.

where: Wрпr (S) is the transfer function of the position controller along the R coordinate; Fрпr - nonlinearity of the coordinate position controller; Wpc (S) - transfer function of the speed regulator according to R; Fpc - nonlinearity of the speed regulator; Km - ASD torque transmission ratio; Tm is the time constant of the ASD torque loop; J is the moment of inertia of the ASD motor; Wдr (S), Wдсr (S) - transfer functions, respectively, of position and speed sensors along the R coordinate; Uзr (S), Uдr (S) - Laplace images, respectively, of signals for setting signals for setting the position and from position sensors along the R coordinate; Uзvr (s) ¬ Laplace images from speed sensors along the R coordinate; Uzm (S) - Laplace image of the torque setting signal in ASD; Md (S) - Laplace image of the torque developed by the ASD electric motor; Fтх (S), Fcx (S) - Laplace image, respectively, of thrust and drag force along the X coordinate; KT, K2r - respectively the coefficients of thrust, counteraction along the X coordinate and along the R coordinate, $K4 = \frac{1/К2}{1+(\frac{1}{К2})}$.

In order to reduce the influence of the wind load on the QC movement (reflected on the model by the force Ff), when determining the transfer function Wрпr (S), the position contour should be adjusted to a symmetric optimum [6]. In this case:

$$\text{Wрпr(S)} \times \text{Wкs(S)} \times \text{К}_\text{Т} \times \frac{\text{К4}}{S} \times \frac{1}{s} \times W\text{дпr}(S) = \frac{1+4\times\tau1\times S}{8\times\tau1^2\times S^2\times(1+\tau1\times S)} \quad (3)$$

Where Wкс (S) is the closed-loop transfer function for the coordinate velocity R. When using a low-inertia position sensor, Wкс (S) $= \frac{\frac{1}{\text{Кдс}}}{1+1{,}5\times\text{Тм}\times S}$ and $\tau1 = 1{,}5 \times \text{Тм}$ [9]

Then:

$$\text{Wрп}r\text{(S)} = \frac{0{,}05\times\text{Кдс}\times(1+6\times\text{Тм}\times\text{S})}{\text{Кдп}\times\text{Кт}\times\text{К4}\times\text{Тм}^2} = \text{Крп}r(1 + \text{Трп}r \times \text{S}) \quad (4)$$

This is the transfer function of the PD controller. In the case of using an inertial position sensor, we have:

$$\text{Wкс}(S) = \frac{\frac{1}{\text{Кдс}}\times(1+\text{Тдс}\times S)}{1+1{,}5\times\text{Тм}\times S} \tag{5}$$

$$\text{Wрпr}(S) = \frac{0{,}05\times\text{Кдс}\times(1+6\times\text{Тм}\times S)\times(1+\text{Тдп}r\times S)}{\text{Кдп}\times\text{Кт}\times\text{К4}\times\text{Тм}^2} = \text{Крпr}(1+\text{Трп}r\times S)(1+\text{Тдп}r\times S) \tag{6}$$

This is the transfer function of the PD controller and PD controller combination, which is practically impossible to implement. Hence follows a practically important conclusion: in the control loop along the R coordinate, it is desirable to use a low-inertia position sensor.

In order to verify the results obtained, mathematical modelling of the control loop of the IMS QC with inertial and low-inertia sensors along the R coordinate was carried out with the following parameters: Крп =250000; Трп =0,006 с; Кдс=0,08; Тм=0,001 с; Кт=0,005 $\frac{\text{Н}\times\text{м}\times\text{с}^2}{rad^2}$; К4 = 0,5 $\frac{\text{рад}}{\text{с}^2}$; Кдп =1; Тдп=0,03 с. Simulation results for positional control, i.e. at Uзпr = 1(t) and disturbing influence Fcr = 10(t-1) are presented in figures 4 and 5.

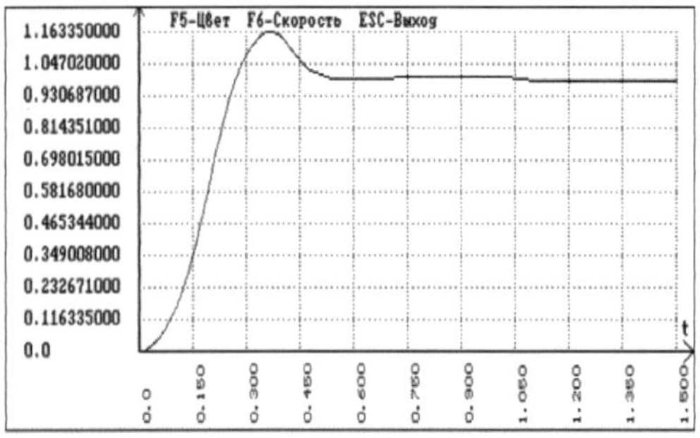

Figure 4. Transient processes in the control loop of the IMS QC along the R coordinate with a low-inertia position sensor.

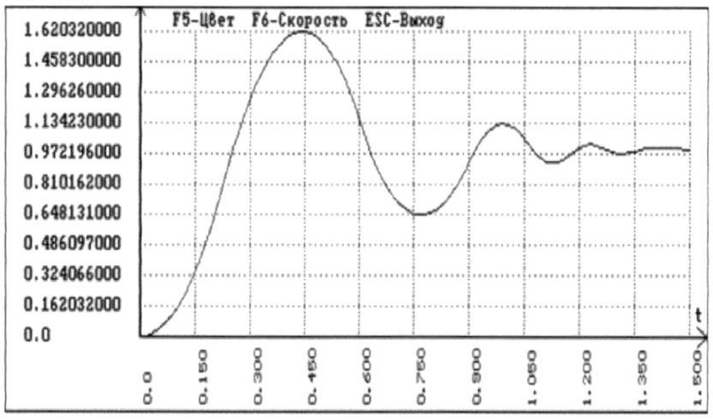

Figure 5. Transient processes in the control loop of the IMS QC along the R coordinate with an inertial position sensor.

Analysis of the obtained transient processes shows that the control loop with an inertial position sensor can be described by a transfer function of the form:

$$\text{Wкr}(S) = \frac{U\text{дr}(S)}{U\text{зr}(S)} \approx= \frac{\text{Кr}}{1+Tr\times S} \tag{7}$$

The transfer function Wкr (S) is obtained to allow synthesizing the control loop along the X coordinate. In order to reduce the influence of the wind load Fc on the motion of the QC, when determining the transfer function Wрпr (S), the position contour should also be adjusted to the symmetric optimum [9]

$$\text{Wрпх}(S)\times\text{Wkr}(S)\times\text{K1}\times\frac{\frac{1}{\text{К2}}}{1+\left(\frac{1}{\text{К2}}\right)\times\text{S}}\times\frac{1}{\text{S}}\times\text{Wдпх}(S) = \frac{1+4\times\tau6\times\text{S}}{8\times\tau6^2\times\text{S}^2\times(1+\tau6\times\text{S})} \tag{8}$$

When using a low-inertia position sensor (Wдп(S) =Кдп) we have ($\tau6 = \text{T}r$):

$$\text{Wрпr}(S) = \frac{0{,}12\times\text{К2}\times(1+4\times\text{Т}r\times S)\times\left(1+\left(\frac{1}{\text{К2}}\right)\times S\right)}{\text{Кдп}\times\text{К}r\times K1\times\text{Т}r^2\times S} = \frac{\text{Крп}r\times(1+\text{Трп}r1\times S)\times(1+\text{Трп}r2\times S)}{S} \tag{9}$$

This is the transfer function of the PID controller.

In order to check the results obtained, mathematical modelling of the control loop of the IMS QC was carried out along the X coordinate when moving by 1m with the following parameters: Крп= 400; Трп1=0,16 с; Трп2 = 0,1 с; К2=10 1/с; К1=2 $\frac{м}{рад}$; Кдп =1; Тдп=0,03 с.; Крм = 30 $\frac{rad \times м}{rad^2}$. The simulation results for positional control, i.e., with Uзпг = 1(t) и Fc = 10(t – 1) are presented in figures 6 and 7.

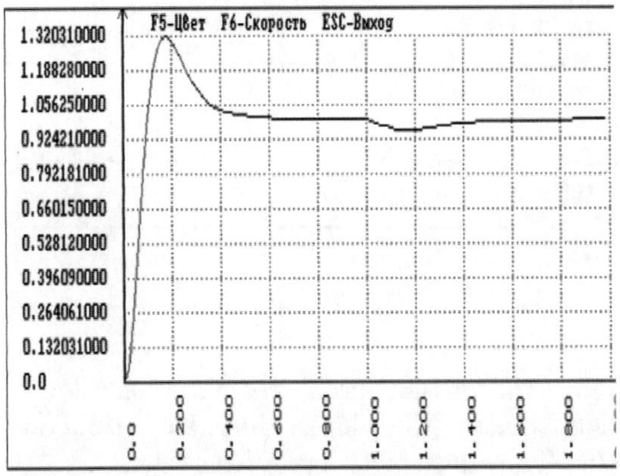

Figure 6. Transient processes in the control loop of the X coordinate in the IMS QC with a low-inertia position sensor.

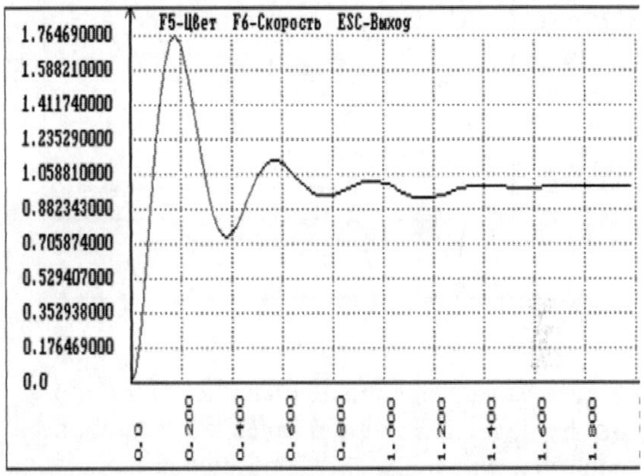

Figure 7. Transient processes in the control loop of the X coordinate in the IMS QC with an inertial position sensor.

It can be seen that the inertia of the position sensor along the X and R coordinates significantly worsens the quality of transient processes - the oscillation increases from 30% to 76%. Consider a control loop with a linearized parabolic controller shown in figure 8.

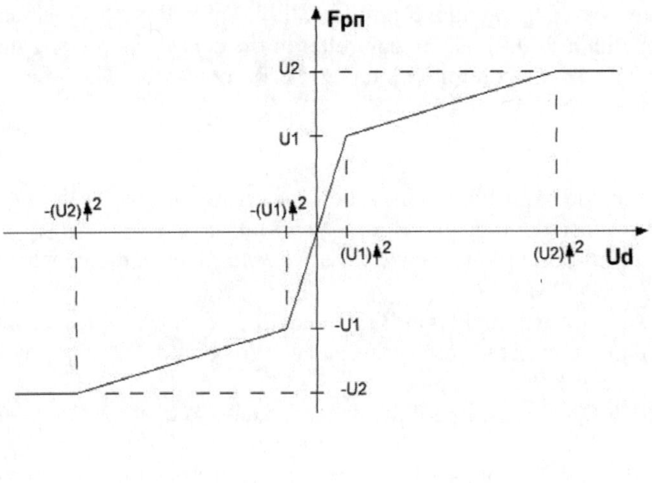

Figure 8. Static characteristic of the linearized parabolic position controller.

IOP Conf. Series: Materials Science and Engineering 1155 (2021) 012038 doi:10.1088/1757-899X/1155/1/012038

The simulation results for positional control, i.e., with Uзпi = 1(t) and Fc = 10(t – 1) with a linearized parabolic controller are presented in figures 9 and 10.

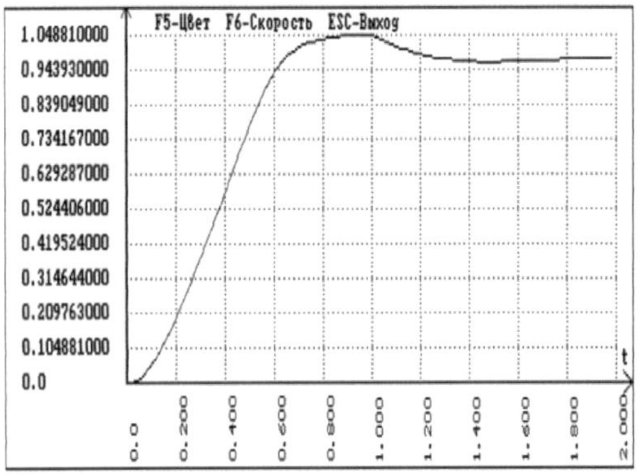

Figure 9. Transient processes in the control loop in the IMS QC with a linearized parabolic controller with a low-inertia position sensor.

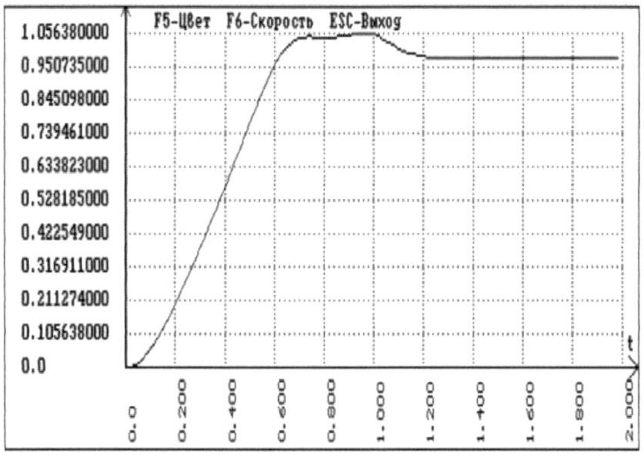

Figure 10. Transient processes in the control loop in the IMS QC with a linearized parabolic controller with an inertial position sensor.

In this case, it can be seen that a control loop with a linearized parabolic controller has good transients..

3. Conclusions

The inertial position sensor significantly degrades the characteristics of the control loop of the spacecraft along the X coordinate and along the R coordinate, making it ineffective, especially under external disturbances.

In the X control loop, fast position and speed sensors must be used in the ASD.

In the control loop along the R coordinate, it is necessary to use low-inertia position and speed sensors in the ASD.

The use of a control loop with a linearized parabolic controller allows providing the necessary characteristics and application of systems with a PID - position controller in the control loop along the X coordinate and with a PD - position regulator in the control loop along the R coordinate. This allows to ensure an acceptable quality of transient processes.

References

[1] Logachev V G and Minin I V 2015 Method of position stabilization and control of spacecraft in space using data from inertial and visual sensors *Fundamental Research* **11-1** 85-91

[2] Epov M I and Zlygostev I N 2012 Application of unmanned aerial vehicles in airborne geophysical reconnaissance *Interexpo Geo-Siberia* **2(3)** 22-7

[3] Rokachevsky O A, Sleptsov V V, Ablaeva A E and Dinh Ba Phuong 2021 Synthesis of the control loop for the altitude coordinate of the information-measuring and control system of the quadcopter *Scientific and Technical Volga Region Bulletin* **1** 38-41

[4] Kolotov M E and Smirnova T A 2016 Decomposition of the linear model of a quadrocopter *Young Scientist* **13(117)**

[5] Sleptsov V V, Grishnyaev I A and Dinh Ba Phuong 2020 The main tasks of designing information-

measuring and control systems of quadcopters *Proc. Sci. and Practical Conf. on Actual Problems of Science and Technology* **VII** (Samara) 34-7

[6] Sleptsov V V and Nasir Ruabhi 1999 Application of state identifiers in tracking systems of industrial robots *Instruments Instrumentation and Automation and Control Systems* (Moscow) 29-35

[7] Yushkin D A and Evdokimov S A 2015 Development of an adaptive fuzzy PID controller for an automatic control and stabilization system for a multi-rotor UAV of the quadcopter type *Actual Problems of Modern Technology Science and Education* **2(1)** 194-8

[8] Afonin V L, Podzorov P V and Sleptsov V V 2006 *Machining Equipment Based on Mechanisms of Parallel Structure* (Moscow: MSUT "STANKIN", Janus-K) p 452

[9] Madani T and Benallegue A 2006 Backstepping control for a quadrotor helicopter *IEEE/RSJ Int. Conf. on Intelligent Robots and Systems* 3255-60

Development and verification of a mathematical model of heating after the metal sample transition to a liquid-solid state in the turbine blades casting process

R A Vdovin[1] and R R Kyarimov

Samara University, 34, Moskovskoe sh., Samara, 443086, Russia

[1]E-mail: vdovin.ssau@gmail.com

Abstract. The paper presents the results of the development of a mathematical model of the melting zone of a discrete-continuous casting installation, and also shows the verification of the results of computer modelling obtained using the developed mathematical model. An important feature of the developed mathematical model is the ability to calculate heat fluxes and temperature fields for all elements of the technological system furnace-flask-filler-mold-metal.

1. Introduction

Blanks of turbine rotor blades of gas turbine engines (GTE) and gas turbine plants (GTP) made of heat-resistant alloys, due to their developed geometric shape, high requirements for the accuracy of geometric dimensions, roughness and quality of the surface layer, can be obtained exclusively by casting with directional crystallization (DC). This method makes it possible to manufacture high-precision blades with a complex spatial shape and developed internal channels of the cooling system; with a minimum allowance or no allowance along the feather of the blade blank due to possible volumetric shrinkage and subsequent machining operations [1].

At the same time, the technological process of casting turbine rotor blades is one of the most complex, time-consuming and expensive methods. The technology of casting blades with a monocrystalline structure significantly increases the cost of the product and complicates the process of manufacturing the blades. Casting with directional crystallization either forms a columnar directional structure of grains, which excludes the appearance of transverse grain boundaries - stress concentrators, or forms a single monolithic crystal, increasing the mechanical characteristics of the blade.

Currently, the use of the DC method in the manufacture of GTE turbine rotor blades is one of the most effective methods. The use of this method makes it possible to improve the main performance characteristics on traditional high-temperature casting alloys of the ZhS class by eliminating the transverse grain boundaries and improving the metallurgical quality of the castings of the turbine blades [2, 3].

Thus, the method of directional solidification, due to the elimination of grain boundaries oriented perpendicularly to the main acting stresses in the blades, provides an increase in the main operational characteristics of heat-resistant alloys, first of all, long-term strength, ductility, and heat resistance.

In most cases, in the production of castings of GTE turbine rotor blades, vacuum melting units are used, heating in which is performed in electric resistance or induction furnaces.

Equiaxial, more precisely, equiaxially textured, cast structure of blades is obtained using vacuum induction furnaces semi-continuous melting of molds (class UPPF-3), and blanks of blades with a longitudinally oriented columnar structure are obtained by casting with directional crystallization in continuous continuous-flow method vacuum furnaces (class PMP-2 and PMP-4), in installations of vacuum directional solidification (class UVNK-8P) or in the VIP-NK vacuum melting unit [4 - 7].

For the serial production of turbine blades, productive designs of continuous casting installations with a horizontal crystallization chamber are used, which have a capacity of about 350 blades per day. Such installations include continuous-acting PMP-1 and PMP-2 furnaces. Such high productivity is achieved by changing the technological system of heating and directional cooling of the mold.

In the central part of the PMP-2 installation, there is a working chamber consisting of three conventional zones: heating, melting and crystallization. On both sides of the working chamber, there are lock chambers that serve for automatic loading and unloading of graphite flasks without breaking the vacuum in it. According to the operating conditions of the foundry, the graphite flask is heated to a temperature of 1500–1520 °C throughout the entire volume until the metal crystallizes, being constantly heated to the same temperature. Therefore, the use of stubs made from a working alloy is excluded, since its melting point is lower than the heating temperature of the ceramic mold (1350 °C). For the operating conditions of the PMP-2 installation, stub crystals with a melting point above the melting point of the alloy are used.

2. Description of research models

The stage of the system heating process after the transition of the upper half of the metal sample to the liquid-solid state begins when the upper half of the metal sample reaches the temperature of its transition to the liquid-solid state and ends when the lower half of the metal sample reaches the same temperature. At the initial moment of this stage, the liquid-solid metal flows into the gap between the lower half of the sample and the bowl. Due to the small size of this gap, it is assumed that the temperature of the metal in the lower half of the bowl is equalized and becomes equal to $T^{0}_{17,0}$ [8 - 10].

In connection with the filling of the gap between the bowl and the initial sample with metal, the heating scheme takes the form shown in the figure 1. At this stage of the process, the calculation formulas for the following heat fluxes and temperatures are changed.

In the calculations, the following assumptions were made.

It is assumed that the temperature of the furnace changes in time according to a linear law, which is determined by the temperature in the corresponding furnace hones and the speed of the train of the flasks. In principle, the intermittent movement of the flask train leads to an abrupt change in the furnace temperature at each moment of time when the flask is pushed. However, taking into account the relatively small value of each movement of the flask, the unevenness of the furnace temperature along the flask and the flow of heat over the flasks, this assumption gives a relatively small error. It is assumed that all thermophysical characteristics of the system are equal to their average value in the calculated temperature range and are constant values.

It is assumed that the initial temperature of the flask-form-metal-stopper system is 20 °C.

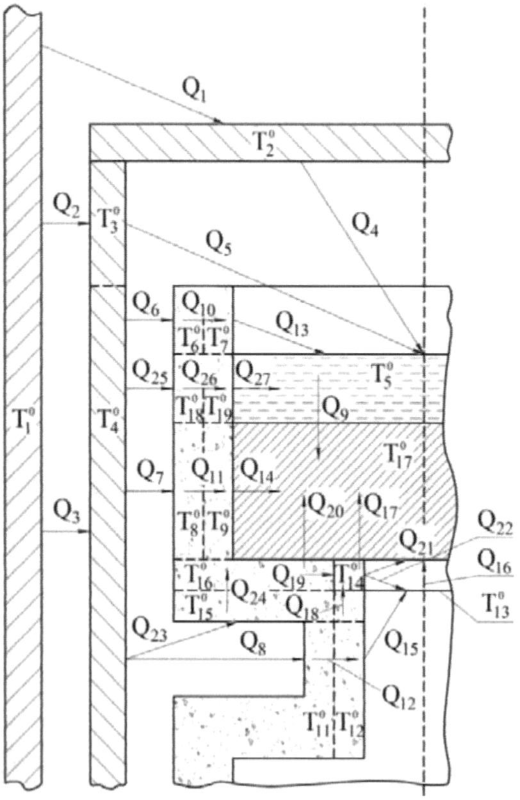

Figure 1. Scheme for calculating the heating of the flask-mold-metal-stopper system in the non-stationary temperature field of the furnace.

Based on preliminary calculations, a calculated time interval of $6 \cdot 10^{-1}$ seconds was adopted, which provides a sufficiently small value of the calculation error.

It is assumed that with contact heat transfer between different temperature zones of the system, the thermal resistance between these zones is determined by the distance between the centres of these zones, and the temperature between the centres of these zones changes according to a linear or logarithmic law (depending on the geometric shape of the calculated zones). However, if any of the zones is heated due to radiation coming from its surface, then the thermal resistance of this zone is determined by its entire thickness. When heat transfer between zones by radiation, it is assumed that the temperature in the entire volume of these zones is the same. Additional calculations have shown that with the available geometric and thermophysical characteristics of the system and a very small (only $6 \cdot 10^{-1}$ seconds) value of the calculated time interval, this assumption also provides a sufficiently small value of the calculation error.

Based on preliminary calculations, it was found that the heat fluxes along the side walls of the flask and the ceramic mold are much less than the heat fluxes going across these walls; therefore, the heat fluxes along the side walls were not taken into account in the calculations.

It is assumed that the vertical heaters of the furnace form a continuous plane, and that the heating of the flask lid due to radiation from the upper thermal insulation is much less than direct heating from the heater.

It is assumed that the transition of a metal from a solid-liquid to a liquid-solid state occurs in the middle of the crystallization temperature range (at 1330 ^{0}C). It is assumed that in the temperature range between the solidus and liquidus of the alloy, the heat spent on its melting is taken into account in the calculations as an additional heat capacity of the alloy.

Calculation of heat fluxes is carried out according to dependencies (1) - (15).

$$Q_{4,i} = \varepsilon_{П2} \cdot C_0 \cdot \left[\left(\frac{T_{2,i}^0}{100} \right)^4 - \left(\frac{T_{5,i}^0}{100} \right)^4 \right] \cdot H_4^/ \cdot (\tau_{i+1} - \tau_i) \tag{1}$$

$$Q_{5,i} = \varepsilon_{П2} \cdot C_0 \cdot \left[\left(\frac{T_{3,i}^0}{100} \right)^4 - \left(\frac{T_{5,i}^0}{100} \right)^4 \right] \cdot H_5^/ \cdot (\tau_{i+1} - \tau_i) \tag{2}$$

$$Q_{6,i} = \varepsilon_{П3} \cdot C_0 \cdot \left[\left(\frac{T_{4,i}^0}{100} \right)^4 - \left(\frac{T_{6,i}^0}{100} \right)^4 \right] \cdot H_6^/ \cdot (\tau_{i+1} - \tau_i) \tag{3}$$

$$Q_{9,i} = \frac{T_{5,i}^0 - T_{17,i}^0}{0,5 \cdot h_{ME} \cdot \frac{D_{ME}^2}{D_Ч^2}} \cdot \lambda_{ME} \cdot F_Ч \cdot (\tau_{i+1} - \tau_i) \tag{4}$$

$$Q_{10,i} = \frac{T_{6,i}^0 - T_{7,i}^0}{\ln \frac{d_2}{d_1}} \cdot 2 \cdot \pi \cdot \lambda_{KEP} \cdot h_{ME} \cdot \left(1 - \frac{D_{ME}^2}{D_Ч^2} \right) \cdot (\tau_{i+1} - \tau_i) \tag{5}$$

$$Q_{11,i} = \frac{T_{8,i}^0 - T_{9,i}^0}{\ln \frac{d_2}{d_8}} \cdot 2 \cdot \pi \cdot \lambda_{KEP} \cdot 0,5 \cdot h_Ч \cdot (\tau_{i+1} - \tau_i) \tag{6}$$

$$Q_{13,i} = \varepsilon_{П4} \cdot C_0 \cdot \left[\left(\frac{T_{7,i}^0}{100} \right)^4 - \left(\frac{T_{5,i}^0}{100} \right)^4 \right] \cdot H_{13}^/ \cdot (\tau_{i+1} - \tau_i) \tag{7}$$

$$Q_{14,i} = \frac{T_{9,i}^0 - T_{17,i}^0}{\ln \frac{d_8}{d_1} \cdot \frac{1}{\lambda_{KEP}} + \ln \frac{d_1}{d_9} \cdot \frac{1}{\lambda_{ME}}} \cdot 2 \cdot \pi \cdot \frac{h_{ME}}{2} \cdot (\tau_{i+1} - \tau_i) \tag{8}$$

$$Q_{16,i} = \varepsilon_{П6} \cdot C_0 \cdot \left[\left(\frac{T_{13,i}^0}{100} \right)^4 - \left(\frac{T_{17,i}^0}{100} \right)^4 \right] \cdot H_{16}^/ \cdot (\tau_{i+1} - \tau_i) \tag{9}$$

$$Q_{17,i} = \frac{T_{14,i}^0 - T_{17,i}^0}{\frac{0,5 \cdot h_{CT}}{\lambda_{KEP} \cdot F_{CT}} + \frac{0,25 \cdot h_{ME}}{\lambda_{ME} \cdot F_{ME}}} \cdot (\tau_{i+1} - \tau_i) \tag{10}$$

$$Q_{20,i} = \frac{T_{16,i}^0 - T_{17,i}^0}{\frac{0,25 \cdot h_{ДН}}{\lambda_{KEP} \cdot F_{ДН}^/} + \frac{0,25 \cdot h_{ME}}{\lambda_{ME} \cdot F_{ДН}^/}} \cdot (\tau_{i+1} - \tau_i) \tag{11}$$

$$Q_{21,i} = \varepsilon_{\Pi 4} \cdot C_0 \cdot \left[\left(\frac{T^0_{14,i}}{100} \right)^4 - \left(\frac{T^0_{17,i}}{100} \right)^4 \right] \cdot H^{/}_{21} \cdot (\tau_{i+1} - \tau_i) \tag{12}$$

$$Q_{25,i} = \varepsilon_{\Pi 3} \cdot C_0 \cdot \left[\left(\frac{T^0_{4,i}}{100} \right)^4 - \left(\frac{T^0_{18,i}}{100} \right)^4 \right] \cdot H_{25} \cdot (\tau_{i+1} - \tau_i) \tag{13}$$

$$Q_{26,i} = \frac{T^0_{18,i} - T^0_{19,i}}{\ln \frac{d_2}{d_8}} \cdot 2 \cdot \pi \cdot \lambda_{KEP} \cdot h_{ME} \cdot \left(\frac{D^2_{ME}}{D^2_{Ч}} - \frac{1}{2} \right) \cdot (\tau_{i+1} - \tau_i) \tag{14}$$

$$Q_{27,i} = \frac{T^0_{19,i} - T^0_{5,i}}{\ln \frac{d_8}{d_1} \cdot \frac{1}{\lambda_{KEP}} + \ln \frac{d_1}{d_9} \cdot \frac{1}{\lambda_{ME}}} \cdot 2 \cdot \pi \cdot h_{ME} \cdot \left(\frac{D^2_{ME}}{D^2_{Ч}} - \frac{1}{2} \right) \cdot (\tau_{i+1} - \tau_i) \tag{15}$$

The calculation of temperature fields is carried out according to dependencies (16) – (21).

$$T^0_{5,i+1} = \frac{Q_{4,i} + Q_{5,i} + Q_{13,i} + Q_{27,i} - Q_{9,i}}{C_{ME} \cdot \rho_{ME} \cdot F_{Ч} \cdot h_{Ж}} + T^0_{5,i} \tag{16}$$

$$T^0_{6,i+1} = \frac{Q_{6,i} - Q_{10,i}}{C_{KEP} \cdot \rho_{KEP} \cdot F^{/}_{Ф} \cdot h_T} + T^0_{6,i} \tag{17}$$

$$T^0_{7,i+1} = \frac{Q_{10,i} - Q_{13,i}}{C_{KEP} \cdot \rho_{KEP} \cdot F^{//}_{Ф} \cdot h_T} + T^0_{7,i} \tag{18}$$

$$T^0_{17,i+1} = \frac{Q_{9,i} + Q_{14,i} + Q_{20,i} + Q_{21,i} + Q_{17,i} + Q_{16,i}}{C_{ME} \cdot \rho_{ME} \cdot F_{Ч} \cdot h_{Ж}} + T^0_{17,i} \tag{19}$$

$$T^0_{18,i+1} = \frac{Q_{25,i} - Q_{26,i}}{C_{KEP} \cdot \rho_{KEP} \cdot F^{/}_{Ф} \cdot h_{Ж}} + T^0_{18,i} \tag{20}$$

$$T^0_{19,i+1} = \frac{Q_{26,i} - Q_{27,i}}{C_{KEP} \cdot \rho_{KEP} \cdot F^{/}_{Ф} \cdot h_{Ж}} + T^0_{19,i} \tag{21}$$

3. Results

As a result of the calculation of the heating of the technological system, the following results were obtained: the first melting point of nickel (1453 ^{0}C) was reached by the fusible diaphragm. This happened 3597 seconds after the start of heating the flask and furnace. In this case, the distance from the mold box to the refrigerator is 1.07 m (a distance that is not a multiple of 0.2 m (the length of the mold box) is obtained because the movement of the mold box is considered continuous in the calculation). The furnace temperature in this place is 1472^0 C. At this moment, the temperature of the retaining ring is $T^0_K = 1441\,^0$C, the average temperature of the metal is $T^0_{CP} = 1445\,^0$C, the temperature difference between the upper and lower surfaces of the metal $\Delta T^0_{\Pi} = 16$ K. The heat flux between the diaphragm and the metal is directed from the diaphragm to the metal. The total amount of heat received by the metal sample through the upper and lower ends and the side surface is distributed as follows: 33.5 % flows through the upper end, 61.3 % through the side surface, and 5.2 % through the lower end. It should be borne in mind that the surface area of the end face of a metal sample with a diameter of $6 \cdot 10^{-2}$ m and a height of $1 \cdot 10^{-1}$ m is $28,2 \cdot 10^{-4}$ m^2, and the area of the lateral surface of this sample is $188,5 \cdot 10^{-4}$ m^2. Since the area of the lateral surface of the sample is 6.68 times greater than the area of its end, the amount of heat entering the sample through the unit area for the upper end is 3.68 times greater than for the lateral surface.

4. Conclusions

Thus, based on the calculations performed, the following conclusions can be drawn. In order for the temperature of the metal during draining to exceed the melting temperature of the diaphragm, it is necessary that the heating rate of the diaphragm and the retaining ring be less than the heating rate of the metal. The solution to this problem is possible in four main ways: the first is to reduce heat transfer to the diaphragm and retaining ring. The second way is to increase the heat capacity of the diaphragm and retaining ring. The third method is to increase the intensity of metal heating. The fourth method consists in the selection of the corresponding geometric and physical parameters of the form-metal-stopper system [11 - 12].

Acknowledgment

The research was supported by a grant of Russian Science Foundation (project № 20-79-00254).

References

[1] Takeshi N 2014 Development of CMC turbine parts for aero engines *Engineering Review* **47(1)** 29-32

[2] Vieira L F, Paggiand R A and Salmoria G V 2012 Innovative developments in virtual and physical prototyping *Proceeding* 385-8

[3] Khaimovich I N and Demianinko E G 2015 *Mathematical Material and Process Simulation* (Samara: SSAU) p 82

[4] Szeliga D P 2014 Modeling of directional solidification of columnar grain structure in CMSX–4 nickel-based superalloy *Materials Engineering and Performance* **23(3)** 1088-95

[5] Szeliga D P 2015 The unidirectional crystallization of metals and alloys (turbine blades) *Rzeszow University of Technology* **1** 45

[6] Chubin Y 2013 Competitive grain growth mechanism in three dimensions during directional solidification of a nickel-based superalloy *Journal of Alloys and Compounds* **578** 557-84

[7] Dai H J, Dong H B and D'Souza N 2011 Grain selection in spiral selectors during investment casting of single-crystal components: **II** *Numerical modeling Metallurgical and Materials Transactions* **42(A)** 3439-46

[8] Vdovin R A, Smelov V G, Sufiiarov V S and Borisov E V 2018 Designing of the digital casting process for the gas turbine engine blades with a single-crystal structure *IOP Conf. Ser.: Mater. Sci. Eng.* **441(1)** 9

[9] Vdovin R A 2019 Improving the quality of the manufacturing process of turbine blades of the gas turbine engine *Journal of Physics: Conf. Ser.* **1399(4)** 6

[10] Vdovin R A 2020 Designing a duplicate model of the GTE turbine blade casting process *IOP Conf. Ser.: Mater. Sci. Eng.* **709(3)** 5

[11] Bolotov M A, Pechenin V A and Ruzanov N V 2016 Uncertainties in measuring the compressor-blade profile in a gas-turbine engine *Russian Engineering Research* **36(12)** 1058-65

[12] Pechenin V A, Bolotov M A and Rusanov N V 2014 Method of evaluation of profile form and shaped surfaces with application of wavelets *Research Journal of Applied Sciences* **9(11)** 820-4

The study of the welding relief shape influence on the linear friction welding process

I M Gatiatullin[1], V R Galimov[1], A U Medvedev[1] and D V Karetnikov[2]

[1] Ufa State Aviation Technical University, K. Marx Str.,12, Ufa, 450008, Republic of Bashkortostan, Russia
[2] Department "Equipment and technologies for welding and control", Ufa State Petroleum Technological University, Kosmonavtov Str.,1, Ufa, 450062, Republic of Bashkortostan, Russia

E-mail: medvedev.otsp@yandex.ru

Abstract. The study of the welding process of asymmetric parts made of Ti-6A1-4V alloy was made to expand the field of application of the process of linear friction welding and to improve the use of technology of existing assemblies. During of the study, it was found that the process of welding T-joints has features in a longer duration of the unsteady stage and an increase in the amplitude value of the shear force during the steady state. Welded joints do not contain seam flaws over the area of the welded joint, which it made unnecessary to assign an allowance along the perimeter of the welded section. The change in conditions at the joint is expressed in an increase in the microhardness of the welded seam in comparison with the Ti-6A1-4V joints, which are produced by the traditional technology more than 50 HV.

1. Introduction

Linear friction welding (LFW) is a relatively young method of friction welding among other ones, actively used in the manufacture of blisk (monowheels) of gas turbine engine (GTE) compressors. An example, the construction of a blisk using LFW is shown in figure 1, a. Modern and advanced compressors of GTE include blades, disks and blisks of two-phase titanium alloys, the most common is Ti-6Al-4V (Russian VT-6 alloy).

The LFW technology requires symmetric relief of the interface joint on the disk and blades (figure 1, a) for creating the identical heating and deformation conditions for each of the parts, but it reduces the material utilization factor, increases the labor intensity of processing one blank, resists to the development of a technology for repairing blisks by welding repaired blades, due difficulty to make the required relief on the repaired assembly.

Apply the scheme of LFW, shown in figure 1, b, welding of blades to an unprepared surface of the disk will expand the technological capabilities in the manufacture and repair of blisks.

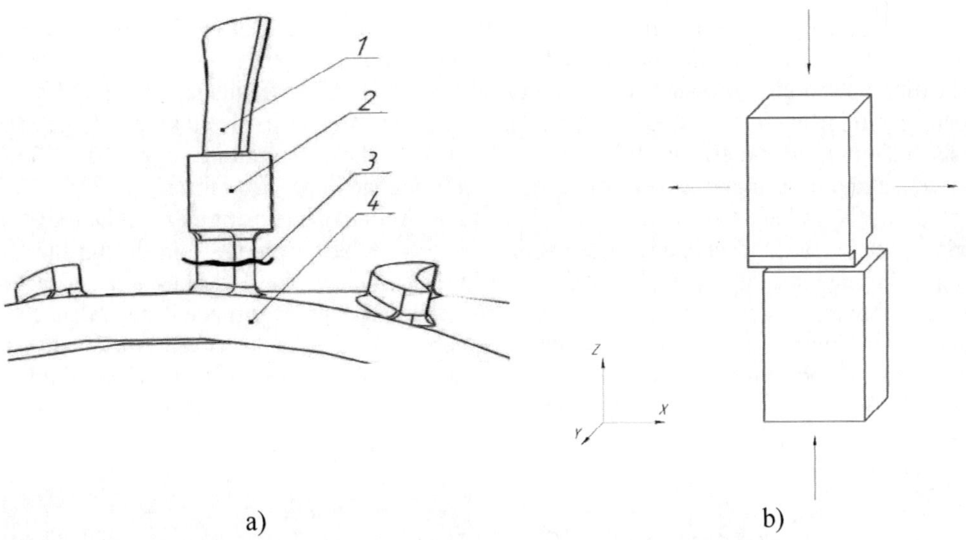

Figure 1. Partially bladed blisk (a) and an LFW process diagram (b). 1 - blade airfoil portion, 2 - place for setting the blade in the welding machine, 3 - bead squeezed out during welding, 4 – disc.

Available scientific material contains articles of development of the LFW technology of foreign scientists which shows interest in welding outside the border. In the paper [2], a two-dimensional model of the LFW of a T-joint was developed in the Deform, which demonstrated the irregularity extrusing a flash, the different rates of extrusing the flash at the edges and in the middle part of the section are a consequence of the lack of symmetry of the thermal fields relative to the interface joint. In the papers [3,4] a decrease in the width of the relief during welding of the Ti6Al4V alloy increases the time of the steady-state upsetting stage, its rate and heat release. Metallographic studies and measurements of the microhardness of T-joints of aluminum alloy AA6082 were reported in [5] also showed the absence of symmetry in the structure of welded joints. The above results indicate that a change in the design of a T-joint affects the processes of heating and plastic deformation during the welding process, however, the questions of the influence of the peculiarities of the formation of T-joints on the welding technology and the properties of welded joints remain unexplored, which makes it difficult to use such joints for the current main LFW application areas - blisk manufacturing [6].

In the Russian Federation, experimental research of LFW had not been made; the available mathematical models of the LFW process are focused on the analysis tasks of symmetric interface joint. The model was described in [7], which allows one to take into account the absence of symmetry in the heat distribution during LFW, was based on a one-dimensional scheme, and it is not obviously to applicate for calculating the temperature field during T-joint welding.

The purpose of this study is to assess the effect of the absence of a relief on the flange in linear friction welding of VT6 alloy T-joints on the welding process and the structure and structure of welded joints.

2. Methodology

Specimens for welding were made of VT-6 alloy, and had the shape of rectangular parallelepiped measuring 13*26*35 mm. To ensure the conditions corresponding to T-joint welding without preparing a relief on the flange, a relief 4 mm wide was formed on the welded surface of one specimen from a pair. Welding of the specimens was made on a equipment with hydraulic drives for reciprocating motion and upsetting. The control system of the hydraulic drives of the installation ensured the maintenance of oscillatory movements with a given frequency and amplitude, and the possibility of forging. The duration of the process was set upon reaching a predetermined value of the heating upset. During the welding process, the forces generated by the forging and oscillation drives, as well as the movements of both plunger pins with a sampling rate of 1 kHz, were recorded.

The heating and forging pressure was taken equal to 100 MPa, the frequency and amplitude of the reciprocating motion were taken equal to 50 Hz and 2 mm, respectively, taking into account the previously accumulated experience [8, 9] on welding two-phase titanium alloys VT6, VT8-1 and VT8M-1. The reciprocating movement was carried out along the larger (26 mm) size.

The study of the welded specimens included visual inspection, analysis of data on the movement of the forging plunger pin and changes in the force on the oscillation plunger pin during the welding process, metallographic analysis and measurement of the microhardness of various areas of the joint.

Specimen preparation for studying the structure and microhardness was made in accordance with the scheme shown in figure 2, a. The microhardness was measured according to GOST R ISO 6507-1-2007 by the Vickers method in the directions shown in figure 2, b. A tetrahedral diamond pyramid with an apex angle of 136° was used as an indenter. The measurements were carried out with a force of 0.9807 N with a step of 0.01 mm.

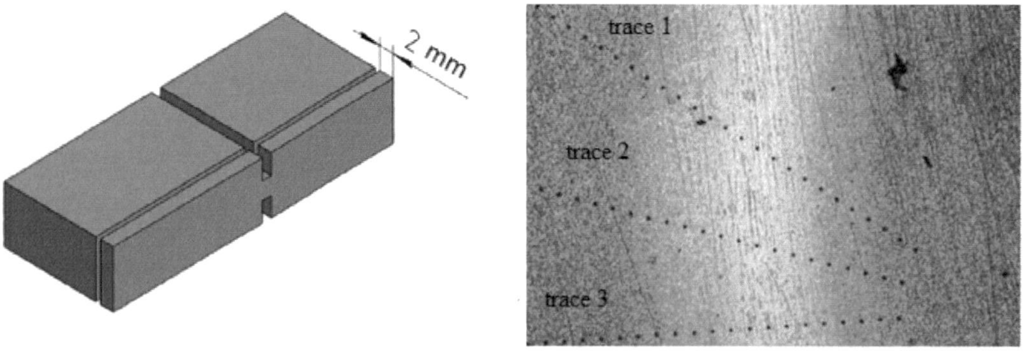

a) b)

Figure 2. Specimen preparation schemes for metallographic examination (a) and microhardness measurement (b).

Investigations of the microstructure were carried out using an optical inverted microscope Carl Zeiss Axio Observer.A1m with a magnification range of 5x - 100x; an instrumental microscope VMM 150 with a magnification of 20X was used to assess the size of welded joint areas on thin sections.

3. Results and discussion

For all specimens, the welding process was completed normally - the required upsetting value was reached, and, as can be seen from figure 3, a flash was formed along the entire perimeter of the welded section as a thin strip. The largest amount of metal was extruded along the larger side of the welded section.

Analysis of the cyclograms of the welding process was showed despite the general similarity between the welding cycles of symmetric and asymmetric specimen, there are also have significant differences. While welding asymmetric specimens, it is also possible to distinguish unsteady and steady stages - sections 1 and 2 in figure 4. The duration of the unsteady stage in welding asymmetric specimens reaches 0.6 s, which is comparable to the analogous parameter for symmetric specimens with a threefold larger cross section. The steady-state stage in the welding of symmetrical specimens is characterized by the constancy of the upsetting rate and shear force [8, 9], and in the welding of asymmetric specimens in this time interval, an increase in the amplitude value of the shear force takes place at a rate of 4.8 kN/s (or 16% per second) at a constant rate of upset.

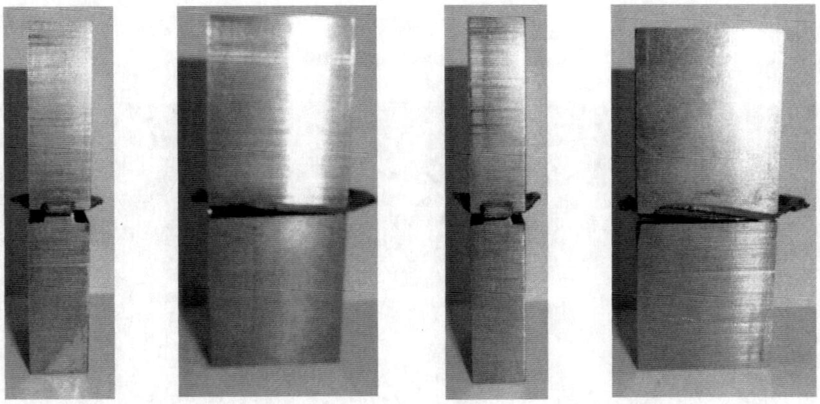

Figure 3. Appearance of samples after the welding.

The metallographic study was showed the formation a solid weld in all cases, immersed in a specimen which didn't have a relief (figure 5).

It is interesting to note the differences in the structure of the near-surface sections of the weld seam of specimens which had a symmetrical relief. In welded joints of specimens with a symmetrical relief (figure 6), the flash thickness is greater than the seam thickness in the middle part, and in the near-surface layer a "trumpet" is formed, often containing discontinuities, which necessitates the appointment of an allowance along the entire perimeter of the welded section. Flash formation in T-joints occurs under slightly different conditions, and as can be seen from figure 5, the width of the weld is practically unchanged, and its peripheral part hasn't flaws.

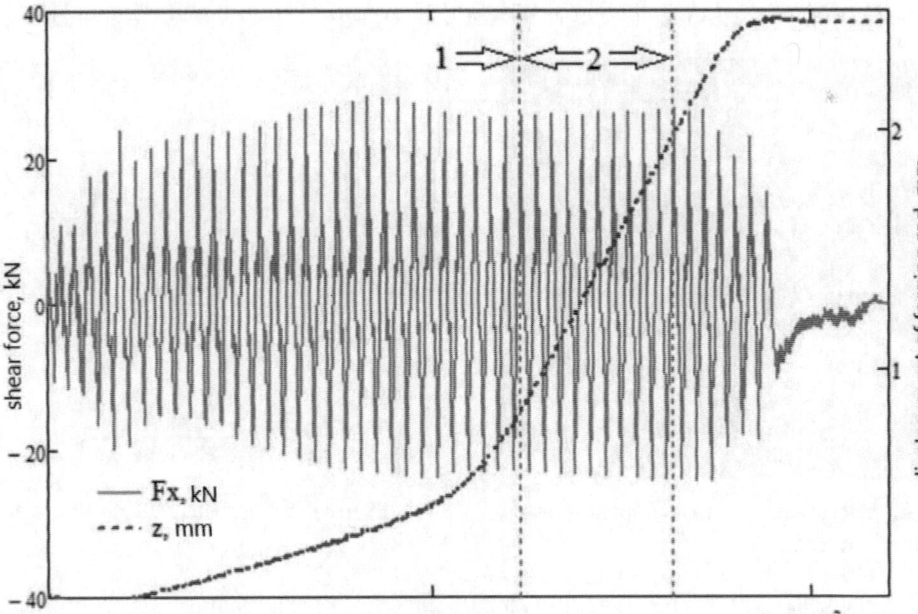

Figure 4. Cyclogram of the welding process.

The formation of a trough-shaped seam demonstrates the lack of symmetry about the plane of the joint in the distribution of temperature, stresses and strain rates. In the middle part of the section the upsetting occurs due to the deformation of both parts, and with distance from the axis of the weld, the deformation of the more massive part decreases, and at the surface of the flash should be formed practically only due to the deformation of the thinner specimen.

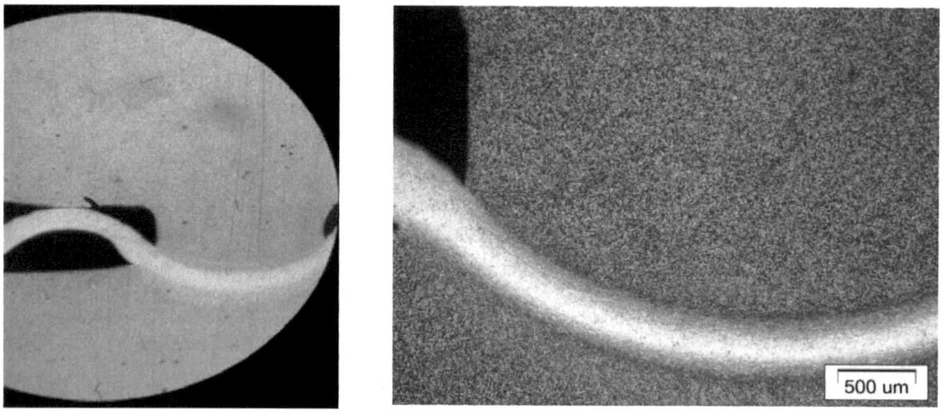

Figure 5. Macro and microstructure of T-joint.

The immersing of the weld into the body of a more massive specimen is accompanied by an increase in its area, which is possibly due to an increase in the amplitude value of the shear force at the steady stage of the process.

Traditionally, the seam and the heat-affected zone, which are distinguished in the butt welded joint, are clearly observed in T-welded joints. According to modern concepts, welded joints of VT6 alloy contain a γ' phase with a microhardness in the range of 400-450 HV, it was formed due to large plastic deformation in the β region at the stage of heating, subsequent recrystallization of the β phase and martensitic transformation during cooling. [**Ошибка! Источник ссылки не найден.**, [11], [12]].

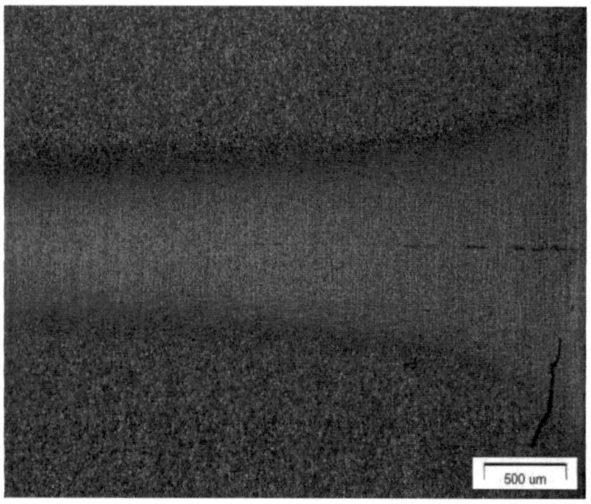

Figure 6. Microstructure of the near-surface area of the butt joint.

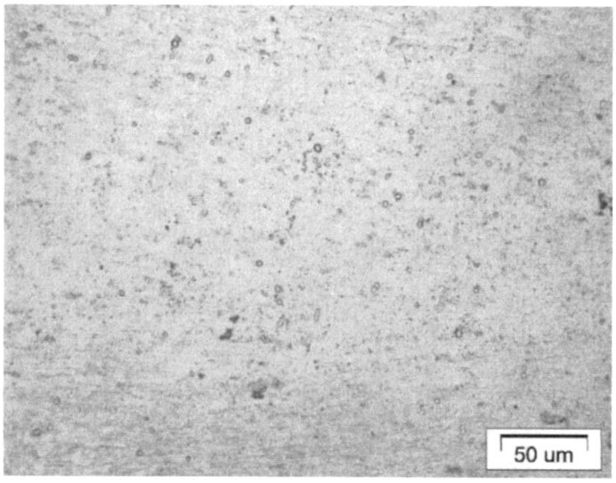

Figure 7. Microstructure of the weld.

In the microstructure of the T-joint welded joint (figure 7), neither traces of the primary microstructure nor traces of recrystallization in the β region were revealed. The noted features indicate that the formation of flash, as in the compounds obtained by the traditional technology, most likely proceeds at temperatures above the polymorphic transformation, but due to the smaller sizes of the heated zones, different heat removal conditions and higher cooling rates, the recrystallization processes practically don't have time to develop.

The termomechanical-affected zone (TMAZ) is a transition region, one of the boundaries of which corresponds in structure to the base material, and the other to the weld. Within the TMAZ, there is a region with a texture formed during metal deformation in the two-phase region. On both sides of the welded seam, the structure and microstructure of the TMAZ are the same, the length of the TMAZ itself is rather small.

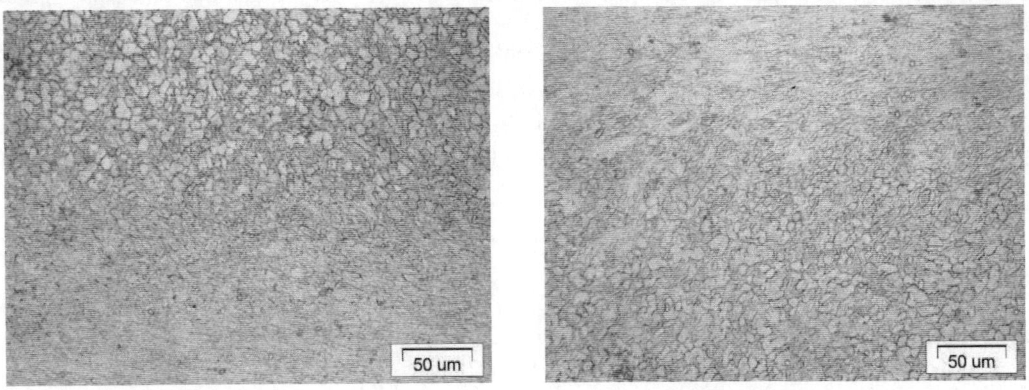

Figure 8. Microstructure of the TMAZ from the side of the wall (left) and from the side of the shelf (right).

Another confirmation of the specificity of the microstructure of the investigated welds and the conditions for their formation are the results of measuring the microhardness. The values of this indicator within the seam and the TMAZ were in the range of 337 ... 500 HV. As can be seen from figure 8, the distribution of microhardness has an extremum at the joint of the specimens to be welded, while on both sides from the seam border there are areas where the microhardness was lower than the values 1of this indicator for the base material (corresponds to the line marked as OM in figure 8). The length of the areas with reduced microhardness is extremely low, and is comparable to the step taken when measuring the microhardness.

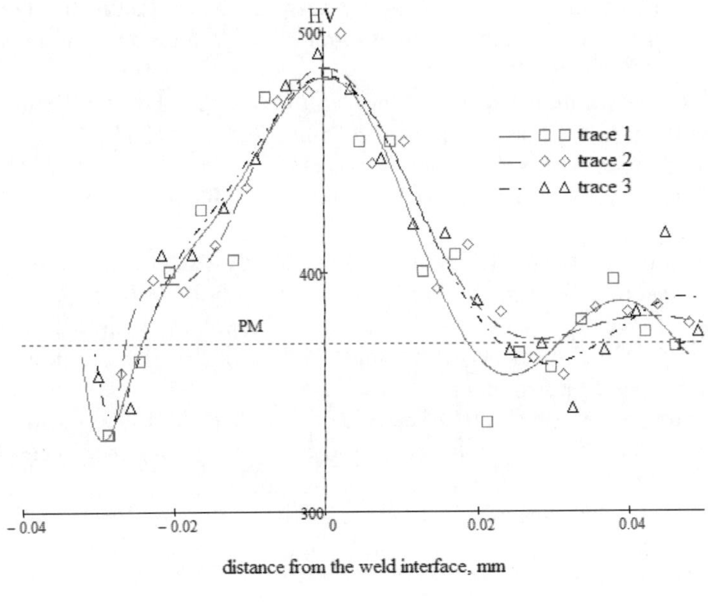

Figure 9. Microhardness distribution across the weld in accordance with the scheme on figure 2.

4. Findings

It has been shown experimentally that in the modes used for linear friction welding of joints with symmetric relief, T-joints can be obtained without preparing a relief on the shelf without flaws over the entire section of the weld, including near-surface regions.

The welding cycle of such joints is distinguished by a longer duration of the unsteady stage and an increase in the amplitude value of the shear force, which is most likely associated with an increase in the interface joint during the welding process.

The differences in the microstructure and microhardness of the weld metal of T-joints from the previously investigated joints of symmetric parts with a cross section of 13*26 indicate the specificity of the thermal regime and the stress-strain state of the metal in the zone of the forming T-joint, and the need for further studies of both the process of welding processes and properties welded joints.

Acknowledgement

The metallograthuc analysis was performed at the Centre for Collective Use "Nanotech", Ufa State Aviation Technical University.

References

[1] Mateo A 2014 On the feasibility of blisk produced by linear friction welding *Revista De Metalurgia* **50(3)** 23

[2] Lee L A, McAndrew A R, Buhr C, Beamish K A and Colegrove P A 2015 2D linear friction weld modelling of a Ti-6Al-4V T-Joint *Engineering Science and Technology Review* **8(6)** 44-8

[3] McAndrew A R, Colegrove P A, Addison A C, Flipo B C, Russell M J and Lee L A 2015 Modelling of the workspecimen geometry effects on Ti-6Al-4V linear friction welds *Materials and Design* **87** 1087-99

[4] McAndrew A R, Colegrove P A, Flipo B C D and Bühr C 2017 3D modelling of Ti–6Al–4V linear friction welds *Science and Technology of Weldingand Joining* **22(6)** 496-504

[5] Ingarao G, Baffari D, Campanella D, Valvo E L , Buffa G and Livan Fratini 2019 A novel linear friction welding based approach for sheet-bulk joining *AIP Conference Proceedings* **2113** 050023

[6] McAndrew A R, Flipo B C D 2018 Linear friction welding for near net shape manufacturing of titanium alloy Ti-6Al-4V aerospace components *9th International Conference on Mechanical and Aerospace Engineering (ICMAE)*

[7] Medvedev A U, Galimov V R, Gatiyatullin I M and Murugova O V 2020 Finite difference model of temperature fields in linear friction welding *Solid State Phenomena* **303** 175-80

[8] Medvedev A Y, Bychkov V M, Selivanov A S, Pavlinich S P, Dautov S K and Supov A V 2015 Linear friction welding of two-phase titanium alloys VT6 and VT8-1 *Welding International* **29(1)** 66-9

[9] Medvedev A Y, Vairis A, Nikiforov R V and Supov A V 2012 Energy balance of the linear friction welding process *Engineering Science and Technology Review* **5(3)** 20-4

[10] Karavaeva M V, Kiseleva S K, Bychkov V M, Medvedev A Y, Supov A V, Musin F F, Aleksandrov I V and Latysh V V 2012 Influence of the value of upsetting on the formation of a welded joint in linear friction welding *Letters on materials* **2** 40-4

[11] Medvedev A Y, Astanin V V and Semenova I P 2019 The role of nanostructured superplasticity in the manufacture of a monowheel of a gas turbine engine compressor *Nanoindustry* **12(3-4)** 220-7

[12] Wanjara P, Booth-Morrison C, Hsu E and Jahazi M 2005 Process optimization for linear friction welding of Ti6Al4V *Proceedings of the 7th International conference on trends in welding research* (Georgia, USA) 860-85

Theoretical aspects of the implementation of remote monitoring and control of technological processes controlled using SoftPLC

P A Nikishechkin and A A Akimov*
MSUT "STANKIN", Moscow, Russian Federation

*E-mail: akimov2804@gmail.com

Abstract. The presented article discusses the theoretical aspects of the implementation of remote monitoring and control of technological processes controlled using the SoftPLC. The developed software environment must meet the requirements for visualization systems: speed, ease of understanding the control process, reliability and information content.

1. Introduction

The introduction of new techniques and information technologies in the management of technological processes, as well as the tasks of increasing the information content and providing opportunities for remote monitoring of technological processes, are currently relevant in the management of heterogeneous technological processes [1-3]. When building systems for monitoring and visualizing technological processes, the main tasks are:

- data exchange with "devices for communication with the object" in real time through drivers, information processing in real time;
- logical control;
- displaying information on the monitor screen in a convenient and human-readable form;
- maintaining a real-time database with technological information;
- alarm management;
- preparation and generation of reports on the progress of the technological process;
- implementation of network interaction between SCADA-stations (computers);
- providing communication with external applications (DBMS, spreadsheets, word processors, etc.);
- the ability to develop an automated process control system in a client-server or in a distributed architecture.

The requirements for visualization systems include: speed, ease of understanding the control process, reliability and information content.

The article proposes a solution to the problem of increasing the information content and providing opportunities for remote monitoring of technological processes by developing an interactive environment for visualization and control of heterogeneous technological processes controlled by the SoftPLC [2,4-7].

2. Structure of SoftPLC with a built-in development environment and visualization of interfaces

SoftPLC developed at the department of CNC MSUT "STANKIN" is a consequence of the development of technologies in the field of PLC: the core of the SoftPLC is implemented in software, and the input / output of control signals is carried out through hardware expansion modules. The computational module of the controller is fully integrated into the core of the CNC system. The terminal editor is implemented as an external application, which allows the PLC to be used both as part of a CNC system and as a stand-alone product. The PLC program editor is intended for creating, editing, configuring, debugging, starting / stopping an electroautomatic program, as well as its verification [8-10]. PLC development is implemented in the Functional Block Diagram (FBD) language of the IEC 61131 standard. SoftPLC have a number of advantages over classical PLCs, which lead to the emergence of this direction in the development of intelligent automation tools:

- no need for additional equipment;
- low cost compared to PLC;
- the ability to quickly upgrade by installing updates, since SoftPLC is a software implementation;
- the ability to remotely diagnose and resolve error situations on the Internet.

The structure of the SoftPLC client application consists of:

- PLC programming environment;
- configurator of hardware input / output devices;
- PLC variable manager.

Variables can be bound to either hardware I / O modules or be part of the internal variables of the PLC program.

Figure 1 shows the developed structure of SoftPLC with a built-in development environment and visualization of user interfaces.

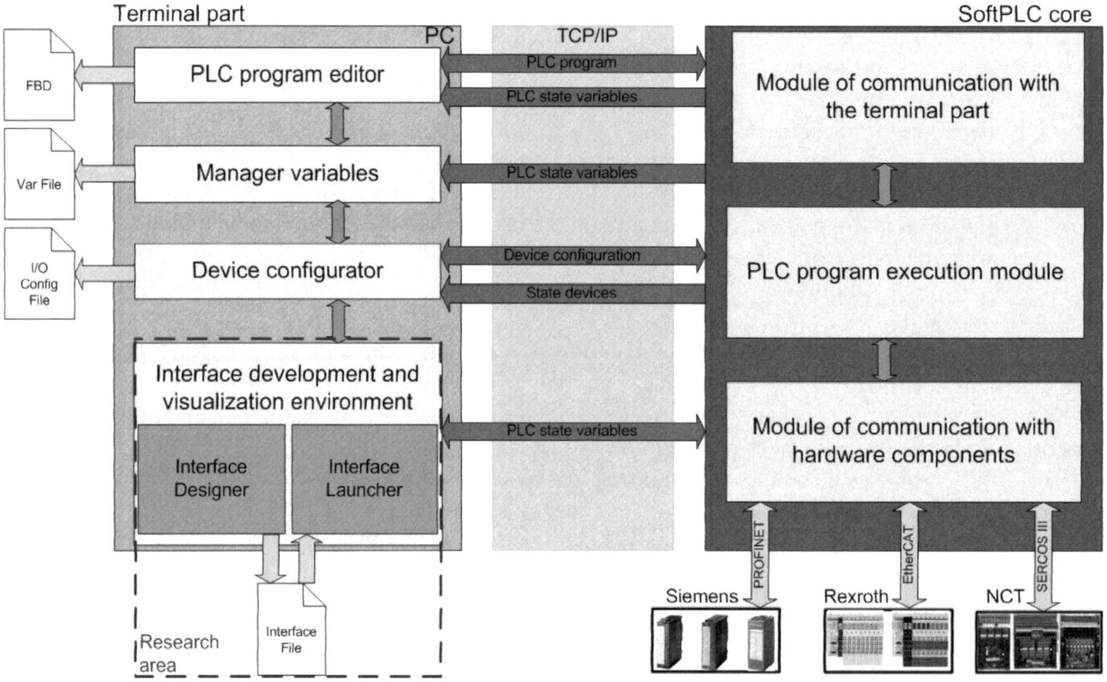

Figure 1. Structure of SoftPLC with a built-in development environment and visualization of user interfaces.

The organization of communication between the terminal part and the kernel based on the TCP / IP protocol stack allows separating the main data streams of the control system and the data streams of the logical task. The SoftPLC core can be divided into: a component that implements the SoftPLC operation logic and a configuration component [11-15].

State machine SoftPLC is an automaton graph with a finite number of states, the set of which is determined at the design stage. The transition from one state to another occurs when the values of the set of system variables that characterize the system as a whole change.

Shared I / O memory defines the underlying I / O mechanism. The shared memory area is located in the system area of the core - the area of control drivers and allows you to synchronize work with the SoftPLC hardware and its software implementation.

The level of control and data transmission contains computing resources with installed software and mathematical software SoftPLC, which are located in the control cabinets of electrical equipment of the production system, and carry out centralized control of production processes. To interact with remote I / O modules, specialized boards are used that implement work based on the selected industrial protocol. The control and diagnostic terminals implement the possibilities of visualizing the technological process of the production line and their parameters [2,6].

SoftPLC client application has a component architecture and is used for:

- design, development (programming), debugging, testing and maintenance of software products created for the implementation of algorithms for the operation of SoftPLC, which is part of the CNC systems;
- for setting up, checking and supervising monitoring and control systems;
- configuring the user interface, client and server applications;
- to visualize the work of SoftPLC, both target and WEB-visualization.
-

To implement the above functions, the architecture of the SoftPLC client application consists of the following components:

- PLC programming environment, responsible for the implementation and maintenance of programs that implement the algorithms of SoftPLC;
- monitoring and control tools are built into the programming environment, they can also be implemented using custom applications using open programming interfaces.

The developed software environment consists of a designer of interfaces, consisting of visual components with the ability to bind them to the PLC with variables defined in the PLC program. Also, the software environment includes interface launcher, which allows you to emulate a controlled process, providing ample opportunities for debugging PLC programs, as well as monitoring and controlling a real technological process [15-17].

3. Conclusions

This work presents a solution to the actual problem associated with the possibility of remote monitoring of technological processes controlled by the SoftPLC [3,18].

Developed a structure model for building interface designer to visualize and control heterogeneous technological processes, characterized by the communication with the controller core using the mechanism of shared memory.

A method of interaction between the interface launcher and the SoftPLC core has been implemented, allowing you to display up-to-date information about the technological process and manage it.

Acknowledgments
Acknowledgments. This work was supported by the Ministry of Education and Science of the Russian Federation as a state program of grants of the President of the Russian Federation for state support of young Russian scientists (grant MK-4469.2021.4).

References
[1] Martinova L I and Martinov G M 2019 Prospects for CNC machine tools *Russian Engineering Research* **39(12)** 1080–3
[2] Martinov G M, Nikishechkin P A, Grigoriev A S and Chervonnova N Yu 2019 Organizing interaction of basic components in the CNC System AxiOMA Control for integrating new technologies and solutions *Automation and Remote Control* **80(3)** 584–91
[3] Martinov G M, Sokolov S V, Martinova L I, Grigoryev A S and Nikishechkin P A 2017 Approach to the diagnosis and configuration of servo drives in heterogeneous machine control systems *8th International Conference (ICSI)* 586-94
[4] Yuqian L, Xun X and Lihui W 2020 Smart manufacturing process and system automation – a critical review of the standards and envisioned scenarios *Journal of Manufacturing Systems* **56** 312-25
[5] Nikishechkin P A, Chervonnova N Y and Nikich A N 2020 An approach of developing solution for monitoring the status and parameters of technological equipment for the implementation of Industry 4.0 *IOP Conference Series: Materials Science and Engineering* **709(4)** 044065
[6] Chao L, Hrishikesh V, Yuqian L and Xun X 2019 A cyber-physical machine tools platform using OPC UA and MTConnect *Journal of Manufacturing Systems* **51** 61–74
[7] Wang M, Cao P, Song K, Zhu G, Zhang W and Zhao Q 2018 Information interaction model of digital workshop based on OPC UA for CNC machine tools manufacturing *Beijing Gongye Daxue Xuebao/Journal of Beijing University of Technology* **44** 1040-6
[8] Martinov G M, Grigoryev A S and Nikishechkin P A 2015 Real-time diagnosis and forecasting algorithms of the tool wear in the CNC systems *Advances in Swarm and Computational Intelligence* **9142** 115-26
[9] Armin V, Martin O and Thomas H 2017 OPC UA integration for field devices *IEEE 15th International Conference on Industrial Informatics (INDIN)* 419-24
[10] Martinov G M, Issa A and Martinova L I 2019 Controlling CAN servo step drives and their remote monitoring by using protocol OPC UA *2019 International Multi-Conference on Industrial Engineering and Modern Technologies* (Vladivostok: IEEE) pp 1-5
[11] Martinov G M, Al Khoury A and Issa A 2018 An approach of developing low cost ARM based CNC systems by controlling CAN drive *International Conference on Modern Trends in Manufacturing Technologies and Equipment (ICMTMTE 2018)* vol 224 (Sevastopol)
[12] Pushkov R L, Martinova L I and Evstafieva S V 2018 Extending functionality of control system by adding engraving capabilities *International Russian Automation Conference* (Sochi) pp 1-5
[13] Kovalev I A, Nikishechkin P A and Grigoriev A S 2017 Approach to programmable controller building by its main modules synthesizing based on requirements specification for industrial automation *2017 International Conference on Industrial Engineering, Applications and Manufacturing* (Saint-Petersburg: IEEE Xplore) 1-4
[14] Sang Z and Xu X 2017 The framework of a cloud-based CNC system *The 50th CIRP Conference on Manufacturing Systems* **63** 82-8
[15] Martinova L I, Kozak N V, Nezhmetdinov R A, Pushkov R L and Obukhov A I 2015 The russian multi-functional CNC system AxiOMA control: practical aspects of application *Automation and Remote Control* **76(1)** 179-86
[16] Grigoriev S N, Sinopalnikov V A and Tereshin M V 2012 Control of parameters of the cutting process on the basis of diagnostics of the machine tool and workpiece *Measurement techniques* **55(5)** 555-8

[17] Obukhov A I, Evstafieva S V and Martinova L I 2020 Real-time cutting simulation based on voxel model *IOP Conference Series: Materials Science and Engineering* **709** 044062
[18] Grigoriev S N, Gurin V D, Volosova M A and Cherkasova N Y 2013 Development of residual cutting tool life prediction algorithm by processing on CNC machine tool *Materialwissenschaft und werkstofftechnik* **44(9)** 790-6

Changing properties of mechanically activated water suspensions

S D Rudnev[1], A I Krikun[2] and V V Feoktistova[1,3]

[1] Department of Machinery and Technological Systems, Institute of Engineering Technologies, Kemerovo State University, 6 Krasnaya Street, Kemerovo, 650000, Russia
[2] Department of Technological Machinery and Equipment, Institute of Food Production, Far Eastern State Technical Fisheries University, 52B Lugovaya Street, Vladivostok, 690087, Russia

[3] E-mail: feonika13@mail.ru

Abstract. The authors carried out a series of experimental studies to research the changes of properties in mechanically activated water and suspensions. This article presents a part of the combined temperature and hydrogen exponent studies carried out at two different universities, using two separate approaches to mechanical dispersion. Study I used a planetary whisk mixer, whereas study II used a vibrating mechanical activator, which is a modern alternative to dispersion devices used at food production facilities. The research of the properties of mechanically activated water and liquid suspensions using these methods was conducted due to both the importance of these substances for all areas of the food production industry, and the modern development of technologies and techniques used at production facilities, as well as the lack of unity among the numerous research studies conducted by leading scientists specializing in the study of the properties of water, liquid suspensions and emulsions in resting and activated states. The results of the experimental studies conducted directed authors to conduct further research of mechanically activated suspensions during the filtration process.

1. Introduction

Water is the most common binary inorganic component on the planet, as it plays a key role in various processes of flora and fauna, inanimate nature and in human production activities (it is the main part of the hydrosphere, is found in the compositions of earth rocks and minerals, can be found in soil and in the atmosphere, occupies 50÷99% of the animal and plant mass, and it is a fundamental link in the maintenance of most of technological operations etc.) [1-3].

Currently, water is the most well-studied substance in the world (by composition, structure, properties, purpose, abnormal properties, changes upon activation by various methods, etc.). Leading scientists have proposed a myriad of theories that can explain most of water's properties, however, the area of mechanical activation of water and liquid suspensions has not been fully researched. Understanding of the structure and properties of water, as well as the possibility of its correction using mechanical activation, carries a practical importance when used in technological processes, especially in food production industries, which explains the goal and purpose of jointly conducted experiments [1-6].

2. Literary overview

It is known that under the activation by external thermal, mechanical, and other influences, solid matters and liquid states are able to change its properties. Mechanical activation plays a paramount role in preparation of raw food materials for various production processes [7-19].

The process of mechanical activation of solid matters (shock, shock-abrasive or abrasive effects leading to structural, chemical or state changes [12]) is fragmentarily researched and generally recognized. However, despite the extensive number of scientific studies conducted in regard to this topic, the area of mechanical activation of liquids requires further research.

At this time, leading scientists researching the mechanical activation of liquids [1, 2, 5, 6, 8-19] have already established that the given process is capable of inflicting complex effects on the matter, triggering chemical reactions. Direct mechanical effects on the molecular formation led to a disordered state, as well as the weakening and rupture of intermolecular bonds and their self-organization until an energy equilibrium can be reached (giant heteropatric water clusters break into many smaller clusters), thus reducing viscosity (μ) and surface tension (σ). One of the modern methods of mechanical activation in the food industry is vibration activation, and the leading devices of our time are the devices of vibration activation [2, 5, 9, 11, 13, 14, 16, 19].

Vibration activation of water and liquid solutions is a complex, multi-stage process of altering the energy state of the matter under the influence of vibrations. Vibration effects are able to speed up many production processes and obtain numerous additional effects. The effects that arise from the action of vibration of nonlinear mechanical systems are discussed in detail in source [19]. The reversible change of some properties of water and dispersed systems containing liquid state matters are discussed in sources [6, 15].

The interest in the study of changes in temperature *(t)* and hydrogen exponent, the activity of hydrogen ions *(pH)* properties during the process of mechanical activation (vibration activation) of water and liquid suspensions is driven by the fact that the said indicators are the most significant factors in the processes of food production. Additionally, *pH* is linearly connected to the hardness of water. This indicator leads to premature corrosion and equipment failure and is not applicable in a number of technological processes [3, 6-8].

3. Research methods and materials

The research of changes of properties of mechanically activated water and liquid suspensions was carried out in several stages on the premises of two different universities: stage I was conducted at Kemerovo State University and stage II was conducted at the Far Eastern State Technical Fisheries University.

3.1. *Stage I of the experimental study*

At this stage, distilled water (hereinafter, water) was exposed to mechanical activation. A planetary rim mixer with volume $(V)=4\cdot10^{-3}$ m^3 and number of revolutions of the activation device $(n)=10^2\div320$ min^{-1} was used as the activation device. All measurements of *t* of the air in the research room and in the water (during the process of mechanical activation), as well as the *pH* level, were done electrometrically, in accordance with the current standards, using a universal pH-meter (graduation: 0.1 and 0.01). The process was recorded using photo and video footage [1].

3.2. *Stage II of the experimental study*

During stage II, water and liquid suspension (hereinafter, suspension) where subjected to mechanical activation (vibration and vibration mechanical activation). A vibrating mechanical activator was used as the activating device ($n=10^4$ min^{-1}, amplitude of vibration (A)=2.5÷5.3 mm, time (τ)=180 s). All measurements of *t* in air in the laboratory, water and in the suspension (before and after mechanical activation) were carried out in accordance with the current standards, using measurement tools with graduation: 1, 0.1 and 0.01°C). The *pH* levels of the water and suspension were established using indicators, which were determined through the comparison of samples under study to the standard scale. The process was recorded using photo and video footage.

4. Results and discussion

This section presents a summary of the general results of two independent experiments, depicted through the use of graphics (figures 1, 2).

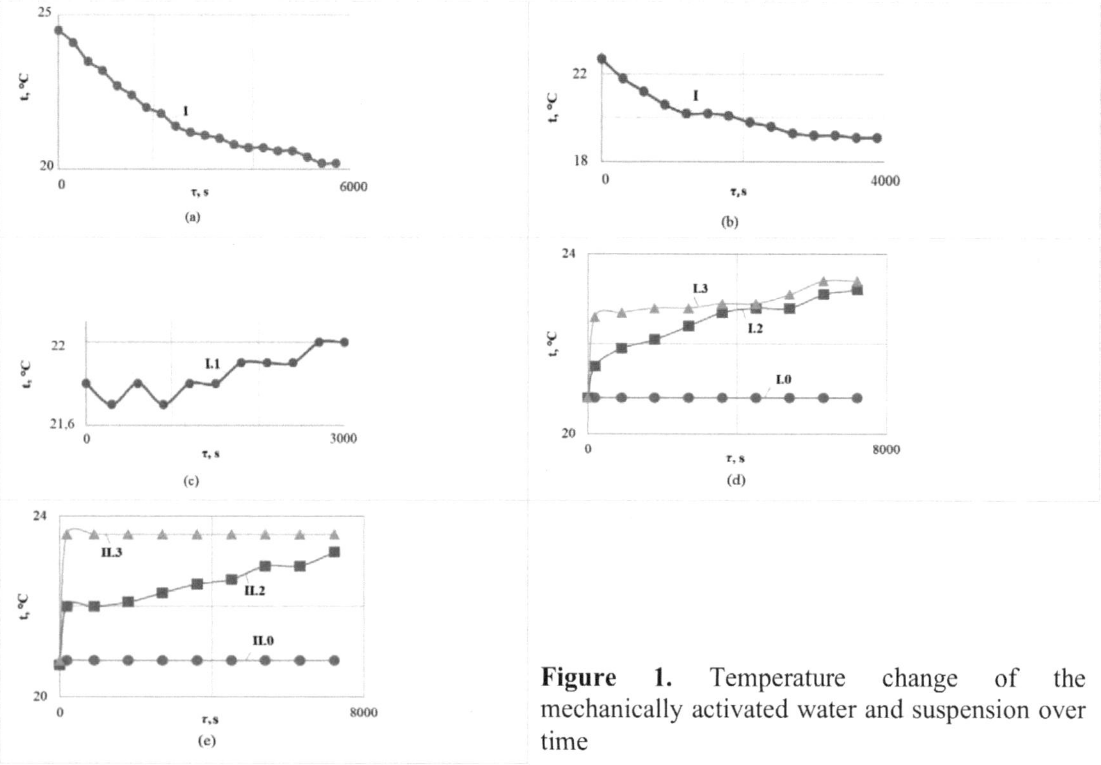

Figure 1. Temperature change of the mechanically activated water and suspension over time

The results obtained through the tracking of t during τ during the process of mechanical activation of water (I) using a mixer (at n=10^2 min^{-1} и n=180 min^{-1}) are shown on figures 1(a), 1(b); after mixing – shown in figure 1 (denoted I.1); before and after vibration activation of water (I) and suspension (II) using the vibrating mechanical activator in the «vibration» mode (at n=10^4) in figure 1(c) (I.2 and II.2); before and after vibrating activation of water – in figure 1(d) (I.3 and II.3); and with no activation (I.0 and II.0). Figure 2 demonstrates the results of changing the pH level of mechanically activated water (I) and suspension (II). All designations are duplicated in figure 1.

The analysis of figures 1(a), 1(b) demonstrates that when water is stirred, a decrease in the t typical to this process can be observed (by 11.54÷17.96%.). According to Newton's law of viscosity, the μ of water must increase as the temperature decreases, however, the literary overview and previous studies [1, 9] indicate its decrease upon mechanical activation. The unusual behaviour of μ can be explained using the cluster theory [6].

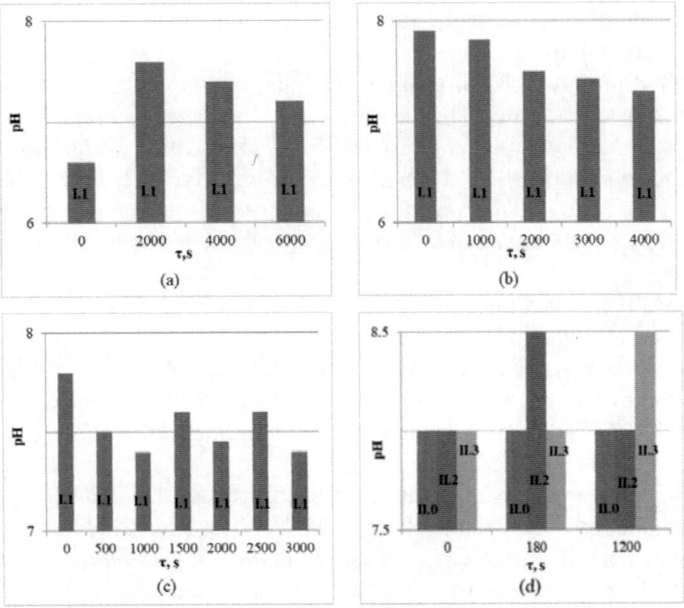

Figure 2. Changes in pH level of mechanically activated water and suspension

This also explains the growth of *t* when using a mechanically activated water mixer over time (figure 1(c)). During the vibration and vibration activation of water and suspension (figures 1(d), 1(e)), a steady growth and maintenance of *t* level can be observed (by 12.08÷13.46%.). Going forward, a study is planned to find the levels of n at which the temperature of water and suspension would not decrease under when put through the action of mechanical activation, but rather increase instead.

The analysis of hyetographs (figure 2) demonstrates that during the process of mechanical activation (vibration and vibration mechanical activation) with various n, the pH level may decrease both below its neutral state (pH<7), assuming an acidic state, or increase above the neutral state (pH>7), assuming an alkaline state [1]. However, the following indicators remain within the acceptable limits for drinking water and technological needs (pH=6÷9).

5. Conclusion

The literarily review on the methods of mechanical activation of liquids, in combination with the results derived from previous and current studies, conclude that the vibration mechanical activation of water, suspensions, and emulsions presents a great interest for further research of various technological processes. Particularly the filtration process, as it retains the activate state for the longest duration of time with a minimum time required for activation.

References

[1] Rudnev S D, Shevchenko T V and Sergeeva E Y 2020 Changing properties of water under mechanical activation *Modern directions of scientific research: technology, quality and safety: materials N(V)K* 121-124

[2] Shevchenko T V, Ustinova Y V and Teplova E Y 2016 Features of the structure of the water molecule and methods of change of its properties *Innovations in the food industry: education, science and production: materials of VNPK* 147-149

[3] Kulsky L A, Gornovsky I T, Koganovsky A M and Shevchenko M A 1980 *Directory for Properties, Analysis Methods and Water Purification: in two Parts* (Kiev: Naukova Dumka) p 1206

[4] Nikolaev A F 2007 A modern look on the structure of the water *News SPbGTI (TU)* **1(27)** 110-5

[5] Blekhman I I, Blekhman L I, Vaisberg L A, Vasilkov V B and Yakimova K S 2008

«Abnormal» properties in liquids under vibration *Property of the Academy of Sciences. Mechanics* **422(4)** 470-4
[6] Goncharuk V V, Smirnov V N, Siroeshkin A V and Malyarenko V V 2007 Clusters and giant heteropatric clusters of water *The chemistry and technology of water* **29(1)** 3-17
[7] Bilenko L F *New handbook for chemists and technologists. 8.5 Mechanical activation when dispersing solid materials. 8.5.1 The essence of mechanical activation of materials and the area of its use Retrieved from: http://chemanalytica.com/book/novyy_spravochnik_khimika_i_tekhnologa/09_protsessy_i_apparaty_khimicheskikh_tekhnologiy_chast_I/5252*
[8] Vorobiev Y V 2013 Fundaments of the theory of mechanical activation of liquid mediums *Bulletin of* TSTU **3** 608-613
[9] Rudnev S D, Markov A S, Ivanov V V, Meshchaninov A V and Prilipko A V 2017 The intensification of the process of obtaining homogenous highly concentrated mixtures with a liquid state in bread production technologies *International scientific research journal* **07(61)** 72-78
[10] Geller S V 2005 Vortex heaters of liquid: cavitation, creation *Youth technology* **11** 56-59
[11] Zinovieva E V, Fedosov S V and Akulova M V 2015 Effects of mechanical activation on structural properties of stirred cement paste *Bulletin of Civil Engineers* **3(50)** 159-162
[12] Gusev A I 2007 *Nanomaterials, Nanostructures, Nanotechnologies* (Moscow: Fizmatlit) p 416
[13] Skosar V Y 2014 Possible mechanisms of energy release in cavitation devices *Aviation and space machinery and technology* **7** 27-33
[14] Efremov I M and Figura K N 2010 The process of vibrations distribution under mixing conditions *Systems. Methods. Technologies* **2(6)** 42-45
[15] Kosincev V I 2010 Water purification using mechanical activation *Advances in modern natural science* **3** 128-129
[16] Kolpakov A V 2012 Research of the process of dispersing of liquid matters through the use of impulse hydrodynamic method *Technology and machinery in livestock farming* **3** 122-128
[17] Vorobiev Y V and Kuzmon A P 2012 Chemical processes of organic liquids, initiated by a hydrodynamic activator *Bulletin of TSTU* **4** 905-911
[18] Kalacheva L P, Koryakina V V and Fedorova A F 2009 Hydrogen production by means of mechanical activation of condensed phases of water *Electronic scientific journal of Oil and Gas Business* **2** 12
[19] Blekhman I I 2017 *What can Vibration do? About «Vibration Mechanics» and Vibration Technolog*y (Moscow: Lenand) p 216

IOP Conf. Series: Materials Science and Engineering 1155 (2021) 012043 doi:10.1088/1757-899X/1155/1/012043

Ensuring the collection of mechanical engineering products, taking into account the tolerances for computer-aided design in "GePARD.3D"

D Zhuravlev[1], L Khvaschevskaya[1], A Shabalin[1] and M Gaer[2]

[1]Irkutsk State National Research Technical University, 83 Lermontov st, Irkutsk 664074, Russian Federation
[2]Irkutsk State University, 1 Karl Marx st, Irkutsk, 664003, Irkutsk, Russian Federation

E-mail: xvlf@mail.ru

Abstract. The article discusses a geometric approach to ensuring the assembly of mechanical engineering products, taking into account the spatial permissible deviations of components in computer-aided design using configuration manifolds and biquaternions. The main advantage of this approach is the modeling of assemblies taking into account the tolerances with the possibility of direct access to the tree for constructing the 3D model of the product, which greatly facilitates the user's work. The developed configuration model of the assembly and the technology for analyzing spatial dimensional relationships in the product, taking into account the tolerances and using biquaternions, allow obtaining analysis results with predictable accuracy already at the stage of geometric modeling.

1. Introduction

Analysis of the current stage of development of machine-building production shows that work to ensure the quality of products is carried out in the following main areas:

1. In the field of machine design - along the line of creating technological designs of parts, assemblies and machines, along the line of improving mathematical models and methods for conducting spatial dimensional-accuracy analysis, focused on computer-aided design.
2. In the field of technology development - along the line of the correct choice of bases and obtaining the required accuracy of processing parts, complete processing of parts in mechanical shops, creating rational technological processes for assembling machines.
3. In the field of production - in the line of strict adherence to technology, obtaining high quality manufacturing of parts, blanks of its production, as well as units from cooperative factories.
4. In the field of assembly design automation - along the line of creating CAx-systems in the concept of "digital enterprise".

The purpose of design automation is to improve the quality of products with minimal labor, material and energy resources.

Despite the tremendous advances made in recent years in the development of CAD systems, modeling methods, assembly structure representation and data analysis still require further improvement and development.

The accuracy of a product is the most important indicator of its quality, which is formed at the stage of product design and is implemented during the assembly process. The main task of the assembly is to ensure the mating of components with the required geometric accuracy parameters that will best ensure the functioning of the product. Components of high quality and recognized as suitable may either not form the required tolerances of the assembly unit, or they may not be assembled at all.

Accuracy problems arising in the process of creating complex science-intensive products are relevant not only in Russia, but throughout the world. A set of problems related to tolerances (in particular, the problem of the generalized mathematical representation of tolerances) arose in the new conditions of production with the use of computers. Research A. Reykvich [1] gave impetus to the beginning of active research in this area. The research results of a group of scientists led by A. Clement [2] contributed to the beginning of the development of a new generation of ISO standards related to tolerances [3]. Representation of geometric variations using matrix transformations is considered by D. Whitney, O. Gilbert, M. Yastrebski in [4]. Conversions represent both nominal relationships between parts and deviations due to geometric tolerances. Desrochers A, Béron V, Laperrière L in [5] use screw parameters to simulate flat tolerance zones.

In the early stages of design, there is a problem of synthesizing a product design, expressing its function as fully as possible, taking into account many restrictions. Since at the stage of conceptual design, about 60% of the cost of the life cycle is laid, the main task is to find geometric shapes that are optimally combined with each other through assigned sets of tolerances that will best describe the design idea.

A full-fledged analysis of spatial dimensional-accuracy relationships in nodes is one of the most important tools for creating high-quality products that can function with high reliability throughout the entire service life, therefore, the CAD system's functionality must support the assigned tolerances as a natural part of the geometric 3D model of the product.

One of the important stages of technological preparation of production is design, 3D modeling and manufacture of various types of equipment. Moreover, its dimensions, geometric parameters and manufacturing accuracy are determined by the values of the geometric parameters of the corresponding components of the product. Therefore, the accuracy of the design and manufacture of tooling is not guaranteed, which leads to further overhead if the technological processes and machining programs on CNC machines of the product components and related tooling are developed on the basis of nominal 3D models. To understand the scale of losses, it is enough to say that for industries associated with a large number of complex bypass-forming equipment, the process of its creation is 60-80% of labor and funds from the total volume of the Chamber of Commerce and Industry and approximately 10-15% of the cost of the product.

Further development of the electronic model of the product involves numerous analyzes and simulations carried out at the technical design stage. Therefore, it is possible to reflect the real situation that will develop during the assembly process or during the operation of the product, to identify the accumulated errors of the real geometry of the product, if already at the stage of conceptual design the problem of creating a 3D model of the product is solved, in which tolerances and information about them are its integral part.

Today, well-known CAD systems (SolidWorks, Autodesk, 3DCS, etc.) solve the problem of architecture, laid down in the last century, by "total product development anew".

A feature of the modern stage of development of computer-aided design systems is the emphasis on the "intellectual level" of systems.

2. Geometric approach to computer-aided design and analysis of assemblies taking into account tolerances and using biquaternions in the GePARD.3D system

When specifying the surface of a part, it is necessary to take into account many different factors that affect its shape and parameters. According to the main theorem of the theory of surfaces (Bonnet's theorem), any surface is determined by uniquely specifying a pair of quadratic forms in the domain U on the plane, provided that the initial conditions are specified. For a part with an elementary topological structure, the real position of the surface can be obtained by varying the coefficients of the quadratic forms. In the case of non-elementary surfaces, it is possible to use atlases consisting of several maps, the number of which is determined by the requirement of minimality and clarity.

When designing a mechanism, the designer must assign different types of tolerances. In our approach, tolerances are represented by three main groups:

- location tolerances (they are characterized by bivectors and biquaternions that set the connection between the reference points of the initial conditions);
- bending tolerances (characterized by a change in the coefficients of the second quadratic form while keeping the first quadratic form unchanged;
- metric distortion tolerances (characterized in terms of variations of the first quadratic form).

Each part included in the product, as well as its geometric elements, for the geometric characteristics of which tolerances are assigned, are uniquely associated with configuration manifolds [6]. Each point of such a variety sets the value of those parameters that can change. Information about unchanged product parameters does not affect the design of the manifold, which makes it possible to significantly reduce the number of unknown quantities. Clear rules are set for determining all parameters of a part (product) at a given point of the configuration manifold. The variety is chosen by the designer. To obtain a new position of the geometrical element of the part with the assigned tolerance for the geometrical characteristic, we set the marked reference, the configuration manifold and its own normalized biquaternion of displacement.

The relationship between parts, with assigned tolerances and included in the product, is described by the assembly graph. To formalize these connections, we use biquaternions.

The assembly is divided into levels, each of which corresponds to a so-called C-part in the same order as the assembly level. In other words, a level k assembly is the assembly that produces a C-part of order *k*. So a monolithic component that cannot be divided into individual parts is called a C-part of order zero. Then a C-part of order *k* is a C-part obtained by assembling C-parts of order not greater than (*k*-1). Accordingly, the (part or assembly) tolerance is called the (part or assembly) level k tolerance if the result of the level *k* assembly depends on it, and *k* is the smallest such value.

If the assembly consists of the *(n+*1*)-th* level. Let us denote by m_j the number of C-parts of level j (j = 0, 1, ..., *n*), by q_p - the number of surfaces of the *i-th* C-part of level *j* (j = 0, 1, ..., n; i = 0, 1 ,..., *m*), for which the tolerances of level j are specified, and is the number of tolerances assigned to the surface. The configuration space of the level *j* , and p_{ji}^{k} of the surface S_{ki}^{j} is obtained in the form:

$$K_{ji}^{k} = \prod_{\alpha}^{p_k^i} K_{ji}^{k\alpha} , \tag{1}$$

where: $K_{ji}^{k\alpha}$ — configuration spaces S_{ki}^{j} characterizing the deviation of the surface from the nominal values relative to one of the tolerances of level *j*.

The configuration manifold of the assembly is obtained as follows:

$$K = \prod_{j=0}^{n}\prod_{i=1}^{m_j}\prod_{k=1}^{q_k}\prod_{\alpha=1}^{p_{ji}^{\alpha}} K_{ji}^{k\alpha} . \tag{2}$$

Using biquaternions and configurational manifolds, a mathematical model of the spatial configurational dimensional chain (CDC) of the product has been developed. The developed technology of nonlinear analysis of spatial dimensional-accuracy links in assembly units to ensure collection is directly related to the technology of configurational direct modeling [7].

3. Conclusion

The research results were implemented in an experimental version of a new generation CAD system "GePARD. 3D ". Figure 1 shows an example of a spatial analysis of a node, taking into account the spatial tolerances assigned to its components, in "GePARD. 3D ".

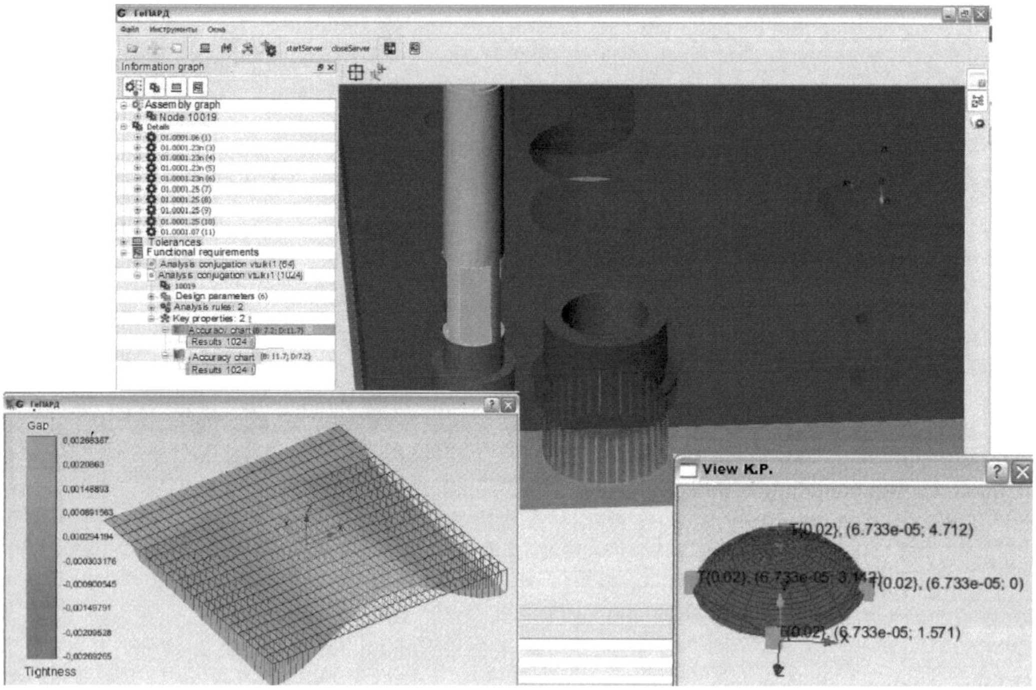

Figure 1. Spatial analysis of tolerances in GEPARD. 3D ".

Many well-known companies are developing various approaches that are designed to provide the user with direct modeling. In this case, direct modeling is carried out in separate applications without direct access to the tree of its construction. The technology of configurational direct modeling of product geometry developed by us is free from this drawback. In addition, since the surfaces are defined by their first and second quadratic forms, it is equally convenient to work with both analytically described surfaces and free-form surfaces: import, create, edit and carry out a collection analysis taking into account the assigned tolerances.

The developed technology for ensuring the collection of mechanical engineering products, taking into account the spatial permissible deviations of components, based on new approaches to the design and manufacture of mechanical engineering products, makes it possible to ensure the required accuracy of the main geometric characteristics of the product, both by the statistical method and by the worst-case method. Its implementation is closely related to the reengineering of business processes for the provision of life cycle products at the present stage of development of mechanical engineering.

References

[1] Requicha A 1983 *International Journal of Robotics and Automation* **2(4)** 45-60
[2] Mathieu L, Clement A and Bourdet P 1997 Modeling, Representation and Processing of Tolerances, Tolerance Inspection: a Servey of Current Hypotesis. *Computer Aided Tolerancing 5th CIRP Seminar* (Toronto, Canada: Springer) 1-38
[3] Srinivasan V 2001 An integrated view of Ceometrical Product Specifications and Verification *7th CIRP Seminar on Computer-Aided Tolerancing* (France,Cachan:Springer) 7-18
[4] Whitney D, Gilbert O and Jastrzebski M 1994 Representation of geometric variations using matrix transforms for statistical tolerance analysis in assemblies. *Research in Engineering*

Design. **6** 191

[5] Desrochers A, Béron V and Laperrière L 2003 Revisiting screw parameter formulation for accurate modeling of planar tolerance zones. *Proceeding of the 8th international CIRP seminar on computer aided tolerancing* (Charlotte, NC) 239-48

[6] Khvaschevskaya L and Zhuravlev D 2018 Modeling of spatial tolerances of assembly units using biquaternions. *Bulletin of Irkutsk State Technical University* **22 (11)** 71-88

[7] Gaer M and Zhuravlev D 2012 Configuration direct modeling technology. *Bulletin of Irkutsk State Technical University* **70(11)** 44-7

Binary control characteristics of resistive models of the induction devices

A A Tyapin[1], V I Panteleev[1], E S Kinev[2,4], S N Efimov[3] and S S Bezhitsky[1,3]

[1]Siberian Federal University, 79 Svobodny Ave., Krasnoyarsk, 660041, Russia
[2]Thermal Electrical Systems LLC, 12 Spandaryan St., Krasnoyarsk, 660020, Russia
[3]Reshetnev Siberian State University of Science and Technology, 31 Krasnoyarsky Rabochy Ave., Krasnoyarsk, 660037, Russia

[4]E-mail: Kinev_ES@ontecom.com

Abstract: The article presents an approach to the construction of parametric models of power circuits for metallurgical induction devices. The parameters of the equivalent circuit of inductors intended for mixing the aluminum melt or heating it before extrusion have components determined by the properties of the secondary element. The properties of the metal change during heating, therefore, in order to create adequate mathematical models, the simulation systems should take into account the change in reactivity and losses. Staged control of reactivity using controlled keys is considered traditional. Adjustable resistive components of equivalent circuits are rarely used. As a means of controlling the mode of equivalent resistive two-terminal devices, a code impulse action on the switches is proposed, which changes the impedance in dynamics. Controlling the step of the impulse action on two-pole models with their serial, parallel and mixed connection, can provide the required smooth nature of the resistive conductivity change. To carry out a numerical experiment, two-pole switchable models of different bit sizes have been built, which have the properties of analog-digital elements. The study of typical circuits of parametric links was carried out using a circuit simulator, while examples of adjusting characteristics of operating and system parameters were obtained. In the course of a numerical experiment, a solution was obtained for a set of operating parameters of an induction device, reflected by a family of vector diagrams.

1. Introduction

Induction electric heating is considered the most effective means of heating aluminum before pressing [1]. Heating inductors are made of a copper tube using liquid cooling [2]. Heating of aluminum ingots is carried out until a temperature of 530 ° C is reached. To compensate for reactive power and a local increase in the voltage of individual sections of inductors, capacitor banks are widely used [3]. The value of the capacitance of the capacitor banks is changed by connecting capacitors in parallel. The capacitors are switched by power switches. Control of switching elements is performed automatically, using a microcontroller or relay-contactor control system [4, 5]. When the ingot is heated, the electrophysical parameters of the heating complex change smoothly [6]. The equivalent load impedance contains resistive components, the change of which is not fully taken into account in the control modes of the regulator [7].

General view of methodical and periodic electromagnetic inductors for heating aluminum ingots is shown in Figure 1. The designs of the inductors are quite diverse, and the sections can be connected to a three-phase or single-phase network. In each specific case, the problem of optimal power supply of the induction complex is solved. This task depends on the technological and technical requirements, as well as the circuitry of the windings.

a b c

Figure 1. General view of different types of induction heaters.

An example of an equivalent circuit for a single-phase induction unit with parametric elements is shown in Figure 2. The circuit implements a balancing device for a three-phase inductor. The balancing device consists of a capacitor bank and a power choke. This is a typical solution for a powerful single-phase load in the form of an ID inductor designed for heating aluminum poles (Figure 1, a), weighing up to 1 ton [8]. The power capacitors are connected to the voltage U_{ab}, and the electromagnetic choke with taps is connected to the voltage U_{bc}. Parallel to the load, an additional adjustable capacitor bank is installed, at line voltage U_{ca}. Such a capacitors battery is designed to maintain the resistive resistance of the power branch by creating a resonance of currents [9]. In the course of balancing to the initial parameters of the inductor, the impedance drift is not taken into account.

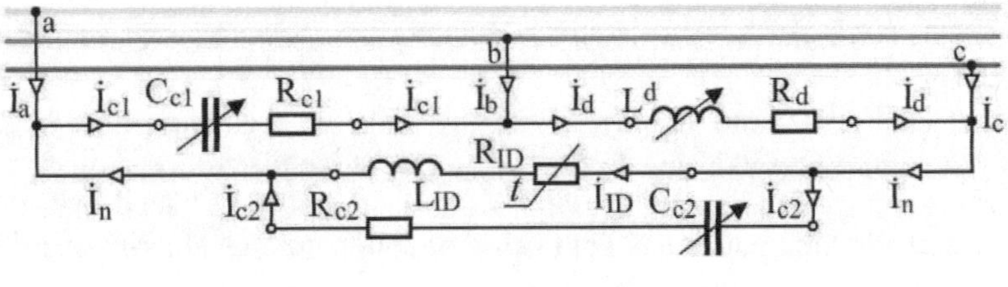

Figure 2. Schematic of an induction installation with parametric models.

In the simulated device, standard regulators of the capacitance of the capacitor banks and the inductance of the choke are used. Switching is performed with gate keys operating under the control of the controller [5, 10]. Traditional discrete controllers are usually built using thyristors or triacs. Modeling such elements in an arbitrary simulator, as a rule, does not cause difficulties.

2. Problem statement

Judging by the diagram (Figure 2), the RID element is a resistive two-port element with variable parameters. Modeling power induction devices with adjustable resistive elements contains certain difficulties. Analog control of two-pole elements in a numerical experiment is not possible. Therefore, to simulate a parametric two-port element, discrete control with the use of idealized keys should be used. Discrete control characteristics can be different. In the simplest case, you can use binary code.

Dynamic control of two-terminal elements has a number of features that are fully manifested in the time domain. Certain difficulties arise when optimizing the switching intervals, since the simulation uses computational methods that are critical to the choice of the iteration step. In addition to simulating dynamic modes, the use of discretely controlled elements makes it possible to automate a numerical experiment in the field of steady-state modes. Using the functions of impulse control of switched models, it is possible to program multivariate studies of a set of quasi-stationary states of a model of a power supply system, a rather complex induction complex.

3. Solution

In a numerical experiment, the dependence of resistance (conductivity) on temperature must be taken into account. At the same time, the element temperature changes much more slowly than the transient mode lasts. This feature makes it possible to overcome the difficulties with the selection of switching intervals. Nevertheless, it is the parametric nature of the two-terminal element that allows us to talk about impulse control, since in fact the impedance changes over time. Thus, at the stage of modeling in the simulator, the two-terminal element must be replaced by a parametric model.

The degree of complexity of the structure of the parametric model is determined by the level of correspondence of the synthesized control characteristic to the conditions for regulating the curve of changing the operating parameters [6, 11]. The characteristics of the change in the impedance of a parametric two-terminal network in time are monotonous. The impedance variation range does not exceed 20% of the nominal value. When modeling the power circuits of induction complexes, models created from elements with lumped parameters are used. Thus, the studied circuitry becomes digital-analog.

To change the parameters of two-pole elements, you can use a code pulse control [11, 12]. The state of the electric key of each switched element of the model is determined by a binary code. At the same time, in pulsed parametric models, you can apply different schemes for connecting elements, with a number of ideal keys. Usually, the number of keys corresponds to the number of bits of the regulated model [11, 13].

Depending on the selected switching time intervals $\Delta t = t_{k+1} - t_k$, the model can be dynamic, or go to the region of stationary modes. Low switching frequency provides a set of quasi-stationary modes of inductive load. For the simulation of transient modes, it provides a choice of small time intervals between commutations [4, 14]. This problem is solved when overcoming conflicts, between the time step and the key switching intervals. In the software used, this problem is solved automatically, at the algorithmic level. This increases the complexity of the program code, but excludes crashes during the numerical experiment.

An example of circuitry of the simplest switching resistive regulators of a serial and parallel structure is shown in Figure 3, a, b. Regulated elements can be identified in the Z-region as resistors R (Ohm), and also in the Y-region as conductivity G (Cm). It should be noted that in most practical circuit solutions, especially taking into account the use of computational algorithms for nodal analysis, the conductivity representation is preferable.

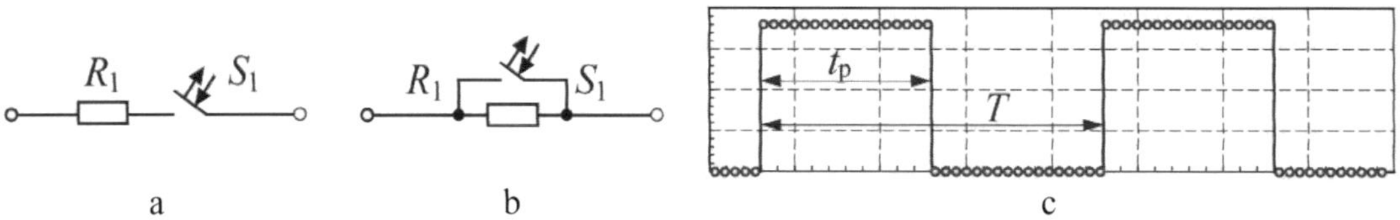

Figure 3. Fragments of switching regulators circuitry.

In the figures, the following designations are adopted: R_1 is the base value of the resistor, S_1 is an ideal controlled switch, then $k_f = t_p/T$ is the duty cycle of the pulses, for the control sequence shown in Figure 3, c. On the basis of linear resistive two-terminal networks, more complex circuit models of controlled elements are created. With the help of impulse action, it is possible to realize both linear and non-linear transient characteristics. The undoubted advantage of such models can be considered the linearity of their parameters and the relatively low order of the matrix equations obtained in the nodal basis. The use of such models is preferable in conjunction with typical multipoles elements of circuit theory, called controlled voltage and current sources [15, 16].

The dynamic state of a branch with pulse-controlled conductivity can be described by the expression:

$$g(t, k_f) = \begin{cases} G, \text{ if}: nT \le t \le [T + t_p]; \\ 0, \text{ if}: [nT + t_p] < (n+1)T, \end{cases} \tag{1}$$

where $n = 1, 2, 3, \dots, \infty$.

Applying an expansion in a trigonometric series, for a function $g(t, k)$, that is symmetric with respect to the middle of the switching interval, it is possible to rewrite the expression for the dynamic conductivity of a resistive link as a sum of constant and variable components:

$$g(t, k_f) = g(k_f) + \tilde{g}(t, k_f) = G k_f + \frac{2G}{\pi} \sum_{n=1}^{\infty} \frac{\sin n \pi k_f}{n} \cos \frac{2n\pi}{T} t. \tag{2}$$

In the case of using a binary control code, it is convenient to use a weighting factor for an adjustable resistance matrix, according to Figure 4, a, b, equal to two $k = R_{n+1} / R_n = 2$.

An example of pulse-controlled resistive elements of a more complex, parallel and serial structure is shown in Figure 4, a, b.

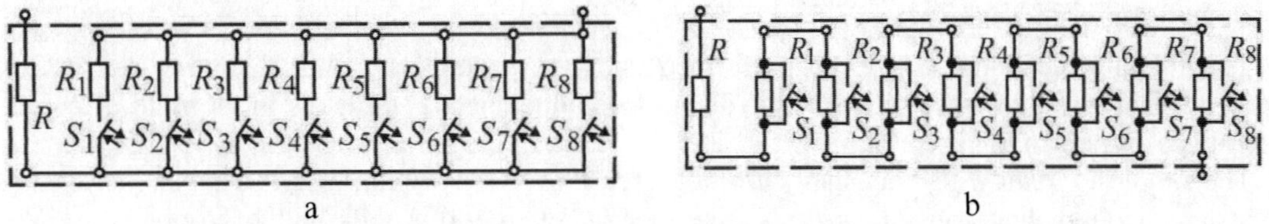

Figure 4. Variants of circuit models of parallel (a) and serial (b) regulator.

In the notation adopted in the figure, R_n – are the values of the adjustable resistors of the matrix, taken with a given weighting factor. An averaging capacitance can be installed in parallel to each switched element [11, 14]. At the same time, the modulating effect of the step process decreases and, accordingly, the intensity of the higher harmonics of voltage and current, caused by switching, decreases. The value of the averaging capacitance is selected from the condition of the minimum possible inertia of the impulse link. Characteristics of binary code control of resistive switches of a five-bit parametric model are shown in Figure 5. In the general case, the number of bits can be increased.

However, it should be understood that when constructing models of 8, 16 or 24-bit controlled elements, it may be necessary to increase the bit depth of the mathematical kernel used by the calculator [16, 17]. This is due not so much to the condition of expanding the bandwidth of the simulated induction device, with pulse-controlled elements, as to the provision of appropriate conditions for the convergence of the iterative computational process. By the way, it can be noted that, according to an approximate estimate, to expand the signal bandwidth above 10 Hz, the frequency of the control pulses of the transient mode should be selected with a value greater than 10 kHz [11].

Therefore, to control elements in industrial installations, the model operates at a frequency of more than 50 kHz. In this case, the error in setting the resistance values during pulse regulation of a five-bit resistive matrix is determined by the ratio of the value of the least significant digit to the senior one: $\varepsilon_R = R_1 / R_5 = 6.25\%$.

To achieve greater accuracy in resistive analog-digital models, it is preferable to operate with eight-bit matrix constructions, for which the equality is fulfilled:

$$\varepsilon_R^{\text{H}} = R_1 / R_8 = 2/256 = 0.78\,\%. \tag{3}$$

Characteristics with pulse sequences of control keys S_1 - S_5, without delay at zero time, are shown in Figure 5, a. The characteristics of key management, with a delay at the zero point in time, are shown in Figure 5, b. In this context, the quality of electronic switches, for which MOSFET components are usually used, is not discussed here. Keys are considered perfect. Moments of switching on the graphs are designated by the corresponding letters [t_1, t_2, t_3, t_4, t_5, t_6], switching occurs instantly [16, 18].

The pulse graphical patterns presented in Figure 5 determine the fundamental difference between characteristics 1 and 2 shown in Figure 6, a.

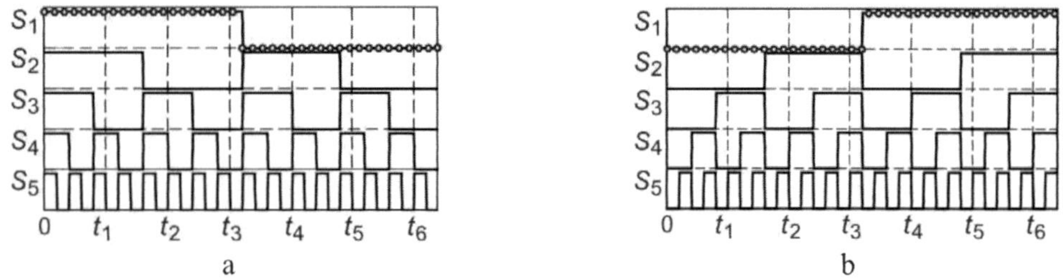

Figure 5. Control characteristics of the switches of the parametric model.

The tuning of parametric models is performed using the circuitry of standard current sources J and EMF E. An example of circuit designs for diagnosing adjustment parameters in the field of currents and voltages is shown in Figure 6.

The simplest circuit with a constant current source J, shown in Figure 6, a, is designed to diagnose the instantaneous voltage on a pulsed resistive model, relative to a node with zero potential. When applying control pulses according to Figure 5, a, to the switches S_n, inside the model $R(t)$, the instantaneous current $i(t)$ of the model takes on values that change over time, according to Figure 7, a (curve 1).

Figure 6. Test circuits for tuning parametric models.

The following designations are adopted in the diagrams: J, E - direct current and EMF sources, $R(t)$ - time-controlled resistive model, Out - instantaneous voltage sensor in the circuit node, EI - current controlled voltage source (CCSV). In the diagram of Figure 6, b, the current change in time is already controlled. The controlled source EI converts the instantaneous current of the primary short-circuited branch into an output voltage measured by the Out sensor.

Since the transfer coefficient EI-source is equal to unity, the output voltage actually reflects the characteristic of the change in the current of the primary branch [18 - 20].

4. Methods and materials

Calculation of circuits with parametric elements and controlled sources according to Kirchhoff's laws is difficult, therefore, it is preferable to use calculations in a software simulation environment using a typical description of all elements [16, 20, 21]. In fact, the construction of circuit models corresponds to a numerical experiment. The model is created in a certain sequence, in accordance with a previously thought-out program of the experiment. The description of the model is formed using a modeling language similar to some versions of the Ansys software environment.

Fragments of the matrix description of EMF and current sources are shown below. The current of the EMF source is directed from node *m* to node *n*, the voltage is directed towards.

$$\mathrm{E_k} = V_\mathrm{n}(t) - V_\mathrm{m}(t) = U \,. \qquad \begin{array}{c|ccc} & \mathrm{m} & \mathrm{n} & i_\mathrm{E} \\ \hline \mathrm{m} & 0 & 0 & 1 \\ \mathrm{n} & 0 & 0 & -1 \\ i_\mathrm{E} & -1 & 1 & 0 \end{array} \begin{bmatrix} V_\mathrm{m}(t) \\ V_\mathrm{n}(t) \\ i_\mathrm{E}(t) \end{bmatrix} = \begin{bmatrix} 0 \\ 0 \\ \mathrm{E_k} \end{bmatrix}. \tag{4}$$

The current of the current source is directed from node *m* to node *n*, the voltage is directed towards. When forming a system of modified nodal equations, for small circuits, use the stamp of the current source given below, obtained from the component equation of the Z-branch.

$$J_\mathrm{k} = \mathrm{I}\,, \qquad \begin{array}{c|ccc} & \mathrm{m} & \mathrm{n} & i_\mathrm{J} \\ \hline \mathrm{m} & 0 & 0 & 1 \\ \mathrm{n} & 0 & 0 & -1 \\ i_\mathrm{J} & 0 & 0 & 1 \end{array} \begin{bmatrix} V_\mathrm{m}(t) \\ V_\mathrm{n}(t) \\ i_\mathrm{k}(t) \end{bmatrix} = \begin{bmatrix} 0 \\ 0 \\ \mathrm{J_k} \end{bmatrix}. \tag{5}$$

Let us consider the description of mathematical models used to construct an algorithm for analyzing circuits with a controlled source of EMF [17, 18, 22]. The input current is directed from node *i* to node *j*, the input voltage is zero, the output voltage is directed from node *m* to node *n* against the source, the output current and EMF source are directed from node *n* to node *m*.

The matrix description of the controlled CCSV source is formed on the basis of component equations, which have the form:

$$u_2(t) = k_R \cdot i_1(t), \quad V_\mathrm{n}(t) - V_\mathrm{m}(t) = k_R \cdot i_1(t),$$

$$V_\mathrm{m}(t) - V_\mathrm{n}(t) + k_R \cdot i_1(t) = 0\,, \quad V_\mathrm{j}(t) - V_\mathrm{i}(t) = 0\,,$$

where: $u_2(t)$ – is the output voltage of the source, $i_1(t)$ – is the input current of the source, $k = R_\mathrm{T}$ – is the transfer coefficient of the source (transition resistance), $V(t)$ – are the potentials of the nodes.

The system of extended nodal equations for a controlled source presented below is compiled taking into account the control and controlled branches.

$$\begin{array}{c|cccccc} & \mathrm{i} & \mathrm{j} & \mathrm{n} & \mathrm{m} & i_1 & i_2 \\ \hline \mathrm{i} & 0 & 0 & 0 & 0 & 1 & 0 \\ \mathrm{j} & 0 & 0 & 0 & 0 & -1 & 0 \\ \mathrm{n} & 0 & 0 & 0 & 0 & 0 & -1 \\ \mathrm{m} & 0 & 0 & 0 & 0 & 0 & 1 \\ i_1 & -1 & 1 & 0 & 0 & 0 & 0 \\ i_2 & 0 & 0 & -1 & 1 & k_\mathrm{R} & 0 \end{array} \begin{bmatrix} V_\mathrm{i}(t) \\ V_\mathrm{j}(t) \\ V_\mathrm{n}(t) \\ V_\mathrm{m}(t) \\ i_1(t) \\ i_2(t) \end{bmatrix} = \begin{bmatrix} 0 \\ 0 \\ 0 \\ 0 \\ 0 \\ 0 \end{bmatrix} \tag{6}$$

The above relations are automatically integrated into the expression of the main computational method. In the software environment for mathematical modeling of induction devices, a modified nodal analysis is programmed [16 - 18].

The generalized expression of the computational method, converted to the time domain, is shown below:

$$\begin{bmatrix} \begin{bmatrix} g_{11} & & \cdots \\ & \cdots & \\ \cdots & & g_{kk} \end{bmatrix} & \begin{bmatrix} b_{kk+1} & & \cdots \\ & \cdots & \\ \cdots & & b_{nn} \end{bmatrix} \\ \begin{bmatrix} a_{11} & & \cdots \\ & \cdots & \\ \cdots & & a_{kk} \end{bmatrix} & \begin{bmatrix} r_{kk+1} & & \cdots \\ & \cdots & \\ \cdots & & r_{nn} \end{bmatrix} \end{bmatrix} \cdot \begin{bmatrix} \begin{bmatrix} V_1^g(t) \\ \cdots \\ V_k^g(t) \end{bmatrix} \\ \begin{bmatrix} i_1^r(t) \\ \cdots \\ i_n^r(t) \end{bmatrix} \end{bmatrix} = \begin{bmatrix} \begin{bmatrix} \cdots \\ \sum j_k^g(t) \\ \cdots \end{bmatrix} \\ \begin{bmatrix} \cdots \\ \sum e_n^r(t) \\ \cdots \end{bmatrix} \end{bmatrix}, \tag{7}$$

where $[g_{kk}]$ – is the matrix of nodal admittances of the regular part of the circuit, $[r_{nn}]$ – is the resistance matrix, $[a_{kk}]$, $[b_{nn}]$, – are matrices with unit and zero coefficients taking into account the controlled elements, $[\Sigma j_k]$ – is the matrix of nodal currents, $[\Sigma e_n]$ – matrix array of EMF sources.

The characteristics of the controlled mode parameters corresponding to the code binary control of the five-bit resistive model are shown in Figure 7. Judging by the characteristics of Figure 7, a, b, the stepwise change in the mode parameters is modulated by the switching frequency.

The smallest step size is typical for the S_5 key of the least significant bit. The set of commutations of all switches at each moment of time is determined by the instantaneous value of the controlled function without averaging [11, 23].

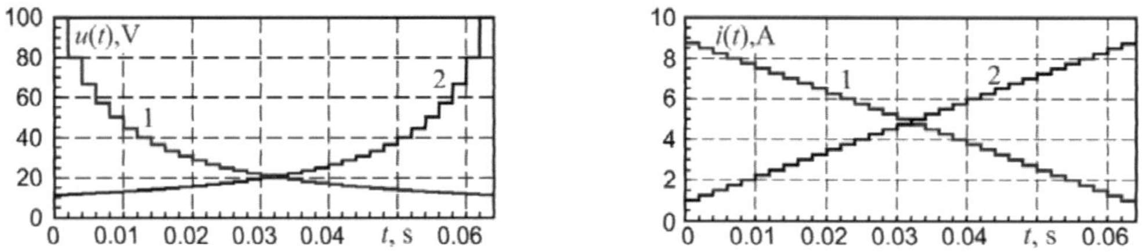

a b

Figure 7. Dynamics of changes in the operating parameters of the resistive model.

Graphs with nonlinear dependences of parameters (1) and (2) (Figure 7, a) correspond to voltage. Graphs with linear characteristics (1) and (2) (Figure 7, b) correspond to the current. Both voltages and currents are determined by the same pulse control sequences (Figure 5, a, b).

Depending on the task of modeling, it is possible to recommend the use of the combined inclusion of models according to Figure 4, a, b. Moreover, to simulate nonlinear curves, it is convenient to use inversion of characteristics using voltage-controlled voltage sources (CVSV) and current controlled current sources (CCSC). Such sources in a relatively simple way provide a mode of galvanic isolation of different parts of the circuit [11, 14, 24].

5. Results and discussion

Taking into account the possibilities of using parametric models, one can proceed to the analysis of transient modes of induction complexes, taking into account the switching of capacitor banks and relatively simple resistive models of the inductor. The shown features of the control characteristics

(Figure 7) require a separate study for the stage of constructing more complex, serial-parallel models. In addition, additional study requires the issue of choosing the control ranges of parametric models, as well as the construction of methods for the synthesis of controlled elements. An example of a vector diagram of an induction unit obtained using switched models is shown in Figure 8. Taking into account the depth of load change, the control characteristics take on a more complex form [25, 26].

The diagram shows that balancing the electromagnetic mode of the ID by the Steinmetz circuit is effective only if the resistive nature of the load is maintained. The degradation of the power factor of the inductor with the exit from the resonance mode immediately leads to a distortion of symmetry, accompanied by a divergence of the current vectors. That is why, in practice, it is necessary to have precise measuring equipment and a microcontroller control device that automatically regulates the electromagnetic state of the ID. A very significant change in the position of the vectors on the diagram is characteristic of a possible uncoordinated change in the parameters of balancing devices, with the admissibility of a simultaneous drift of the electrophysical parameters of the loaded inductor.

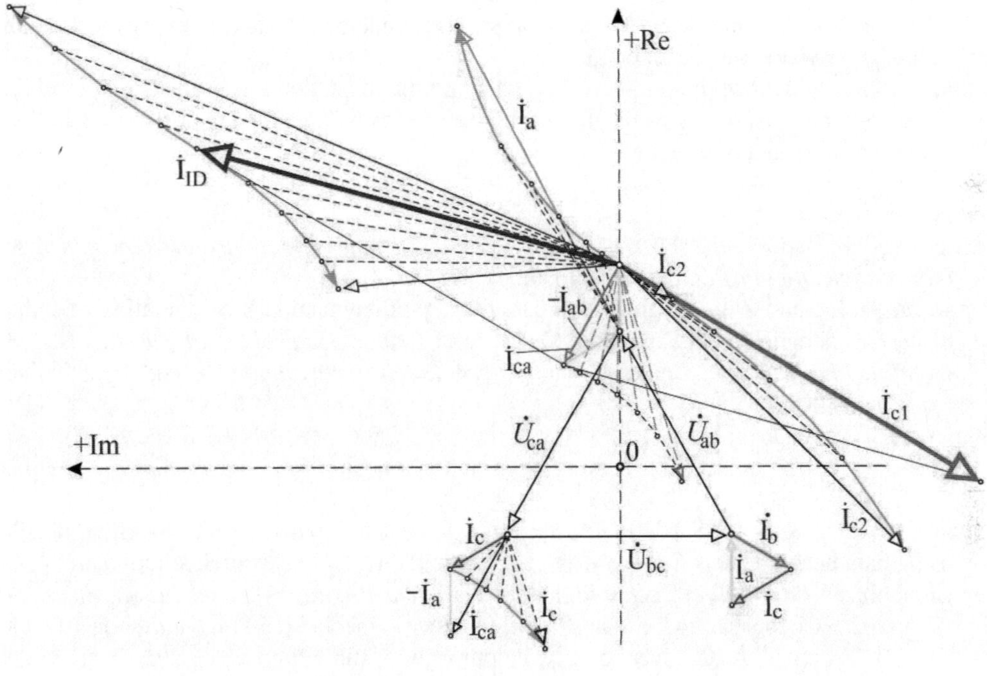

Figure 8. Vector diagram of an induction unit with switchable models.

When constructing a vector diagram, the results of modeling modes for a depth of regulation of an inductive load of 15% were used. In addition, in the program of the numerical experiment, the range of regulation of each of the balancing elements, extended to ± 25% with respect to the symmetric mode, is specified. It can be noted that practical coincidence of the given conditions is unlikely. However, the diagram clearly shows the application of inductor currents (3-4 kA). Naturally, the currents of consumption from the network with such a deep mismatch go beyond the permissible limits, limited by the resonant characteristics of the parallel circuit of the inductor.

The capabilities of the impulse regulation of the models can be used in the calculation of the stationary regime. To do this, it is necessary to draw up a program for a numerical experiment and describe the simulation sequence in the format of a program code. In this way, a control command file is formed, which makes it possible to carry out multivariate calculations according to the global description of the induction device.

As a result of obtaining a set of operating characteristics, when changing the load parameters, presented in the form of "fan" vector diagrams. The problem of modeling the dynamics of induction heaters with the construction of phase trajectories and portraits for the mode of characteristics [27] is programmed similarly to the problem of analyzing vector diagrams [28]. In view of the large volume of calculations and the cumbersomeness of the problem, the described fragments of the numerical experiment are separated into separate projects and will be considered separately.

6. Conclusion

The models of resistive pole networks proposed in two parameters can be used when setting up induction installations with variable functions. Pulse control of a resistive unit subsystem, as part of complex RLC models in a numerical experiment, allows the use of a matrix mathematical apparatus and algorithms for iterative analysis of analog systems. When constructing models, the elemental basis of controlled voltage and current sources is widely used, as applied to the theory of circuits. The use of the simulator makes it possible to exclude the formation and solution of complex systems of equations, passing directly to the understandable and obvious circuitry of the modeled system. In this case, the synthesized models turn out to be analog-digital, and the structure of the obtained control characteristics can become nonlinear. With an appropriate choice of control intervals, the control problem can be both parametric and stationary.

Additional conveniences of the proposed approach are manifested in the form of the possibility of obtaining an evolutionary set of vector diagrams for several steady-state modes of the modeled system during the tuning of the models.

References

[1] Aliferov A I, Lupi S and Forzan M 2017 *Electrotechnological installations and systems. Induction heating installations* (Novosibirsk: NSTU)

[2] Danilushkin A I and Zhivotyagin D A 2019 Energy-efficient modes of operation of induction heaters in a metal processing line *Collection of scientific articles of the 4th All-Russian scientific and technical conference with international participation* (Kursk: JSC "University Book") pp 82-5

[3] Kinev E S, Tyapin A A, Litovchenko A V, Efimov S N and Bezhitskiy S S 2020 Energy modes of a three-section inductor for heating aluminum *Journal of Physics: Conference Series* **1679** 52048

[4] Tyapin A A and Kinev E S 2020 Numerical analysis of the modes of an induction installation using parametric models *Collection of reports of the XII All-Russian Scientific and Technical Conference "Information Technologies in Electrical Engineering and Power Engineering" ITEE-2020* (Cheboksary: Chuvash State University named after I N Ulyanov) pp 107-11

[5] Kinev E S, Tyapin A A, Golovenko E A, Avdulov A A and Efimov S N 2020 Universal Mhd Device For Automation Of Casting Control Of Aluminum *IOP Conference Series: Materials Science and Engineering* 32019

[6] Bazarov A A and Pavlova O E 2019 Electrotechnical complex for heating ingots with uniform energy consumption *Modern Russian equipment to improve the reliability, efficiency and safety of the energy complex of Russia* (Samara: SamSTU) pp 6-9

[7] [Tyapin A A, Kinev E S, Baykova K A and Litovchenko A V 2020 Modeling the energy characteristics of induction heaters *Priority areas of innovation in industry: Collection of scientific articles of the VI International scientific. conf.* (June 29-30, 2020 Part 1. Kazan: LLC "Envelope") pp 83-8

[8] Tyapin A A and Kinev E S 2020 Application of the inductor modeling technique for designing a series of induction devices *The scientific heritage* **50(1)** 61-5

[9] Tyapin A A and Kinev E S 2020 Fundamentals of PWM Inverter Control Strategy of Linear Metallurgical MHD Machine *The scientific heritage* **51(1)** 63-7

[10] Malyshev I V 2014 *Circuitry of pulse and digital devices* (Taganrog: SFedU Publishing House)

[11] Zakharov V K and Lypar Yu I 1984 *Electronic devices for automation and telemechanics* (L.: Energoatomizdat)

[12] Tyapin A A and Kinev E S 2020 Method of modeling and calculation of modes of induction heaters *Collection of scientific works of the 66th International Scientific Conference of the Eurasian Scientific Association* **8 (66)** 119-23
[13] Williams B W 2006 Principles and Elements of Power Electronics Devices, Drivers, Applications and Passive Components (Glasgow. United Kingdom)
[14] Wilamowski B M and Irwin J D 2011 *The Industrial Electronics Handbook. Power electronics and motor drives* (Taylor and Francis Group, LLC. Boca Raton, London, New York. USA)
[15] Baskakov S I 2005 *Radio circuits and signals* (M.: Higher school)
[16] Chua L O *et al.* 1980 *Machine Analysis of Electronic Circuits: Algorithms and Computational Methods* (M.: Energiya)
[17] Chua L O, Desoer C and Kuh E 1987 *Linear and nonlinear circuits* (McGraw-Hill, New York)
[18] Zeveke G V, Ionkin P A, Netushil A V and Strakhov S V 1989 *Fundamentals of circuit theory* (M.: "Energoatomizdat")
[19] Horowitz P and Hill W 1998 *The art of circuitry* (M.: "Mir")
[20] Bychkov Yu A, Zolotnitskiy V M, Chernyshev E P and Belyanin A N 2008 *Fundamentals of theoretical electrical engineering* (SP-b.: Publishing house "Lan")
[21] Karpov E A, Timofeev V N, Perfiliev Yu S, Khatsyuk M Yu and Pervukhin M V 2019 *Modeling of transient processes in linear and nonlinear electrical circuits* (Krasnoyarsk: Publishing house of the Siberian Federal University)
[22] Karpov E A, Timofeev V N and Khatsyuk M Yu 2017 *Theoretical foundations of electrical engineering. Fundamentals of nonlinear electrical engineering in exercises and tasks* (Krasnoyarsk: Publishing house of the Siberian Federal University)
[23] Shaffer R 20017 *Fundamentals of Power Electronics with MATLAB* (Charles River Media, Boston, Massachusetts)
[24] Kinev E, Tyapin A and Litovchenko A 2020 Analysis of the closed circuit network in the power supply system of a metallurgical enterprise *The Scientific Heritage* **49-1** 69-74
[25] Kinev E S, Tyapin A A, Golovenko E A, Avdulov A A and Efimov S N 20202 Universal Mhd Device For Automation Of Casting Control Of Aluminum *IOP Conference Series: Materials Science and Engineering* 32019
[26] Frizen V E and Nazarov S L 2018 *Calculation and selection of electrical equipment for low-voltage distribution networks of industrial enterprises* (Yekaterinburg: Federal State Educational Institution Ural Federal University named after Boris Yeltsin)
[27] Kinev E and Tyapin A 2020 Three-phase inductor performance correction using local resonances *The Scientific Heritage* **48-1** 36-44
[28] Kinev E S, Tyapin A A and Efimov S N 2020 Combined connection on of windings when upgrading an induction heater *Energy security and energy saving* **4** 39-48

Methods and models of traffic analysis

J I Shamlitskiy, D V Rogova, A S Polyakova, A A Popov and L V Lipinskiy

Reshetnev Siberian State University of Science and Technology, 31, Krasnoyarskiy rabochiy pr., Krasnoyarsk, 660037, Russia

E-mail: Tolynbms@yandex.ru, dasha_28_05@mail.ru

Abstract. This article analyzes the traffic situation, and also discusses the methods of collecting data and how they are applied on the road. In addition, the advantages and disadvantages of the methods studied in the work are given. There is a separate talk about the Floating Car Data (FCD) method, what is the essence of the method, its advantages and disadvantages. In turn, the paper presents a difference in the approaches to using data collection methods, namely the Floating Car Data method on the use of the Global Positioning System in cars and the method based on telephone devices with a Global Positioning System signal. The average annual daily traffic is considered as one of the most important traffic data. Describes a way to implement the Floating Car Data method, the advantage in using and improving this method and the prospects of the method in the future.

1. Introduction

Today, for several years now, there has been a trend of increasing demand for improved traffic management, the development of new methods for collecting data, as well as the development of ways to access traffic information in real time. The use of traditional road sensors (for example, based on the use of inductive loops) for data collection is necessary but insufficient due to their limited coverage and expensive implementation and maintenance costs. In recent years, we have witnessed the emergence of alternative data sources that are promising, cost-effective solutions to overcome the limitations inherent in stationary detectors. The introduction of new methods for assessing traffic information is an important task to improve the efficiency of the use of the road network. Today, information about road accidents, traffic jams or other difficulties on the roads is often not received promptly enough. Changing the existing systems for collecting information is a way that can correct the current situation. The very essence of the FCD method is that the traffic situation is simulated in real time. Even though the idea of collecting data from devices "in the car" via mobile phones or GPS is not new, the FCD market has only recently started to develop and provide users with a wide range of applications and benefits. The main task in this work is to study data collection methods, namely the complexity of the implementation of the FCD method, with the aim of improving this method in future developments [1].

2. Intrusive data collection methods

Methods for calculating the traffic situation can be divided into two categories: intrusive (that is, requiring the physical presence of a fixed vehicle in a certain area where this fixation is performed, as well as its interaction with certain systems) and non-intrusive methods (which do not require the interaction of a vehicle with intermediate links in the transmission of location information. The intrusive method is that the data logger and sensor are placed on or in the road. This is the most common and most important approach [2, 3].

2.1. Pneumatic Travel Tubes

Rubber tubes are placed across road lanes to detect vehicles by measuring pressure changes as the vehicle's tire passes through the tube. The generated air impulse is recorded and processed by a meter located on the side of the road. The main disadvantage of this technology is that it has limited lane coverage and its effectiveness is dependent on weather, temperature and road conditions. This system can also be ineffective when measuring low-speed streams [4, 5].

2.2. Piezoelectric sensors

The sensors are placed in a groove along the road surface. The principle is to convert mechanical energy into electrical energy. Mechanical deformation of a piezoelectric material changes the surface charge density of the material so that a potential difference arises between the electrodes. The amplitude and frequency of the signal are directly proportional to the degree of deformation. This system can be used to measure weight and speed.

2.3. Magnetic loops

This method is the most common technology used to collect traffic data. The hinges are embedded in roads in a square formation that generates a magnetic field. The information is then transferred to a counting device located on the side of the road. It has a generally short lifespan because it can be damaged by heavy vehicles but is not affected by bad weather conditions. Over the past decades, this technology has become widespread in Europe. However, implementation and maintenance costs are expensive [6].

In addition to the commonly used methods, recently, due to the development of technology, new, promising methods of collecting information have appeared.

2.4. Passive magnetic method

Magnetic sensors are mounted under or over the road surface. They count the number of vehicles, their type and speed. However, under operating conditions, the sensors have difficulty distinguishing between closely spaced vehicles. Microwave Radar: This technology can detect moving vehicles and measure their speed. The radar records counting data, speed and simple vehicle classification.

2.5. Ultrasonic and passive acoustic methods

These devices emit sound waves to detect vehicles by measuring the return time of the signal to the device. Ultrasonic sensors are located above the lane.

3. Non-intrusive data collection methods

The principle of FCD is to collect traffic data in real time by locating the vehicle's location using mobile phones or GPS across the entire road network. Data like this are sent anonymously to the central processing center [7].

After collecting and processing information, it can be transferred to drivers to obtain the traffic situation on the road. Rather, the FCD is a source of high-quality data that complements existing technologies. They will help to improve safety, the efficiency of obtaining traffic conditions and the reliability of the transport system. They become central to the development of a new transport system [8].

3.1. Floating Car Data

There are two main types of traffic data acquisition, namely Global Positioning System (GPS) and Cellular Systems (GSM).

Figure 1 shows an example of a GPS based network.

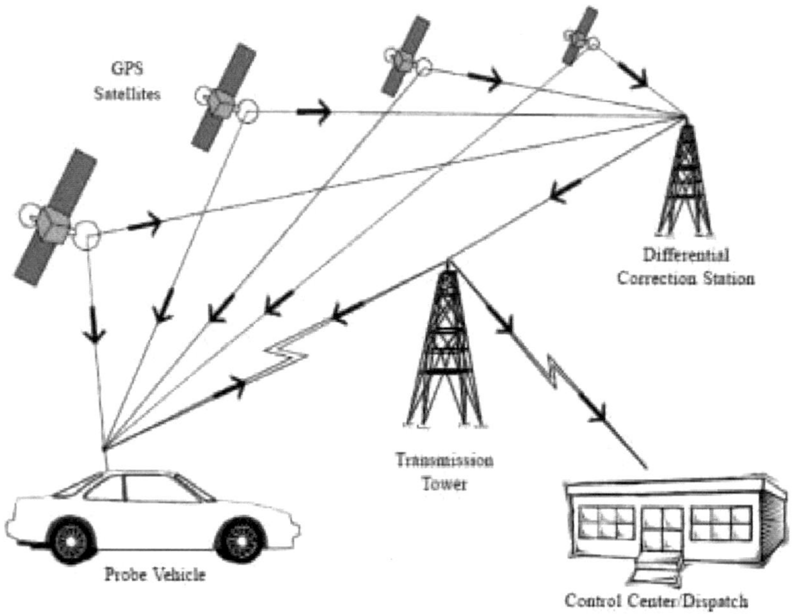

Figure 1. GPS based method.

Figure 2 shows an example of a cellular-based FCD network (GSM).

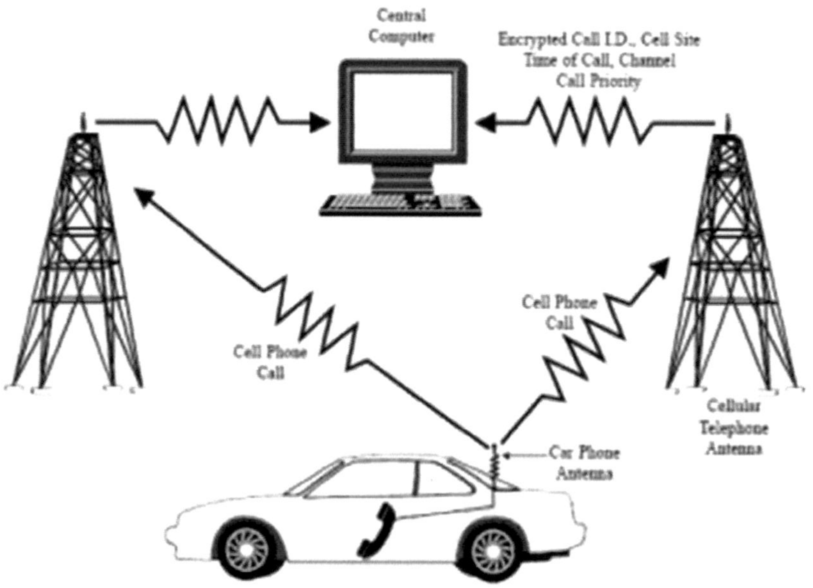

Figure 2. FCD Cellular Systems (GSM).

Although GPS is becoming more popular and affordable, only a limited number of vehicles are equipped with this system. As a rule, vehicles providing transportation services are equipped with them. The vehicle positioning accuracy is relatively high, usually less than 30 m [8, 9].

More often than not, traffic data from private vehicles or trucks is more suitable for highways and rural areas. In the case of urban traffic, taxi fleets are especially useful due to their large number and the already existing on-board communication systems. Currently, GPS sounder data is widely used as a source of real-time information by many service providers, but it suffers from a limited number of equipped vehicles and high equipment costs compared to floating cellular data [1, 9].

Currently, most vehicles are equipped with one or more mobile phones, which makes it possible to use mobile phones as anonymous traffic probes. The positioning of the mobile phone is regularly transmitted to the network using triangulation or other methods, and then, travel time and additional data can be estimated over a number of road segments before being converted into useful information by road centers [3, 10].

Unlike stationary road detectors and GPS systems, cars do not require special devices, equipment, and there is no need to build a special infrastructure along road networks. Therefore, it is cheaper than conventional detectors and offers wider coverage options. This method is easier to set up, requires less maintenance, but requires complex algorithms to retrieve and process data and therefore requires skilled personnel to predict traffic conditions and send to users.

The types of data for assessing the traffic situation provided by transport centers around the world are quite important, such as: Annual average daily traffic and Vehicle Kilometers Traveled. These two raw traffic datasets are mainly derived from fixed sensor measurements. They play a key role in traffic engineering.

Annual average daily traffic (AADT) is the average annual number of vehicles passing a point in a given section of the count every day (usually expressed in vehicles per day) [1, 11]. It simply represents the flow of vehicles along a section of road (for example, a highway) on an average day of the year. AADT is regarded as one of the most important raw traffic datasets, where it provides the necessary input for developing traffic models and calibration exercises that can be used to plan new road construction, determine the geometry of the roadway, and congestion. management, road surface design and much more. AADT is generally available on most European road networks. This data is collected by traffic management centers, updated and disseminated to users by traffic information centers in most countries. On the one hand, AADT is calculated annually for all motorway or road segments. On the other hand, real-time traffic streams can be provided every minute or hour. This data is measured using the traffic recorders for some highways. AADT calculation methods are usually based on data information obtained from two types of counts: continuous automatic traffic count and short-term traffic count. A combination of these two measurements is commonly used to obtain an AADT estimate and transmit it to the user in the form of traffic conditions.

One of the most common augmentation methods for assessing AADT is the factoring method [2, 12]. In this case, the permanent sections of traffic are first manually classified into different groups (known as seasonal categories) based on the similarity of the road traffic characteristics. Each permanent station is then assigned a seasonal factor category according to the site location, assuming that the seasonal variability and movement characteristics in the short and long term are similar in the same geographic area. However, the most important questions are the optimal number of groups and the way in which short samples are assigned to groups of seasonal factors. Although this method is arguably the simplest and most used worldwide, it is still limited in terms of accuracy. More complex mathematical methods such as linear regression, neural network, genetic algorithm, etc. Have been designed to provide more accurate AADT values.

As you can see from the block diagram in Figure 3, data processing flows are a complex system.

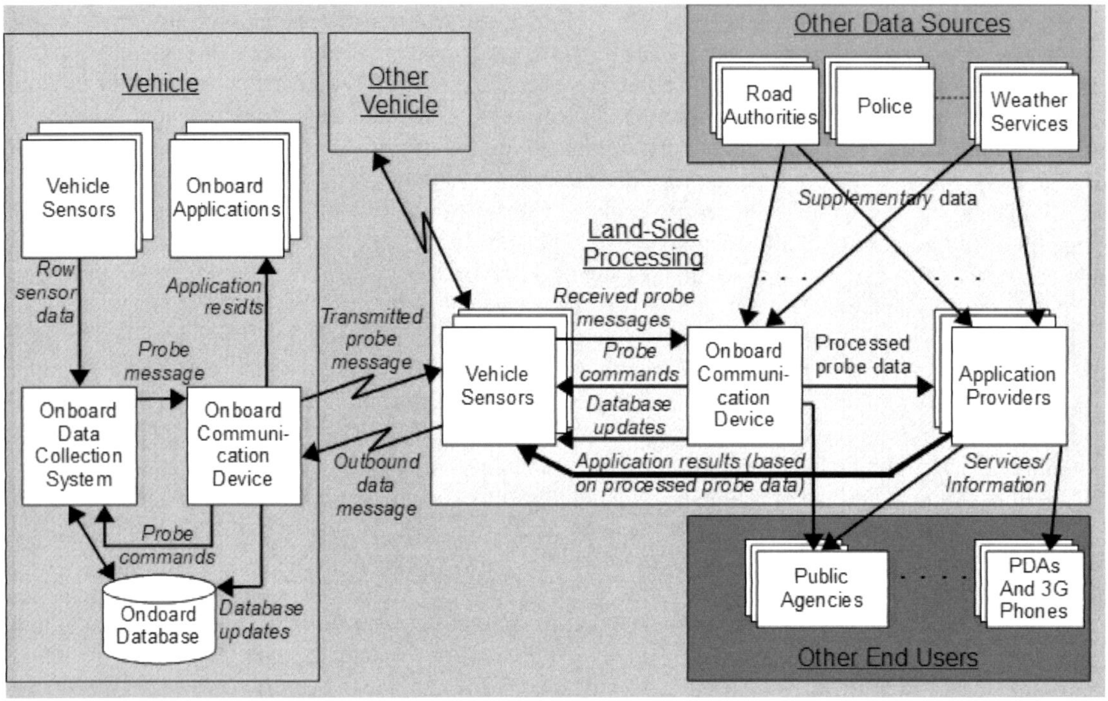

Figure 3. Data processing scheme.

The method considered in the study consists of two stages, namely, transferring data from vehicles (for example, via GSM) and transferring data back to drivers, as an end user. Trips made without vehicles should be filtered using pattern matching. As with GPS point data, map matching is required to map point locations to specific roads. One of the disadvantages of this technology is that the continuous transmission of a large amount of information from vehicles creates an excessive load on the transmission channels and represents a significant cost factor when using a paid communication system [1, 3]. For this reason, it would be preferable to transfer compressed data rather than individual values. For example, the average speed of a given vehicle can be transmitted to a center responsible for collecting and processing traffic data at chronological intervals. However, it would be much more efficient to filter the data according to its relevance. For example, the transmission may be limited to information about detected traffic congestion only.

4. Results and discussions

The main question about FCD is not whether the technology is an effective way of collecting traffic information in real time, but rather an assessment of what applications and benefits it can provide in the short to medium term. More accurate and up-to-date traffic information in real time can lead to significant improvements in areas such as: reduced congestion, improved O-D matrices, better monitoring of traffic conditions.

Optimizing the existing infrastructure by making better use of the existing road network will reduce driving times, thereby lowering costs for drivers and road network service providers, so it is expected that these improvements will affect all participants in the transport process, albeit to varying degrees. For example, road users will receive reliable information in real time on the best routes from their location to their destination.

The above software and hardware solutions can also be useful for identifying time dependences of traffic.

These are daily, weekly and seasonal fluctuations in traffic intensity. For example, in urban traffic conditions, the intensity of traffic flows can vary by more than 50%.

The intelligent combination of FCD with road sensors is the ideal input for dynamic traffic patterns. New data fusion algorithms will take advantage of each technology, leading to optimal solutions to traffic management problems. Thus, the traffic model can provide short-term traffic forecasting (15 to 30 minutes). For example, accidents (collisions) can be automatically detected thanks to dynamic models fed with high quality real-time data (i.e., through fusion techniques that can also include meteorological data and other parameters).

In recent years, the number of sources for collecting traffic data has increased significantly. The combination of traditional road sensors with floating vehicle data can provide high accuracy and real-time transmission of road traffic. Stationary detectors are difficult to implement for the following reasons: the need for large financial costs, the need for qualified personnel for their installation and maintenance, the dependence of such sensors on weather conditions. FCD is not aimed at replacing existing sensors, but rather to supplement their information with more accurate data. If current trends continue, transport operators can reap significant benefits from the combination of fixed and mobile traffic measurements across a wide range of areas. Due to favorable market prospects, private suppliers are currently developing all the necessary conditions for the deployment of new systems in the traffic accounting market.

5. Conclusion

This article discusses the benefits of implementing the FCD principle into existing traffic information collection systems. The advantages of this principle, its advantages and disadvantages are substantiated.

Examples of different ways to implement the FCD principle, their differences and possible implementation difficulties are given. The effectiveness of non-intrusive methods has already been proven by various road navigation systems.

The implementation of the FCD principle will significantly increase the efficiency of the use of transport infrastructure, as well as reduce the costs of end users.

References

[1] Bishop R, Shah T, Pelle R, Hoyle D, Pearson T, Haines L, Brass A, Hulme H, Graham S P, LN Taracha E, Kanga S, Lu C, Hass B, Wortman J, White O, Gardner M J, Nene V, de Villiers E P 2005 Analysis of the transcriptome of the protozoan Theileria parva using MPSS reveals that the majority of genes are transcriptionally active in the schizont stage *Nucleic acids research* **33(17)** 5503-11

[2] Bishop R 2005 *Arizona I-19 Wi-Fi Corridor: Assessment of Opportunities for Probe Data Operations* vol Report TRQS-02 (Phoenix, Arizona Department of Transportation) p 36

[3] Dolgikh I Yu, Korolev A N and Zakharov V M 2014 Mathematical modeling of electromagnetic and thermal processes during induction heating *Electrical engineering. Energy. Mechanical engineering* 85-8

[4] Granato S, Linn County Regional Planning Commission 1998 The impact of factoring traffic counts for daily and monthly variation in reducing sample counting error *Transportation Conference Proceeding* **8(1)** 122-5

[5] Toppen A, Wunderlich D 2004 Travel time data collection for measurement of advanced traveler information systems accuracy *At the Crossroads: Integrating Mobility Safety and Security. ITS America 2004, 14th Annual Meeting and ExpositionITS America* **2(3)** 92-117

[6] Höpfner M, Ehrenpfordt I, Nicolay E, Maier B, Brockfeld E, Krajzewicz D 2007 Simulation und Feldtest für die Validation mobilfunkbasierter Verkehrsdaten *CeBIT in Motion-Forum for Telematics and Navigation* **5(1)** 7-38

[7] Kaack L H, Chen G H, Morgan M G 2019 Truck traffic monitoring with satellite images *Proceedings of the 2nd ACM SIGCAS Conference on Computing and Sustainable Societies* **1130(12)** 155-64

[8] Hajibabai L, Bai Y, Ouyang Y 2014 Joint optimization of freight facility location and pavement

infrastructure rehabilitation under network traffic equilibrium *Transportation Research Part B: Methodological* **63(1)** 38-52
[9] Manual H C 2000 Highway capacity manual (Washington, DC) 2(1) p 1134
[10] Teply S 1995 Canadian Capacity Guide for Signalized Intersections Committee Canadian Capacity Guide for Signalized Intersections 117 p
[11] Levashev A G, Mikhailov A Yu 2004 Basic parameters for assessing the throughput of controlled intersections *VINITI* **1(3)** 14-9
[12] Silyanov V V 1977 *The theory of traffic flows in the design of roads and traffic management* Moscow Transport 303 p

Temperature stabilization of the angular velocity measuring device

A V Golikov[1,4], M A Barulina[1,3], E V Pankratova[1] and M V Efremov[2]

[1] Institute of Precision Mechanics and Control, Russian Academy of Sciences, Saratov, 410028, Russia
[2] Research and production enterprise «Antares», Saratov, 410600, Russia

E-mail: barulina@iptmuran.ru[3], golikov@iptmuran.ru[4]

Abstract. The article examines the possibility of using an active two-circuit thermal control system to reduce the influence of temperature disturbances on the components of the angular velocity measuring device. Special attention is devoted to the problem of temperature stabilization in fiber-optic gyroscopes mounted in the measuring device. The study of the inhomogeneous nonstationary temperature field of the angular velocity measuring device with a two-circuit reversible thermal control system on Peltier thermal batteries is carried out. Thermal and mathematical models of the considered dynamical system are constructed. Specialized software for numerical simulation has been developed, which allows calculating and visualizing real-time inhomogeneous, three-dimensional, non-stationary temperature fields of the device under conditions of complex non-stationary thermal effects of internal and external sources. It is shown that the use of a two-circuit reversible thermal control system can help to significantly reduce the influence of external non-stationary temperature disturbances on the temperature stability of fiber-optic gyroscopes in various operating conditions, such as vacuum and weightlessness. The energy consumption and the increase in the weight and size characteristics of the device are estimated.

1. Introduction

In this article, the problem of ensuring temperature stabilization of the angular velocity measuring device (BIUS) on fiber-optic gyroscopes (FOG) with a two-circuit reversible thermal control system is solving. Fiber-optics gyroscopes are the most sensitive for temperature gradients of the BIUS components.

Among gyroscopic devices of the various operating principles, the FOG stands out with the relatively low cost, reliability, low power consumption, absence of mechanical parts, and many other advantages. At the same time, such a sensor has some disadvantages, for example, sensitivity to various external influences – mechanical, electromagnetic, temperature and others, which cause inaccuracies of the measured angular velocity.

The modern high-precious FOG has strict requirements for temperature [1-8] because temperature disturbances are one of the significant factors, which is affecting the FOG precision. Most of the FOG errors are associated with the fiber coil – sensitive element of the device [2,5,6,7,13]. Over the past decades, the efforts of Russian and other countries' scientists have achieved significant results in improving the precision and stability of FOG [9-17]. Many engineering solutions have been found and applied to ensure the maximum possible protection of the FOG from external disturbing influences and to increase its accuracy. There are various methods for minimization of the temperature errors in FOG, used in navigation systems, such as special methods of winding a fiber on a coil (for example, quadrupole winding), algorithmic compensation of the thermal drift, etc. [2,4,8,16,17]. The most accurate instruments have an accuracy of 0.01 deg/hour and better.

However, it's not enough to use only passive methods to achieve inertial accuracies for FOG <0.01 deg/hour. Thermoelastic deformations in micrometers and percentage of micrometers have a significant impact on the performance of high-precision devices. Also, the temperature instability of the radiation source impact on the scale factor (SF) of the FOG. For example, to achieve an accuracy of the FOG <0.01 deg/hour, the relative error of SF [1,7,9,10,11,13], should be less than 0.01%. In contemporary advanced navigation-class navigation systems, an erbium superluminescent optical radiation source can be used, but studies shown [12, 17, 18] that the temperature instability of the radiation source at high rotational speeds has a significant effect on the scale coefficient.

Thus, the problem of minimizing errors caused by temperature perturbations remains relevant.

The purpose of this work is to study the possibility of using an active two-circuit thermal control system (TCS) based on the use of Peltier thermoelements to minimize the influence of external and internal temperature disturbances on the FOG sensor element and the laser radiation source. The object of the study is the angular velocity measuring device (BIUS-M-1) with fiber-optic gyroscopes, designed to use as part of the strapdown navigation system of the spacecraft, which is being developed at the Antares Research and Production Enterprise in Saratov.

To evaluate the efficiency of the TCS, it is necessary to study the non-stationary, inhomogeneous, three-dimensional temperature field of the BIUS-M-1, which contains three fiber-optic gyroscopes, three electronic boards with heat sources of a given power, the internal circuit of the TCS of one FOG, and the external circuit of the TCS of the entire measuring device.

2. Model description

The measuring device is based on three orthogonally arranged FOGs and a service electronics unit. BIUS-M-1 has a modular structure and contains elements (gyroscopes, electronic boards, etc.) with heat sources of a given power. The unit provides for thermal shunting of the heat-generating elements of individual boards to the base through the heat sink racks, shunts, and radiator plates. The guaranteed temperature range of the mounted plane (thermoplastics) of the device during testing and operation mode is in (-10 ÷ 40) °C.

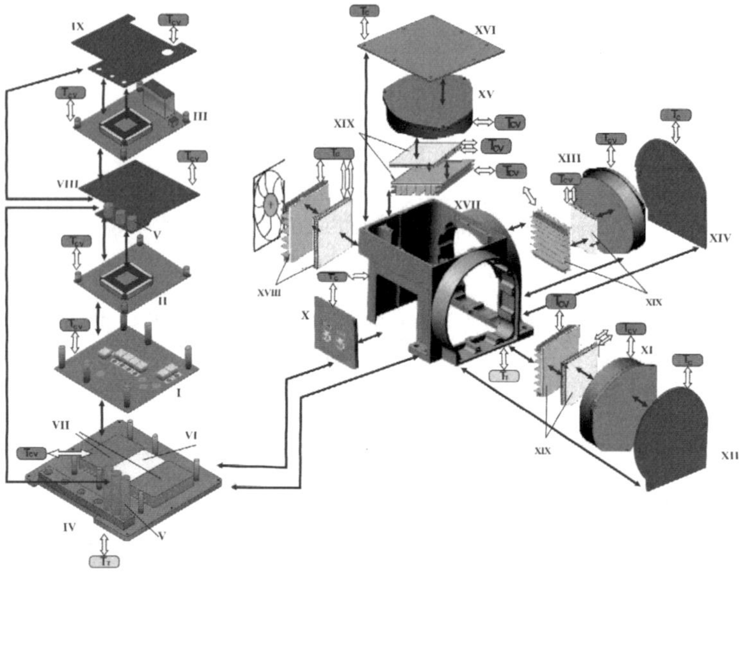

Figure 1. BIUS thermal model. The arrows indicate the thermal connections.

For temperature stabilization, we assumed that the Peltier thermal battery is placed on the device case, which formed the external contour of TCS, and on every FOG included in BIUS (the second and internal contours of TCS).

To construct a thermal model following the general method of thermal modeling [2, 20], the BIUS has to be decompose into separate structural elements. Figure 1 shows the thermal model of the BIUS with an indication of the thermal connections between the structural elements and the external environment.

Each element, if it consists of smaller non-uniform parts and contains heat sources, has also to be divided into smaller elements to obtain a more accurate view of the temperature field. In our case, each of the boards of the service electronics unit containing heat sources is decomposed into 25 elements to obtain a detailed picture of the temperature field.

3. Mathematical model and the temperature regulation low

The mathematical model of the non-stationary inhomogeneous temperature field for the considering problem is presented in the form of a finite difference algorithm and have the following view [2,20]:

$$T_i(t+\Delta t)=\left[1-\frac{\Delta t}{c_i}\left(\sum_{j=1}^{N} q_{ij}+q_{ic}\right)\right]T_i+\frac{\Delta t}{c_i}\left(\sum_{j=1}^{N} q_{ij}T_j+q_{ic}T_{ci}+Q_i\right), \quad (1)$$

where: $T_i(t)$, $T_i(t+\Delta t)$, c_i $(i=1,\dots,M)$ – temperatures of the i-th element in the current and subsequent moment and his thermal capacity; q_{ij} - thermal conductivity between elements i,j (j=1,…,N); q_{ic} - thermal conductivity between i-th element and the environment; T_{ci} - environment temperature; Q_i - heat source power; M – the number of elements; N - the number of elements that have thermal contact with i-th element; Δt - calculation step.

Thermal conductivity coefficients included in the expression (1), are determined by known formulas [2,20] for every type of heat exchange.

The main supposed difference of the device design from the base one is the presence of TCS actuators, namely, three Peltier thermoelements for each FOG with radiators.

The equations describing the thermoelectric characteristics of Peltier batteries have the form:

$$Q_1=\left[-\varepsilon_{\mathrm{T}}J(T_{\mathrm{in}}+273)+\frac{J^2R}{2}\right]nm,\quad Q_2=\left[\varepsilon_{\mathrm{T}}J(T_{\mathrm{ext}}+273)+\frac{J^2R}{2}\right]nm, \quad (2)$$

where Q_1 - the power of the heat dissipation (cooling facilities) on the working junction; Q_2 - the power of the heat dissipation (cooling facilities) on the external junction; ε_{T} – coefficient of thermoelectromotive force in thermoelements; J – current flowing through the thermocouple; R – electrical resistance; n, m - the number of thermoelements in one Peltier module and the number of modules.

The most widely used systems in aerospace instrument engineering are proportional and positional temperature control systems, which provide sufficient accuracy of temperature control and at the same time are relatively simple, cheap, and reliable in operation. The expression describing such a Peltier thermoelectric current control law has the form:

$$J=\begin{cases} J_{\max}, & \text{for } T_{\mathrm{d}}-T_{\mathrm{Z}}\ge T_{\mathrm{L}};\\ k(T_{\mathrm{d}}-T_{\mathrm{Z}}-T_{\mathrm{N}}), & \text{for } T_{\mathrm{N}}\le T_{\mathrm{d}}-T_{\mathrm{Z}}\le T_{\mathrm{L}};\\ 0, & \text{for } -T_{\mathrm{N}}\le T_{\mathrm{d}}-T_{\mathrm{Z}}\le T_{\mathrm{N}};\\ k(T_{\mathrm{d}}-T_{\mathrm{Z}}+T_{\mathrm{N}}), & \text{for } -T_{\mathrm{L}}\le T_{\mathrm{d}}-T_{\mathrm{Z}}\le -T_{\mathrm{N}};\\ -J_{\max}, & \text{for } T_{\mathrm{d}}-T_{\mathrm{Z}}\le -T_{\mathrm{L}}; \end{cases} \quad (3)$$

where J_{max} – maximum current value; T_L, T_N – linearity and insensitivity zones; $k = \mathrm{tg}\,\alpha = J_{max}/(T_L - T_N)$ – slope of characteristics of the temperature control law; T_Z – preset temperature of thermal stabilization; T_d – the temperature of the temperature sensors installed in the necessary places of the BIUS and FOG.

The graph describing the control law (3) is shown in figure 2.

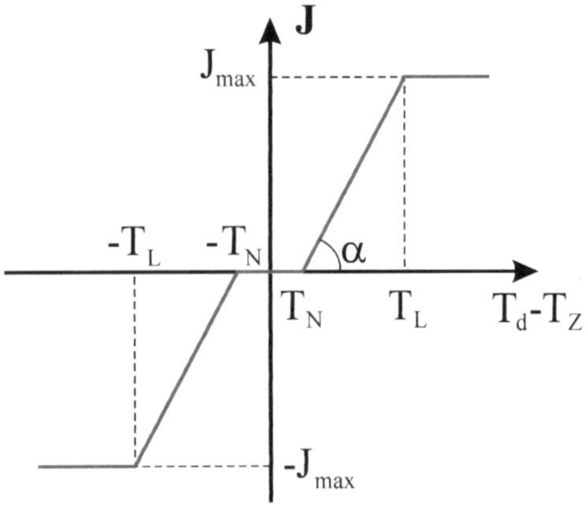

Figure 2. The law of temperature control in the Peltier module.

The theory and practice of the use of TCS on Peltier thermoelements [17-19] show that while choosing the optimal parameters of the TCS in terms of energy consumption and the efficiency of maintaining a given thermal regime, the following rules should be followed:

- initial temperature T_0 should be selected as close as possible to the preset temperature of thermal stabilization T_Z. The stabilization temperature for BIUS is assumed to be equal to $T_Z = +30$ °C from the maximum range $T_c = (-10 \div +40)°$ C
- preferred locations of the thermal sensors are on the BIUS case and the FOG cases.
- some kinds of thermal insulation of the device and FOGs are necessary for the effective operation of the TSC.

4. Computer simulation

All constructed mathematical models were implemented in specialized software for the numerical simulation. The specialized software is developed on the C++ programming language and allows visualizing non-stationary inhomogeneous three-dimensional temperature fields of the device under consideration in dynamics.

With the help of the developed software, a series of computer experiments was carried out to simulate thermal processes in the BIUS in the main thermal modes defined in previous works [3], simulating the Earth conditions and the conditions of outer space (vacuum and weightlessness). At the same time, a comparative analysis of temperatures was carried out when the TCS was turned on and off.

Below are color topograms of the temperature field of the BIUS-TCS dynamic system (figure 3), and graphs of the current temperatures of the parameters of the TCS (figure 4), obtained as a result of computer modeling in terrestrial conditions (1 - thermal mode) [3].

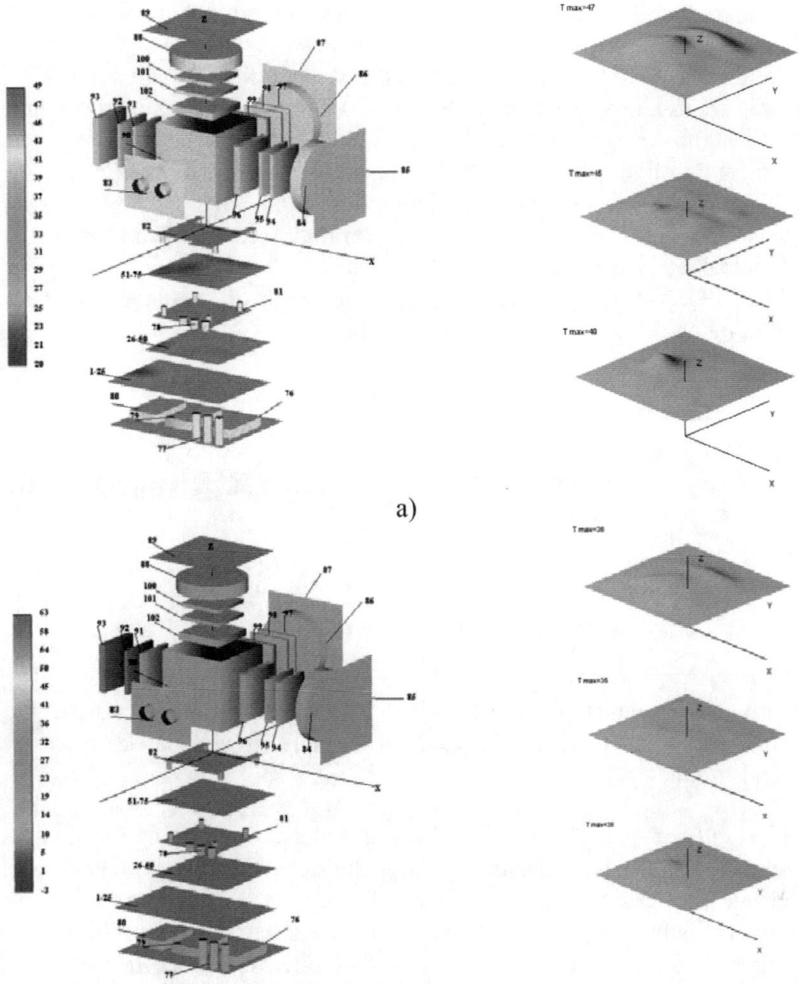

a)

b)

Figure 3. Temperature fields of the BIUS-TCS dynamic system and temperature topograms on electronics boards in steady-state mode at $T_c = T_T$ =+40 °C: a) when the TCS is turned off; b) when the TCS is turned on.

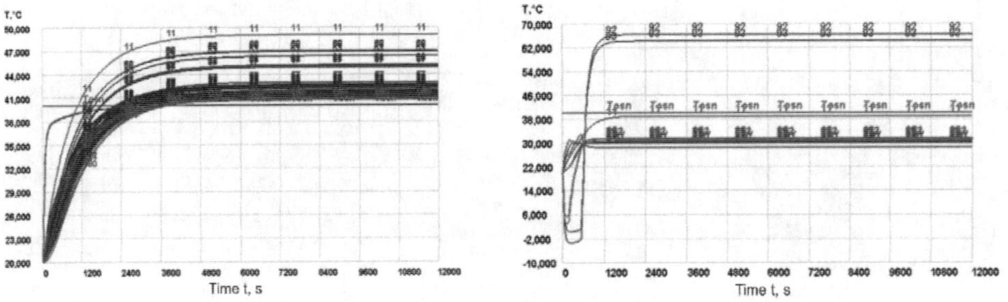

a) b)

Figure 4. Current temperatures in the BIUS-TCS dynamic system at Tc = TT =+40°C in first thermal mode: a) when the TCS is turned off; b) when the TCS is turned on;

An analysis of the results shows the following. The nature of thermal processes when the TCS supply system is switched off and running is significantly different.

The behavior of the BIUS-TCS dynamic system changes significantly when the TCS is enabled. By T_c = +40°C the maximum temperature of the hottest BIUS element (on the power supply board) has decreased by 11 °C and do not exceed 38°C (then the TCS turned off it was 49°C). The temperature of the remaining elements takes the value, close to T_Z c accurate to tenths of a degree (0.3°C), and for gyroscopes and the body equal to T_Z accurate to hundredths of a degree (0.02°C).

The next stage is the analysis of the BIUS temperatures in the second thermal mode, which simulates the conditions of outer space (vacuum and weightlessness).

The graphs of the BIUS temperatures obtained from the results of calculations in the second thermal regime are shown in figure 5.

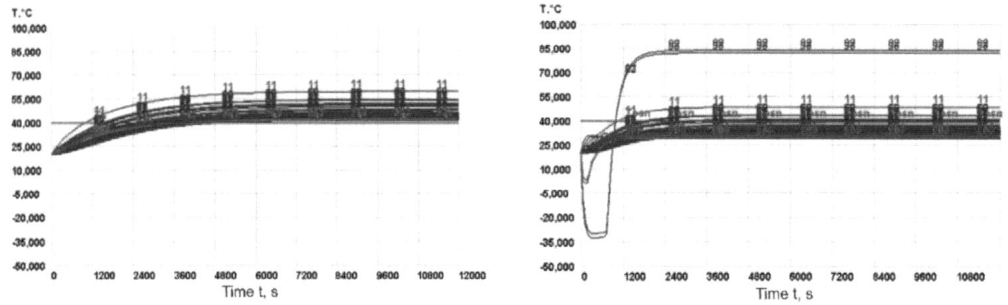

a) b)

Figure 5. Current temperatures in the BIUS-TCS elements at the temperature of the medium and the thermal board +40°C in zero gravity and vacuum a) when the TCS is turned off; b) when the TCS is turned on.

Comparing the results of simulations of the 1st and 2nd thermal heating modes of the BIUS-M-1 with the operating TCS, we see that the maximum superheat over the thermal board in zero gravity and under vacuum outside and inside the unit have naturally increased and reached the values of ΔT_{max} = 8,5°C on boards, which is in 2 times more than overheating in terrestrial conditions. However, this is still more than 2 times less than in the basic design without the TCS under the same conditions (ΔT_{max} = 20°C). In the other elements of the BIUS, the temperature is close to the set one T_Z.

The simulation of the non-stationary inner temperature disturbance based on the given cyclogram simulating the tests of a measuring unit in a thermal chamber is particularly indicative from the point of view of the effectiveness of the use of the TCS.

Figure 6 shows the graphs of the current temperatures at all calculated points of the device, obtained as a result of computer modeling.

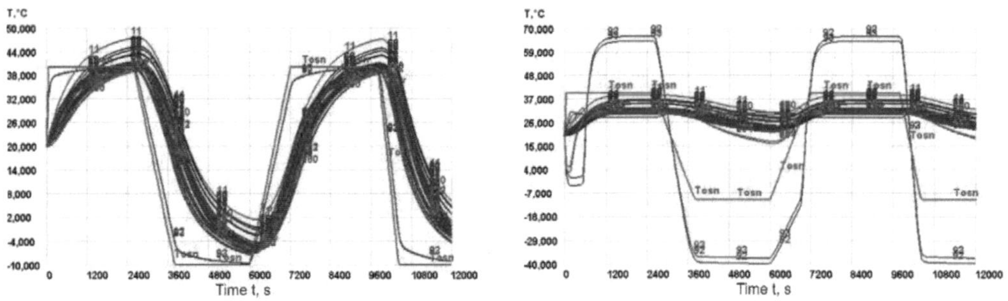

a) b)

Figure 6. Current temperatures in the BIUS-TCS dynamic system at a temperature perturbation according to a given cyclogram: a) when the TCS is turned off; b) when the TCS is turned on.

From the presented graphs, the effect of the working TCS is obvious. Due to thermal stabilization, the temperature situation is significantly improved by the maximum overheating of the elements with internal heat sources over the environment (with the TCS running, the maximum temperature difference over the medium temperature decreased by 9°C). The total power consumption of the TCS, takes the value <2.5 W under normal conditions.

When the temperature of the medium and the housing changes from the "shelf" to the "shelf" according to the linear law (-10 ÷ +40) °C the temperatures of the main elements of the BIUS vary slightly around the set temperature T_Z (figure 6b), and the temperatures of the gyroscopes are equal to the set temperature with an accuracy of 0.2°C.

Peak power consumption of the external circuit TCS (at maximum differences between T_c and T_Z), in the thermal stabilization mode < 12 W under normal and extreme conditions.

5. Conclusion

The article examines the possibility of using an active two-circuit thermal control system to reduce the influence of temperature disturbances on the components of the angular velocity measuring device on the example of the BIUS-M-1 developed at the Antares Research and Production Enterprise in Saratov.

Mathematical models of thermal processes occurring in the BIUS-M-1 dynamic system with fiber-optic gyroscopes and a two-circuit reversible TCS are constructed and implemented in specialized software. The specialized software allows calculating and visualizing inhomogeneous, three-dimensional, non-stationary temperature fields of the system under consideration under conditions of complex non-stationary thermal effects of internal and external character.

Computational experiments were carried out using the developed software and a detailed thermal analysis of the operation of the BIUS-M-1 device under thermal conditions was performed. Qualitative and quantitative estimates of the parameters of thermal processes are obtained.

It is shown that in the case of the use of TCS, we can expect a decrease in the maximum overheating over the temperature of the medium to -1°C. At the same time, the calculations show that the accuracy of temperature stabilization of the external circuit is not worse than 0.32°C, and in FOG area (inner circuit) is not worse than 0.02°C. The peak power and current consumption of both TCS circuits are 3 W and 0.9 A, respectively, and in steady-state mode, the total power consumption does not exceed 3 W, and the current is 0.2 A. According to calculations, the influence of external temperature changes on the temperature of the BIUS and its components can be weakened by about 250 times.

The "price" for such an improvement in the temperature situation in the BIUS with the TCS is the power consumption of the thermal control system up to 13 W in steady-state mode at normal gravity and normal pressure of the medium and the accepted degree of thermal insulation of 0.25; an increase in weight and dimensions due to the executive elements of the TSC and the use of thermal insulation by about 5-10% compared to the basic version of the BIUS without the thermal control system.

Thus, it can be expected that the use of the connected two circuits of the TCS will significantly reduce the influence of external and internal thermal influences on the gyroscopes and other elements of the BIUS. However, due to the contemporary strict requirements for energy consumption and the overall mass characteristics of space-based devices, in the case of the implementation of the design solution for the implementation of the considered TCS in the BIUS, there may be restrictions on its application due to the above reasons.

Acknowledgments
The work was supported by an RFBR grant 19-08-00839.

References

[1] Lefevre H 1993 *The Fiber-Optic Gyroscope* (MA: Artech House, Norwood) p 24
[2] Dzhashitov V E and Pankratov V M 2005 *Sensors, devices and systems of aerospace and marine instrumentation under thermal influences* (St. Petersburg: State Research Center of the Russian Federation TsNII «Elektropribor») p 404

[3] Golikov A V, Pankratov, V M, Barulina M A, Pankratova E V and Efremov M V 2019 Investigation of temperature fields in complex navigation devices intended for operation in space *Innovations and investments* **11** 167-75
[4] Dzhashitov V E, Pankratov V M, Golikov A V, Gubanov A G and Efremov M V 2011 Ensuring the thermal invariance of a fiber-optic gyroscope *Gyroscopy and Navigation Publishing House of the SSC RF «Central Research Institute «Electropribor»* **4(75)** 42-56
[5] Shupe D M 1980 Thermally induced non-reciprocity in the fiber-optic interferometer *Appl. Opt.* **19(5)** 654-5 doi: 10.1364/AO.19.000654
[6] Schadt F and Mohr F 2011 Error Signal Formation in FOGs Through Thermal and Elastooptical Environment Influence on the Sensing Coil *Inertial Sensors and Systems – Symp. Gyro Technology* September 2011 (Karlsrue: Germany) p 20-1
[7] Esipenko I A and Lykov D A 2017 Mathematical model of thermal drift of a fiber-optic gyroscope and its experimental verification *Herald of the Bauman Moscow State Technical University* **5(116)** 31-46 doi: 10.18698/0236-3933-2017-5-31-46
[8] Vakhrameev E I, Galyagin K S, Ivonin A S and Oshivalov M A 2013 Prediction and correction of the thermal drift of a fiber-optic gyroscope *Izv.VUZov* **5(56)** 79-84
[9] Korkishko Yu N, Fedorov V A and Prilutsky V E 2007 Fiber-optic gyroscope of the navigation class toch-news *XIV St. Petersburg International Conf. on Integrated Navigation Systems* (St. Petersburg: SSC RF «Central Research Institute «Electropribor») p 141-50
[10] Meshkovsky I K, Strigalev V E, Deineka G B and Peshekhonov V G 2011 Three-axis fiber-optic gyroscope *Development results XVIII St. Petersburg International Conf. on Integrated Navigation Systems* (St. Petersburg: State Research Center of the Russian Federation JSC Concern «Central Research Institute «Electropribor») p 8-14
[11] Kolevatov A P, Nikolaev S G, Andreev A G and Ermakov V S 2009 Successes in the development of BINS based on fiber-optic gyroscopes *XVI St. Petersburg International Conf. on Integrated Navigation Systems* (St. Petersburg: SSC RF «Central Research Institute «Electropribor») p 13-20
[12] Blazhnov B A, Volynsky D V, Rupasov AV, Stepanov A P and Untilov A A 2019 High-precision fiber-optic gyroscopes. Current state and prospects of development *XIII All-Russian meeting on problems of VSPU management* (Moscow 17-20 June) 1330-5
[13] Dranitsyna E V, Egorov D A, Untilov A A, Deineka G B, Sharkov I A and Deineka I G 2012 Reducing the effect of temperature changes on the output signal of a fiber-optic gyroscope *Gyroscopy and Navigation* **4(79)** 10-20
[14] Kurbatov A M 2015 Fiber-optic On ways to improve new gyroscopes with open and closed feedback loops *Gyroscopy and Navigation* **1** 43
[15] Kurbatov A M, Kurbatov R A and Goryachkin A M 2019 Improving the accuracy of a fiber-optic gyroscope by suppressing parasitic effects in integrated optical phase modulators *Gyroscopy and Navigation* **2(105)** 52-69
[16] Li Zh, Meng Zh, Liu T and Yao X S 2013 A novel method for determining and improving the quality of a quadrupolar fiber gyro coil under temperature variations *Optics Express* **2(21)** 2521-30
[17] Wu W, Xian T, Hu G and Zhou K 2020 Rapid and precise compensation of scale factor in a fiber-optic gyroscope with a twin-peaks source *Optic Letters* **11(45)** 3107-10
[18] Sharkov I A, Rupasov A V, Strigalev V E and Volkovsky S A 2013 Influence of the temperature instability of the source characteristics on the readings of the fiber-optic gyroscope *Scientific and Technical Bulletin of Information Technologies, Mechanics and Optics* **6(88)** 31-5
[19] Ingberman M I, Fromberg E M and Graboy L P 1979 *Thermostating in communication technology* (Moscow: Svyaz) p 144
[20] Dulnev G N, Parfenov V G and Sigalov A V 1990 *Methods for calculating the thermal regime of devices* (M.: Radio and communication) p 312

IOP Conf. Series: Materials Science and Engineering 1155 (2021) 012047 doi:10.1088/1757-899X/1155/1/012047

Simulation of balanced scorecard for forecasting cybernetic cost collaboration of engineering enterprises

A I Utkin

Ivanovo State University, 39, st. Ermaka, Ivanovo, 153025, Russian Federation

E-mail: alexeyutkin98@mail.ru

Abstract. The article examines the problem of simulation modeling of the relationship between the level of production efficiency of mechanical engineering enterprises and the distribution of cybernetic costs in the process of mechatronic robotic functions collaboration. Within the framework of the balanced scorecard concept, based on the methodology of experimental studies using raster electronic mechatronic modeling of cybernetic contact zones, a comprehensive forecasting of the influence of key leading positive and negative projection factors on the effectiveness of cybernetic cost collaboration is carried out. Through the author's simulation method of criteria-based analysis strategic projections of the balanced scorecard are formed and the simulation model of the resource potential of the cybernetic cost collaboration of engineering enterprises is proposed. As a result of the research, an analytical tool for evaluating and managing the production efficiency is elaborated and it contains the concept of simulation modeling of the interdependence between the cybernetic cost collaboration and the factors of the balanced scorecard projections. The simulation technology of the balanced scorecard ensures the convergence of the processes of functioning of mechatronic robotic factors, components and systems within the framework of optimizing cybernetic costs and identifying prospects for the sustainable development of mechanical engineering production.

1. Introduction

In the conditions of clustering of mechanical engineering production, the development of new mechatronic models of the resource potential of enterprises participating in innovative clusters becomes possible due to the multidimensional collaboration of cybernetic costs for designing an ecosystem of electronic robotic objects. In relation to the machine-building industry, the "leading points" of the emergence of system-forming engineering enterprises are the key cloud procedures for the development and manufacture of products, which form the reference positive and negative factors of electronic projections of the mechatronic balanced scorecard [1]. At the same time, the processes of digital mechanical clustering, based on the information and technological integration of cybernetic cost sources and the complication of automatic simulation tools, determine the negative effect in the form of uneven collaboration segments and a decrease in the production efficiency of engineering enterprises. The inefficiency of the implementation of mechatronic robotic machine-building functions of system-forming enterprises ultimately leads to the lack of prospects for the development of the innovation cluster as a whole [2]. In 2010-2018, low rates of functionality of robotic systems (2.34%), development of mechatronic resources (3.25%), collaboration of enterprise life cycles (3.64%) caused the rate of decline in innovation activity of electronic platforms for optimizing production costs (-4.3%), electronic engineering of cluster processes (-5.9%), description of clouds of technical means dysfunctions (-6.3%), maintenance of electronic profiles of cluster participants (-6.7%) [3]. In turn, for the period from 2019 to the first quarter of 2020, the moderate average annual rate of implementation of mechatronic systems Enterprise Resource Planning (ERP) and Product Data Management (PDM) (7.84%) affected a slight increase in the production efficiency of the cybernetic cost collaboration and electronic platforms (the rate of decline, respectively -3.2%, -4.3%, -5.9%, -5.7%) [3]. The observed changes reflect the decisive role of the collaboration of cybernetic costs in the formation of the resource potential of mechanical engineering enterprises and the possibility of a positive impact of the key collaboration factors on the level of production efficiency of mechatronic objects. In this regard, to manage the effectiveness of cybernetic functions of engineering enterprises, it is necessary to simulate strategic projections of the balanced scorecard that allow forecasting the impact of leading positive and negative projection factors on the effectiveness of cybernetic cost collaboration [4-6].

2. Materials and methods

Modern mechatronic technologies of Industry 4.0 are characterized by the production of new robotic systems that integrate various combinations of electronic objects and cybernetic costs using high-tech equipment, advanced techniques and methods to improve the quality and efficiency of products in machine-building production [7]. Through the constant introduction of technological innovations, the improvement of the most critical phases of the life cycle of technical products ensures the optimization of production costs and design of robotics with the release of 3-4 new unique types of mechatronic products per year. As a result of the most rational application of brain models of robotic processes based on the software product CAMS (delatex), the range of manufactured engineering equipment includes a balanced and optimally selected composition of electronic mechatronic objects, obtained with the constant interaction of key factors of the balanced scorecard projections and parameters of the resource potential of cybernetic cost collaboration [8-10]. The simulation of the balanced scorecard allows us to comprehensively assess and improve the production efficiency of engineering enterprises when constructing a mathematical dependence of the resource potential of the collaboration on the key leading indicators of strategic projections (typical positive and negative factors of the innovation and production clustering mechanism).

The problem of the research is to determine the approach to simulation modeling of the interdependence between the level of production efficiency of mechanical engineering enterprises (the resource potential of cybernetic cost collaboration) and the factors (robotic functions) of the balanced scorecard projections. The relevance of the problem is determined by the need to identify strategic prospects for the effectiveness of the distribution of cybernetic costs in the process of collaboration of mechatronic functions associated with overcoming the decline in innovative activity of machine-building enterprises.

The purpose of the work is to elaborate a simulation forecasting model of the resource potential of the cybernetic cost collaboration for further development of a simulation technology for adapting the balanced scorecard to the convergence of the processes of functioning of mechatronic robotic factors.

The research methods are criteria-based analysis (the method developed by the author), the Delphi method, the method of forecasting scenario, the analysis of causal relationships, empirical analysis, raster electronic mechatronic modeling of cybernetic contact zones using JSM-6490LV, elements of the analysis of ERP and PDM solutions.

To select the key leading factors of the strategic projections of the balanced scorecard, it is proposed to use the author's method of criteria-based analysis through the unified list of criteria for identifying typical factors of production efficiency of cybernetic cost collaboration. Criteria-based analysis within the framework of an innovative approach to simulation modeling of the balanced scorecard will allow us to establish the correspondence of the generated trends in the formation of collaboration with the essence of the identification criteria. As reference indicators based on the selection results, it is assumed to identify factors that meet all the established criteria. The experimental research methodology, based on raster electronic mechatronic modeling of cybernetic contact zones using JSM-6490LV and the analysis of ERP and PDM solutions, becomes the basis for building a simulation model of the resource potential of the collaboration and forming the concept of the interdependence between the cybernetic cost collaboration of enterprises and the factors of the balanced scorecard.

3. Results

Comprehensive forecasting of the obtained positive and negative factors of the balanced scorecard projections, the level of efficiency of cybernetic cost collaboration and the trends in the functioning of mechatronic robotics in machine-building production characterize the potential for improving the existing cluster cloud resources and visualization of electronic production facilities with specific features of production activities (conditions for the occurrence of factors). In this regard, the key reference factors should be considered standard and unified in relation to the simulation of the resource potential of cybernetic cost collaboration of engineering enterprises that have similar specifics and specialize in the production and sale of computer-integrated engineering systems and intelligent robotic systems. A necessary condition for improving the strategic projections of the balanced scorecard and the emergence of a positive effect from the collaboration of cybernetic costs of machine-building enterprises should be a positive relationship between the processes of the emergence of factors and cybernetic costs caused by cybernetic shifts and unified identification criteria. In the case of mechanical engineering enterprises, this interdependence consists in the dual function of factors, in which they act as key projection factors and cost components (electronic and computer integrated resources). At the same time, the unified selection criteria that determine the effectiveness of the use of cybernetic information systems in the production of mechatronic products provide a projection of the influence of factors on the strategic prospects for improving production efficiency (achieving sustainable development of the resource potential of the collaboration). The concept of simulation modeling of the interdependence between the cybernetic cost collaboration of mechanical engineering enterprises and the factors of the balanced scorecard projections based on the author's method of criteria-based analysis is presented in figure 1.

The parameters of cybernetic costs that determine the vectors of the influence of factors (resources) on the indicator of resource potential are multiplicative values. Therefore, it is necessary to take them as scalars that are close in modular value to 1.

Forecasting of production efficiency using the experimental research methodology based on raster electronic mechatronic modeling of JSM-6490LV cybernetic contact zones and ERP and PDM technologies shows that the effectiveness of cybernetic cost collaboration is influenced, in addition to the resource factors reflected in the model, by the vertical compression coefficients of electronic objects and maneuverability coefficients (the functionality of financial resources invested in mechatronic products). Therefore, for a more accurate assessment of the impact of the factors of the balanced scorecard projections, it is necessary to take into account these coefficients.

Based on the double function of the projection factors of the balanced scorecard, we have elaborated the simulation model of the resource potential of cybernetic costs of machine-building enterprises. Parameters (resources) of cybernetic costs related to all stages of improvement of strategic projections are numbered by indexes 1, 2, 3. The general view of the simulation model is given below (equation 1):

$$I_{collab}^{CC} = RS^{RQP} + \alpha_1 RS^{EPO} + \alpha_2 RS^{MI} + \alpha_3 RS^{CEM} + \alpha_4 RS^{ESP} + \varepsilon \quad (1)$$

I_{collab}^{CC} – Indicator of the resource potential of cybernetic cost collaboration;

$\alpha_1 \ldots \alpha_4$ – Regression equation ratios (parameters-ranks of cybernetic costs);

RS^{RQP} – Resource of predominance of high-quality product sales by the main nomenclature group;

RS^{EPO} – Resource of predominance of expensive production objects;

RS^{MI} – Resource of motivational instruments for improving production efficiency;

RS^{CEM} – Combinations of electronic modules for mutually beneficial collaboration;

RS^{ESP} – Supply of expensive software packages without recycling.

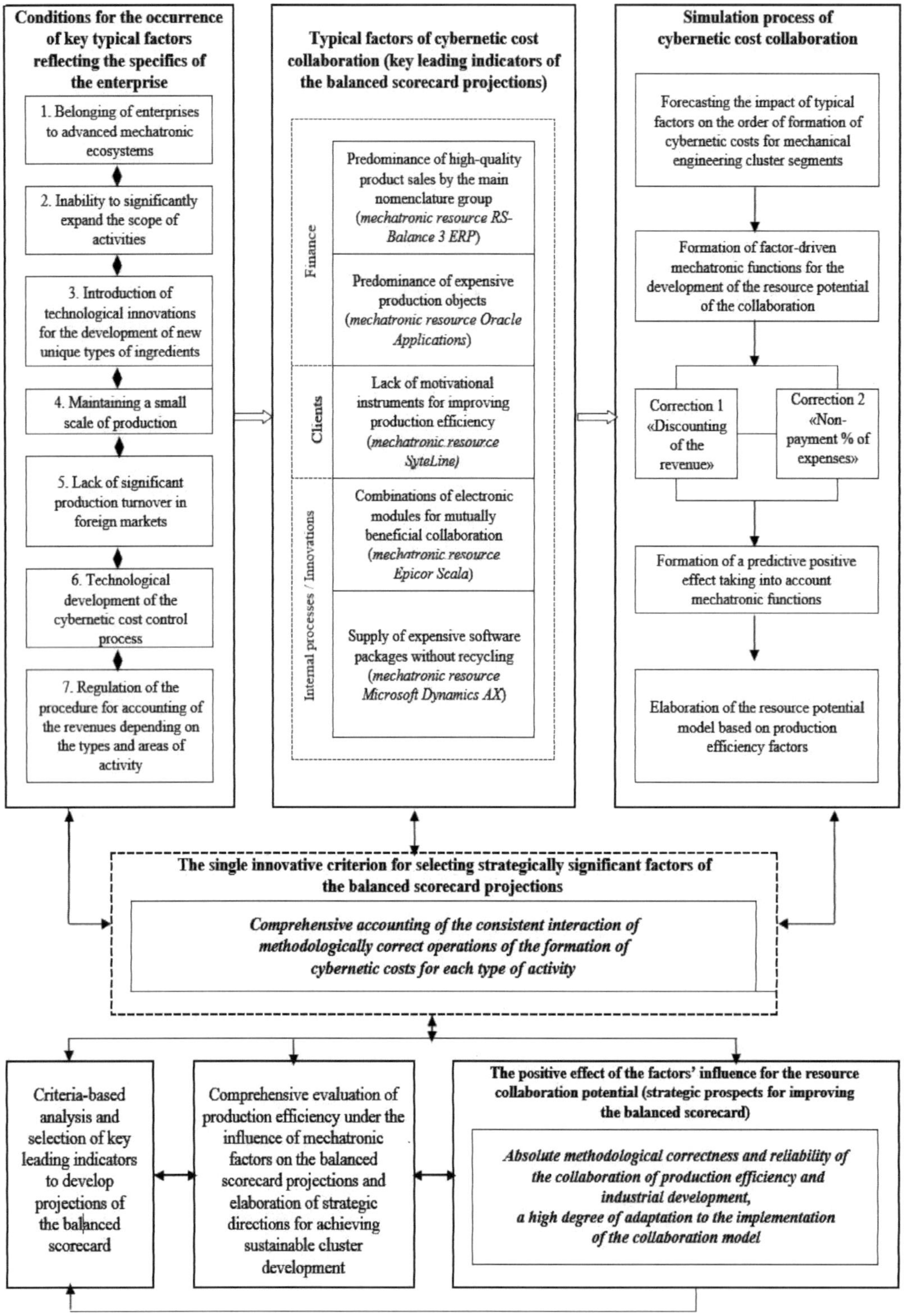

Figure 1. The concept of simulation modeling of the interdependence between the cybernetic cost collaboration of mechanical engineering enterprises and the factors of the balanced scorecard projections.

The level of influence of reference factors of cybernetic costs determines the possibilities and limitations of adapting machine-building production to the adjustment of functional tools aimed at increasing the innovation activity of the subjects of cybernetic cost collaboration using the elaborated simulation model. In the conditions of constant innovation and technological improvement of production output, the forecasting process based on the simulation model of cybernetic cost collaboration allows us to study the surface and structure of mechatronic resources and analyze the prospects for increasing the resolution of electronic objects in order to comprehensively manage the production efficiency of engineering enterprises based on the balanced scorecard.

4. Discussion

The proposed simulation technology for the formation of the balanced scorecard ensures the convergence of the introduced functional adjustments based on the combination of tools for integrated assessment and management of the production efficiency of machine-building enterprises. The unified nature of the innovation criterion, which determines the possibility of using a simulation model of the resource potential of the cybernetic cost collaboration and identifying the prospects for their optimization, is due to the priority influence of mechatronic robotic factors, components and systems on the choice of strategy and mission determining the development of machine-building clustering mechanisms as the main goal of the collaboration. In accordance with the strategy and mission, the developed analytical tools for evaluating and managing production efficiency within the framework of the balanced scorecard concept are aimed at achieving this goal and will always be based on the conducted simulation modeling.

5. Conclusion

The main conclusions of the research are presented as follows:

- The construction of the concept of simulation modeling of the interdependence between the cybernetic cost collaboration of mechanical engineering enterprises and the factors of the balanced scorecard projections allowed us to develop the simulation model of the resource potential of the collaboration and to forecast the efficiency of machine-building production;
- The balanced scorecard and simulation model developed using the author's method of criteria-based analysis allowed us to expand the theoretical understanding of the role and functionality of mechatronic factors and processes in the comprehensive assessment and management of resource potential based on key reference indicators;
- The most important practical significance of the research results is the determination of the innovation mechanism for selecting key leading indicators using a single innovative criterion;
- The proposed tools for improving the strategic projections of the balanced scorecard and forecasting cybernetic cost collaboration of machine-building enterprises reflect the possibilities of their optimization and identification of prospects for sustainable development while transferring positive trends in production efficiency to the field of activity of other organizations.

References

[1] Rogachev A F, Medvedeva L N, Timoshenko M A, Tokarev K E and Shiro M S 2019 Using Computer Modeling during Evaluation of Socio-Labor Potential of the Region *Advances in Intelligent Systems and Computing* **726** 989-1004
[2] Odei S A, Prokop V and Stejskal J 2020 Innovation Collaborations of Firms: the Case of Hungarian Multinational Companies *Economy of Region* **16(1)** 257-67
[3] Burkaltseva D D, Boychenko O, Sivash O S, Mazur N M, Zotova S A and Novikov A V 2017 The Construction of the Digital Organizational, Social and Economic Production Mechanism in the Agro-Industry *European Research Studies Journal* **XX(4B)** 350-65
[4] Utkin A I and Speransky S N 2019 Income Potential Management of the Cluster-Forming

Enterprises of Ivanovo Region *Proceedings of Higher Education Institutions. Textile Industry Technology* **3(381)** 14-20

[5] McCann P and Ortega-Argiles R 2013 Modern Regional Innovation Policy *Cambridge Journal of Regions, Economy and Society* **6(2)** 187-216

[6] Tushavin V A 2013 Analysis of the Quality of IT-Services using Classification Trees *Economics and Management of Control Systems* **10(4.1)** 211-7

[7] Habib A 2011 Management of Textile and Apparel Industry by PDM and ERP Software *Cotton Bangladesh* **4** 33-6

[8] Mironov V V and Konovalova I D 2019 On the Relationship of Structural Changes and Economic Growth in the World Economy and Russia *Voprosy Ekonomiki* **1** 54-78

[9] Utkin A 2020 Elaboration of the Balanced Scorecard for the Assessment of Regional Cluster System Development *The Scientific Heritage* **56-6(56)** 54-6

[10] Kobushko Ia, Kobushko I, Starinskyi M and Zavalna Zh 2020 Managing Team Effectiveness based on Key Performance Indicators of its Members *International Journal for Quality Research* **14(4)** 1245-60

Automated control system functions for complex structured technological processes

A M Korneev, T A Smetannikova, T V Lavrukhina and F A Al-Saeedi

Lipetsk State Technical University, Lipetsk, Russia

E-mail: lavrukhina_tv@mail.ru

Abstract. A system for determining the optimal technology and process control has been developed. The system provides identification of global and local technologies, flexible management of end-to-end technology with the issuance of appropriate recommendations about the further progress of the production process. The data obtained during the operation of the system is recorded by the report generation subsystem for further analysis. The developed system allows you to implement optimal process control, both at the level of an individual product and for a batch of products. The system provides the ability to adapt to changing production conditions. A procedure for selecting the most effective optimization method for managing local technology has been developed.

1. Introduction

The approach to process of modeling and management focuses on building models of quality indicators, determining the optimal values of technological factors that provide the highest level of output properties. In this paper we consider an automated system that allows to identity global and local technologies, as well as to control the progress of the production process [1-9].

The main features of this automated system are:

- the ability to analyze end-to-end technology;
- universal system of modeling;
- independence from the type and structure of the technological process for which statistical information is available;
- finding the optimal technology and monitoring its implementation, and when the factors deviate from the recommended values, a forecast of the quality characteristics for this technology is made, and its adjustment is made in order to increase the level of output properties.

Proposed automated system consists of several subsystems. The information about technology is included into data base. Moreover, relevant characteristics of finished product quality are also included into it. Then all information goes to the subsystem of pre-processing of data, where it is analyzed. The obtained data is used in the identification subsystems of global and local technologies.

Identification of the global technology consists in choosing the optimal technological modes that ensure obtaining the required properties with the maximum probability [2, 4-8]. After identifying the global technology, the output properties prediction subsystem determines the values of the quality indicators of the finished product, provided that rational boundaries are observed.

2. Research method

The process modeling subsystem provides the construction of mathematical models of metric and non-metric quality indicators. For this purpose, models of the dependence of quality characteristics on technological factors are built on the basis of regression analysis. At the same time, the researcher can control the complexity of the resulting models at his own discretion. The following types of models can be constructed:

- a linear model, which can be written in the general form as follows: $f(x) = a_0 + a_1x_1 + \cdots + a_nx_n$, there $a_0, \ldots, a_n$ - model parameters;
- a nonlinear model: $f(x) = a_0 + \sum_{k=1}^{K} a_k f_i(x_j)$, where $(K + 1)$ -the number of parameters of the model, $j = 1, \ldots, n$ - the number of the factor, $i = 1, \ldots, r$ - the number of the function, and there is a linear function in the set of functions *f*;
- with using the products of linear functions of factors: $f(x) = a_0 + \sum_{k=1}^{K_1} a_k x_j + \sum_{k=1}^{K_2} a_{K_1+k} x_{j_1} x_{j_2}$;
- with using products and nonlinear functions of factors: $f(x) = a_0 + \sum_{k=1}^{K_1} a_k\, f_i(x_j) + \sum_{k=1}^{K_2} a_{K_1+k} f_{i_1}(x_{j_1}) f_{i_2}(x_{j_2})$.

The subsystem for modeling technological processes allows you to obtain mathematical models of individual quality indicators. However, when managing the technology, the problem arises of determining the optimal values of technological factors that ensure the minimum deviation of all quality indicators from the required values [2]. The local technology identification subsystem allows us to obtain point optimal values of technological factors that provide the highest level of output properties, based on the obtained models. For optimization, a weighted complex criterion is used – the root-mean-square convolution function. The system provides for the use of expert evaluation methods to determine the weight coefficients of quality indicators, which allows you to get the smallest deviation of the most important indicator from the optimal value.

The resulting nonlinear least squares problem in the system is solved by the following methods: Gauss-Newton, Levenberg-Marquardt, DFP and BFGS, which are described in detail in [2]. Then the problem of choosing a solution method occurs that allows to get the most accurate approximation to the optimum for certain class of models. The researches help come to the conclusion that it is impossible to uniquely determine the most optimal procedure among the listed ones for the considering types of models. That's why the automatic selection of the most rational method in each specific case is used in the developed system. Below you can see the results of calculation for 4 models of quality indicator, reflecting the dependence on six technological factors. The following models of the form were obtained:

- a linear model $f(x) = 7.383 + 11.077x_1 + 9.496x_2 - 4.449x_3$;
- a nonlinear model $f(x) = 9.003 + 1.282x_1^3 + 18.672\sqrt{x_2} - 3.723x_2^3 - 5.737\sqrt{x_3}$;
- with using the products of linear functions of factors $f(x) = 7.477 + 10.681x_1 + 1553.591x_2x_4 - 12598.549x_4x_5 - 34.004x_2x_6$;
- with using products and nonlinear functions of factors $f(x) = 7.932 + 20.441\sqrt{x_2} - 3.987x_2^3 + 1.291x_1^3 - 293.477x_3x_4^2$.

Table 1 shows the average values of relative deviations of quality indicators from the specified optimal values.

Table 1. Example of forming subsets of complex shapes.

Model	Method Gauss-Newton	Levenberg-Marquardt	DFP	BFGS
linear	0.23146	0.02384	0.22555	0.22556
nonlinear	0.12870	0.14912	0.17970	0.05133
products of linear functions	0.31628	0.17437	0.46323	0.42733
products and nonlinear functions	0.13694	-0.04750	0.16580	0.17691

The given example demonstrates that for different types of models, different methods show the best results. In this case, these are the BFGS methods for nonlinear models and the Levenberg-Marquardt methods in the other cases. This allows us to conclude that the chosen strategy for finding the optimal method in each particular case is correct. After the identification of the local technology, the forecast of the values of the quality characteristics is carried out on the basis of the constructed models for the obtained optimal values of the technological factors.

3. Results and analysis

The problem of choosing optimization boundaries becomes an important step, because they determine the possibility of finding acceptable values of technological parameters. Consider an example of the influence of factors on the properties of products. An example of such a model:

$$y_1 = 0.027 + \frac{16311.738}{x_1^2} - \frac{85.295}{x_2^2} + \frac{0.0003}{x_3^2} + 0.033x_4 +$$
$$+1.73 * 10^{-12}x_5^3 + 2.64 * 10^{-8}x_6^3 + \frac{528.898}{x_7^2} - \frac{437.947}{x_8^2}$$

As constraints, we take: $\begin{cases} 1310 \le x_1 \le 1465 \\ 236 \le x_2 \le 294 \\ 0.4 \le x_3 \le 1.03 \\ 0.006 \le x_4 \le 0.088 \\ 1580 \le x_5 \le 1710 \\ 18 \le x_6 \le 23.67 \\ 550 \le x_7 \le 1290 \\ 250 \le x_8 \le 1250 \end{cases}$

Let's set the point value that we need to get at the output to 0.048. Then the optimization problem will look like this:

- find the minimum value of the function

$$y_1 = 0.027 + \frac{16311.738}{x_1^2} - \frac{85.295}{x_2^2} + \frac{0.0003}{x_3^2} + 0.033x_4 +$$
$$+1.73 * 10^{-12}x_5^3 + 2.64 * 10^{-8}x_6^3 + \frac{528.898}{x_7^2} - \frac{437.947}{x_8^2}$$

- under the following conditions

$$g_1(x) = 1310 - x_1 \le 0, g_2(x) = x_1 - 1465 \le 0,$$
$$g_3(x) = 236 - x_2 \le 0, g_4(x) = x_2 - 294 \le 0,$$
$$g_5(x) = 0.4 - x_3 \le 0, g_6(x) = x_3 - 1.03 \le 0,$$
$$g_7(x) = 0.006 - x_4 \le 0, g_8(x) = x_4 - 0.088 \le 0,$$
$$g_9(x) = 1580 - x_5 \le 0, g_{10}(x) = x_5 - 1710 \le 0,$$

$$g_{11}(x) = 18 - x_6 \leq 0, g_{12}(x) = x_6 - 23.67 \leq 0,$$
$$g_{13}(x) = 550 - x_7 \leq 0, g_{14}(x) = x_7 - 1290 \leq 0,$$
$$g_{15}(x) = 250 - x_8 \leq 0, g_{16}(x) = x_8 - 1250 \leq 0,$$
$$g_{17}(x) = 0.021 - \frac{16311.738}{x_1^2} + \frac{85.295}{x_2^2} - \frac{0.0003}{x_3^2} - 0.033x_4 -$$
$$-1.73 * 10^{-12}x_5^3 - 2.64 * 10^{-8}x_6^3 - \frac{528.898}{x_7^2} + \frac{437.947}{x_8^2} \leq 0$$

The restriction $g_{17}(x)$ shows that the range of acceptable values from the bottom is limited by the number 0.048, reaching it during optimization, we will find the necessary values of the technological parameters.

1) We will set $x^0 = (1310; 294; 0.4; 0.088; 1710; 18; 550; 1250)^{\mathrm{T}}$; $f(x^0) = 0.05$

$$\varepsilon_1 = -0.01; \varepsilon_2 = 0.0001; M = 5.$$

2) Set $k = 0$
3) Check the condition $k \geq M: 0 < 5$.
4) Calculate $g_j(x^0), j = 1,2,\ldots,17$: $g_1(x) = 0, g_2(x) = -155, g_3(x) = -58, g_4(x) = 0,$

$$g_5(x) = 0, g_6(x) = -0.63, g_7(x) = -0.082, g_8(x) = 0, g_9(x) = -130, g_{10}(x) = 0,$$
$$g_{11}(x) = -0.032, g_{12}(x) = -5.638, g_{13}(x) = 0, g_{14}(x) = -740, g_{15}(x) = -1000,$$
$$g_{16}(x) = 0, g_{17}(x) = -0.005$$

5) Check $\varepsilon_1 \leq g_j(x^0) \leq 0$

Restrictions with numbers 1,4,5,8,10,13,16,17 fall under this condition. Table 2 shows components of the gradient vector.

Table 2. Components of the gradient vector.

$\frac{df}{dx_1}$	$\frac{df}{dx_2}$	$\frac{df}{dx_3}$	$\frac{df}{dx_4}$	$\frac{df}{dx_5}$	$\frac{df}{dx_6}$	$\frac{df}{dx_7}$	$\frac{df}{dx_8}$
-1.451E-05	6.713E-06	-8.556E-03	3.300E-02	1.520E-05	2.572E-05	-6.358E-06	4.485E-07

6) Calculate $\Delta x^k = -\left[E - A_k^T\left(A_k A_k^T\right)^{-1} A_k\right] \nabla f(x^k)$, where A_k presented in the table 3.

Table 3. Matrix A_k.

-1	0	0	0	0	0	0	0
0	1	0	0	0	0	0	0
0	0	-1	0	0	0	0	0
0	0	0	1	0	0	0	0
0	0	0	0	1	0	0	0
0	0	0	0	0	-1	0	0
0	0	0	0	0	0	1	0
1.45E-05	-6.71E-06	8.56E-03	-3.30E-02	-1.52E-05	-2.57E-05	6.36E-06	-4.48E-07

7) Check the condition $\|\Delta x^k\| \leq \varepsilon_2$: the condition is met $\Delta x^k = 0$
8) Calculate $\lambda^0 = -\left(A_k A_k^T\right)^{-1} A_k \nabla f(x^k)$ the components of which are presented in the table 4.

Table 4. The components λ^0.

-1.43E-11	-6.60E-12	-8.41E-09	-3.24E-08	-1.49E-11	2.53E-11	6.25E-12

We remove the constraint number 8 from the list of active constraints, because the largest element of the negative values by modulus corresponds to it.

9) Calculate $\Delta x^k = -\left[E - A_k^T\left(A_k A_k^T\right)^{-1} A_k\right] \nabla f(x^k)$, where A_k presented in the table 5 and the components of $\Delta x^k$are presented in the table 6.

Table 5. Matrix A_k.

-1	0	0	0	0	0	0	0
0	1	0	0	0	0	0	0
0	0	-1	0	0	0	0	0
0	0	0	0	1	0	0	0
0	0	0	0	0	-1	0	0
0	0	0	0	0	0	1	0
-2.38E+06	-1.10E+06	-1.40E+09	-5.41E+09	-2.49E+06	4.22E+06	1.04E+06	-1.64E+11

Table 6. The components Δx^k.

-1.0E-14	0	-2.5E-17	-3.3E-02	0	0	0	1.09E-03

Calculate t_k^* , in this case, this parameter can not be obtained the function increases monotonically, the components of Δx^k will be in this case presented in the table 7.

Table 7. The components Δx^k.

№ conditions	difference	t_k^j
2	-155	negative
3	Not influence	
6	-0.63	negative
7	0	2.49
9	Not influence	
11	Not influence	
12	Not influence	
14	Not influence	
15	-1000	negative

As you can see, the single increment can be estimated as $t_k^j = 2.49$. Get a new point and check it $x^0 = (1310; 294; 0.4; 0.06; 1710; 18; 550; 1250)^{\mathrm{T}}$; $f(x^0) = 0.048$. As you can see, we have reached the optimal point and these parameters are optimal.

The functions of monitoring and correcting the technology are performed by tracking technology execution subsystems for optimal ranges and values of factors, issue recommendations on the continuation of the process and predict the output properties. By means of these subsystems, it is possible to adapt the technology to the changing conditions of the production process.

In the production of a batch of products, for some reason, the values of technological factors at one of the stages may go beyond the calculated rational limits, which will entail a change in the predicted level of output properties. If it turns out to be significant, the system issues recommendations on the transfer of products to another grade or defects. The adjustment of the technology is carried out as following. At the *k-th* step of the used iterative method, the technological space, from which the part corresponding to the already implemented stages of the production process is excluded, is divided into k_n subspaces, where n is the number of factors. For each subspace, the value of the criterion for the amount of shared information between factors and quality indicators is calculated. The area with the highest criterion value is selected. The procedure continues for $k = 1,2,3...$, until the amount of information begins to decrease. As a result, we get rational limits of changes in the still unrealized technological factors.

After adjusting the global technology, new optimal "point" values of factors and corresponding values of quality indicators are determined. In this case, the existing values of quality indicators corresponding to the previous stages are substituted into the statistical models of quality indicators. This allows you to reduce the forecast error for models, and therefore determine the values of technological factors that will improve the quality of this product.

In the conditions of real production, several classes of situations may arise that lead to the task of correcting the found optimal technology. Let's consider several similar situations and illustrate their solution in the framework of an automated process control system. If one of the following cases occurs, the system adjusts the found technological limits and finds the optimal continuation of the production process.

1. Violation of technology, which corresponds to a situation where the values of technological factors do not correspond to the calculated optimal values.

The values of the technological factors obtained after the violation of the technology and further adjustment are shown in figure 1 and in table 8.

Note. In the figures: ● - before the state change; ■ - after the state change; the left rectangle is the original optimal technology; the right rectangle is the changed ranges.

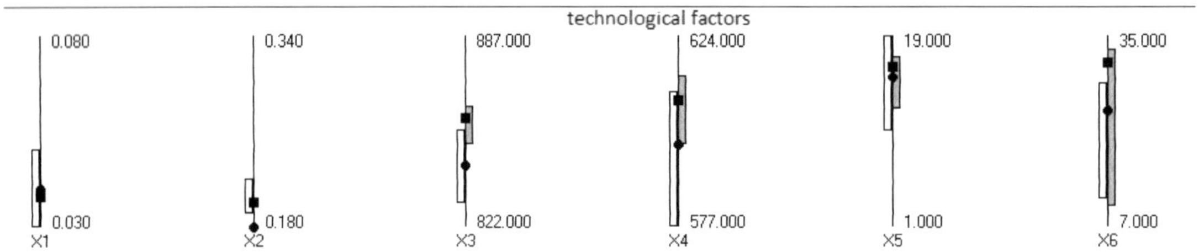

Figure 1. The values of technological factors obtained after the technology violation and adjustment.

Table 8. The values of technological factors obtained after the technology violation and adjustment.

		technological factors					
		X1	*X2*	*X3*	*X4*	*X5*	*X6*
before adjustment	min	0.03	0.19	830	577	10	11
	max	0.05	0.22	855	610	19	28
	opt	0.038	0.18	843	597	15	24
after adjustment	min	0.03	0.19	850	597	12	10
	max	0.05	0.22	863	614	17	33
	opt	0.04	0.2	859	608	16	31

2. Change in production conditions. After changing the temperature range of the end of rolling in the hot rolling mill, an adjustment was made and the following values of the factors were obtained (figure 2 and table 9).

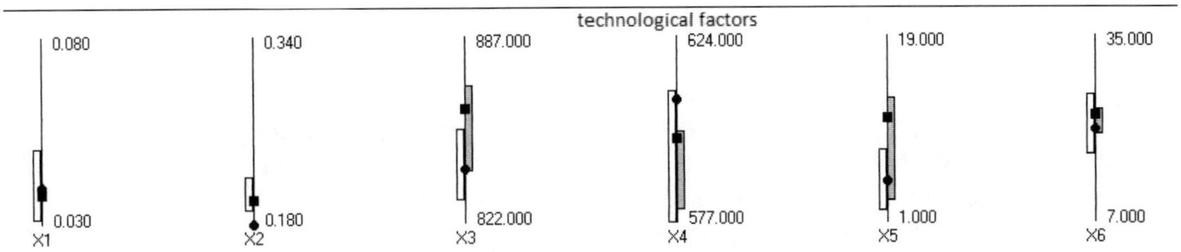

Figure 2. The values of technological factors obtained when changing production conditions.

Table 9. The values of technological factors obtained when changing production conditions.

		technological factors					
		X1	*X2*	*X3*	*X4*	*X5*	*X6*
before adjustment	min	0.03	0.19	830	577	2	17
	max	0.05	0.22	855	610	8	26
	opt	0.038	0.18	841	608	5	21
after adjustment	min	0.03	0.19	840	580	3	20
	max	0.05	0.22	870	600	13	24
	opt	0.04	0.2	862	598	11	23

3. Raw material drift. The following results were obtained when the properties of raw materials drifted (figure 3).

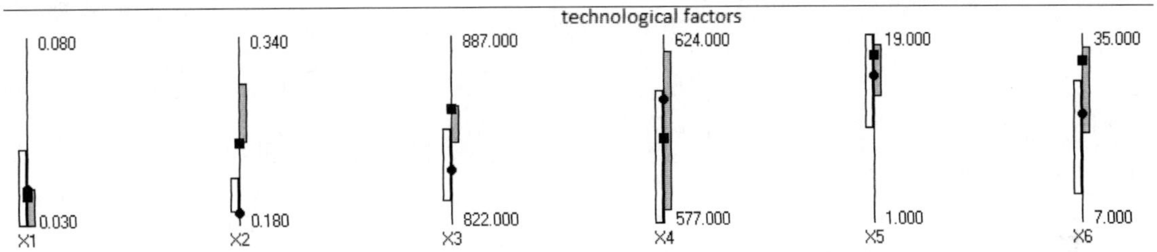

Figure 3. The values of technological factors obtained when changing production conditions.

4. Changing the requirements for the quality indicators of finished products. In fact, this change corresponds to the transition to a different type of product.

4. Conclusion

The data obtained during the operation of the system is recorded by the report generation subsystem for further analysis. The developed system allows you to implement optimal process control, both at the level of an individual product and for a batch of products. The system provides the ability to adapt to changing production conditions. A procedure for selecting the most effective optimization method for managing local technology has been developed.

References

[1] Draper N and Smith G 1991 *Applied regression analysis* **1** (Moscow: Finance and Statistics) p 365
[2] Draper N and Smith G 1991 *Applied regression analysis* **2** (Moscow: Finance and Statistics) p 349
[3] Recleitis G, Ravindran A and Ragsdel K 1986 *Optimization in technology* Book 1 (Per. s engl. - M.: Mir) p 349
[4] Korneev A M, Lavrukhina T V and Smetannikova T A 2013 Description of the technological process with a finite state machine *Proceedings of the Workshop on Computer Science and Information Technogies CSIT'2013* **1** (Vienna-Budapest-Bratislava) p 155-9

[5] Korneev A M, Salfetnikov M V, Glazkova Y A, Lavrukhina T V and Abdullakh L S 2018 Mathematical modelling and optimization of risk analysis in emergencies *International Journal of Engineering and Technology(UAE)* **7(4.7(7)** 209-11
[6] Zhigljavsky A and Zilinskas A 2008 *Stochastic Global Optimization* (Berlin: Springer) p 262
[7] Mishachev N M, Shmyrin A M and Suprunov I I 2020 simulation of sequential processing of a moving extended object *International Transaction Journal of Engineering, Management and Applied Sciences and Technologies* **11(7)** 11A07T
[8] Yartsev A G and Shmyrin A M 2017 Identification of neighborhood system taking into account restrictions for coefficients and parameters *In the book: INTERACTIVE SYSTEMS: Problems of Human-Computer Interaction. collection of scientific papers* 124-7
[9] Shmyrin A M and Yartsev A G 2018 Damping procedure for mixed control with variable coefficients of the neighborhood model *International Journal of Engineering and Technology(UAE)* **7(3)** 21-3

IOP Conf. Series: Materials Science and Engineering 1155 (2021) 012049 doi:10.1088/1757-899X/1155/1/012049

Mathematical description of elastic mechanisms of force compensating manipulator

G Ya Pyatibratov, L L Altunyan and A A Danshina

Department of Electric Power Supply and Electric Motor Drive, Platov South-Russian State Polytechnic University, Prosveshcheniya street, Novocherkassk, 346428, Russia

E-mail: x5551@bk.ru

Abstract. Improvement of control systems for manipulators and robots was required implementation complex automation of production processes. It is necessary to have a reliable mathematical description of elastic mechanisms interacting during operation to solve problems of analysis and synthesis of control devices of force compensating manipulators. The work goal is the development of the mathematical description of elastic executive mechanisms of the ShBM-150M balanced manipulator. Studies have shown that low-frequency components of moved weight elastic vibrations are due to flexural deformation of a manipulator's shoulder and arm. Inertia equivalent values of manipulator links are changed by 2.2 and 37 times during manipulator operation with a constant load value. Total flexural deformation of shoulder and arm deflection is 25 mm during changing mass per 100 kg. And parameters of the mathematical model can be changed by 11 times, which needs adaptive control use.

1. Introduction

Currently, in different industries, manipulators and robots are the basis for complex automation of production processes [1]. Improvement of modern industrial manipulators (IM) is required to solve tasks of mechanization and automation of production processes. Analysis of the technical characteristics and capabilities of modern IM showed that the negative effect of elastic structures and mechanical links increases on quality and productivity of work because of complications of tasks and new approaches to design focused on reducing materials [2-4]. The dynamic loads of mechanisms and electric drives are increased during IM work stimulation. The vibrations of controlled axes are grown [2].

The requirements of manipulators control systems are raised at implementation of modern intelligent production technology. It is necessary to have a real mathematical description of multilink elastic mechanisms in realization.

2. Setting of the problem

The operation of modern manipulators and robots has negative effect by the flexibility of structures and mechanical transmissions.

Because vibrations of executive mechanism (EM) are increased, quality of IM work is degraded, the accuracy of weight position is lowed.

Recently, force compensating manipulators (FCM) become a frequent practice. The electric drives (ED) control systems compensating weight of links and moved load are used in them. This allows a worker to move and position objects with minimal force values between 2 and 5 % of the weight. It is necessary to create systems with high-precision force control in elastic manipulators EM.

It is necessary to have a reliable mathematical description allowing to study power energy and kinematic effects in links, mechanisms, and gears of a manipulator for the design of FCS electromechanical system. The elasticity of mechanisms and structures, nonlinearity of characteristics, changes in a structure, and parameters values of control object are needed to use in FCM mathematical description [5].

A difficult task is the development of FCM mathematical models with multilink elastic mechanisms with cross internal feedbacks and nonlinear characteristics. Therefore, it is important to substantiate using hypotheses, formulate correct assumptions and determine rational approaches and research methods during solving development of a mathematical description of complex manipulator mechanisms and perform its correct simplification at the conceptual stage of FCM study.

3. Papers' review determining relevance

Various approaches and methods are used during developing FCS mathematical description [6]. At the same time, a difficult task is creation of real mechanisms mathematical description with an elastic multilink structure resulting in large elements deformations [7].

Recently, a mathematical description with Hamilton principle is used for study of energy power and kinematic mechanisms action of multilink manipulators. Based on its method Lagrange equations of second kind are obtained:

$$\frac{d}{dt}\left(\partial T / \partial \dot{q}_i\right) - \partial T / \partial q_i + \partial D_M / \partial q_i + \partial U / \partial q_i = Q_i, \; i = 1, 2, n, \tag{1}$$

where n is number of degrees of system freedom; q_i, $\dot{q}_i$ are generalized coordinates of speed; Q_i is generalized force acting on system.

Influence of inertia, stiffness coefficients and mechanisms damping on motion of FCS EM are determined by analysis of Lagrange equations components describing change in kinetic (T) and potential (U) energy of studied system [8]. Damping mechanisms properties are considered using the dissipative dispersion function (D) determined from results of experimental studies. It allows a demonstration that real values of internal viscous friction forces in system with structural damping [9].

Following assumptions were made during developing a mathematical description of FCM [10]:

- wave processes were not taken into account in study of mechanisms deformations, it possible to use discrete mathematical mechanisms models in which concentrated masses are connected by weightless elastic links;
- deformations and stresses in elements of mechanical transmissions and structures of a manipulator are followed Hooke's law;
- forces and torques acting on mechanical system elements exerted on non-deflecting point mass;
- elastic elements (EE) have constant stiffness characterized proportionality constant between force and deformation;
- twisting of motor shaft and gearbox is neglectfully small and it is neglected;
- stiffness of a manipulator stand fixing to a base is neglected.

4. Goal and tasks of paper

The goal of the work is justification of rational approaches and methods for completing a reliable a mathematical description of FCM with elastic mechanical transmissions and parameters changing during operation. The solved problem is focused on developing a real mathematical model of manipulator multilink mechanisms, making analysis and synthesis of control systems requiring their functioning quality.

5. Approaches and methods of theoretical study

Justification of approaches and methods used in development of the FCM mathematical description will be considered by example of the ShBM-150M type manipulator. Its construction consists of two pantograph mechanisms [11]. The main advantage of such mechanisms is performing large vertical and horizontal motions of unit with minimal moving actuator providing movement with saved required orientation [12]. However, IM mathematical description is heavily complicated with of such mechanism.

Features of the ShBM-150M manipulator construction will be considered affecting on development of a mathematical description of elastic mechanisms. The kinematic structure of the ShBM-150M manipulator mechanisms is shown in figure 1.

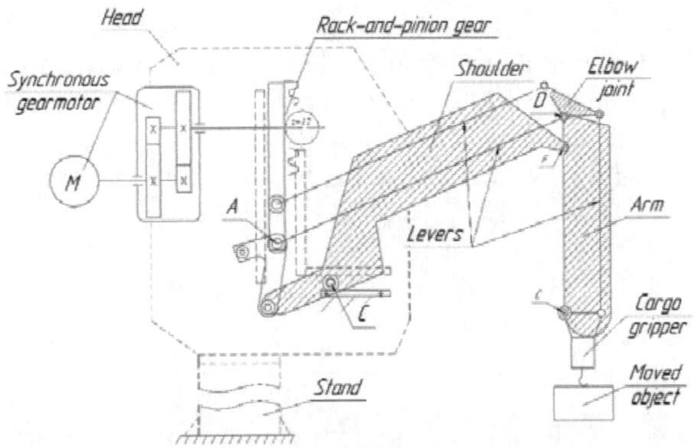

Figure 1. Kinematic structure of the ShBM-150M manipulator mechanisms.

A lift actuator is mounted in the manipulator head. It consists of an electric motor compensating required load weight during vertical movement. Sew-Eurodrive frequency-controlled electric drive consists of Movitrac B frequency converter (0.75 kW, 4.2 A), a DRS80S4 synchronous gearmotor (0.75 kW; 5.1 Nm; 1400 rpm) [13].

The gearmotor is jointed to two gear wheels rotating a rack through a helical gearbox. Required object movement is provided rack motion through the lever system by shoulder and arm. Structurally, the levers are located inside squaring shoulder and arm. Structural elements and mechanical transmissions of a multilink manipulator are got complex elastic deformations by Hooke law during operation. It possible to analyze their components separately.

A moved cargo orientation is provided by two parallelogram mechanisms with a rack, levers, shoulder and arm. The parallelogram mechanism structure is led by working levers in tension and compression, and shoulder and arm of the ShBM-150M manipulator are mainly got bending deformations.

Studies have shown that deflection value is greater than tension or compressive deformations of same elements with same forces values acted to long levers and structures of the FCM [12]. This is due to the following main factors:

a) the tensile stretch of a long structural element is determined by sectional area, and the deflection is dependent on value of axial moment of inertia;
b) for structural steels, Young's modulus determining tension and compressive deformation has many folds larger value than the shear modulus.

For the ShBM-150M manipulator, studies of structural elements and transmission mechanisms using the Ansys Mechanical software package [14] showed that tensile-compression deformation of levers is 0.04 mm, the box-type arm is 0.012 mm, and the arm is 0.034 mm with an acting force of 1000 N. The deflection (bend) of the shoulder is 6.9 mm, and the arm is 2 mm with same force value. Thus, arm deflection is 166 times greater than stretching deformation, and shoulder deflection are 202 times greater than tension deformation. Studies have shown that equivalent deformation of mechanical transmissions and manipulator structure have resulted in cargo shift in the vertical position by 25 mm with a change in cargo mass per 1000 N.

A collaborative Ansys Mechanical software analysis of the ShBM-150M manipulator mechanism driving forces from a motor to a cargo showed that equivalent stretching mechanism elements can be ignored during studying low-frequency IM vibrations. And vibration processes can be studied with only deformations of the shoulder and arm. In this case, the shoulder bend must be studied about the C point, and the arm bend will be studied about the F point with levers action of parallelogram mechanisms.

A design model is importantly drawn up at a conceptual stage of development of a mathematical model of the manipulator ShBM-150M mechanisms. This helps to determine elastic influence of mechanisms. Mechanisms design models are developed with various levels of detail depending on solved tasks.

The construction analysis of the ShBM-150M manipulator showed that maximal inertia of a synchronous motor with a gearbox, the shoulder, arm and a cargo is had. Minimal stiffness coefficients are determined by bending deformations of the shoulder and arm. Therefore, mechanisms design model characterizing bending deformations of the shoulder and arm should be presented in a three-mass model form showed in figure 2.

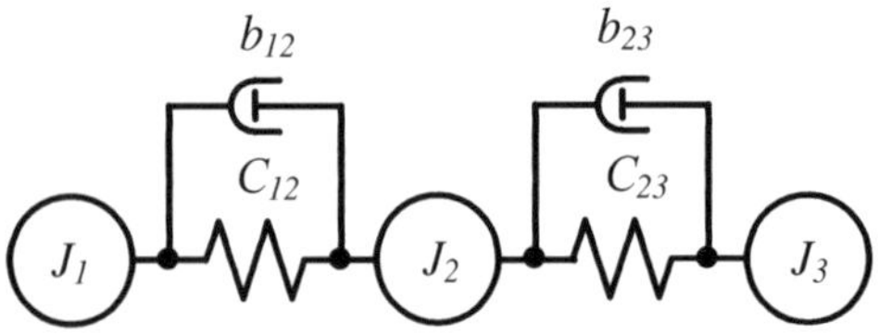

Figure 2. The three-mass design model of the ShBM-150M manipulator mechanism.

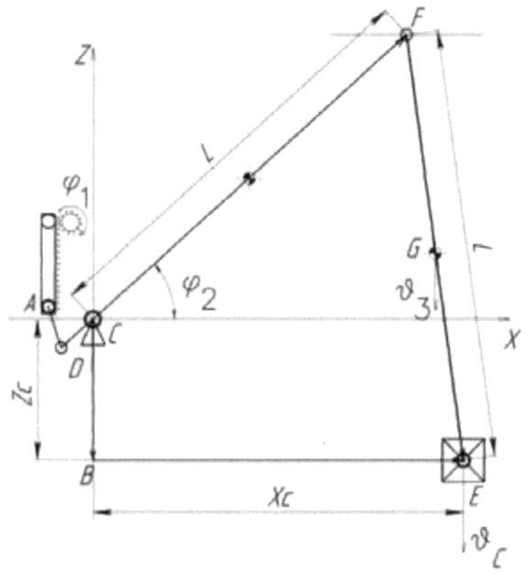

Figure 3. The construction of the manipulator executive mechanism.

In figure 2 J_1 inertia consists of inertial properties of motor rotor, gearbox and rack gear, J_2 is shoulder inertia, J_3 is arm and cargo with gripper inertia. Stiffness coefficients C_{12}, C_{23} characterize equal bending deformation of the shoulder and arm. Damping coefficients b_{12}, b_{23} describe dissipative properties of the shoulder and arm.

The inertia values of structural elements of the ShBM-150M manipulator were refined using the SolidWorks Motion software [15]. And the stiffness coefficients values of mechanical transmissions were refined using the Ansys Mechanical software.

The mathematical description is developed to study the electromechanical system of the ShBM-150M manipulator. Therefore, the parameters of design model must put into motor shaft. The reduction of masses and inertia is completed with constant kinetic energy, and equivalent values of elements stiffness coefficients are determined from constant potential energy condition of the system [16]. Studies have shown that the parameters of the design model of the manipulator can change with changing cargo weight and shoulder and arm configuration. The range of changing inertial properties and elastic deformations values of mechanical transmissions of the ShBM-150M manipulator is determined by cargo weight value also values of masses and inertia of levers, shoulder and arm put into motor shaft.

The gear ratio and reduction radius are must to know for putting parameters of design model into motor shaft. When determining these dependencies, shoulder rotates the center of mass about the C point in vertical cargo movement, and arm complex moves in linear direction with cargo and rotation about the *F* point.

The method of parameters putting of design model into motor shaft is explained in figure 3. In the figure, the radial direction of cargo movement X_C coincides with the *X*-axis, and its vertical movement Z_C coincides with the *Z*-axis.

The generalized basic parameter of the φ_2 angle was chosen at determination of manipulator links configuration influence in different cargo positions. The angle determines shoulder down about horizontal position and it uniquely determines position of moved shoulder, arm and cargo.

In figure 3, the distance between the rotation axis of the shoulder point *C* and horizontal and vertical move of cargo is characterized by variables X_C and Z_C. The length of arm and shoulder is determined by the parameter $L = 1.5$ m.

The analytical equations for reduction radius ρ_C and gear ratio i_3 describing motion of a mass centers of a cargo, arm, and shoulder are got nonlinear. It is far more complex study.

For example, the gear ratio putting rotational arm movement into a motor shaft is:

$$i_3 = \left[L\cos\varphi_2 - \frac{L\sin\varphi_2\left(\cos\varphi_2 - \frac{X_C}{L}\right)}{\sqrt{1-\left(\cos\varphi_2 - \frac{X_C}{L}\right)^2}} \right] \left[\rho_C \frac{\sin\varphi_2}{\sin\left(\arccos\left(\frac{X_C}{L} - \cos\varphi_2\right)\right)} \right]^{-1} \tag{2}$$

They were taking a Taylor series expansion for linearization of obtained nonlinear dependences. The simplified linearized analytical equations were got using only linear components of a Taylor series.

The linearized gear ratios of the shoulder and arm, the radius of linear arm move and putting angle of the shoulder rotation into a motor shaft are determined by the following equations:

$$i_2 = (\gamma + \beta\varphi_2)^{-1};\ i_3 = (\theta\varphi_2)^{-1};\ \rho_3 = \rho_C + \lambda\varphi_2;\ \varphi_2 = \Lambda + \gamma\phi_2, \tag{3}$$

where $\gamma = \rho_C / L$; $\beta = (\rho_C(L - X_C))\left(L\sqrt{2LX_C - X_C^2}\right)^{-1}$; $\theta = -\rho_C\left(2LX_C - X_C^2\right)^{-1/2}$; $\lambda = \rho_C(L - X_C)\left(2\sqrt{2LX_C - X_C^2}\right)^{-1}$; $\Lambda = \sqrt{2LX_C - X_C^2} L^{-1/2}$.

The analysis showed that the range between nonlinear and linearized characteristics does no more than 10%. At the same time, it was found that the greatest linearization errors are got in out points of a manipulator working area, and minimal values are got in area middle.

The dependences of the equivalent total inertia of the shoulder, arm, and cargo putting into a motor shaft were determined due to the linearized equations. The equations were got in changing configuration of the manipulator links determined by cargo position.

Studies have shown that parameters of the mathematical model putting into a motor shaft with moved cargo are changed in the following ranges: shoulder inertia $J_2 = (4.2\cdot10^{-7} - 1.5\cdot10^{-5})$ kg·m^2, arm inertia at vertical cargo motion with 10 kg mass $J_3 = (1.1\cdot10^{-5} - 2.5\cdot10^{-5})$ kg·m^2 and changing mass between 10 and 150 kg $J_3 = (1.1\cdot10^{-5} - 1\cdot10^{-5})$ kg·m^2. Changing stiffness of shoulder is $C_{12} = (0.03 - 0.15)$ Nm/rad and arm is $C_{23} = (0.3 - 0.8)$ Nm/rad. Damping coefficients b_{12}, b_{23} change in range between 0.00008 Nm/rad and 0.00025 Nm/rad. Notice that the inertia, stiffness and damping coefficients are constants for the elements. The parameters change is due to give into a motor shaft.

The analysis showed that mechanism parameters can change of ten times. Must be factored into development of control systems for the manipulator [17].

The mathematical model was developed differential equations form describing the movement of the mechanism manipulator and cargo. The model was got with considered approaches and assumptions based on Lagrange equations, and variable values of mass reduction radius and stiffness coefficients given to a motor shaft in equation (4).

The block diagram of three-mass mathematical model of mechanical part of the ShBM-150M manipulator was obtained based on the equations of cargo motion (4). This model makes allowances for analyzed correlation and changes in the mechanism parameters. The simplified block diagram is shown in figure 4. Only basic correlations and parameters presented in more complex model are given in it.

$$\begin{cases} \ddot{\phi}_2 J_2\left(\gamma+\beta(\Lambda+\gamma\phi_2)\right)^2+\dot{\phi}_2^2 J_2\rho_C\beta\left(\gamma+\beta(\Lambda+\gamma\phi_2)\right)- \\ -\dot{\phi}_3^2\left[J_{arm}\gamma\theta^2(\Lambda+\gamma\phi_2)+m_{arm}\gamma\lambda\left(\rho_C+\lambda(\Lambda+\gamma\phi_2)\right)\right]- \\ -C_{23}\left(\phi_2-\phi_3\right)\theta^2\left[\gamma\left(\Lambda+\gamma\phi_2\right)\left(\phi_2-\phi_3\right)+\left(\Lambda+\gamma\phi_2\right)^2\right]+ \\ +C_{12}\left(\phi_1-\phi_2\right)\left[\left[\gamma+\beta(\Lambda+\gamma\phi_2)\right]^2+\gamma\beta\left[\gamma+\beta(\Lambda+\gamma\phi_2)\left(\phi_1-\phi_2\right)\right]\right]=0; \\ \ddot{\phi}_3\left[m_{arm}\left(\rho_C+\lambda(\Lambda+\gamma\phi_2)\right)^2+m_C{\rho_C}^2+J_{arm}\theta^2\left(\Lambda+\gamma\phi_2\right)^2\right]+ \\ +2\dot{\phi}_2\dot{\phi}_3\left[J_{arm}\gamma\theta^2(\Lambda+\gamma\phi_2)+m_{arm}\gamma\lambda\left(\rho_C+\lambda(\Lambda+\gamma\phi_2)\right)\right]- \\ -C_{23}\left(\phi_2-\phi_3\right)\theta^2\left[\left(\Lambda+\gamma\phi_2\right)^2\right]=F_W\rho_C-m_C g\rho_C; \\ \ddot{\phi}_1 J_1+C_{12}\left(\phi_1-\phi_2\right)\left[\gamma+\beta(\Lambda+\gamma\phi_2)\right]^2=M_1-M_L. \end{cases} \quad (4)$$

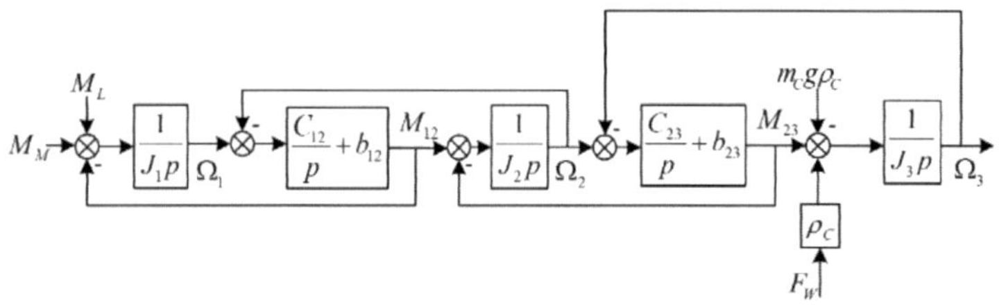

Figure 4. The block diagram for study of the ShBM-150M manipulator mechanism.

The M_W component of motor torque compensating a weight of the shoulder and arm, the force F acted by a worker for cargo are marked in figure 4. The mathematical model validity of the ShBM-150M mechanisms has been confirmed by large experimental studies with real manipulator. For example, experimental and simulated oscillograms are presented in figure 5. The "Tenzo-M" type load cell [18] with 2500 N capacity was jointed between the gripper and cargo for experimental studies of elastic vibrations of the manipulator mechanisms. In figure 5 the force oscillogram was got at dead stop of a cargo with 100 kg mass.

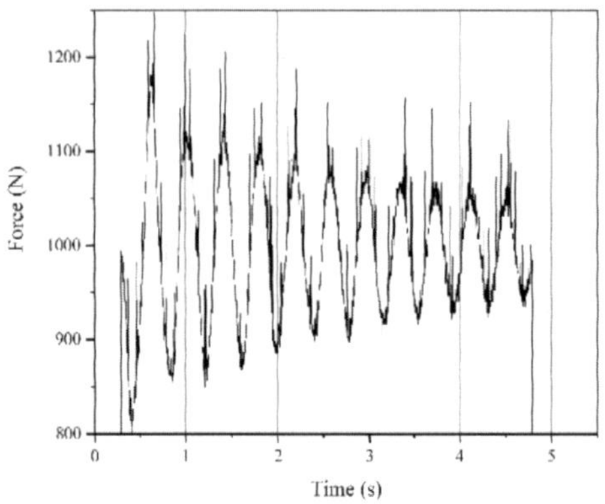

Figure 5. The force oscillogram in manipulator mechanism at shock pulse input on cargo with 100 kg mass with maximum go out of cargo from the stand.

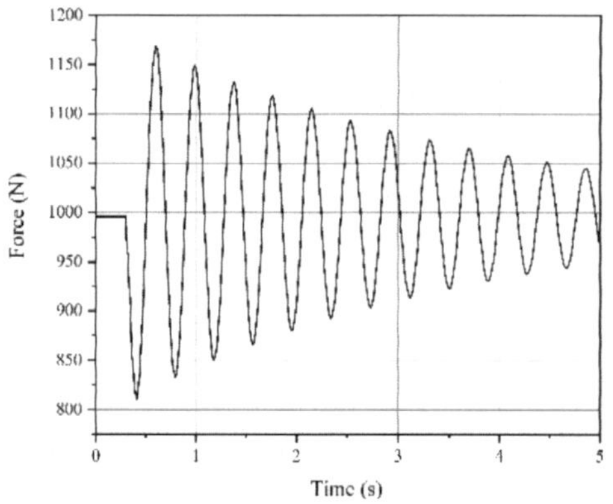

Figure 6. The force oscillogram in manipulator mechanism at shock pulse input on cargo with 100 kg mass with at simulating manipulator operation.

In figure 5, there is the oscillogram with maximum go out of cargo from the stand. Its analysis showed that the cargo frequency vibration is 2.5 Hz. In this case, the amplitude of first force vibration over the final value is 176 N, the damping decrement is 0.13, the vibration decay time is more than 5 s. The force oscillograms taken in other positions showed the damping decrement because of position varies in the range between 0.12 and 0.44. Then if cargo is placed farther from a stand the damping decrement is smaller.

In figure 6 there is the oscillogram at simulating manipulator operation with similar conditions of experiment with the given parameters: $J_1 = 0.000178$ kg·m², $J_2 = 8.4 \cdot 10^{-6}$ kg·m², $J_3 = 6.39 \cdot 10^{-5}$ kg·m², $C_{12} = 0.07$ Nm/rad, $C_{23} = 0.56$ Nm/rad, $b_{12} = 0.00008$ Nms/rad, $b_{23} = 0.0001$ Nms/rad.

The simulation was made with a cargo mass of 100 kg and maximum distance of a cargo from a stand. Into the system, pulsed disturbance was inputted every 0.3 seconds with 200 N amplitude. In this case, the cargo vibration frequency is 2.52 Hz, the amplitude of first force vibration over the final value is 181 N, damping decrement is 0.11, vibration decay time is 6.3 s. The imprecision between compared research results was: 0.8% for frequencies, 2.8% for maximum amplitude over final value, 18% for damping decrements of the transient.

6. Analysis of research results

The effect research of changing parameters of mechanical transmissions on manipulator operation was obtained using the mathematical model of the ShBM-150M. In figure 7 bode plots of changing torque in mechanism EE of the ShBM-150M were shown. Bode plots were given with different cargo positions about stand with cargo weight of 100 kg.

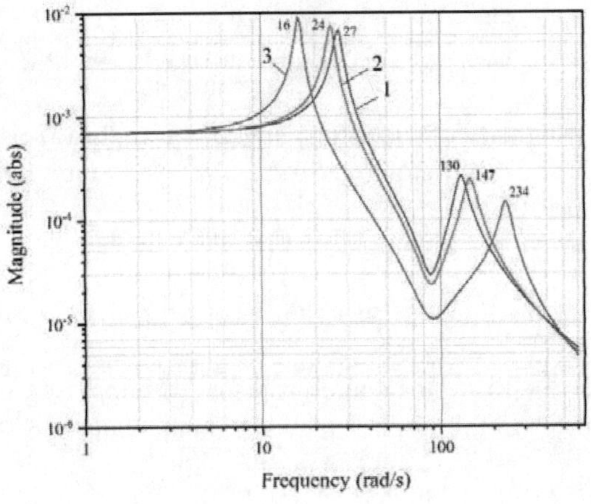

Figure 7. Bode plot of lift mechanism of the ShBM-150M.

Line 1 is links position in minimum distance of a cargo about stand. It is stiff configuration of mechanisms. Line 2 is position at middle distance of a cargo with vertical position of the arm. Line 3 is position at maximum distance of cargo about the stand. In figure 7 values of resonant frequencies are marked about maximum values of all bode plots. Values are different at cargo position. They differ 1.67 times for low frequency, and 1.75 times for high frequency. Research has shown that low frequency resonances are primarily determined by tension shoulder, and high frequency resonances are determined by tension arm. The difference between low frequency and high frequencies is from 4.8 to 14.6 times depending on a cargo position. However, if a cargo is placed near a stand (line 1) then difference is smaller.

The study of changing parameters of mechanisms for joint balanced manipulator operation showed that it is necessary to take on change of mechanisms parameters and complete adaptive control with the mathematical model of the ShBM-150M during solving problem of synthesizing a control system [19].

7. Conclusion

The dynamic loads of mechanisms are increased, there are large EM vibrations, the transients properties are degraded, the accuracy of cargo positioning is decreased because of elastic mechanical transmissions. This is because the FCM is realized with light-weight construction and high drive velocity.

- A valid mathematical description of multilink elastic mechanisms in various conditions and operations is importantly got during solve problems of analysis and synthesis of force control systems. It is necessary to improve the quality of the FCM function.
- Studies have shown that for the ShBM-150M manipulator, the low-frequency components of cargo vertical elastic vibrations are mainly due to bending deformations of the shoulder and arm. They are 25 mm in cargo weight changing by 100 kg.
- The parameters changing of the mathematical model of the ShBM-150M manipulator mechanisms is dependent on configuration of the shoulder and arm of the manipulator, and weight value and cargo position.
- At the ShBM-150M manipulator operation the equivalent value of the shoulder inertia putted into a motor shaft can change by 37 times. The arm inertia can change by 2.2 times. A cargo inertia can change by 11 times. Therefore, using of an adaptive control system is needed.

Acknowledgment
The reported study was funded by RFBR, project number 19-38-90047.

References

[1] Vladov N L, Danilevskiy V N, *et al.* 1988 *Balanced Manipulators* (Moscow: Machinery Press) p 264
[2] Pyatibratov G Ya, Danshina A A, *et al.* 2016 Methods of efficient parameters multifactorial determination of industrial manipulators gears and electric drives *Proc. 2nd Int. Conf. on Ind. Eng. (ICIE-2016)* **150** (Chelyabinsk: SUSU) p 1403-9
[3] Christen G 1989 Simulation of operating behavior of computer-controlled device drives *Feingeratetechnik* **4** 156-8
[4] Kaneko K and Ohnishi K 1990 A drive control of flexible joint *Proc. Int. Power Electron. Conf* **1** (Tokyo) p 442-9
[5] Altunyan L L, Danshina A A and Pyatibratov G Ya 2019 Goals and problems of improving of electromechanical control system of ShBM-150M type balanced manipulator *Rus. Electromechanics* **62(5)** 54-61
[6] Yurevich E I 2000 *Control of Robots and Robotic Systems* (St. Petersburg: SPSIT Press) p 171
[7] Li M, Wu H, Handroos H, Skilton R, Keep J and Loving A 2018 Comparison of deformation models of flexible manipulator joints for use in DEMO *IEEE Trans. on Plasma Sci.* **46(5)** 1198-204
[8] Lure A I 1961 *Analytic Mechanics* (Moscow: Fizmatgiz Press) p 824
[9] Veyts V L and Tsarev G V 1992 *Dinamics and Simulation of Electromechanical Actuators* (Saransk: Mordov University Press) p 228
[10] Kshirsagar P, Jiang D, Zhang, *et al.* 2014 Implementation and evaluation of online system identification of electromechanical systems using adaptive filters *IEEE Trans. Energy Conver. Congress and Exposition (ECCE)* (Pittsburgh, PA, USA) p 2477-84
[11] *Website of Public company "BOMZ"* Retrieved from: http://bomz.su/catalog/shbm-150m/
[12] Andre P, Kofman Zh-M, *et al.* 1986 *Robot Construction* (Moscow: Mir Press) p 360
[13] *Website of Public company "Sew-Eurodrive"* Retrieved from: https://www.sew-eurodrive.ru/
[14] Fedorova N N, Valger S A, *et al.* 2017 *Basic Work in ANSYS 17* (Moscow: DMK Press) p 210
[15] Zinovev D V 2017 *Base of Simulation in SolidWorks* (Moscow: DMK Press) p 240
[16] Filonov I P, Antsiporovich P P, *et al.* 1998 *Mechanism, Machinery and Manipulator Theory*

(Moscow: Dizayn PRO Press) p 656
[17] Pyatibratov G, Danshina A, *et al.* 2020 Efficiency determination of vibrations damping of elastic mechanisms with electric drive *IEEE Transl. 27th Int. Workshop on Electric Drives: MPEI Department of Electric Drives 90th Anniversary (IWED)* (Moscow: MPEI) p 1-8
[18] *Website of Public company "Tenso-M"* Retrieved from: https://www.tenso-m.ru
[19] Lozynskyy A, Marushchak Y, *et al.* 2020 Synthesis of combine feedback control of electromechanical system by feedback linearization method *IEEE Transl. Problems of Aut. Electrodrive. Theory and Practice (PAEP)* (Kremenchuk) p 1-6

Machine learning in the development of regulators: the specifics of data preparation and the choice of the ML model algorithm

N A Staroverova[1], M L Shustrova and M Z Fatykhov

Kazan National Research Technological University, 72 Karl Marx St, Kazan, 420015, Russian Federation

[1]E-mail: nata-staroverova@yandex.ru

Abstract. The article considers the stages of regulators development based on machine learning. Machine learning, like a fuzzy logic, is used in PID regulators in two ways: it is able to build the regulator or a unit for adjusting its coefficients. A feature of machine algorithms is the "learning" ability which allows you to transfer the experience of an expert to the system. The presence of a learning procedure distinguishes the configuration of regulators based on machine algorithms from the classical approach. The publication analyzed the possibility of using regression machine algorithms (models) of the Scikit-learn library (Python) to solve control problems using the example of a level regulator. The search for an effective data processing model led to the construction of ensemble algorithms based on stacking. The high degree of compliance between predicted and test data confirms the accuracy and stability of the results. The results presented in the article can be used in a simulating or in developing a controller based on a machine algorithm.

1. Introduction

The economic development of modern society is impossible without the development of industrial production. It is important to ensure an increase in the volume of manufactured products, an increase in their quality characteristics, as well as ensuring the safety of the operation of enterprises.

A significant contribution to the achievement of these indicators is made by the development of production technologies, hardware and process control technologies.

An important role in ensuring the quality of control of technological processes is played by the laws of regulation, which are the basis for the functioning of control systems. Historically, proportional controllers were the first to be created. Soon it was supplemented with integral and differential components and PID controllers were appeared. Despite the fact that operation principles of PID controllers were developed at the beginning of the twentieth century, this type of regulator is still the most widely used in automated control systems. The popularity of PID controllers is due to their reliability, functionality, and simplicity. Other advantages are the small amount of overshoot and the relatively high speed of reaching the set point. However, there are also unresolved problems. It is necessary to eliminate the integral saturation. Controlling objects with hysteresis and nonlinearities is not always effective. The issues of automatic tuning and adaptation are worthy of special attention [1-3].

The growing level of development of modern technologies forms the trend of introducing intelligent technologies into production management processes. With an increasing in the volume of information flows, the degree of complexity of their processing are also increasing. The dynamics of modern control systems are often unknown. Processes requiring regulation cannot be called independent. The measurement results of key technological parameters are often very noisy, and the load on the system is not constant, which additionally causes certain difficulties. Often there are tasks that require non-standard solutions. These problems cannot always be solved by conventional algorithmic methods, and it is often impossible to find a clear solution algorithm. It is advisable to solve problems such as monitoring, control and prediction of process parameters, the direct measurement of which turns out to be expensive, difficult to implement or impossible. These factors determine the relevance of the development of adaptive regulators. Adaptive control can provide monitoring of regulator settings of continuously changing properties of control object, which allows improving quality of control. For example, in highly inertial processes, it is advisable to carry out operational correction of control signals long before changing process parameters.

Artificial intelligence technologies (machine algorithms, artificial neural networks, deep learning) are one of the key areas of development in this context. These technologies have become widely used both for the implementation of predictive analysis functionality and for ensuring optimal regulation of process parameters [4-5].

The relevance of a regulator based on artificial intelligence lies in its ability to provide maximum peak performance, high readiness for operation, the ability to predict events, as well as convenience in operation and maintenance, which is a significant advantage. One of the difficulties in developing adaptive regulators is the high degree of individualization of each regulator. The creation of an adaptive regulator includes the stages of determining a data set and parameters for conducting training, selecting a data processing model, conducting direct training, testing and implementing the regulator in a functioning system.

The purpose of this work is to determine an effective machine learning algorithm, as well as a data set and parameters for investigating the possibilities of machine learning in regulating technological processes.

2. Selecting and preparation data for learning

Regulator is designed for automatic level control in the separator condensate collection compartment (figure 1), which is a part of the complex gas treatment unit. Hydrocarbon liquid is fed to separator R-2 where it is separated into weathering gas, light phase flow and water-methanol solution. The adjustable parameter is the level in the separator controlled by the level meter LT-1 [5].

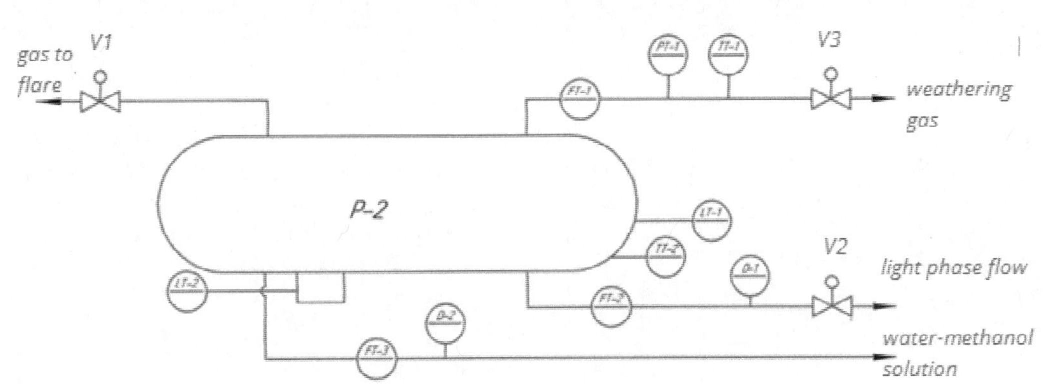

Figure 1. Regulation object scheme.

The analysis is carried out on data accumulated in the specified area during the year. The data for analysis is divided into two sets - for training and for testing.

According to classical control theory, level control is calculated based on the mismatch signal between the regulator set point and the current level value. However, in order to implement the machine learning strategy, it is advisable to consider at the initial stage all the technological parameters available for measurement in this area.

Training the model on a primary experimental data often does not produce the expected result. This fact can be explained by several reasons. The data for learning may contain inconsistent experimental values, i.e. the same values of independent parameter vectors have different values of dependent parameter vectors. Measurement errors and subjective errors can also lead to inconsistencies [6-8].

Another cause is a model parameter selection error if the model does not include meaningful input parameters. Such data strings affect learning speed, the magnitude of learning errors and generalizations, and may ultimately lead to inoperability of the constructed machine learning model. Therefore, this data should be excluded from the training sample. The values in the source database can be divided into three components: noise, non-informative values, and datum objects. The main task in data pre-processing is to eliminate noise and non-informative values and bring the reference parameters into normal form.

The process of excluding parameters is divided into two stages: the elimination of noise emissions and the elimination of uninformative peripheral objects. The process stops when only datum objects remain, the exclusion of which significantly increases the functionality. The correlation method allows you to clearly determine the dependence of the input, transition, and output values. it can be used to estimate the influence of parameters. The primary data analysis made it possible to determine the effect of each of the analyzed parameters on the valve opening degree. Diagrams of some parameters are shown in figure 2.

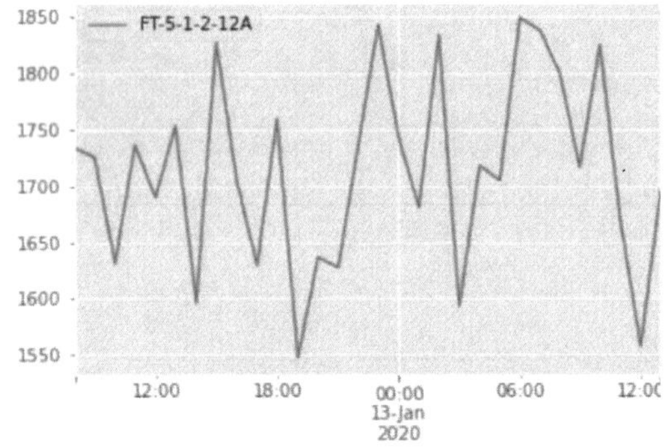

a

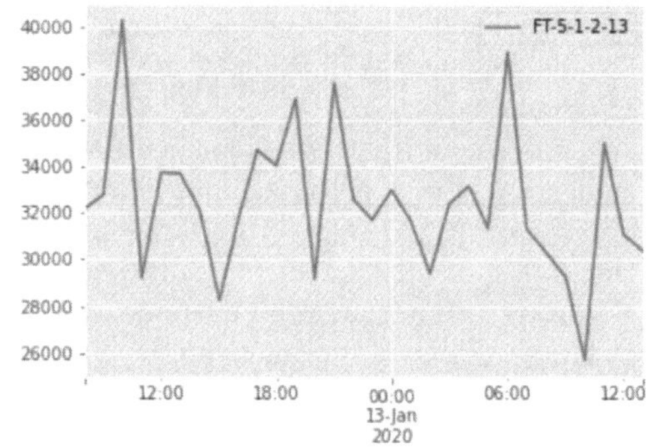

b

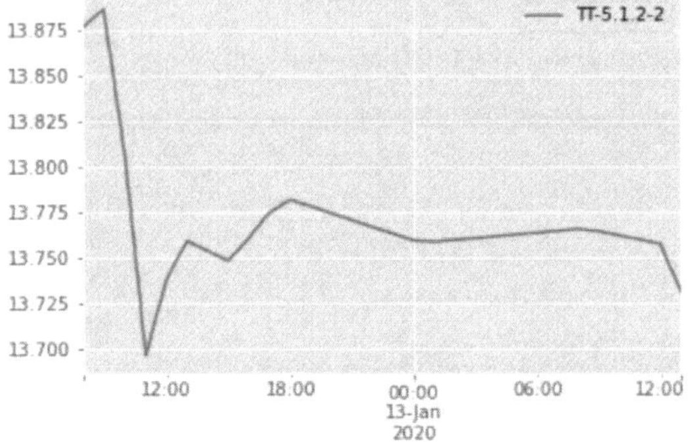

c

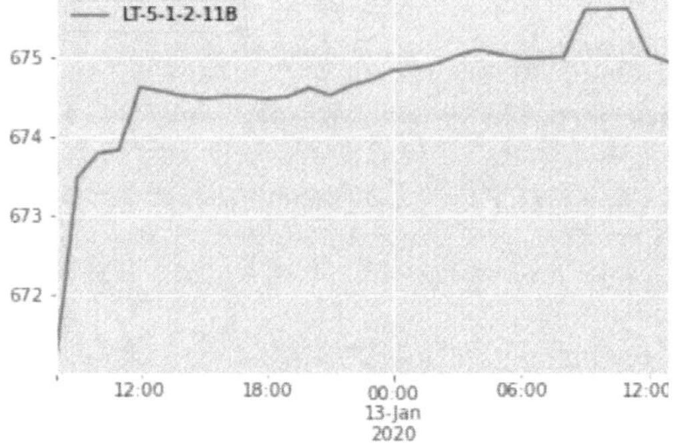

d

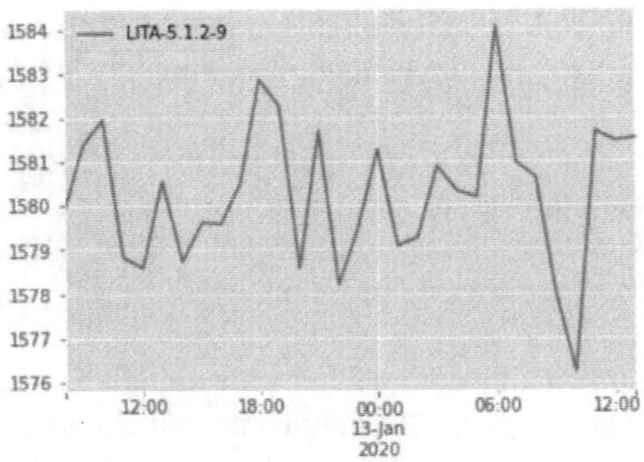

e

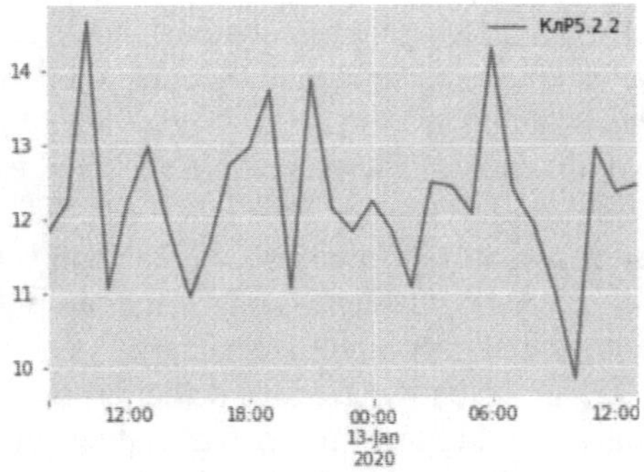

f

Figure 2. Evolution of key process parameters over time: a - gas flow rate in the line after the separator, FT-1, m3/h; b - hydrocarbon condensate flow rate, FT-2, m3/h; c- temperature in the separator, TT-2 0C; d - level in heavy phase collection compartment, LT-2, mm; e - level in the light phase compartment of the separator, LT-1, mm; f -valve-regulator, V-1 %.

Several dataframes with different sets were formed to analyze the joint effect of parameters on the output. In the next step this data frames were tested on regression models suitable for the forecasting problem. The influence of the analyzed technological parameters on the resulting variable (the state of the control valve) was assessed by analyzing their correlation (table 1). A correlation coefficient below 0.5 indicates almost no influence of the parameters on each other [9-10].

The process state of the valve depends directly on the level reading; the maximum level value corresponds to the maximum degree of valve opening. However, the heat map shows that this relationship is not so obvious between the valve and the level readings. However, there is a high correlation with hydrocarbon condensate flow from the separator. This is natural: this flow rate determines the increase in the level.

Table 1. Correlation between valve condition and sensor readings.

	LT-1	TT-2	FT-2	FT-3	FT-1	TT-1	V-1
LT-1	1.0000	-0.1519	0.6333	0.3282	0.5994	-0.2509	0.6150
TT-2	- 0.1519	1.0000	-0.4830	-0.5075	-0.2395	0.8295	-0.5683
FT-2	0.6333	-0.4830	1.0000	0.6293	0.4743	-0.5370	0.9663
FT-3	0.3282	-0.5075	0.6293	1.0000	0.0613	-0.4642	0.6548
FT-1	0.5994	-0.2395	0.4743	0.0613	1.0000	-0.3645	0.4978
TT-1	-0.2509	0.8295	-0.5370	-0.4642	-0.3645	1.0000	-0.5947
V-1	0.6150	-0.5683	0.9663	0.6548	0.4978	-0.5947	1.0000

3. Instruments and methods

Before starting work, it is advisable to normalize the data and bring them to a single dimensionless scale. There are several methods of normalization: decimal scaling, minimax normalization, and normalization using standard deviation.

Decimal scaling means dividing the original numbers by the maximum number of digits from the sample. The disadvantage of this method is the absence of any clear range of numbers. It will always spread up or down.

Minimax normalization eliminates the drawback of the previous example by bringing it to a certain interval [a, b]. Basically, intervals [0, 1], [-1, 1] are taken for its further use, for example, in activation functions in neural networks. This method is effective with relatively uniform value distributions. The presence of abnormal emissions in the sample can affect the entire distribution, even if these emissions do not significantly affect the dataset.

Normalization by standard deviation solves this problem by introducing an estimate of mean and variance. In this case, the average value and standard deviation of the sample are calculated, and then the data is converted for each individual sample value.

In the work, all three methods considered were analyzed to determine the method that is most effective within the given task.

The tools for solving the prediction problem are regression algorithms of the Scikit-learn library of the Python development environment. Linear algorithms were used in the task are Linear Regression (‘LR’), Ridge Regression (‘R’), Lasso Regression (‘L’) (LASSO — Least Absolute Shrinkage and Selection Operator), Elastic Net Regression (‘ELN’), Least Angle Regression (LARS) (‘LARS’), Bayesian ridge regression (‘BR’).

Non-linear algorithms were also used. There are k-nearest neighbors regressor (‘KNR’), Decision Tree Regressor (‘DTR’), Linear Support Vector Machine – Regression / (‘LSVR’), Epsilon-Support Vector Regression (‘SVR’) [9-13]

Ensemble algorithms considered in the research are AdaBoost Regressor (‘ABR’) (AdaBoost = Adaptive Boosting), Bagging Regressor (‘BR’) (Bagging = Bootstrap aggregating), Extra Trees Regressor (‘ETR’), Gradient Boosting Regressor (‘GBR’), Random Forest Classifier (‘RFR’).

Each model was tested on four datasets. Each dataset was used both without normalization and with three methods of normalization. Testing was carried out to determine an algorithm that would show consistently accurate results and be resistant to data noise.

Model testing results were performed by estimating the determination coefficient. For acceptable models, it is assumed that the determination coefficient should be not less than 50%. In this case, the absolute value of the multiple correlation coefficient exceeds 70%.

Models with a determination coefficient above 80% can be considered statistically valid. The variable is exactly described by the model if the determination coefficient is close to one.

4. Results

The efficiency of each algorithm is estimated by cross-validation. During testing, it was revealed that almost all models show accuracy above 0.8. The exception was Least Absolute Shrinkage and Selection Operator)/Lasso Regression ("L") and Elastic Net Regression ("ELN"). The Normalizer method, in turn, is not effective for Lasso Regression ("L") and Elastic Net Regression ("ELN"). It shows satisfactory results for other models.

The algorithms selected for further work meet a number of requirements: a stable and high result greater than 0.9, both on raw data and with all methods of normalization, and with any set of parameters. These features indicate a good generalizing ability of algorithms and their resistance to noise and outlier, which is an important property.

Such algorithms include linear algorithms - Least Angles Method ("LARSCV"), Bayesian ridge regression ("BR"), nonlinear algorithms - k-closest neighbour method (regression) ("KNR"), Regression trees ("DTR"), Reference vector method (regression) ("SVR") and ensemble algorithms - ("BR") (Bootstrap aggregating), Extra Trees Regressor, Gradient Boosting Regressor ("GBR"), Random Forest Classifier ("RFR").

The accuracy in 0.9 is considered quite sufficient. Therefore, an ensemble algorithm Random Forest Classifier ("RFR"), which showed an average determination coefficient of 0.97, could be recommended to predict the degree of valve opening in this process scheme. However, in addition to accuracy, it is advisable to achieve greater stability of the prediction process in critical situations, for example, at outlier points in parameter readings. Stacking allows to combine several weaker algorithms and train a final estimator on their results. Any algorithm, or neural network, can be used as a final estimator. Linear and nonlinear regression algorithms ('LARSCV', 'BR', 'KNR', 'DTR', 'SVR') were used as an "ensemble of weak students. As a final estimator, ensemble algorithms (table 2) with the largest determination coefficients are adopted. Standard error (mse) is used to estimate the response to outliers, mae is an estimate of absolute prediction error. The correspondence of the predicted data and the original information is estimated by the determination coefficient.

Table 2. Staking test results with different final estimators.

final estimator	determination coefficient	mae; mse
RandomForestRegressor (max_depth = None, n_estimators = 700), figure 3 - a	0.988	0.2677; 0.1675
BaggingRegressor (n_estimators=700), figure 3 –b	0.987	0.2694; 0.1753
ExtraTreesRegressor (n_estimators=700), figure 3 –c	0.988	0.2704; 0.1686
GradientBoostingRegressor (n_estimators=700) figure 3 –d	0.988	0.2695; 0.1648

All ensemble sets were analyzed with different levels of noise, i.e. with 3 sets of parameters tied to one valve and showed satisfactory results (figure 3), not much different from the results in table 1. The influence of the "set" of parameters on the accuracy of the results will be discussed in more detail in a separate article.

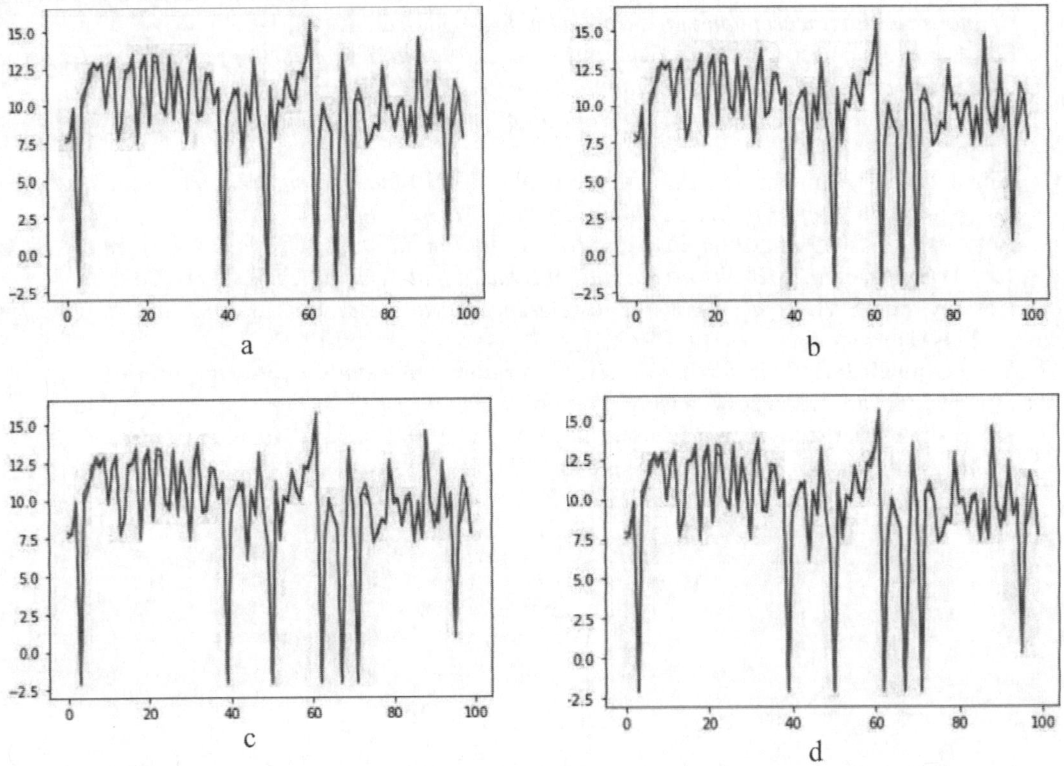

Figure 3. Staking test results with different final estimators: agreement of the predicted (red) and test (green) predicate.

The coefficient of determination of the obtained ensemble sets is 0.98, which shows the sufficient predictive ability of the model. In addition, the absolute and root-mean-square errors are small, which proves that most of the model's predictions correspond to the actual values of the test sample. This effect is also noticeable in figure 3: small deviations are present only at the peak values of the predicted and testing data, in other positions there is a coincidence of parameters.

5. Conclusion

The article provides analysis of annual cycle data of gas condensate treatment shop separators unit. The data were cleared and prepared for machine learning. The results obtained by testing 12 regression algorithms conclude that machine regression algorithms show good accuracy and reproducibility in predicting the degree of valve opening within the set task.

In this example, ensemble stacking with linear and nonlinear regression algorithms ('LARSCV', 'BR', 'KNR', 'DTR', 'SVR ') turned out to be the most effective. Among final estimators, the best results were demonstrated by a final estimator including ‘Random Forest Regressor’, ‘Bagging Regressor’, ‘Extra Trees Regressor’, ‘Gradient Boosting Regressor’. In each case, we have a determination coefficient of 0.98 These results will be used to analyze the effect of various parameters on the stability of the model forecast, as well as to build regulators based on them.

References

[1] Jimmi W Key 2016 *Control engineering Russia* **63(3)** 62-6
[2] Jimmi W Key 2016 *Control engineering Russia* **64(4)** 106-10
[3] Gaynullin R, Zatsarinnaya Y and Staroverova N 2019 *IOP Conference Series: Materials Science and Engineering* **537(3)** 032019
[4] Mukhamadiev R, Staroverova N and Shustrova M 2020 *International Multi-Conference on Industrial Engineering and Modern Technologies (FarEastCon)*, 1-8
[5] Shustrova M, Fafurin A and Baytimirov A 2019 *International Conference on Industrial Engineering Applications and Manufacturing ICIEAM* 8743033
[6] Trawiński B, *et al.* 2012 *International Journal of Applied Mathematics and Computer Science* **22** 867-81
[7] Gibadullin R, Baimukhametova G and Perukhin M 2019 *International Conference on Industrial Engineering Applications and Manufacturing ICIEAM* 8743064
[8] Botchkarev A 2019 *Interdisciplinary Journal of Information, Knowledge & Management* **14** 35
[9] Järvelä Sanna, *et al.* 2020 *British Journal of Educational Technology* **51** 2391-406
[10] Kosarevskaia S A, *et al.* 2021 *IOP Conference Series: Materials Science and Engineering* **1047(1)**
[11] Yang Qijiu 2020 *IOP Conference Series: Materials Science and Engineering* **782(3)**
[12] Ohsugi Hideharu, et al. 2017 *Scientific reports* **7(1)** 1-4
[13] Kangash A, Maryandyshev P, Zatsarinnaya Y and Volkova M 2019 *IOP Conference Series: Materials Science and Engineering* **643(1)** 01215

Direct numerical simulation of anisotropic turbulent flow for incompressible fluid

A M Balonishnikov[1,3] and J V Kruchkova[2]

[1] Department of Higher Mathematics and Mechanics, Saint Petersburg University of Aerospace Instrumentation, 67 Bolshaya Morskaya, SPb 190000, Russia
[2] Department of Foreign Languages, Saint Petersburg University of the Ministry of Internal Affairs of Russia, 1 Pilot Pilyotov str., SPb 198206, Russia

[3] E-mail: balonishnikov@mail.ru

Abstract. Equation for small-scale velocity or pulsation is a starting point to build almost all models of turbulence. Transport equation for Reynolds stresses, dissipation rate, kinetic energy are derived from this equation, adding some assumptions about structure of terms, including in these equations. Equation for small-scale velocity can be simplified, if we assume that large – scale velocity and its gradients are constants instead of linear profile for large-scale velocity. We implemented the direct numerical simulation of this equation under the simple shear, leading to anisotropy. The nonlinear helicity terms were computed in spectral space, using the three-dimensional Fast Fourier transformation, then, the inverse Fast Fourier transformation was used to return in physical space. Aliasing terms were not removed. Four – order Runge- Kutta method was used for integration in time. Evolution of Reynolds stresses in time were computed.

1. Introduction

A basis for direct numerical simulation of fully developed turbulence was laid down in [1, 2]. Numerical solutions of Navier - Stokes equations for incompressible fluids with more powerful computers have been investigated in [3-5]. The pumping of energy in the isotropic flow was implemented by some random force with known statistical properties.

There is a more realistic picture, when energy from large-scale flow is transferred by an anisotropic shear term to small-scale eddies with a size less than the integral scale of turbulence. There are the results of the simplified models of anisotropic cascades [6] and the direct numerical simulation of anisotropic turbulence [7]. The prevailing assumption of the anisotopic turbulence simulation is a linear profile of large-scale or mean velocity [7]. A simplified form of the small-scale velocity transport equation, when large-scale velocity and its gradients were considered as constants, was proposed in [8]. The simplicity of this equation should lead to a better analytical analysis. Thus, the motivation of our simulation is to reveal unexplored properties of a new transport equation.

What is known about the simulation of small-scale velocity, when large-scale velocity has linear profile? The next system of differential equations was derived from Navier –Stokes equations [7]:

$$(\partial_t + x_2\partial_1)u_1 + u_2 - H_1 + \partial_1 p = \varepsilon\Delta u_1,$$

$$(\partial_t + x_2\partial_1)u_2 - H_2 + \partial_2 p = \varepsilon\Delta u_2,$$

$$(\partial_t + x_2\partial_1)u_3 - H_3 + \partial_3 p = \varepsilon\Delta u_3,$$

$$\partial_j u_j = 0,$$

where $H_i = -u_j\partial_j u_i$ - nonlinear terms, p – kinematics pressure, lower index corresponds to three spatial directions, the summing over repeating indexes is assumed. The equations are transformed to dimensionless form by means of the velocity tensor – gradient invariant S and the length L , which is equal to integral scale. Large-scale velocity is $\mathbf{U} = (x_2, 0, 0)$ in dimensionless variables. Flow depends on the Rossby number,

$$Ro = \frac{u}{SL},$$

where u is the small-scale velocity scale and Rossby number

$$\varepsilon = \frac{\nu}{SL^2},$$

where ν is viscosity, $\mathrm{Re} = \frac{1}{\varepsilon}$ is Reynolds number.

Unfortunately, results were given only for two -dimensional flow. In paper [9], the system of ordinary differential equation for the velocity Fourier - components was solved at the grid with $512\times512\times512$ points under the action of simple shear. The integration in time was carried out up to $St = 2$.

An agreement with Kolmogorov energy spectrum ($\propto k^{-5/3}$) and Lumley spectrum ($\propto k^{-7/3}$) for off-diagonal components E_{ij}($i \neq j$) was achieved. Recently, an experimental paper was published with a measurement of all components E_{ij} [10], a comparison with the dns results is not available.

This system of differential equation was numerically solved also in [11], aperiodic oscillations of energy and Reynolds's stress were revealed. An anisotropy, caused by the gradients of mean velocity, was investigated in [12].

Velocity $\mathbf{v}$ was represented as a sun of small-scale velocity $\mathbf{u}$ and large-scale velocity $\mathbf{U}$:

$$\mathbf{v} = \mathbf{u} + \mathbf{U}.$$

We obtain the next equation from Navier- Stokes equation for incompressible fluid:

$$\partial_t u_\alpha + u_\beta\partial_\beta u_\alpha + U_\beta\partial_\beta u_\alpha + u_\beta\partial_\beta U_\alpha = -\partial_\alpha p + \nu\Delta u_\alpha$$

Authors of the paper [12] considered shear $\partial_\beta U_\alpha$ as anisotropic random forces:

$$\mathbf{f} = (0, 0, f_z(x)),$$

where $f_z(x) = F_1\cos(2\pi x/L_x + \phi_1(t)) + F_2\cos(2\pi x/L_x + \phi_2(t))$, F_1 and F_2 - are constants, random fazes $\phi_1(t)$ and $\phi_2(t)$ are mutually independent, homogeneous distributed, δ - correlated in time. We can transit to the coordinate system, moving with local velocity $\mathbf{U}$, and nullify corresponding term in the transport equation. The inverse transformation in the initial coordinate system, being at rest, is given by Galileo transformation [13]:

$$u_i(\mathbf{k},t)_U = \exp[-ik_j U_j(t - t_0 0]u_i(\mathbf{k},t)]$$

The results of simulation are in agreement with Kolmogorov's and Lumley's spectra.

2. Transport equation for polarization components of small-scale velocity

Using the approach [3], let us transform the small-scale velocity $\mathbf{v}(\mathbf{x},t)=(v_1,v_2,v_3)$ in Fourier series in spatial variables:

$$\mathbf{v}(\mathbf{x},t)=\sum_{\mathbf{k}}\mathbf{v}_{\mathbf{k}}(t)\exp(i\mathbf{k}\cdot\mathbf{x}), \tag{1}$$

where $i=\sqrt{-1}$, $\mathbf{k}=(2\pi/L)\mathbf{n}$ - is the wave vector, $\mathbf{n}=(n_1,n_2,n_3)$ - is a vector with the integer components, here $L-$ is the integral scale of turbulence. We should consider $\mathbf{n}$ in numerical simulation as finite values:

$$-\tfrac{N}{2}\le n_i<\tfrac{N}{2}, \quad -(2\pi/L)\tfrac{N}{2}\le k_i<(2\pi/L)\tfrac{N}{2}, \tag{2}$$

where $i=1,2,3$ and N is a sufficient large number, being a positive integer power of 2 for using Fast Fourier Transform. These inequality can be rewrote as $|\mathbf{n}|<\tfrac{N}{2}\equiv K$ or $|\mathbf{k}|<k_{\max}\equiv(2\pi/L)K$.

We seek the solution as:

$$\mathbf{v}(\mathbf{x},t)=\sum_{|\mathbf{k}|<k_{\max}}\mathbf{v}_{\mathbf{k}}(t)\exp(i\mathbf{k}\cdot\mathbf{x}) \cdot \tag{3}$$

Let us define the collocation points as the following ones:

$$\mathbf{x}_{\mathbf{j}}=\frac{L}{N}(j_1,j_2,j_3), \quad |\mathbf{j}|<N/2. \tag{4}$$

Then, an approximate solution (3) of the small-scale transport equation in the collocation points shall be:

$$\mathbf{v}(\mathbf{x}_{\mathbf{j}},t)=\sum_{|\mathbf{k}|<k_{\max}}\mathbf{v}_{\mathbf{k}}(t)\exp(i\mathbf{k}\cdot\mathbf{x}_{\mathbf{j}}). \tag{5}$$

The inverse Fourier transform with finite number of terms shall be:

$$\mathbf{v}_k(t)=\frac{1}{N^3}\sum_{|\mathbf{j}|<K}\mathbf{v}(\mathbf{x}_j,t)\exp(-i\mathbf{k}\cdot\mathbf{x}_{\mathbf{j}}) \tag{6}$$

Due to the wave number cutoff, the nonlinear terms of the transport equation will have the wave numbers out of the range (2), it is so-called aliasing error. Since the small-scale velocity $\mathbf{v}(\mathbf{x},t)$ has the real components, there is an equality:

$$\mathbf{v}_{\mathbf{k}}=\mathbf{v}^*_{-\mathbf{k}} \tag{7}$$

Vorticity is defined as

$$\omega=\nabla\times\mathbf{v}(\mathbf{x},t), \tag{8}$$

kinetic energy of small-scale eddies is equal to:

$$E=\tfrac{1}{2}\mathbf{v}(\mathbf{x},t)\cdot\mathbf{v}(\mathbf{x},t), \tag{9}$$

subgrid Reynolds tensor is defined as:

$$R_{ij}=v_i(\mathbf{x},t)v_j(\mathbf{x},t) \tag{10}$$

Vorticity ω has an expansion into discrete Fourier series as:

$$\omega(\mathbf{x}_j,t)=\sum_{\mathbf{k}} i\mathbf{k}\times\mathbf{v}_{\mathbf{k}}\exp(i\mathbf{k}\cdot\mathbf{x}_j) \tag{11}$$

Kinetic energy (9) and subgrid Reynolds tensor (10) have the similar expansion into Fourier series.

Combining the results of [3,8], transport equations for two polarization Fourier components of the small-scale velocity will be for anisotropic turbulence as:

$$\partial_t u_1 + iU_j k_j u_1 = a_{1\mu}u_\mu - \frac{\varepsilon^1\cdot[\mathbf{k}\times(\mathbf{k}\times\mathbf{s}_{\mathbf{k}})]}{k^2} - \nu k^2 u_1 \tag{12}$$

$$\partial_t u_2 + iU_j k_j u_2 = a_{2\mu}u_\mu - \frac{\varepsilon^2\cdot[\mathbf{k}\times(\mathbf{k}\times\mathbf{s}_{\mathbf{k}})]}{k^2} - \nu k^2 u_2, \tag{13}$$

where summation over repeating indices is assumed $(j;\, j=1,2,3)$, $(\mu;\, \mu=1,2)$, $u_1(\mathbf{k},t),\, u_2(\mathbf{k},t)$ are two polarization components of the small-scale velocity:

$$\mathbf{v}(\mathbf{k},t)=u_1\mathbf{e}^1+u_2\mathbf{e}^2;\quad u_1=\mathbf{v}\cdot\mathbf{e}^1,\, u_2=\mathbf{v}\cdot\mathbf{e}^2\ , \tag{14}$$

directing along the unit polarization vectors $\mathbf{e}^1$ and $\mathbf{e}^2$, these are orthogonal to each other and to vector $\mathbf{k}(k_1,k_2,k_3)$, its components vary from $k_{\max}=2\pi/L$ ($L-$ integral scale of turbulence, separating large-scale motion and small-scale eddies) up to some $k_{\min}$ (a theoretical value $k_{\min}=\infty$ and $k_{\min}$ in computations is limited by the computer capacity), $i=\sqrt{-1}$, helicity $\mathbf{s}=\mathbf{u}\times\omega$, $\omega=\nabla\times\mathbf{u}$ is vorticity, $\mathbf{U}$ is large-scale velocity, in moving coordinate system $\mathbf{U}=0$, four coefficients are defined as $a_{\lambda\mu}=-e_j^\gamma e_m^\mu \partial_m U_j$, ν - molecular viscosity. An option of unite vectors $\mathbf{e}^1$ and $\mathbf{e}^2$ is sufficiently arbitrary.

The components of strain rate tensor are defined as:

$$S_{ij}=\frac{1}{2}(\partial_i U_j+\partial_j U_i),$$

one of its invariant $S=\sqrt{2S_{ij}S_{ji}}$.

Let us define the unit vector $\mathbf{e}$, as:

$$\mathbf{e}=(\sqrt{2},\quad \pi,\quad 0)\ \cdot\frac{1}{\sqrt{2+\pi^2}}. \tag{15}$$

After that, we define the unit vectors $\mathbf{e}^1$ and $\mathbf{e}^2$:

$$\mathbf{e}^1=\frac{\mathbf{e}\times\mathbf{k}}{|\mathbf{e}\times\mathbf{k}|}, \tag{16}$$

(vector $\mathbf{e}$ in (15) are chosen as nonparallel to vector $\mathbf{k}$, given in three-dimensional rectangular grid, used for numerical simulation)

$$\mathbf{e}^2=\frac{\mathbf{k}\times\mathbf{e}^1}{|\mathbf{k}\times\mathbf{e}^1|}=\frac{\mathbf{k}\times\mathbf{e}^1}{k}. \tag{17}$$

It is clear that the denominator in (16) and (17) are not zeros. We have a relation, following from (11), that $\omega(\mathbf{k},t) = i\mathbf{k}\times\mathbf{v}(\mathbf{k},t)$. Also, we obtain from formulas (14)-(17), that:

$$\omega(\mathbf{k},t) = ik(u_1\mathbf{e}^2 - u_2\mathbf{e}^1) \tag{18}$$

We have the only nonzero component of velocity tensor- gradient for simple shear:

$$\partial_1 U_2 = S \neq 0$$

It can be easily obtained for this case: $a_{11} = -e_2^1 e_1^1 S$, $a_{12} = -e_2^1 e_1^2 S$, $a_{21} = -e_2^2 e_1^1 S$, $a_{22} = -e_2^2 e_1^2 S$.

The system of equations (12), (13) turns into the next system:

$$\partial_t u_1 = -e_2^1 e_1^1 S u_1 - e_2^1 e_1^2 S u_2 - \frac{\mathbf{e}^1\cdot[\mathbf{k}\times(\mathbf{k}\times\mathbf{S}_k)]}{k^2} - \nu k^2 u_1 \tag{19}$$

$$\partial_t u_2 = -e_2^2 e_1^1 S u_1 - e_2^2 e_1^2 S u_2 - \frac{\mathbf{e}^2\cdot[\mathbf{k}\times(\mathbf{k}\times\mathbf{S}_k)]}{k^2} - \nu k^2 u_2 \tag{20}$$

Using well-known formula for double vector product: $\mathbf{A}\times(\mathbf{B}\times\mathbf{C}) = \mathbf{B}(\mathbf{A}\cdot\mathbf{C}) - \mathbf{C}(\mathbf{A}\cdot\mathbf{B})$, where $\mathbf{A},\mathbf{B},\mathbf{C}$ - arbitrary vectors.

We deduce: $\mathbf{k}\times(\mathbf{k}\times\mathbf{S}_k) = \mathbf{k}(\mathbf{k}\cdot\mathbf{S}_k) - \mathbf{S}_k k^2$.

In result, we obtain the next system of equations from the system (19) and (20):

$$\partial_t u_1 = -e_2^1 e_1^1 S u_1 - e_2^1 e_1^2 S u_2 - \frac{(\mathbf{e}^1\cdot\mathbf{k})(\mathbf{k}\cdot\mathbf{S}_k) - k^2(\mathbf{e}^1\cdot\mathbf{S}_k)}{k^2} - \nu k^2 u_1 \tag{21}$$

$$\partial_t u_2 = -e_2^2 e_1^1 S u_1 - e_2^2 e_1^2 S u_2 - \frac{(\mathbf{e}^2\cdot\mathbf{k})(\mathbf{k}\cdot\mathbf{S}_k) - k^2(\mathbf{e}^2\cdot\mathbf{S}_k)}{k^2} - \nu k^2 u_2 . \tag{22}$$

Using expressions (3)-(5), we can write down the explicit expressions for the unit vectors in Cartesian coordinate system:

$$\mathbf{e}^1 = \frac{(e_2k_3 - e_3k_2)\mathbf{i}_1 - (e_1k_3 - e_3k_1)\mathbf{i}_2 + (e_1k_2 - e_2k_1)\mathbf{i}_3}{\sqrt{(e_2k_3 - e_3k_2)^2 + (e_1k_3 - e_3k_1)^2 + (e_1k_2 - e_2k_1)^2}} \tag{23}$$

and, taking into account (15), we have:

$$\mathbf{e}^1 = \frac{e_2k_3\mathbf{i}_1 - e_1k_3\mathbf{i}_2 + (e_1k_2 - e_2k_1)\mathbf{i}_3}{\sqrt{k_3{}^2 + (e_1k_2 - e_2k_1)^2}}, \tag{24}$$

for unit vector $\mathbf{e}^2$, taking into account the mutual orthogonality vector $\mathbf{e}^1$ and vector $\mathbf{k}$, we obtain the relation:

$$\mathbf{e}^2 = \frac{(k_2e_3^1 - k_3e_2^1)\mathbf{i}_1 - (k_1e_3^1 - k_3e_1^1)\mathbf{i}_2 + (k_1e_2^1 - k_2e_1^1)\mathbf{i}_3}{k} \tag{25}$$

Also, taking into account orthogonality of $\mathbf{e}_1$ and $\mathbf{e}_2$ to vector $\mathbf{k}$: $\mathbf{e}_1\cdot\mathbf{k} = \mathbf{e}_2\cdot\mathbf{k} = 0$, system of equation (21), (22) simplifies:

$$\partial_t u_1 = -e_2^1 e_1^1 S u_1 - e_2^1 e_1^2 S u_2 + \mathbf{e}^1\cdot\mathbf{S}_\mathbf{k} - \nu k^2 u_1 \tag{26}$$

$$\partial_t u_2 = -e_2^2 e_1^1 S u_1 - e_2^2 e_1^2 S u_2 + \mathbf{e}^2 \cdot \mathbf{S}_\mathbf{k} - \nu k^2 u_2 \tag{27}$$

Thus, for given vector $\mathbf{k}(k_1, k_2, k_3)$ we calculate vector $\mathbf{e}^1$, using (16), and vector $\mathbf{e}^2$, using (17).

Let us make a transition to dimensionless variables, using time scale $(1/S)$ and length scale (L):

$$tS \to t, \quad kL \to k, \quad u_{\lambda\mu}/(LS) \to u_{\lambda\mu}, \quad \mathbf{S}_\mathbf{k}/(L S^2) \to \mathbf{s}_\mathbf{k}, \quad \nu \to 1/\mathrm{Re}_s.$$

where $\mathrm{Re}_s = \dfrac{SL^2}{\nu}$ is subgrid Reynolds number, the only parameter in the system.

The system equation (26) and (27) in dimensionless variables becomes as the following:

$$\partial_t u_1 = -e_2^1 e_1^1 u_1 - e_2^1 e_1^2 u_2 + \mathbf{e}^1 \cdot \mathbf{s}_\mathbf{k} - \frac{1}{\mathrm{Re}_s} k^2 u_1 \tag{28}$$

$$\partial_t u_2 = -e_2^2 e_1^1 u_1 - e_2^2 e_1^2 u_2 + \mathbf{e}^2 \cdot \mathbf{s}_\mathbf{k} - \frac{1}{\mathrm{Re}_s} k^2 u_2 \tag{29}$$

It seems that we obtained the simplest description of anisotropic turbulence here. All nonlinearity is in Fourier components of helicity: $\mathbf{S} = \mathbf{v} \times \omega$.

3. Choice of equation's parameters

Parameter L is a spatial scale, dividing large and small fluid motion, $k_0 = 2\pi / L$; $L = 1 -$ expressed in dimensionless variables. The dissipation rate ε is part of Kolmogorov's energy spectrum $E(k)$:

$$E(k) = C_K \varepsilon^{2/3} k^{-5/3}, \tag{30}$$

where $C_K \approx 1.5$ [14].

The formula (30) is used for setting up the initial conditions for small – scale polarization components of velocity ($\varepsilon = 1$ as dimensionless variable):

$$< |u_1|^2 + |u_2|^2 >= E(k)/(4\pi k^2)$$

A computational grid in space of the wave numbers $\mathbf{k}$ is defined as: $(k_{\max}/k_0) \times (k_{\max}/k_0) \times (k_{\max}/k_0)$, where the choice of $k_{\max}$ is limited of our computer capacity.

The fast Fourier transform is usually used for $(k_{\max}/k_0) = 2^n$.

The setting up of initial condition can be defined, following to [15] as:

$$u_\alpha(\mathbf{k}.0) = (\delta_{\alpha\gamma} - k_\alpha k_\gamma / k^2) r_\gamma(\mathbf{k},0),$$

$$\text{or } u_1 = \pm u_2 \quad u_2 = \pm \tfrac{1}{2} E(k)/(4\pi k^2),$$

where $r_\gamma(\mathbf{k},0)$ - statistically independent random variables with Gaussian distribution with zero mean and dispersion, being proportional to energy spectrum (30).

Following to [3], to compute $\mathbf{s}_\mathbf{k}$ and $E_\mathbf{k}$ (which are Fourier Images of $\mathbf{u} \times \omega(\mathbf{x}_\mathrm{j})$ and $\frac{1}{2}\mathbf{u} \cdot \mathbf{u}(\mathbf{x}_\mathrm{j})$ respectively) we compute Fourier vorticity: $\omega_\mathbf{k} = i\mathbf{k} \times \mathbf{u}_\mathbf{k}$, which transforms in physical space to obtain $\mathbf{u}(\mathbf{x}_\mathrm{j})$ and $\omega(\mathbf{x}_\mathrm{j})$. Then, we compute variables $\mathbf{s}(\mathbf{x}_\mathrm{j}) = \mathbf{u}(\mathbf{x}_\mathrm{j}) \times \omega(\mathbf{x}_\mathrm{j})$ and $E(\mathbf{x}_j) = \frac{1}{2}\mathbf{u}(\mathbf{x}_\mathrm{j}) \cdot \mathbf{u}(\mathbf{x}_\mathrm{j})$

. At last, we come back in wave-numbers space with help of Inverse Fourier transform. The Fast three dimensional Fourier Transform is used. The integration in time is implemented with help of the explicit 4-th order Runge - Kutta method.

4. The computational results and discussion

Here, we give preliminary results of dns for anisotropic incompressible turbulence, using a new simplified transport equation for the small-scale velocity [8]. We used 8x8x8 grid of the wave numbers. Subgrid Reynolds number was $\mathrm{Re}_s = 10$. The integration in time was stopped at $St = 3$. The averaging over the random initial conditions is absent. We give a typical evolution of the polarization Fourier velocity components.

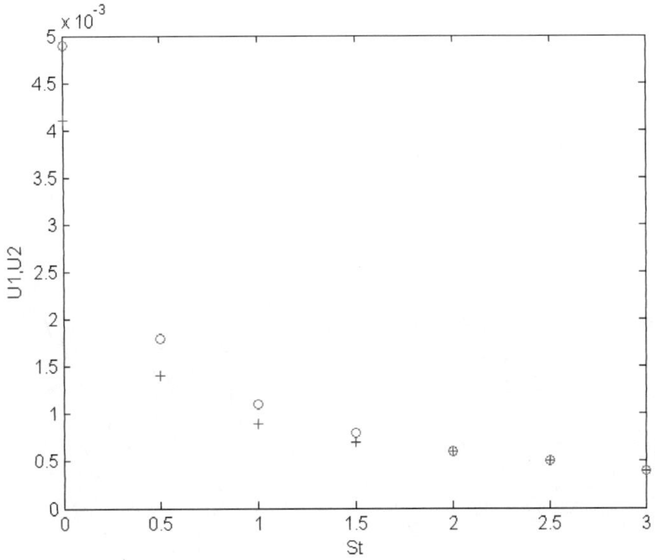

Figure 1. Evolution of the polarization Fourier velocity components in time; $+, U1 = \sum_k |u_1|^2$; $o, U2 = \sum_k |u_2|^2$; St - dimensionless time.

We see from figure 1 a tendency to isotropy. Also, we can describe the process as self- sustained one.

References

[1] Patterson G S and Orszag S A 1971 *Physics of Fluids* **14** 2538-41
[2] Patterson G S and Orszag S A 1972 *Physical Review Letters* **28** 76-9
[3] Schumann U, Grotzbach G and Kleiser L 1980 *Predicted methods for turbulent flow* ed W Kollman (New –York: Hemisphere publ. corp.) p 103-227
[4] Kaneda Y and Ishihara T 2006 *Journal of Turbulence* **7** 20
[5] Ishihara T, Gotoh T and Kaneda Y 2009 *Annu. Rev. Fluid Mech.* 165-80
[6] Bolotnov I A, Lahey Jr R T , Drew D A, Jansen K E and Oberai A A 2008 *Journal of Turbulence* **8(26)**
[7] Jimenez J 2007 Simulation of homogeneous and inhomogeneous shear turbulence *Center for Turbulent Research Annual Research Briefs*. 367-78
[8] Balonishnikov A M 2003 *Technical Physics* **73(10)** 36-9
[9] Ishihara T, Yoshida K and Kaneda Y 2002 *Phys. Rev. Lett.* **88(15)** 154501
[10] Elsinga G E and Marusic I 2016 *Physics of Fluids*. **28** 011701

[11] Mamatsashvili G, Khujatze G, Chagelishvili G, Dong S, Jimenez J and Foysi H 2016 *Phys. Rev. E* **94** 023111
[12] Biferale L, Daumont I, Lanotte A and Toschi F 2004 *European Journal of Mechanics B/Fluids* **23** 401-14
[13] Pandya R V R 2013 A new Eulerian theory of turbulence constrained by random Galilean invariance arxiv 1307.2324 16
[14] Sreenivasan K P 1995 *Physics of Fluids* **7** **(11)** 2778-84
[15] Orszag S A 1977 *Handbook of turbulence (Fundamentals and Application* **1**) ed W Frost and T H Moulden (New -York: Plenum Press) p 311

Simulation modelling of the filling oil products process in the arctic zone using neural network forecasting

Ya V Grebnev[1,2] and A K Moskalev[1]

[1] Siberian Federal University, 79 Svobodny pr., 660041 Krasnoyarsk, Russia
[2] Emergency Forecasting Department, EMERCOM of Russia for the Krasnoyarsk Territory, 68 Mira Ave., Krasnoyarsk, 660049, Russia

E-mail: yaroslav.grebnev@gmail.com, ak_moskalev@mail.ru

Abstract. The intensification of the extraction of natural resources in the Arctic leads to an increase in the anthropogenic load on the environment. Due to wear and tear of equipment, corrosion of metals, mechanical damage, due to lack of proper maintenance and prolongation of the operating time in difficult climatic conditions, the operation of equipment used at enterprises leads to an increase in the risk of accidents. On the territory of the Russian Arctic, there are a number of large potentially dangerous objects, which have over 60 tanks for storing petroleum products. In 2020, there were technological accidents related to the bottling of petroleum products, which actualized the problem of taking prompt measures to prevent such emergencies. The methods for assessing the area of bottling of petroleum products currently used, especially in the Arctic zone, have a number of limitations. In the work with the use of the software product Toxy + risk, the risk of an emergency at one of the enterprises of the Russian Arctic was calculated. Neural network prediction of the area of pollution of the surface of the Kheta River was carried out using the NeuroPro neural network simulator. Comparison of the simulation results with the data obtained earlier in the analysis of the accident in the tank farm located beyond the Arctic Circle in the city of Norilsk.

1. Introduction

The Arctic territory of the Krasnoyarsk Territory is a territory of intensive extraction of natural resources, metal production and other types of industrial activities that threaten the natural ecosystem of this unique region. One of the most dangerous and widespread pollutants in the ecosystem of the North is oil products used for heating, transport and industrial equipment. Despite the measures taken in the field of technosphere and industrial safety, accidents continue to occur, and their scale is increasing every year. So in 2020, in the Arctic zone of the Krasnoyarsk Territory, there were three technogenic accidents associated with the bottling of oil products, which caused irreparable damage to the ecology of the region and affected the production processes at enterprises.

Routine operations at the oil depot with petroleum products are as follows:

- pumping from petroleum products from a barge to storage tanks;
- storing petroleum products in tanks;
- supplying petroleum products to tank trucks;
- bunkering of vessels, loading oil products into tankers and barges;

- intra-base pumping;
- accounting of received and released oil products.

Storage of oil products is carried out in vertical steel tanks, separate for each brand of oil products. The release of oil products is carried out in tank trucks, tankers and barges, as well as the tank farm carries out bunkering of ships.

The most dangerous places of accidents are the points of loading of oil products and unloading of tankers and places with navigational hazards (shallow water, ridging and compression of ice, etc.).

Figure 1 shows various scenarios leading to an accidental oil spill from a reservoir.

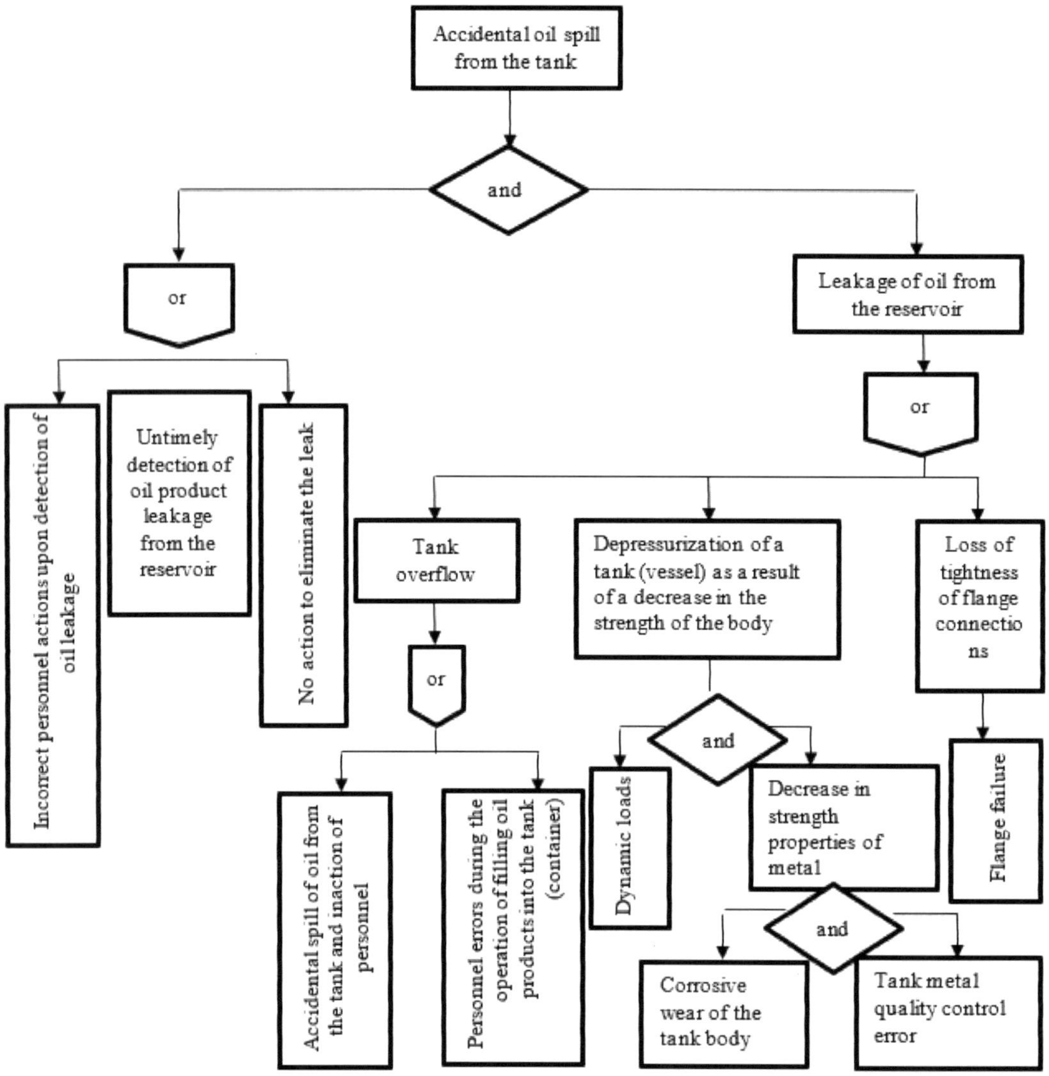

Figure 1. Fault tree.

The most dangerous scenarios for an accident development at a tank farm are accidents in a tank farm for storing petroleum products. The probable causes of emergencies include:

- incidents in production processes, problems with power supply, hydraulic shocks, etc.;

- human factor, violation of labor protection rules, errors in the operation of equipment, etc.;
- failures of technological equipment, these accidents can occur as a result of wear and tear of metal products, factory defects, poor-quality construction and installation work, including defective welding;
- factors of a natural and man-made nature, earthquakes and landslides that violate the tightness of tank equipment, thunderstorms, strong winds and hurricanes, the impact of flood waters, the effect of damaging factors from accidents that occurred at objects and installations located nearby;
- deliberate illegal actions of people.

A large number of works are devoted to solving the problem of bottling oil products on soil and in water bodies [1-3]. However, a universal approach to solving this problem has not yet been developed [4-6]. As part of the implementation of the program for the sustainable development of the Arctic region and the adoption of preventive measures to prevent man-made accidents in this region, the assessment of the risk of emergencies associated with the bottling of oil products remains an urgent task. One of the effective means of solving the complex problem of risk forecasting is the widespread implementation of monitoring and data processing systems. One of the effective means of solving this problem is the use of big data technologies and neural network forecasting, which are successfully used in various branches of science and technology. In this work, we set the task: based on the data obtained in the analysis of the consequences of the accident that occurred in 2020 at one of the enterprises of the Arctic zone of the Krasnoyarsk Territory, to assess the risk of a man-made accident and, using neural network algorithms, to simulate its consequences.

2. Materials and methods

The tasks set in the work: assessment of the likelihood of emergencies in the tank farm of the enterprise and modelling the situation with the ingress of oil products into the Kheta River are solved using the TOXI + Risk software [7] and the neuroimitator NeuroPgo 0.25 developed at the Federal Research Center of the KSC SB RAS [8]. The data array for risk assessment and modelling was obtained from the Main Directorate of the Ministry of Emergency Situations of Russia for the Krasnoyarsk Territory, the agency for civil defense, emergency situations and fire safety of the Krasnoyarsk Territory on the basis of emergencies that occurred in 2020.

Initially, for the purposes of predicting an emergency spill of oil products, the scale of the calculated plan was set and the area of objects with the personnel present was calculated. Also, the areas of objects were determined, which indicated the number of people falling into the risk zone, as well as the coefficient of their presence. The number of personnel falling into the risk zone corresponds to the maximum number of people who can simultaneously be at the facility. People who can be in the industrial area of the enterprise and can get into the dangerous affected area belong to the group of personnel.

The use of the TOXI + Risk software product allows you to analyze the risk of a probable emergency and assess the consequences of potentially dangerous facilities for the economy enterprises. To use the software product, the following actions were analysed:

- organization of work on risk analysis;
- identification of hazards;
- value of possible irrecoverable human losses, determined by the number of fatalities as a result of an accident;
- value of possible sanitary human losses, defined as the number of victims requiring hospitalization;
- the expected frequency of the accident.

The following scenarios of the occurrence and development of an emergency are possible in the company's tank farm:

1. Tank destruction
   1.1 Instant ignition causing spill fire
   1.2 No multi-link ignition
      1.2.1 Delayed ignition and spill fire
      1.2.2 Non-ignited dispersion of oil products

The initial data for the calculations were the following indicators:

- density of the oil product p_{oil} =850 km/m^3;
- coefficient of distribution of oil products on the concrete pavement f_p=30 m^{-1};
- temperature of the oil product obtained by calculation t_p=39°C;
- limit of concentration of flame propagation $\varphi_f = 0{,}9\%$;
- molecular weight of oil M= 50 kg/mol;
- molar volume of steam V_0=22,41 m^3/kmol;
- gravitational acceleration g=9.8 m/s^2;
- atmospheric pressure p_0=101,325 kPa;
- RVS height, H_p=18 meters;
- RVS diameter, D_p=40 meters;
- RVS filling degree ε =0,9;
- number of RVS in embankment n=1;
- size of RVS depressurization hole d_{hole}=0,05 meters;
- coefficient of consumption of oil products during depressurization of the RVS $\mu = 0{,}62$;
- time of blocking the flow of liquid from the hole at local depressurization of the RVS $\tau_{blocking} = 3600$ seconds.

Data obtained at the protected object:

1. saturated oil vapor pressure p_s=60 kPa
2. saturated oil vapor concentration:

$$\varphi_s = \frac{p_s}{p_0} \cdot 100\% = 59{,}215\% \tag{1}$$

3. evaporation rate:

$$W = 10^{-6}\sqrt{Mp_s} = 4{,}24 \cdot 10^{-4} kg/(m^2 \cdot s) \tag{2}$$

4. oil vapor density at operating temperature and atmospheric pressure:

$$p_{steam} = \frac{M}{V_0(1+0.00367t_p)} = 2{,}15\ km/m^3 \tag{3}$$

To solve the problem of modelling the ingress of oil products into the Yenisei River, multilayer neural networks with a sigmoid activation function of the following form were used:

$$f(x) = \frac{x}{c+|x|} \tag{4}$$

where c is the characteristic of the neuron; x is the set of input parameters.

The learning algorithm of the neural network was reduced to training and testing the network with predetermined characteristics. The size of the area of bottling of oil products in a reservoir depends on many factors that should be taken into account when modelling and compiling feature vectors:

1. the volume of oil product entering the reservoir;
2. protective structures and their type on the path of distribution of oil products;
3. terrain;
4. the level of watering of the area;
5. the presence of vegetation: woody, shrub, herbaceous;
6. ambient temperature;
7. spill containment time;
8. rainfall;
9. mass of oil dispersed in water, tons;
10. wind speed, m /s;
11. water flow rate, m /s;
12. type of oil product;
13. direction of the wind.

The input parameters were selected empirically. The dimension of the vectors of the features noted above was established empirically, was thirteen. In the calculation, a multilayer neural network with 12 layers and the number of neurons in hidden layers equal to 204 was used, which shows good results in practice. The percentage of reliability of the results obtained was in the range of 70-78%.

3. Results and discussion

Based on the methodology described in section 2 of this work with the use of the TOXI + Risk software product, the following results were obtained for the risk of an accident at the Taimyr Fuel Company, table 1.

Table 1. The results of calculating the probability of an emergency at the Taimyr Fuel Company.

The source of the accident	Alarm triggering event	Outflow hole diameter, mm	Depressurization probability, year ^{1}
Pipelines	Depressurization with subsequent outflow of liquid or two-phase medium	5	$2.8 \cdot 10^{-3}$
		12.5	$3.6 \cdot 10^{-4}$
		25	$5.2 \cdot 10^{-4}$
		50	$3.0 \cdot 10^{-4}$
		Complete destruction	$2.0 \cdot 10^{-4}$
Pumps	Failure of pumping equipment	-	$4 \cdot 10^{-4}$
Storage tanks at near atmospheric pressure	Depressurization with subsequent outflow of liquid into the embankment	All types	$2.0 \cdot 10^{-4}$
	Quasi-instant destruction	-	$6.0 \cdot 10^{-6}$
Fixed roof tanks	Pressure vent valves fire	-	$10.0 \cdot 10^{-5}$
	Fire across the entire surface of the tank	-	$10.0 \cdot 10^{-5}$
Filling devices (tips, hoses, sleeves)	Flute crack	Up to 5	$1.8 \cdot 10^{-3}$
	Blowout	-	$4.3 \cdot 10^{-5}$

As follows from the above data, the greatest probability is an emergency situation associated with depressurization of process pipelines, and the greatest danger is the destruction of a reservoir with an oil product.

The main causes of emergencies are risks of production processes, such as metal corrosion, mechanical damage, shutdown electricity, natural and man-made hazards; dangers of the human factor (wrong actions or inaction of the personnel and management of the facility, malicious intent, terrorist acts).

Technological processes of storage, reception, bunkering of ships, filling oil products into tankers and barges, pumping oil products, violation of operating rules and during repair work, creates the risk of emergencies associated with depressurization of equipment and pipelines and bottling of oil products into the soil and water bodies.

The solution to the second task, namely neural network forecasting of the oil slick area, was carried out using the technique described earlier in the materials and methods section. Using the monitoring data on the accident on 12.07.2020 at the oil depot of the Taimyr Fuel Company, we performed a simulation. The obtained computational results were compared with the accident prevention action plan in the event of an emergency and real data on the accident.

Comparison of the results obtained using the neural network and real values of the oil slick area are presented in table 2.

Table 2. Data on the real and estimated area of the oil slick.

Date	Actual area with a cumulative total, m^2	Calculated area with a cumulative total, m^2	Deviation
12.07.2020	15200	14800	400
13.07.2020	22600	21200	1400
14.07.2020	29300	33350	-4050
15.07.2020	32200	35100	-2900
16.07.2020	35500	36100	-600
17.07.2020	36800	36500	300
18.07.2020	38300	38500	-200

Discrepancies in the first days of real data and simulated ones associated with the adoption of operational measures to contain and eliminate the oil spill were carried out in 3 stages. At the first stage, measures were taken to localize oil spills by installing booms and using the "Unisorb" sorbent. The use of booms was reflected in the results obtained during modelling by means of neural network forecasting (in table 1 starting from 07.14.2020), since in fact, predicting the time of application and the number of booms and sorbent substances is a separate difficult task, which is rather difficult to estimate numerically. In fact, with the accumulation of oil at the booms, the spill area was reduced, and there was also a partial pollution of the coastal strip.

At the second stage of work, measures were taken to collect the bulk of the fuel-water mixture and contaminated surface soil. At the third stage, measures were taken to pump the fuel-water mixture from the area of the reservoir mouth to the temporary storage warehouse in the area of the tailing dump. There are 84 tanks installed at the temporary storage warehouse, with a total volume of 21,000 cubic meters.

As a result of taking prompt measures, it was possible to avoid spilling of oil products over a large area, which resulted in a significant discrepancy between the model data; the data differed by the amount of error.

As a result of taking prompt measures, it was possible to avoid the spill of oil products over a large area, which resulted in a significant discrepancy between the model data; later, after the bottling was localized, the model data and the real data differed by the amount of error.

4. Conclusion

An important and urgent problem of man-made accidents associated with the bottling of oil products in the Arctic zone of the Krasnoyarsk Territory, which has become aggravated due to the increase in the number of accidents at enterprises, is considered in the work. One of the effective ways to minimize the losses of enterprises and prevent man-made accidents is the use of modern means of monitoring and forecasting emergency situations, drawing up models of the development of the situation and calculating the risks of probable emergency situations. The development of modern technologies for neural network forecasting and modelling makes it possible to automate the processes of predicting the risks of emergencies, which was directly demonstrated in this work. The results obtained can be applied in the development of plans for the prevention and elimination of emergency spills of oil and oil products.

Acknowledgement
The reported study was funded by Krasnoyarsk Regional Fund of Science, the research project name is "Development of a set of necessary preventive measures to protect the population and the Arctic the territory of the Krasnoyarsk Territory from natural and man-made emergencies based on neural network cause".

References

[1] Fentiman A and Zabbey N 2015 Environmental degradation and cultural erosion in Ogoniland: a case study of the oil spills in Bodo *Extr Ind Soc Int J* **2** 615-24
[2] Li X, Chen G and Zhu H 2007 Modelling and assessment of accidental oil release from damaged subsea pipelines *Marine Pollution Bulletin* **123** 133-41
[3] Grebnev Ya and Moskalev A 2020 Influence of hydroelectric power station on flood situation and modeling of hydrological characteristics *E3S Web Conf.* **221** 1-7
[4] Guandalini R, Agate G and Moia F 2017 Numerical 3D modelling of oil dispersion in the sea due to different accident scenarios *Energy Procedia* **125** 161-9
[5] Grebnev Ya, Moskalev A, Vershkov A and Gazizulina A 2019 The practice of connectionist model for predicting forest fires in the Arctic zones of the Krasnoyarsk Territory *International Journal of System Assurance Engineering and Management Int J Syst Assur Eng Manag* 1-9
[6] Grebnev Ya and Yarovoy A 2018 Monitoring and forecasting floods in the Krasnoyarsk Territory using neural network algorithms *Scientific and analytical journal "Siberian Fire and Rescue Bulletin"* **3** 13-6
[7] Website of STC "Industrial Safety" CJSC *Automated Environment Modelling System Toxy + Risk* Retrieved from: https://toxi.ru/
[8] Gorban A *et al.* 1998 *Neuroinformatics* (Novosibirsk: Nauka, SB RAS) p 296

Methods of diagnostics of electrical equipment by fiber-optic sensors on the Faraday Effect

T M Levina and D N Bilalova[1]

Ufa State Petroleum Technological University, Branch in Salavat, 22B, Gubkin st., Salavat, Russia

[1]E-mail: dinabil2015@yandex.ru

Abstract. One of the important problems in the operation and creation of a complex technical system is the provision of diagnostics of electrical equipment. This is especially important when diagnosing electrical equipment at an industrial complex belonging to the class of spark - explosive production facilities. The task of ensuring the operability of electrical equipment and preventing emergency situations due to timely detected breakage or violation of the insulation properties of the equipment includes a whole range of certain measures. In the process of operation, technical diagnostics of electrical equipment occupies a special place among complex technical devices at oil refineries. One of the most modern sensors for diagnosing electrical equipment is multiplicative sensors, which use optical fiber. Fiber-optic sensors have special proper-ties. The Faraday Effect is the basis for the construction of fiber-optic sensors for diagnostics of electrical equipment. Methods for diagnosing electrical equipment with fiber-optic sensors based on the Faraday Effect depend on the principles of the sensors themselves. This article presents the results of the study of the creation of experimental fiber-optic sensors based on the Faraday Effect. In this paper, it is proposed to carry out diagnostics of electrical equipment at an industrial complex belonging to the class of spark - explosive production facilities, such as oil refineries, using the use of fiber-optic sensors based on the Faraday Effect.

1. Introduction

Providing diagnostics of electrical equipment is one of the most important problems in the operation and creation of a complex technical system. This is especially important when diagnosing electrical equipment in an industrial complex belonging to the class of spark-explosive production facilities, such as oil refineries.

The task of ensuring the operability of electrical equipment and preventing emergencies due to timely detected breakage or violation of the insulation properties of the equipment includes a set of measures starting from:

- determination of the location of electrical equipment depending on the influence of aggressive media on it;
- dimensions of electrical equipment;
- technical performance characteristics;
- operational performance characteristics;
- economic calculations of power supply or distribution systems, etc.

In addition, the performance characteristics of electrical equipment are inextricably linked with such objects of oil refineries as:

- overhead power lines;
- cable power lines;
- step-down substations and power transformers;
- cells, switches and disconnectors;
- current and voltage measuring transformers;
- switchgears;
- switches;
- static capacitors;
- current limiting reactors;
- synchronous and asynchronous motors of control systems and machine tools, installations of submerged centrifugal pumps;
- electric motor control stations;
- grounding devices;
- pumps of the supply system of reagents;
- other auxiliary equipment.

2. Theoretical and experimental methods of diagnostics of electrical equipment

Technical diagnostics of electrical equipment during operation occupies a special place among complex technical devices at oil refineries [1]. Due to the fact that electrical equipment is subject to the risk of defects, malfunctions and failures, equipment control is used for the equipment in the form of a diagnostic system GOST 20911-89. "Technical diagnostics. Terms and definitions" uses the phrase The System of Technical Diagnostics as "A set of means, object and performers necessary for carrying out diagnostics (control) according to the rules established in the technical documentation" in relation to the electrical equipment of oil refining enterprises. It is necessary to pay special attention to the means (sensors) used for diagnostics (control), taking into account their current state and improved technical and operational characteristics [2].

One of the most modern sensors for diagnosing electrical equipment is multiplicative sensors, which use optical fiber.

Fiber-optic sensors have special properties such as: high throughput, lack of electrical conductivity, small dimensions, weight, spark-explosion safety, etc.

Currently, a number of domestic and foreign scientists are engaged in the theory, calculation and design of fiber-optic sensors for diagnostics of electrical equipment: Busurin V. I., Burkov V. D., Vitsinsky S. A., Isakov V. N., Kazachkov Yu. P., Kirin I. G., Kuznetsova V. I., Lovchy I. L., Pimenov A.V., Williams P, Rose A, Day G, Milner T, Deeter M.

The development of such sensors is carried out by such companies as: the Canadian company NxtPhase T&D Corporation, the Swedish company PowerSense, the American companies OptiSense Network, Inc., Airak, InoFieldMetrics, Inc. (FMI). In Russia, the development and research of such sensors is carried out by LLC "Unique Fiber Devices". Similar studies are conducted at the Ufa State Aviation Technical University, St. Petersburg ITMO University under the guidance of academician, Doctor of Technical Sciences, Professor Meshkovsky I. K. and other scientists of the University [4-5].

The Faraday Effect is the basis for the construction of fiber-optic sensors for diagnostics of electrical equipment. It consists in turning the plane of polarization of linearly polarized light in a sensitive element in the form of a coil of optical fiber with linear double refraction, under the action of a magnetic field created by a conductor with a measured current, the magnetic lines of force of which coincide with the direction of light [3].

Quantitatively, the effect is described by the formula:

$$\varphi = V \cdot H \cdot L,$$

where φ is the angle of rotation of the plane of polarization (or often called Faraday rotation); L is the optical path of a plane-polarized wave in a medium whose magnetic rotational capacity is characterized by the Verde constant V. The Verde constant depends on the properties of the substance, the wavelength of light, and temperature.

Methods for diagnosing electrical equipment with fiber-optic sensors based on the Faraday effect depend on the principles of construction of the sensors themselves (table 1).

Table 1. Principles of construction of fiber-optic sensors based on the Faraday Effect and the measurement range.

The principle of construction by fiber-optic sensors on the Faraday effect	**Measuring range**
Rotation of the plane of polarization of light propagating in an optical fiber under the influence of a magnetic field (the angle of rotation of the plane of polarization of light is ±45°)	at rated current values, the measurement range is from 200 to 20 kA
Splitting the light beam from the output of the optical fiber coil into two beams	measurement of current in a small specified range of currents-usually up to 120% of its rated current value
Interweaving of input and output optical fibers of an optical fiber turn.	high-voltage power engineering (> 10 kV), requires breaking the power line
Serial connection between each other (n-1) sensitive elements of the light guide	registers current pulses with a duration of not more than 30 ns and not less than 1 ns
Serial connection of an optical fiber with a mirror piezo ceramic membrane	DC analog output signal range (20-4=16 m A)
Rotation of the polarization plane in a Faraday element with plane anisotropy	current measurement limits from 10^2 to 10^5 A
Light input into the optical fiber of the branched part in the first arm	dynamic range of magnetic field measurements: from 50 μT (40 A / m) to 150 mT (120 ka / m)

The paper presents the results of a study of the creation of experimental fiber-optic sensors based on the Faraday Effect.

The construction of such sensors can be based on the following example.

Taking into account GOST 20911-89, we will consider the construction of an experimental diagnostic system for monitoring the magnitude of electric current with a fiber-optic sensing element based on the Faraday Effect (figure 2). The system consists of three parallel fiber-optic chains [6]. And it works as follows: when a controlled electric current I flows from a controlled power source 1, a magnetic field of intensity H is created on the first fiber-optic chain. This field acts on the sensing element (SE) 4 (twisted optical fiber). The coherent monochromatic radiation J1 generated by the laser diode 2 in the polarizer 3 is converted into a linearly polarized light wave J2. In the sensing element 4, under the action of an external magnetic field of intensity H, the plane of polarization of light propagating along the direction of the magnetic field rotates. The light flux J3 from the output of the sensing element 4 passes through the analyzer 5 and gets to the photodiode 6, then to the amplifier 7 and the analog-to-digital converter 8, which transmits the output signal from the first fiber-optic chain to the computer 15.

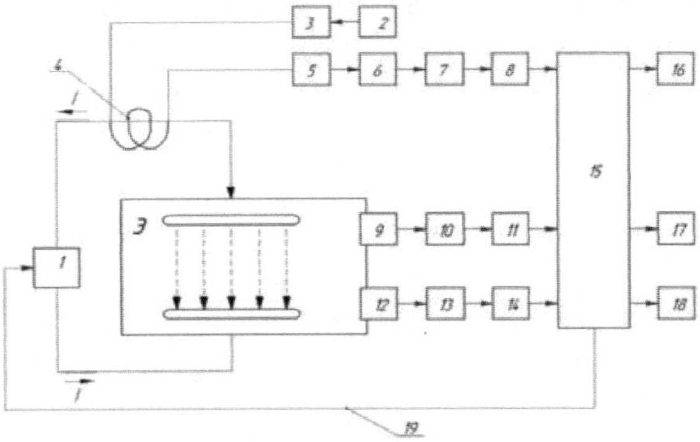

Figure 1. Functional diagram of the information and measurement system for monitoring the electric current of an electric desalting plant with a built-in digital system

In the second fiber-optic chain, the signal from the sensitive element that controls the temperature T 9, in the form of an electrical signal, enters the amplifier 10 and also enters the ADC 11, which transmits the output signal from the second fiber-optic chain to the computer 15.

The third fiber-optic chain works in the same way as the second only through the sensing element 12-the pressure control P, the amplifier 13 and the ADC 14, which transmits the output signal from the third fiber-optic chain to the computer 15.

The outputs of the ADC (8, 11, 14) of the three fiber-optic circuits are connected to the computer 15, which is connected to the liquid crystal display (LCD) 16, the data output device for printing 17 and the data input device in the computer 18, the output of the computer by means of a control signal transmission device is connected to an adjustable power source 1.

In addition, fiber-optic sensors based on the Faraday effect for the diagnosis of electrical equipment can be easily interfaced into information and measurement systems to obtain information about various parameters of the controlled object, which must be processed, i.e., compare the obtained parameters with the parameters set as reference, determine the signs and values of the difference in the measurement parameters, calculate some derived parameters, etc. (figure 2).

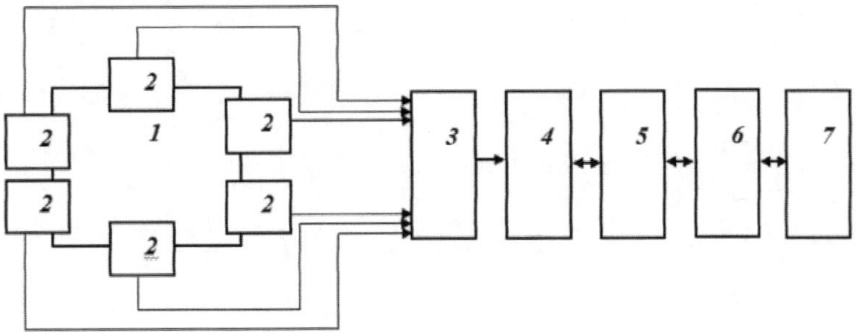

Figure 2. Block diagram of the information and measurement system for monitoring the magnetic field.

IOP Conf. Series: Materials Science and Engineering 1155 (2021) 012053 doi:10.1088/1757-899X/1155/1/012053

Information from the object of research 1, goes to the primary measuring transducers 2 ... N, then the information goes to a single data collection point 3 and along the highway 4, the information goes to the computer 5, from where through the interface 6 to the device for forming control actions 7.

Here it is necessary to pay attention to one of the significant shortcomings. The measurement of a controlled physical quantity strongly depends on the quality of the information obtained at the initial stage of measurement. Thus, the highest requirement should be placed on the quality of information [7-9]. In addition, it is necessary for the measured means to ensure the mutual conjugation of its elements with each other. The solution of this problem can be achieved through consistency of information obtained in a form suitable for direct perception of a human operator and information in the fiber-optical sensors the Faraday Effect is presented in the form of a specific physical quantity, convenient for transmission, and further transformations in automation systems. This value is called a signal, and it is uniquely related to the controlled physical quantity or parameter of a particular technological process [10].

Modern state systems of industrial devices cover only a part of the controlled values that are most often used in production. These values are classified in five groups:

- heat and power (temperature, pressure, differential pressure, level and flow rate);
- electric power (DC and AC currents and voltages, power, power factor, frequency and insulation resistance);
- mechanical quantities (linear and angular displacement, angular velocity, force, hardness, vibration);
- chemical composition (concentration, chemical properties, composition);
- characteristics of physical properties (humidity, density, electrical conductivity, viscosity, illumination, etc.).

Fiber-optic sensors on the Faraday Effect for diagnostics of electrical equipment can be connected to form structural schemes, for example, a single direct conversion, with feedback, sequential direct conversion [11, 12].

The experimental fiber-optic sensors based on the Faraday Effect and systems based on them (figures 1 and 2) presented as a result of the study consist of a single sensing element. In the case of serial connection of several fiber-optic sensors on the Faraday effect, the output value of the previous one is the input value of the subsequent one (figure 3 (a,b). The scheme of fiber-optic sensors based on the Faraday Effect for diagnostics of electrical equipment with feedback (figure 3 (c)) is characterized by high accuracy, versatility and low dependence of the measurement coefficient on external disturbances.

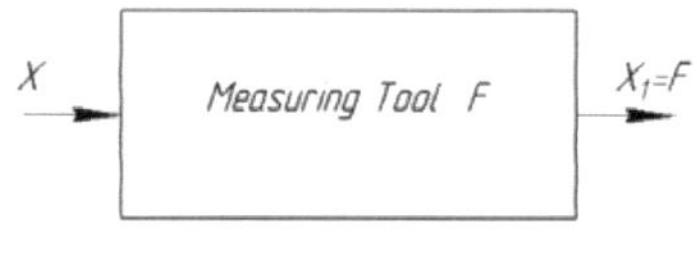

a)

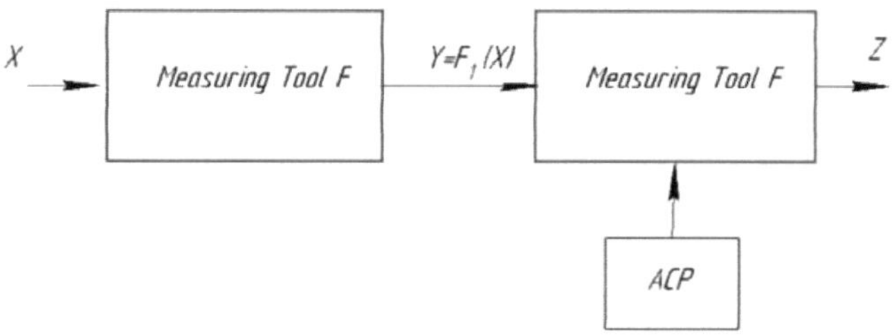

b)

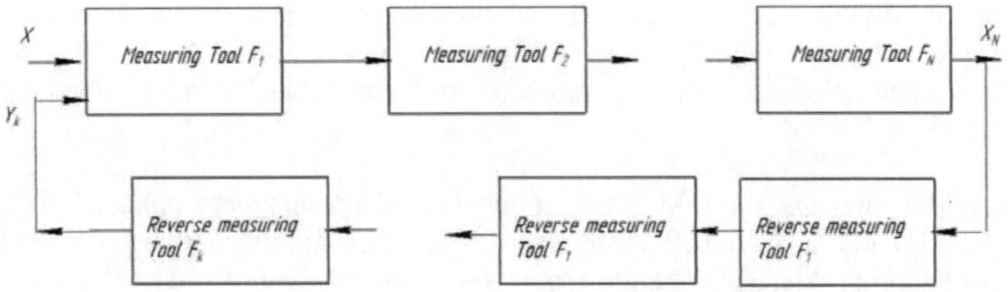

c)

Figure 3. Serial connections of several measuring instruments

Thus, it is necessary to apply a systematic approach, which in this case may mean that when solving the problem of creating fiber-optic sensors based on the Faraday effect for the diagnosis of electrical equipment, there is a solution to the complex of works aimed at:

- agreeing on the maximum level of unification and standardization of fiber-optic means of measuring the magnetic field;
- establishment of nomenclature requirements and administrative documents that ensure compliance with the requirements for modern measuring instruments, a certain number of modifications of the measuring instrument of the same type (from the point of view of the totality of coordination of two different subject areas in the design, manufacture and operation);
- deepening of the subject and technological specification and specialization of enterprises-designers and manufacturers.

Conducting this complex of works gives the creation of an integrated complex-a fiber-optic means of measuring the magnetic field [13].

3. Conclusion

In this paper, it is proposed to carry out diagnostics of electrical equipment at an industrial complex belonging to the class of spark - explosive production facilities, such as oil refineries, using fiber-optic sensors based on the Faraday Effect.

It is particularly important to note that it is necessary to take into account the system approach to the development of functional schemes of such sensors, which, due to their construction principles, can diagnose such parameters as electric current or magnetic field in the places of insulation breakage at power cables.

The application of a systematic approach in this case may indicate that the development and design of fiber-optic sensors on the Faraday Effect reduces the problem of the following parameters:

- coordination of the maximum level of unification and standardization of the fiber-optic magnetic field measuring device;
- establishment of nomenclature requirements and administrative documents that ensure compliance with the requirements for modern measuring instruments, a certain number of modifications of the measuring instrument of the same type (from the point of view of the totality of coordination of two different subject areas in the design, manufacture and operation);
- deepening of the subject and technological specification and specialization of enterprises-designers and manufacturers.

IOP Conf. Series: Materials Science and Engineering 1155 (2021) 012053 doi:10.1088/1757-899X/1155/1/012053

Conducting this complex of works gives the creation of an integrated complex-a fiber-optic means of measuring the magnetic field, which is required according to GOST 20911-89 "Technical diagnostics. Terms and definitions" in the formulation of the system of technical diagnostics.

References

[1] Urakseev M A and Levina T M 2008 Application of the magneto-optical Faraday element in information and measurement systems for monitoring the magnetic field and electric current Caspian Journal: Management and High technologies **2 (2)** pp 24-31

[2] Rogers A J 1995 Optical fiber current measurement *Optical Fiber Sensor Technology* eds K. T. V. Grattan and B. T. Meggitt (London: Chapman and Hall) pp 421-39

[3] Day G W 1989 Recent advances in Faraday effect sensors Springer Proceedings in Physics *Optical Fiber Sensors Springer Berlin* **44** p 250

[4] Levina T M, Zharinov Yu A and Shamaev F F 2012 Application of electro-safe sensors in the control system of the technological process of oil desalination *Caspian Journal: management and high technologies Astrakhan: ASU Publishing House* **4** pp 18-23

[5] Levina T M and Zharinov Yu A 2013 Methods of controlling spark and explosion-proof sensors on the CDU *Proc. Int. Baku forum of young scientists on new challenges in the european area* (Azerbaijan: Eurodoc) pp 53-55

[6] Evdokimova N G, Luneva N N and Levina T M 2020 Experience in the application of investment controlling inllc "Gazprom Neftekhim Salavat" *Directions of improvement Journal of Advanced Research in Dynamical and Control Systems* **12(2)** pp 371–379

[7] Rogers A J 1973 Optical technique for measurement of current at high voltage *Proc. IEE* **120 (3)** pp 261-67

[8] Ning Y N, Chu B C B and Jackson D A 1991 Miniature Faraday current sensor based on multiple critical angle reflections in a bulk-optic ring *Opt. Lett* **16** pp1996-98

[9] Okosi T, Otsu M, Nisihara H, Kyuma K and Hatate K 1991 *Fiber-optic Sensors* (Leningrad: Energoatomizdat) p 255

[10] Urakseev M A and Levina T M 2008 Application of the magneto-optical Faraday element in information and measurement systems for monitoring the magnetic field and electric current *Caspian Journal: Management and High technologies* **2 (2)** pp 24-31

[11] Urakseev M A, Levina T M and Galiullin I V 2007 Pat. of the Russian Federation No. 62712 G01R29/00 appl. 11.12.2006, publ. 27.04.2007

[12] Shamaev F F and Levina T M 2013 Pat. of the Russian Federation No. 136190 G01R 33/032.

[13] Vildanov R G, Khismatullin A S and Luneva N N 2019 The investigation of magnetization reversal loss sensor *IOP Conf. Series: Materials Science and Engineering* **560(1)** 012109

Model reproduction of non-equilibrium thermodynamics principles as a means to provide software self-development

E A Kol'chugina

Department of software and computer application, Penza State University, Krasnaya Street 40, Penza, 400026, Russian Federation

E-mail: kea_sci@list.ru

Abstract Further evolution of technologies in material sciences, hardware and software demands reproduction of self-organization as a manifestation of non-equilibrium thermodynamics principles in new artificial environment. This is especially actual for software engineering, where reproduction of non-equilibrium thermodynamics principles will allow enhancing speed and quality of development process. In this article we reveal, list and analyze principles to be reproduced in artificial systems to provide capabilities for self-development, self-healing and self-improvement. It is shown that self-development process is impossible without reproducing the metaphor of electric charge with its abilities for attraction and repulsion. But this metaphor is usually avoided in computer models. We propose and consider a method to reproduce the metaphor of electric charge in computer models by means of UNIX/Linux operating system tools. The obtained results provide the basis to establish an artificial physics, a new research area dedicated to adequate reproduction of principles of physics in the artificial environment of computer models.

1. Introduction

The reproduction of non-equilibrium thermodynamics principles in computer models became of importance not only for physics, material technologies, natural sciences, but for cybernetics and software engineering also.

In the end, software development has become an industry where it is necessary to engage developers' teams to implement any more or less competitive project. The epoch of successful individual developers has gone.

Modern software projects involve huge investments of money and human labour. Many of such projects are dedicated for distributed or mobile systems. But the extensive way to achieve high computational performance and powerful functionality faces with two challenges.

The first is a debugging problem. It became impossible to test and debug applications completely during their life cycle.

The second is a problem of adaptation and self-improvement. Typical software is rigid and incapable to interact with its environment for adaptation and improvement. So, changes of environment, e.g. updates of operating system, can lead to destruction of the software due to arising conflicts.

Thus, there is occurring a need of endless revision and release of new software versions.

The solution is to make software more reasonable, if not intelligent, and to endow it with abilities to improve, heal and teach itself. This demands creation of new nonstandard programming paradigms.

The most widely known paradigm to create intellectual software and hardware systems is neural networks. But this paradigm is too complicated to be used for design of simple service software, e.g. utilities. The resulting software should not be too clever and complicated, and, as a result, expensive.

A shorter and simpler way to make software reasonable, but not intelligent, is to reproduce the principles observed in nature and stated in mathematics, physics, chemistry and biology. These principles allow the creation of self-organizing systems, and they were formulated primarily for non-equilibrium thermodynamics of irreversible processes [1-3].

Self-organizing systems exist due to and through the interactions with their environment by energy-matter exchange. Self-organizing systems are capable for self-assembly and self-ordering in response of environmental changes. So, reproduction of non-equilibrium thermodynamics principles in virtual cybernetic medium can endow program systems with abilities for self-development throughout their entire life cycle.

One of scientific branches aimed to reproduce non-equilibrium dynamics of biological systems in cybernetic medium is artificial life [4]. A subdivision of artificial life closely connected with origin of life theory, molecular biology and biochemistry is artificial chemistry [5-21,23,24]. Software artificial chemistry models allow considering programs as artificial molecules. Respectively, the interactions between programs are considered as analogues of chemical reactions which lead to construction and destruction of artificial molecules, i.e. programs. This point of view opens the way to endow programs and groups of interacting programs with abilities for self-ordering and reordering, self-design, adaptation and self-improvement, i.e. continued self-development.

Many software models of artificial chemistry try to imitate the principles of self-organization and non-equilibrium thermodynamics in software systems [5-7]. An overview and analysis of such models was given by the author in [7].

But in [7] it was shown also that attempts to replicate non-equilibrium thermodynamics principles in software systems are imperfect. The reason is that basic principles of physics of the real world are not reproduced in computer models adequately.

In [7] it was revealed the necessity to formulate and study principles of deeper level of reality simulation in the form of artificial physics.

Here in this article, we make a first step towards forming these principles of artificial physics. We start from analysis of basic categories and principles of the physics of the real world, which are connected with non-equilibrium thermodynamics processes. We are also giving analysis of methods of reproduction of these categories and principles in artificial life and artificial chemistry models.

2. Methods

To analyze the basic categories and principles necessary for reproducing non-equilibrium dynamics in computer models, we need firstly to name and list such categories and principles.

According to the analysis of artificial chemistry models given in [7], these categories, which must be reproduced to ensure the presence of processes with non-equilibrium dynamics, include space, time, energy, entropy, and electric charge. Here we consider different ways of representations of such categories in virtual medium.

2.1. Space and time

In many models of artificial life and artificial chemistry related to the theory of origin of life, such as “Venus” and “Luna” [8], “Tierra” [9-11], “Avida” [12-17], “Amoeba” [18-20], the model space is understood as a memory space of some virtual computer. That space is usually one-dimensional, more rarely it has two or more dimensions. An interesting remark given in [10] is that model space can be considered as non-Euclidian space, since every cell can be reached directly from any other cell by jump performed by central processor unit (CPU).

An analogy between computer memory and model space establishes another important quality of space defined in abovementioned artificial chemistry models: the space in such models is always discrete or cellular.

The model time is organized in more simple and obvious way: it is linear and unidirectional. Really the model time is a CPU time.

2.2. Energy and dissipation
It is known that time and space of the real world possess their own energy. The same is true for the artificial worlds in "Venus" and "Luna" [8], "Tierra" [9-11], "Avida" [12-17], "Amoeba" [18-20], and other similar models of artificial chemistry. In these models, energy is the amount of available CPU time. This establishes a direct relationship between model time, space and energy: the more time the analogue of non-equilibrium process is emulated, the more energy it consumes; the more model energy a computer analogue of non-equilibrium process retains, the more time it is executed; the wider model space the computer analogues of non-equilibrium processes occupy, the more energy they possess [7,9-17].

According to this introduced analogy, the matter in computer worlds is represented by bits and bytes of information stored in memory.

But computer models have an important restriction that seriously limits similarity between physics of artificial and real worlds. Physics of the real world allows possibility of mutual transformations between energy and matter. In artificial computer worlds energy (CPU time) could be transformed into matter (information stored in memory). The converse is not true: information stored in memory can be traded for processor time, but not directly converted to it.

This makes possible to talk about energy dissipation in models of non-equilibrium processes of artificial life and artificial chemistry. An energy, or CPU time, being spent once on changing the state of the memory cannot be restored back.

So software artificial chemistry models trying to reproduce non-equilibrium thermodynamics of real-world self-organizing systems are inevitably dissipative and suffer energy loss.

2.3. Entropy
Definitions of non-equilibrium, non-equilibrium state, and non-equilibrium process are usually given using the term entropy [1-3]. But in computer models, the situation is different. Usually entropy looks like a secondary (compared to energy) principle, which is synonymous with disorder and heterogeneity.

The problem is that it is not clear what entropy itself is. Entropy can be defined in different ways, as shown in [7]. When applied to models of artificial chemistry, entropy can be defined, for example, as a measure of the absence of structures representing data or programs in the memory of a virtual computer [21], or by analyzing the frequency of bit sequences and the speed of their replication by means of statistical mechanics [13].

Thus, to be used in computer models, the definition of entropy should be refined, primarily in physics and natural sciences.

2.4. Self-organization as one of the key principles of constitution of matter
In our opinion, the best definition of self-organization was given by A. P. Rudenko [22]. According to this definition, the main condition for self-organization is non-equilibrium, the main reason is useful work against equilibrium, and the driving force is a part of free energy of exchange process spent on doing internal useful work [22].

In [22] self-organization, along with organization, is understood as one of the key principles of constitution and existence of matter. These principles determine existence of matter in dynamic and static forms. The organization is entropic principle, while self-organization is anti-entropic principle.

In artificial chemistry models, the metaphors for non-equilibrium state and free energy are program execution and CPU time, respectively. Useful work against equilibrium is work that provides and maintains self-ordering. In artificial chemistry models, useful and rewarded calculations can be considered as such.

Current modern software is built on the principle of organization: software execution leads to the loss of the model analogue of energy (CPU time) and subsequent termination. After that, complex structures representing code and data are destroyed.

On the contrary, reproduction of self-organization principle in virtual medium implies that:

- software tends to perpetual execution as analogue of non-equilibrium state;
- perpetual execution requires an unlimited source of analogue of energy, i.e. CPU time;
- CPU time should be provided to the software primarily as a reward for performing useful operations.

At the first glance, the means to ensure self-organization in computer medium are obvious. The question is how to represent an analogue of work against equilibrium adequately. In the real physical world, such work is done naturally, spontaneously, and without compulsion. But in [23,7], it was found that the von Neumann architecture is not suitable for reproducing spontaneity, since the concept of this architecture requires that all interactions between programs must be authorized by the central processor unit.

In [7] it was also stated that spontaneous interactions should arise independently, in decentralized manner and inevitably, according to the properties of interacting components and the strength of their mutual attraction and repelling.

And here we face the most important problem: how to ensure decentralized spontaneity. The most typical examples of spontaneous interactions in the real world are those based on an electric charge.

Thus, to solve the problem excavated, it is necessary to imitate in computer models analogue of electric charge.

2.5. Electric charge, attraction and repelling

The basic law concerning to electricity says that like charges inevitably repel and unlike charges inevitably attract. The reproduction of this law in virtual computer medium is of immense necessity for software engineering: this is a key to achieve spontaneous self-assembly and self-development of programs.

Despite the fact that electricity has been actively used by mankind in the last two centuries, it is still not clear what it is. To understand something completely means to be able to reproduce it, but the electric charge and field have never been artificially recreated from scratch until now.

The lack of complete understanding of nature of electric charge in physics makes simulation of electric charge in computer models a great difficulty.

Electric forces of attraction can be seen as a special kind of "glue" that combines and unites atoms and molecules. According to this point of view, there are two ways can be offered to simulate forces of attraction in computer models of artificial chemistry. Any of these methods implies that artificial atoms and molecules can be attracted to each other through something common that can be shared between them.

The first and obvious way to implement such shareable medium is to use common data structures, e.g. variables situated in shared memory. This solution was considered and experimentally studied in [24]. In the conducted experiments it was revealed that common variables can serve as a base for spontaneous formation of dynamical cyclic structures. These structures were formed by simplest functions that run as independent processes and use the same shared variables. The result of simulation in general looked like a spontaneous formation of high-level programs form the variety of simpler ones, where shared variables played the role of "glue" or a means to simulate attraction and repelling forces.

But the relations formed between functions in [24] were weak, and they were easily destroyed by invading functions. The reason is that relations were indirect and mediated. Other drawbacks of approach proposed in [24] were lack of selectivity and asymmetry on the set of possible relations.

The second way to create "glue" between programs is to use common interactions, not objects. These interactions can mimic shared electron clouds that combine atoms into molecules. To implement this, a process executing program unit destined to be combined with other units must be a system of "sinks" and "sources".

The "sources" play role of extra electrons on external orbitals, while "sinks" represent unfilled orbitals. The process with excess of "sources" is an analogue of negative ion, while the process with excess of "sinks" is a model analogue of positive ion.

These "sinks" and "sources" can be organized by means of inter-process communication (IPC) system in UNIX/Linux operating systems. Sockets are the most probable candidates, but other IPC tools are also possible to use.

Such approach is similar to those in physics, where interconnections and shared particles are understood as vortexes [25]. We don't discuss the ideas proposed in [25] from the point of view of physics, but they can serve as useful abstractions from the point of view of cybernetics to imitate particles and their interconnections in impoverished computer medium.

The proposed approach will be studied in details in our further works. It is expected that interactions between programs and corresponding processes will be direct, stronger and more stable than it was observed in [24]. But the problem of representing asymmetry on the set of possible interactions remains.

3. Results, discussion and further works

The biggest drawback of any computer model that tries to reproduce self-organization during non-equilibrium processes is the lack of spontaneity and inevitability of interactions. To overcome this, it is necessary to reproduce physical structure of matter in computer models properly.

Here, in this article, we have detected, listed and analyzed physical categories and principles to be simulated, which are necessary to reproduce self-organization artificially. We found out that some terms used in studies of self-organizing systems, e.g. "entropy", need clearer definition to be represented in computer models (see table 1).

Table 1. Representation of basic physical categories and principles in computer models (a summary).

Category or principle	Form of representation	Method of realization
Time	Explicitly	Usually represented as linear and unidirectional; identical to CPU time
Space	Explicitly	Memory space of the real or virtual computer
Matter	Explicitly or implicitly	Information stored in memory
Energy	Explicitly or implicitly	An amount of available CPU time
Entropy	Explicitly or implicitly	As the definition of entropy is not univocal, this category represented in different ways
Self-organization	Explicitly (as a goal of simulation) or implicitly (as a side effect)	Dynamically through intra system interactions that arising due to modeling rules
Electric charge	Not represented	Possibly can be represented by means of sockets

According to the summary given in table 1, most of the considered categories and principles are represented explicitly or implicitly in artificial chemistry models with the exception of electric charge. But here we found that this basic principle is essential for reproducing self-organization in a virtual medium. Electric charge simulation allows running programs to be "glued together" into dynamical structures without central control of the processor unit. We proposed that electric charge, electric attraction, and electron clouds can be represented using a system of "sinks" and "sources".

We also found candidate tools to simulate electric charge in computer models. These can be sockets in UNIX/Linux operating systems. In our further works we shall study this approach in details.

4. Conclusion

Development of artificial intelligence gave impact to studies in artificial life, a new research area to achieve efficient technologies based on metaphor of self-organization systems in biology. The necessity of deep understanding of the nature of self-organization in biological systems required a new level of research in the form of artificial chemistry models.

In [7] and here it was shown that further research in the field of artificial chemistry and the need for effective reproduction of self-organization in computer models require an adequate representation of physical categories and principles in such models, since the physical level of matter organization is the first where self-organization is observed.

The proposed approaches and the results achieved in this article can become the basis for the development of a new direction of self-organization research in the field of computer science: artificial physics.

The studies of this field are extremely necessary for computer science and technique to endow hardware and software with new abilities for self-organization manifested in forms of self-development, self-healing and self-improvement.

References

[1] Prigogine I 1955 *Introduction to Thermodynamics of Irreversible Processes* (Springfield: Thomas)

[2] Gyarmati I 1970 *Non-Equilibrium Thermodynamics Field Theory and Variational Principles* (Berlin-N.Y: Springer)

[3] Ziegler H 1963 Some extremum principles in irreversible thermodynamics with application tocontinuum mechanics *Progress in Solid Mechanics* **4** ed. I. N. Sneddon and R. Hill (Amsterdam, The Netherlands: North-Holland Publishing Company) **2** p 91-193

[4] Langton C G 1989 Artificial life *Artificial Life: Proc. of an Interdisciplinary Workshop on the Synthesis and Simulation of Living Systems (September 1987, Los Alamos, New Mexico) (SFI Studies in the Sciences of Complexity* **VI**) ed. C. G. Langton (Redwood City, CA: Addison-Wesley) p 1-47

[5] Dittrich P, Ziegler J and Banzhaf W 2001 Artificial chemistries – a review *Artificial Life* **7(3)** 225-75

[6] Banzhaf W and Yamamoto L 2015 *Artificial Chemistries* (Cambridge, Massachusetts; London, England: The MIT Press)

[7] Kol'chugina E A 2019 *Self-organization and self-development of programs in artificial chemistry models* (in Russian) (Kazan: OOO "Buk")

[8] Rasmussen S, Knudsen C and Feldberg R 1991 Dynamics of programmable matter *Artificial Life II: Proc. of Workshop on Artificial Life (February 1990, Santa Fe, New Mexico) (SFIStudies in the Sciences of Complexity* **X**) ed. C. G. Langton, C. Taylor et al (Redwood City, CA: Addison-Wesley) p 211-54

[9] Ray T S 1991 An approach to the synthesis of life *Artificial Life II: Proc. of Workshop on Artificial Life (February 1990, Santa Fe, New Mexico) (SFI Studies in the Sciences of Complexity* **X**) ed. C. G. Langton, C. Taylor et al (Redwood City, CA: Addison-Wesley) p 371-408

[10] Ray T S 1997 An evolutionary approach to synthetic biology: zen and the art of creating life *Artificial Life: an overview* ed. C. G. Langton (Cambridge, Massachusetts: The MIT Press) p 179-209

[11] Thearling K and Ray T S 1994 Evolving multi-cellular artificial life *Artificial Life IV, Proc. of the 4th International Workshop on Synthesis and Simulation of Living Systems (6-8 July 1994, the Massachusetts Institute of Technology)* ed. R. A. Brooks and P. Maes (Cambridge, MA: MIT Press) p 283-8

[12] Adami C and Brown C T 1994 Evolutionary learning in the 2D artificial life system «Avida» *Artificial Life IV, Proc. of the 4th International Workshop on Synthesis and Simulation of*

Living Systems (6-8 July 1994, the Massachusetts Institute of Technology) ed. R. A. Brooks and P. Maes (Cambridge, MA: MIT Press) p 377-81

[13] Adami C 1994 On modelling life *Artificial Life IV, Proc. of the 4th International Workshop on Synthesis and Simulation of Living Systems (6-8 July 1994, the Massachusetts Institute of Technology)* ed. R. A. Brooks and P. Maes (Cambridge, MA: MIT Press) p 269-76

[14] Adami C 2002 Ab inito modeling of ecosystems with artificial life *Natural Resource Modeling* **15** 133-45

[15] Adami C 2003 Sequence complexity in darwinian evolution *Complexity* **8(2)** 49-56

[16] Orfia C and Adami C 2002 Evolution of genetic organization in digital organisms *Evolution as Computation; DIMACS Workshop (January 1999, Princeton)* ed. L. F. Landweber and E. Winfree (Berlin Heidelberg : Springer-Verlag Berlin Heidelberg) p 296-313

[17] Orfia C, Bryson D M, Wilke C 2009 A software platform for research in computational evolutionary biology *Artificial Life Models in Software, 2nd edition* ed. M. Komosinski and M. Adamatzky (London : Springer) p 3-35

[18] Pargellis A N 1996 The spontaneous generation of digital "Life" *Physica D* **91(1-2)** 3-35

[19] Pargellis A N 1996 The evolution of self-replicating computer organisms *Physica D* **98(1)** 111-27

[20] Pargellis A N 2003 Self-organizing genetic codes and the emergence of digital life *Complexity* **8** 69-78

[21] Kol'chugina E A 2009 Non-equilibrium programming *Izv. Vyssh. Uchebn. Zav. Povolzh. Region. Techn. Nauki* **3(11)** 25-31

[22] Rudenko A P 2013 Self-organization and synergetics *Complex Systems* **2(7)** 4-39

[23] Kol'chugina E A 2016 Self-organization of program systems based on artificial chemistry's models *Neurocomputers* **2** 77-83

[24] Kol'chugina E A 2018 Spontaneous emergence of programs from "primordial soup" of functions in distributed computer systems *Automatic Control and Computer Sciences* **52(1)** 40-8

[25] Kasterin N P 1937 *Generalization of the basic equations of aerodynamics and electrodynamics: Report at a Special meeting at Acad. Sciences of the USSR 9 December 1936* (Moscow: Academy of Science of USSR Publishing Office)

Hardware-software complex of a speech-like interference generator for speech information protection

V V Alekseev, A V Yakovlev, M V Moiseeva and A A Tikhomirova

Tambov State Technical University, Sovetskaya street, 106, Tambov, 392000, Russia

E-mail: mariyamoiseeva@mail.ru

Abstract. The article describes a model of a speech-like interference generator based on a system analysis of ways to solve the problem of protecting speech information in office-type premises. The results of the effectiveness of the model application are presented.

1. Relevance

Many companies suffer significant financial losses every year due to leakage of confidential information. To combat this problem, they allocate significant financial resources, take various protective measures, but often do not pay due attention to the technical channels of speech information leakage, which is a significant mistake.

In acoustic information leakage channel, the technical unmasking (intelligence) features of the objects of protection are acoustic waves. Such leakage channels are typical for acoustic verbal intelligence (for intercepting speech information from places of human communication) and acoustic signal intelligence (for obtaining intelligence information about acoustic «portraits» of various acoustic devices, whose work is accompanied by acoustic fields).

Speech is a source of acoustic information, the carrier of acoustic information is acoustic signals. In air, the signal propagates in the form of a longitudinal elastic wave, which is the vibration of air particles along the direction of wave propagation. The acoustic signal can be intercepted by an intruder using a microphone, which converts the signal into electrical and allows to transmit the signal outside the monitored room.

The main means of protecting speech information from leakage through these channels are soundproofing of premises, search for «bugs» and active acoustic masking. The basis of masking means are audio jamming devices. In practice, noise generators have found the most widespread use. Application of this method makes it possible to reduce the signal-to-noise ratio at the input of the intelligence equipment by increasing the level of interference.

However, acoustic interference created by technical means of information security should not bring significant discomfort to the participants in the dialogue present in the room. The problem arises of bringing the power of acoustic interference to an optimal level that meets the requirements of both the room security and comfortable conversation.

2. Analysis of ways to solve the problem

The process of speech perception in noise is accompanied by the loss of the components of the speech message. The intelligibility of a speech message is characterized by the number of correctly received words that reflect the qualitative area of intelligibility, which is expressed in the categories of the details of the certificate of the intercepted conversation compiled by the attacker.

As a solution, it was proposed to use an instrumental-computational method for assessing speech intelligibility, which allows you to select the level of noise interference of the protection means in such a way that the radiation of the generator is minimal, but ensures the protection of speech information from leakage. The power that meets these criteria is proposed to be considered optimal. Thus, it is necessary to measure the parameters set by the method, calculate the verbal intelligibility and adjust the masking tool to the optimal radiation level.

The method of forming speech-like interference correlated in level, spectrum and time with a hidden signal is the most effective way to actively protect speech information.

The use of speech-like interference in the information protection system against leaking over an acoustic channel is not only an effective method of preventing speech information leakage, but also is a «softer» method in relation to people in the room. A lower noise level required to achieve the set level of protection (relative to the use of other types of generators), a signal spectrum close to human speech – these factors make working in the space to be protected more comfortable and calm.

In systems of active speech information protection in rooms for negotiations, it is possible to use speech-like signals and speech sequences formed taking into account the linguistic features of the language and the statistical characteristics of the occurrence of phonemes in this language, as well as the length of words and sentences as masking signals. An overview of the active protection system against unauthorized eavesdropping is shown in figure 1.

The formation of speech-like signals can be performed by the compilation method based on the structural units of speech. As a result, the speech-like signals formed in this way retain all the shades of the speech of a certain speaker, and it is very difficult to distinguish them from the information signals of the same speaker.

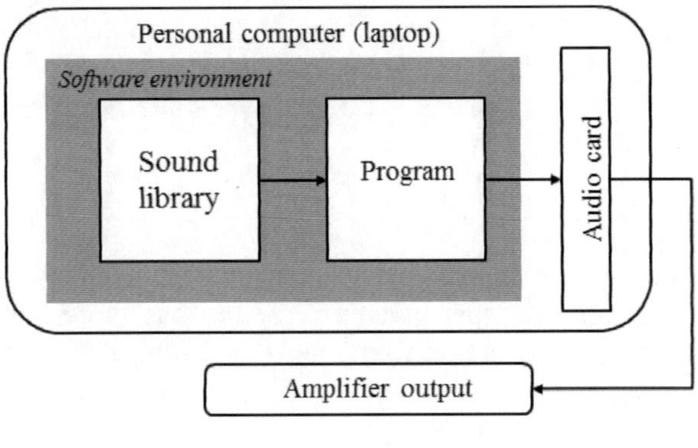

Figure 1. Overview of the active protection system against unauthorized eavesdropping.

To improve the quality of synthesized speech, you can use exponential spline functions at the boundaries of the transition from one phonemic structure to another and by superimposing the end of one phonemic structure on the beginning of the second phonemic structure. In this case, there will be some faster attenuation of the oscillation amplitudes of the end of one phonemic structure and an increase in the amplitude of the beginning of the second phonemic structure. This mechanism of compilation speech synthesis will eliminate signal jumps at the boundaries of phonemic structures. To perform such a compilation speech synthesis, you need a database of phonemic structural units of speech with slightly enlarged segments in the time domain, since during the synthesis the end of one phonemic structure is superimposed on the beginning of the second phonemic structure.

The analysis of speech segmentation methods showed that for the formation of the base of phonemic structural units of speech for speech synthesis by the compilation method, the segmentation method using dynamic programming is the most convenient. To do this, you must have a phonetic record of continuous speech, labeled manually into phonemic structural elements and containing all phonemic

IOP Conf. Series: Materials Science and Engineering 1155 (2021) 012055 doi:10.1088/1757-899X/1155/1/012055

structural units necessary for the base. Usually these are 300-400 allophones for Russian, Kazakh, Belarusian speech and about 1200 phonemic structural units.

Since the phonetic bases of the structural elements of speech at the beginning and at the end contain transition areas, splines for the transition areas should be used in speech synthesis. It is recommended to use the so-called «stitching» of allophones when compiling speech synthesis. The transition section of the end of the previous allophone, multiplied by a decreasing function varying from 1 to 0, is superimposed on the transition section of the subsequent allophone, multiplied by an increasing function from 0 to 1. If the lengths of the overlapping transition sections are not equal, then the length of the formed transition region is chosen equal to the length of the longer transition section.

Table 1. Values of signal-to-noise ratios at which the required efficiency of protection of acoustic (speech) information is provided.

Type of interference	Verbal intelligibility W, %	Signal-to-noise ratio q_i in octave bands					Signal-to-noise ratio in the frequency band 180 ... 5600 Hz
		250	500	1000	2000	4000	
«White» noise	20	+ 0.8	- 2.2	- 10.7	- 18.2	- 24.7	- 10.0
	30	+ 3.1	+ 0.1	- 8.4	- 15.9	- 22.4	- 7.7
	40	+ 5.1	+ 2.1	- 6.4	- 13.9	- 20.4	- 5.7
«Pink» noise	20	- 5.9	- 5.9	- 11.4	- 15.9	- 19.4	- 8.8
	30	- 3.7	- 3.7	- 9.2	- 13.7	- 17.2	- 6.7
	40	- 1.9	- 1.9	- 7.4	- 11.9	- 15.4	- 4.9
Noise with the decline in the spectral density of 6 dB per octave	20	- 14.1	- 11.1	- 3.6	- 15.1	- 15.6	- 13.0
	30	- 12.0	- 9.0	- 11.5	- 13.0	- 13.5	- 10.8
	40	- 10.0	- 7.2	- 9.7	- 11.2	- 11.7	- 9.0
Noise speech-like interference	20	- 3.9	- 7.9	- 12.9	- 15.9	- 16.9	- 9.0
	30	- 1.7	- 5.7	- 10.7	- 13.7	- 14.7	- 6.8
	40	+ 0.1	- 3.9	- 8.9	- 11.9	- 12.9	- 5.0

Acoustic and vibroacoustic masking systems use interference of both "white" and "pink" noises, as well as speech-like interference. In protection complexes, interference is used to mask speech, which is similar in structure to masked speech. This can be interference from an external source or interference from a speech-like noise synthesizer by a phonemic cloner. The interference generated by such a synthesizer is not just speech-like, the phonemic cloner ensures the formation of such interference that maximally matches the sounds of the speech of a particular person or group of people whose conversations are protected from eavesdropping.

3. Model description

A model of a speech-like noise generator based on the Matlab mathematical package is proposed. The model is built on the basis of standard objects of the Simulink visual design environment. In the presented model, it is proposed to multiply the transition areas by linear functions varying from 1 to 0 and from 0 to 1. However, due to the fact that hearing sensitivity is nonlinear, it is more efficient to use spline functions of a higher order (up to the third degree).

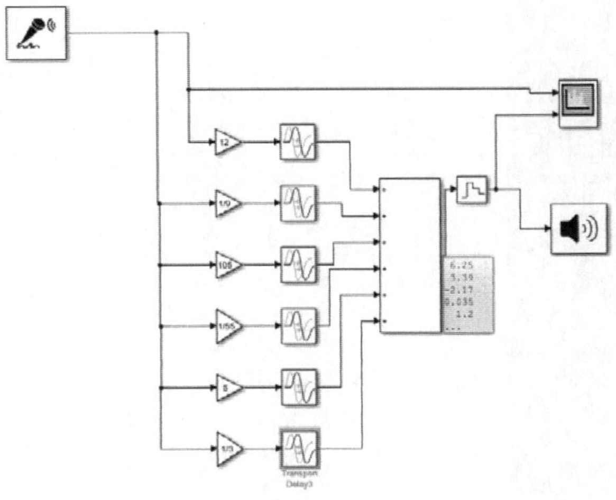

Figure 2. Model of a speech-like interference generator with 6 shaper stages.

Studies of the model have shown that an increase in the number of allophone cascades leads to a change in both the spectrum of the generated speech-like interference and its amplitude. These changes are clearly visible in the transition from a three-stage scheme to a four-stage one.

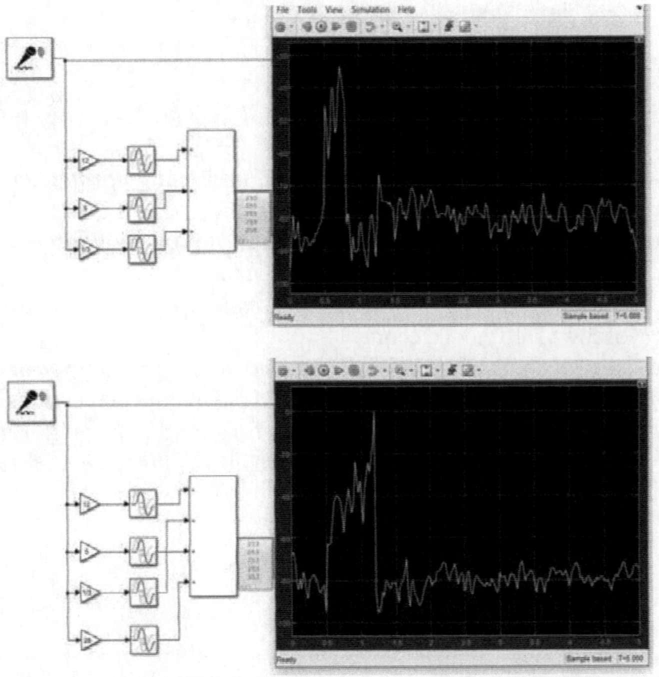

Figure 3. Simulation results of a speech-like interference generator with a different number of shaper stages.

Figure 4 shows the spectra of the speech message of the interlocutors and the speech-like interference synthesized by the generator. The simulation result shows a sufficient similarity of these spectra.

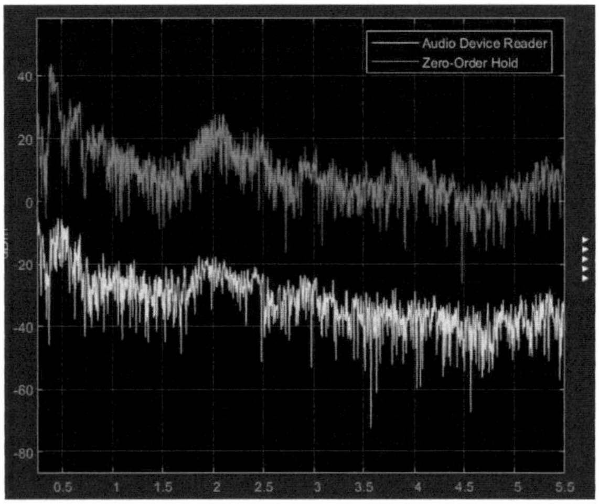

Figure 4. Spectra of the speech message of the interlocutors and the speech-like interference synthesized by the generator.

4. Conclusion

Thus, a model of a speech-like interference generator based on the Matlab mathematical package has been synthesized. The model is built on the basis of standard objects of the Simulink visual design environment. The simulation result shows a sufficient similarity of the spectra of the speech message of the interlocutors and the speech-like interference synthesized by the generator.

References

[1] Tsaregorodtsev A V and Taraskin M M 2017 *Methods and means of information protection in public administration* (Moscow: Prospect)

[2] Lynkov L M, Golikov V F and Borbotko T V 2013 *Fundamentals of information protection and intellectual property management* (Minsk: BSUIR)

[3] Crow V A, Vorona V A and Kostenko V O 2016 Methods and means of protecting information from leakage through technical channels, *Computational nanotechnology*

[4] Buzov G A, Kalinin S V and Kondratiev A V 2005 *Protection against information leakage through technical channels* (Moscow: Hotline-Telecom)

[5] Zheleznyak V K, Makarov Yu K and Khorev A A 2000 *Some methodological approaches to assessing the effectiveness of protection of speech information* (Moscow: Special equipment)

[6] Khorev A A 2013 *Methods of protecting the allocated premises from the leakage of speech (acoustic) information through technical channels: vibroacoustic protection systems* (Moscow: Special equipment)

IOP Conf. Series: Materials Science and Engineering 1155 (2021) 012056 doi:10.1088/1757-899X/1155/1/012056

Recognition of handwritten MNIST digits on low-memory 2 Kb RAM Arduino board using LogNNet reservoir neural network

Y A Izotov[1,3], A A Velichko[1], A A Ivshin[2] and R E Novitskiy[2]

[1]Institute of Physics and Technology, Petrozavodsk State University, 33 Lenin str., 185910, Petrozavodsk, Russia

[2]K-SkAI LLC, premises 20, 17 Naberezhnaya Varkausa, 185031, Petrozavodsk, Russia

[3]E-mail: izotov93@yandex.ru

Abstract. The presented compact algorithm for recognizing handwritten digits of the MNIST database, created on the LogNNet reservoir neural network, reaches the recognition accuracy of 82%. The algorithm was tested on a low-memory Arduino board with 2 Kb static RAM low-power microcontroller. The dependences of the accuracy and time of image recognition on the number of neurons in the reservoir have been investigated. The memory allocation demonstrates that the algorithm stores all the necessary information in RAM without using additional data storage, and operates with original images without preliminary processing. The simple structure of the algorithm, with appropriate training, can be adapted for wide practical application, for example, for creating mobile biosensors for early diagnosis of adverse events in medicine. The study results are important for the implementation of artificial intelligence on peripheral constrained IoT devices and for edge computing.

1. Introduction

Neural networks occupy a leading position in the processing of large amounts of information in various fields of science and technology, for example, in recognition of medical images [1], modelling [2] and control [3]. The particular importance of the analysis of data sets in real time using artificial intelligence technologies can be illustrated by the early diagnosis of adverse events in medicine [4], where the information about the patient's condition is transmitted from biosensors to receiving devices using the concept of "Internet of Things" (IoT) [5] and data transmission networks between physical objects. The rapid growth of the global market for the Internet of Things of Medical Devices (IoMT) [6], improved confidentiality, high speed of real time data processing, significant reduction of technology cost, and improved quality of outpatient medical services drive a global paradigm shift in healthcare. Decentralization of medical care and a shift from "cloud" technologies for processing medical data to "Edge computing" and "Fog computing" [7] are expected in the near future. The integration of IoT and artificial intelligence (AI) will allow for intelligent pre-processing of data on the peripheral devices without sending information to the cloud, reflecting the key area of the contemporary scientific research in the field of the Internet of things [8] and edge computing [9]. The significant research task is the development of new neural network technologies that could be successfully implemented in the IoT

[10], widely used in daily activities, in healthcare [11], in designing robots [12] and in a smart homes [13], while consuming less computing resources and RAM memory.

The research on deep learning networks based on complex schemes using the convolutional filters is actively progressing [14]. Such neural networks require significant computing resources not only during training, but also during operation. An alternative for deep learning neural networks are neural networks based on reservoir computations [15]. The main idea of the reservoir computing is to use the recurrent neural network as a reservoir with rich dynamics and powerful computing capabilities. In this case, the reservoir is formed randomly, and it eliminates the need to conduct its training the in most cases. Only the output neural network, the classifier, is trained. The most popular networks are echo state networks (ESN [16]) and liquid state machines (LSM [17]). Until recently, reservoir networks did not find practical implementation on devices with low computational resources and memory, as storing a significant array of weights was a major problem. To address this issue, we have proposed the architecture of the neural network LogNNet [18], which uses a recurrent logistic mapping that describes a model of deterministic chaos. The chaotic filters, which affect the incoming information, are generated during the operation of the algorithm. As a result, the information is chaotically mixed, however, the valuable signs, which are not visible from the beginning, are extracted from the data. Reservoir neural networks use a similar mechanism. With a simple feedforward network architecture, where signals propagate exclusively from input to output, LogNNet is easy to program and install on hardware platforms. Chaotic filters are created during program operation and consume an insignificant amount of RAM.

Compact algorithms are usually tested on well-known databases such as MNIST-10 (the database of handwritten numbers [19]), CIFAR10/100 (the database of photographs of various objects [20]), Chars4K (the database of graphic symbols [21]). LogNNet was tested on MNIST-10 database and demonstrated a recognition accuracy of ~ 96.3% [18], whereas the LogNNet architecture used no more than 29 Kb of RAM. Other compact algorithms that operate on peripheral devices with 2-16 Kb RAM are algorithms based on the GBDT decision tree [22], Bonsai [23], and ProtoNN [24], which uses the kNN method. It is worth to mention the algorithm that implements a convolution neural network and uses only 2 Kb RAM [25], demonstrating a classification accuracy of ~ 99.15% on MNIST-10 database. Algorithms based on the architecture of the recurrent network Spectral-RNN [26] and FastGRNN [27] achieve ~ 98% classification accuracy on MNIST-10 with a model size of ~ 6 Kb. All mentioned algorithms have a complex structure, and their adaptation for the user's tasks requires significant efforts.

Arduino boards are effectively used to automate processes in many areas, such as education [28], medicine [29], chemical production [30], and resource monitoring in the IoT [31]. A project on an Arduino board for MNIST-10 recognition is described in the literature [32], which uses a linear classifier with two hidden layers and achieves the recognition accuracy of ~ 96%. This result is comparable to the results for the LogNNet network 784:100:60:10 [18]. However, the method in project [32] applies preliminary down sampling of images, and it requires third-party computing resources in addition to the Arduino board.

In the current study, we investigate the capabilities of the LogNNet neural network algorithm for recognizing handwritten digits from the MNIST-10 database on an Arduino UNO board with an extremely low ~ 2 Kb RAM (in datasheet SRAM). The images are received in the original form, and the image recognition happens entirely on the board. The algorithm is simple to implement and can be used for scientific and educational purposes. The research results have the application potential for the tasks of integrating IoT and AI.

2. Architecture of the neural network LogNNet

The LogNNet architecture developed in [18] is demonstrated in figure 1. Its configuration is designed to recognize handwritten digits with a size of 28 × 28 pixels from the MNIST-10 database (available on Yan LeCun's Internet page [19]). An image of a digit is fed into the network input, and then the image is transformed into a one-dimensional array Y using the T-pattern transformation. We used the simplest template shown in figure 1 with per-column image input. An array Y is fed to the input of the reservoir.

The array contains the values of the pixel intensities normalized to 1, has 784 values in total, and the zero offset element is $Y[0] = 1$.

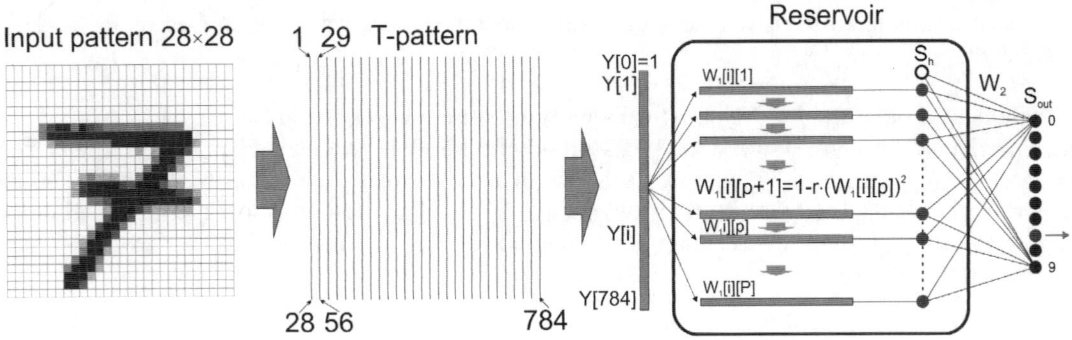

Figure 1. LogNNet neural network architecture.

The reservoir and the output neural network operate as feedforward networks, according to the equations

$$S_h = f_h(Y \cdot W_1)\ , \tag{1}$$

$$S_{out} = f_{out}(S_h \cdot W_2)\ . \tag{2}$$

In equations (1) and (2), S_h are the neurons of the hidden layer located in the reservoir, S_{out} are the neurons of the output layer, Y is the array of input data, W_1, W_2 are the matrix of weight coefficients, from the input to the hidden layer and from the hidden layer to the output layer, respectively, f_h is the identical activation function, with subsequent normalization of values in the range from -0.5 to 0.5, and f_{out} is the logistic activation function.

The matrix W_1 is set using the equations (3-4):

$$W_1[i][1] = A \cdot \sin\left(\frac{i}{784} \cdot \frac{\pi}{B}\right)\ , \tag{3}$$

$$W_1[i][p+1] = 1 - r \cdot W_1[i][p] \cdot W_1[i][p], \tag{4}$$

where r, A and B are constants, i and p are variables, $i = 0\text{-}784$, $p = 1..P$, P is the number of neurons in the hidden layer.

Equation (4) is a logistic mapping where the next value is calculated based on the previous one. The equation is commonly used in population biology, and is an example of a simple equation for calculating a sequence of chaotic values. Equations (3-4) can fill the matrix W_1 of any dimension with chaotic numbers. The elements of the W_1 matrix are sequentially calculated during the network operation, and occupy a small amount of RAM memory. This approach saves device memory; however, it additionally loads the processor with calculations and increases the network operation time.

The matrix of coefficients W_2 is trained using the backpropagation method on the MNIST-10 database. The database contains 60,000 training handwritten images and 10,000 test images of numbers from 0 to 9, in the form of a matrix of 28×28 pixels. The brightness of a pixel is represented by a value from 0 to 255.

The normalization coefficients of the function f_h are calculated from the values of S_h using the first 1000 elements of the training set. The maximum, minimum, and weighted mean values of S_h for each neuron of the hidden layer were used as normalization coefficients.

The constants *r*, *A* and *B* were selected by optimization methods using the particle swarm method, where the fitness function was the function of the neural network recognition accuracy. The values of these constants are given in the appendix (figure A2).

After determining all the parameters, the trained neural network was programmed on the Arduino UNO board.

3. Implementation of the LogNNet network on the Arduino UNO board

The objective of the study is to test the LogNNet architecture for recognizing handwritten digits from the MNIST-10 database on a microcontroller with an extremely low amount of RAM. For the implementation of the LogNNet neural network and its subsequent testing, a low-memory 2 Kb RAM board Arduino UNO with an ATmega 328P microcontroller is chosen. Table 1 shows the main characteristics of the board.

Table 1. Characteristics of the Arduino UNO board.

Characteristic	Value
SRAM	2048 bytes
Flash memory	32 Kbytes
Microprocessor	ATmega 328P
Clock frequency	16 MHz
Weight	46 g
Board dimensions	69×53×20 (mm)

In [18], we demonstrated that with an increase in the number of neurons in the hidden layer of the *P* reservoir, the recognition accuracy of MNIST-10 digits increases. The main limiting factor for the number of neurons in the hidden layer, when implementing LogNNet on the ATmega 328P microcontroller, is the amount of static RAM. The mathematical calculations for the network operation require a large number of variables of various types that take up RAM memory. The optimal configuration of the neural network LogNNet 784:20:10 was selected, where the numbers indicate the parameters of the network: $S = 784$ - the size of the input data array Y, $P = 20$ - the number of neurons in the hidden layer S_h, and $M = 10$ - the number of neurons in the output layer S_{out} that determine the result of the network operation.

A fragment of the executable program code describing the main “void loop” is presented in figure 2a, and the functions and procedures used are shown in figure A1 in Appendix. The program is written in the C programming language in the Arduino IDE development environment, and has a compact form in the editor, occupying 62 lines and 5934 bytes of the device's flash memory. All necessary constants (S, P, M), values of the weights of the W_2 arrays and normalization parameters are contained in the library file LogNNet.h (see figure A2, Appendix) and are loaded into the RAM memory at the start of the program. At the start of the program, the serial port of the board is initialized by the procedure Serial.begin(115200) in the lines 45-47 of the “void setup” code block, as demonstrated in figure 2a.

The testing process is performed by transferring images to the board one by one through the serial port. The transferred images have the original scale of 28 × 28 pixels, and are not pre-processed.

The algorithm of the program's operation checks in a loop the presence of data in the input buffer via the Serial.available function and reads all 784 bytes of the Y array. Next, the neural network response is calculated.

At the first stage, a preliminary calculation of the neurons of the hidden layer takes place, the code for the Hidden_Layer_Calculation procedure is shown in figure A1 (a) (see Appendix, lines 7-19). The procedure calculates the weight coefficients W_1 using Equaitons (3, 4), then the arrays Y and W_1 are multiplied in accordance with Equation (1). The values of the hidden layer neurons are averaged and normalized for the final calculation. The normalization procedure is called

Hidden_Layer_Normalization and its code is demonstrated in figure A1 (a) (see Appendix, lines 21-27). The procedure uses the minimum (*minH*), maximum (*maxH*) and weighted average (*meanH*) parameters from the LogNNet.h library file. These values are pre-determined during network training, and they are floating point numbers. The values are written to the microcontroller memory in integer form (to save RAM), and they are divided by a scale factor for the reverse conversion to the "float" type. For arrays *minH* and *maxH*, the scale factor is 100, and for the *meanH* array, the coefficient is 10000. The final values of the neurons of the hidden layer are stored in the S_h array. The length of the S_h array is 21, and it contains 20 hidden layer neurons calculated in the Hidden_Layer_Normalization procedure and 1 bias neuron.

```
#include "LogNNet.h"

float Sh[P+1];
int i = 1;
byte Y[S+1];
⋮

void setup() {
  Serial.begin(115200);
}

void loop() {
  if (Serial.available() > 0) {
    Y[i] = Serial.read();
    i++;
    if (i == S+1) {
      Y[0] = 255;
      i = 1;
      Hidden_Layer_Calculating(Y);
      Hidden_Layer_Normalization();
      byte Digit = Output_Layer();
      Serial.print(String(Digit));
    }
  }
}
```

(a)

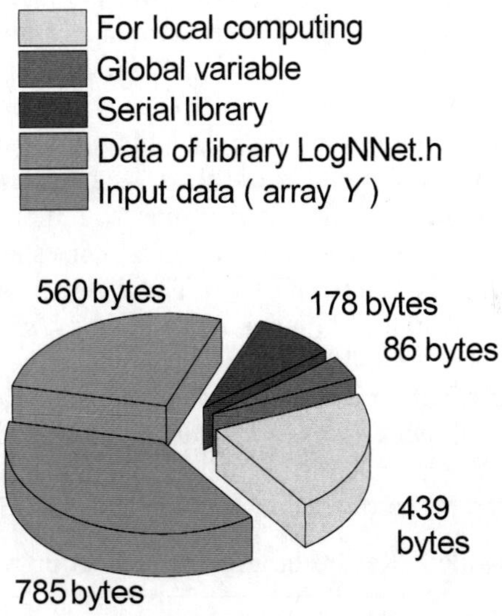

(b)

Figure 2. A fragment of the executable program for the Arduino UNO (a), and the allocation of RAM memory of the microcontroller board for the configuration LogNNet 784:20:10 (b).

The neurons of the output layer are calculated in the Output_Layer function, and the result of the neural network operation is obtained; the code is demonstrated in figure A1 (b) (see Appendix, lines 29-43). The function calculates the values of the S_{out} array and outputs the most probable digit corresponding to the recognized image, in the range from 0 to 9 (variable *Digit*). The function uses an array of weighting coefficients W_2, which was obtained during preliminary network training on a computer. The W_2 array is stored in the controller RAM in integer form and is loaded from the LogNNet.h library. The scale factor of the W_2 array equals to 100.

3.1. Memory allocation

The allocation of the microcontroller's RAM is presented in figure 2b. Global variables *Sh* and *i* occupy 86 bytes in RAM (4.2%). The data of the LogNNet.h library are loaded at the start of the program and occupy 560 bytes (27.34%). The program uses a library for working with a serial port (Serial), which occupies 178 bytes (8.68%). Input data from the serial port is written to array *Y*, which is 785 bytes (38.33%). As a result, with 20 hidden layer neurons, the volume of permanently used RAM is 1609 bytes, and the microcontroller has 439 bytes left for mathematical calculations.

3.2. Neural network training and testing

The training of the neural network LogNNet is performed on a computer.

The process of network testing consists of sequentially downloading each of the 10,000 pictures of the MNIST-10 database test set through the serial port, calculating S_{out} and receiving the *Digit* result back to the computer. This technique allows full test of the network on the entire test set, without loading the set into the board's memory.

4. Results and discussion

Table 2 presents the results of the MNIST-10 image recognition accuracy using the LogNNet network with a different number of neurons *P* in the reservoir, and the processing time of one image by the Arduino UNO board. The image recognition accuracy on a personal computer and on the Arduino UNO board differ after the 2nd decimal place of the result number. It may be explained by the method of storing the W_2 weights and normalization parameters in integer form, when significant digits of numbers are lost. In addition, Arduino UNO uses the float type numbers with 8 significant digits for calculations, while the computer software (Delphi programming language) operates with the real type numbers with 16 significant digits. Nevertheless, such good accuracy is a significant result and allows preliminary testing of the network on a computer with more processing power.

The highest recognition accuracy corresponds to the network configuration with the number of neurons in the reservoir $P = 20$. The accuracy reaches values of ~ 82% and is comparable, for example, with the NeuralNet Pruning network, which has an accuracy of ~ 81% [33].

An increase in the number of neurons in the reservoir ($P > 20$) does not give a noticeable increase in the classification accuracy and requires additional expansion of RAM memory. For each additional neuron, 30 bytes of RAM are added to increase the arrays S_h, *minH*, *maxH*, *meanH*, and W_2. A further accuracy increase can be achieved by internal preprocessing of the array *Y* or by increasing the number of hidden layers of the output classifier. These ideas can be as a subject for future research.

Table 2. Results of LogNNet operation with different number of neurons P in the reservoir.

Configuration of network LogNNet *S*:*P*:*M*	Recognition accuracy of MNIST-10 image on a computer	Recognition accuracy of MNIST-10 image on Arduino UNO board	Recognition time of one image on Arduino UNO board
LogNNet 784:20:10	82.04 %	82.03 %	7.11 s
LogNNet 784:18:10	80.46 %	80.45 %	6.01 s
LogNNet 784:15:10	76.89 %	76.94 %	4.06 s
LogNNet 784:10:10	70.00 %	70.01 %	2.06 s

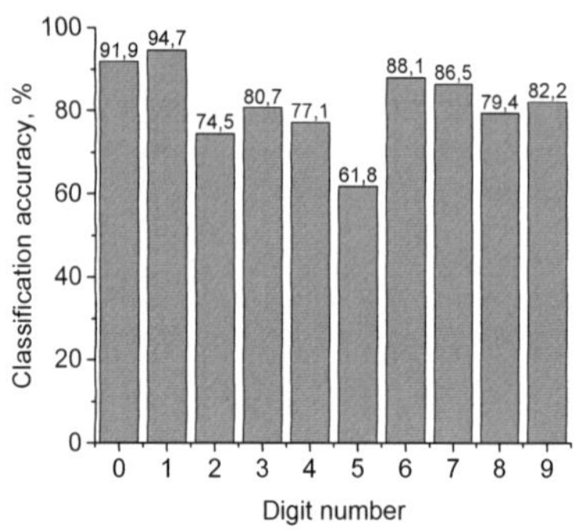

Figure 3. Distribution of classification accuracy by digits for LogNNet 784:20:10.

The distribution of the classification accuracy of individual digits from the MNIST-10 base for LogNNet 784:20:10 is shown in figure 3. The neural network recognizes digits with different probability. For the digit "1", the probability is the highest and equals to ~ 94.7%, or, in other words, 769 images out of 1032 possible images are correctly recognized. The network showed the worst result when recognizing the digit “5”, when the probability was ~ 61.8% out of 892 images. The similar reduced classification accuracy of the digit “5” was observed in other studies, for example, in Hopfield network [34]. The reason for this may be the similarity of the digit “5” to digits “3”, “6” and “9”.

The recognition time for one image on the board is several seconds. The time is mainly determined by the duration of the sequential calculation of the W_1 array elements in the Hidden_Layer_Calculation procedure.

The recognition time can be reduced by applying the faster Algorithm 2 described in [18], but algorithm implementation requires a microcontroller with a large amount of RAM memory. Such processing speed of input data is acceptable for private practical applications that do not require a quick response, for example, in environmental monitoring systems or medical monitoring. The range of tasks to be solved by LogNNet is not limited to recognizing the MNIST database images, and can be expanded to other areas after corresponding retraining.

5. Conclusion

The study presents a compact algorithm for recognizing handwritten digits of the MNIST-10 database, created on the basis of the LogNNet reservoir neural network. The algorithm is tested on the low-memory Arduino UNO board, which has a low-power microcontroller and 2048 bytes of RAM. Classification accuracy of ~ 82% was achieved with a small ($P = 20$) number of neurons in the reservoir. The program code is compact, simple in structure and can be used in the scientific and educational processes. With appropriate training, the presented LogNNet algorithm can be adapted for wide practical applications, for example, for creating mobile biosensors for early diagnosis of adverse events in IoMT, and implemented on a wide range of hardware and software, for example, Raspberry PI, Orange Pi, Arduino, Microduino and Femtoduino. The results facilitate the creation of prototypes of peripheral devices with low RAM, performing required functions using own artificial intelligence in the space of the Internet of Things.

Acknowledgment

The authors express their gratitude to Dr. Andrei Rikkiev for the valuable comments in the course of the article translation and revision.

Appendix A

```

void Hidden_Layer_Calculating(byte Y[S+1]) {
  float W1 = 0;
  Sh[0] = 1;
  for (int j = 1; j <= P; j++) {
    Sh[j] = 0;
    for (int i = 0; i <= S; i++) {
      W1 = A * sin(i/float(S)*PI/B);
      for (int k = 2; k <= j; k++)
        W1 = 1 - (r * W1 * W1);
      Sh[j] = Sh[j] + Y[i]/255.0 * W1;
    }
  }
}

void Hidden_Layer_Normalization() {
  Sh[0] = 1;
  for (int j = 1; j <= P; j++)
    Sh[j] = ((Sh[j] - minH[j-1]/100.0) /
      (maxH[j-1]/100.0 - minH[j-1]/100.0))
        - 0.5 - meanH[j-1]/10000.0;
}
```

(a)

```

byte Output_Layer() {
  float Sout[M];
  for (int j = 0; j < M; j++) {
    Sout[j] = 0;
    for (int i = 0; i <= P; i++)
      Sout[j] = Sout[j]+Sh[i]*W2[i][j]/100.0;
    Sout[j] = 1 / (1 + exp(-1*Sout[j]));
  }
  byte digit = 0;
  for (int i = 0; i < M; i++) {
    if (Sout[i] > Sout[digit])
      digit = i;
  }
  return digit;
}

```

(b)

Figure A1. The procedures Hidden_Layer_Calculation, Hidden_Layer_Normalization (a) and the Output_Layer function (b) used for calculating the LogNNet network on the Arduino UNO.

```
#pragma once
#include <Arduino.h>

#define S 784
#define P 20
#define M 10

#define A 0.672304710111946
#define B 1.57740808024101
#define r 1.86376633343066

int maxH[P] = {16458, 13985, 16194, 4846, 9755, 12284, 5592, 5826, 9223,
              6653, 3504, 11273, 3993, 5204, 7919, 3889, 9294, 6141, 5063, 5334};
int minH[P] = {1043, 944, 900, -2416, -1539, -2423, -2509, -816, -761, -2896,
              -2205, -125, -2282, -457, -17, -2211, -112, -276, -858, 37};
int meanH[P] = {-2188, -2301, -2094, -595, -1707, -1204, -667, -1419, -1395,
             -846, -1025, -1562, -451, -1382, -1471, -1016, -1929, -1348, -1137, -1274};
int W2[P+1][M] = {
                {-491, -915, -453, -353, -455, -419, -516, -584, -324, -431},
                {211, -522, 120, -463, -509, -115, 96, -412, 344, 16},
                {17, -1044, 775, -133, -1325, -149, -105, -806, -539, -741},
                {304, 56, -933, -890, 535, -152, 283, 229, 1362, 965},
                {-95, -38, -2354, -4, 965, 219, -1500, 1769, 1154, 1590},
                {-620, 4, -1564, -710, 2701, -1338, 1485, -657, 331, 897},
                {-1012, 1187, 1633, -1741, -1798, -1191, 126, 667, 2425, 46},
                {-644, -242, 391, -1942, -334, -64, 745, -131, 914, 838},
                {-317, -523, -594, -825, -988, 991, 1085, -257, 373, 99},
                {105, 565, -1483, 1147, 614, 1539, -364, -508, -389, -1703},
                {-819, 331, 196, 2209, -900, -340, -471, 118, -781, -684},
                {1250, -2406, -340, 925, -580, 416, -98, -850, -727, -664},
                {-200, -1120, -480, 1527, -516, 245, -685, 68, -1278, -331},
                {137, 542, -203, 464, -111, 323, -479, -399, -237, -77},
                {7, -352, 312, -839, 778, 836, -501, -154, 1087, 220},
                {712, 1207, 1556, -872, 431, 501, -752, -1718, 668, -1003},
                {602, -996, 745, -123, 965, -1148, -1411, 1678, -923, -888},
                {1558, -856, 14, 410, -1222, -1103, 752, -14, -1345, 602},
                {-305, -712, 645, 913, -46, -1199, -969, -469, 86, 727},
                {-588, -72, 655, 733, -20, -930, -197, 501, -1088, -244},
                {-541, -1298, -427, 564, 599, 342, 218, 997, -947, -444}
};
```

Figure A2. LogNNet.h library source code.

The source code version is available in the GitHub repository:
https://github.com/izotov93/LogNNet_Arduino

References

[1] Yamashita R, Nishio M, Do R K G and Togashi K 2018 Convolutional neural networks: an overview and application in radiology *Insights Imaging* **9** 611-29

[2] Mohandes S R, Zhang X and Mahdiyar A 2019 A comprehensive review on the application of artificial neural networks in building energy analysis *Neurocomputing* **340** 55-75

[3] Sohangir S, Wang D, Pomeranets A and Khoshgoftaar T 2018 Big Data: Deep Learning for financial sentiment analysis *J. Big Data* **5**

[4] Obermeyer Z and Emanuel E J 2016 Predicting the Future — Big Data, Machine Learning, and Clinical Medicine *N. Engl. J. Med.* **375** 1216-9

[5] Hamdan S, Ayyash M and Almajali S 2020 Edge-Computing Architectures for Internet of Things Applications: A Survey *Sensors* **20**

[6] Al-Turjman F, Nawaz M H and Ulusar U D 2020 Intelligence in the Internet of Medical Things era: A systematic review of current and future trends *Comput. Commun.* **150** 644-60

[7] Klonoff D C 2017 Fog Computing and Edge Computing Architectures for Processing Data From Diabetes Devices Connected to the Medical Internet of Things *J. Diabetes Sci. Technol.* **11** 647-52

[8] Ghosh A, Chakraborty D and Law A 2018 Artificial intelligence in Internet of things *CAAI Trans. Intell. Technol.* **3** 208-18

[9] Osia S A, Shamsabadi A S, Taheri A, Rabiee H R and Haddadi H 2018 Private and Scalable Personal Data Analytics Using Hybrid Edge-to-Cloud Deep Learning *Computer (Long. Beach. Calif).* **51** 42-9
[10] Chawla A, Babu P, Gawande T, Aumayr E, Jacob P and Fallon S 2021 *Intelligent Monitoring of IoT Devices using Neural Networks*
[11] Arora M, Dhawan S and Singh K 2021 *Deep Learning in Health Care: Automatic Cervix Image Classification Using Convolutional Neural Network* 145-51
[12] Liao J-C, Chen S-H, Zhuang Z-Y, Wu B-W and Chen Y-J 2021 Designing and Manufacturing of Automatic Robotic Lawn Mower *Process* **9**
[13] Siswipraptini P, Aziza R, Sangadji I, Indrianto I and Siregar R 2019 *Automated Smart Home Controller Based on Adaptive Linear Neural Network*
[14] Yu H, Li H, Shi H and Hua G 2020 Any-Precision Deep Neural Networks *Artif. Intell. Eng.* 93-113
[15] Tanaka G, Yamane T, Héroux J B, Nakane R, Kanazawa N, Takeda S, Numata H, Nakano D and Hirose A 2019 Recent advances in physical reservoir computing: A review *Neural Networks* **115** 100-23
[16] Jaeger H 2001 The "Echo State" Approach to Analysing and Training Recurrent Neural Networks *GMD-Report 148, Ger. Natl. Res. Inst. Comput. Sci.*
[17] Maass W, Natschläger T and Markram H 2002 Real-Time Computing Without Stable States: A New Framework for Neural Computation Based on Perturbations *Neural Comput.* **14** 2531-60
[18] Velichko A 2020 Neural Network for Low-Memory IoT Devices and MNIST Image Recognition Using Kernels Based on Logistic Map *Electronics* **9** 1432
[19] Baldominos A, Saez Y and Isasi P 2019 A Survey of Handwritten Character Recognition with MNIST and EMNIST *Appl. Sci.* **9**
[20] Chai F and Kang K-D 2021 Adaptive Deep Learning for Soft Real-Time Image Classification *Technol.* **9**
[21] Fedorov I, Adams R, Mattina M and Whatmough P 2019 SpArSe: Sparse Architecture Search for CNNs on Resource-Constrained Microcontrollers *NeurIPS*
[22] Friedman J H 2002 Stochastic gradient boosting *Comput. Stat. Data Anal.* **38** 367-78
[23] Kumar A, Goyal S and Varma M 2017 Resource-Efficient Machine Learning in 2 KB RAM for the Internet of Things *Proceedings of the 34th International Conference on Machine Learning* **70** ICML'17 (JMLR.org) 1935-44
[24] Gupta C, Suggala A S, Goyal A, Simhadri H V, Paranjape B, Kumar A, Goyal S, Udupa R, Varma M and Jain P 2017 ProtoNN: Compressed and Accurate kNN for Resource-scarce Devices *Proceedings of the 34th International Conference on Machine Learning* Proceedings of Machine Learning Research **70** ed D Precup and Y W Teh (International Convention Centre, Sydney, Australia: PMLR) 1331-40
[25] Gural A and Murmann B 2019 Memory-Optimal Direct Convolutions for Maximizing Classification Accuracy in Embedded Applications *Proceedings of the 36th International Conference on Machine Learning* Proceedings of Machine Learning Research **97** ed K Chaudhuri and R Salakhutdinov (Long Beach, California, USA: PMLR) 2515-24
[26] Zhang J, Lei Q and Dhillon I 2018 Stabilizing Gradients for Deep Neural Networks via Efficient {SVD} Parameterization *Proceedings of the 35th International Conference on Machine Learning* Proceedings of Machine Learning Research **80** ed J Dy and A Krause (Stockholmsmässan, Stockholm Sweden: PMLR) 5806-14
[27] Kusupati A, Singh M, Bhatia K, Kumar A, Jain P and Varma M 2018 FastGRNN: A Fast, Accurate, Stable and Tiny Kilobyte Sized Gated Recurrent Neural Network. *Advances in Neural Information Processing Systems*
[28] López-Belmonte J, Marín Marín J, Costa R and Moreno Guerrero A 2020 Arduino Advances in Web of Science. A Scientific Mapping of Literary Production *IEEE Access* **PP** 1
[29] Wu H, Dyson M and Nazarpour K 2021 Arduino-Based Myoelectric Control: Towards

Longitudinal Study of Prosthesis Use *Sensors* **21**

[30] Wijaya N, Novela D, Shahu N and Sattar M 2020 Arduino-based Mini Shaker for Automatic Chemical Solution Mixer *J. Robot. Control* **1**

[31] Mozaffari Legha M and Farjah E 2020 IoT Based Load Management of a Micro-Grid Using Arduino and HMAS TT - *IUST* **16** 228-34

[32] Chovanec M Deep neural network on arduino - MNIST handwritten | Hackaday.io

[33] Han S, Mao H and Dally W 2016 Deep Compression: Compressing Deep Neural Network with Pruning, Trained Quantization and Huffman Coding *CoRR* **abs/1510.0**

[34] Belyaev M A and Velichko A A 2020 Classification of handwritten digits using the Hopfield network *IOP Conf. Ser. Mater. Sci. Eng.* **862** 52048

Face recognition using multitasking cascading convolutional networks

A G Musikhin[1] and S Yu Burenin

MIREA-Russian Technological University, Moscow, 119454, Russia

[1]E-mail: mag.iit.kvt@gmail.com

Abstract. The face recognition has many applications in such areas like intelligent security and access control, biometrics, safeguard, verification, attendance accounting, machine vision, etc. Identification of a personality by the face has a number of advantages over other methods: the person does not need to be contacted physically, which is the most acceptable way for mass applications and no complicated or expensive equipment is required. This article discusses the problem of recognition and identification of a person's face using convolutional neural networks that process frames from a camera in real time or from a recorded video file with the subsequent entry of the identified person into the database. Multitasking Cascade Convolutional Neural Network (MTCNN) has three convolutional networks (P-Net, R-Net, and O-Net) and is capable of outperforming many face detection tests while maintaining real-time performance. The proposed method for human face recognition was developed as a software product, tested and showed the probability of correct recognition in real time 96.02%.

1. Introduction

There is a growing interest in technology face recognition, whose task is to automatically localize faces on video and person identification. The concept of "recognition" can be determined as the assignment of the studied object to one of the known classes. "Identification" is the establishment of identity of the unknown object to the known one basing on the coincidence of recognized features.

The face recognition has many applications in such areas like intelligent security and access control, biometrics, safeguard, verification, attendance accounting, machine vision, etc. Traditional ways of identification such as keys or passwords do not provide high degree of reliability, therefore, the biometric identification systems are being developed dynamically. Identification of a personality by the face has a number of advantages over other methods:

- the person does not need to be contacted physically, which is the most acceptable way for mass applications;
- no complicated or expensive equipment is required.

Various techniques are used to solve the recognition problem, such as based on neural networks, on Karhunen–Loève decomposition, on Haar cascades, on the equal intensity lines, on the algebraic moments, on the elastic (deformable) samples.

The main problem of the most face recognition methods is the dependence of the angle, position, lighting conditions, etc. This problem can be minimized using the neural networks based methods.

For producing an automated system of the person identification and the attendance accounting Chintalapati et al. [1] in 2013 proposed a method based on the Viola Jones method [2] for face detection followed by the equalization histogram for feature extraction and the support vector machine (SVM) classifier for face recognition. Later in 2017 Rathod et al. [3] proposed an automated system of the same technique for face detection and a classifier for face recognition. However, these methods were based on classical machine algorithms. Arsenovic et al. [4] in 2017 proposed a modern FaceTime method using of cascaded convolutional neural network for face detection and convolutional neural network for facial features generation, which were then used for recognition. In this article proposed the structure of multi-task cascaded convolutional networks (MTCNN) [5] for face recognition and Siamese neural network [6] to extract the traits of the faces to create a 128- element vector representing these traits. Most commercial automated systems software use images of faces that should be taken close-up distance and which main condition for the successful work is an open, well-lit face. Image processing is carried out sequentially, which is a rather lengthy process.

2. Materials and methods

This article discusses the problem of recognition and identification of a person's face using convolutional neural networks that process frames from a camera in real time or from a recorded video file with the subsequent entry of the identified person into the database.

One of the main advantages of the proposed solution is robustness of the algorythm against common problems such as occlusion (partially visible / obscured faces), orientation of the face and the influence of room lighting.

The software processes the video file (frames from CCTV cameras) and detects faces using a new method based on the MTCNN [5], recognizes the detected faces using the Siamese neural network [6], which learns to transform facial images into a compact Euclidean space, where the distance corresponds to the measure similarity of faces, and adds the result of face recognition to the database.

Multitasking Cascade Convolutional Neural Network (MTCNN), shown in figure 1, has three convolutional networks (P-Net, R-Net, and O-Net) and is capable of outperforming many face detection tests while maintaining real-time performance. An image is taken and resized to different scales to build a pyramid of images, which is the input to the next three-stage cascading network.

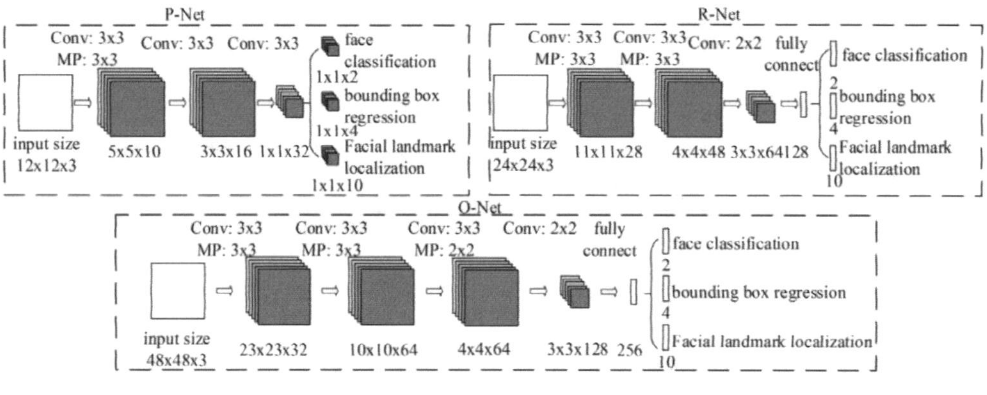

Figure 1. MTCNN architecture.

Three stage of the MTCNN:

- The Proposal Network (P-Net) This first stage is FCN (Fully Convolutional Network). The difference between CNN (Convolutional Neural Network) and FCN is that a fully convolutional network does not use the dense layer as part of the architecture. P-Net is used to obtain candidate windows and their bounding box regression vectors.

- The Refine Network (R-Net) All candidates from P-Net go to R-Net. This network is a CNN and not an FCN like the previous one, since there is a dense layer at the last stage of the network architecture. R-Net outputs the decision of whether the image is a face or not, in the form of a 4-element vector, which is the bounding box for the face, and a 10-element vector for localizing the facial landmarks.
- The Output Network (O-Net) This step is similar to R-Net, but this output network aims to describe the face in more detail and output the positions of the five facial landmarks for the eyes, nose, and mouth.

The MTCNN's job is to deduce three things: classification (face/non-face), bounding box coordinates, and facial landmarks location. The Siamese network shown in figure 2 is a type of a neural network architecture that learns to differentiate input data. That is, it allows you to learn to understand which images are similar and which are not. Siamese networks are made up of two identical neural networks, each with the same exact weights. Each network takes one of two images as input. The outputs of the last layers of each network are then sent to a function that determines if the images contain the same IDs. This is done by calculating the distance between the two outputs. The Siamese network is used to extract quality facial features and create a 128-element vector to represent those features.

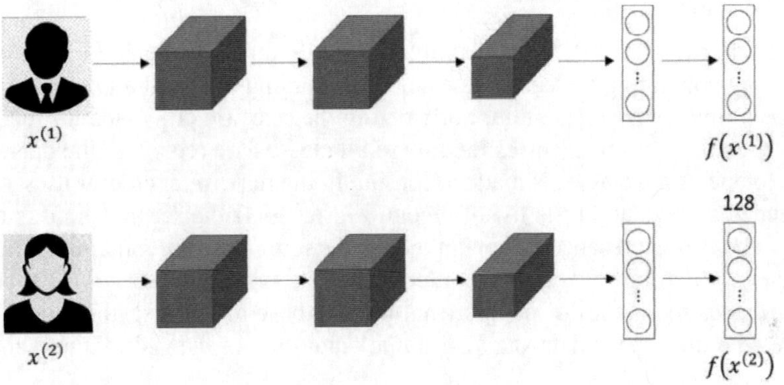

Figure 2. Siamese network.

The operation of the software includes a classifier training process and an identification process. When trained (figure 3), the input video is passed through the MTCNN, which gives the coordinates of a bounding box around the face. With the help of coordinates, the face is cut out of the frame and transferred to the change, including flipping, blurring, adding noise, horizontal displacement, vertical displacement, increasing and decreasing the brightness of the image. The resized frames are sized to 160 by 160 pixels and are saved to the folder corresponding to the person ID. After creating the database, it is necessary to train the Siamese network on it. The Siamese network model is trained until the classification accuracy is 99%. After training, the data file is saved and used later for identification tasks.

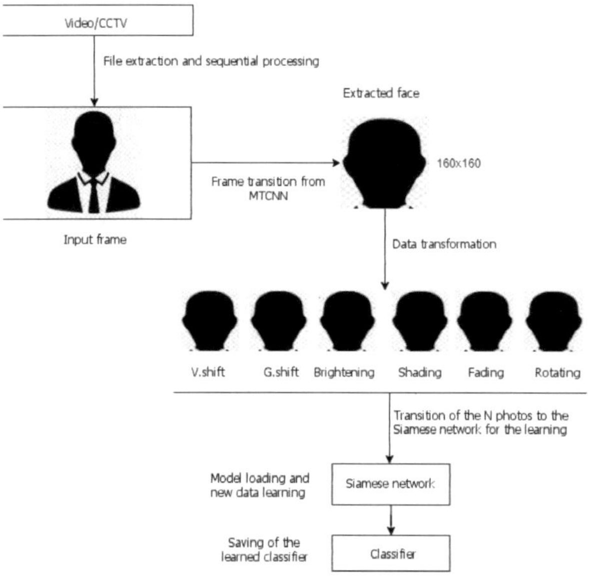

Figure 3. The learning process scheme.

During identification (figure 4) the video frame passes through the MTCNN, which returns a twodimensional array containing the coordinates of the bounding boxes of all detected faces. Then the faces are extracted from the frame and, after normalizing the size, are processed for recognition using a trained classifier. Confidence scores corresponding to all classes are recorded, the decision on whether a person belongs to one of the classes is made according to the maximum credibility score when the set threshold is exceeded and added to the list of recognized faces. If the set threshold is not exceeded, a decision is made about the presence of an unknown person. The list containing the names of the recognized persons is then transferred to the attendance accounting module. This module creates an attendance file and adds the names of people from the database to it. Next, the module creates a new column in the file with the current date and adds a mark and time of the visit of the recognized person.

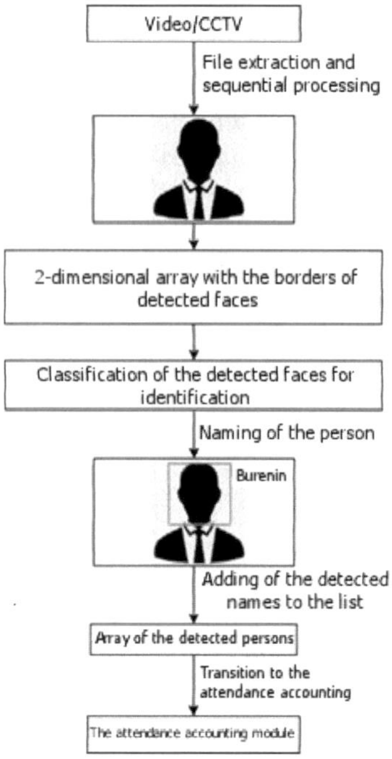

Figure 4. The identification and the attendance registration process scheme.

3. Results and discussion

The proposed method for human face recognition was developed as a software product, tested on a dataset of 10,000 images of 20 people, that is, 500 images per person, and showed the probability of correct recognition in real time 96.02% with the probability of false recognition less than 10%. This result exceeds indicator of other automated attendance recording systems used in practice (table 1).

Table 1. Comparison with other methods.

Number	Method	Accuracy
1	Chintalapati et al. [1]	95.00%
2	FaceTime [4]	95.02%
3	Proposed method	96.02%

4. Conclusion

The proposed system is sensitive to the distance to the faces in the video, as well as to video with low resolution. Further work may include improved face detection and recognition independent of video resolution by preprocessing with an image resolution enhancer.

References

[1] Chintalapati S and Raghunadh M V 2013 Automated attendance management system based on face recognition algorithms *IEEE International Conference on Computational Intelligence and Computing Research.* Enathi, India 1-5

[2] Paul V and Jones M J 2004 Robust real-time face detection *International Journal of Computer Vision* **57(2)** 137-54

[3] Rathod, Hemantkumar, Yudhisthir Ware, Snehal Sane, Suresh Raulo, Vishal Pakhare and Imdad A Rizvi 2017 Automated attendance system using machine learning approach *IEEE Explore*

[4] Arsenovic M, Sladojevic S, Anderla A and Stefanovic D 2017 FaceTime – Deep learning based face recognition attendance system *Conference: – IEEE 15th International Symposium on Intelligent Systems and Informatics (SISY) At* Subotica, Serbia

[5] Zhang Kaipeng, Zhanpeng Zhang, Zhifeng Li and Yu Qiao 2016 *Joint face detection and alignment using multitask cascaded convolutional networks* [Electronic resource] 1499-503 Link : https://arxiv.org/ftp/arxiv/papers/1604/1604.02878.pdf

[6] Koch G, Zemel R and Salakhutdinov R 2015 Siamese Neural Networks for One-shot Image Recognition *Proceedings of the32nd International Conference on Machine Learning* Lille France (JMLR: W&CP **37**) p 3-6

Concealment of sensor network node interaction

E A Kushko and N Y Parotkin

Institute of Informatics and Telecommunications, Reshetnev Siberian State University of Science and Technology, Krasnoyarsk, Russian Federation

E-mail: evgeny.kushko@gmail.com

Abstract. The aim of this study is to increase the security of sensor network hosts by means of node interaction concealment. This paper outlines existing methods and approaches to information security of sensor network and internet of things systems. The authors developed a new secure data communication method based on dynamic network topology. This method allows to conceal sensor network data communication from an intruder breached an outer security perimeter. The new method secures data transfer and differs from the existing switching approaches that conceal sensor network hosts.

1. Introduction

The Industrial Internet of Things (IIoT) is a multi-level system that consists of sensors, various industrial nodes and unit controllers, means of collecting, transmitting and processing data, as well as analytical tools, data visualization tools and other components. This system is designed for offline data collection and real time data communication and enables remote monitoring and management of industrial facilities with no (or minimal) human involvement. Industry 4.0 assumes widespread introduction of industrial Internet of Things in various economic sectors since it aims to create automated cyber-physical systems collecting, processing, visualizing and systematizing data, and thus implementing and controlling various industrial processes [1].

The lower layer of the industrial Internet of things is a wired or wireless sensor network. The sensor network is a dynamic, distributed, self-organizing network of sensors and execution units, which is designed to accomplish automation, diagnostics, telemetry and machine-to-machine interaction tasks. The following requirements are considered when building a sensor network: ease of deployment and operation, no need for frequent maintenance, high fault tolerance and reliability, as well as low performance and bandwidth requirements [2]. The industrial Internet of Things has a positive economic impact in various industries, so the number of such systems is increasing. However, with the spread of the industrial Internet of Things, the number of attacks on these systems is also increasing.

2. Sensor network security measures

Usually, one can protect traditional local networks with layered defense, network segmentation, monitoring and managing network traffic, concealing the architecture and configuration of the information system, encrypting transmitted data, providing trusted channels and routes, ensuring the authenticity of connections and building a security perimeter. These measures require qualified specialists to study and analyze data, network traffic and log files received from various information security tools and services in order to detect abnormal behavior, since an attacker who has overcome the outer network perimeter, is almost free in his actions.

Sensor networks, in turn, require autonomy, reliability, fault tolerance and scalability, so the sensor network protection mechanisms are primarily aimed at availability: ensuring stable communication channels, building optimal routes, protecting against DoS attacks. In addition, sensor networks are usually run on low-performance devices and sometimes have strict latency limits (real time data collection and analysis). Given all this, developers of sensor network systems create their own protocols and technologies, which make it impossible to use traditional local network security measures [3].

These factors produce a number of vulnerabilities that an attacker can exploit. First of all, the lack of intruder detection, authentication and encryption mechanisms — an attacker can impersonate a legitimate node of the sensor network, redirect traffic, hack specific nodes, intercept, analyze and modify network traffic, affects the security of sensor networks [4].

As additional protection measures, researchers offer solutions aimed at network exploration protection, for example network steganography and moving target technology. Network steganography conceals the fact of data transfer. The data is hidden in the headers and payload of network packets, the packet-transmission order changes, and some packets are deliberately lost. Network steganography methods do not change the data and do not interfere the functionality of network services. However, network steganography is difficult for implementation and limited in use, because of the limitations on the amount of transmitted data [5].

Moving target technology complicates network exploration by periodically changing the structure and configuration of the network over time. The functioning of the information system is not disrupted, so the attacker has no limits in his actions, but he cannot obtain long-term information about the network [6]. The most effective systems are of the decentralized type, since there is no vulnerable central node. This technology is at the initial stage of development: theoretical solutions and some technical prototypes have been proposed, but researchers still face difficulties with their implementation and application.

These solutions have been tested only for the protection of traditional computer network systems, as a result, they can be inapplicable for sensor networks due to their characteristics.

3. Solution description

The developed method of secure data communication for dynamic network topology is based on moving target technology. Nodes participating in secure data communication move in multicast groups and transmit data using group broadcasting. Each node is simultaneously connected to several multicast groups of secure data communication and redirects the received data to all connected groups, i.e. avalanche routing is used.

When data is transmitted this way, an attacker who intercepts and analyzes the data cannot identify receivers or senders, since the data packets do not explicitly contain their IP addresses. The data packet, regardless of the amount of data transmitted, has a fixed size and is encrypted. Furthermore, participants of the secure data communication switch between multicast groups and the logical structure of the system has a dynamic topology. Retransmission in combination with a fixed packet size does not allow to determine whether the packet is the request or response, and which node is the sender or receiver. As a result, an attacker cannot identify data flows or determine relationships between nodes, as well as cannot obtain long-term information about the logical structure of the system.

At the stage of initialization, each participant node forms a pool of secure data communication multicast group addresses, through which the data is transmitted, according to an algorithm depending on the current date and time. This pool of addresses changes at regular intervals. Next, the node selects two multicast groups and connects to them. At the end of another time interval, which is less than the interval for re-forming the multicast group address pool, the node again randomly selects the multicast groups. The general algorithm for initializing secure data communication is shown in figure 1.

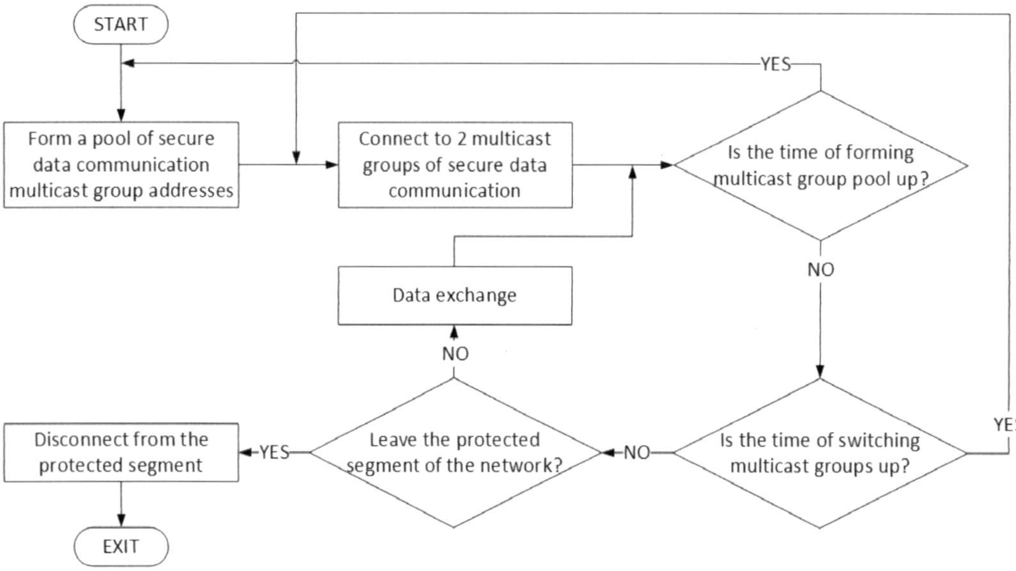

Figure 1. Secure data communication initialization algorithm.

In order to transmit data, each node participating in secure data communication generates a data packet of the structure shown in figure 2, and transmits it to all currently connected multicast groups of secure data communication. All data packets are of the same size, and if the size of transmitted data exceeds the size of the packet, the data is split into several packets. A packet containing the receiver and sender IDs is encrypted with the receiver's public key using a lightweight cryptography algorithm [7]. Nodes connected to receiving multicast groups of secure data communication retransmit this packet to another multicast groups and so on. It should be noted that during retransmission, the packet is modified so that the initiator of data communication could not be identified. It is crucial because otherwise listening to all the network traffic, it would be possible to determine original data sender. Each retransmission generates different packet, and as a result, the sender cannot be identified.

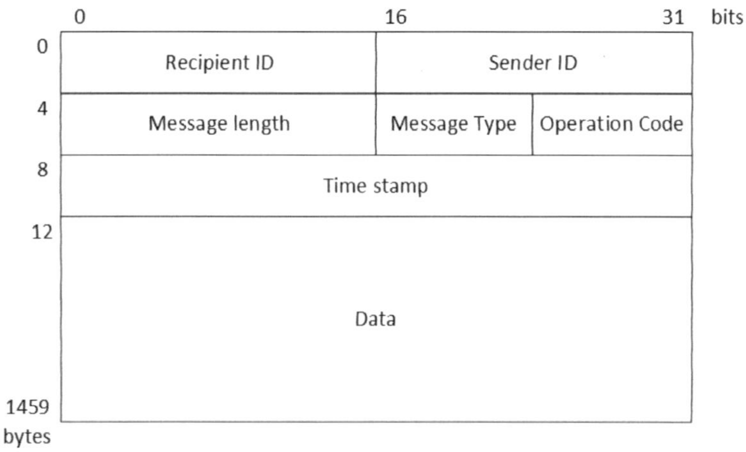

Figure 2. Data packet structure.

After the retransmission, the node attempts to decrypt the packet header and extract its ID. If the data packet is successfully decrypted, the node processes the packet, otherwise discards it. To ensure that the retransmission is not infinite, each data packet has a lifetime. In addition, the packet is discarded if the maximum data packet size is exceeded, or the data packet was previously processed by the node, regardless of whether it is the intended receiver or not.

4. Implementation of the developed solution

The practical implementation of secure data communication based on a dynamic network topology is a daemon that receives UDP datagrams from the MQTT-SN client, encapsulates them in a data packet and transmits them further over the network. MQTT-SN v1.2 is chosen as the application layer protocol, since it supports UDP and is designed to function precisely in a wireless sensor network, in conditions of low bandwidth, poor quality communication channel and low performance of end devices, in contrast to the usual MQTT [8]. NSP Proxy is a software implementation of the solution being developed.

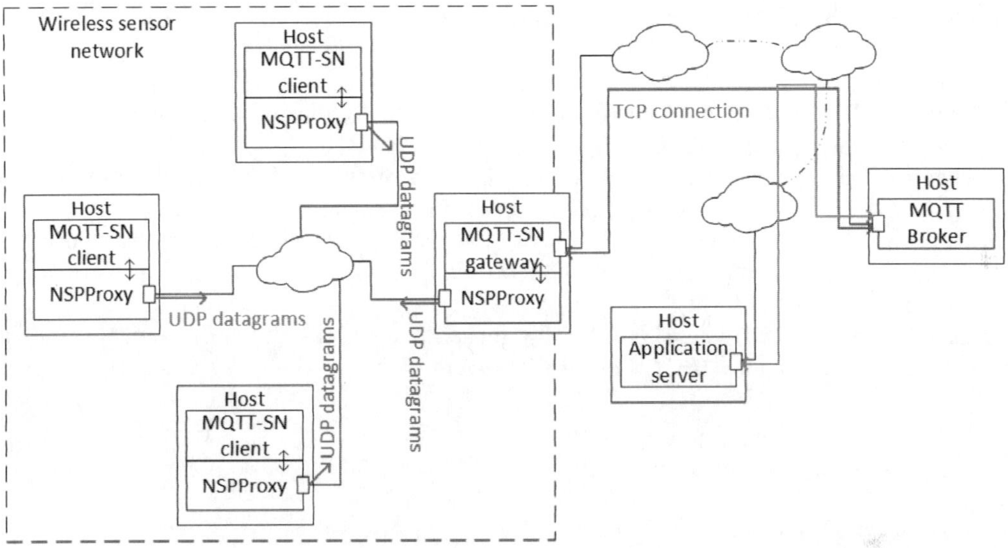

Figure 3. Sensor network diagram.

Figure 3 shows the sensor network diagram. The MQTT-SN network consists of clients, translators and gateways, which in turn are connected to the MQTT broker, normally situated outside the sensor network, by means of a stable and optionally secure TCP connection via MQTT protocol.

The sensor network based on MQTT-SN does not support secure connections: data is transmitted in unencrypted UDP datagrams, there are no authentication mechanisms. The method of secure data communication involves the protection of data by a cryptosystem with a public key. It should be noted that the MQTT-SN client (sensor) only needs to know the public keys of the translator or gateway that forward the data to the MQTT broker. Accordingly, the administrator registers the public and private keys in the device, as well as the public keys necessary for a particular device to receive and transmit data from specific nodes.

Performance characteristics of the developed system: the average bandwidth is 320 kbit/s and the delay is 28 ms. These values were obtained by generating a substantial amount of traffic on a router with a maximum speed of 300 Mbit/s through a wireless connection, the channel width was shared between all participants of the experiment. In a real-world sensor network, such performance characteristics are considered acceptable, a large amount of data is usually not transmitted in a short period of time.

5. Conclusion

The developed method of secure data communication based on a dynamic network topology does not involve the presence of central nodes. It is built on the basis of standard technologies that are supported by most devices operating over a WLAN network (including LAN). The use of this method does not affect other services or security requirements. The presented method allows to protect from analysis not only the transmitted data, but also the sides of network interaction, and the whole sensor network from the exploration.

References

[1] Boyes H, Hallaq B, Cunningham J and Watson T 2018 The industrial internet of things (IIoT): An analysis framework *Computers in industry* **101** 1-12

[2] Singh P K 2020 *Handbook of wireless sensor networks: issues and challenges in current Scenario's* ed Bhargava B K *et al.* (Cham: Springer Nature Switzerland AG)

[3] Finogeev A G, Nefedova I S, Finogeev E A, Kuang V T and Botvinkin P V 2014 Analysis and classification of attacks via wireless sensor networks in SCADA systems *Caspian Journal: Management and High Technologies* **1(1)** 12-23

[4] Kumar H K and Kar A 2009 Wireless sensor network security analysis *Int. journal of computer science & information technology (IJCSIT)* **1** 1-10

[5] Mazurczyk W, Smolarczyk M and Szczypiorski K 2011 Retransmission steganography and its detection *Soft Comput.* **15** 505-15

[6] Zhuang R, DeLoach S A and Ou X 2014 Towards a theory of moving target defense *Proc. of the 1st ACM Workshop on Moving Target Defense* (Scottsdale: ACM) 31-40

[7] Bhardwaj I, Kumar A and Bansal M 2017 A review on lightweight cryptography algorithms for data security and authentication in IoTs *4th Int. Conf. on Signal Processing, Computing and Control (ISPCC)* (Waknaghat: IEEE) 504-9

[8] Stanford-Clark A and Truong H L 2013 Mqtt for sensor networks (mqtt-sn) protocol specification *Int. business machines (IBM) Corporation version* **1(2)**

Designing an on-board computer for an armored personnel carrier

S N Kostarev[1,2,3], S N Goryachev[1], D V Miromanov[1], T G Sereda[2,5] and O V Kochetova[2,4]

[1] Perm Military Institute of National Guard Troops of the Russian Federation, 1, Gremjachij log St., Perm, 614030, Russia
[2] Perm State Agro-Technological University named after Academician D N Pryanishnikov, 23, Petropavlovskaja St., Perm, 614990, Russia
[3] Perm National Research Polytechnic University, 29, Komsomol'skij Avenue, Perm, 614990, Russia
[4] Perm Institute of the FPS of Russia, 125, Karpinskogo St., Perm, 614012, Russia

[5] E-mail: iums@dom.raid.ru

Abstract. Timely and accurate solution of the tasks facing the Russian Guard is impossible without a good state of military equipment. The Kupol monitoring system used by the Russian Guard is intended only for monitoring the location of mobile objects using GLONASS / GPS satellite navigation systems, for transmitting navigation information over mobile radio communication networks. Currently, the condition of the armored personnel carrier (BTR-80) is monitored periodically, in accordance with the maintenance regulations. However, the condition of an individual vehicle may differ from the average armored personnel carrier, and therefore it became necessary to re-equip the BTR-80 with an on-board computer that accurately diagnoses the wear (serviceability) of equipment and the combat capability of the vehicle's units. There is a functional connection between the structural parameters and the parameters of the output processes, due to which, according to the parameters of the latter, it is possible to adequately assess the technical state of military equipment (unit), as well as the quality of its work. The nominal parameters of the structural parameters correspond to the nominal parameters of the parameters of the output processes. As the technical state of military equipment (unit) shifts to the worse, the parameters of the output processes can increase (fuel consumption, vibration) or decrease (oil pressure in the engine). The designed monitoring systems of the BTR-80 are grouped by the type of tasks performed or the drive mechanism: engine and transmission, radio communication, electrical equipment, pneumatic systems, hydraulics, and others. The design of the on-board computer is performed on a programmable logic device using the international standard IEC61131-3. The simulation of the relay-contact circuit showed positive results.

1. Introduction

The history of the first on-board computers goes back to the 1980s [1-3]. Currently, on-board computers and computer technologies are widely used in many transport areas [4-7]. The implementation of automation tools for military equipment is also an urgent task. Currently, the rolling stock of Rosgvardiya is being modernized. The problem of control and combat readiness of military equipment is acute in all units of Rosgvardiya. For use in Rosgvardiya, a monitoring system "Dome" has been developed, which serves as a database and analysis of the combat capability of military equipment. But the current state of each individual unit of military equipment needs additional detail. Periodic technical inspection of military equipment is insufficient, and therefore the development of an on-board computer for an armored personnel carrier is an urgent task. To transmit readings from individual components and mechanisms, various mechanical sensors are used, which very quickly become worn out, and some equipment does not have monitoring tools. For monitoring the condition of military equipment, it is necessary to develop tools for analyzing the components and mechanisms of the machine. On the object under study (an armored personnel carrier) the transmitted information is the parameters of the power unit (temperature, engine oil pressure, turbocharger operation, the condition of the transmission shafts), the operation of the pneumatic and hydraulic systems, the battery voltage, the operation of the generator unit, and other parameters. There is a need to improve the transmission of data from the nodes and mechanisms of the equipment to the control and measuring devices.

2. Equipment and devices used in studies

The research used the theory of finite automata, the theory of constructing relay-contact circuits. The software and hardware is implemented on the Omron PLC and the NT-21 touch screen monitor.

3. The results of the study and their discussion

3.1. Purpose of automated monitoring means of the military equipment condition

During diagnostics, to assess the technical state of military equipment (unit), the output processes of the functioning mechanism are used. Work output processes are distinguished (fuel consumption, power consumption or output, heat exchange with the external environment) and related ones (vibration, noise, light phenomena, etc.). All output processes are quantitatively assessed using the appropriate parameters (power output can be assessed by the corresponding value and the rate of its increase). Depending on the amount of data that contains the parameters of the output processes, they can be private or generalized. The former characterize the technical condition of a particular unit of the system (knocks in the engine crank mechanism, steering wheel play, etc.), the generalized ones characterize the technical condition of military equipment (unit) as a whole (fuel consumption per 100 km of track, path and acceleration time of military technique up to a given speed, etc.). The parameters of the output processes, in contrast to the structural ones, are most often calculated on operating military equipment and are used to identify its technical condition without disassembly.

3.2. Composition of diagnostic modules

At the object under study (armored personnel carrier (APC), the transmitted information is the engine temperature, the operation of the turbocharger, the condition of the gearbox shafts, the voltage of the battery, the operation of the generator unit, the operation of special equipment and other parameters. There is a need to improve the transmission of data from units and mechanisms of equipment to the control and measuring devices. We divide the technological parameters of the equipment into the next groups. The group (Engine, transmission) will be described by the function MT (P, T, Tax, TK, 1D, 2D, C, F). The group (Electrical equipment) will be described by the array E (U, I, O1, O2, 03, O4). The group (Pneumatic system) will be described by the function PS (P1, P2, P3). The group (Radio communication unit) will be described by R (R1–R6). The values of the variables are given in table 1.

Table 1. Monitoring of the current values of the measured parameters and limit values was recorded in the Data Memory cells.

N	Bit address CIO	Value	Memory address DM current	max	min	Monitoring object
			Engine, transmission			
1	100.00	P	100	105	104	Oil pressure
2	100.01	T	90	95		Antifreeze temperature
3	100.02	Tax	80	85		Tachometer
4	100.03	TK	70	75	74	Turbocharger
5	100.04	1D	60			Transmission Input shaft sensor
6	100.05	2D	55			Transmission Output shaft sensor
7	100.06	C	54			Clutch malfunction
8	100.07	F	50			Injectors
9	100.10	MT				General section diagnostics
			Electrical equipment			
10	100.11	U	40		44	Battery voltage
11	100.12	I	30	35	34	Recharging current
12	100.13	O1	20			Dimensions
13	100.14	O2	15			Low beam
14	100.15	O3	14			High beam
15	100.16	O4	16			Stop-signal
16	101.13	E				General section diagnostics
			Pneumatic system			
17	101.00	P1	10	15	14	Total pressure
18	101.01	P2	110	115	114	Tire pressure
19	101.02	P3	120			Brake system
20	101.12	PS				General section diagnostics
			Radio communication unit			
21	101.03	R1	130			Receiver-transmitter
22	101.04	R2	140			High-frequency pulse generator
23	101.05	R3	150			Antenna filter block
24	101.06	R4	160			Laryngophone amplifier
25	101.07	R5	170	175	174	RF cable
26	101.08	R6	180	175	174	Voltage transformer
	101.14	R				General section diagnostics

3.3. Compilation of logical equations and development of software for the equipment operation

Logical equations for monitoring the "Engine, transmission" section are shown in table 2.

Table 2. Logical equations for monitoring the "Engine, transmission" section.

Logical equations	Equipment
$(DM100 < DM105) \cdot (DM100 > DM104) = P$	*Oil pressure*
$DM90 < DM95 = T$	*Antifreeze temperature*
$DM80 < DM85 = Tax$	*Tachometer*

IOP Conf. Series: Materials Science and Engineering 1155 (2021) 012059 doi:10.1088/1757-899X/1155/1/012059

$(DM70 < DM75) \cdot (DM70 > DM74) = TK$	*Turbocharger*
$DM60 > 0 = 1D$	*Transmission Input shaft sensor*
$DM55 > 0 = 2D$	*Transmission Output shaft sensor*
$\overline{DM54} = \overline{C}$	*Clutch malfunction*
$\overline{DM50} = \overline{F}$	*Injectors*
$P \cdot T \cdot Tax \cdot TK \cdot 1D \cdot 2D \cdot C \cdot F = DT$	*General section diagnostics*

Using the example of table 2, logical equations were developed for all sections (table 1). Researching the model for monitoring and controlling the equipment condition were carried out using the OMRON CX-Programmer and CX-Designer software [8,9]. The programming language of the OMRON CP1L controller was ladder-contact logic (the IEC 61131-3 standard). A simulation of the operation of the circuit using the example of monitoring oil pressure and engine speed is shown in figures 1 and 2.

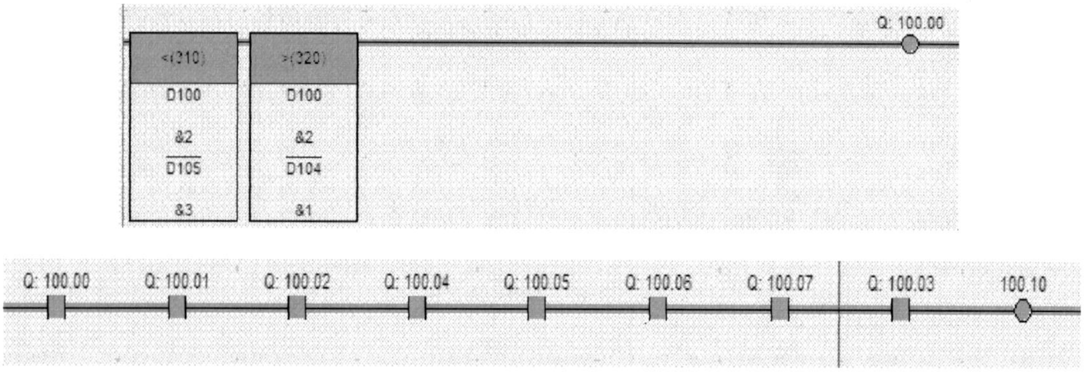

Figure 1. Simulation of the circuit. Engine oil pressure is within normal limits. All monitoring units of the "Engine, transmission" operate normally.

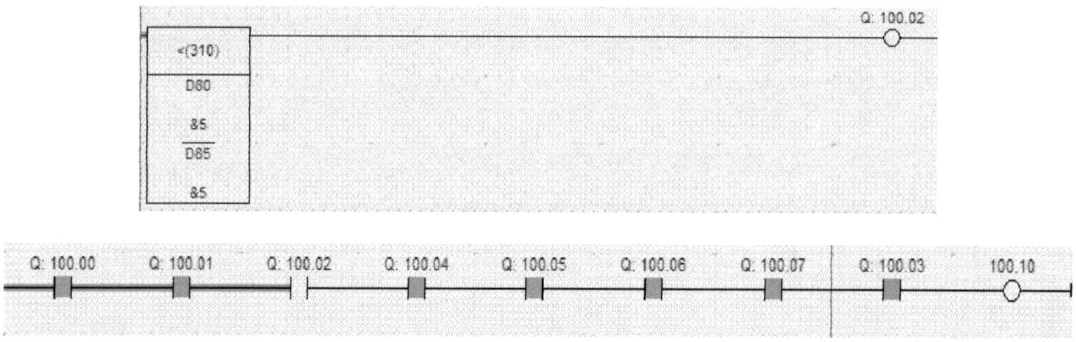

Figure 2. The tachometer showed dangerous engine speed, the general circuit of the section went into a warning state.

4. Conclusion

For monitoring the military equipment condition, using the BTR-80 as an example, an on-board computer based on the OMRON programmable logic controller has been developed. The equipment monitoring is shown, divided into 4 sections, including about 30 positions. In the future, it is possible to expand the monitoring nodes and refine the software. A simulation of the circuit has shown satisfactory results.

References

[1] Sharma U K and Mc Tamaney L S 1987 Real time system architecture for a mobile robot *Proceedings of SPIE - The International Society for Optical Engineering* **852** 25-31

[2] Harland D B 1983 A simple model of a car suspension system using an analogue computer *Teaching Mathematics and its Applications: an international journal of IMA* **2(2)** 79

[3] Oliveira A G *at el.* 2021 A Look at the Evolution of Autonomous Cars and Its Impact on Society Along with Their Perspective on Future Mobility *Smart Innovation, Systems and Technologies* **201** 583-94

[4] Zhou L, Dang J and Zhang Z 2021 Research on fault diagnosis for on-board equipment of train control system based on imbalanced text classification *Journal of Applied Science and Engineering (Taiwan)* **24(2)** 167-75

[5] Samoylenko A P, Panychev A I and Panychev S A 2017 Technologies of aggregation for integral criteria of evaluating the efficiency of on-board computer systems *RPC 2017 - Proceedings of the 2nd Russian-Pacific Conference on Computer Technology and Application* 157-61

[6] Popkov A and Bekher S 2018 Use of computer software for technical training at rail car repair shops *MATEC Web of Conferences. 10th International Scientific and Technical Conference "Polytransport Systems"* 04001

[7] Bukowski M, Luckner M and Kunicki R 2020 Estimation of free space on car park using computer vision algorithms *Advances in Intelligent Systems and Computing* **920** 316-25

[8] Kostarev S N and Sereda T G 2017 *IOP Conf. Ser.: Earth Environ. Sci.* **87** 032016

[9] Kostarev S N *et al* 2019 *IOP Conf. Ser.: Mater. Sci. Eng.* **537** 032030

IOP Conf. Series: Materials Science and Engineering 1155 (2021) 012060 doi:10.1088/1757-899X/1155/1/012060

Modelling of power consumption for Advanced Encryption Standard and PRESENT ciphers

E Maro

Southern Federal University, Chehova str., 2, Taganrog, Rostov-on-Don region, 347922, Russia

E-mail: eamaro@sfedu.ru

Abstract. Paper presents a result of modelling by ELMO tool and measuring on hardware stand power consumption traces for encryption algorithms Advanced Encryption Standard and PRESENT. Power consumption traces for different version of AES (basic one-round AES algorithm are compared with AESMasked_R1 and MBedAES versions) and PRESENT (with various round numbers) are generated. Assessment of instructions with leaks were made by comparison power consumption traces for fixed-input and power consumption traces for random-inputs (total quantity of traces is 10^4). Numbers of total and potentially leakage instructions for each examined version of AES and PRESENT ciphers implementation are estimated.

1. Introduction

The cryptographic algorithms used are mathematically (computationally) secure schemes for current computation resources of the attackers and existing approaches of ciphers analysis. At the same time analyzing the security of cryptographic systems is important to consider their implementation in software and hardware tools of information protection. Thus, computationally secure cryptographic algorithms could be vulnerable to attacks using additional information about the data encryption process in real physical devices. The method of analyzing the side channels of cryptographic devices is based on search a mathematical correlation in parameters of encryption device functioning, for example, such parameters as computation time, power consumption, electromagnetic and acoustic emitting, etc.

The research topic is focused on theoretical and experimental analysis of the power consumption traces during execution of cryptographic algorithms and assessment of instructions with identified leaks. To collect measurements a hardware stand was assembled on Arduino Nano board, which performs AES and PRESENT encryption algorithms in different versions, and a high-precision oscilloscope RIGOL DS1054. Experimental hardware stand is shown on figure 1. To point encryption area's during measurements of power consumption an output signals were sent to fixed pin. An independent power supply source (4 cells, 1.5 V nominal) was used (to minimize possible voltage fluctuations from an external device). The measurements were carried out through a shunt resistor.

Power consumption traces were modeled using ELMO software [1, 2]. The ELMO simulator consists of two main components: a device emulator (a modified version of the Thumbulator emulator is used) and a set of leak detection algorithms. During constructing a model of hardware equipment in ELMO bit values and bit's changes at inputs and outputs of the arithmetic logic unit (ALU), as well as in commands for working with memory, are considered. Each operand is compared with the corresponding operand of the previous instruction. The resulting power consumption is modeled as a linear combination of processed bit values or bit changes. ELMO provides the ability to simulate 21 instructions, which are often always used in cryptography tasks: ldr, ldrb, ldrh, str, strb, strh, lsls, lsrs, rors, muls, eors, ands, adds, adds #imm, subs, subs #imm, orrs, cmp, cmp #imm, movs and movs #imm.

All instructions are modeled independently and divided into 5 separate groups:

- ALU instructions (adds, adds #imm, ands, eors, movs, movs #imm, orrs, subs, subs #imm, cmp, cmp #imm).
- Shift instructions (lsls, lsrs, rors).
- Load instructions (ldr, ldrb, ldrh)
- Save instructions (str, strb, strh)
- Multiplication instruction (muls).

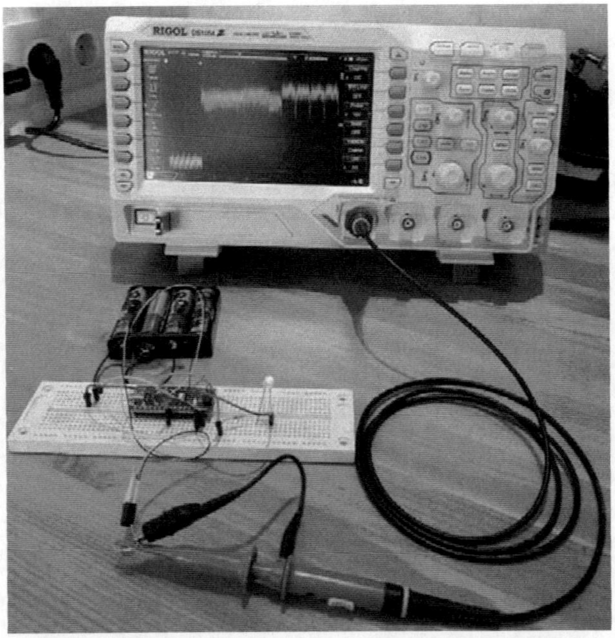

Figure 1. Experimental hardware stand based on Arduino Nano board and oscilloscope RIGOL DS1054.

The data for generating power consumption model is collected after processor executes a sequence of three instructions. Each constructed power consumption trace is preprocessed to find a point-of-interest, after which linear regression is applied to find the coefficients of the model.

The general device model consists of 19 main components, each of which simulates a fixed part of the architecture:

- A linear combination of bit state change between each operand of the current instruction and the corresponding operand of the previous and subsequent instructions.
- A linear combination of the bit values of the operands of the current instruction.
- Groups of the previous and subsequent instructions.

2. AES and PRESENT ciphers description

Advanced Encryption Standard (AES) [3] is symmetric block cipher with 128-bits block size and 128, 192 or 256-bits key length. Round transformation consists of 4 different operations (figure 2): substitutions in eight 8-bits length blocks (S-boxes), state shift ShiftRows() for a different bytes positions, shuffle columns MixColumns() by multiplication of state (considered as polynomials over GF

(2^8)) to polynomial $g(x)=\{03\}x^3+\{01\}x^2+\{01\}x+\{02\}$ by modulo x^4+1, addition modulo two AddRoundKey() with round keys. AES general scheme is presented at figure 2.

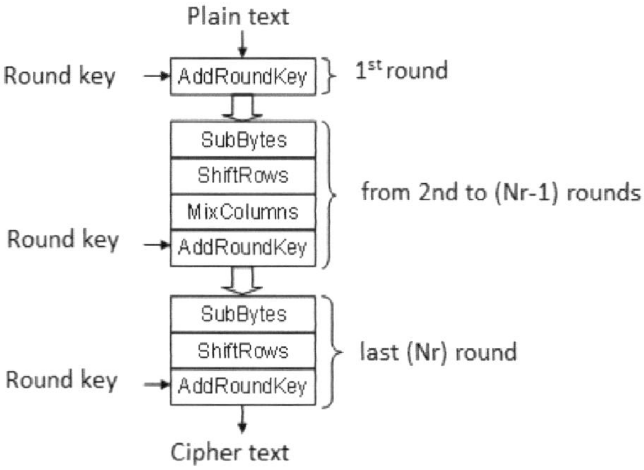

Figure 2. AES cipher scheme.

PRESENT cipher [4] (figure 3) is based on substitution-permutation network (SP-network) and contains 31 rounds with three operations: addition modulo 2 with round key, substitutions in sixteen 4-bits length blocks (S-boxes) and permutation (H).

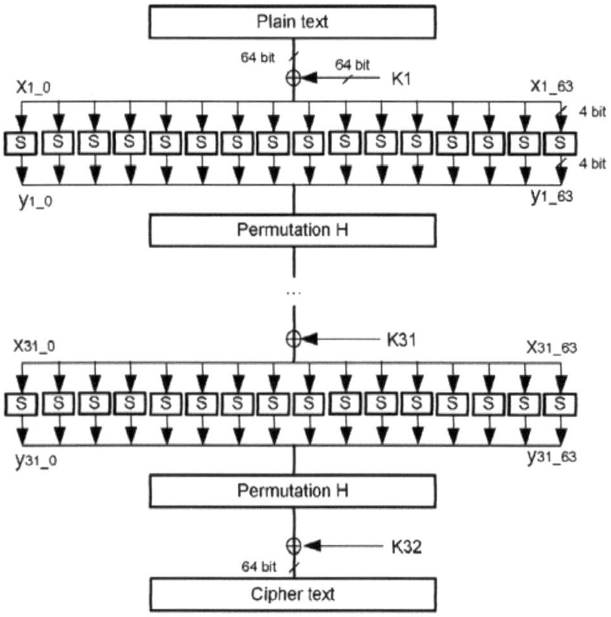

Figure 3. PRESENT cipher scheme.

3. AES and PRESENT power traces modelling and measurement

Model of power consumption trace was generated by ELMO tool during AES encryption. Figure 4 presents power measurement trace for one-round AES encryption. We use version of AES algorithm with directly implemented encryption scheme [5].

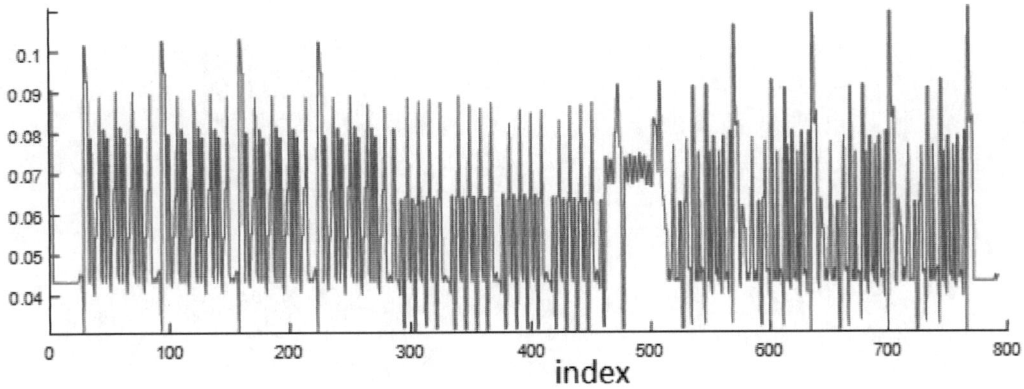

Figure 4. A model of power consumption trace during one-round AES encryption in ELMO.

Modeling of power consumption measurements by ELMO tool allows to estimate number of instructions containing information about the protected data (instruction with leaks) based on execution of t-test (with a threshold value of | 4.5 |) for fixed and random plaintexts. Based on the results of comparing power consumption traces of one-round AES cipher (10^4 traces), a graph Fix vs Random power consumption test for instructions is presented at figure 5.

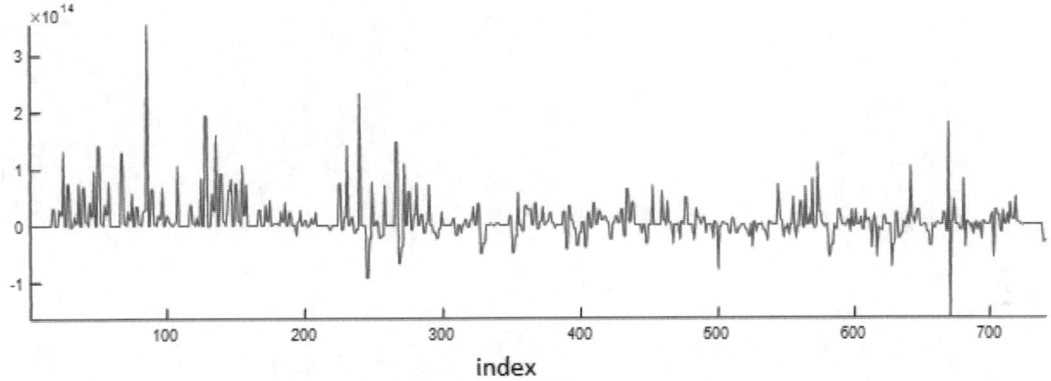

Figure 5. Fix vs Random test result for one-round AES instructions.

Table 1 shows results of simulating first-order side-channel leaks for several implementations of the AES cipher.

Table 1. Simulation results for side-channel leaks for various AES cipher implementations.

AES cipher implementation	Total number of instructions	Number of instructions with identified power consumption leaks
AES (one-round)	741	514
AESMasked_R1	451	83
MBedAES	2402	1755

We simulated power consumption traces for PRESENT cipher [6] by ELMO tool. Example of generated trace is shown on figure 6. After that we compare traces for fix and random input values (plain texts) and evaluate count of instruction with leaks. Result of leakage assessment for different number of rounds is presented at table 2.

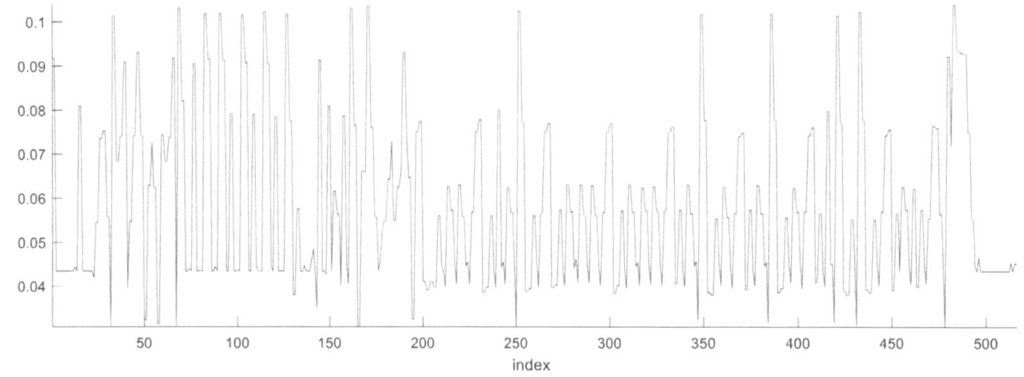

Figure 6. A model of power consumption trace during one-round PRESENT encryption in ELMO.

Fix vs Random instruction graph for one-round PRESENT is shown at figure 7. This graph allows to mark instructions with leaks (evaluate peak points).

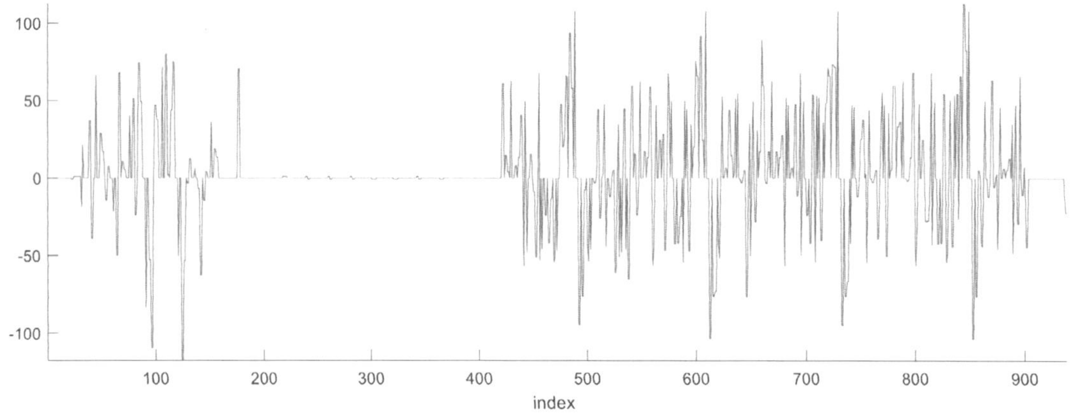

Figure 7. Fix vs Random test result for one-round PRESENT instructions.

Table 2. Simulation results for side-channel leaks for various rounds of PRESENT cipher.

Number of rounds	Total number of instructions	Number of instructions with identified power consumption leaks
1 round	938	379
2 rounds	1771	774
10 rounds	8435	3929
20 rounds	16765	7773
31 rounds (full-round)	26047	12137

Acknowledgments
This research was funded by RSF research project No 19-71-00041.

References

[1] McCann D, Oswald E and Whitnall C 2017 Towards Practical Tools for Side Channel Aware Software Engineering: Grey Box Modelling for Instruction Leakages *USENIX Security*

Symposium 199-216
[2] *ELMO: Evaluating Leaks for the ARM Cortex-M0 Instructions of Use and Documentation* (The ELMO Developers https://raw.githubusercontent.com/scaresearch/ELO/master/ELMODocumentation.pdf)
[3] *Advanced Encryption Standard (AES)* (Federal Information Processing Standards Publications (FIPS PUBS) 197, https://nvlpubs.nist.gov/nistpubs/FIPS/NIST.FIPS.197.pdf)
[4] Bogdanov A, Knudsen L, Leander G, Paar C, Poschmann A, Robshaw M, Seurin Y and Vikkelsoe C 2007 PRESENT: an ultra-lightweight block cipher *Lect Note. Comput. Sci.* **4727** 450-66 doi: 10.1007/978-3-540-74735-2_31
[5] *C++ AES(Advanced Encryption Standard) implementation* (GitHub, https://github.com/SergeyBel/AES)
[6] *PRESENT-C* (GitHub, https://github.com/bozhu/PRESENT-C)

Smart manufacturing control system based on deep reinforcement learning

E I Nikolaev[1], N I Zakharova[1] and V V Zakharov[2]

[1] North-Caucasus Federal University, 2, Kulakov Ave., Stavropol, 355029, Russia
[2] Stavropol State Agrarian University, 12, Zootekhnichesky lane, Stavropol, 355017, Russia

E-mail: notdeveloper@gmail.com

Abstract. The base of smart manufacturing contains two main components: cyber-physical systems and the Internet-of-Things (IoT). But the progress in machine learning, Big Data and deep learning technologies have a great impact on all areas of the modern economy, including information technology, industry. The capabilities of deep learning models associated with the ability of intelligent systems to make decisions significantly transform the key paradigms of intelligent production systems. The gap between the principles of human decision-making and intelligent systems is constantly closing. New approaches to decision-making based on deep models allow bringing the benefits of human control into production processes, and, at the same time, getting rid of errors associated with the human factor. The paper proposes an architecture for building multipurpose manufacturing systems based on deep machine learning models and reinforcement learning technologies.

1. Introduction

The research presented in this paper is interdisciplinary in nature. First, it offers a deep model architecture that allows control actions to be generated sequentially. The similar nature of the generation of control actions is inherent in humans. Secondly, the architecture of the control system for intelligent production is investigated, possible options for implementation are analyzed, the main theses of the construction of intelligent production for a wide variety of purposes. Combining smart manufacturing technologies with the latest advances in deep learning can improve the efficiency of smart manufacturing control systems.

The basic principles of building human-cyber-physical systems (HCPs) and cyber-physical systems (CPs) are presented in the works of many researchers: in [1] discussed the basic terminology of the subject area under consideration and the main differences between smart manufacturing and intelligent manufacturing; in [2] the evolutionary footprint of intelligent manufacturing is reviewed from the perspective of HCPs; in [3] discussed the principles of building CPs from an Industry 4.0 perspective.

Along with the development of intelligent manufacturing technologies, there is a rapid growth in theoretical and applied research in the field of deep learning, generative adversarial network (GAN), and effective methods of deep model training. The combination of the latest achievements from the considered areas allows us to conclude that the prerequisites for the transition to a qualitatively new level of smart manufacturing and intelligent manufacturing have been created. This work proposes the architecture of smart manufacture by example, proposes the main patterns of interaction of the control system of production equipment with actuators. The proposed architectures were tested on a CNC machine for engraving with a laser module.

2. Background and related work

There is a lot of existing work on smart manufacturing and intelligent manufacturing, circular economy, predictive maintenance, quality control and zero-defect manufacturing. The review of related work will thus be narrowed down to the most interesting and referenced papers since 2015. In [x00], the main structural elements of cyber-physical systems are shown, their role in the industry of the future is described, the place of CNC machine tools in intelligent production systems is shown. The various architectures of production systems, which are the core of the fourth industrial revolution (Industry 4.0), is described in [4]. The principles of building a CNC machine control system based on deep learning and reinforcement learning are shown in [5]. In [6], the principles of training a deep model using competitive learning technologies are considered: the developed neural network is used as a control system of a CNC machine. Research aimed at solving problems in the field of smart manufacturing is presented in [7-10].

The construction of intelligent production management systems that are significantly superior in efficiency to existing Smart Manufacturing technologies has become possible thanks to the development of deep learning, reinforcement learning and adversarial networks technologies. Deep neural networks have shown high efficiency in the tasks of classification [11-13], object detection [14-16] and segmentation [17-19]. Generative adversarial networks [20] has demonstrated promising prospects in image synthesis [21-24]. The problems solved by GANs are not limited by tasks related to the synthesis of new graphic content. In the framework of the probabilistic approach to the design and training of GAN, the problem of generating samples from the target distribution is solved. The many proposed applications of GAN were summarized in [25], devoted to the problem of cross-domain adaptation. In this research, an application of GAN technologies for the synthesis of control commands for the intelligent control system for laser engraver is proposed.

3. Proposed smart manufacturing architecture

Intelligent control is understood as such a sequence of commands to the executive device (machine tool), which leads to the solution of the problem and is as close as possible to the sequence of actions that a person would perform in solving this problem.

3.1. The structure and principles of operation of the executive device

An executive device means a computer numerical control machine (CNC), the schematic diagram of which is shown in figure 1.

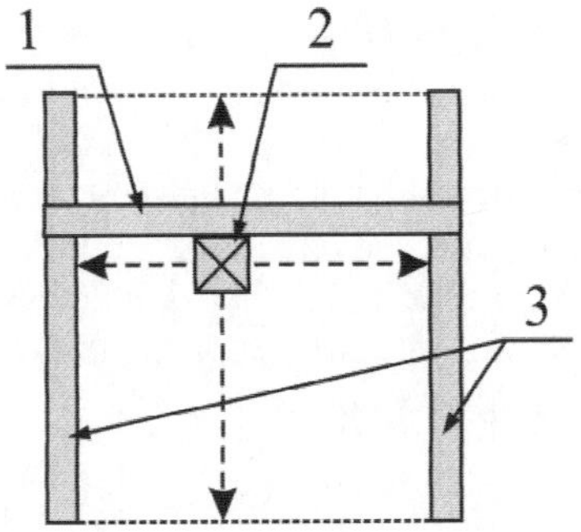

Figure 1. CNC machine structure: laser engraver scheme.

The two-axis CNC machine is controlled by a microcontroller that executes the G-code. The sequence of G-code commands is formed in advance on the computer and loaded onto the microcontroller. The machine is drawing the image (obviously, it is drawing the image in grayscale format). The microcontroller, moving the working tool (2, figure 1) along the axis Ox (3, figure 1) and

Oy (1, figure 1) axes, draws the original image. A pencil, laser module, filament, cutter, etc. can be used as a working tool. The machine in figure 1 draws the image line by line, while the image consists of dots or lines. Figure 2 shows the source image and the result obtained by CNC.

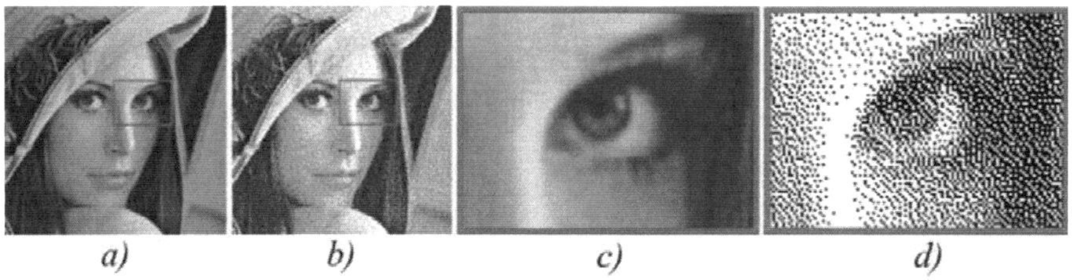

Figure 2. The original image and the resulting synthesized image: a) input image; b) synthesized image; c) enlarged fragment of the original image; d) enlarged fragment of the synthesized image.

In the practical applications, the following problems come to the fore:

- The operating time of the machine. Existing modifications of machine tools require a significant amount of time to build an image, therefore, most often machines are used that allow synthesizing an image of a small format.
- The quality of the synthesized image. Figure 2(d) shows what effect the technological characteristics of the working tool have on the quality of the synthesized image. In most implementations, such machines synthesize dotted images. In rare cases, there are methods for synthesizing images using line segments rather than points. The choice of the template (points or segments) strongly depends on the characteristics of the original image.

3.2. The control system and proposed format of the control commands

The main idea of the research is to analyze the feasibility, prove the feasibility and develop a prototype of an intelligent control system for synthesizing images through a single control command or a sequence of control commands. At the same time, the control commands are intended for a two-axis CNC machine with a wide range of working tools (further presentation of the research progress is made for this executive device).

Having analyzed the actions of a person when drawing an image, we find that the nature of his actions differs significantly from the sequential point-by-point reproduction of the target image inherent in machine methods. Simplified, you can imagine the process of drawing by a person as a sequential application of simple graphic primitives. This process can be formalized as a sequence of drawing individual points or segments. For an actuator, this is optimal from the point of view of control, but this is too rough an approximation to the human method of image synthesis.

Therefore, curves will be used in the work as the graphic primitives that make up the target image, and specifically their simplest form – Bezier curves. Bezier curves are convenient for control systems since they are a type of graphic primitive that is constructed algorithmically, that is, they are vector curves. It should also be noted that when using Bezier curves, different variations are available, specified by the order of the Bezier curve. Figure 3 shows Bezier curves of different orders.

IOP Conf. Series: Materials Science and Engineering 1155 (2021) 012061 doi:10.1088/1757-899X/1155/1/012061

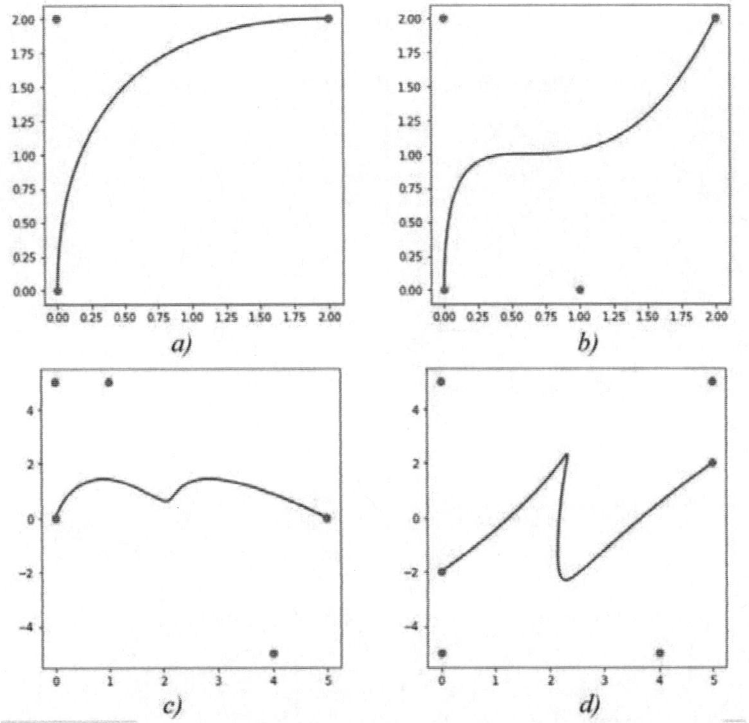

Figure 3. Bezier curves of various orders and its control points: a) 2-order curve; b) 3-order curve; c) 4-order curve; d) 5-order curve.

Another advantage of using Bezier curves is the simplicity and strict sequence of their construction:

$$P(t) = tP_1 + (1-t)P_2 \tag{1}$$

$$P(t) = t^2P_1 + 2t(1-t)P_2 + (1-t)^2P_3 \tag{2}$$

$$P(t) = t^3P_1 + 3t^2(1-t)P_2 + 3t(1-t)^2P_3 + (1-t)^3P_4 \tag{3}$$

$$P(t) = t^4P_1 + 4t^3(1-t)P_2 + 6t^2(1-t)^2P_3 + 4t(1-t)^3P_4 + (1-t)^4P_5 \tag{4}$$

Expressions (1) – (4) present equations for constructing Bezier curves from the first to fourth order, respectively. P – the coordinate value (x or y); P_i, $i=\overline{1,5}$ – control points which define the Bezier curve. The advance of Bezier curves for machine learning is that a complex curve (3-order Bezier curve) is constructed from four points. Four points on the plane are presented as a vector of length 8, that is, this is the output (tensor) of the neural network.

3.3. Control system architecture

The proposed control system is discussed in detail in [26]. As a result of an experimental study of the processes of interaction between an intelligent control system and an executive device (laser engraver), this work proposes an architecture of an intelligent production system that can be applied to production systems for various purposes.

Figure 4 shows a diagram of the production process and the place of the developed intelligent control system in this process.

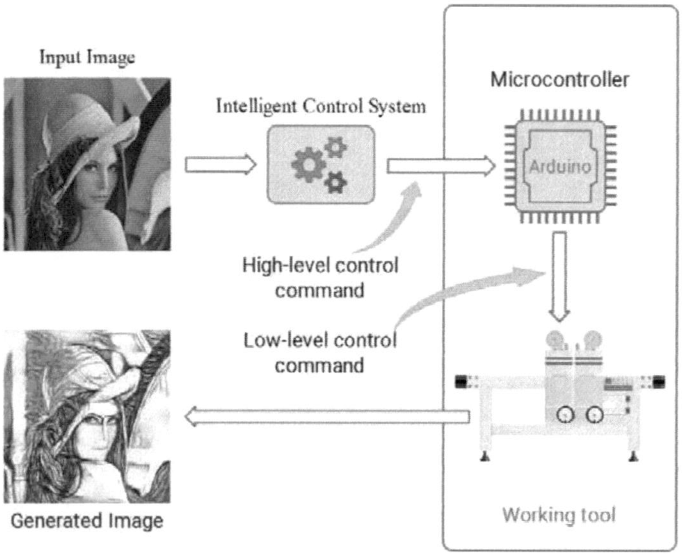

Figure 4. Diagram of the production process using an intelligent control system.

During the design of an intelligent control system, an important aspect of its functioning was revealed (figure 5).

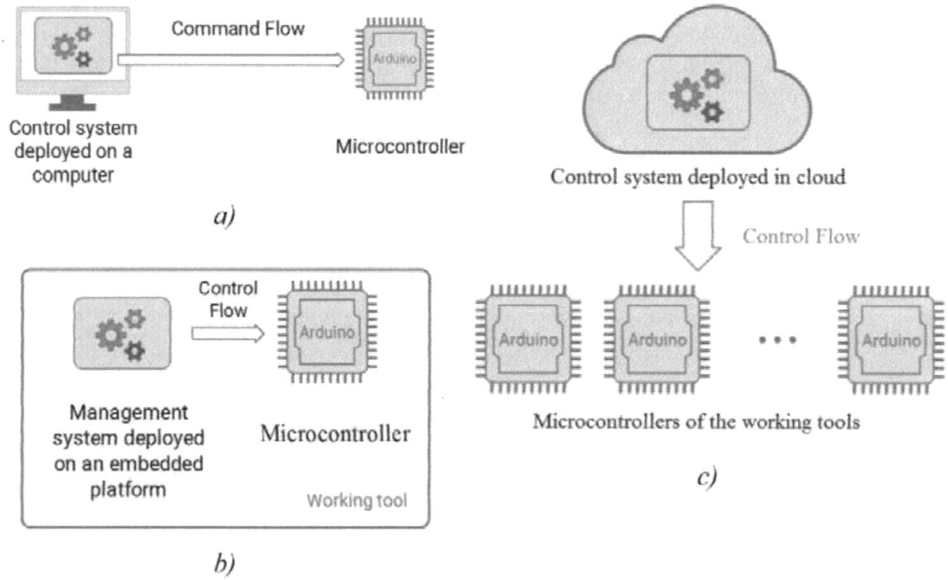

Figure 5. Different strategies for the interaction of the software control system and the actuator: a) the control system is deployed on a personal computer; b) the control system is deployed on an embedded graphics computing platform; c) the control system is deployed in the cloud.

Figure 5 shows that the control system is weakly connected to the control object, that is, it can be easily separated from the control object (actuator). When using a deep model or even multiple deep models as the core of the control system, high-performance GPU or TPU hardware may be required, so it is recommended to place the control system in the cloud. Through cloud architecture, many executive devices can be controlled at the same time.

4. Experimental result

In [26] the detailed experiment conditions are described. After training deep generative model which is a core of the control system, the proposed architectures can be realised on the real actuators. The result of the manufacturing process is shown in figure 6.

Figure 6. The examples of synthesized images at the different stages of the training process.

The actuator (laser engraver) repeats control commands and performs visualization by burning material. Additionally, the proposed architecture (figure 5) can be applied to various executive devices or groups of devices. The sequential manner of control, implemented in the proposed control system, makes it possible to control not only production equipment, but can be applied in the field of unmanned vehicles, unmanned aerial vehicles and in other areas where control actions are required that are comparable to human actions.

5. Conclusion

As part of this experimental study, an intelligent generative model of sequential synthesis of control commands was developed. Second, the possibility of generating control commands for a CNC machine like human actions was shown.

Further, the proposed intelligent control system is deployed on the laser engraver. When using the developed control system, a reduction in the material processing time of the CNC machine is achieved compared to the use of the GRBL software.

Finally, the ideas and results from efforts like those presented in this paper are necessary for many smart manufactures to reach intelligent and sustainable manufacturing/production and stay competitive and profitable.

References

[1] Yao X, Zhou J, Zhang J and Boër C 2017 From Intelligent Manufacturing to Smart Manufacturing for Industry 4.0 Driven by Next Generation Artificial Intelligence and Further *5th Int. Conf. on Enterprise Systems* 311-8

[2] Zhou Y, Yu F, Chen J and Kuo Y 2020 Cyber-Physical-Social Systems: A State-of-the-Art Survey, Challenges and Opportunities *IEEE Communications Surveys & Tutorials* **22(1)** 389-425

[3] Wan J, Li X, Dai H, Kusiak H, Martínez-García M and Li D 2021 Artificial-Intelligence-Driven Customized Manufacturing Factory: Key Technologies, Applications, and Challenges *Proc. of the IEEE* **109(4)** 377-98

[4] Lee J, Bagheri B and Kao H 2015 A Cyber-Physical Systems architecture for Industry 4.0-based manufacturing systems *Manufacturing Letters* **3** 18-23

[5] Nikolaev E I 2019 Towards intelligent control system for computer numerical control machines *IOP Conf. Ser.: Materials Science and Engineering* **537(3)** 032085

[6] Nikolaev E I, Zakharov V V and Zakharova N I 2019 Style Transfer for CNC Machine Input Data Preprocessing *IOP Conf. Ser.: Materials Science and Engineering* **582(1)** 012013

[7] Grangel-Gonzalez I, Halilaj L, Coskun G, Auer S, Collarana D and Hoffmeister M 2016 Towards a semantic administrative shell for industry 4.0 components *IEEE 10th Int. Conf. on Semantic*

Computing (ICSC) 230-7

[8] Datta S K, Da Costa, Harri J and Bonnet C 2016 Integrating connected vehicles in Internet of Things ecosystems: Challenges and solutions *IEEE 17th int. symp. on a world of wireless, mobile and multimedia networks* 1-6

[9] Chen B T, Wan J F, Celesti A, Li D, Abbas H and Zhang Q 2018 Edge computing in IoT-based manufacturing *IEEE Communications Magazine* **56(9)** 103-9

[10] Wang S, Wan J F, Zhang D, Li D and Zhang C 2016 Towards smart factory for industry 4.0: a self-organized multi-agent system with big data based feedback and coordination *Computing Networks* **101** 158-68

[11] Simonyan K and Zisserman A 2015 Very deep convolutional networks for large-scale image recognition *3rd Int. Conf. on Learning Representation*

[12] Perez L and Jason W 2017 The effectiveness of data augmentation in image classification using deep learning *arXiv preprint* arXiv:1712.04621

[13] Wang F, Mengqing J, Chen Q, Shuo Y, Chenxi L, Honggang Z, Xiaogang W and Xiaoou T 2017 Residual Attention Network for Image Classification *IEEE Conf. on Comp. Vision and Pattern Recognition (CVPR)* 6450-8

[14] Chen X, Ma H, Wan J, Li B and Xia T 2017 Multi-view 3d object detection network for autonomous driving *IEEE Conf. on Comp. Vision and Pattern Recognition (CVPR)* 6526-34

[15] Redmon J, Divvala S, Girshick R and Farhadi A 2016 You only look once: Unified, real-time object detection *IEEE Conf. on Comp. Vision and Pattern Recognition (CVPR)* 779-88

[16] Yang F, Choi W and Lin Y 2016 Exploit all the layers: Fast and accurate cnn object detector with scale dependent pooling and cascaded rejection classifiers *IEEE Conf. on Comp. Vision and Pattern Recognition (CVPR)* 2129-37

[17] Lin G, Shen C, Hengel A and Reid I 2016 Efficient piecewise training of deep structured models for semantic segmentation *Conf. on Comp. Vision and Pattern Recognition (CVPR)* 3194-203

[18] Harley A W, Konstantinos G D and Iasonas K 2017 Segmentation-Aware Convolutional Networks Using Local Attention Masks *IEEE Int. Conf. on Computing Vision (ICCV)* 5048-57

[19] Ronneberger O, Fischer P and Brox T 2015 U-Net: Convolutional Networks for Biomedical Image Segmentation *Proc. of the Int. Conf. on Medical Image Computing and Computer-Assisted Intervention* **9351** 234-41

[20] Goodfellow I, Pouget-Abadie J, Mirza M, Xu B and Bengio Y 2014 Generative adversarial nets *Proc. of the 27th Int. Conf. on Neural Inf. Proc. Systems* (Cambidge, MA, USA) MIT Press 2672-80

[21] Isola P, Zhu J-Y, Zhou T and Efros A 2017 Image-to-Image Translation with Conditional Adversarial Networks *Proc. of the IEEE Conf. on Comp. Vision and Pattern Recognition (CVPR)* 5967-76

[22] Salimans T, Goodfellow I, Zaremba W and Cheung V 2016 Improved techniques for training GANs *Proc. of the 30th Int. Conf. on Neural Information Processing Systems* 2234-42

[23] Tung F, Harley A, Seto W and Fragkiadaki K 2017 Adversarial inverse graphics networks: Learning 2d-to-3d lifting and image-to-image translation from unpaired supervision *Proc. of the Int. Conf. on Computing Vision* 4364-72

[24] Gulrajani I, Ahmed F, Arjovsky M, Dumoulin V and Courville A 2017 Improved training of wasserstein GANs *Proc. of the 31st Int. Conf. on Neural Information Processing Systems* 5769-79

[25] Ganin Y, Lempitsky V 2015 Unsupervised domain adaptation by backpropagation *Proc. of the 32nd Int. Conf. on Machine Learning* **37** 1180-9

[26] Nikolaev E I 2019 Laser Engraver Control System based on Reinforcement Adversarial Learning *Proc. of the Int. Russian Automation Conf.* (Sochi, Russia) 1-5

Modelling and optimization of information systems with distributed databases

I Ya Lvovich[1], Ya E Lvovich[2], A P Preobrazhenskiy[1], Yu P Preobrazhenskiy[1] and O N Choporov[2]

[1] Information Systems and Technologies Department, Voronezh institute of high technologies, 73a, Lenina st, Voronezh, 394043, Russia
[2] Information Security Department, Voronezh state technical university, 14, Moscow dist., Voronezh, 394026, Russia

E-mail: komkovvivt@yandex.ru

Abstract. The paper deals with the problems of modern corporate information systems. It is necessary to carry out the structural synthesis of their components. It is important that there is a choice of the required number of parallel processing in the databases, as well as the assignment of priorities for the execution of the corresponding processing conditions. On the basis of enumeration schemes, alternative variables will be introduced. When calculations are carried out using the developed models, it is possible to calculate the values of predictive estimates for each of the search steps. That is, many promising options for information and telecommunication systems are selected, with the additional inclusion of procedures for taking into account restrictions and multi-criteria.

1. Introduction

When designing wide classes of corporate information systems (CIS), it is necessary to correctly select the components that make up their composition. One of the central components is the database. Recently, they have been designed in a distributed manner. It is required to solve problems associated with the structural synthesis of such components. Why is this necessary? When implementing the synthesis procedures, it is necessary to carry out work on the construction of a multi-alternative optimization model. On its basis, it is possible to build interesting predictive assessments from the point of view of practice. There are some peculiarities during their construction. They are connected with the fact that when the objective function is formed in the optimization problem, then its values that were determined for each of the search steps in the calculated predictive estimates determine the need to implement a simulation experiment using transactional calculations. This makes it possible to refine the characteristics of the analyzed information systems [1].

These models have specific features. This leads to the fact that modified algorithms are used for selecting a set of promising options for information and telecommunication systems and additional inclusion of procedures for taking into account restrictions and multicriteria. There may be several options. If a multicriteria choice is carried out according to a rational option, then the process of combining assessment according to fuzzy requirements and a randomized search for a compromise will take place [2].

2. Characteristics of multi-alternative optimization based on a simulation model of a complex information and telecommunications system

When we carry out optimization using procedures for structural synthesis of such a component of the CIS as a distributed database, then it is important to solve the following problem. On different workstations, we can see some parts of the distributed database. It is part of the larger CIS. In a large corporate information system, there are sections of the local computer network. It can be divided into many computers $M=\{M_i\}, i=\overline{1,I}.$, which are considered as having similar characteristics [3]. Many files $F\{f_j\}, j=\overline{1,J}.$ are contained within a distributed database. Interaction within the analyzed information and telecommunication systems occurs using network equipment [4].

It is important to distribute the database files among the components of information and telecommunication systems. This should be done in such a way as to minimize the processing time of requests. How can you ensure this minimization? An optimization model is being built. To create it, we introduce alternative variables:

$$x_{ji}=\begin{cases}1, if\ j-th\ file\, is\ on\ i-th\ computer, \\ 0, otherwise,\end{cases}, \quad j=\overline{1,J}, \quad i=\overline{1,I}. \tag{1}$$

Complex processes are taking place in the information and telecommunication system. There is a dependence of the processing time of requests on the corresponding values of the variables $x_{ji}, j=\overline{1,j}$, $i=\overline{1,I}$. To build it, we need to use simulation modeling. The model will be like this $T=\Psi(x_{ji})$. The objective function should be minimized. Appropriate restrictions are created for it:

$$x_{ji}=\begin{cases}1, \\ 0\end{cases}, \quad j=\overline{1,J}, \quad i=\overline{1,I}. \tag{2}$$

1) For any fileм f_1 with data, only one of the analyzed computers can be associated at the moment

$$i=1\sum_{i=1}^{I} x_{ji}=1, \quad j=\overline{1,J};$$

2) There are limitations for a computer in terms of the total amount of information that will be for it ... It is bounded from above by the value V_i

$$\sum_{j=1}^{J} v_j x_{ji} \le V_i, i=\overline{1,I} \tag{3}$$

here V_i - it will be the amount of information inside the j-th file.

After taking into account these constraints, we can create a multi-alternative optimization model. On its basis, a structural synthesis of databases takes place [5, 6], which will be used inside information and telecommunication systems

$$T = \Psi(x_{ji}) \to min, \ \sum_{i=1}^{I} x_{ji} = 1, \ j = \overline{1,J},$$

$$\sum_{j=1}^{J} v_j x_{ji} \le V_i, \qquad i = \overline{1,I}, \ x_{ji} = \begin{cases} 1, \\ 0 \end{cases}, \quad j = \overline{1,J}, \ i = \overline{1,I}. \tag{4}$$

It is required for practical applications to determine the dependency $T = \Psi(x_{ji})$. This can be done, for example, based on a transaction hierarchy approach. In this case, a simulation model is applied.

Interconnected structures should be formed in the information and telecommunication system. For this purpose, modeling of the CIS architecture is applied. Modeling transactions by architectural features must be carried out in the correct way. Therefore, it is necessary to structure structures at the architectural level when designing, it is required to correlate with various types of links ..

When transactional computations are organized [7], then objects will be considered in the form of resources, correlated with the transactions being carried out. Then we can identify key links between transport and logical structures (figure 1):

- resources can be shared using transactions [8].
- object methods will influence how transaction bodies will be formed.
- if objects interact, then integrity constraints will be allocated for them using transactions.

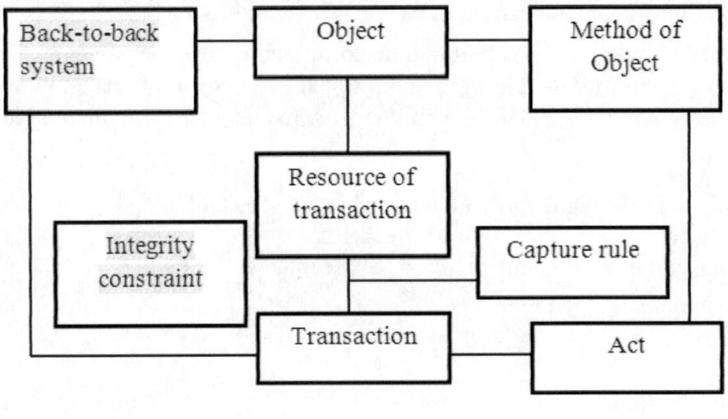

Figure 1. Illustration of the relationship between logical and transactional structures.

It is important to note that the flow of transactions in relation to objects acts like a work flow. What are the features of this? There are still additional constraints on the integrity of the computations performed [9].

Components need to be labeled appropriately. We consider control flows as components in the process structure. In this regard, transactions show how the workflows will occur. At the same time, we support the condition that integrity will be ensured in the calculations that are performed.

Modular and transactional structures will be linked by means of objects. In this case, the transition from one holistic state to another will be within the framework of the transaction.

The required modular structure depends on what work should be done in order for the transaction to be tested appropriately [10].

Due to the physical structure of the information component, telecommunication systems will be mapped accordingly to the hardware components.

What needs to be done in order to correctly describe the components of transactional structures?

It is required that links with existing workflows, particularities of hierarchies in transactions, characteristics of related objects, how the objects will be associated, in which way will be to a consistent state using the appropriate sequence of transactions. To use the hierarchy of transactions that is required to support organizational and structural relationships among transactions, it is necessary to use class diagrams. As a result, based on the above components, there are opportunities to build a simulation model. On its basis, key parameters in the designed information and telecommunication systems can be shown. Modeling of a set of options for information and telecommunication systems is carried out within the framework of simulation. The results obtained can be used in order to obtain a multi-criteria assessment on them. The next step is how decisions will be made [11].

3. Carrying out a multi-criteria choice according to options subject to unclear requirements

We have several options for information and telecommunication systems. To make a choice on them, you can use a method in which there is a multi-criteria choice for alternatives of options. In this case, a compositional rule is used, which allows aggregation by descriptions of alternatives. We proceed from the assumption that information about the preferences of persons who will make decisions is represented as fuzzy judgments. The main problem that will be in multi-criteria choice when fuzzy models are applied is how the assessments and related criteria. Why does this problem arise? This is due to the fact that different results will be obtained for the same problems, but using different methods. To give a solution to such problems, you can use a heuristic approach. We believe that many solutions [12] can be described using a set of criteria $x_l,...,x_p$. Then we say that linguistic variables are applied. They will be set on the basis of base sets $u_l,...,u_p$. Let's give a specific example. We proceed from the fact that for the variable x_l "transaction time" the value can be LOW, and the variable x_2 "cost" - the value "GOOD", and so on. In a set that meets several criteria that will be correlated with some values, we can describe the perceptions of persons related to how decisions will be made. These solutions show how satisfactory the alternative would be. We will consider the variable S "satisfactory" as linguistic. Next, we consider an example of a statement:

$$d_1 \text{ : "If } x_l = \text{"LOW" and } x_2 = \text{"GOOD", then S = "HIGH".} \quad (5)$$

When we consider the general case, then the statement d_i is represented as follows:

$$d_i\text{: "If } x_l = A_{li}\, u\, x_2 = A_{2i}\, u \ldots x_p = A_{pi},\ mo\, S = B_i\text{''}. \quad (6)$$

We introduce the notation of intersection $(x_l = A_{li} \cap x_2 = A_{2i} \cap \ldots \cap x_p = A_{pi})$ in the form $x = A_i$.

Operations associated with the intersection of fuzzy sets will correspond to the definition of the minimum by their membership functions: $\mu_{Ai}(\upsilon) = \min(\mu_{A_{il}}(u_l),...,\mu_{Aip}(u_p))$. In the specified expression $V = U_l \times \ldots \times U_p; v = (u_l,...,u_p), \mu_{Ail}(u_j)$ shows what the value of the element's membership u_i in a fuzzy set A_{il} will be. After that, we will write expression (6) as follows:

$$d_i\text{: "If } x = A_i, mo\, S = B_i\text{."} \quad (7)$$

Base sets U and W we will denote V as W. In this case A_i will be a fuzzy subset of W. When constructing a model B_i is considered as a fuzzy subset in the unit interval I.

When developing a model, it is necessary to present the rules. In this case, we rely on the operation of implication. Lukasiewicz's fuzzy implication will be as follows:

$$\mu_H(w,i) = \min(1,(1-\mu_A(w)+\mu_B(i))), \tag{8}$$

In this case, H - will be a fuzzy subset on $W \times 1, w \in W, i \in I$.

Similarly to how it was done above, we can make transformations of statements $d_l,...,d_q$ to sets $H_l,...,H_q$. The set D is considered as their intersection: $D = H_l \cap ... \cap H_q$

For any of $(w,i) \in W \times I$ $\mu_D(w,i) = \min(\mu_{Hi}(w,i)), i = \overline{1,q}$.

To what extent we satisfy the alternative associated with a fuzzy subset of A and W, we can deduce using of the compositional inference rule $G = A \circ D$, in it, G is considered as a fuzzy subset over the interval I.

Summarizing, we get $\mu_G(i) = \max(\min \mu_A(w)\mu_D(w,i))$. Alternatives will be generated by generating point estimates. For a fuzzy set $C \in I$ we can find α -level set $(\alpha \in [0,1])$:

$$C_\alpha = \{i | \mu_C(i) \geq \alpha, i \in I\}. \tag{9}$$

For any C_α there is a possibility to determine the average number of elements $M(C_\alpha)$:

In a set $M(C_\alpha) = \sum_{j=1}^{n} i_j / n, i_j \in C_\alpha$ that consists of n elements if $C_\alpha = \{a \leq i \leq b\}$ $M(C_\alpha) = (a+b)/2$; i

$C_\alpha = \bigcup_{j=1}^{n} \{a_j \leq i \leq b_j\}$, $M(C_\alpha) = \frac{\sum_{j=1}^{n} 0,5(a_j+b_j)(b_j-a_j)}{\sum_{j=1}^{n}(b_j-a_j)}$, when $0 \leq a_l \leq b_l \leq ... \leq a_n \leq b_n \leq 1$.

In this case, we can record a point value over a set C in this way:

$$F(C) = \frac{1}{\alpha_{\max}} \int_0^{\alpha_{\max}} M(C_\alpha) \partial a, \tag{10}$$

$\alpha_{\max}$ is considered as the maximum value over set C.

When alternatives are chosen, then it is important for any of them to find satisfaction. The corresponding point estimate is calculated. Which alternative would be considered the best? The whichever has the largest value [13]. A random enumeration is performed based on the index i_1. It demonstrates what the numbers will be in the criteria. We introduce a discrete random variable $\tilde{\eta}$ tat taking values $1,2,...,I_1$.. The initial distribution is uniform

$$P(\tilde{\eta} = i_1) = \frac{1}{I_1} \forall i_1 = \overline{1, I_1}. \tag{11}$$

When changing the distribution (11) using information from the decision maker, we rely on an algorithm that has such steps.

1 step. We implement the random variable $\tilde{\eta} = \alpha$ in a machine way. Then we conduct an experiment using the criterion $\Psi_\alpha(x)$, $\alpha \in \overline{1, I_1}$. It is important to follow the procedures. Values are

calculated based on other criteria $\Psi_{i1}(x),(i_1 = \overline{1, I_1},\ i_\alpha \neq \alpha\ )$ *if* $x = x_\alpha^k$. Using the obtained values, we will be able to construct a vector z^k:

$$z^k = \begin{vmatrix} \Psi_1\left(x_\alpha^k\right) \\ \cdot \\ \cdot \\ \cdot \\ \Psi_{I1}\left(x_\alpha^k\right) \end{vmatrix}. \tag{12}$$

Step 2. It is necessary to involve the person who will make the decision. For such a person, the appropriate question is formulated: "Will we get satisfactory values for all the components in the vector z^k?". We will complete the solution of the problem under consideration if there is a satisfactory answer. Otherwise, an analysis is required regarding the possibilities for folding local criteria to the global one [14]. It will look like this:

$$F(x) = \sum_{i=1}^{I_1} P_{i1}^* \hat{\Psi}_{i1}(x), \tag{13}$$

In the specified expression P_{i1}^* - will be the steady-state values of probabilities

$$P^k(\tilde{\eta} = i_1), \left(i_1 = \overline{1, I_1}\right); \quad \hat{\Psi}_{i1}(x) = \frac{\Psi_{i1}(x) - \Psi_{i1}^{\min}(x)}{\Psi_{i1}^{\max}(x) - \Psi_{i1}^{\min}(x)}, \tag{14}$$

$\Psi_i^{\max}, \Psi_i^{\min}$ - for local criteria the maximum and minimum values. In this case, they will lie inside in the areas of feasible solutions. How does this happen? Uniform distribution is considered for the first step

$$P^1(\tilde{\eta} = i_1) = \frac{1}{I_1}, \left(i_1 = \overline{1, I_1}\right) \tag{15}$$

The preferences of the decision-makers are changing. Therefore, there will be changes in the values of the probabilities. But their total sum will be as follows

$$\sum_{i=1}^{I_1} P^k(\tilde{\eta} = i_1) = 1. \tag{16}$$

When the process of learning situations is carried out, then there will be a tendency of the value P_{i1}^*

of probabilities (figure 2) to certain values that are steady-state [15].

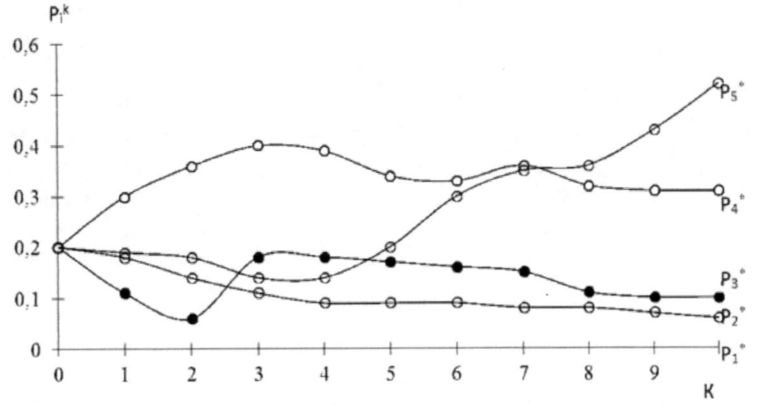

Figure 2. Illustration of a diagram of the settings of probabilistic estimates associated with preferences by local criteria.

A visual registration of the person who makes the decision takes place. It will be indicated to them as a function to be optimized by machine experiment. In this case, we rely on a procedure when criterion (16) is applied. When the decision-maker believes that the values $P^k(\tilde{\eta} = i_1)$, $(i = \overline{1, I_1})$ will not be outside the ranges of steady-state values, then he marks the criterion $\Psi_\beta(x_\alpha^k), (\beta \in \overline{1, I_1})$ in which the least satisfactory value will be. We can interpret his answer in the k - th step in the form of signed estimates for the values

$$\Delta_{i1}^k F(x_\alpha^k) = siqn\left(\frac{\partial F(\Psi)}{\partial \Psi_{i1}}\right) = \begin{cases} 1, i_1 = \beta, \\ -1, i_1 \neq \beta. \end{cases} \quad (17)$$

Step 3. The settings process will be carried out according to the values of the probabilities $P^{k+1}(\tilde{\eta} = 1_1),$. They correlate with how the generation of random discrete variables $\tilde{\eta}$ for the $(k+1)$-th step will proceed:

$$P^{k+1}(\tilde{\eta} = i_1) = \frac{P^k(\tilde{\eta} = i_1) + \varepsilon^{k+1} æ\left(\Delta_{i1}^k F(x_\alpha^k)\right)}{1 + \varepsilon^{k+1}}, \quad (i_1 = \overline{1, I_1}), \quad (18)$$

In this case ε^{k+1} - is considered in the form of a step size correlated with an iterative procedure

$$(a) = \begin{cases} I, \ if \ a \rangle 0, \\ 0, \ if \ a \leq 0, \end{cases} \quad (19)$$

in the numerator we mean an iterative gradient procedure. The expression written in the denominator shows how the condition is satisfied

$$\sum_{i_1=1}^{I_1} P^{k+1}(\tilde{\eta} = i_1) = 1. \quad (20)$$

When the step size ε^{k+1} is chosen, then we are based on heuristic considerations. When a person who will make a decision on two adjacent steps $(k-1) and (k)$ will make a choice according to the same criterion β, then we can assume that there will be an increase in its significance according to experimental research. In such cases, the $P^{k+1}(\tilde{\eta} = \beta)$ will be growth. We can ensure this by increasing in steps ε^{k+1}. In cases where the preferences are not the same, it is required to achieve opposition relative ε^{k+1}. How can this be explained? Then the persons associated with how decisions are made are not characterized by a stable opinion. There will be an adjustment of opinions. And this leads to the fact that there is no need to carry out a noticeable correction with respect to the distribution over a discrete random variable $\tilde{\eta}$. That is, when a choice is made according to the step size, we rely on the rule

$$\varepsilon^{k+1} = \varepsilon^{k} \exp\left[\gamma \Delta_{\beta}^{k} F \cdot \Delta_{\beta}^{k-1} F\right] \quad (21)$$

In the expression γ - considered as a constant factor. After setting the probabilities, return to the first stage.

4. Conclusion

Distributed database must undergo a process of structural synthesis. This is necessary to improve the characteristics of the CIS. Although optimization procedures are performed, the values of the objective function are calculated. For this, a simulation model is involved. It provides opportunities to formalize the dependence of the query processing time on alternative variables. The use of alternative variables is necessary in tasks related to the synthesis of the number of parallel processing and priorities for fulfilling certain processing conditions. When the process of selection of the final options for the design is carried out according to a variety of promising options, then the involvement of a multi-criteria assessment is required. Formalization of the specified requirements in a fuzzy way allows improving its efficiency. This introduces linguistic variables. The algorithm uses a randomized scheme associated with the search for a compromise solution.

References

[1] Stankovic J A 2014 Research directions for the internet of things *IEEE Internet of Things Journal* **1** 3-9

[2] Rios L M and Sahinidis N V 2013 Derivative-free optimization: a review of algorithms and comparison of software implementations *Journal of Global Optimization* **54** 1247-93

[3] Lutakamale A S, Kaijage S 2017 Wildfire Monitoring and Detection System Using Wireless Sensor Network: A Case Study of Tanzania *Wireless Sensor Network* **9** 274-89

[4] Minerva R, Biru A and Rotondi D 2015 Towards a definition of the Internet of Things (IoT) *IEEE Internet Initiative* Torino, Italy Retrieved from: https://iot.ieee.org/images/files/pdf/IEEE_IoT_Towards_Definition_Internet_of_Things_Revision1_27MAY15.pdf

[5] 2016 Broadband Internet Technical Advisory Group *Internet of Things (IoT) Security and Privacy Recommendations* Retrieved from: https://www.bitag.org/documents/BITAG_Report_-_Internet_of_Things_(IoT)_Security_and_Privacy_Recommendations.pdf

[6] Odu G O and Charles-Owaba O E 2013 Review of Multi-criteria Optimization Methods *Theory and Applications* **3** 1-14

[7] Shah A and Ghahramani Z 2015 Parallel predictive entropy search for batch global optimization of expensive objective functions *Advances in Neural Information Processing Systems* 3330-8

[8] Neittaanmki P, Repin S and Tuovinen T (Eds.) 2016 *Mathematical Modeling and Optimization of Complex Structures; Series: Computational Methods in Applied Sciences* Springer International Publishing AG, Switzerland

[9] Morais H, Kádár P, Faria P, Vale Z A and Khodr H M 2010 Optimal Scheduling of a Renewable Micro-Grid in an Isolated Load Area Using Mixed-Integer Linear Programming *Elsevier*

Editorial System(tm) for Renewable Energy Magazine **35(1)** 151-6
[10] Orlova D E 2018 Stability of solutions in ensuring the functioning of organizational and technical systems *Modeling, Optimization and Information Technologies* **6(1)** 325-36
[11] Talluri S, Kim M K and Schoenherr T 2013 The relationship between operating efficiency and service quality: are they compatible? *Int. J Prod Res* **51** 548-2567
[12] Groefsema H, Beest N R T P 2015 Design-time compliance of service compositions in dynamic service environments *Int. Conf. on Service Oriented Computing & Applications* 108-15
[13] Lvovich I Ya, Lvovich Ya E, Preobrazhenskiy A P, Preobrazhenskiy Yu P, Choporov O N 2019 Modelling of information systems with increased efficiency with application of optimization-expert evaluation *Journal of Physics: Conf. Ser. Krasnoyarsk Science and Technology City Hall of the Russian Union of Scientific and Engineering Associations; Polytechnical Institute of Siberian Federal University. Bristol, United Kingdom* 33079
[14] Yao Y and Chen J 2010 Global optimization of a central air-conditioning system using decomposition-coordination method *Energy and Buildings* **5** 570-83
[15] Lvovich I, Preobrazhenskiy A, Preobrazhenskiy Y, Lvovich Y, Choporov O 2019 Managing developing internet of things systems based on models and algorithms of multi-alternative aggregation *Int. Seminar on Electron Devices Design and Production, SED-2019 - Proceedings* 8798413

Cloud computing in industrial automation systems

A U Mentsiev[1,3], Y A Kulpeiis[2] and K K Smagulova[2]

[1]Faculty of information technology, Chechen State University, 32 Sheripov Street, Grozny, 364024, Russia

[2]Production Processes Automation Department, Faculty of Energy, Automation and Telecommunication, Karaganda Technical University, 56 N, Nazarbayev av., Karaganda, 100027, Republic of Kazakhstan

[3]E-mail: a.mentsiev@chesu.ru

Abstract. Automated systems in industries increase production and make things more accurate. These both things give very much profit to the manufacturer. No doubt these automated systems of different types reduced the rate of error and increase efficiency but it was facing some very serious issues like loss of data, security of data, and maintenance of the automated system. These are issues where cloud computing involves resolution and provides solutions to overcome these issues. This paper describes the problems of automated systems in industry and the help of cloud computing for the mitigation of the issues faces by automated systems. In this paper discussion about the future role of cloud computing is described for enhancing the automation in industry.

1. Introduction

Terms like "cloud" and "cloud computing" have become more noticeable in recent times. Cloud computing is the new centralized online data repository tool that provides services related to servers, networking, databases, storage, analytics, and whatnot. The use of this technology makes things faster, easier and economical. The users only pay for the cloud services they use. The rest is free. This reduces the operating cost and helps in running the system both effectively and efficiently. Cloud computing is a term that is becoming very familiar now a day. It is providing solution of the issue which are creating hurdles for installation and working of automated system in the industry. Term cloud computing is defined as access of network on demand for sharing of different computing resources like applications, storage, services, servers, and networks. It has some specific properties. Cloud system provides services like storage on server and network response automatically without human interaction to each user. In cloud computing networks and servers can be accessed through many different platforms like mobiles, computers, laptops, tablets, and other resources by using web services or another standard way to access the cloud. Resources in the cloud are kept in a pool to provide services to different consumers at a time. Assignment of resources is made by using a model called multi-tenant. This model dynamically assigned the resources and reassigned them on user demand. In cloud computing consumer is not aware of the physical location of resources that are providing him services [1]. Cloud provides elasticity in services. Cloud can automatically increase the storage and computing capacity according to the needs of the user. Users can get unlimited storage space and computing power. While the use of control systems like robots and automated computers for different process handling instead of humans is called industrial automation. This automation in industry increased the manufacturing and production of industries. The operating cost of these systems is also low. It also eliminates the cost of healthcare. It also increases the working time. It also saves the salary of workers. Maintain ace cost is much lower than the salaries of workers. This system can work 365 days of the year which results in an increase in productivity. An automated system also reduced the chances of error because machines work on fixed patterns and according to commands given to them [2].

2. Discussion

Industrial automation is defined as the use of computer systems and artificial intelligence in place of humans in different industries. It provides high quality with flexibility. It enhances the productivity as well as the safety of the system. [3] Cloud computing has made its mark in industrial automation systems as well. It not only saving time but it's also a great source of saving money. Above that it reduces the chance of error and organizations can create reliable workflows. This plays directly on better IT and corporate management.

2.1. Cloud computing in Finance

The first and most benefit of a cloud financial system is maintenance cost is very low. There are a lot of features the financial industry could use to their advantage. SaaS programs are being adopted by many banks and insurance in order to handle records and back-office tasks. Cloud-powered AI and machine learning is being used in capital markets to help with financial data management and investment strategies [4].

2.2. Cloud computing in automotive

For the automotive industry, the best technology solutions are keeping up with data on their supply. This helps them improve their communication with their clients, to provide them with exactly what customers want. It is much easier to keep track of what parts are available for both manufacturers and automotive suppliers by keeping data in a common cloud environment. With smart cars becoming more readily available, auto companies can use cloud-based analytics solutions to use a large amount of data produced by the car on the road [4].

2.3. Cloud computing in manufacturing

Every part of a manufacturing process can be integrated with cloud solutions. It is much faster to filter different size orders through the same channels and provide products according to their need. Automation software programs can be used to automatically manage the installation, management, and configuration of cloud-based computing systems. Using cloud-based resources more efficiently. Most companies use the cloud for essential services and applications. Cloud automation is a time saver and improves cloud management, becoming more noticeable in IT companies [4].

3. The issue with industrial automation system

Industrial automation is usually adopted for the reduction of accidents industry and it is used for increasing efficiency. An automated system like robots, equipment that works autonomously, and automated conveyor system are used in the industry which is increasing efficiency and production. There are different types of industrial automation systems are working in the industry. These systems are producing data and worked through different kinds of automation like fixed, programmable, and integrated automation. As described above the goal of an automated system is to improve efficiency than it is very important to make an analysis of data which is produced by these systems. This analysis can be used for making actionable intelligent systems. Cloud computing makes manipulation of data much easier which is produced by these automated systems. Data manipulation is not the only issue that automated systems are facing. There are some other issues in automated systems in the industry are existed. Automation in industry and system control is still relied on legacy security for dealing with threats that are creating hurdles. It was a good policy in the past and it was working. But after introducing IP-based communication in the automated system of industries relying on these outdated and traditional security tools is increasing risks for data leakage [6]. Because automated cyber-attacks can easily found vulnerabilities in the systems which are working on IP-based communication. Unplanned failure of automated systems has also affected the system and it is also a big challenge for the automation of industry. According to an estimation unplanned downtime causes the loss of $260000 per hour for the manufacturers who are using automated systems in their companies. This estimation is taken from all the businesses. Cloud computing provides solutions for all these issue and addition of internet of things devices in the automation of industry make this mitigation process very easy [7].

4. The role of cloud computing for mitigation of automated systems issues in the industry

The centralization feature of cloud computing provides a very good solution to issues facing by automated systems of industry. Role cloud computing in mitigation of issue related to automated systems of industry are given below.

4.1. Enhance analysis of data

All the automated systems used in the industry are controlled through data and its manipulation. Such scheduled operations that pass the instruction to the function for executing on time are created from previous data. Cloud computing allows the automated systems in industries to perform the scheduled operation in real-time and provide all the required computing resources to handle big data sets [8]. In cloud computing, it does not matter that data is complex or large because cloud computing provides more computing resources for complex data which can be easily handled for generating analytics of data.

4.2. Enhancement in measures for cybersecurity

Cybersecurity is always a big issue for automated systems in industries. Because due to automation IoT devices usage is increasing. These devices need many communication protocols which left many vulnerabilities which can easily be exposed by cybersecurity attacks. A centralized system of industrial cloud computing helps in overcoming this issue and provides storage for complex and large data sets. We can say that technologies used in automated completely depend on cloud computing for protection from cyberattacks. In the area of cybersecurity cloud computing technology is excelling day by day. It is due to the continuous updating of cloud computing involved in the industry. This is also due to options of using SIEM tools and integration of third-party security information for enhancing security [9]. Cloud computing also provides easy data recovery in case of data loss from the data which was stored as backup.

4.3. Define the future of automation systems in industries

The future of automated industrial systems is based on emerging technology such as the internet of things, Artificial intelligence, edge computing, and Machine learning. Integrating these technologies with the current system needs simulation to see the effects before implementation. Cloud computing provides an environment for testing and doing analysis for integration [10].

4.4. Enhance consolidation of workload

Centralization of cloud used in the industry plays a vital role for effective automation of industrial systems. It is the industrial cloud which helps in combining different industrial automation system in a single secure and measurable platform [9]. In this way, the network and IT infrastructure involved in handling the operation of the industry can be reduced. This system also reduced the maintenance and management cost of the system because it is easy to handle all the resources in one place instead of handling them at many different platforms [11].

5. Conclusion

Many fields have already adopted cloud computing in their manufacturing, supplying, or selling businesses. These businesses are prime examples to see how beneficial it can be to integrate and implement cloud computing in industrial automation systems. These examples can also be used for analysis and learning of better usage of cloud computing in a future implementation. The benefits of cloud computing are still being measured against the risks of output loss, protection, availability, and unknown server locations in the process control industry. Virtualized servers, however, are becoming increasingly more attractive to end-users with the bevy of technological advances. For older, space-constrained plants that need these latest developments in equipment, where few applications can be sustained in a small, cloud environment, the rise of cloud computing eventually requires survival. This may also contribute to the transfer of secondary monitoring and expert systems to the cloud as well. We will possibly see more and more of these systems being virtualized, providing plant operators with more cost and space advantages.

References

[1] Amirova E F, Petrova L I, Ziuzya E V, Sleptsov V V, Krishtaleva T I and Kuznetsova M V 2019 Import substitution as an economic incentive mechanism for Russian commodity producers *International Journal of Civil Engineering and Technology* **10(2)** 926-31

[2] Breivold H P 2020 Towards factories of the future: migration of industrial legacy automation systems in the cloud computing and Internet-of-things context *Enterprise Information Systems* **14(4)** 542-62

[3] Lavrov E A et al. 2021 *IOP Conf. Ser.: Mater. Sci. Eng.* **1047** 012117

[4] Magomadov V S 2020 *IOP Conf. Ser.: Mater. Sci. Eng.* **862** 032080

[5] Magomadov V S 2020 *IOP Conf. Ser.: Mater. Sci. Eng.* **862** 032101

[6] Magomedov I A et al. 2020 *IOP Conf. Ser.: Mater. Sci. Eng.* **862** 052071

[7] Magomedov I A 2020 *IOP Conf. Ser.: Mater. Sci. Eng.* **734** 012155

[8] Pop P and Raagaard M L 2018 Enabling Fog Computing for Industrial Automation Through Time-Sensitive Networking (TSN) *IEEE Communications Standards Magazine* **2(2)** 55-61

[9] Rostova O, Shirokova S and Sokolitsyna N 2019 *IOP Conf. Ser.: Mater. Sci. Eng.* **497** 012017

[10] Zharova M, Shirokova S and Rostova O 2019 Management of pilot IT projects in the preparation of energy resources *E3S Web of Conferences* **110** 02033

[11] Zolkin A L, Losev A N, Gridina D V and Aygumov T G 2021 *IOP Conf. Ser.: Mater. Sci. Eng.* **1047** 012106

On the application of a model for assessing the transactional reliability of information systems of industrial enterprises

I V Kovalev[1,2,3,4], D I Kovalev[2], N A Testoyedov[4,5], D V Borovinskii[6], A A Koltashev[5], A K Shatrov[5] and V A Bartenev[5]

[1]Siberian Federal University, 79, Svobodny pr., Krasnoyarsk, 660041, Russia
[2]Krasnoyarsk State Agrarian University, 90, Mira pr., Krasnoyarsk, 660049, Russia
[3]China Aviation Industry General Aircraft Zhejiang Institute Co., Ltd, China
[4]Reshetnev Siberian State University of Science and Technology, 31, Krasnoyarsky Rabochy Av., Krasnoyarsk, 660037, Russia
[5]JSC "Academician M F Reshetnev Information satellite systems", 52 Lenin street, Zheleznogorsk, Krasnoyarsk region, 662972, Russia
[6]FSBEE HE Siberian Fire and Rescue Academy EMERCOM of Russia, Severnaya street, Zheleznogorsk, Krasnoyarsk region, 662972, Russia

E-mail: kovalev.fsu@mail.ru

Abstract. The article examines an example of the application of a model for assessing the transactional reliability of information systems of industrial enterprises. This transactional reliability determines the correctness of the calculations performed and the integrity of the data, and, consequently, the characteristics of the quality of software tools that are used in information systems of industrial enterprises to perform control and data processing functions. It is shown that all the parameters of this model can be obtained from the results of testing the system, which makes it possible to construct adequate estimates of the reliability parameters. The calculated indicators obtained using the developed model differed from the real ones by no more than 5%, which indicates the sufficient adequacy of the model. The modern system Microsoft Dynamics 365 for Finance and Operations (Dynamics AX/Axapta) was used as an environment for modeling and analyzing the transactional reliability of software tools.

1. Introduction

When developing critical projects of information systems of industrial enterprises, developers try, to one degree or another, to obtain an assessment of the reliability of software. This is usually done based on the final test results of the system. However, modern software used in complex information systems uses a colossal amount of data in its work, passing through standard modules and functions. Therefore, it is almost impossible to identify all the connections and ways of information processing, even for a fairly simple program. Based on this, the detailing of the elements (conditionally called software modules) of the reliability calculation should be limited to complete software components. These components, interacting with each other, constitute a more complex combination, for which we need to calculate the operational (hereinafter referred to as "transactional") reliability.

Currently, transactional information processing is becoming one of the most important aspects that determine the correctness of calculations and data integrity, and, consequently, the characteristics of the quality of software used in control and information processing systems [1-4]. The large number and ever-increasing complexity of software tools for transactional information processing require increased attention to transaction modeling throughout the development process. The transactional computing model implemented in a software tool essentially determines whether the software tool will be in a consistent state and maintain the required level of system reliability.

The main advantage of this model is that all the parameters of this model can be obtained from the data after the system test, which makes it possible to construct estimates of the reliability parameters [5-7].

2. Materials and methods

2.1. Information systems of industrial enterprises

Such a model is useful for evaluating information systems for processing and storing data, for example, for modern Enterprise Resource Planning (ERP) systems [8]. These systems allow not only to maintain operational (goods, services, production), financial, tax accounting at the enterprise, but also allow the management, the planning department to build budgets and monitor their implementation, plan the activities of the enterprise, manage the enterprise, analyzing the planned and actual indicators in the system.

Systems of this class are more focused on working with financial information for solving problems of managing large corporations of various types of activities with geographically separated resources. This includes everything that is necessary to obtain resources, manufacture products, transport them and settlements for customer orders. In addition to the listed functional requirements, ERP implements new approaches to the use of graphics, the use of relational data, CASE-technologies for their development, architecture of computing systems of the "client-server" type and their implementation as open systems.

ERP is a methodology aimed at efficient management of all production resources of an enterprise [9]. It provides a solution to the problems of planning the activities of the enterprise in physical and monetary terms, modeling the capabilities of the enterprise. This methodology is based on a number of large interrelated functionalities, including:

- Business Planning (BP);
- Sales and Operations Planning (S&OP);
- Production Planning (PP);
- Master Production Scheduling (MPS);
- Material Requirements Planning (MRP);
- Capacity Requirements Planning (CRP);
- Shop Floor Control (SFC);
- Just-in-Time (JIT);
- Forecasting (F);
- Project and program management (PPM);
- Maintaining information about the range and composition of products (MIR&CP);
- Maintaining information on technological routes, supply chains (MITR&SC);
- Cost management (CM);
- Financial management (FM);
- Human resources management (HRM).

An enlarged ERP management structure is shown in figure 1. The presence of a powerful infrastructure and methodology for building systems contributes to the achievement of a high level of efficiency in the implementation of management systems such as ERP in industrial enterprises. According to some estimates, the introduction of such systems can lead to a reduction in stocks up to

30%, an increase in labor productivity up to 25%, an increase in the number of orders completed on time, up to 20%.

2.2. Dynamics AX / Axapta
The modern system Microsoft Dynamics 365 for Finance and Operations (Dynamics AX/Axapta) was chosen as an environment for the development of a system of model-algorithmic support for analyzing the transactional reliability of software tools. This is an automated enterprise management system that provides the management of the company and its employees with the most complete information necessary for a successful business.

Dynamics AX is a scalable system for midsize and large enterprises, corporations and holding structures, providing a single, integrated solution aimed at increasing business manageability and increasing enterprise profits.

Dynamics AX is a powerful tool that realizes opportunities for both operational operation of the enterprise and for the formation of a long-term strategy, which allows you to lay the foundation for successful operations.

The uniqueness of the Dynamics AX system is in its modern advanced technology, which provides a single information space for the enterprise. Powerful functionality, optimal price / quality ratio, ease of learning, ease of use and configuration of the system make it possible to build an information management system at an enterprise that corresponds to its industry and individual characteristics.

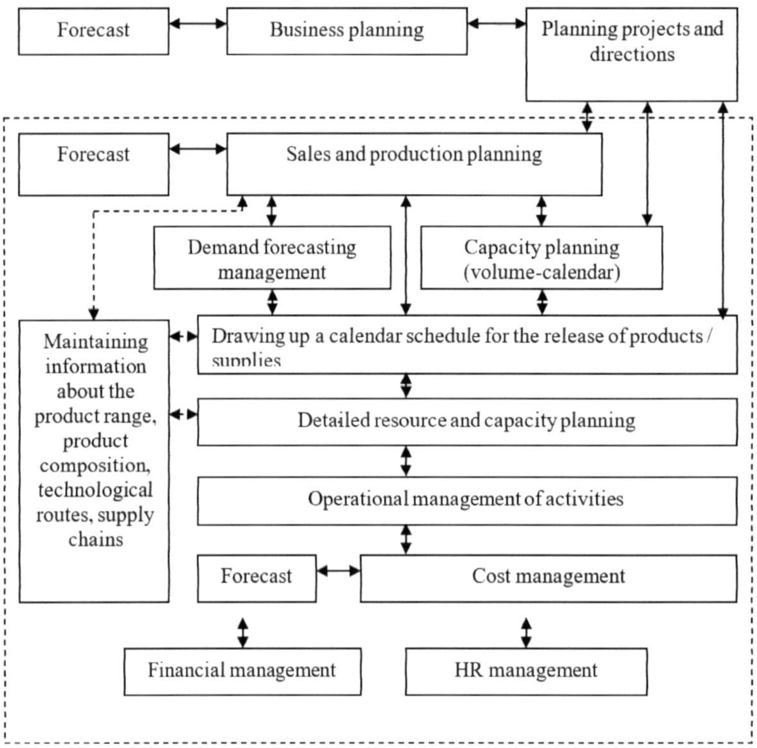

Figure 1. General scheme of the ERP.

Facts in favor of choosing Dynamics AX:

- Dynamics AX system is written in the object-oriented language X ++, similar in syntax to C ++ and Java;
- Dynamics AX system has an open source code available for modifications;

- Dynamics AX system has built-in visual programming tools;
- Dynamics AX system has about 2000 classes that implement the business logic of the system;
- Dynamics AX system has a built-in code profiler with run-time granularity down to a line of source code;
- Dynamics AX system has built-in tools for monitoring and tracing errors;
- Dynamics AX system has the ability to organize distributed computing.

The system of model-algorithmic support for the analysis of transactional reliability of software is implemented as a module "Reliability Models" of the ERP-system Dynamics AX.

The logic of the system of model-algorithmic support for the analysis of transactional reliability of software is written in X ++ in the form of a class hierarchy and is integrated into the logic of Dynamics AX [8, 9]. Final and intermediate results are saved in the database for further analysis. The module is built as a set of dialog boxes and reports for interactive interaction with the user [10]. The user can create new parameters for the model, adjust existing parameters, make calculations of reliability parameters, save results and build reports on already performed and saved calculations. After opening the module, the user can configure the main reference books of the module: parameters of class descriptions, descriptions of transactions. After choosing a model for assessing transactional reliability, the user must fill out reference books about the classes used or transactions included in the calculations, as well as operational profiles based on the test results. Calculations are performed automatically and controlled by the user. The entered and calculated data are saved in the database. The results of past calculations can be printed using reports.

2.3. Model for assessing the transactional reliability of information systems

Reliability of software is defined as the probability of functioning without errors for a certain period of time in a particular operating system. For most developers, reliability is defined as the correctness of the software, i.e. the number of errors to be corrected during the test.

Transactional reliability differs in its meaning from the classical understanding of reliability. It can be used to assess the reliability of software for processing and storing data, where the logical unit of work is a transaction [11].

Transactional reliability depends not only on the reliability of the components, but also on the specific problem area for which the software was developed, formally - on the set of operational profiles of the components. An operational profile is a set of disconnected and non-overlapping input ranges and the probabilities of their use by components. Profiles may differ for different operating systems. Obtaining an operational profile requires dividing the input range space into subsets (leaves), and then estimating the probabilities of using each subset (leaf). A subset with a sufficiently high probability of use can be divided into smaller zones. Profiles are tables that include information about the ranges of the input data of the module and the probability of failure on a given range of data [2-4].

To calculate the overall transactional reliability R_{tr}, it is necessary to compile a complete list of all transactions and calculate the reliability of each transaction using the operating profiles.

A transaction is a sequence of actions that is considered completed successfully if there were no failures at any step. A transaction can be either a single transaction of information systems, or a certain, logically completed, independent sequence of user actions in the system. For correct calculations, you must select one type of transaction.

Let's define:

m – total number of classes;

n – total number of transactions;

K_i – class, i=1,…,m;

O_i – operating profile (multiple input ranges) i-th class, i=1,…,m;

F_i – vector representing the probabilities of failure i class for each input range, i=1,…,m;

f_{il} – l-th element vector F_i , l – vector dimension F_i;

T_j – transaction, j=1,…,n;

D_j – many classes, belonging to the j-th transaction, j=1,…,n;
H_{ij} – vector of probabilities of using the range of values of the i-th class in the j-th transaction;
h_{ijk} – k-th vector element H_{ij}.
Wij – weight (reliability) i-th class j-th transaction;
PU_j – the probability of using the j-th transaction, where $\sum_{j=1}^{n} PU_j = 1$;
R_{tj} – reliability of the j-th transaction;
R_{tr} – transactional reliability of the system.

The weight (reliability) of the i-th class of the j-th transaction is determined by the formula:

$$W_{ij} = \sum_{k=1}^{l} h_{ijk}(1 - f_{ik}). \tag{1}$$

The reliability of a transaction can be calculated as:

$$R_{tj} = \prod_{k=1}^{n} W_{jk} \; . \tag{2}$$

After calculating the reliability of all transactions, the total transactional reliability of the system is calculated:

$$R_{tr} = \sum_{j=1}^{m} PU_j \times R_{tj} \tag{3}$$

3. Results

As an example, consider the "Master Planning" module of the Dynamics AX system, implemented at the enterprise of the Krasnoyarsk Territory in the «FORS» company [6].

As a result of testing, 4 transactions were identified, which in total use 18 classes with a maximum number of input ranges of 4.

Table 1. Operating profiles of classes.

Classes (K_i)	Failure probabilities in the range			
	1	2	3	4
K1	0.000010	0	0.000070	0
K2	0	0	0.000010	0.000010
K3	0	0.000040	0.000550	0
K4	0	0	0	0.000050
K5	0	0	0.000020	0
K6	0.000340	0	0	0
K7	0	0	0.001100	0
K8	0	0	0	0.000090
K9	0	0.000001	0	0
K10	0	0.000440	0	0
K11	0.000670	0	0	0
K12	0	0	0	0.000010
K13	0.000180	0	0	0
K14	0	0.000580	0	0
K15	0	0	0.000001	0
K16	0.000010	0	0	0
K17	0	0	0.000050	0
K18	0	0.000010	0	0

The probabilities of using transactions are:

- PU_1=0.330000;
- PU_2=0.250000;
- PU_3=0.150000;
- PU_4=0.270000.

Let's describe subsets of classes included in transactions:

- D_1= {1, 2, 4, 5, 6, 8, 9, 11, 13, 14, 15, 17, 18};
- D_2= {1, 2, 3, 4, 5, 6, 16, 17, 18};
- D_3= {2, 3, 5, 7, 8, 10, 12, 15, 16, 17};
- D_4= {6, 7, 8, 15, 16, 17, 18}.

The vectors of the probability of using the ranges and the vectors of the weights are as follows (see table 2).

The vectors of the probabilities of using the ranges and vectors of weights of transactions T2, T3 and T4 are formed in a similar way.

Table 2. Vectors of the probabilities of using the ranges and of the weights of the T1 transaction.

Classes (K_i)	Ranges use probabilities				
	1	2	3	4	W_{1i}
K1	0	0.100000	0.900000	0	0.999937
K2	0.500000	0.500000	0	0	1
K4	0	0.140000	0.030000	0.830000	0.999959
K5	0.150000	0.250000	0	0.6	1
K6	0.005000	0.044000	0.456000	0.495000	0.999998
K8	0	0.550000	0.262000	0.188000	0.999983
K9	0.999000	0.001000	0	0	1
K11	0	0.727000	0.005000	0.268000	1
K13	1	0	0	0	0.999820
K14	0	0.441000	0.334000	0.225000	0.999744
K15	0	0.456000	0	0.544000	1
K17	0.745000	0.111000	0	0.144000	1
K18	0	0.498000	0.229000	0.273000	0.999995

After that, having received all the necessary data, we can calculate the reliability of each transaction using the methodology presented in the article:

- Rt_1 = 0.999436;
- Rt_2 = 0.999855;
- Rt_3 = 0.999662;
- Rt_4 = 0.999710.

As a result, we get the final transactional reliability of the system R_{tr} =0.999649.

4. Discussion

Analyzing the calculation results, we conclude that when working with these types of documents (four types of transactional data processing), the system will fail in 351 cases out of 106, provided that the operator does not make mistakes.

As a result of the real test, according to which the above calculation was made, out of 106 transactions, 338 transactions were unsuccessful.

The value of the experiment lies in the fact that both classes and their input ranges were identified that contribute the most to the probability of failure. This allowed us to pay more attention to testing and identifying errors in the code and in the "blockage" of the system under test.

The greatest weight in the probability of failure is made by the classes: K7, K10, K14. By reducing the probability of failure on input ranges in these classes by half, you can achieve the following results:

- $Rt_1 = 0.999564$;
- $Rt_2 = 0.999855$;
- $Rt_3 = 0.999784$;
- $Rt_4 = 0.999860$.

As a result, we get a new transactional reliability of the system $R_{tr} = 0.999750$.

5. Conclusion

During the full system testing phase, you can find the blockage in the system - those components that contribute the most to the likelihood of failure.

By knowing exactly which operating profiles and input ranges are most likely to fail, we spend less time finding and fixing bugs in our code.

For a ready-made system for processing and storing information with specified operating profiles, it is possible to calculate the final transactional reliability and system availability, as well as track their changes in the course of eliminating errors in the blockage.

As a result of using the system in a real project, it was possible to identify the blockage in the system and improve the reliability indicators. The calculated indicators obtained using the developed models differed from the real ones by no more than 5%, which indicates the sufficient adequacy of the model.

References

[1] Kovalev I V *et al* 2020 *IOP Conf. Ser.: Mater. Sci. Eng.* **862** 062105
[2] Kovalev I V *et al* 2019 *IOP Conf. Ser.: Mater. Sci. Eng.* **537** 052038
[3] Petrosyan M O *et al* 2016 *IOP Conf. Ser.: Mater. Sci. Eng.* **122** 012026
[4] Tsarev R Yu *et al* 2016 *IOP Conf. Ser.: Mater. Sci. Eng. 155* 012041
[5] Antamoshkin O A *et al* 2016 *IOP Conf. Ser.: Mater. Sci. Eng.* **155** 012043
[6] Kovalev I V *et al* 2015 *IOP Conf. Ser.: Mater. Sci. Eng.* **70** 012008
[7] Kuznetsov P A *et al* 2015 *IOP Conf. Ser.: Mater. Sci. Eng.* **94** 012019
[8] NetU: Business solutions Retrieved from: https://www.netugroup.com/category/our-services/business-solutions/
[9] Kozhuhova O A and Kukartsev V V 2011 *Actual problems of aviation and astronautics* **1(7)** 448-9
[10] Selezneva S V 2010 *Science and modernity* **1-2** 121-3
[11] Demin A M 2007 *Bulletin of the Ivanovo State Power Engineering University* **2** 60-8
[12] Kukartsev V V *et al* 2011 *Reshetnev readings* **2(15)** 703-5

Flow model in the impeller of a centrifugal pump

V V Chernenko and D V Chernenko

Reshetnev Siberian State University of Science and Technology, 31, Krasnoyarsky Rabochy Av., Krasnoyarsk, 660037, Russian Federation

E-mail: g_ramzes@mail.ru

Abstract. The calculation of the fluid flow in the interscapular channel of a centrifugal impeller with a finite number of blades, the construction of the energy characteristics of the impeller and its optimization by the number of vanes are considered. The calculation consists of two parts: firstly, the determination of the theoretical pump head taking into account the effect of a finite number of vanes based on the analysis of the force interaction, and, secondly, the determination of hydraulic losses in the impeller by integrating the friction stresses over the limiting surfaces. The results from both parts are used to optimize the number of vanes in the pump impeller.

Based on the results of the analysis of the experimental visualization of the flow in the impeller (Figure 1) [1], it can be concluded that the difference between the circumferential component of the fluid velocity and the circumferential velocity of the leading disk at the unpressurized side of the vane, and their equality on the pressure side.

Considering that the force interaction in a liquid propagates only in the form of a longitudinal elastic wave, it can be concluded that the instantaneous direction of the liquid velocity should coincide with the rectilinear direction of wave propagation (force interaction). If we neglect the dissipation of the motion energy and the damping of the wave on the scale of the considered geometric forms, we can assume that the value of the instantaneous velocity along the line of elastic force interaction is a constant value. The initial (generating) point of the line of elastic force interaction is located on the surface of the vane and sets the value of the velocity along the line. Therefore, it is possible to define a family of characteristic lines for the transferable movement, along which the value of the transferable velocity is determined and constant.

Based on the foregoing, it is possible to write down an expression for the pump head at each point at the outlet of the impeller for various types of vanes.

The calculation scheme for a wheel with rectilinear tangential vanes is shown in Figure 2.

In this case, the expression for the pump head at an arbitrary point j takes the form:

$$H_j = U_2 C_{2uj} = U_2 U_{transf} \cos\gamma = U_2 U_i \cos\gamma . \quad (1)$$

The circumferential velocity is determined by the expression:

$$U = R\omega \quad \Rightarrow \quad H_j = R_i R_2 \omega^2 \cos\gamma . \quad (2)$$

The radius at any point of the vane is determined by the expression:

$$R_i = R_2 \cos\gamma . \quad (3)$$

Figure 1. Visualization of the flow in the impeller of a low-flow pump with rectilinear vanes: $\dot{V}=50\cdot10^{-6}$ m^3/s; D_2=0,0405; b_2=0,003; D_1=0,0155.

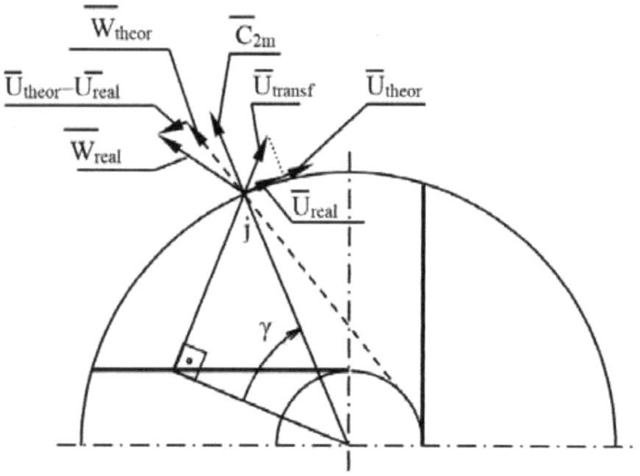

Figure 2. Calculation scheme for an impeller with rectilinear tangential vanes.

The final expression for the pump head will take the form:

$$H_j = R_2^2 \omega^2 cos^2 \gamma \tag{4}$$

The theoretical head in case a finite number of vanes is determined step by step according to the following algorithm. At each step, the increment of angle γ is set. The head value at the current point is determined by the equation (4). The theoretical head is defined as the averaging of the obtained values:

$$H_T = \frac{\sum_{j=1}^{n} H_j}{n}. \tag{5}$$

Influence coefficient of a finite number of vanes:

$$k_z = \frac{H_T}{H_{T\infty}}. \tag{6}$$

The value of the theoretical head in an impeller with an infinite number of vanes is determined by the classical expression of Euler's streamjet theory [2].

Figure 3 shows the change in the theoretical head for a finite number of vanes, depending on the number of vanes.

Given the complexity of viscous flows in the impellers, the real characteristics differ from the theoretical ones by the amount of hydraulic losses:

$$H = H_T - \Delta H = \eta_{\text{г}} H_T \tag{7}$$

Hydraulic losses depend on the magnitude and direction of the relative velocities in the flow part of the impeller and are determined by the integral of the friction stresses along the limiting surfaces: the cylindrical surfaces of the vanes and the end surfaces of the cover disks.

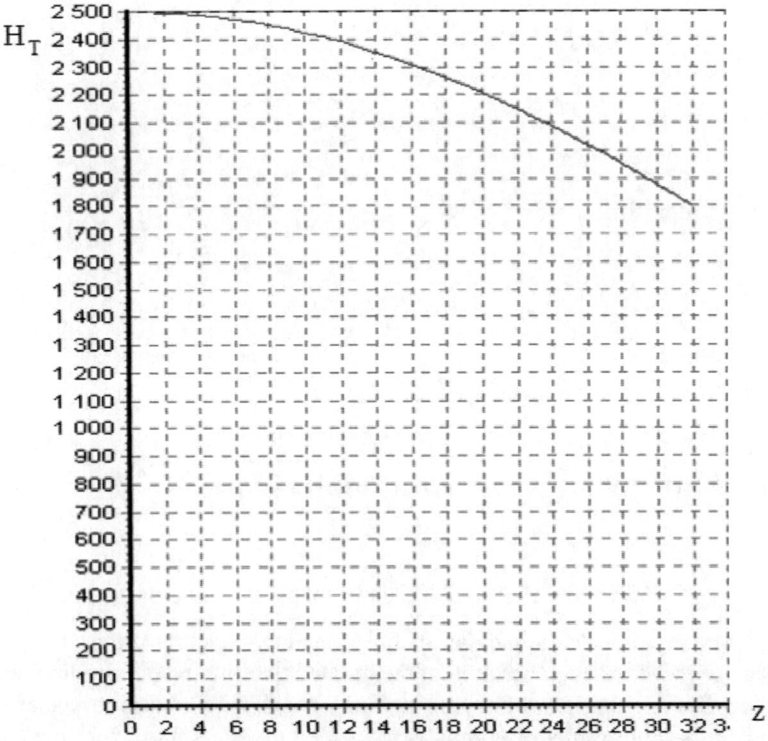

Figure 3. Theoretical head at a finite number of vanes.

To date, the assessment of the actual values of the head depending on the main geometric and operating parameters of the impeller in most practical applications is based on semi-empirical expressions obtained as the result of generalization and analysis of experimental data [3; 4; 5; 6].

A method for determining hydraulic losses, which is the integration of friction stresses along the limiting surface of a curved channel, is presented in the work [1]. Here, when integrating the finite-difference analogs of the equations of impulses of the spatial boundary layer, the thickness of the impulse loss δ_{φ}^{**} and the bevel angle of the bottom streamline ε are calculated at each step.

The total head drop is written as an integral [7]

$$\Delta H = \frac{\Delta P}{\rho} = \frac{P_2 - P_1}{\rho} = \frac{1}{\rho}\int_s \tau_0 dS = \frac{1}{\rho}\int_s \tau_0 d\varphi d\psi ; \tag{8}$$

Expanding τ_0 into projections, we obtain

$$\tau_0 = \sqrt{\tau_{0\varphi}^2 + \tau_{0\psi}^2} ; \tag{9}$$

$$\Delta H = \frac{1}{\rho}\int_s \sqrt{\tau_{0\varphi}^2 + \tau_{0\psi}^2} \cdot d\varphi d\psi . \tag{10}$$

Considering $\tau_{0\psi} = \varepsilon\tau_{0\varphi}$, we get

$$\Delta H = \frac{1}{\rho}\int_S \sqrt{\tau_{0\varphi}^2 + \varepsilon^2 \tau_{0\varphi}^2}\, d\varphi d\psi = \frac{1}{\rho}\sqrt{\left(1+\varepsilon^2\right)}\int_S \tau_{0\varphi} d\varphi d\psi . \tag{11}$$

From the law of friction [8]

$$\tau_{0\varphi} = 0{,}01256\, \rho U^2 \left(\frac{U\delta_\varphi^{**}}{\nu}\right)^{-0.25} . \tag{12}$$

Substituting in (11), we obtain the expression for the head loss

$$\Delta H = \frac{1}{\rho}\sqrt{\left(1+\varepsilon^2\right)}\int_S \tau_{0\varphi} d\varphi d\psi = 0{,}01256\sqrt{\left(1+\varepsilon^2\right)}\int_S U^2 \left(\frac{U\delta_\varphi^{**}}{\nu}\right)^{-0.25} d\varphi d\psi . \tag{13}$$

Substituting the found values of ε and δ_φ^{**}, we obtain the value of the head losses.

The presented method for calculating the spatial boundary layer is quite simple and intuitive. However, there is a need for further elaboration of the method to bring it to a form that makes it possible to calculate the three-dimensional flow of a working fluid in a channel of arbitrary shape.

It should be noted that the results of comparison of the numerical and experimental visualization show a similar flow pattern, which indirectly confirms the validity of the assumptions and the correctness of the conclusions when constructing a method for determining the field of transferable velocity in the interscapular channel of the impeller.

Based on the results of theoretical studies, a algorithm was developed, which allows calculating local values of head along the length of the interscapular channel using the known geometric parameters of the wheel, angular velocity and flow rate.

Sample calculation results are presented in Figures 4 and 5.

Figure 4 shows the hydraulic losses of the head along the channel length.

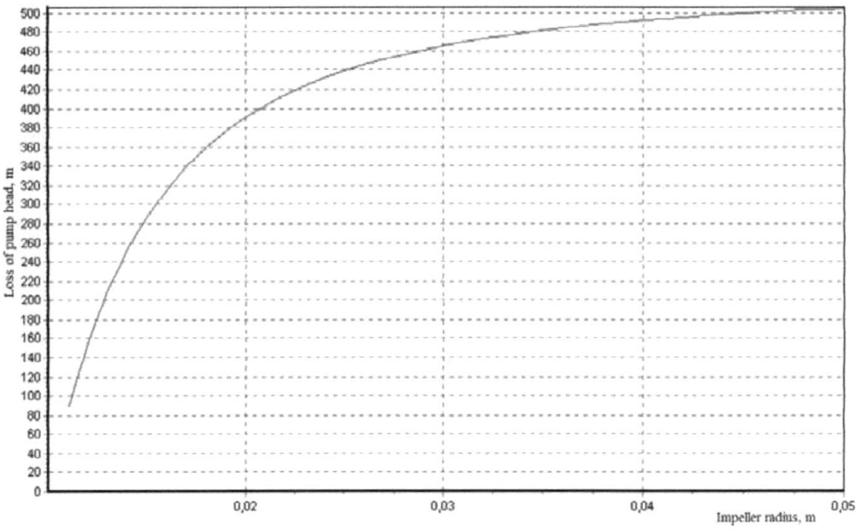

Figure 4. Hydraulic head losses along the channel length.

Figure 5 shows the dependences of the coefficient of influence of the finite number of vanes k_z, hydraulic efficiency $\eta_г$ and optimization parameter $k_z \cdot \eta_г$ on the number of impeller vanes.

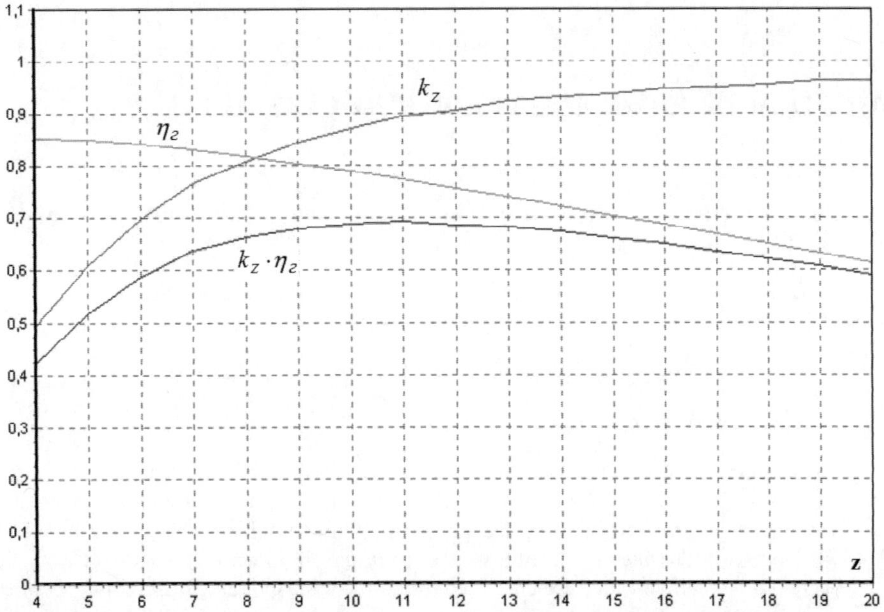

Figure 5. Dependences of k_z, $\eta_г$ and ($k_z \cdot \eta_г$) on the number of vanes.

Having carried out a series of calculations for a different number of vanes, it is possible to select the most optimal option according to the criterion $k_z \cdot \eta_г \rightarrow max$.

References

[1] Chernenko D V 2005 *Hydrodynamics of centrifugal vane superchargers of power plants of aircrafts* (Krasnoyarsk) p 167
[2] Kirillov I I 1972 *Turbomachine theory* (Leningrad: Mashinostroyeniye) p 536
[3] Stodola A 1945 *Steam and gas turbines* (New York: P.Smith) p 736
[4] Eckert B 1959 *Axial and centrifugal compressors* (Moscow: Mashgiz) p 680
[5] Proskura G F 1954 *Hydrodynamics of turbomachines* (Moscow: Mashgiz) p 417
[6] Pfleiderer K L 1960 *Vane machines for liquids and gases* (Moscow: Gostechizdat) p 684
[7] Schlichting G 1969 *Boundary layer theory* (Moscow: Nauka) p 744
[8] Stepanov G Yu 1962 *Hydrodynamics of turbomachine grids* (Moscow: Phismatgiz) p 512

IOP Conf. Series: Materials Science and Engineering 1155 (2021) 012066 doi:10.1088/1757-899X/1155/1/012066

Determination of natural gas content of coal

A A Kenetayeva[1,3]**, Sh E Usupayev**[2]**, V V Zhurov**[1]**, M Rabatuly**[1]**, M K Baymuldin**[1] **and G M Balniyazova**[1]

[1] Karaganda State Technical University, Karaganda, Kazakhstan
[2] Central-Asian Institute for Applied Geosciences, Bishkek, Kyrgyzstan

[3] E-mail: aigul_tate@bk.ru

Abstract: The main provisions of the laboratory work are as follows. Purpose - to determine the gas content of coal, which characterizes the actual gas content in coal, and is also the main parameter for predicting gas release into existing mine workings. The gas content is determined in laboratory conditions, and the gas pressure can be determined indirectly on the basis of sorption isotherms, which are characteristic of the corresponding different types of coal. Terms and Definitions. Gas bearing capacity, gas content is the volume of gas contained in the mass [m^3/t] of coal. Depending on the composition of the gas, the gas content can be divided, for example, into methane content and carbon dioxide content. Since methane is the main constituent of gas in most cases, the term "gas content" is usually used to mean "methane content". The value of gas content is usually taken in normal conditions (273 K and 1013.25 hPa).

In practice, the gas content of coal seams can be determined by sampling from the coal massif, as well as by recalculating the gas content of the extraction area during the development of seams.

In the "Spetsshakhtomontazhdegazatsiya" CD JSC AMT department there is a laboratory for determining the gas content of coal seams, equipped with equipment from the German company DMT and using its method of sampling and cutting samples [1].

The main provisions of the laboratory work are as follows.

Purpose - to determine the gas content of coal, which characterizes the actual gas content in coal, and is also the main parameter for predicting gas release into existing mine workings. The gas content is determined in laboratory conditions, and the gas pressure can be determined indirectly on the basis of sorption isotherms, which are characteristic of the corresponding different types of coal.

Terms and Definitions. Gas bearing capacity, gas content is the volume of gas contained in the mass [m^3/t] of coal. Depending on the composition of the gas, the gas content can be divided, for example, into methane content and carbon dioxide content. Since methane is the main constituent of gas in most cases, the term "gas content" is usually used to mean "methane content". The value of gas content is usually taken in normal conditions (273 K and 1013.25 hPa).

The total gas content q is the sum of the gas content in normal conditions and the desorbed gas content. The total gas content includes both adsorbed gas and free gas.

The q_{1bar} value is defined as the gas content retained in the methane atmosphere under normal conditions (at equilibrium partial pressure).

Desorbed gas content q_{des} refers to the maximum part of the gas content that is desorbed in the methane atmosphere under normal conditions at a balanced partial pressure.

Ash content *A* is defined as the residue after coal combustion, measured in % of mass at the design moisture content.

Moisture content is defined as the decrease in mass of a sample after heating to 106 °C and is measured as % of mass.

The volatile content *F* is determined from the decrease in the mass of the sample after heating in a hermetically sealed environment, measured in % from the mass. When determining the gas content, the content of volatile content is determined by the dry ashless mass of coal.

Sorption isotherm is related to the volume of gas adsorbed on a solid surface as a function of pressure, which is valid for a specific temperature of gas and solid material.

The work to determine the gas content is carried out as follows. With a special drilling tool with a pneumatic drive (figure 1), a well is drilled along the formation.

Every 2m the coal samples are taken using a special ejector (figure 2) and placed in a sealed container.

The sampling is carried out as follows. The valves are closed during drilling. When it is necessary to take a sample, the valves are opened, and due to the ejection, the coal particles are sucked into a special device, from where the sample is placed into a sealed vessel.

Under standard drilling conditions, the well length is 20 - 22m. Coal samples are delivered to the laboratory and, using a gas analyzer, the amount of gas released from one container is determined [2].

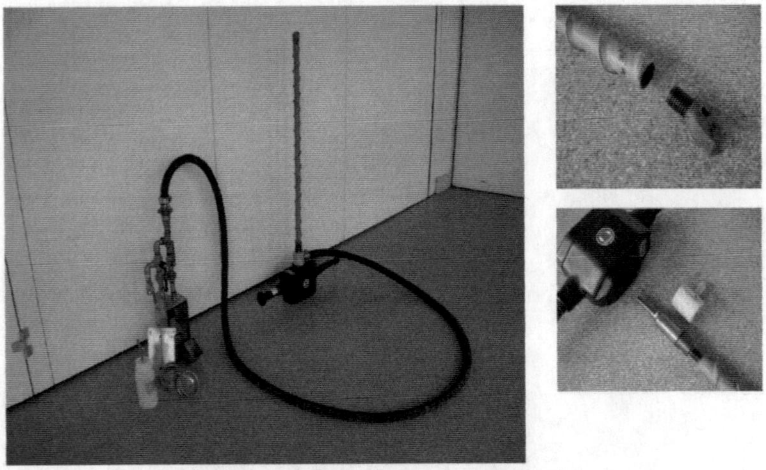

Figure 1. Drilling tool Turmag IV.

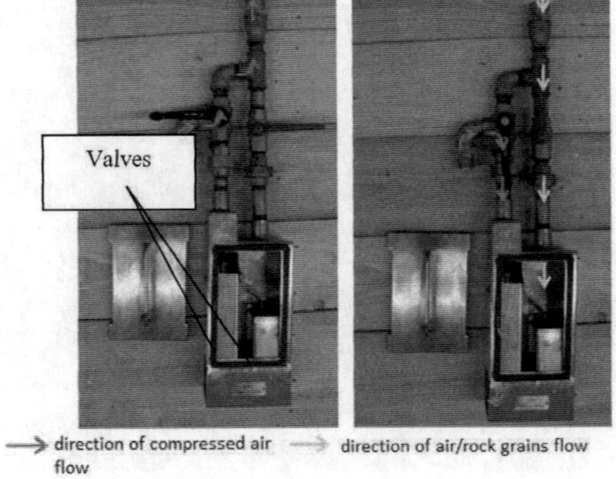

Figure 2. The principle of ejector coal sampling.

The adsorbed gas content fraction is the sum of three elements:

- gas released during drilling, q_1
- gas released into the sample container, q_2
- residual gas in the coal sample, q_3

q_1 - calculated based on q_2

q_2 - is determined by measuring the concentration of methane in the container with the sample by the gas analyzer (figure 3).

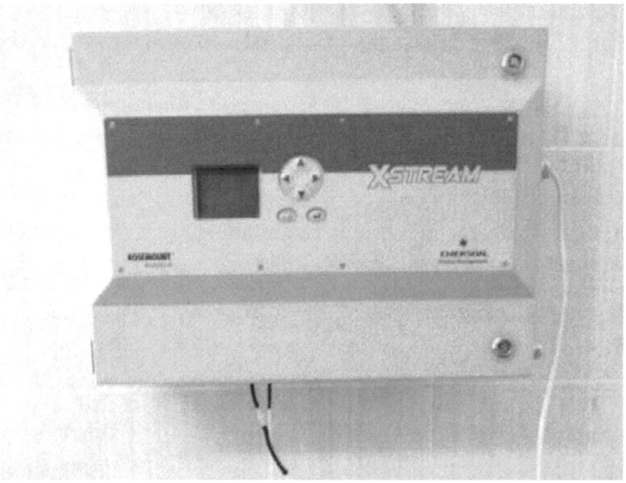

Figure 3. Gas analyzer.

Then, using a mill, grind the sample. The q_3 value (residual gas) is determined by the volume of gas released during grinding in the mill (figure 4).

Figure 4. Sample grinding mill.

After that, ash content and adsorbed gas content are determined by dry ash-free mass using a furnace (figure 5).

In addition to the methane content in coal samples, the laboratory determines its moisture content, ash content, and the release of volatiles [3,4].

Based on the methane content from several samples taken from different depths, the resulting coal methane content curve is plotted. In theory, the curve should reflect the properties of the coal mass in terms of stress distribution. That is, the massif immediately adjacent to the walls of the mine was unloaded and divided into blocks with the formation of significant cracks. In this zone, the formation gas content is minimal. With an increase in the distance from the working wall, the effect of unloading decreases, the zone of block division is replaced by a zone of inelastic deformations, with a gradual decrease in the number of cracks in the rock mass. In this zone, gas content should increase with increasing well depth. At a distance of 10 - 15 m from the working wall, the effect of unloading on the massif is minimal, there is a zone of elastic deformations, the gas content reaches a value close to natural.

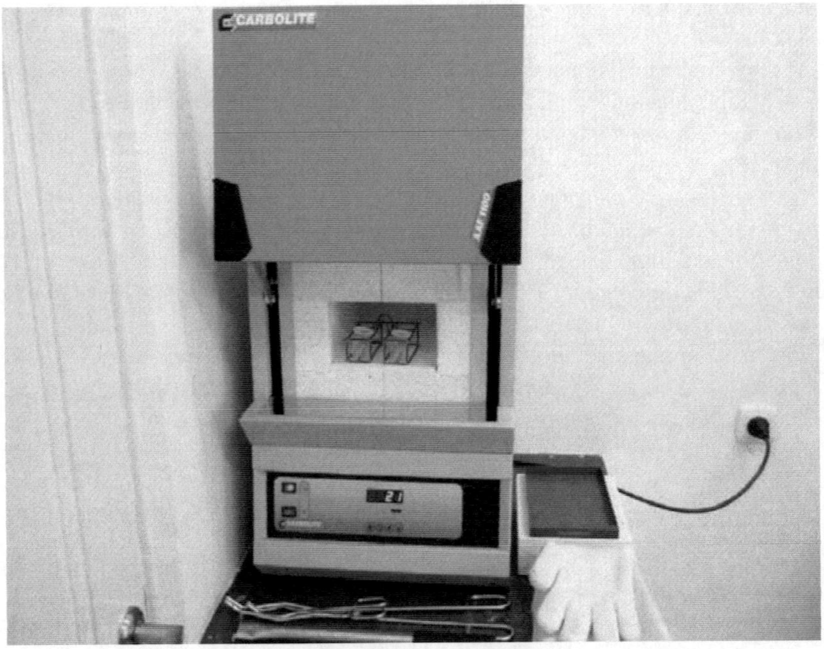

Figure 5. Coal samples combustion furnace.

Indeed, the results of work on coal sampling from wells drilled from the 33 K_{10}-c conveyor industrial drift of the Abayskaya mine at a distance of 55 and 65 m from the installation chamber of 33 K_{10}-c showed that the nature of the change in gas content looks as follows.

From 2 to 4m, the gas content increases at a slow rate, apparently due to the formation of significant cracks formed as a result of unloading (block division). Then, from 4 to 8 m, its sharp increase is observed (the zone of inelastic deformations, the number of cracks is sharply reduced). And from 8m, the growth curve practically flattens, reaching its maximum (elastic deformation zone). Thus, the gas content of the K_{10} formation in the area of the installation chamber of 33 K_{10}-c of the Abayskaya mine at a depth of 520 m is 18.9 - 19.3 m^3/t.

At the Abayskaya mine, the work was carried out to determine the gas content of the K_{10} formation in the southern block. The formation depth at the sampling area was 435 m. Samples were taken from the well drilled by the SBG-1M machine at depths of 38, 40, 42 m.[5].

The materials show that in a practically intact massif the gas content is constant, although its fluctuations are possible in zones of geological disturbances, increased rock pressure, etc.

IOP Conf. Series: Materials Science and Engineering 1155 (2021) 012066 doi:10.1088/1757-899X/1155/1/012066

Practice shows that the curves obtained can be very different in appearance. The change in the gas content of the seams with the distance from the walls of the workings can have a different character, which is influenced by such properties of the coal mass as plasticity, moisture, fracturing, degree of unloading, etc.

For example, at the Kuzembayev mine, coal samples were taken from the K_{10} seam from the pair working of the $37K_{10}$-в ventilation drift. The processing results showed that at a depth of 570 m, the gas content of the K_{10} formation changes.

The block division in this case is practically absent or is located within 2 m from the wall of the working. The zone of inelastic deformation extends from 2 to 10 m. The zone of elastic deformations is located from 10 to 21 m. No cracking occurs in this zone, and the gas content of the coal massif remains practically unchanged [6].

References

[1] Ayruni A T 1981 *Theory and practice of mine gases control at great depths* (Moscow: Science) p 332

[2] Kolmakov V A 1981 *Methane emission and its control in mines* (Moscow: Science) p 236

[3] Filimonov E N, Portnov V S, Egorov V V, Stefliuk Iu Iu and Kenetayeva A A 2016 *Some aspects of the K_{10} formation gas content study in the conditions of the Abayskaya mine of CD AMT JSC* p 11

[4] Zenkovich L M 1976 *Natural gas permeability and specific intensity of gas recovery from working coal seams of the main deposits of the USSR* vol 17 (Occupational safety and health, labor protection and mine rescue work) p 10-11

[5] Ayruni A T 1990 *Gas content of the USSR hard coal mines. Integrated development of gas-bearing coal deposits* ed G D Lidina (Moscow: Science) p 213

[6] 2002 *Problems of the development of methane-bearing coal seams, industrial extraction and use of mine methane in the Karaganda basin* ed A T Ayruni (Moscow: Academy of Mining Sciences of Russia Press) p 346

Methods of improving the power supply reliability of industrial site

A S Khismatullin and M G Bashirov

Branch of Ufa State Petroleum Technological University in Salavat, Salavat, the Republic of Bashkortostan, Russia

E-mail: hism5az@mail.ru

Abstract. Ensuring reliability is one of the most important problems in the creation and operation of any technical system. It is especially relevant for complex systems such as power supply systems. A modern industrial site from the power supply point of view is a complex technical system of hazardous industrial facilities, the technological processes continuity and safety of which directly depends on the power supply reliability. Electricity receivers of the considered industrial site in terms of power supply reliability belong to the second category. The task of ensuring the power supply reliability includes a whole range of technical, economic and organizational measures aimed at reducing damage from disruption of the electricity consumers' normal operation. In addition, in modern market conditions, the power supply reliability is inextricably linked with the economic performance and industrial enterprises energy security. The paper proposes to modernize the industrial site power supply circuit, using the existing standby diesel power plant, which would automatically provide the industrial site power supply in automatic mode in case of disappearance, decrease, increase in voltage level, phase failure, and the phase sequence violation of the three-phase supply voltage on the busbar sections TP-1, TP-2 site. It is obvious that with the existing power supply scheme of the industrial site, the manual switching sequence, control over the backup diesel power plant operation, the normal power supply scheme restoration, it takes an average of 2.5 hours, which is not desirable, especially in winter. Therefore, it is necessary to introduce at the industrial site production control devices for the reserve automatic transfer, which ensure the power supply restoration to the industrial site in automatic mode.

1. Introduction

Ensuring reliability is one of the most important problems in the creation and operation of any technical system. It is especially relevant for complex systems such as power supply systems [1-10]. The task of ensuring the power supply systems reliability includes a whole range of technical, economic and organizational measures aimed at reducing damage from disruption of the electricity consumers normal operation, such as:

- selection of criteria and quantitative characteristics of reliability;
- reliability tests and predicting the operating equipment reliability;
- the optimal structure selection of the designed (reconstructed) power supply systems according to the reliability criterion;
- ensuring the specified technical and operational characteristics of consumers' work;

- the development of the most efficient in terms of reliability, operating system software (substantiation of preventive maintenance modes, norms of spare elements and methods of troubleshooting).

In addition, in modern market conditions, the power supply reliability is inextricably linked with the economic performance and industrial enterprises energy security.

2. Theoretical, experimental, technical and technological methods

All Union State Standard 27.002-95 «Reliability in technology. Basic concepts, terms and definitions» defines reliability as «an object's ability to keep the values of all parameters characterizing the ability to perform the required functions in specified modes in time». With regard to electric power systems and, in particular, to power supply systems, it is necessary to take into account their large dimension (in terms of the number of elements and interconnections between them), dependence on adjacent technical systems (fuel and energy complex and technology of the consumer enterprise) and the continuity in time of production processes, distribution and consumption of electricity.

Therefore, the power supply reliability should be understood as the continuous electricity provision to consumers of a given quality in accordance with the power consumption schedule and according to the scheme that is provided for long-term operation.

Electricity receivers of the considered industrial site in terms of power supply reliability belong to the second category. The industrial site is a structural unit of a large industrial enterprise in the city of Salavat.

The industrial site contains:

- a radio relay station that receives and transmits data to the CDS (central dispatch service) in terms of volume, temperature, pressure, etc. pumped gas in main gas pipelines (MG);
- technological communication center (TCC) performing maintenance, repair of telephone communication equipment between the enterprise division facilities, gas distribution stations (GDS), automobile gas filling compressor stations (CNG filling stations), etc .;
- seasonal gas boiler house providing heat supply, hot water supply of the site buildings;
- gas station, mechanical repair shops, garages, warehouses, welding and turning areas;
- security service (SS) performing control functions inside the access mode, the state of the fire and security alarm system, video surveillance.

Let ua consider the power supply of an industrial site according to the scheme shown in Figure 1.

It is seen from Figure 1, the main step-down substation-1 110/35 kV power supply is carried out by the local power authority (LPA) through a double-circuit overhead line 110 kV from the regional substation located at a distance of 20 km, and the main step-down substation-2 35/10 kV power supply is carried out by the local power authority through a double-circuit overhead line 35 kV at a distance of 10 km from the main step-down substation-1 110/35. Moreover, double-circuit overhead lines 110 kV and overhead lines 35 kV are on the same power transmission line support.

The industrial site power supply from the main step-down substation-2 35/10 kV is carried out through two separate power lines on reinforced concrete vibrated racks (VR 110) with aerial non-insulated cables. The industrial site is located at a distance of 0.5 km from the main step-down substation 35/10 kV.

It is obvious that LPA double-circuit overhead lines (110/35 kV and 35/10 kV), located on one power transmission line support, cannot be independent, mutually redundant.

If an LPA double-circuit power transmission line has been damaged, requiring the disconnection of both circuits (for example, a break in a lightning protection cable, simultaneous destruction of circuit insulators), then the power supply to the consumer is interrupted during the repair work, which is unacceptable. An accident with one source of power supply (PS) should not in any way affect the PS provision from another source.

IOP Conf. Series: Materials Science and Engineering 1155 (2021) 012067 doi:10.1088/1757-899X/1155/1/012067

If an accident occurs, e.g., a breakage of the lightning protection cable of a 110 kV power transmission line during non-working hours, there is a power outage at the site. At most industrial sites of the main gas pipelines line production management; at least a standby diesel power plant (DPP) operating in manual mode is installed. There is no permanent engineering staff at the site. Production start-up, DPP switch-on, power supply restoration to the site during off-hours, weekends, holidays is carried out by an emergency recovery team, and assembled according to an emergency response plan. According to the emergency response plan the emergency recovery team collection time is 2.5 hours.

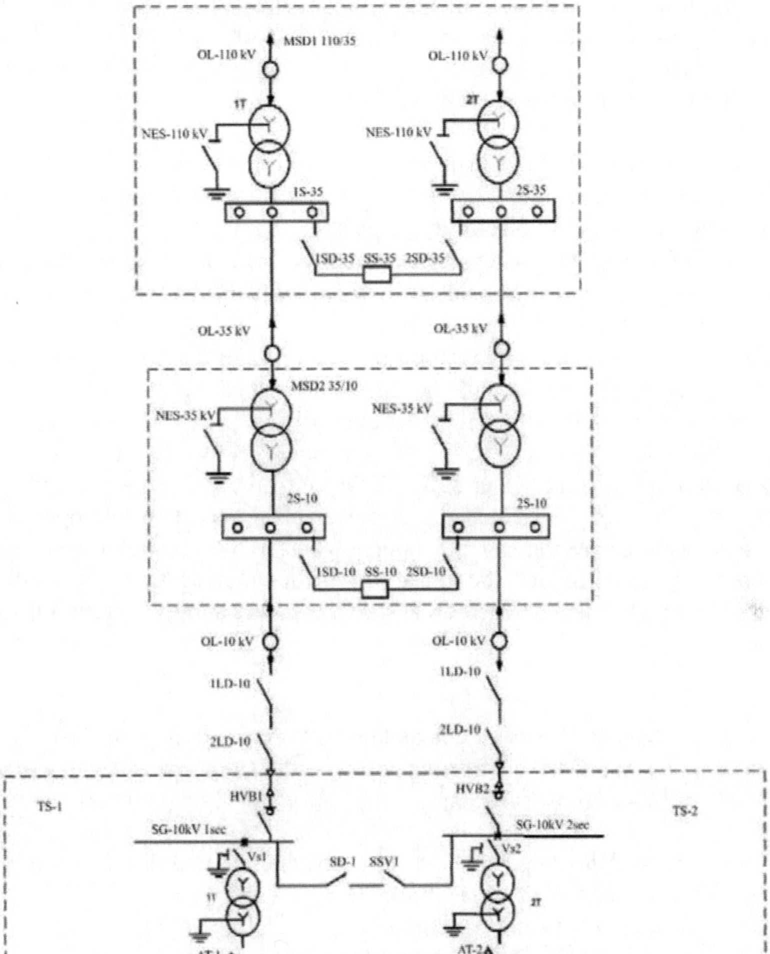

Figure 1. Industrial site power supply diagram.

Under adverse weather conditions in winter a sustained interruption in site power supply can lead to defrosting of heating networks, heat consumption networks, water heaters of supply ventilation units located in industrial site buildings and workshops, a break in telephone communication, lack of video surveillance, control over fire and security alarms of buildings, etc.

3. Conclusion

The paper proposes to modernize the industrial site power supply circuit, using the existing standby diesel power plant, which would automatically provide the industrial site power supply in automatic mode in case of disappearance, decrease, increase in voltage level, phase failure, and the phase sequence violation of the three-phase supply voltage on the busbar sections TP-1, TP-2 site.

Currently, in the event of an emergency power outage during non-working hours, weekends, holidays emergency power supply is carried out as follows:

- In the event of a voltage failure at both high-voltage inputs TP-1 and TP-2, the dispatcher of the technological communication center and the security service report to dispatchers in the directions about the absence of electricity at the site
- The collection of operational - repair personnel is carried out. The power plant is being launched; the personnel are acting in accordance with the instructions for transferring the industrial site power supply to a backup source.
- The control over the power supply restoration according to the normal scheme is carried out by light signaling in RU-10 kV. in the first and second busbar sections or specified by telephone from the dispatcher of local power authority.

It is obvious that with the existing power supply scheme of the industrial site (Fig. 1), the manual switching sequence, control over the backup diesel power plant operation, the normal power supply scheme restoration; it takes an average of 2.5 hours, which is not desirable, especially in winter. Therefore, it is necessary to introduce the linear production management of main gas pipelines automatic transfer switch (ATS) devices, which ensure the power supply restoration to the industrial site in automatic mode.

References

[1] Alabugin A A, Alabugina R A and Shchelkonogov A E 2018 Methods for improving the reliability of power supply of enterprises based on the autonomous cogeneration sources *Int. Multi-Conf. on Industrial Engineering and Modern Technologies, FarEastCon 2018.* 8602674

[2] Khismatullin A S 2018 Automated software to determine thermal diffusivity of oil-gas mixture *Journal of Physics: Conf. Series* **1015(5)** 052013

[3] Mullakaev M S , Abramov V O and Abramova A V 2015 Development of ultrasonic equipment and technology for well stimulation and enhanced oil recovery *Journal of Petroleum Science and Engineering* **125** pp 201-208

[4] Khismatullin A S 2018 Method for increasing oil resources transformers with long-term operation *IOP Conf. Series: Materials Science and Engineering* **327(2)** 022058

[5] Vildanov R G, Khismatullin A S and Luneva N N 2019 Economic aspects of reactive power compensation at gas-chemical plant *IOP Conf. Series: Materials Science and Engineering* **560(1)** 012108

[6] Vildanov R G, Khismatullin A S and Luneva N N 2019 The investigation of magnetization reversal loss sensor. *IOP Conf. Series: Materials Science and Engineering* **560(1)** 012109

[7] Mullakaev M S, Wexler G B and Mullakaev R M 2019 Mobile sonochemical complex of oil sludge processing. *SOCAR Proceedings* **3** 88-96

[8] Surakov M R *et al.* 2020 Software and Hardware System Development for Determining Pump-and-Compressor Equipment Technical State *Proceedings - 2020 Int. Ural Conf. on*

Electrical Power Engineering, UralCon 2020. pp 215-219

[9] Evdokimova N G, Luneva N N and Levina T M 2020 Experience in the application of investment controlling inllc "Gazprom Neftekhim Salavat *Directions of improvement Journal of Advanced Research in Dynamical and Control Systems* **12(2)** pp 371–379

[10] Filatova G, Petrov A and Batmanov M 2020 Research of the algorithm and the prototype device for fault location based on synchronized two-side measurement *Proc. of the 2nd 2020 Int.Youth Conf.on Radio Electronics, Electrical and Power Engineering, REEPE 2020*. 9059140

Analysis of causes of failures of process equipment of oil refining and petrochemistry

M V Vakulenko*, N V Zharinova and A M Zabaturin

Ufa State Petroleum Technological University, Ufa, 450062, Russian Federation

*E-mail: d_margarita@mail.ru

Abstract. The development of modern industry is associated with an increase in the total scale of production and processing of fire and explosion hazardous and toxic materials, a significant increase in the single capacity of plants and apparatus, and a complication of technologies and regimes for managing the production process. As a result, there is a steady trend in industry to increase the number of accidents with increasingly severe social, environmental and economic consequences. Despite significant efforts in the development of technical safety systems, accident rates (number of accidents or injuries per unit of production) in our country have increased significantly in recent years. The most dangerous situation has developed in the oil refineries of Bashkiria, Samara, Irkutsk and Perm regions where the depreciation depreciation of some technological equipment is 70%. Thus, safety is one of the main characteristics of industrial facilities.

1. Introduction

The following factors explain the relevance of improving fire and explosion safety of hydrocarbon processing plants:

- concentration of hydrocarbon systems with higher combustion heat. High pressure of saturated vapors and, as a result, explosion hazard of hydrocarbon systems and increased ability to contaminate the atmosphere with hazardous emissions;
- the presence of potential hazards (volumetric fires and explosions) causing material and human losses, which follows from the properties listed above;
- faster development of production volumes and lagging pace of development of environmental events, which is typical for modern production;
- the emergence of hard-to-dispose production wastes and new types of wastes, the application and processing of which has not yet been found;
- extremely high energy saturation of oil refining facilities. Thus, a typical oil refinery (refinery) with a capacity of 10-15 million tons/year focuses on territory from 200 to 500 thousand tons of hydrocarbon fuel, the energy content of which is equivalent to 2-5 megatons of TNT;
- intensification of technology, increase of unit capacity of devices, as a result of which parameters such as temperature, pressure, content of fire and explosion hazardous substances rise and approach critical ones;
- imperfect technology for collection and disposal of fire and explosion hazardous components of petroleum products caught in the environment;

- excessive concentration of oil refining, petrochemicals and chemistry in individual regions [1-4].

2. Experimental procedures

A systematic analysis of the main causes of emergency situations occurring at refining and petrochemistry plants makes it possible to distinguish the following:

- organizational and technical (lack of raw materials and auxiliary materials, lack of energy resources, late shipment of finished products, error or inadequate competence of decision-making persons);
- technological (change of temperature, pressure and flow from regulated values, frequent change of quality of raw materials supplied for processing, sometimes of very low quality);
- mechanical (failure of process equipment, equipment of pipeline systems) [5-8].

The structure of the causes of equipment failures in percent is shown in Figure 1. Any of these causes or their totality determines the degree of risk, danger of emergency situations and their consequences, as well as the influence on the associated process units or production as a whole. On the other hand, accidents caused by one of the reasons may cause failures for other reasons or other reasons.

From the above data it can be seen that 67.5% of failures occur due to organizational and technical reasons. The shortage of own working capital, which outpaces the growth rate of prices for machine-building products of basic petrochemical equipment (column vessels, reactors, heating furnaces, compressor equipment, automation and parameter control equipment) in comparison with the growth rate of prices for products of the energy complex does not allow technically re-equipping or updating existing obsolete fixed assets. The physical and moral wear and tear of most production equipment and in general process units that have been in operation for more than 35 years (depreciation is 70%) requires radical reconstruction and technical re-equipment. Mainly due to the lack of progressive equipment and poor equipment with automation and microprocessor equipment, many oil refining and petrochemical synthesis enterprises are forced to extensively expand most of the production on an old basis [9-13].

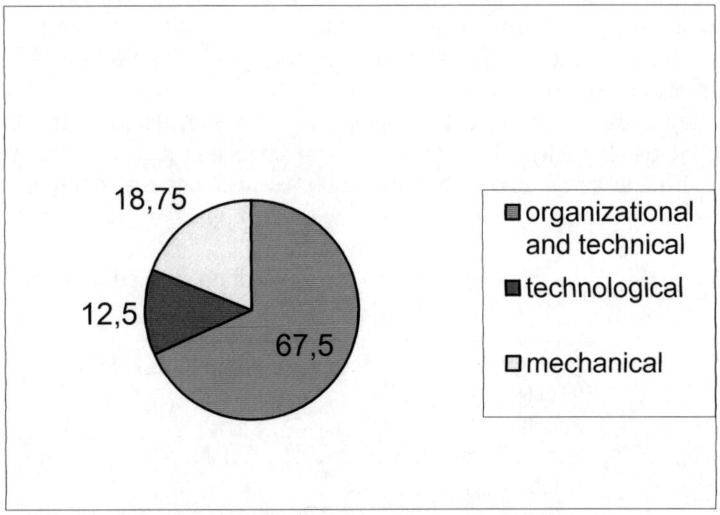

Figure 1. Failure cause structure (%).

The most dangerous plants are: rectification units, catalytic cracking, raw material treatment.

Table 1 shows the power potential of the plants reduced to the TNT equivalent of a modern refinery.

Complex situations arise during the operation of the main equipment of process plants and industries - heating furnaces, reactors, separation (rectification) columns, settling tanks, tank farm.

The distribution of the number of accidents by types of process equipment, % at the plant EDP-AVT6 is given in table 2.

Table 1. Power potential of plants reduced to TNT equivalent of modern refineries.

Process Unit Name	TNT equivalent W_T, t
EDPAVT 6 (column K-2)	95.6
AVT-3 (column K-2)	55.9
Absorption cleaning (capacity E-3)	14.8
Reflux Cleaning (Column K-3)	3.9
Catalytic cracking (reactor R-1)	120.2
EDP 1,2 (electric dehydrator E-3)	6
Hydrotreating LCh-24-2000 (absorber K-202)	1.1
Reforming L-35-11/300 (column K-1)	21.1
Stabilization 22-4 (ColumnK-1)	8.5
Feedstock Treatment (Column)	107.7

Pumping equipment contains only about 0.6% of the total amounts. This type of equipment is "dangerous" in terms of the frequency of malfunctions and failures associated with the release of explosive substances into the environment.

Possible scenarios for the accident of electric dehydrators, heat exchangers may be: depressurization due to corrosion or personnel error with oil spill; mechanical destruction.

The main reasons for exit of the furnaces are the possibility of oil ignition in small leaks in the heat exchange surfaces due to close fire.

The characteristic "hazards" of the columns may be corrosion damage, mechanical destruction, depressurization of the flange joints. The risk is due to the concentration of a large number of explosive and fire hazardous substances in high temperatures and pressures, any malfunction or defect associated with the depressurization of column vessels can lead to catastrophic consequences [14-16].

Table 2. Distribution of the number of accidents by types of process equipment, % at the unit EDP-AVT6.

Equipment	Number of accidents, %
Process piping	31.5
Pump stations	18.9
Capacitive devices (heat exchangers, dehydrators)	15.0
Furnaces	11.4
Fractionation, vacuum and other columns	11.2
Promkanalization	8.5
Tankfarms	3.8

Fires and explosions at open process oil refineries are usually the result of emergency situations, developing according to approximately the following scheme:

- flammable superheated liquids and gasoline vapors are leaked as a result of tightness violation or destruction of pipelines, valves and process devices;
- spill products are either ignited or create a vast zone of gas-gas-air mixture of explosive concentration;
- the factors of the resulting fire strongly affect the device or pipeline from which the leakage occurs, as well as neighboring equipment, as a result of which the pressure in it increases above the design one, it loses strength and breaks down or its depressurization occurs due to destruction from the fire of sealing devices.

3. The results of studies and their discussion

Various methods can be used for hazard assessment, such as preliminary hazard analysis, error tree analysis, accident impact analysis, and risk assessment. The purpose of the hazard assessment, regardless of the method chosen, is to determine potential causes of failure or accidents at an industrial facility.

The risk analysis methodology includes the calculation of the probability of an accident and the assessment of its consequences (economic, environmental and social). When assessing the risk of product outflows from the plant, it is important to keep in mind the ambiguity and inconsistency of the initial information on the repeatability of the initial events. This is reflected primarily in the fact that from all available risk assessments, its upper limit is chosen - integrated risk - a complex safety indicator expressed in a single value equivalent.

To calculate the integrated risk, a possible accident scenario is drawn up, which is a sequence of physical phenomena occurring one after another as a result of an emergency.

The scenario should be divided into structural elements based on the following principles:

1) each element should contain a specific physical phenomenon, which can be represented in the form of a mathematical or empirical model;

2) the element should be isolated, ready for use in other scenarios of accident development. For example, an element (thermal damage) is a consequence of both a gas explosion and the combustion of the basin. Therefore, it does not make sense to include estimates of thermal damage in the elements for each of these phenomena, but to separate it into a separate element. The same requirement ensures interchangeability of elements. For example, depending on the type of liquid, both a neutral buoyancy gas cloud and a heavy gas cloud may be formed. In both cases, different physical and mathematical models are used to estimate cloud scattering. At the same time, the main input and output parameters of these models are the same. Therefore, the consequences can be estimated according to the same scenario, using the same elements (techniques), with the exception of methods describing scattering, which for each case will be their own;

3) the possibility of having several techniques describing the same process. For example, the process of scattering heavy gases can be calculated both from complex three-dimensional codes requiring several hours of work, and from simplified empirical models. At the same time, the input parameters for these models are the same. Depending on the evaluation objectives, timeliness and accuracy in the system, you can use a particular type of element. Here, there is also the possibility of replacing the outdated method with a new one, which does not require the processing of the entire system of methods.

In case of emergency depressurization of any of the units of EDP-AVT unit (EDP (electric desalination and electrical dehydration), AD (atmospheric distillation of oil to fuel oil), VD (vacuum distillation of fuel oil to tar with production of oil distillates), gasoline stabilization and secondary distillation unit, flue gas heat recovery unit) and emission to adjacent surface. In case of ignition source - spill fire or explosion of fuel-air mixture followed by fire on spill mirror.

4. Conclusions

As a result of the work done, it is shown that fires and explosions at the EDP-AVT plant occur only in a situation that is characterized by three factors at the same time: leakage (spilling) of combustible liquids; evaporation and formation of fire-and-explosion mixtures of hydrocarbon vapors with air; presence of ignition source.

It should be noted that in order to develop any accident scenario, it is necessary to know how the hazardous substance is distributed to the equipment, the amount of equipment in the units, the number of unit units. Further, fire or explosion frequency tables are compiled. After that, an assessment of the integrated risk of causing social, material and environmental damage to the accident at the plant is carried out.

References

[1] Vakulenko M V and Zharinova N V 2020 *IOP Conf. Ser.: Mater. Sci. Eng.* **862** 062022
[2] Zharinova N V and Vakulenko M V 2020 *IOP Conf. Ser.: Mater. Sci. Eng.* **862** 032014
[3] Nasibullina OA, Gareev A G and Rizvanov R G 2018 *Solid State Phenomena* **284** 1302-6
[4] Gaysin E Sh, FrolovY A and NasibullinaO A 2020 *IOP Conf. Ser.: Earth and Environ. Sci.* **459** 032055
[5] Abdullin I G and Gareev A G 1993 *Physico-chemical mechanics of materials* [in Russian – Fiziko-Khimicheskaya Mekhanika Materialov]**29(5)**97-8
[6] Nasibullina O A and Tyusenkov A S 2020 *IOP Conf. Ser.: Mater. Sci. Eng.* **862** 022007
[7] KuzeevI R, Ibragimov I G, BayazitovM I, Davydov S NandKhairudinov I R 1985 *Chem. and Technology of Fuels and Oils* **22(3)** 111-3
[8] Nasibullina O A and Gareev A G 2019 *Materials Science Forum* **946** 20-4
[9] Tyusenkov A S, Rubtsov A V and Tlyasheva R R 2017 *Solid State Phenomena* **265** 868-72
[10] Nasibullina O A and Rizvanov R G 2020 *Materials Science Forum* **992** 695-9
[11] Karetnikov D V, Rizvanov R G, Fairushin A M and Kolokhov K S 2013 *Welding International* **27** (7)557-60
[12] Nasibullina O A and Rizvanov R G 2020 *Materials Science Forum* **989** 28-2
[13] Zaripov M Z, Fairushin A M and Karetnikov D V 2019 *Materials Science Forum* **946** 883-8
[14] Tokarev AS,Karetnikov DV,Rizvanov RG, FayrushinAMandZaripov MZ 2020 *Materials Science Forum* **991** 70-6
[15] Nasibullina O A and Rizvanov R G 2020 *IOP Conf. Ser.: Mater. Sci. Eng.* **969** 012004
[16] Yakhin A V, Karetnikov D V, Rizvanov R G, Abakacheva E M and Gareev A G 2019 *Chemical and Petroleum Engineering* **54(11-12)** 801-5

Methods of calculation of standard parameters in the system of maintenance and repair of the atmospheric column

M V Vakulenko* and N V Zharinova
Ufa State Petroleum Technological University, Ufa 450062, Russian Federation

*E-mail: d_margarita@mail.ru

Abstract. The modern theory of repairability of machines and mechanisms has received significant development. Today there are already theoretical concepts and prerequisites for creating a scientifically sound system of maintenance and repair of technical products of various branches of technology. From these positions the review of various repair strategy in aspect of their applicability to an atmospheric column of K-2 of installation of primary oil refining of EDP-AVT-6 is interesting.

1. Introduction

By maintenance and repair system, is meant a set of interconnected tools, documentation and executors necessary to maintain and restore the quality of products included in this system [1]. The objectives of the repair system are:

1) prevention of equipment breakdowns and emergency stops;
2) maintenance of equipment operability during the entire service life;
3) ensuring reliable operation of the equipment;
4) ensuring the quality of products;
5) ensuring safety during equipment operation and environmental protection.

Organization of the enterprise maintenance system is carried out in the following matters:

a) selection of the type of repair organization;
b) development of a strategy and formation of conditions providing repair system.

The repair system strategy means the goals by coordinating and allocating the relevant resources of the enterprise, a generalizing model of actions necessary to achieve the set [2].

The presence of chlorides and water in the refined oil contributes to the hydrochloric corrosion of the equipment, leads to long downtime of process plants, reduces the service life of expensive catalysts used in secondary processes, and degrades the quality of commercial oil products.

Deep demineralization of oil ensures reduction of corrosion and reduction of deposits in equipment, increase of inter-repair runs of plants (especially AVT, visbreaking, thermal cracking and coking), improvement of quality of raw materials for catalytic processes, as well as commercial products - fuels, bitumen and electrode coke [3].

EDP-AVT - combined atmospheric-vacuum tube with oil desalination and dehydration unit. The purpose of the plant is the preparation and primary processing of oil. The plant includes EDP unit (electric desalination and electrodeposition), AT (atmospheric distillation of oil to fuel oil), VT (vacuum distillation of fuel oil to tar to obtain oil distillates), gasoline stabilization and secondary distillation unit, flue gas heat recovery unit. The plant is designed to process unresolved West Siberian oils, their mixtures with coal-bearing oils, gas stable condensates (Yamburg, Urengoysky, Karachaganaksky, etc.), or other oils close in chemical composition and physical properties.

The choice of column K-2 justified by the fact that the column is the largest on the plant, the high temperature and nature of the processes makes it the main source of danger. Also, the column K-2 has a large destruction radius, the equipment on the plant is located at a short distance from each other, which, during depressurization and subsequent explosion of the column, will lead to subsequent resolution of the equipment [4-7].

2. Experimental procedures

The idea of the first strategy is that a piece of technological equipment is considered as a single object, the failure of any part of which leads to a "failure." A probability curve of failure distribution over time is constructed for this piece of equipment. Such a curve can be of the following types:

a) with a period of maximum number of failures clearly expressed in time;

b) with almost uniform distribution of failures;

c) with the distribution of the maximum number of failures in the initial period of time (start-up, adjustment and run-in period).

The corresponding calculations show that for equipment having a fault distribution pattern corresponding to a curve of type "a," there is a certain time of forced repair τ_1, at which the equipment idle time will be minimal. For other fault allocation cases, no such time exists and forced scheduled repairs only result in increased downtime [8-10].

This strategy has the following drawbacks: not applicable to all types of equipment; does not take into account the possibility of changing the nature of the failure distribution curve over time, which leads to a change in the optimal repair period; does not take into account the uneven labor intensity of various repairs, which does not allow planning downtime; requires a very large amount of operational observations.

The second repair strategy is as follows. A piece of equipment is considered as an object consisting of many separate parts with different service lives. Repair is carried out after a period of time τ_{1h}, equal to the smallest service life of parts assigned to scheduled repair. Parts with service life less than τ_1 are repaired as exits fail during overhaul. All other parts are divided into groups so that the first group includes parts whose service life is more than τ_1 but less than $2\tau_1$; in the second group - more than $2\tau_1$, but less than $3\tau_1$ in the third group - more than $3\tau_1$, but less than $4\tau_1$, etc. The following assumptions are accepted:

a) fluctuation of service life of parts does not exceed τ_1;

b) labor intensity of repair of any group of parts is approximately the same.

Accepted assumptions are statistically valid for a number of equipment types studied. The number of part groups requiring scheduled repair is graphically displayed.

The idea of the third strategy is that the optimal periodicity of prevention (small repairs) only for assembly units of the product is established on the basis of the condition of achieving their maximum reliability in the interprophylactic period with a minimum of labor costs for preventive work. This condition depends on the minimum labor costs (hours) at a given level of reliability or the maximum reliability at a given level of labor costs.

The expression for optimal periods of interprophylactic service (days, hours) is as follows:

$$\tau_{opt} = \frac{1}{\lambda_1}\sqrt{\frac{\tau_2 \tau \lambda}{\tau_3}} \tag{1}$$

where λ_1 - is the failure rate of the assembly unit of the product for the interprophylactic period;
λ - product failure rate in general for the same period;
τ - the effective periodicity of prevention prior to optimization;
τ_2 - labour input of preventive maintenance of assembly unit;
τ_3 - labour intensity of elimination of sudden failures of the assembly unit.

This strategy allows you to assign separate periods of prevention (minor repairs) for three periods of the product's life: running in, normal operation and increased wear and tear.

The number of all prophylaxis is

$$n = \frac{\tau_{\max}}{\tau_{\text{opt}}} \quad (2)$$

where $\tau_{\max}$ - all service life of a product.

The fourth strategy is as follows. Separately, for each period of the product operation, for each assembly unit, by statistical processing, the value λ_1 and the number of failures n_1 are then determined also for each period, and for each assembly unit a ranked number of numbers from the number of failures is made $n_1 = 0$; $n_1 = 1, \ldots, n_2 = m$.

By resorting to probabilistic Poisson distribution, you can get the probability of failure detection per operating time of the article T_x at the number of failures n_1

$$P_1 = (\lambda_{1j}\tau_{1j})^{n_{1j}} e^{-\lambda_{1j}\tau_{1j}} \quad (3)$$

where j- is the number of the assembly unit of the product.

For each assembly unit and a specific period of operation, you can plot the dependencies:

$$P_{1j} = \varphi(\tau_1 j) \quad (4)$$

Size $\tau_{1\text{opt}}$ is on crossing of a curve with number of refusals of $n_1 = 2$ with abscissa axis

$$\tau_{1\text{opt}} = \frac{\lambda_1\tau_1}{2} \quad (5)$$

since this case corresponds to the probability of no more than one failure in the assembly unit equal to 0.01.

If the curve does not match the number of failures $n_1 = 2$ with the abscissa axis, it can be extrapolated to this axis. As with the previous strategy

$$n = \frac{\tau_{\max}}{\tau_{1\text{opt}}} \quad (6)$$

When such a strategy is applied, the number of necessary input data decreases and is reduced only to the definition of λ_1. Events corresponding to the scope of prevention of each assembly unit are assigned specifically.

The third and fourth strategies are not applicable to the equipment of the primary oil processing plant EDP-AVT-6, since they evaluate the repair system of only the assembly units of the product.

In the case of the fifth strategy, all work related to preventive maintenance of equipment is divided into two separate subsets: M_1 and M_2.

In the first subset, all preventive work of i_1 and direct costs for their carrying out y_1 (rubles) are placed. Thus, the direct costs of y_{1i} and the maximum permissible periodicity of prevention of x_{1i} correspond to each i_1 preventive work. In the second subset, all preventive work of ¼ 2 and their corresponding unit costs y_{2i}, (RUR/h of work) are placed. And in this case, the maximum permissible frequency of prevention of x_{2i} is also known.

It is assumed that the following data are known: ω_{h} - the average annual operating time of the product, y_n - losses from one hour of the product downtime, B_1 - the average value of the downtime when performing each i -th preventive work.

The fifth strategy can be applied to the construction of the maintenance system of the equipment of the crude oil processing unit EDP-AVT-6, since all the necessary data are available [11-15].

The sixth strategy is based on the normal density distribution of failure probabilities of assembly units. At the same time, the overhaul life of these assembly units may be

$$\tau = \tau_{\text{av}} - 3\sigma \tag{7}$$

where τ_{av} - average development on refusal of the considered assembly unit;

3σ –third - sigma deviation from τ_{av}.

According to this strategy, intervals can be assigned between the current repairs of all assembly units of this type of equipment, which requires the collection and systematization of statistical material and is very laborious.

The total number of average repairs is:

$$n = \frac{\tau_{\max}}{\tau} \tag{8}$$

The latter strategy is based on an analysis of the costs associated with the increasing wear and tear of the equipment. The fact is that as wear increases, equipment parameters deteriorate. Hence, it is necessary to assign economically feasible repair dates to restore deteriorating parameters. The comparison of depreciation costs with the costs caused by the wear and tear itself is the basis for the timing between the necessary repairs (average or capital).

The overhaul period cannot be longer than the established duration of operation of the main assembly units of equipment. It can be less if economically feasible.

3. Conclusions

The analysis of repair strategies makes it possible to justify and develop a calculated scheme of maintenance periodicity, number of overhaul periods and intervals between current and overhaul for atmospheric column K-2 and other equipment of the crude oil refinery EDP-AVT-6.

References

[1] Vakulenko M V and Zharinova N V 2020 *IOP Conf. Ser.: Mater. Sci. Eng.* **862** 062022
[2] Zharinova N V and Vakulenko M V 2020 *IOP Conf. Ser.: Mater. Sci. Eng.* **862** 032014
[3] Nasibullina O A and Gareev A G 2019 *Materials Science Forum* **946** 20-4
[4] Rizvanov R G, Mulikov D Sh, Karetnikov D V, Fairushin A M and Tokarev A S 2018 *IOP Conf. Ser.: Mater. Sci. Eng.* **317** 012077
[5] Nasibullina O A and Tyusenkov A S 2020 *IOP Conf. Ser.: Mater. Sci. Eng.* **862** 022007
[6] Gaysin E Sh, FrolovY A and NasibullinaO A 2020 *IOP Conf. Ser.: Earth and Environ. Sci.* **459** 032055
[7] Nasibullina O A and Rizvanov R G 2020 *Materials Science Forum* **992** 695-9
[8] Karetnikov D V, Rizvanov R G, Fairushin A M and Kolokhov K S 2013 *Welding International* **27** (7)557-60
[9] Nasibullina O A and Rizvanov R G 2020 *Materials Science Forum* **989** 28-2
[10] Zaripov M Z, Fairushin A M and Karetnikov D V 2019 *Materials Science Forum* **946** 883-8

[11] Tyusenkov A S and Nasibullina O A 2019 *IOP Conf. Ser.: Mater. Sci. Eng.* **687** 066016
[12] Nasibullina O A and Tyusenkov A S 2019 *IOP Conf. Ser.: Mater. Sci. Eng* .**537** 022023
[13] Yakhin A V, Karetnikov D V, Rizvanov R G, Abakacheva E M and Gareev A G 2019 *Chemical and Petroleum Engineering* **54(11-12)** 801-5
[14] Nasibullina O A and Rizvanov R G 2020 *IOP Conf. Ser.: Mater. Sci. Eng.* **969** 012004
[15] Tokarev A S, Karetnikov D V, Rizvanov R G, Fayrushin A M and Zaripov MZ 2020 *Materials Science Forum* **991** 70-6

Automated diseases detection of plant diseases in space greenhouses

A F Aleynikov[1,2] and V I Toropov[2]

[1]Siberian Federal Scientific Centre of Agro technology, Russian Academy of Science, Krasnoobsk, Russia
[2]Novosibirsk State Technical University, Novosibirsk, Russia

E-mail: fti2009@yandex.ru

Abstract. To detect plant diseases in a space greenhouse, a computer vision method is proposed by counting image pixels in the space of color channels of red, green and blue. An algorithm and a program have been developed for the automated determination of the degree of fungal diseases affecting garden strawberry leaves from the transmitted images of plant leaves. The research results can be used for early diagnosis of diseases and monitoring of the functional state of greenhouse plants.

1. Introduction

Long-term presence of man in space requires providing the crew with adequate nutrition with minimal replenishment from Earth. This is due to the reduced load on consumables for future space missions. Therefore, it becomes necessary to produce fresh, functional plant-based foods using greenhouses. In addition, the presence of greenhouses reduces emotional instability and mitigates the stress load on the crew [1]. For example, garden strawberry is nutritious, has an abundant biochemical composition and has a high taste. It has valuable medicinal properties and an attractive appearance. However, garden strawberries infect over 30 fungal, viral and bacterial diseases. Most diseases (about 80%) are caused by fungi [2]. Considering that the vegetation of plants in space conditions occurs at low gravity and illumination levels, the lack of necessary air convection, the problem of early detection of plant diseases is very urgent.

2. Problem definition

Diagnostics of diseases and their causative agents is an important link in the system of protection of plants growing in extreme conditions. Currently, a large number of special methods for detecting diseases with their inherent advantages and disadvantages are used in diagnostics. Diagnostic methods based on polymerase chain reaction, enzyme immunoassay and immune-chromatographic analysis are highly sensitive, but they are lengthy, laborious and invasive [2]. Visual diagnostics, which are widespread in practice, do not lose their relevance, as they are often preliminary for other diagnostic technologies. The main disadvantages of the visual method can be considered unreliability due to the coincidence of external signs of infectious plant lesions and physiological disorders caused by adverse external influences, the need for a highly qualified expert and the late detection of diseases.

The concept of using informative images of plant health monitoring is widespread, based on the development of a normalized difference vegetative index and the determination of the reflectivity of plant leaves in the near infrared range [3]. But this concept is not acceptable for localized health assessments of plants grown in confined spaces. We propose to install color cameras of technical vision systems with high optical resolution to assess the health of plants in greenhouses. Then the images are transferred to the ground segment for its subsequent processing. For residential greenhouses, it is proposed to carry out early diagnosis of plant diseases using smartphones using computer vision methods, for example, based on counting image pixels in the space of color channels of red, green and blue (R, G, B).

3. Materials and methods

Portable electronic equipment is an indispensable companion in the life of a modern person. For example, all manufactured smartphones are endowed with the functions of a telephone, a pocket personal computer, a music player and a camera. The smartphone is able to replace a light source, a control panel for various electrical engineering, a building level, a GPS navigator, etc. The evolution of portable electronic equipment is at a high pace and is developing through the creation of special software applications to meet the needs of the population in various areas of its activity.

A promising direction is the use of smartphones as measuring and diagnostic tools.

Currently, smartphones are being developed and are already being used as a "diagnostic center" for individual health by transforming it into a blood pressure monitor, thermometer, glucometer, electrocardiograph and other useful medical devices and devices. For example, thanks to the development of innovative digital technologies and artificial intelligence, smartphones have found wide application in the traditional diagnosis of human diseases [4].

The device in real time, through an application installed on the smartphone of the person caring for the patient, transmits data about his health: heart rate and breathing parameters, stress level and sleep cycles, etc.

To implement the diagnosis of plant diseases in the form of an application to a smartphone, computer vision methods are promising by counting image pixels in the space of color channels of red, green and blue (R, G, B) [5-10].

A commonly used indicator of plant health is leaf color, which is related to their chlorophyll content.

The development of an algorithm for the program for determining the degree of damage to plant leaves by diseases in the form of applications for a smartphone was based on the following methodological aspects.

Imaging is the first step for any vision system prior to performing image analysis. The smartphones digital camera is used to capture images at the required resolution. To improve image quality, it is necessary to maintain an equal angle and illumination. When approaching a leaf, the camera focuses on the leaf, which is positioned on a substrate with a uniform white or black background to create a contrasting image of the affected plant leaf.

The purpose of image preprocessing is to guarantee the following condition: the extraction of informative parameters does not affect the background, size and shape of the leaf, the intensity of the light source, and the characteristics of the camera for diagnosing plant diseases [11]. Image preprocessing is also used to highlight certain features and reveal details in an image. In this case, various methods are used, such as image filtering, resizing, segmentation, morphological and other operations [12]. In addition, the captured images may contain some noise. Noise removal is performed prior to image analysis using high pass, low pass, median and linear filters, etc. The image can also be enhanced to distinguish between subject and background. Once captured, images are converted to a spatial representation of a different color if required for further analysis. In some cases, masking and pixel removal is required to detect diseases on plant leaves. Masking is setting the pixel value in an image to some other background value, or zero. At this stage, the highly colored pixels need to be identified. For example, when identifying disease, green pixels represent a healthy area of a leaf. Therefore, it is preferable to remove green pixels and save pixels from the infected part of the study area. Masking is performed based on the specified threshold. The red, green, and blue component of a pixel is set to zero if the green component of the pixel intensity is less than a pre-calculated threshold. Masking significantly reduces processing time because disease segmentation is obtained by setting the non-disease portion to zero and «1» to the diseased portion of the leaf.

Image segmentation is the division of an image into an object and an area or background.

In disease identification, image segmentation is used to separate the image between areas with necrotic spots and healthy areas in leaves [13].

In some cases, the infected part, after it has been removed, is segmented into several spots of the same size.

4. Experimental results

The algorithm for the implementation of the proposed method of computer vision was originally implemented on a personal computer using previously obtained color images of leaves of garden strawberry (figure 1).

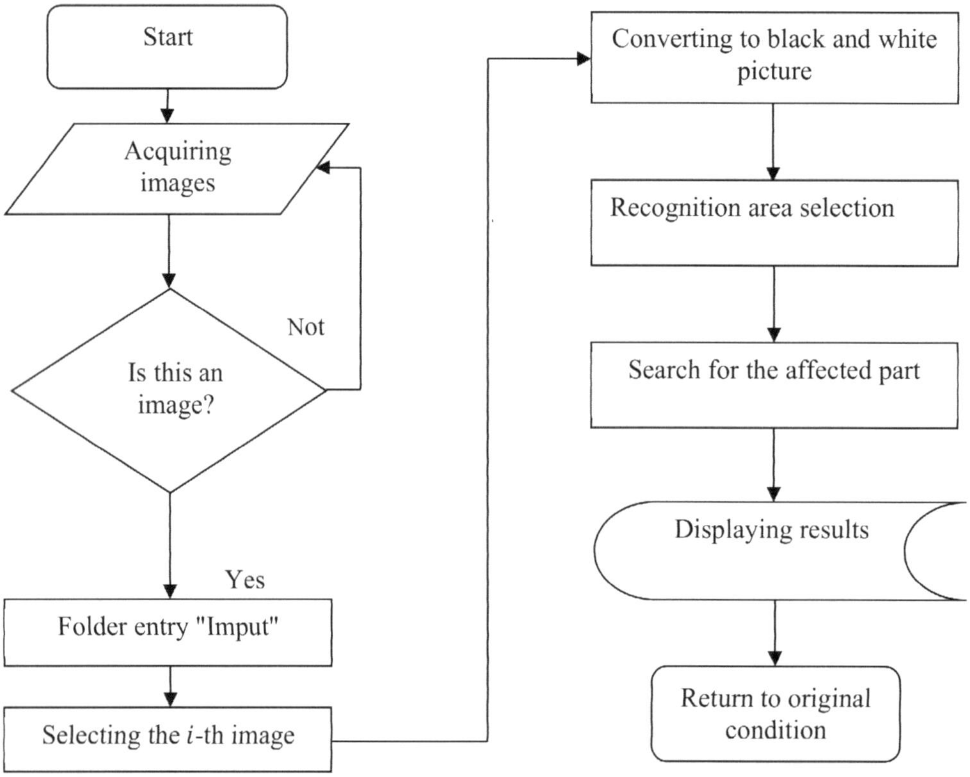

Figure 1. Algorithm of the program for determining the degree of damage to the leaf of garden strawberry.

This is due to the need to thoroughly refine the program algorithm, since a computer display has a higher resolution and a larger program window display area than a smartphone display. In addition, when analyzing a sufficient number of images using the created program, it will be possible to identify inaccuracies in its operations and to correct or modify the program according to the data obtained using the experimenter's experience.

The Java programming language and the Android Studio 3.4.1 operating system were used as a language for the development of the logical part of the information system. To build the graphical interface, software was used to facilitate the development and integration of various modules of the LibGDX software project. The LibGDX project is a cross-platform game development and visualization framework based on the Java programming language with some components written in C and C ++ to improve the performance of certain code. Currently supports Windows, Linux, Mac OS X, Android, iOS and HTML5 as target platforms [14].

With the help of a camera, a color image of a plant leaf is formed. Then the resulting image is copied to the "Imput" folder. The choice of an image from a folder is carried out using vertical sliders and a mouse click (figure 2).

Then the image is converted to the basic format in pixels so that it fits completely on the screen of the LibGDX program.

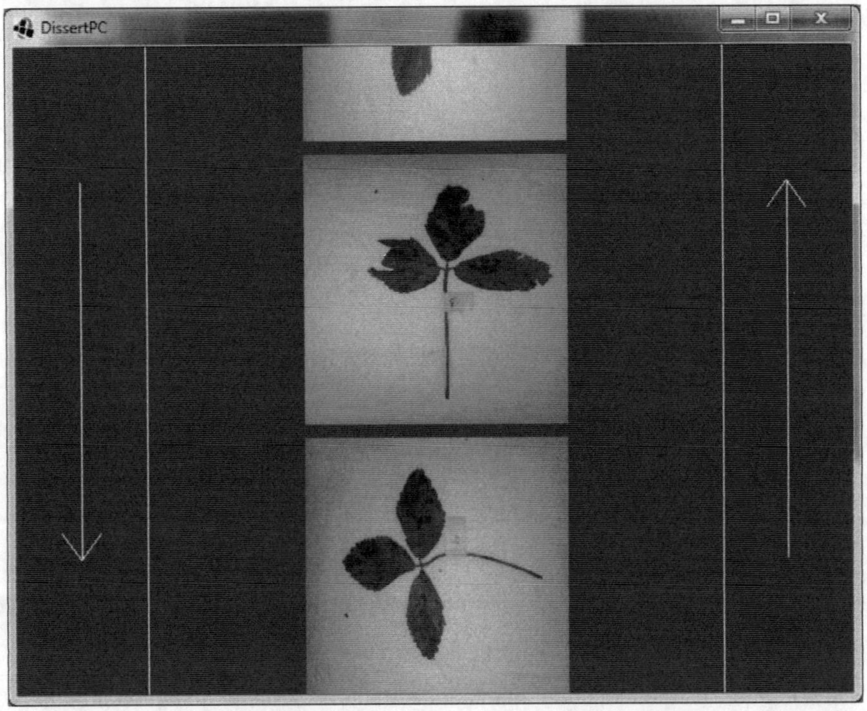

Figure 2. Selecting an image for recognition.

Next, the process of converting a color image to black and white occurs. The image transformation is segmented by analyzing the black and white color intensity distribution on a histogram to match the requirements of the plant disease dataset. A histogram is a graph with brightness located on the x-axis with a maximum size of 256 pixels. The y-axis is a sequence of pixels from 1 to 200 with a corresponding brightness level.

Once the image has been segmented, the extracted area is processed to remove pixilated areas that are dominated by green, that is, places where the leaf is considered to be unambiguously healthy.

After that, using the slider under the graph, we select the pixels of the desired brightness from 0 to 255, where 0 is black, and 255 is absolutely white. At this stage, select the pixels of the plant leaf area, without the background (figure. 3).

Then the degree of damage to the plant is determined (figure. 4). This procedure is performed by analyzing each pixel by comparing its color signatures, for example, comparing red to green and blue to green.

For a more accurate assessment, two histograms-graphs were created with the x-axis from 1 to 500 pixels and with the y-axis to display the red / green and blue / green pixel ratios vertically on them, which are responsible for removing the pixel areas in which dominated red or blue color As the slider moves to the right along the x-axis, the number of pixels on the screen decreases in proportion to the peak on the histogram (figure. 4).

Figure 3. Selecting the sheet area.

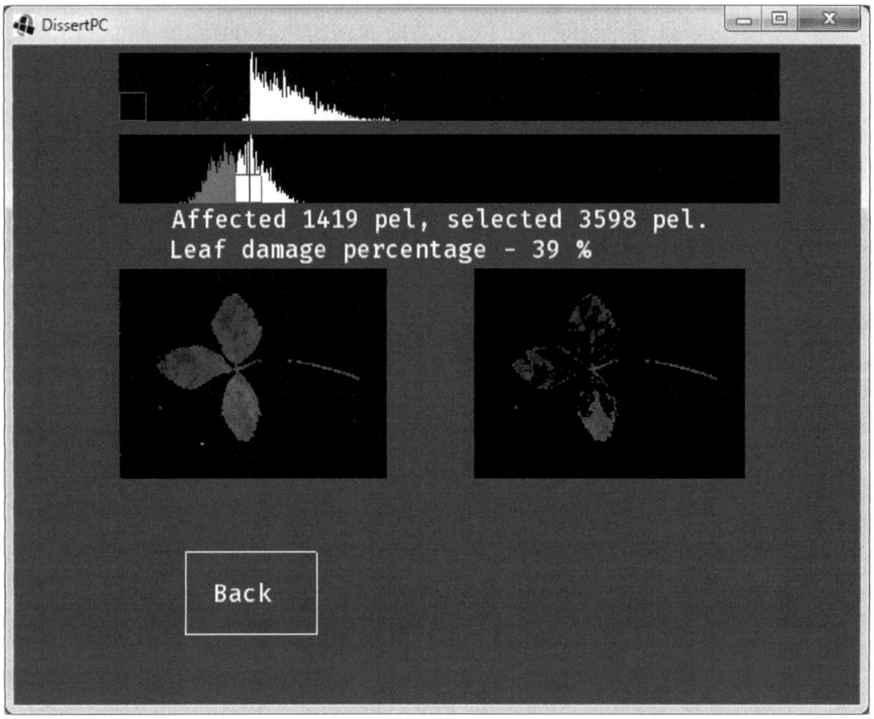

Figure 4. Display of the result of the lesion assessment.

Here is an example of calculating a peak on a histogram. Take a pixel with values red = 70, green = 50, blue = 60. This pixel will be added to the element of the peak (red / green) array equal to (70/50) × 100 = 140, and the peak with the x-axis coordinates equal to 140 will be higher by one.

Histograms of pixel array element values have peaks. The higher the peak on the histogram, the more pixels it has with this ratio. Thus, we reduce the number of pixels close to green and, therefore, find the affected areas. Then remove pixels where the green value is greater than the red and blue value. These questionable pixels with array values from 0 to 499 under the histogram are removed with a slider. After removing the questionable pixels, the affected part is calculated and displayed on the screen.

5. Results and discussion

The introduction of the aforementioned operations using two histograms and sliders will allow diagnosing diseases when its symptoms are still little noticeable to the observer. It should be noted that the above algorithm and software are presented in the form of a basic block, which will be improved in the process of further research and experimental work.

When testing the basic block of the program in the field, it will be possible to take into account the observations and wishes of plant protection experts, agronomists, producers and other specialists in the cultivation of specific crops and its further adjustment and transformation.

For example, due to the importance of such a plant trait as area, in the photosynthetic process and regulation of water balance, it becomes necessary to determine its size.

Of course, in the future, it is necessary to solve other more complex problems, for example, determining the dominant fungal plant disease, the problem of classifying diseases [15].

6. Conclusion

On the basis of one of the computer vision methods, which counts image pixels in the space of red, green and blue color channels (R, G, B), an algorithm and a program have been developed for the automated determination of the degree of damage to plant leaves by fungal diseases.

In this case, images from an uninhabited space greenhouse can be automatically transmitted to Earth with the required period of time.

The algorithm and software are presented in the form of a basic block, which can be supplemented and improved in the course of experimental work and solve related problems to determine other important plant traits.

The research results can be implemented in the form of a software application that can be autonomously installed on the researcher's Smartphone and will be used by him for monitoring the sanitary state of the space greenhouse.

Acknowledgments

This work was supported by the budget project of the Siberian Federal Scientific Centre of Agro-biotechnology of the Russian Academy of Sciences No. 0533-2021-0007.

References

[1] Berkovich Yu A, Konovalova I O, Smolyanina S O, Erokhin A N, Avercheva O V, Bassarskaya E M, Kochetova G V, Zhigalova T V, Yakovleva O S and Tarakanov I G 2017 LED crop illumination inside space greenhouses *Reviews in Human Space Exploration* **6** 11-24 DOI 10.1016/j.reach.2017.06.001

[2] Alejnikov A F, Cheshkova A F and Mineev V V 2020 Choice of impedance parameter of strawberry tissue for detection of fungal diseases *IOP Conference Serie: Earth and Environmental Science* **548** 032005 DOI 10.1088/1755-1315/548/3/032005

[3] Tucker R, Callaham J A, Zeidler C, Paul A L and Ferl R J NDVI imaging within space exploration plant growth modules – A case study from EDEN ISS Antarctica 2020 *Life Sciences in Space Research* **26** 1-9 DOI 10.1016/j.lssr.2020.03.006

[4] Kanchi S and Sabela M I Bisetty Smartphone based bioanalytical and diagnosis applications: A

review 2018 *Biosensors and Bioelectronics* **102** 136-49 DOI 10.1016/j.bios.2017.11.021
[5] Camargo A and Smith J S An image-processing based algorithm to automatically identify plant disease visual symptoms 2009 *Biosystems Engineering* **102 (1)** 9-21 DOI 10.1016/j.biosystemseng.2008.09.030
[6] Gong A, Wu X, Qiu Z and He Y A handheld device for leaf area measurement 2013 *Computers and Electronics in Agriculture* **98** 74-80 DOI 10.1016/j.compag.2013.07.0132
[7] Li Z, Paul R, Tis B T, Saville A C, Hansel J C, Yu T, Ristaino J B and Wei Q Non-invasive plant disease diagnostics enabled by smartphone-based fingerprinting of leaf volatiles 2019 *Nature Plants* **5 (8)** 56-66 DOI 10.1038/s41477-019-0476-y
[8] Padhye P and Rajani K Machine vision guided system for classification and detection of plant diseases using support vector machine 2014 *International Journal of Computational Science and Engineering* **5** 249-54
[9] Raut S and Fulsunge A Plant Disease Detection in Image Processing Using MATLAB 2017 *International Journal of Innovative Research in Science, Engineering and Technology* **6 (6)** 10373 DOI 10.15680/IJIRSET.2017.0606034
[10] Kumar S and Raghavendra B K Diseases Detection of Various Plant Leaf Using Image Processing Techniques: A Review 2019 *5-th International Conference on Advanced Computing & Communication Systems (ICACCS)* 313-16 DOI 10.1109/ICACCS.2019.8728325
[11] Patil J K and Kumar R Advances in image processing for detection of plant diseases 2011 *Journal of Advanced Bioinformatics Applications and Research* **2 (2)** 135-41
[12] Renugambal K and Senthilraja B Application of image processing techniques in plant disease recognition 2015 *International Journal of Engineering Research & Technology* **4 (3)** 919-23
[13] Tichkule S. K. and Gawali D. H. Plant diseases detection using image processing techniques 2016 *Online International Conference on Green Engineering and Technologies (IC-GET)* DOI 10.1109/get.2016.7916653
[14] 2020 LibGDX – cross-platform framework for game development and visualization *On-line developer guide* http://www.libgdx.ru/p/guide.html Date of treatment
[15] Khan M A Detection and Classification of Plant Diseases Using Image Processing and Multiclass Support Vector Machine 2020 *International Journal of Computer Trends and Technology (IJCTT)* **68 (4)** 5-11 DOI 10.14445/22312803/IJCTT-V68I4P102

Phase equilibrium and synthesis in ionic melts of the system Li, Na, Pb || WO_4

J A Kochkarov[1], A A Baysangurova[2] and R A Bisergaeva[2]

[1]Kabardino-Balkarian State University, 173, st. Chernyshevsky, Nalchik, 360004, Russia
[2]Faculty of Biology and Chemistry, Chechen State University, 32 A, Sheripova street, Grozny, Russia

E-mail: chemest2@rambler.ru, radimaa-2009@mail.ru

Abstract. The phase diagram of a three-component system has been studied for the first time by the method of differential thermal analysis and XRF Li,Na,Pb//WO_4. The coordinates of the invariant points are determined (42,5% Na_2WO_4, 47,5% Li_2WO_4, 10% Pb_2WO_4, 450°C), areas of crystallization of the initial components. The possibility of chemical synthesis of lead oxide tungsten bronzes in ionic melts of the studied system is shown.

1. Introduction

Interest in a comprehensive study of lead tungstate single crystals is due to its scintillation properties [1-5]. It was found that lead tungstate occupies an exceptional position in the tungstate family with a sheltie structure. Lead tungstate single crystal is a scintillation material [1], used in the Large Hadron Collider electromagnetic calorimeter and the photon detector in the ALICE experiment at CERN [1, 2]. Now we can say unequivocally that lead tungstate is the most promising scintillation material in the next decade.

The scintillation properties of lead tungstate single crystals are due to their ability to glow under the influence of ionizing radiation, which is of great theoretical and practical interest from the point of view of the possibility of registering ionizing radiation (charged particles, gamma quanta, etc.). As it turned out, due to the large charge of the lead and tungsten nuclei, and the high density (8 g / cm^3), lead tungstate effectively absorbs gamma quanta.

At present, studies are continuing on the possibility of improving the scintillation properties of single crystals of d-element tungstate's [3,4].

In the modern world, scientific and practical research based on accelerators is impossible without the use of highly efficient detecting elements capable of maintaining a high stability of their parameters under the influence of radiation for a long period of time.

That is why, in the early 1990s, the LHC program initiated a number of research projects to develop advanced detectors with improved performance. As part of the functioning of this project, the scintillation properties of lead tungstate were revealed, thanks to the joint work of scientists in the field of materials science, technologists and physicists specializing in high energy physics.

Over the past five years, the technology for the production of single crystals has grown from the release of individual samples to mass production with precisely specified characteristics and properties.

Until recently, the Bogoroditsk Plant of Technochemical Products in Russia carried out mass production of lead tungstate single crystals by the Czochralski method. Small quantities of lead tungstate single crystals are produced by the Shanghai Institute of Ceramics (China) using the modified Bridgman method [6], Institute of Single Crystals (Kharkiv, Ukraine) [7].

In addition, the luminescent properties of lead tungstate single crystals were discovered. [8].

At the same time, the existing methods of synthesizing lead tungstate are far from perfect, require a lot of labor and energy, and pollute the environment.

Oxide tungsten bronzes are promising for the manufacture of anodes of chemical current sources, cathodes of electrolysis baths, are used as catalysts in organic synthesis, to obtain high-quality printing inks, materials for semiconductor diodes and pressure sensors. They are also used for the manufacture of electrodes used in redox titration, since they have high activity and selectivity and in some processes successfully replace platinum metals. In the molten state, they are very strong reducing agents and are used for etching laser rods. Serves as a protective coating for some metal parts, prevents oxidation of the titanium base of iridium oxide anodes.

Thus, the development of an optimal technology for the synthesis of lead tungstate and oxide tungsten lead bronzes in ionic melts of multicomponent systems are of great theoretical and practical interest.

It seems to us that the role of phase diagrams of multicomponent systems in solving the problems of finding optimal conditions for the synthesis of substances is extremely great.

The development of complexation and the presence of solid solutions of varying degrees of stability complicate the study of the system. This is of great methodological value, since allows you to study real systems in their morphological development and mutual influence.

Three-component systems $(NaCl)_2$-Na_2CO_3-$(NaF)_2$, $(KCl)_2$-K_2CO_3-K_2MoO_4 and $(KCl)_2$-K_2CO_3-K_2WO_4belong to the simple eutectic type. Each of them implements a single triple eutectic.

In the systems $NaCl)_2$-$(NaF)_2$-Na_2WO_4, $(NaCl)_2$-$(NaF)_2$-Na_2MoO_4, $(KCl)_2$-$(KF)_2$-K_2WO_4 [9], $(KF)_2$-$(KCl)_2$-K_2WO_4, $(KCl)_2$-$(KF)_2$-K_2CO_3, $(NaCl)_2$-Na_2CO_3-Na_2WO_4 [10], $(NaCl)_2$-Na_2CO_3-Na_2MoO_4, $(NaF)_2$-Na_2CO_3-Na_2MoO_4, $(NaF)_2$-Na_2CO_3-Na_2WO_4, $(KF)_2$-K_2CO_3-K_2MoO_4, $(KF)_2$-K_2CO_3-K_2MoO_4, $(KF)_2$-K_2CO_3-K_2WO_4, in addition to the liquidus surfaces of the initial components, crystallization regions of double complex compounds $NaCl \cdot Na_2WO_4$, $NaCl \cdot Na_2MoO_4$, $KF \cdot K_2WO_4$, $KF \cdot K_2MoO_4$, $KF \cdot K_2CO_3$, 2 $NaF \cdot Na_2MoO_4$, $2NaF \cdot Na_2WO_4$.

Table 1. The phase diagrams of these systems are triangulated into phase unit blocks.

	FEBi	HBT	System
I	$(NaCl)_2$-$(NaF)_2$-$NaCl \cdot Na_2WO_4$	E_{613}	Na//F, Cl, WO_4
II	$(NaF)_2$-$NaCl \cdot Na_2WO_4$-$2NaF \cdot Na_2WO_4$	empty	-"-"-"-"-"-"-"-
III	Na_2WO_4-$NaCl \cdot Na_2WO_4$-$2NaF \cdot Na_2WO_4$	E_{606}, P_{627}	-"-"-"-"-"-"-"-
IV	$(NaCl)_2$-$(NaF)_2$-$NaCl \cdot Na_2MoO_4$	E_{586}	Na//F, Cl, MoO_4
V	$(NaF)_2$-$NaCl \cdot Na_2MoO_4$-$2NaF \cdot Na_2MoO_4$	empty	-"-"-"-"-"-"-"-
VI	Na_2MoO_4-$NaCl \cdot Na_2MoO_4$-$2NaF \cdot Na_2MoO_4$	E_{580}, P_{592}	-"-"-"-"-"-"-"-
VII	$(KCl)_2$-$(KF)_2$-$KF \cdot K_2WO_4$	E_{505}	K//F, Cl, WO_4
VIII	$(KCl)_2 \cdot K_2WO_4$-$KF \cdot K_2WO_4$	E_{584}	-"-"-"-"-"-"-"-
IX	$(KCl)_2$-$(KF)_2$-$KF \cdot K_2MoO_4$	E_{549}	K//F, Cl, MoO_4
X	$(KCl)_2$-K_2MoO_4-$KF \cdot K_2MoO_4$	E_{558}	-"-"-"-"-"-"-"-
XI	$(KCl)_2$-K_2CO_3-$KF \cdot K_2CO_3$	empty	K//F, Cl, CO_3
XII	$(KCl)_2$-$(KF)_2$-$KF \cdot K_2CO_3$	E_{528}, P_{532}	-"-"-"-"-"-"-"-
XIII	$(NaCl)_2$-Na_2CO_3-[$NaCl \cdot Na_2WO_4$]	E_{588}	Na//Cl, CO_3, WO_4
XIV	Na_2CO_3-Na_2WO_4-[$NaCl \cdot Na_2WO_4$]	E_{569}	-"-"-"-"-"-"-"-
XV	$(NaCl)_2$- Na_2CO_3-[$NaCl \cdot Na_2MoO_4$]	E_{572}	Na//Cl, CO_3, MoO_4
XVI	Na_2CO_3-Na_2MoO_4-[$NaCl \cdot Na_2MoO_4$]	E_{548}	-"-"-"-"-"-"-"-
XVII	$(NaF)_2$-Na_2CO_3-$2NaF \cdot Na_2MoO_4$	empty	Na//F, CO_3, MoO_4
XVIII	Na_2CO_3-Na_2MoO_4-$2NaF \cdot Na_2MoO_4$	E_{544}, P_{570}	-"-"-"-"-"-"-"-
XIX	$(NaF)_2$-Na_2CO_3-$2NaF \cdot Na_2WO_4$	empty	Na//F, CO_3, WO_4

XX	Na_2CO_3-Na_2WO_4-$2NaF·Na_2WO_4$	E_{552}, P_{580}	-"-"-"-"-"-"-"-"-
XXI	$(KF)_2$-$KF·K_2CO_3$-$KF·K_2MoO_4$	E_{624}	K//F, CO_3, MoO_4
XXII	K_2CO_3-$KF·K_2CO_3$-$KF·K_2MoO_4$	E_{622}	-"-"-"-"-"-"-"-"-
XXIII	K_2CO_3-$KF·K_2MoO_4$-K_2MoO_4	E_{630}	-"-"-"-"-"-"-"-"-
XXIV	$(KF)_2$-$KF·K_2CO_3$-$KF·K_2WO_4$	E_{623}	K//F, CO_3, WO_4
XXV	K_2CO_3-$KF·K_2CO_3$-$KF·K_2$WO4	E_{621}	-"-"-"-"-"-"-"-"-
XXVI	K_2CO_3--K_2WO_4-$KF·K_2WO_4$	E_{630}	-"-"-"-"-"-"-"-"-

FEBi II, VI, XII, XVII and XIX are empty, due to the migration of triple eutectics into neighboring adjacent FEBi with inversion into triple peritectics.

In the systems Na_2CO_3-Na_2MoO_4-Na_2WO_4, K_2CO_3-K_2MoO_4-K_2WO_4, belt solid solutions based on Na_2MoO_4 and Na_2WO_4, K_2MoO_4 and K_2WO_4 do not decompose.

Three-component reciprocal systems Na, K//F, Cl и Na, K//Cl, CO_3 irreversible reciprocal diagonal type, without complexation.

Systems Na, K//F, CO_3, Na, K//F, MoO_4 и Na, K//F, WO_4 are irreversibly reciprocal diagonal-adiagonal type with complexation.

Table 2. Their phase complexes are triangulated by stable diagonals and adiagonals to the corresponding FEBi.

	FEBi	HBT	System
I	$(KF)_2$-$(NaF)_2$-$KF·K_2CO_3$	empty	Na, K//F, CO_3
II	$(NaF)_2$-K_2CO_3-$KF·K_2CO_3$	E_{558}, P_{584}	-"-"-"-"-"-"-"-"-
III	$(NaF)_2$-K_2CO_3-Na_2CO_3	solid solution	-"-"-"-"-"-"-"-"-
IV	$(NaF)_2$-$(KF)_2$-$KF·K_2MoO_4$	E_{648}	Na, K//F, MoO_4
V	$(NaF)_2$-K_2MoO_4-$KF·K_2MoO_4$	E_{672}	-"-"-"-"-"-"-"-"-
VI	$(NaF)_2$-K_2MoO_4-$Na_2MoO_4·K_2MoO_4$	P_{645}	-"-"-"-"-"-"-"-"-
VII	$(NaF)_2$-Na_2MoO_4-$Na_2MoO_4·K_2MoO_4$	E_{593}, R_{599}	-"-"-"-"-"-"-"-"-
VIII	$(NaF)_2$-$(KF)_2$-$KF·K_2WO_4$	E_{645}	Na, K//F, WO_4
IX	$(NaF)_2$-K_2WO_4-$KF·K_2WO_4$	E_{664}	-"-"-"-"-"-"-"-"-
X	$(NaF)_2$-K_2WO_4-$Na_2WO_4·K_2WO_4$	empty	-"-"-"-"-"-"-"-"-
XI	$(NaF)_2$-$Na_2WO_4·K_2WO_4$-$2NaF·Na_2WO_4$	empty	-"-"-"-"-"-"-"-"-
XII	Na_2WO_4-$Na_2WO_4·K_2WO_4$-$2NaF·Na_2WO_4$	E_{591}, P_{596}, P_{610}	-"-"-"-"-"-"-"-"-

A distinctive feature of the Na, K // F, MoO_4 system is that the compound $2NaF·Na_2MoO_4$ "wedges out" and is not included in the composition of the completely crystallized alloy, due to its decomposition below the temperature of HBT R. Thus, the pole of this compound cannot be triangulating, and the point R does not belong to FEB VII. Note also that HBT E_{645} migrates from FEB VII to VI, to which it belongs.

In the Na, K // F, WO_4 system, HBT P_{610} and P_{596} are localized in FEB XII, but belong to FEBi X and XI, respectively. Migration processes do not take place here.

The systems Na, K // Cl, MoO_4 and Na, K // Cl, WO_4 belong to reversibly reciprocal systems of the adiagonal type with complexation. In them, the chemical processes of mutual exchange are suppressed by the processes of complex formation.

Table 3. The compounds $NaCl·Na_2MoO_4$ and $NaCl·Na_2WO_4$ are the poles of three-beam triangulating stars, which divide the corresponding squares of the compositions by FEBi.

	FEBi	HBT	System
I	$(NaCl)_2$-$(KCl)_2$-$NaCl·Na_2MoO_4$	empty	Na, K//Cl, MoO_4
II	$(KCl)_2$-K_2MoO_4-$NaCl·Na_2MoO_4$	P_1, E, P_2	-"-"-"-"-"-"-"-"-
III	K_2MoO_4-$NaCl·Na_2MoO_4$-$Na_2MoO_4·K_2MoO_4$	P	-"-"-"-"-"-"-"-"-

IV	Na_2MoO_4-$NaCl{\cdot}Na_2MoO_4$-$Na_2MoO_4{\cdot}K_2MoO_4$	empty	-"-"-"-"-"-"-"-"-
V	$(NaCl)_2$-$(KCl)_2$-$NaCl{\cdot}Na_2WO_4$	empty	Na, K//Cl, WO_4
VI	$(KCl)_2$-K_2WO_4-$NaCl{\cdot}Na_2WO_4$	P, E_{508}	-"-"-"-"-"-"-"-"-
VII	K_2WO_4-$NaCl{\cdot}Na_2WO_4$-$Na_2WO_4{\cdot}K_2WO_4$	E_{559}	-"-"-"-"-"-"-"-"-
VIII	Na_2WO_4-$NaCl{\cdot}Na_2WO_4$-$Na_2WO_4{\cdot}K_2WO_4$	E_{590}	-"-"-"-"-"-"-"-"-

HBT P1, E, P2 of the Na, K//Cl, MoO_4 systems are localized in FEB II, but HBT P_1 and P_2 belong to FEBs I and III, respectively, and HBT P belongs to FEBu IV.

The topology of the Na, K//Cl, WO_4 system differs significantly. Here, in FEB VI, only two NVT E_{508} and P are localized, where P belongs directly to FEB V.

Systems Na, K//CO_3, MoO_4, Na, K//CO_3, WO_4 и Na, K//MoO_4, WO belong to the irreversible reciprocal diagonal type with complex formation. In them, binary solid solutions based on Na_2CO_3 and K_2CO_3, Na_2WO_4 and Na_2MoO_4, K_2WO_4 and K_2MoO_4, $Na_2WO_4{\cdot}K_2WO_4$ and $Na_2MoO_4{\cdot}K_2MoO_4$ do not decompose and form solutions of the belt type.

Table 4. The squares of the compositions are divided by triangulating sections into FEBi.

FEBi	HBT	System
Na_2CO_3-K_2CO_3, K_2MoO_4	absent	Na, K//CO_3, MoO_4
Na_2CO_3-K_2MoO_4-$Na_2MoO_4{\cdot}K_2MoO_4$	E_{605}	-"-"-"-"-"-"-"-"-
Na_2CO_3-Na_2MoO_4-$Na_2MoO_4{\cdot}K_2MoO_4$	E_{570}	-"-"-"-"-"-"-"-"-
Na_2CO_3-K_2CO_3-K_2WO_4	absent	Na, K//CO_3, WO_4
Na_2CO_3-K_2WO_4-$Na_2WO_4{\cdot}K_2WO_4$	E_{586}	-"-"-"-"-"-"-"-"-
Na_2CO_3-Na_2WO_4-$Na_2WO_4{\cdot}K_2WO_4$	E_{556}	-"-"-"-"-"-"-"-"-
K_2WO_4-K_2MoO_4-$Na_2WO_4{\cdot}K_2WO_4$-$Na_2MoO_4{\cdot}K_2MoO_4$	absent	Na, K//MoO_4, WO_4
Na_2WO_4-Na_2MoO_4-$Na_2WO_4{\cdot}K_2WO_4$-$Na_2MoO_4{\cdot}K_2MoO_4$	absent	-"-"-"-"-"-"-"-"-

Thus, the analysis of the literature review on the systems under study allows us to conclude that the main experimental method used to study systems with two or more components is the method of cross sections based on the VPA data. Since the 70s, research has been intensively conducted using DTA and XRF.

The aim of this work is to study phase equilibria and the possibility of chemical synthesis of lead tungstate in ionic melts of a three-component system Li,Na,Pb//WO_4.

This goal was achieved by solving the following tasks:

- Using differential thermal analysis to reveal the coordinates of invariant points, crystallization areas of the initial components and binary compounds and build the phase tree and crystallization tree.
- Show the possibility of chemical synthesis of oxide tungsten lead bronzes in ionic melts of the studied system.

2. Experimental part

The object of research is the phase diagram of the three-component system Li, Na, Pb || WO_4 (table1).

Experimental study of the phase diagram of the three-component Li, Na, Pb // WO4 system was carried out by differential thermal analysis (DTA).

Used Pt-Pt / Rh thermocouples and platinum microcrucibles with a capacity of 0.5 g. The qualification of the starting salts is not lower than "chemically pure". The crystallization (melting) temperature of each sample was measured twice, the difference between the crystallization and melting temperatures was no more than 3-4 ° C, the measurement error of the crystallization (melting) temperature was ± 2 ° C, the compositions of the eutectics were determined with an absolute accuracy of ± 0,5% by each component. All compositions are expressed in mol. %, and the temperature - in oC. We used the techniques of the projection thermographic method.

The chemical composition of the obtained oxide tungsten bronzes was determined by X-ray phase analysis (XRD).

In the three-component system Na2WO4-Li2WO4-Pb2WO$_4$, only one eutectic with a melting point of 450 ° C and concentration coordinates was revealed: 42,5% Na_2WO_4, 47,5% Li_2WO_4, 10% Pb_2WO_4. Component liquidus determined Na_2WO_4, Li_2WO_4 и Pb_2WO_4 (surface of primary crystallization of componentspositions of monovariant lines (lines of joint crystallization of two phases - lines of double eutectics e_1-E, e_2-E и e_3-E). Consequently, there is no chemical interaction between the components, and the system under study is a simple eutectic one.

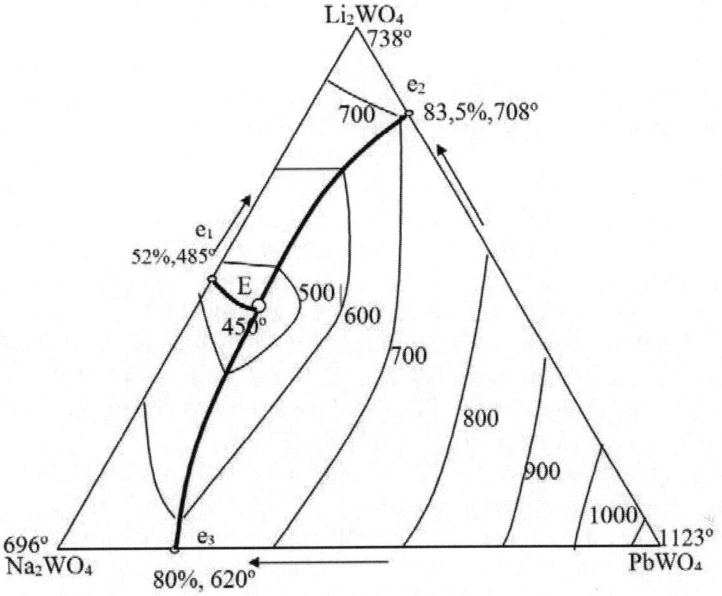

Figure 1. Composition triangle of the ternary system Na_2WO_4-Li_2WO_4-Pb_2WO_4.

Synthesis of oxide tungsten lead bronzes in ionic melts of a three-component system Na_2WO_4-Li_2WO_4-Pb_2WO_4.

For us, of particular interest was the question of the possibility of chemical synthesis of oxide molybdenum lead bronzes in melts of a three-component system Na_2WO_4-Li_2WO_4-Pb_2WO_4.

Synthesis technique. Powders of metallic tungsten and tungsten (VI) oxide were introduced into the initial sample of the eutectic composition based on the percentage of lithium, sodium and lead tungstates in accordance with the reaction equation:

$$3Li_2WO_4 + 2WO_3 + W = 6LiWO_3$$

$$3PbWO_4 + 2WO_3 + W = 6Pb_{0.5}WO_3.$$

$$3Na_2WO_4 + 2WO_3 + W = 6NaWO_3$$

The resulting mixture was thoroughly mixed in a mortar, then dried at a temperature of 150-200 ° C. Then the charge was transferred into a crucible, lowered into a shaft furnace and heated to the melting point. The melt was kept at this temperature for up to 30-45 minutes. Then the melt was poured into a stainless steel cuvette, and after cooling, it was thoroughly ground in a mortar and transferred into boiling distilled water to wash the bronze from salts. After separation from the filtrate, the bronze was dried at 100 ° C, weighed and the product yield was determined, which was 95.6%.

A brown multiphase mixture was obtained, in which the phases were identified by XRF: $Pb_{0,5}WO_3$, $Pb_{0,45}WO_3$, $Pb_{0,46}WO_3$, $Pb_{0,47}WO_3$. $Na_{0,49}WO_3$, $Na_{0,44}WO_3$, $Na_{0,54}WO_3$, $NaWO_3$, $NaWO_3$, $Na_{0,54}WO_3$, $Na_{0,55}WO_3$, $LiWO_3$, $Li_{0,51}WO_3$.

3. Conclusions

The phase diagram of the three-component system Li, Na, Pb // WO_4 and the possibility of chemical synthesis of oxide tungsten bronzes in ionic melts of the three-component system have been studied Li, Na, Pb//WO_4.

It follows from the data obtained that in the three-component system Na_2WO_4-Li_2WO_4-Pb_2WO_4 there is no chemical interaction between the components, as evidenced by the formation of only one eutectic point.

As the results of studying the possibility of chemical synthesis of lead tungstate show, in ionic melts of a three-component system Li, Na, Pb//WO_4 it is possible to synthesize highly dispersed powders of oxide tungsten bronzes with a high yield of the main product.

References

[1] Bernabei R, Belli P, Cappella F, Caracciolo V *et al.* 2013 Crystal scintillators for background measurements *AIP Conference Proceedings* **1549** 189-96
[2] Belli P, Incicchitti A and Cappella F 2014 Inorganic scintillators in direct dark matter investigation *Int. Journal of Modern Physics A* **29(19)** 144-9
[3] Burachas S, Ippolitov M, Manko V, Nikulin S, *et al.* 2010 Temperature dependence of radiation hardness of lead tungstate scintillation crystals *Rad. Measur.* **45** 83-8
[4] Xie J, Shi Y, Yuan H, *et al.* 2009 Influence of negative ions doping on scintillation properties of PbWO4 crystals *Nucl. Instr. Meth. A.* **604(3)** 527-30
[5] Kochkarov Zh A and Sokurova Z A 2016 Synthesis of lead tungstate in melts of multicomponent systems *Izv. of Doug. state ped. Un* **10(2)** 5-9
[6] Kochkarov Zh A and Sokurova Z A 2016 Synthesis of molybdate and oxide tungsten lead bronzes in melts of multicomponent systems *Izv. of Doug. state ped. Un* **10(3)** 20-6
[7] Kochkarov Zh A, Sokurova Z A and Babaeva L Z 2016 Phase equilibria and synthesis in quaternary reciprocal systems Na, K, Pb // Cl, WO4; Na, K, Pb // Cl, WO4 *Izv. Doug. state ped. Un* **10(4)** 12-8
[8] Kochkarov Zh A, Sokurova Z A, Balkizova A Yu and Tokhov M R 2018 Three-component system KCl-K2WO4-PbWO4 *Journal of Scientific and Technical Bulletin of the Volga Region* **5** 38-40
[9] Vasilchenko L M, Trunin A S and Posypaiko V I 1976 *System K // F, Cl, WO4* (Kuibysh. polytechnic in) pp 3386-76
[10] Kochkarov Zh A, Lutsyk V I, Mokhosoev M V, Garkushin I K and Trunin A S 1987 Liquidus of systems Na // WO4, Cl, J (VO3, CO3) *Zh. of inorganic chemistry* **32(1)** 197-201

Information modernization of the general theory of environmental safety ensuring

I A Dmitrieva[1,4], L P Mileshko[2] O V Begun[3] and A V Berezhnaya[1]

[1] Don State Technical University, 1 Gagarin Sq., Rostov-on-Don, 344000, Russia
[2] Southern Federal University, 105/42 Bolshaya Sadovaya St., Rostov-on-Don, 344006, Russia
[3] Anton Chekhov Taganrog State Institute (branch) Rostov State University of Economics, 48 Iniciativnaya St., Taganrog, 34793, Russia

[4]E-mail: idmitrieva2004@mail.ru

Abstract. A brief analysis of the content of the concept "information" in ecology is given. The purpose of this work is to analyze the features of the information component of the second and third principles of the ensuring environmental safety general theory. It is noted that information interactions in ecological systems play a key role in terms of the possibility of maintaining a balance in them to ensure environmental safety. It is argued that the information components of ecological systems of the environment have a physicochemical meaning, which consists in the reflection of material-energy changes in the form of an information environment (information field) of all subsystems. Information processing can be viewed as a kind of partial modification of self-organization. In open systems (pumped) that give off entropy, information can be acquired. It is the entropy and, of course, the uncertainty of the position of the system that can only increase, i.e. information itself can be nothing more than lost. Thus, the indicated problem seems to be important, essential for the currently achieved level of development of the general theory of ensuring environmental safety and information theory.

1. Introduction

According to the second principle of the general theory of environmental safety (GTES) in any system, quantitative and qualitative aggregates of substances, energy and information are inextricably linked, being in continuous change caused by a change in the characteristics of environmental factors (N.F. Reimers., Yu.N. Kurazhkovsky, L.V.Bobukh, K.A.Bobukh, L.P. Mileshko) [1].

Information here is characterized by an energetically weak effect, a coded message perceived by an organism about the possibility of much more powerful influences on it from other organisms or environmental circumstances and prompting it to respond [2].

In accordance with the third principle of GTES, the energy consumption for information interactions is insignificant and does not lead to imbalance in ecological systems but leads to the coordinated behaviour of subsystems (self-organization), as a result of which the degree of ordering of the system increases, i.e. entropy decreases (L.P. Mileshko, V.V. Kotenko) [3].

The development of GTES from the standpoint of information substantiation of its principles is an important task that is essential for its further improvement.

The purpose of this work is to analyze the features of the information component of the second and third principles of the general theory of ensuring environmental safety.

2. Materials and method

According to N. F. Reimers, according to the law of internal dynamic equilibrium (ISDR) - "matter, energy, information and dynamic qualities of individual natural systems and their hierarchy are so interconnected that any change in one of these indicators causes concomitant functional and structural quantitative and qualitative changes that preserve the total sum of material-energy, informational and dynamic qualities of the systems where these changes take place, or in their hierarchy "[2].

In the interpretation of the ISDR N.F. Reimers refers to other laws, principles and rules that can be used as an algorithm for studying the laws of ecology. For example, the concept of information in environmental management, natural chain reactions, the Le Chatelier-Brown principle, the law of irreversibility of evolution, the law of reducing the energy efficiency of environmental management, the rules of one and ten percent, etc.

The ecological meaning of the bit(s) of N.F. Reimers explains as follows. As you know, a bit(s) of information is a unit of the amount of information that is obtained during the implementation of one of two equally probable events (i.e. a binary unit of information). For example, the information that within the range of a dandelion it grew in this place (it could or may not have grown) contains information in 1 bit.

However, unlike technology, in nature, single events occur very rarely (say, the appearance or not the appearance of a species outside its range).

In the case of the dandelion, the chain of events is actually not exhausted by the alternative "grown - not grown", as in the case of a coin toss - "heads - tails". In the coin variant, all that matters is that it is a coin with two sides and that it is tossed. Where did it fall (if not on an edge into a gap and rolled down the slope and disappeared) what dignity it was, how quickly it fell, how high it was thrown, etc. - all this is irrelevant. In the case of a dandelion, all of the above is important, since, for example, if its seed gets on the asphalt or in a pond, it will not grow, it can be destroyed by pests, etc. [2, p.52].

A legitimate question arises about the role of information in ecology: what is the nature of its relationship with matter and energy according to the ISDR and does this relationship exist at all? And if so, in what form? Is it possible to control information by means of any material-energy influences? If information is an ideal display of material-energy interactions, then how can the environment be reflected - as an independent information "field", or rather, an information environment. And in general, how legitimate is this approach? Can environmental systems be balanced through information flow management?

By definition, an "ecological system" is a spatially defined set of living organisms of different species and their habitat, united by material-energy and information interactions [3].

Apparently, the last question can be answered in the affirmative, if the following provisions put forward by T.A. Akimova and V.V. Haskin V.V. are true [3].

Provided that for each element of the system two states are possible (ground and excited) and they are equally probable (in this case $-log_2 p_i = 1$ bit of information), the ratio $I_S = H/S$ can serve as the *information equivalent of entropy*: $I_S = 4.38 \cdot 10^{23}$ bits per entropy unit or $1.045 \cdot 10^{23}$ bits per J / K. Such a value of I_S is obtained as a quotient from dividing the conversion factor of the logarithm base 2 to the natural logarithm ($log_2 x = 1.4427\ lnx$) by the Boltzmann constant. 1 entropy unit *(1 e.u.)* = 1 cal / deg = 4.18 J/K.

Because the

$$\Delta S = \Delta Q/T, \qquad (1)$$

where Q is the amount of heat (energy);

T is the absolute temperature, the knowledge of which makes it possible to estimate *the information equivalent of energy*:

$$I_E = 1.045 \cdot 10^{23} \frac{1}{T} bit/J \tag{2}$$

This means that, energetically, structural information itself is extremely "cheap". Its inverse dependence on temperature has a hidden universal significance and is of independent interest.

The entropy itself and thus the uncertainty of the state of the system can only increase, i.e. information itself can only be lost. Only in open systems (pumped) that give off entropy can information be acquired. Therefore, information processing can be viewed as a kind of private kind of self-organization.

Thus, the problem indicated in the title of the article seems to be important, essential for the currently achieved level of development of the general theory of environmental safety and information theory.

According to the law of energy maximization (LEM) [2] G. and E. Odumov - in competition with other systems, the one that contributes to the flow of energy in the best way and uses its maximum amount in the most efficient way survives (is preserved). "For this purpose, the system: 1) creates accumulators (storage facilities) of high-quality energy: 2) spends (a certain amount) of accumulated energy to ensure the supply of new energy; 3) ensures the circulation of various substances; 4) creates regulation mechanisms that support the stability of the system and its ability to adapt to changing conditions: 5) establishes the exchange with other systems necessary to meet the need for energy of special types" (Odum G., Odum E.). It should be noted that the LEM is also true in relation to information, therefore it can also be considered as the LEM of information: the system that is most conducive to the flow, production and efficient use of energy and information has the best chances of self-preservation. The maximum intake of a substance as such does not guarantee the success of a system in a competitive group of other similar systems.

According to S.I. Ozhegov SI, balance is "the ratio of mutually related process indicators", in this case - the exchange of matter and energy with the environment in non-equilibrium conditions.

An addition to the second principle of HSEB can be formulated as follows: the information component of the environment contains a physicochemical meaning, which consists in displaying material and energy changes in the form of an information environment (information field) of all subsystems [4-7].

The most common chemical signals in the example of water are changes in: taste, smell, acidity, alkalinity, hardness, etc.

Physical signals include changes in temperature, colour, turbidity, etc.

In the case of atmospheric air, chemical signals are changes in its chemical composition and the chemical composition of atmospheric precipitation and snow cover, etc., and physical signals are background radiation, temperature, pressure, direction and speed of wind in the surface layer.

For soils, chemical signals are, for example, changes in the composition of soil air and the state of water in the soil, while physical signals include changes in soil structure, physical degradation, etc.

The transmission of information in populations occurs through material carriers, making it existing and its reception and processing requires little energy. Information is present in certain structures of organisms and is transmitted by streams of physical and chemical signals.

Reading information includes the reception of physical and chemical signals from the atmosphere, hydrosphere, lithosphere and biota and can be carried out in a contact and non-contact way.

The disadvantage of contact methods is the change in the parameters of environmental elements in the zone of contacting substances, which reduces the reliability of the analysis.

For this reason, the most interesting is the reception of optical signals, especially in the infrared range.

The principle of operation of IR devices is based on the transformation into visible infrared radiation of bodies that is insensitive to the human eye [8].

According to the principle of operation, infrared devices are subdivided into passive ones, which use their own IR radiation of the investigated objects and active ones, with built-in artificial sources of IR radiation, illuminating the analysed bodies [8].

There are two large groups of IR receivers: thermal and photonic [9].

Consequently, to build models of the information environment of ecological systems, a hardware and software complex is required, including monitoring subsystems that read information from the atmosphere, hydrosphere and lithosphere.

At the same time, great attention should be paid to ensuring information security of measurements.

The efficiency of the functioning of the system directly depends on the physicochemical nature of the principles of action of the elements.

A distinctive feature of the designed hardware-software complex is the presence of two arrays of sensors - gas sensors and ion-selective electrodes, which requires parallelization of the processes of reading and processing information.

Software can be developed using the C # language.

To train a neural network, a data array is required that will contain measurement data with a previously known concentration of the measured components in the analyte.

To implement a neural network, you can use the free TensorFlow libraries from Google. There are several reasons: the Python language is simple, there is a large amount of teaching material and there is no need for in-depth study of the theoretical principles of creating neural networks.

Common path:

- creating a model;
- compiling the model;
- loading and preparing data;
- training a neural network;
- using the model to determine the results.

3. Materials and method

Thus, a brief analysis of the content of the concept "information" in ecology is given. It is noted that information interactions in ecological systems play a key role in terms of the possibility of maintaining a balance in them to ensure environmental safety.

Information processing can be viewed as a kind of partial modification of self-organization.

In open systems (with pumping), which give off entropy, information can be acquired.

It is the entropy and, of course, the uncertainty of the position of the system that can only increase, i.e. the information itself can be nothing but lost.

References

[1] Mileshko L P 2018 Modernization of the general theory of ensuring environmental safety *Management of Economic Systems: Scientific Electronic Journal* **6 (112)** 32

[2] Mileshko L P, Kotenko V V and Nestyurina E E 2014 Environmental security of the information environment and the role of information interactions in ecological systems *Information Counteraction to Terrorism Threats* **23** 259-63

[3] Akimova T A 2006 *Ecology. Man - Economy - Biota - Environment* (Moscow: UNITI-DANA)

[4] Kotenko V V, Mileshko L P, Petrova E A and Stadnikova Yu V 2018 Fundamentals of the theory of environmental security of the information environment *Information and Security* **21(1)** 94-95

[5] Dmitrieva I A, Gordienko L V, Mileshko L P and Sakharova O V 2020 Environmental safety systems in breadbaking efficiency analysis *IOP Conf. Series: Earth and Environmental Science* **421** 72010

[6] Dmitrieva I A, Mileshko L P, Sakharova O N and Gordienko L V 2020 Progress in environmental safety (review) *IOP Conf. Series: Earth and Environmental Science* **421** 72013

[7] Mileshko L P, Telesh A D, Dmitrieva I A, Borisova A A and Sakharova O N 2020 Methods of computer thermodynamic analysis of chemical reactions in ecological systems *IOP Conf. Series: Materials Science and Engineering* **862** 62030

[8] Borisov Yu 1976 *Infrared Radiation* (Moscow: Energiya)
[9] Kriksunov L Z 1978 *A Handbook on the Basics of Infrared Technology* (Moscow: Sov. Radio)

Assessment of the contribution of phase transformations to the space-time distributions of the temperature of the filtering oil

G F Efimova[1] and T M Levina

Ufa State Petroleum Technological University, Branch in Salavat, 22B, Gubkin st., Salavat, Russia

[1]E-mail: dohle_e@mail.ru

Abstract. The solution of the problem of filtration of paraffinic oil by asymptotic methods in the form of endless sequence of boundary value problems for the expansion coefficients of the required solution in the Maclaurin series is obtained. Estimation formulas were used to calculate the space-time temperature distributions in the first approximation for the vibrational motion of viscous oil and the contribution of phase transformations of paraffin wax. The results can be used for theoretical substantiation and further development of thermal methods for influencing oil-bearing reservoirs, as well as for predicting temperature processes in the bottomhole zone.

The thermal impact on the layer is one of the most effective and perspective methods of enhancing oil recovery from deposits with a high content of paraffins, resins, and asphaltenes. The dependence of oil recovery from a layer containing high-viscosity oil on temperature is complex.

To construct a mathematical model for the filtration of paraffinic oil and solving the thermodynamic problem, we will take into account the following factors: heat transfer between the components of the medium, convective heat transfer, deceleration of thermal effects due to the high heat capacity of the porous body, friction, heat transfer losses, phase transformations of oil components with temperature changes.

The statement of a thermodynamic problem, describing the process of filtration of structured inhomogeneous liquids - paraffinic oils with complex characteristics, presupposes the consideration of filtration conditions and the introduction of some assumptions. Let us assume there is a layer of thickness H, filled with paraffinic oil, the bottom-hole zone of which is waxed. The movement of a two-phase liquid is considered as plane-parallel, because in the conditions of an oil field, the horizontal length exceeds its thickness in many times. The model for analytical research is a borehole of radius r_0, centrally located in a circular layer of a homogeneous porous medium of radius R (figure. 1). The layer is divided into three areas, perpendicular to the well axis. Let us assume that there is no vertical filtration of a homogeneous liquid: for a flat-radial geometry, the speed of liquid flow through the porous walls of the cylinder in the middle layer depends only on r and t and does not depend on the coordinates φ and z: $\vec{v} = \vec{v}(r,t)$. Thermophysical characteristics along the vertical axis – thermal conductivity λ_z and thermal diffusivity a_z – differ from the corresponding properties along the horizontal axis r – λ_r and a_r accordingly. Impermeable rocks with thermophysical parameters (λ_{z1}, a_{z1}, λ_{r1}, a_{r1}) and (λ_{z2}, a_{z2}, λ_{r2}, a_{r2}) are located in the region $z > h$ and $z < -h$, accordingly.

We will take the oil temperature at the initial moment of time as the reference point, the pressure P_0 of the porous body is considered known. At the interfaces between the regions, the temperature will be assumed to be the same and equal to the temperature of the middle part. Heat fluxes at the boundaries of the division of the regions are equal to each other. The values of the physical parameters of an incompressible fluid in a porous fluid – viscosity, Joule-Thomson coefficient, thermal diffusivity, permeability are considered constant and independent of T and P. Liquid is filtered through the rod with a filtration rate w_0. The time is counted from the moment the pressure drop is applied, i.e. from the beginning of the movement of the liquid.

The estimates of the value of thermal conductivity in the direction of liquid filtration show that thermal conductivity determines only local temperature redistribution, and heat transfer over long distances is associated with convection [1]: the characteristic time of temperature equalization is on the order of $l^2/a \sim 10$ s – much less than the typical time for filtration. Consequently, the temperatures of oil, paraffin and the skeleton of the porous fluid at each point coincide; only the radial coordinate of the convective heat transfer rate $U_r \neq 0,\ U_\varphi = 0, U_z = 0$ is different from zero. Thus, to determine at each point, we introduce a single temperature averaged over an elementary microvolume, like other characteristics of the filtration flow [1].

The problem of changing the temperature of a liquid during oscillatory motion in a porous fluid, taking into account phase transitions, is based on the heat balance equation, continuity equations for each phase of paraffinic oil, the equation of motion, initial and boundary conditions. Because oil filtration leads to the appearance of a barothermal effect – a change in temperature during the flow of fluid in a porous fluid in a non-stationary pressure field [2], then this factor must be taken into account when setting the complete problem.

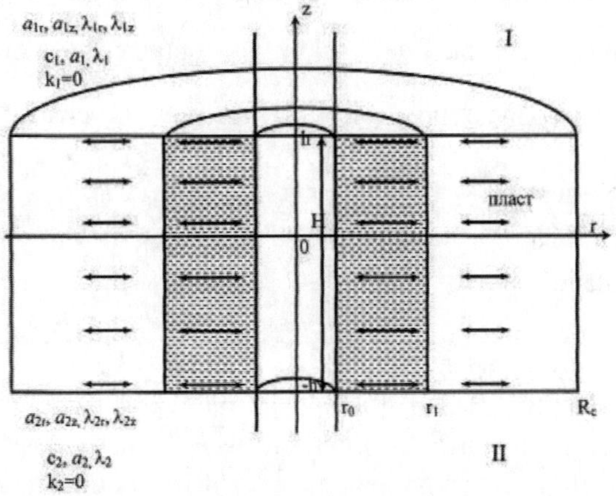

Figure 1. Geometry of the layer containing paraffinic oil in the bottomhole zone

The mathematical formulation of the problem for the first and second regions (subscripts 1 and 2) is represented by the heat conduction equation, and for the middle region – by the equation of the barothermal effect, which takes into account the contribution of phase transitions due to waxing (dewaxing):

$$\frac{\partial T_1}{\partial t} = a_{r1}\frac{1}{r}\frac{\partial}{\partial r}\left(r\frac{\partial T_1}{\partial r}\right) + a_{z1}\frac{\partial^2 T_1}{\partial z^2},\ z > h,\ r > 0,\ t > 0; \quad (1)$$

$$\left[1+\frac{m(1-s)}{c_n}f_1(T)\right]\frac{\partial T}{\partial t}+U_{ef}(r,t)\left[1+\frac{f_1(T)}{c_p\rho}\right]\frac{\partial T}{\partial r}+U_{ef}(r,t)\cdot\varepsilon_{ef}\frac{\partial P}{\partial r}-\eta_{ef}\frac{\partial P}{\partial t}=$$
$$=a_r\frac{1}{r}\frac{\partial}{\partial r}\left(r\frac{\partial T}{\partial r}\right)+a_z\frac{\partial^2 T}{\partial z^2},\quad t>0,\ r>0,\ |z|<h; \tag{2}$$

$$\frac{\partial T_2}{\partial t}=a_{r2}\frac{1}{r}\frac{\partial}{\partial r}\left(r\frac{\partial T_2}{\partial r}\right)+a_{z2}\frac{\partial^2 T_2}{\partial z^2},\ z<-h,\ r>0,\ t>0\,. \tag{3}$$

At the interfaces, the conditions for equality of temperatures and heat fluxes are specified

$$T\big|_{z=h}=T_1\big|_{z=h},\ \lambda_z\frac{\partial T}{\partial z}\bigg|_{z=h}=\lambda_{z1}\frac{\partial T_1}{\partial z}\bigg|_{z=h}, \tag{4}$$

$$T\big|_{z=-h}=T_2\big|_{z=-h},\ \lambda_z\frac{\partial T}{\partial z}\bigg|_{z=-h}=\lambda_{z2}\frac{\partial T_2}{\partial z}\bigg|_{z=-h}. \tag{5}$$

There are no temperature disturbances at the initial moment of time

$$T\big|_{t=0}=T_1\big|_{t=0}=T_2\big|_{t=0}=0\,. \tag{6}$$

We represent the boundary condition in the form

$$\lim T_i\big|_{r+|z|\to\infty}=0\,. \tag{7}$$

On the left side of (2), a thermal function is introduced that describes the intensity of heat sources caused by the phase transformation of paraffin in a certain temperature range [3]:

$$f_1(T)=C^{\bullet}\left(\frac{T-T_{н}}{T_{m}-T_{н}}\right)^2\exp\left(-2\frac{T-T_{н}}{T_{m}-T_{н}}\right). \tag{8}$$

Where $T_н$, T_m are the temperatures of the beginning and maximum of paraffin crystallization, accordingly, the coefficient $C^{\bullet}$ is a constant value for a given oil (for Mangyshlak oil $C^{\bullet}=54\cdot10^{-3}$).

The solution of the hydrodynamic problem for determining the functions of the rate of convective heat transfer $u(r,t)$ and the pressure field $\frac{\partial P}{\partial r}$, it was obtained in work [3]: during the oscillatory motion of a hydrodynamically incompressible fluid, the pressure changes according to the harmonic law

$$P=P_0\frac{\ln r/R}{\ln r_0/R}\sin\omega t,\ \frac{\partial P}{\partial r}=\frac{P_0}{\ln r_0/R}\frac{\sin\omega t}{r},\ \frac{\partial P}{\partial t}=P_0\omega\frac{\ln r/R}{\ln r_0/R}\cos\omega t\,. \tag{9}$$

The source density function has the form [3]:

$$f(r,t)=\left(\frac{\varepsilon}{T_0}\frac{c_{ж}}{c_n}\frac{k}{\mu a}\frac{P_0^2}{\ln^2 r_0/R}\frac{\sin^2\omega t}{r^2}+\eta\frac{m}{T_0}\frac{c_{ж}}{c_n}P_0\frac{\ln r/R}{\ln r_0/R}\cos\omega t\right)\cdot\gamma(r>r_0). \tag{10}$$

The solution is assumed to be bounded at all points $r>0$.

Problem (1) – (7) is nonlinear, since the coefficients at the derivatives of temperature with respect to time and radial coordinate in the central zone $|z|<1$ depend on temperature. The solution of linear and nonlinear boundary value problems for ordinary differential equations and for partial differential equations with a sufficient degree of reliability is carried out by perturbation methods that allow obtaining approximate analytical representations, and the solution is represented by several first terms of the asymptotic expansion, the number of which usually does not exceed two [4]. The asymptotic method presupposes the transition to dimensionless quantities, the choice of the parameter of the asymptotic expansion, the expansion of the sought functions in terms of the parameter of the asymptotic expansion $\hbar$, the formulation and search for a solution to the problem for the zero, first and other approximations. In the particular case of the zero approximation, this method leads to a "lumped capacitance scheme," and the following approximations allow a more detailed study of the temperature field and estimate the magnitude of the resulting error.

According to the above, we introduce the following dimensionless quantities:

$$T=\frac{T_p}{T_0},\ T_1=\frac{T_{1p}}{T_0},\ T_2=\frac{T_{2p}}{T_0},\ z=\frac{z_p}{h},\ r=\frac{r_p}{h},\ t=\frac{a_1 t_p}{h^2},$$

$$\hbar=\frac{\lambda_{z1}}{\lambda_z},\ \Psi=\frac{c_1\rho_1}{c\rho},\ \Lambda=\frac{\lambda_{z2}}{\lambda_{z1}},\ A=\frac{a_r}{a_{z1}},\ A_1=\frac{a_{r1}}{a_{z1}},\ A_2=\frac{a_{r2}}{a_{z1}},\ A_3=\frac{a_{z2}}{a_{z1}}$$

where t – dimensionless time, t_p – dimensional time, $a=\lambda/\rho\cdot c$ – thermal diffusivity, m²/s; λ – coefficient of thermal conductivity, W/(m K); $\varepsilon=1/(\rho c_p)-\eta$ – Joule-Thomson coefficient, K/Pa; $\eta=\alpha VT/c_p$ – adiabatic coefficient, K/Pa; α – coefficient of thermal expansion of the liquid, K^{-1}; $V=1/\rho$ – specific volume of liquid, m³/kg; T is the magnitude of the temperature effect, K; r_0 – radius of the hollow cylinder, m; m – porosity, k – fluid permeability, m², Darcy; μ – dynamic coefficient of viscosity, cP, $c_п$ – volumetric heat capacity of a saturated porous fluid, J/m³ K, $c_ж=c_p\rho$ – volumetric heat capacity of a liquid at constant pressure, J/m³ K; – density of the fluid coming from the layer, kg/m³; $u(r,t)$ – the rate of convective heat transfer. For the parameter of the asymptotic expansion, we formally take

$$\hbar=\lambda_{z1}/\lambda_z.$$

With this choice of the parameter of the asymptotic expansion in the zero approximation, in a particular case, the problem can be reduced to the well-known "lumped capacity scheme" [5].

It is expedient to seek a solution to the problem in the form of Maclaurin series using the expansion in powers of the parameter $\hbar$:

$$T_i=T_i^{(0)}+\hbar T_i^{(1)}+\hbar^2 T_i^{(2)}+\cdots,\ i=0,1,2. \tag{11}$$

The subscripts at the temperature T refer to the number of the region, and the upper indices correspond to the ordinal number of the approximation.

In the surrounding rocks, the zero approximation always gives an excessive temperature value. The following approximations provide more detailed study of the temperature field and allow you to achieve the maximum error. The solution of the temperature problem for the middle part of the layer in the zero approximation, described in detail in [6], has the form:

$$T^{(0)}=\varepsilon_\sigma\beta\left[t-\frac{\sin 2\omega t}{2\omega}\right]+\eta m\frac{c_ж}{c_п}\frac{P_0 x}{l}\sin\omega t, \tag{12}$$

where $\varepsilon_\delta=\varepsilon+\eta m\frac{c_ж}{c_п}$, $\beta=\frac{c_ж}{2c_п}\frac{k}{\mu}\frac{P_0^2}{l^2}$.

After averaging over the period

$$\left\langle T^{(0)}\right\rangle=\frac{c_ж}{2c_п}\frac{k}{\mu}\frac{P_0^2}{l^2}\varepsilon_\sigma t. \tag{13}$$

The complete formulation of the problem for determining the temperature T in the interval of the layer and the corresponding temperatures at T_1 and T_2 in the top and bottom of the layer for the first approximation is written as:

$$\frac{\partial T_1^{(1)}}{\partial t}-A_1\frac{1}{r}\frac{\partial}{\partial r}\left(r\frac{\partial T_1^{(1)}}{\partial r}\right)-\frac{\partial^2 T_1^{(1)}}{\partial z^2}=0,\quad z>1,\ t>0; \tag{14}$$

$$\Phi_1(T)\frac{\partial T^{(1)}}{\partial t}-U_{ef}(r,t)\Phi_2(T)\frac{\partial T^{(1)}}{\partial r}-\Psi\frac{\partial^2 T^{(2)}}{\partial z}=0,\ |z|<1,\ r>0,\ t>0; \tag{15}$$

$$\frac{\partial T_2^{(1)}}{\partial t}-A_2\frac{1}{r}\frac{\partial}{\partial r}\left(r\frac{\partial T_2^{(1)}}{\partial r}\right)-A_3\frac{\partial^2 T_2^{(1)}}{\partial z^2}=0,\ z<-1,\ t>0; \tag{16}$$

The border conditions:

$$\left.\frac{\partial T_1^{(1)}}{\partial z}\right|_{z=1}=\left.\frac{\partial T^{(2)}}{\partial z}\right|_{z=1},\quad \left.\frac{\partial T_2^{(1)}}{\partial z}\right|_{z=-1}=\Lambda\left.\frac{\partial T^{(2)}}{\partial z}\right|_{z=-1} \tag{17}$$

$$T_1^{(1)}\Big|_{z=1} = T_2^{(1)}\Big|_{z=-1} = T^{(1)}; \tag{18}$$

$$T_1^{(1)}\Big|_{t=0} = T_2^{(1)}\Big|_{t=0} = T^{(1)}\Big|_{t=0} = 0; \tag{19}$$

$$\lim T_{1,2}^{(1)}\Big|_{r+|z|\to\infty} = 0\,. \tag{20}$$

Here it is $\Phi_1(T) = 1 + \dfrac{m(1-s)}{c_{\mathrm{n}}} f_1(T), \quad \Phi_2(T) = 1 + \dfrac{f_1(T)}{c_{\mathrm{p}}\rho}$.

To "decouple" the coefficients and solve the thermodynamic problem in the first approximation, we use the Laplace-Carson transformations, while the Laplace operator has the form

$$\hat{L} = \frac{1}{\Psi}\frac{\partial}{\partial t} - \frac{U_{\mathrm{ef}}(r,t)}{\Psi}\frac{\partial}{\partial r}\,. \tag{21}$$

It is also necessary to introduce an additional condition, averaging the temperature in the interval of the layer $-1 < z < 1$ using the integral:

$$\langle T \rangle = \frac{1}{2}\int_{-1}^{1} T\,dz\,. \tag{22}$$

It is easy to verify that the obtained in this way the problem of determining the average temperature in the layer coincides with the problem in the zero approximation. The uniqueness of the solution to the corresponding problems implies:

$$\langle T \rangle = T^{(0)}, \tag{23}$$

therefore, averaging series (11) for $i = 1$, we obtain:

$$\langle T \rangle = T^{(0)} + \hbar\langle T^{(1)} \rangle + \hbar^2\langle T^{(2)} \rangle + \ldots + \hbar^n\langle T^{(n)} \rangle + \ldots \tag{24}$$

or

$$\hbar\langle T^{(1)} \rangle + \hbar^2\langle T^{(2)} \rangle + \ldots + \hbar^n\langle T^{(n)} \rangle = 0\,. \tag{25}$$

From (25) follows an additional condition imposed on the solutions of the first, second and further approximations:

$$\langle T^{(i)} \rangle\Big|_{r=r_1} = 0, \quad i = 1, 2, \ldots \tag{26}$$

Averaging in condition (26) is carried out in some value $r=r_1$ corresponding to some cylindrical surface. Obviously, the best approximation to the desired solution is obtained when r_1 coincides with the radius of the surface on which the boundary conditions are specified [7].

Taking into account condition (26), problem (14) – (20) has a unique solution that allows one to calculate corrections to the zero approximation. The first approximation $T^{(1)}$ allows one to take into account the dependence of the temperature in the interval of the layer on z. However, taking its accounting should not change the averaged temperature values. Therefore, the first approximation must satisfy the condition of equality to zero of its mean values – (26).

The solution to problem (14) – (20) in the first approximation with respect to the parameter $\hbar$ has the form:

$$\begin{aligned}\langle T^{(1)} \rangle = &-\frac{D(1+\Lambda_0)}{8r^2\Lambda_1}(z^2-1) - \frac{D(1-\Lambda_0)}{4r^2\Lambda_1}\left(z - \frac{1-\Lambda_0}{1+\Lambda_0}\right) + \\ &+\left[\frac{D(1+\Lambda_0)}{12r^2\Lambda_1} + \frac{D(1-\Lambda_0)^2}{2r^2\Lambda_1(1+\Lambda_0)(2\Lambda_1-\Lambda_0-1)}\right]\exp\left(\left(\frac{1+\Lambda_0}{2}\right)^2 t\right) erfc\left(\frac{1+\Lambda_0}{2}\sqrt{t}\right) + \\ &+\left[\frac{D(1+\Lambda_0)}{8r^2\Lambda_1}\left(z^2 - \frac{\frac{2}{3}\Lambda_1 + \Lambda_0 + 1}{2\Lambda_1 - \Lambda_0 - 1}\right) + \frac{D(1-\Lambda_0)}{4r^2\Lambda_1}\left(z + \frac{1-\Lambda_0}{3(2\Lambda_1-\Lambda_0-1)}\right)\right] \times \exp(\Lambda_1^2 t)\, erfc(\Lambda_1\sqrt{t}),\end{aligned} \tag{27}$$

$$T_1^{(1)} = N_1 erfc\frac{z-1}{2\sqrt{t}} - N_2 \exp\left(\frac{1+\Lambda_0}{2}(z-1)+\left(\frac{1+\Lambda_0}{2}\right)^2 t\right) erfc\left(\frac{z-1}{2\sqrt{t}}+\frac{1+\Lambda_0}{2}\sqrt{t}\right) -$$
$$- N_3 \exp\left(\Lambda_1(z-1)+\Lambda_1^2 t\right) erfc\left(\frac{z-1}{2\sqrt{t}}+\Lambda_1\sqrt{t}\right), \quad (28)$$

$$T_2^{(1)} = N_1 erfc\frac{|z-1|}{2\sqrt{t}} - N_2 \exp\left(\frac{1+\Lambda_0}{2}|z-1|+\left(\frac{1+\Lambda_0}{2}\right)^2 t\right) erfc\left(\frac{|z-1|}{2\sqrt{t}}+\frac{1+\Lambda_0}{2}\sqrt{t}\right) -$$
$$- N_3 \exp\left(\Lambda_1|z-1|+\Lambda_1^2 t\right) erfc\left(\frac{|z-1|}{2\sqrt{t}}+\Lambda_1\sqrt{t}\right), \quad (29)$$

where the following notation is introduced:

$$\Lambda_0 = \frac{\Lambda}{\sqrt{A_3}},\ \Lambda_1 = \frac{L}{2}+\frac{\Lambda}{\sqrt{A_3}},\ D = \varepsilon_{ef}\frac{k}{\mu}\frac{c_{ж}}{c_n}\left(\frac{P_0}{\ln r_0/R}\right)^2,\ \Psi = \frac{c_1\rho_1}{c\rho},\ \Lambda = \frac{\lambda_{z2}}{\lambda_{z1}},$$

$$N_1 = \left(\frac{D(1+\Lambda_0)}{8r^2\Lambda_1}+\frac{D(1-\Lambda_0)^2}{4r^2\Lambda_1(1+\Lambda_0)}\right)\left(1+\frac{1+\Lambda_0}{2\Lambda_1-\Lambda_0-1}\right)-\frac{D(1+\Lambda_0)}{24r^2\Lambda_1}\left(1+\frac{1+\Lambda_0}{2\Lambda_1-\Lambda_0-1}\right)-$$
$$-\frac{D(3-\Lambda_0)}{8r^2\Lambda_1}-\frac{D(1+\Lambda_0)}{8r^2\Lambda_1}\left(\frac{1+\Lambda_0}{2}+\frac{1-\Lambda_0}{1+\Lambda_0}-\frac{\Lambda_1}{3}\right)\frac{2}{2\Lambda_1-\Lambda_0-1},$$

$$N_2 = \left(\frac{D(1+\Lambda_0)}{8r^2\Lambda_1}+\frac{D(1-\Lambda_0)^2}{4r^2\Lambda_1(1+\Lambda_0)}\right)\left(1+\frac{1+\Lambda_0}{2\Lambda_1-\Lambda_0-1}\right)+\frac{D(1+\Lambda_0)}{24r^2\Lambda_1}\left(1-\frac{1+\Lambda_0}{2\Lambda_1-\Lambda_0-1}\right),$$

$$N_3 = \frac{D(3-\Lambda_0)}{8r^2\Lambda_1}+\frac{D(1+\Lambda_0)}{8r^2\Lambda_1}\left(\frac{1+\Lambda_0}{2}+\frac{(1-\Lambda_0)^2}{1+\Lambda_0}-\frac{\Lambda_1}{3}\right)\frac{2}{2\Lambda_1-\Lambda_0-1}.$$

The final solution for the two found approximations will be written in the form:

$$T = T^{(0)} + \frac{\lambda_{z1}}{\lambda}T^{(1)},\ T_1 = T_1^{(0)} + \frac{\lambda_{z1}}{\lambda}T_1^{(1)},\ T_2 = T_2^{(0)} + \frac{\lambda_{z1}}{\lambda}T_2^{(1)}. \quad (30)$$

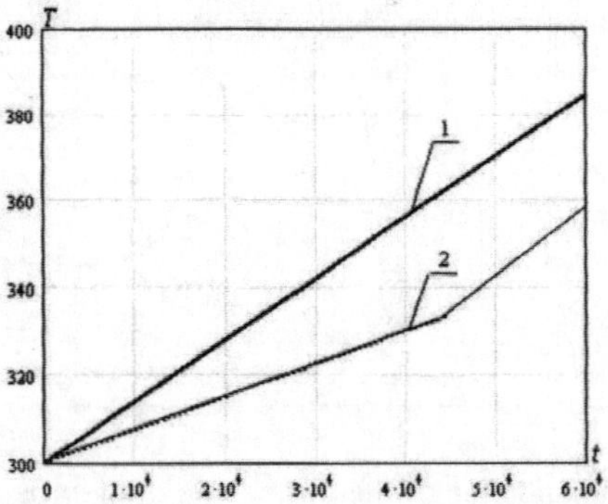

Figure 2. The contribution of phase transitions due to the dissolution of the solid phase of paraffin: 1 - temperature effect without phase transitions, 2 – with phase transitions

The dependence of the temperature of the filtrating paraffinic oil on time, with the accounting of phase transitions, is shown in figure 2. Calculations were carried out for Mangyshlak oil with a paraffin content of 20%.

For the numerical experiment, the following parameter values are accepted: m=0.2, s=0.9, ω=0.26s^{-1}, k=6·10^{-11}m^2, c_n=3·10^{6}J/m^3·K, $c_{жс}$=2.1·10^{6}J/m^3·K, ε=4·10^{7}K/Pa, q=2.3·10^{5}J/кg, C^*=54·10^{-3}, μ=141.7·10^{-3}Pa·s.

Comparison of experimental measurements and theoretical averaged temperature dependence for viscous liquid filtration shows that the calculation results agree with a sufficient degree of accuracy with the actual crystallization curves for any paraffinic oil and petroleum product [3], [8].

Figures 3 – 5 show the calculations of the dimensionless temperature T from the vertical coordinate z for various values of the parameters included in the solution and the dimensionless time. Number 1 denotes the graphs corresponding to the zero approximation, number 2 – the first expansion coefficient, and 3 – the first approximation. The physical and chemical properties of the rocks are the same, $r = r_0$, $r_1 = R_k$.

The temperature is constant in the interval of the layer $-1 < z < 1$ for the zero approximation – curve 1. In the central part of the layer for short times, the zero approximation describes the temperature distribution with a deficiency, and at the edges of the layer with an excess. Taking into account the correction, the solution in the first approximation more realistically describes the temperature distribution in the layer. Zero and first approximations are sufficient for most practical calculations.

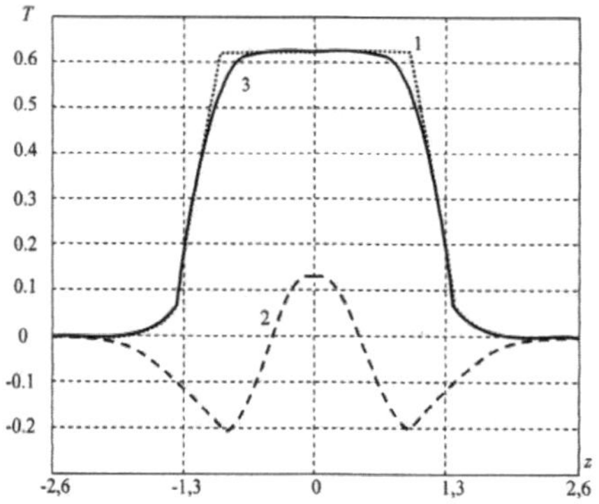

Figure 3. The graph of the dependence of the dimensionless temperature T in the z coordinate in dimensionless coordinates: t=0.1, R_k=60 м, ε=0.04, Ψ=36.841

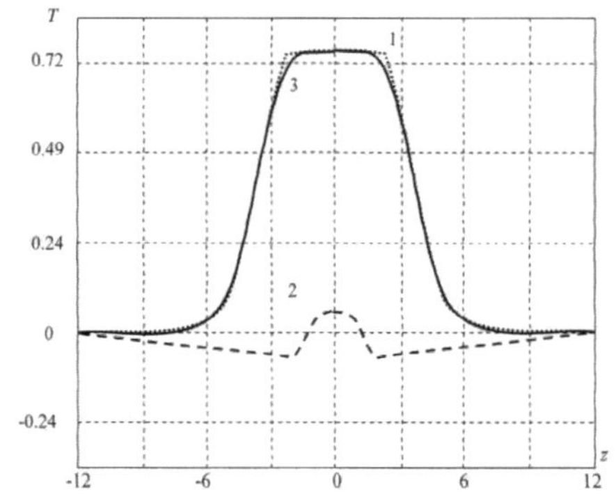

Figure 4. The graph of the dependence of the dimensionless temperature T in the z coordinate in dimensionless coordinates: t=1, R_k=60 м, ε=0.004, Ψ=32.841

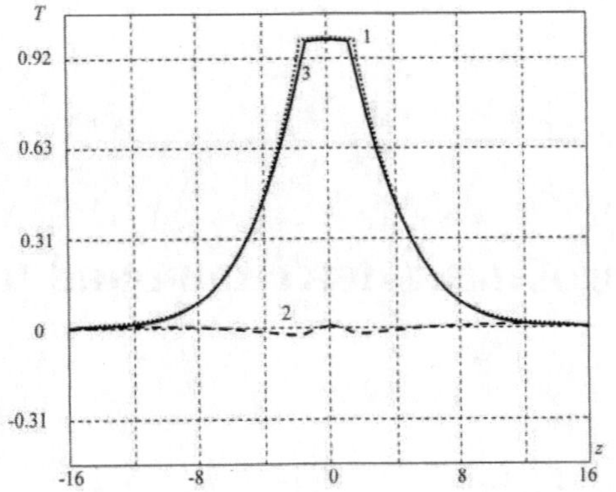

Figure 5. The graph of the dependence of the dimensionless temperature T in the z coordinate in dimensionless coordinates: t=10, R_k=60 м, ε=0.4, L=36.841

The calculations of spatio-temporal temperature distributions during the oscillatory motion of a liquid, taking into account phase transitions, show that during oscillatory motion, a monotonous heating of a porous fluid and a filtering viscous liquid occurs, and due to the dissolution of paraffins, the rate of temperature rise decreases with time – with a paraffin content in oil of about 20%, it occurs reduction of the temperature effect by 1.5 – 2 times.

References

[1] Barenblatt G I, Entov V M and and Ryzhik V M 1972 *The theory of unsteady filtration of liquid and gas* (Moscow: Nedra) p 288
[2] Filippov A I 1989 *Well transient thermometry* (Saratov: Saratov state university) p 116
[3] Efimova G F 2020 *Mathematical modeling of temperature processes in filtration-wave fields* (Ufa: USPTU Publishing House) p 106
[4] Nayfe A Kh 1976 *Perturbation methods* (Moscow: Mir) p 455
[5] Rubinstein L I 1971 *Temperature fields in oil layer* (Moscow: Nedra) p 276
[6] Filippov A I. and Efimova G F 1997 *High Temperature* **35(4)** 549-52
[7] Ishmukhametova A A and Khusainov I G 2008 *Proceedings of the Institute of Mechanics of the Ufa Scientific Center of the Russian Academy of Sciences* **6** 89-94
[8] Bilalova D N, Kireeva N A, Levina T M, Zharinov Y A and Uimanova I P 2020 Digital educational resources in the study of humanities subjects in a technical university *Advances in Social Science Education and Humanities Research Proc. Int. Sci. Conf. "Digitalization of Education: History, Trends and Prospects"* pp 315–19

Selecting laser fluorosensor detection band to monitor oil pipeline leaks

Yu V Fedotov, M L Belov[1], D A Kravtsov, K S Titarenko and V A Gorodnichev

Bauman Moscow State Technical University (BMSTU), 2nd Baumanskaya str., Moscow, 105005, Russia

[1]E-mail: belov@bmstu.ru

Abstract. Detection bands of the laser fluorosensor to detect oil pipeline leaks have been studied at the fluorescence excitation wavelength 355 nm. It is shown that a diversity of the fluorescence spectra of different oils and plant species leads to the need to adapt the number and position of spectral detection bands of the laser fluorosensor for the specific types of the pipeline oil and vegetation along the pipe way. With no interfering factors (that emerge false alarms) to monitor along the pipe way, only one spectral channel (at maximum position of the oil fluorescence spectrum) can be used. However, there can be a need to use both the two spectral channels (the channel 1 at maximum position of the oil fluorescence spectrum, and the channel 2 at maximum position of the vegetation fluorescence spectrum within ~ 680…690 nm or ~ 730…740 nm) and the tree spectral channels (the channel 1 at maximum position of the oil fluorescence spectrum, the channel 2 at maximum position of the vegetation fluorescence spectrum within ~680…690 nm or ~ 730…740 nm, and the channel 3 in the spectral band of 625-650 nm).

1. Introduction

Presently, monitoring of oil pipeline leaks is a critical task because a significant amount of oil, travelling through the pipelines, is lost as a result of their worn-out state and mechanical damages [1,2]. The leak detection systems (LDS) available in the pipelines have LOD (a limit of detection) of a tenth of one percent from the oil-flow and the lower oil leaks are undetectable [3].

A promising operational method for monitoring the low oil-line leaks is to provide remote airborne monitoring of oil pollution on the ground surface along the pipe way.

Currently, the most efficient techniques for remote monitoring of oil pollution on the water and land surface are laser fluorescence methods [4-9].

The paper focuses on the study of possible detection spectral bands of a laser fluorosensor for oil pollution monitoring on the ground surface.

2. Problem description

A problem of LIF monitoring of oil pollution on the ground surface is more complicated than that of LIF oil pollution monitoring on the water surface.

A laser-induced fluorescence (LIF) signal from the oil pollution on the ground surface is less than the LIF signal from the oil spill on the water surface (oil is absorbed by the soil, flows downhill, contaminated areas are covered with dust, etc.). And still, the fluorescence signal from the oil spill on the ground surface can be comparable in magnitude with the fluorescence signal from the landscape elements (generally from vegetation and water objects).

Therefore, to differentiate the oil pollution areas from the landscape elements (preselecting the usable detection spectral bands of the laser fluorosensor) it is necessary to involve data about a LIF spectrum shape both of oil and of vegetation and water objects.

To study possible detection spectral bands of the laser fluorosensor for monitoring the oil pollution on the ground surface, the LIF spectra of oil, vegetation and water objects have been experimentally investigated, and other authors' experimental studies have been analysed [10-23].

A study of possible detection bands (their spectral position and number) for the LIF method to detect oil pipeline leaks using a fluorescence excitation wavelength of 355 nm is given below.

3. Experimental studies of LIF spectra of oil, vegetation, and water objects at excitation wavelength 355 nm

To measure the LIF spectra of oil, vegetation and water objects for the fluorescence excitation at the wavelength of 355 nm we used a laboratory setup (figure 1).

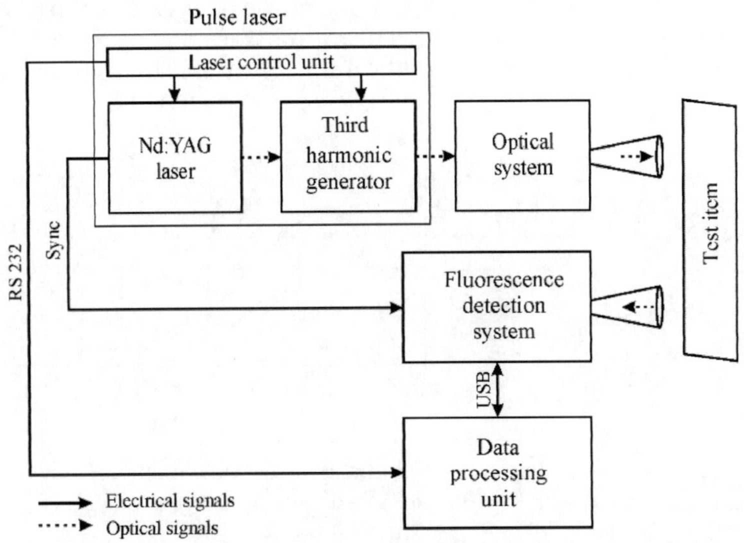

Figure 1. Schematic diagram of the laboratory setup to study the LIF spectra.

The laboratory setup employed a Q-switched diode-pumped solid-state pulse Nd: YAG laser at a wavelength of 1064 nm (pulse length <8 ns, repetition rate up to 500 Hz) and a module of the third harmonic (with the 0.8 mJ laser pulse at the third harmonic). A subsystem to detect fluorescence emission was based on the polychromator and the highly sensitive ICCD and enabled the fluorescence spectra detection in the band of 375 to 790 nm with the spectral resolution of 5 nm. A lens diameter of detector was 15 mm. The distance from the laser source to test item was ~ 1.4 m. A laser exciting beam diameter and detector's field of view in the test item plane were ~ 20 mm.

Equipment calibration involved a wavelength calibration of the polychromator and a detection limit calibration of the fluorescence detection system. To control the experimental laboratory setup calibration, we used a Raman-scattering spectrum of distilled water.

A laboratory setup management was based on software in visual development environment LabVIEW.

Figure 2 illustrates some examples of the LIF spectra of crude oils measured by the laboratory setup.

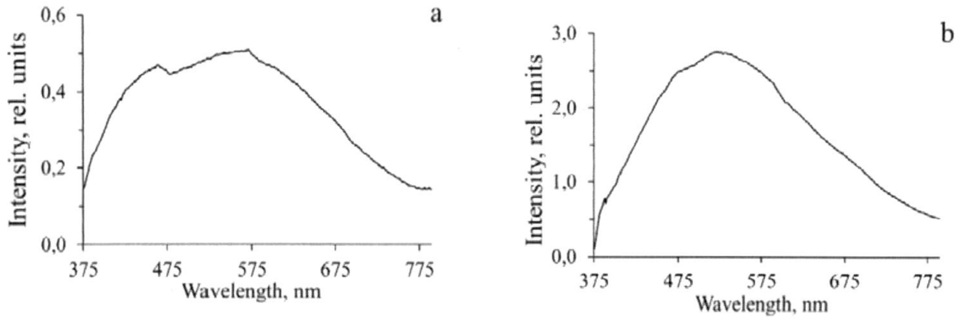

Figure 2. LIF spectra of Almetyevsk oil (2a) and stock tank oil of the Moscow refinery (2b).

It is seen from figure 2 that for oils measured, the LIF spectra are maximum in the ~ 500 - 575 nm spectral band and have a wide spectral width.

Figure 3 (for exciting fluorescence wavelength 355 nm) shows the most typical examples of the experimental LIF spectra of other oils from different oilfields [10-15] (different authors measured spectra using various equipment, so to illustrate them in the same figure we have normalized a value at maximum of each spectrum).

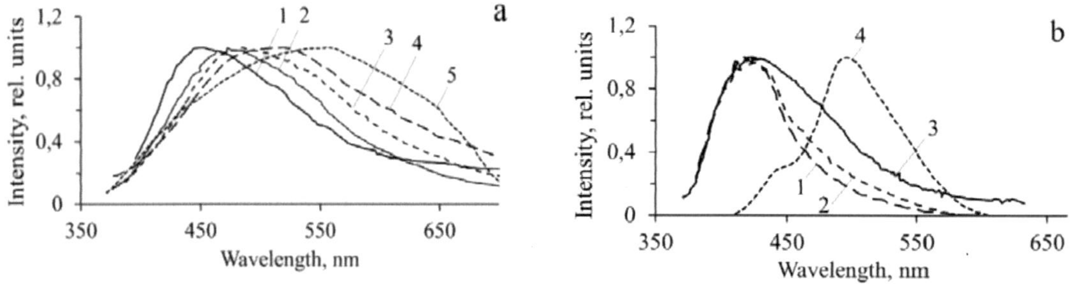

Figure 3. Oil from different oilfields. 3a: 1 – Absheron, 2 –Yakutia, 3 – Western Siberia, 4 – Ust-Kut, 5 – Arabian oil; 3b: 1 – light oil, 2 – heavy oil, 3 – oil from Azerbaijan, 4 – Asgard oil.

Figure 3 shows the LIF spectra of oils, which are similar to those in figure 2 in terms of the maximum position (~ 500 - 575 nm) and the spectral width (curves 4,5). The other spectra in figure 3a (curves 1 – 3) have a slightly less spectral width, and their maximum position is appreciably shifted towards the shorter waves (~450 – 500 nm).

Figure 3b shows the fluorescence spectra, which are far different from those illustrated in figures 2 and 3a. They have a much lesser spectral width, and their maximum position can be within ~ 420 nm or ~ 490 nm.

Of course, for monitoring the oil pipeline leaks it is appropriate to choose the fluorosensor detection band at maximum position of the oil fluorescence spectrum. Because of a large variety of the oil fluorescence spectra from different olifields (figures 2,3) a detection spectral band of the laser fluorosensor for monitoring the pipeline leaks should be adapted to the specific oil transported.

However, this is not the only difficulty to create a laser fluorosensor for monitoring the oil pipeline leaks. The key interfering factor for oil pollution monitoring on the ground surface (due to oil pipeline leaks) is vegetation. For a task of oil pipeline leaks monitoring, this interfering factor leads to the fact that employing the single-channel method to detect LIF (at oil fluorescence spectrum maximum position) can be non-optimum (due to high-risk of false alarms).

Illustrations below show LIF spectra of different species of vegetation measured by the laboratory setup (figure 4).

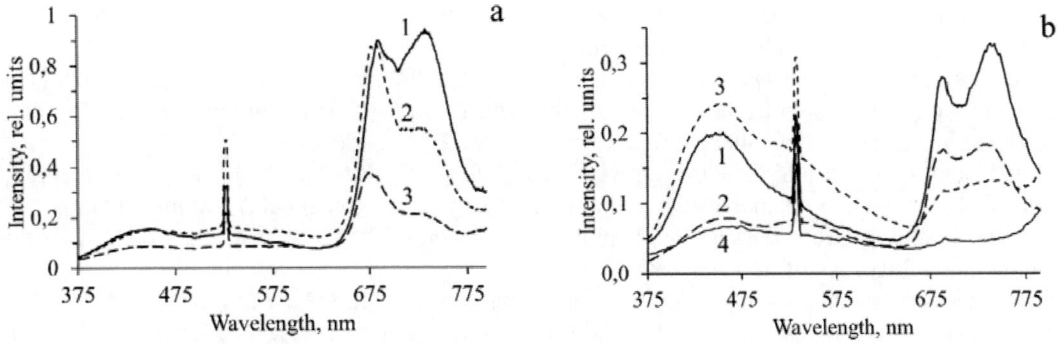

Figure 4. Fluorescence spectra of vegetation. 4a: leaves of hawthorn, 1 – green, 2 – yellow, 3 – brown; 4b: 1 – green grass, 2,3 – cut grass after 7 and 11 days, 4 – dried grass.

It is clearly seen from figure 4 that the LIF spectrum of live vegetation (figure 4a: curves 1-3, figure 4b: curves 1-3) has always typical maxima in the spectral bands of ~ 670…695 nm and ~ 720…745 nm.

The above data also show (that is well known) that in the spectral band of 375 - 650 nm the vegetation fluorescence is sometimes weak and is sometimes significant (i.e. there is fluorescence of vegetation in the spectral area of oil fluorescence). Herewith, in most cases (figure 4a: curves 1-3, 4b: curves 1-2) for live vegetation the fluorescence in the spectral band of ~ 670…695 nm and in the spectral band of 720…745 nm is more than in the spectral band of 375 - 650 nm.

Therefore, to differ the oil fluorescence from the vegetation one, two spectral detection bands (with the same spectral width) can be used: 1 – at maximum position of the fluorescence spectrum of a specific oil, 2 - ~ 670…695 nm or 720…745 nm.

For oils, a fluorescence signal in the spectral band of 670…695 nm or ~ 720…745 nm will be less than a fluorescence signal in the spectral band (with the same width) at maximum position of the oil fluorescence spectrum (figures 2,3).

For vegetation, in the spectral band of 670…695 nm or ~720…745 nm a fluorescence signal will be mostly more than a fluorescence signal in the spectral band (with the same spectral width) at maximum position of the oil fluorescence spectrum (figure 4a - curves 1-3, figure 4b – curves 1,2). The spectral curve 3 in figure 4b for vegetation can be also distinguished from the spectral curve of any kind of oil owing to introducing some threshold ratio of the fluorescence signal in the spectral band of 670…695 nm or ~ 720…745 nm to the fluorescence signal in the spectral band at maximum position of the oil fluorescence spectrum.

Figure 5 shows the most typical illustrations of the experimental spectra of other plants [16-23].

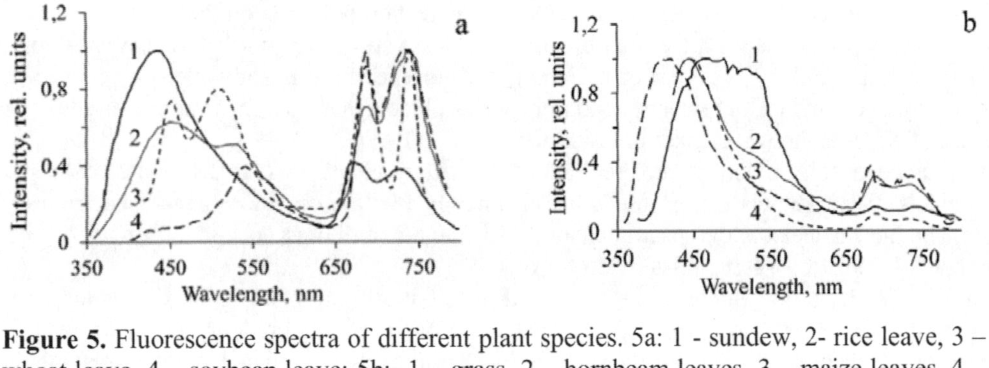

Figure 5. Fluorescence spectra of different plant species. 5a: 1 - sundew, 2- rice leave, 3 – wheat leave, 4 – soybean leave; 5b: 1 – grass, 2 – hornbeam leaves, 3 – maize leaves, 4 – poplar leaves.

It is seen from figure 5 that maximum position of spectrum vegetation (within 375 – 650 nm) can change in the band from ~ 440 to 540 nm (figure 5a).

And herein, there may be situations (like in figure 4) when the typical maximum in the spectral band of 375 - 650 nm can be significantly more than the typical (but weak) maximum in the spectral bands of ~ 680…690 nm or ~ 730…740 nm (figure 5b and curve 1 in figure 5a).

This leads to the fact that for a task of monitoring the oil pipeline leaks a dual-channel method (a channel 1 is at maximum position of the oil fluorescence spectrum, a channel 2 is at maximum position the of vegetation spectrum within ~ 680…690 nm band or ~ 730…740 band) as well can be non-optimum due to high-risk of false alarms.

In this case there may be a need to introduce the third channel in the spectral band of 625-650 nm (between the channels 1 and 2) in order to add-on monitoring the fluorescence spectrum of the area being monitored. The fluorescence signal for all plant species in the spectral band of 625-650 nm will be less or, at least, roughly the same as the signal in the spectral bands of ~ 670…695 nm or ~720…745 nm. At the same time, the fluorescence signal for all oils in the spectral band of 625-650 nm will be more than the signal in the ~ 670…695 nm or ~ 720…745 nm spectral bands.

As to the land-based water objects (rivers, ponds, lakes, puddles, etc.), fluorescence spectra of water objects may also resemble those of some oils. Figure 6 shows illustrations of the LIF spectra of various water objects (figure 6a – spectra of inland fresh water measured by the laboratory setup, figure 6b – spectra of river water [10,13]).

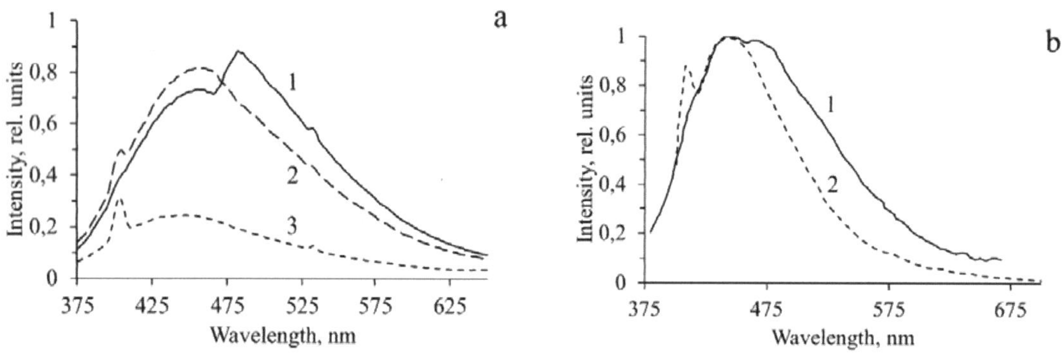

Figure 6. Fluorescence spectra of water objects. 6a: 1 – green pond water, 2 – Moscow Canal water, 3 –Lebedyansky pond water (relatively clear) in Moscow; 6b: 1,2 – river water.

However, a magnitude of the fluorescence signal from the water object (except for heavily polluted or turning green water objects) is significantly less than that of the fluorescence signal from oil (i.e. a threshold algorithm can be used just as for methods to detect oil pollution on the water surface).

Besides, the spectra shapes (of water objects and oil) are still different (different typical degrees of spectra growth and roll-off, position of maxima), and for specific oils and water objects they can be divided (moreover, in the fluorescence spectra of water objects there is a maximum caused by Raman-scattering of water at the wavelength of ~ 403 nm).

For instance, for oils which spectra are illustrated in figures 2 and 3a, to differentiate between oil and water objects, two channels can be used: the channel 1 at of the oil fluorescence spectrum (~500 – 575 nm) and the channel 2 in the spectral band of 625-650 nm (a fluorescence spectrum roll-off between the channels 1 and 2 for water is faster than for oil).

Thus, a diversity of the fluorescence spectra of different oils and plant species leads to the need to adapt the number and position of the spectral detection bands of a laser fluorosensor for oil pollution monitoring on the ground surface for the specific types of the pipeline oil and vegetation along the pipe way.

With no interfering factors (that emerge false alarms) to monitor oil pollution along the pipe way, only one spectral channel (at maximum position of the oil fluorescence spectrum) can be used. However, there can be a need to use both the two spectral channels (the channel 1 at maximum position of the oil fluorescence spectrum, and the channel 2 at maximum position of the vegetation fluorescence spectrum within ~ 680…690 nm or ~ 730…740 nm) and the tree spectral channels (the channel 1 at maximum position of the oil fluorescence spectrum, the channel 2 at maximum position of the vegetation fluorescence spectrum within ~ 680…690 nm or ~ 730…740 nm, and the channel 3 in the spectral band of 625-650 nm).

The number of detection channels, the central wavelengths of the spectral detection channels and their width can be finally selected only from results based on the experimental study of the pipeline oil and vegetation spectra in the oil piping region.

4. Conclusion

Thus, possible detection spectral bands of the laser fluorosensor to detect oil pipeline leaks have been studied at the fluorescence excitation wavelength 355 nm. It is shown that a diversity of the fluorescence spectra of different oils and plant species leads to the need to adapt the number and position of spectral detection bands of the laser fluorosensor for the specific types of the pipeline oil and vegetation along the pipe way. With no interfering factors (that emerge false alarms) to monitor along the pipe way, only one spectral band (at maximum position of the oil fluorescence spectrum) can be used. However, there can be a need to use both the two spectral bands (the band 1 - at maximum position of the oil fluorescence spectrum, and the band 2 - at maximum position of the vegetation fluorescence spectrum within ~ 680…690 nm or ~ 730…740 nm) and the tree spectral bands (the band 1 - at maximum position of the oil fluorescence spectrum, the band 2 - at maximum position of the vegetation fluorescence spectrum within ~ 680…690 nm or ~ 730…740 nm, and the band 3 - in the spectral range of 625-650 nm).

References

[1] Adejoh O F 2014 *World Environment* **4(3)** 93
[2] Sivokon S and Andreev N 2012 *Int. J. Corros. Scale Inhib.* **1(1)** 65
[3] Zhang J, Hoffman A, Kane A and Lewis J 2014 10th *International Pipeline Conference* **1** 1
[4] Pashayev A, Tagiyev B, Allahverdiyev K, Musayev A and Sadikhov I 2015 *Proceedings of SPIE* **9810** 981018-1
[5] Fedotov Yu V, Belov M L, Kravtsov D A and Gorodnichev V A 2019 *Journal of Optical Technology* **86(2)** 23
[6] Sasano M 2012 *Industrial applications of laser remote sensing* ed T Fukuchi, T Shiina (Sharjah: Bentham Science Publishers Ltd) p 89
[7] Brown C E 2011 *Oil Spill Science and Technology* ed M Fingas (Burlington: Elsevier Inc.) p 171
[8] Steffens J, Landulfo E, Courrol L C and Guardani R 2011 *Journal of Fluorescence* **21(3)** 859
[9] Utkin A B, Lavrov A and Vilar R 2011 *Proceedings of SPIE* **7994** 799415-1
[10] Luedeker W, Guenther K P and Dahn H G 1995 *Proc. SPIE* **2504** 426
[11] Malecha M, Bessant C and Saini S 2003 *Appl. Spectrosc.* **57(8)** 1042
[12] Patsayeva S, Yuzhakov V, Varlamov V, Barbini R, Fantoni R, Frassanito C and Palucci A 2000 *EARSeL eProceedings* **1** 106
[13] Karpicz R, Dementjev A, Kuprionis Z, Pakalnis S, Westphal R, Reuter R and Gulbinas V 2005 *Lithuanian Journal of Physics* **45(3)** 213
[14] Palombi L, Lognoli D and Raimondi V 2013 *Proc. of SPIE* **8887** 88870F-1
[15] Hegazi E, Hamdan A and Mastromarino J 2005 *The Arabian Journal for Science and Engineering* **30(1B)** 3
[16] Lichtenthaler H K, Subhash N, Wenzel O and Mieht J 1997 *Intern. geoscience and remote sensing symp.* **4** 1799
[17] Snels M, Guarini R and Dell'Aglio M 2000 *Proc. of SPIE* **4070** 100
[18] Meyer S, Cartelat A, Moya I and Cerovic Z G 2003 *Journal of Experimental Botany* **54(383)** 757

[19] Mishra K B and Gopal R 2005 *Gen. Appl. Plant Physiology* **31(3-4)** 181
[20] Maurya R,and Gopal R 2008 *Applied spectroscopy* **62(4)** 433
[21] Mishra K B and Gopal R 2008 *International Journal of Remote Sensing* **29(1)** 157
[22] Mo C, Kim M S, Kim G, Cheong E J, Yang J and Lim J 2015 *J. of Biosystems Eng.* **40(4)** 335
[23] Yang J, Gong W, Shi1 S, Du1 L, Sun J, Song S, Chen1 B and Zhang Z 2016 *Scientific Reports* **6(1)** 28787

IOP Conf. Series: Materials Science and Engineering 1155 (2021) 012075 doi:10.1088/1757-899X/1155/1/012075

Problems of hydrate formation in production oil products from gas condensate wells

O S Ivanov, V G Afanasenko, R T Zaripov, M A Nigmatullin and A Y Fedorov

Ufa State Petroleum Technological University, 1, Cosmonauts street, Ufa, 450062, Russian Federation

E-mail: olegiv099@mail.ru

Abstract. One of the most important problems in gas production at fields is hydrate formation - a process in which hydrates, settling on the inner walls of technological equipment, reduce their throughput, and therefore it may be necessary to stop the entire technological process for the purpose of unscheduled repairs, which leads to significant economic losses. The article presents the results obtained in the course of research, analytical work, namely: the assessment of the share of hydrocarbon production by region; analyzes of the main methods of combating hydrate formation; development of a methodology for comparing methods of combating hydrate formation among themselves; determining the most acceptable geography for the application of a particular method; forecast of the popularity of the method for 5 years.

1. Introduction

Currently, the oil and gas industry is the leading industry in Russia [1]. Oil and gas are of great importance for the development of the national economy of the country. Natural gas and heavy residues from oil refining are a cheap and convenient form of energy and domestic fuel. Oil serves as a raw material for the production of a large number of types of energy fuel: gasoline, kerosene, jet and diesel fuels for internal combustion engines, fuel oil, various types of lubricants, bitumen, synthetic fatty acids, etc. Today Russia is the largest producer of a driver and supplier of hydrocarbons in the global oil and gas market. Figure 1 shows the distribution of hydrocarbon production volumes by the largest enterprise of PJSC Gazprom in the Russian Federation in 2018 [2].

Figure 1. Volumes of hydrocarbon roduction by PJSC Gazprom [1].

According to figure 1, a hydrocarbon production diagram was drawn up, clearly showing in which regions the largest amount of hydrocarbons is produced (figure 2).

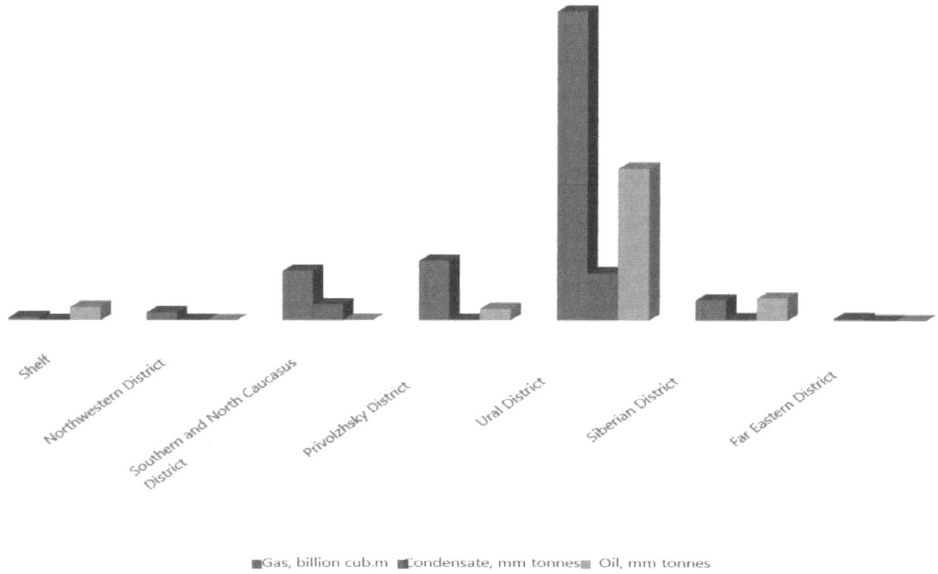

Figure 2. Shares of hydrocarbon production volumes by region.

Technological processes of production, storage, transportation and processing of oil and gas can be complicated by the formation of hydrates [3-4]. Hydrates are solid crystalline com-pounds formed as a result of the interaction of water with hydrocarbons under certain temperature and pressure conditions, mainly due to Van der Waals forces [5-7]. This compound is very unstable and any violation of the equilibrium state of hydrate formation will lead to their decomposition. The phenomenon of hydrate formation is characterized by a sharp decrease in the transfer capability of pipelines, which in turn can lead to an emergency shutdown of pipeline operation. [8].

2. Methods to combat hydrate formation.

To date, to prevent the formation of hydrates and eliminate the already formed hydrate plugs, there are a number of methods [9-10], which are divided into three groups according to the mechanism of action. The classification of existing methods is shown in figure 3.

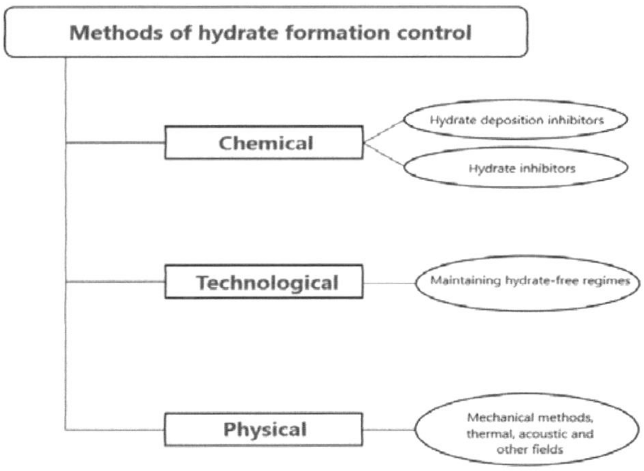

Figure 3. Classification of methods to combat hydrate formation.

The chemical method [9] involves the introduction of inhibitors into the gas flow. Inhibitors are chemicals that slow down or eliminate hydrate formation [11]. Methanol, glycols, calcium chloride solution and others are used as inhibitors. The advantages of this method are applicability in the Far North, high manufacturability, and relative cheapness. The disadvantages are the toxicity of substances, which can lead to poisoning of people, the environment, the possibility of corrosion of equipment and pipelines.

The technological method [9-14] consists in maintaining a hydrate-free mode of the process. This can be achieved by selecting the appropriate working flow rates of the well, at which the temperature at the wellhead will be higher than the equilibrium temperature of hydrate formation. Also, this method can be implemented through the introduction of the so-called "smart" wells - these are wells equipped with sensor systems and/or control valves that allow you to control the operating parameters. This innovation will allow to remotely, timely and quickly respond to all changes in the parameters of the mining process and prevent the emergence of conditions for the formation of hydrates. Smart wells are very widespread in the world, but in Russia these wells are not yet in great demand (figure 4). This method is well applicable in fields with a warm, temperate climate.

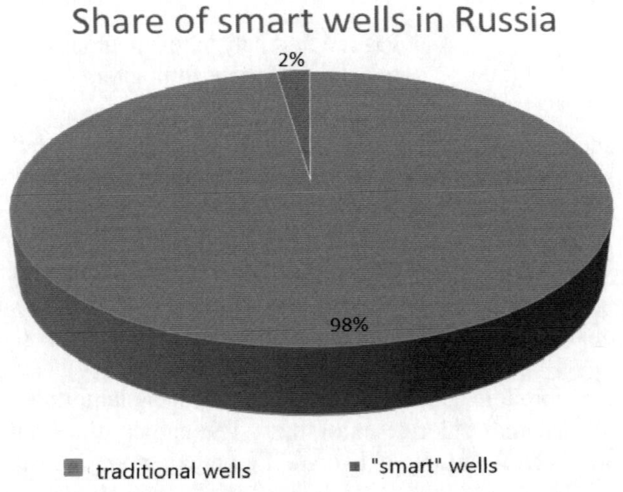

Figure 4. The share of smart wells in Russia and the world.

It is also possible to avoid hydrate formation by heating the gas at specialized stations us-ing steam, or in heat exchangers with other heat carriers. These heaters are installed in places of possible hydrate formation. This method is applicable only for short sections of pipelines and requires significant operating and economic costs.

To destroy the already formed hydrates, the pressure reduction method is used. It is based on a short-term pressure drop below the hydrate decomposition pressure and is applicable only in emergency situations.

The physical method involves the mechanical removal of hydrates by scraping, or by heating the section of the pipeline where hydrates are formed. The application of this method is due to the partial or complete shutdown of equipment, which leads to large economic losses. In this connection, today the physical method is practically not used anywhere.

3. Methodology and comparison of methods to combat hydrate formation.

In the course of the research work, a methodology for comparing the technological and chemical methods with each other was developed. The methodology is as follows: assessment criteria are assigned - economic, impact on human health, environmental (damage to nature as a result of an accident), temperature criterion (the effectiveness of the method under different temperature conditions, from low to high temperatures). In the course of the analysis, it was assumed that all criteria are equal. A comparison diagram of the methods is shown in figure 5.

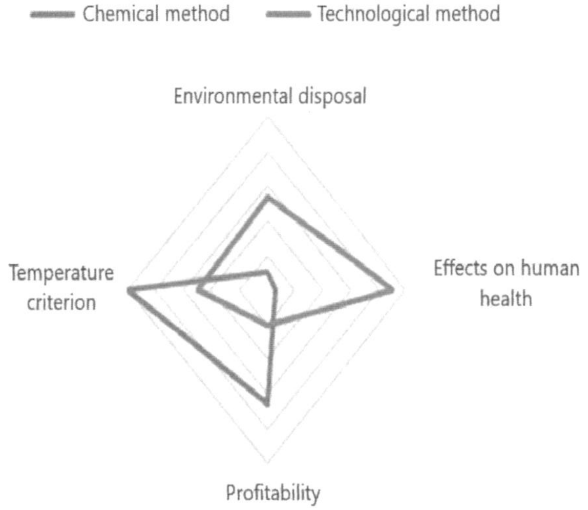

Figure 5. Analysis of methods to combat hydrate formation.

The temperature criterion was formed in such a way that the use of the chemical method is possible both in extremely negative and extremely positive temperatures, and the technological one - only in extremely positive ones.

Based on the comparative analysis, it can be concluded that when all the evaluation criteria are taken into account, these methods are approximately equivalent. However, it should be noted that at the present stage in Russia, the key factor for choosing a method for combating the formation of hydrates is temperature, since most of the production of hydrocarbons occurs in harsh climatic conditions (see figures 1, 2).

Let's carry out a comparative analysis of the inhibitors used in practice. The following inhibitors are mainly used in production: glycols, methanol, calcium chloride and others. The choice of inhibitor depends on its ability to lower the equilibrium temperature of hydrate formation, cost, industrial regeneration, water solubility and toxicity.

Figure 6 shows diagrams of comparative analysis of inhibitors by various parameters. Each parameter was assessed on a 5-point scale: 1 point - unsatisfactory, 5 points - excellent.

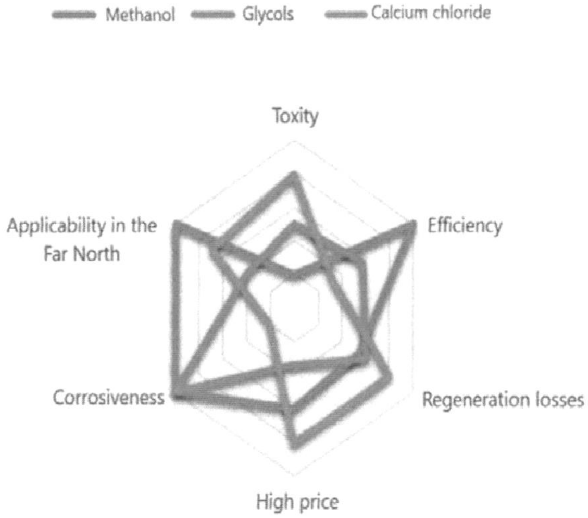

Figure 6. Results of the analysis.

According to figure 6, it can be said that the most preferred inhibitor is methanol [15-17] (largest area of the figure).

Thus, in the course of the research work, a map of the Russian Federation was proposed, which marked the zones where the most optimal use of chemical (zone A) and technological (zone B) methods (figure 7).

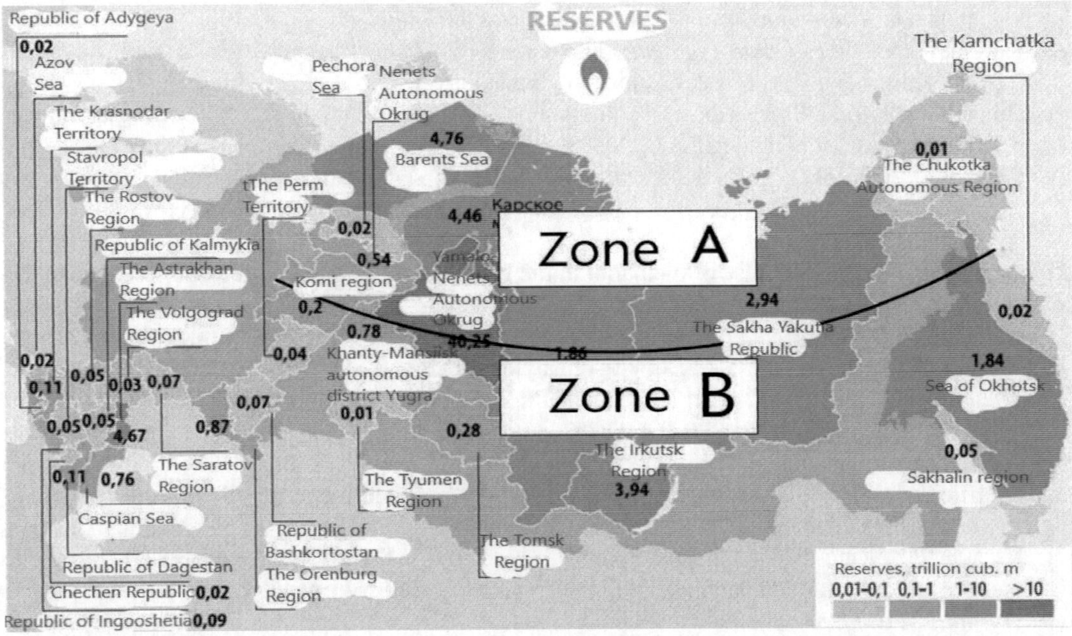

Figure 7. Map of distribution of hydrocarbon production volumes in the Russian Federation.

4. Conclusion

In the conclusion of the review study, the following conclusions were formulated:

- To date, the most optimal methods for preventing hydrate formation are chemical and technological.
- A comprehensive comparison of chemical and technological methods shows that the technological method is the most environmentally friendly, but at the same time, it is much more expensive than the chemical one, and much less effective at low temperatures.
- Comparison of chemical inhibitors showed that it is best to use methanol, but one should take into account its high toxicity (refers to substances of the 3rd hazard class).

Taking into account the above conclusions, a forecast was made according to which, over the next five years, the most common method of combating hydrate formation will be chemical, and the most widely used inhibitor is methanol.

References

[1] Gureeva M A 2012 *Economics of the oil and gas industry: a textbook for students of institutions of sec-ondary vocational education* 2nd ed (Moscow: Publishing Center "Akademiya") p 240

[2] *Gazprom in figures 2014-2018* Retrieved from: https://www.gazprom.ru/f/posts/72/692465/gazprom-in-figures-2014-2018-ru.pdf

[3] Zakozhurnikov Yu A 2010 *Preparation of oil and gas for transportation* (Volgograd: In-Folio) p 176

[4] Katz D L, Cornell D, Kobayashi R, Poettmann F H, Veri J A, Elenbaas J and Wainaug Ch F 1965

Manual for the extraction, transport and processing of natural gas (Moscow: Nedra) p 531
[5] Makogon Yu F 1985 *Gas hydrates, prevention of their formation and use* (Moscow: Nedra) p 232
[6] Karataev K A and Petryakov V A 2011 New technologies for the oil and gas region *Materials of the All-Russian scientific-practical. conf. (*Tyumen: TIU) p 69-71
[7] *Gas hydrates* Retrieved from: https://neftegaz.ru/science/view/706-Gazovye-gidraty
[8] Shagapov V Sh and Urazov R R 2004 Characteristics of a gas pipeline in the presence of hy-drate sediments *Thermophysics of high temperatures* **42(3)** 461-8
[9] Korotaev Yu P 1973 *Control of hydrates during natural gas transportation* ed Yu P Koro-taev, A M Kuliev and R M Musaev (Moscow: Nedra) p 136
[10] Buhgalter E B 1982 Prevention and elimination of hydrate formation during the prepara-tion and transportation of oil and natural gases ed E B Buhgalter *Oilfield business* **10(34)** 41
[11] Skoblo A I, Molokanov Yu K, Vladimirov A I and Shchelkunov V A 2000 *Processes and apparatus of oil and gas processing and petrochemistry* 3rd ed (Moscow: OO "Nedra Businesscenter") p 677
[12] Grunvald A V 2007 The use of methanol in the gas industry as an inhibitor of hydrate formation and the forecast of its consumption in the period up to 2030 *Oil and Gas Business* **2** 35-44
[13] Buhgalter E B 1986 *Methanol and its use in the gas industry* (Moscow: Nedra) p 238
[14] Afanasenko V G, Kulakov P A, Boev E V, Davletov O B and Mazidullin D N 2020 Optimization of ultrasound emulsion under mechanical mixing *Bulletin of the Tomsk Polytechnic University Geo Assets Engineering* **331(4)** 148-55
[15] Afanasenko V G, Kulakov P A, Rubtsov A V, Ivanov O S and Rakhmaeva G I 2020 Research of efficiency of regular separation pack made of corrugated metal sheets *Journal of Physics: Conf. Series* **1515(5)** 052047
[16] Kulakov P A, Rubtsov A V, Afanasenko V G, Sharipova R R and Gudnikova A A 2019 Influence of technical condition parameters on the residual resource of capacitive equipment *Journal of Physics: Conf. Series* **1399(5)** 055052
[17] Naumkin E A, Rubtsov A V, Kulakov P A and Berdin A R 2020 The assessment of the individual resource of the welded joint during repairs of the technological pipeline *IOP Conf. Series: Materials Science and Engineering* **734(1)** 012215

IOP Conf. Series: Materials Science and Engineering 1155 (2021) 012076 doi:10.1088/1757-899X/1155/1/012076

Use of silt sludge after biological wastewater treatment for crop yields increasing

A A Tikhonova[1], I V Vladimtseva[1,2], I V Mogilevskaya[1], O V Kolotova[1], L A Minchenko[2] and M E Spivak[2]

[1]Department of Industrial ecology and life safety, Volgograd State Technical University, 28, Lenin Avenue, Volgograd, 400005, Russia
[2]Department of Chemistry, Food and Sanitary Microbiology, Volgograd State Agricultural University, 26, University Avenue, Volgograd, 400002, Russia

E-mail: alexvlad32@yandex.ru

Abstract. Researching the possible use of the silt sludge after biological wastewater treatment from the industrial enterprises for increasing the agricultural crops' yield was spent. Silt sludge is obtained as a result of biological wastewater treatment in column-type aeration tanks. Founded the concentration's increase of macro - and microelements in the silt sludge, improvement of the sanitary status and decrease of the heavy metals content after its treatment by the enzyme-cavitation method. Low-intensity cavitation with a cavitation number (0.01-0.05) with air oxygen saturation of the sediment (at least 10 mg/m^3) allows for deep organic matter mineralization, dehydration pathogenic microflora's destruction and dehilmintization. Shown the obtained complex fertilization into soil gives it nitrogen, humus and sulfur and increases the winter wheat's yield.

Among the various wastewater treatment methods the most popular is the biological method based on the disposal of pollutants by microbial communities. Biological wastewater treatment is the largest-capacity technology in terms of recyclable volume flows. However, the use of biological treatment facilities with traditional (corridor) type often leads to the problem of a high residual contaminants' concentration in the treated water, so there is a need for post-treatment. The water purified by the biological method needs to be decontaminated with various reagent or non-reagent methods, for example, chlorination, sodium hypochlorite treatment, ultraviolet irradiation or ozonation. Despite the high bactericidal infectability (99.8%), these methods have serious drawbacks – danger to service personnel, the expensive equipment use, utilization needs of used consumables (mercury lamps, electrolysis waste, etc.).

The column-type aeration tanks (tower bioreactors) have been used to improve the wastewater treatment effects recently. These constructions are characterized by the large depth with small cross-section, economy and small building area. The main advantage of the column-type aeration tank is the high air oxygen using degree, and, as a result, the higher cleaning process efficiency.

A huge amount of sludge (excess activates sludge) accumulates at the biological industrial wastewater treatment plants, it contains more than 60% of organic substances. The main problem of further activated sludge using is the pathogenic microflora's and contaminants presence, for example, heavy metals [1]. Most often, the anaerobic methane tanks fermentation is used for the sludge processing. This technology cannot respond to today realities, is characterized by high specific energy consumption, the sludge treatment's length, the methane's, carbon dioxide's and other gases' relasing. After anaerobic fermentation, sediments contain excessive heavy metals' amount, pathogenic micro flora, so they can't be used as fertilizers in agriculture [2].

In recent years, enzymatic cavitation method for processing activated sludge is proposed, it leads to the disinfection and improving the sludge quality through the using of low intensity cavitation and substrate's deep oxygen saturation [3, 4, 5]. Activated sludge (biofilm or raw settling tanks sediment) is saturated with air oxyjets using oxygen. The generation of low-intensity cavitation with a cavitation number of 0.01-0.05 is carried out using turbujets. At the same time, deep mineralization of the sediment, disinfection and deworming are carried out.

Enzymatic cavitation method with a significant energy saving, allows stabilize sludge and obtain a product with the structure is similar to peat without unpleasant odor. Various forms of humic substances are formed in the sludge. These forms promote the heavy metal ions' binding. The resulting sediment is biologically stable, harmless in sanitary because of the pathogenic microflora absence and complete deworming. The sediment contains organic substances in a deeply mineralized form, which are available to plants and soil microflora, so this sludge is promising for use in agricultural production as an organic fertilizer.

The purpose of this work was to study the possibility of using silt sludge obtained in column-type aeration tanks after its processing by the enzyme-cavitation method to increase the agricultural crops yield.

The research was conducted in arid conditions of the Saratov region (Engels) and the Lower Volga region (Volzhsky). The water sludge after aerobic biological treatment of industrial and domestic wastewater in tower bioreactors was treated by the enzyme-cavitation method, including homogenization of the processed sludge mixture and low-intensity cavitation treatment with dissolved oxygen concentration in the mixture of at least 10 mg/m^3 [6, 7]. Technologically, the achievement of low-intensity cavitation with the number G=0.01-0.05, to the necessary for suppressing pathogenic microflora level, preventing stratification of the water-sludge mixture flow and disrupting the pump operation, is achieved by installing eccentrically narrowed section on the pipeline, where profiled resistance elements-Archimedes ' spirals-are placed with a clockwise inclination to the flow [7]. Thanks to this, the movement speed of the cleaned liquid reaches the rotation with pump impeller speed, ensuring the continuity of the flow. Thus, the incoming liquid flow for cleaning "swirls", remaining so after the pump, which contributes to the generation of microcaverns.

The resulting sediment was analyzed for its basic physical and chemical properties. The obtained indicators were compared with the regulatory and technical documentation [8, 9, 10] for determing the possibility of sludge's using as an organic fertilizer. The results of the sediment analysis are in table 1.

Table 1. Physical and chemical parameters of water-silt sediment after treatment by the enzyme-cavitation method.

Monitored indicators	The indicator value in the sludge	Value for regulatory and technical documentation
pH,	6.7	5.5-8.5
Moisture content, %	35	< 82
Organic matter, %	15	> 20
Total nitrogen, %	2.54	> 1.0
Total phosphorus, %	4.2	> 4.0
Total potassium, %	1.25	> 0.3
Mobile sulfur, mg/kg	1950	not normalized
Mobile copper, mg/kg	8.2	not normalized
Mobile zinc, mg/kg	35.0	not normalized
Mobile cobalt, mg/kg	0.18	not normalized
Mobile manganese, mg/kg	56.5	not normalized

Lead, mg/kg	68.0	1000
Cadmium, mg/kg	30.0	30
Zinc, mg/kg	684.3	4000
Copper, mg/kg	137.1	1500
Mercury, mg/kg	0.10	15.0
Cobalt, mg/kg	6.10	not normalized
Mercury, mg/kg	2.3	20.0
Manganese, мg/kg	390.0	2000
Nickel, mg/kg	136.9	400
Fluorine, mg/kg	1.20	10.0
Radium-caesium ($^{134+137}$), Bk/kg	10.8	not normalized
Index of enterococci	not detected	1
Pathogenic bacteria, including Salmonella, helmints eggs and intestinal protozoa	not detected	absense

The data presented in table 1 indicate the sludge after aerobic enzyme-cavitation treatment is characterized by lower amounts of heavy metals: manganese, nickel, copper, zinc, mercury and lead. The amount of organic matter in the sludge (15%) does not formally correspond to the regulatory and technical documentation (RTD) (20%), but in terms of nitrogen, phosphorus and potassium, it significantly exceeds the standards. Sludge treated by the enzyme-cavitation method is actually a complex organo-mineral fertilizer, and RTD is been developed for organic fertilizers with high organic substances and micronutrients deficiency. The complex nature of the fertilizer obtained by authors is confirmed by the presence of the significant amount of microelements. In addition, it should be emphasized that the sludge has the high sanitary and hygienic conditions (the absence of pathogenic microorganisms and helminths). In addition, a feature of the sediment treated by the enzyme-cavitation method is the high content of sulfur (about 2 g/kg). Sulfur is a biogenic element and a part of amino acids, proteins, enzymes and other biologically active substances, sulfate ions are the most important source of mineral nutrition of plants.

In order to test the efficiency of the using silt sludge as a fertilizer, it was applied to light chestnut soil during the cultivation of winter wheat of the Don-93 variety in the Lower Volga Research Institute of Agriculture (Gorodishchensky district of the Volgograd Region). Silt sediment with a moisture content 35% was introduced into the soil in the proportion of 20 t/ha. Two variants of soil treatment were used in the studies:

- deep loosening with a turnover of the topsoil to the depth of 15 cm. In this case, the sediment is sealed to the depth of 10 -15 cm.
- surface (shallow) tillage to the depth of less than 10 cm.

Inoculation of winter wheat made after the autumn introduction of fertilization. In the spring, friendly shoots were observed in the areas with the applied fertilizer, and in the case of the sludge absence in the soil, the shoots appeared only after heavy rains. The highest yield of winter wheat (49.3 kg/ha) were observed after the surface soils' treatment with sludge. The results of the physical and chemical parameters of the soil determining in the autumn and spring periods were shown in table 2.

Table 2. The results of physical and chemical analysis of the soil

Treatment	Sediment presence	Sampling depth, cm	Phosphorus, mg/kg	Potassium, mg/kg	Nitrogen, mg/kg	Humus, %	Iron , mg/kg	Sulphur, mg/kg
			Autumn					
Small	+		264.5	560	414.6	6.78	23150	1860
Deep	+		205.5	618	461.8	6.42	22635	2070

	-		46.3	554	100.8	2.47	20223	69.0
			Spring					
Small	+	0-5	231.6	600	349.8	4.67	23928	1380
	+	10-20	238.0	600	294.0	3.94	24680	1550
Deep	+	0-5	197.5	578	221.2	3.02	21672	800
	+	10-20	205.5	478	182.0	3.11	20867	710
	-	0-5	52.9	614	114.8	2.29	18960	44.5
	-	10-20	51.6	592	98.0	2.20	19493	16.7

The data given in table 2 allow us to conclude the sediment introduction into the soil after its enzyme-cavitation treatment leads to a significant enrichment of its macro-and microelements, especially nitrogen, phosphorus and sulfur. It should be emphasized after the sediment's introduction in the soil the humus amount increases.

To determine the aftereffect of the silt sediment in the third year after its introduction, Prairie barley was sown on the same plots. The yield on the experimental plots was 28 c/ha, it was 73% more than in the control (the same loosening, but without the introduction of sediment).

To study the possibility of using silt sediment to increase crop yields on sandy soils, field tests were conducted on fine-grained polymineral, medium humus sand. The rate of sediment application in this substrate was 20 t/ha. As an agricultural crop, long-term fodder rye was used. Shown during the introducing the sediment into the sandy substrate, it becomes blackened, and the yield of the green rye mass in the experimental plot increased by 23.3% compared to the control one.

Thus, the studies found the aerobic enzyme-cavitation method of treating sludge, obtained during biological industrial wastewater purification, contributes increasing the macro - and micronutrients, improving the sanitary condition and reducing the heavy metals' concentration. Shown the possibility of using the resulting sediment as a fertilizer in the winter wheat cultivation on light chestnut soil. The introduction of sediment enriches the soil with phosphorus, nitrogen, humus and sulfur. The highest yield of winter wheat (49.3 c/ha) was observed after surface tillage with sludge.

References

[1] Stepkina Yu A 2007 Problems of Sludge Formation at Wastewater Treatment Plants. *Treatment and disposal methods Construction Engineering* **10** 15-8
[2] Zykova I V 2008 *Neutralization of excess activated sludge and wastewater sludge from heavy metals* (St. Petersburg) p 384
[3] Pirsol I S 1975 *Cavitation* (Moscow: Mir) p 95
[4] Stepkin A A and Stepkina Yu A 2003 *Pat. of the Russian Federation* No2210550 appl.24.12.2001, publ. 20.08.2003
[5] Stepkina Yu A 2007 Integrated wastewater treatment in vertical design using low-intensity cavitation **1** 44-49
[6] Tikhonova A A, Chernova Yu A and Vladimtseva I V 2016 Application of low-intensity cavitation to suppress pathogenic microflora in biological treatment towers *Natural and technical sciences* **5** 19-22
[7] Stepkin A A and Stepkina Yu A 2005 Pat. of the Russian Federation No226716 appl. 24.12.2003, publ. 20.09.2005
[8] GN 2.1.7.2041 -06 2006 Maximum permissible concentration (MPC) of chemical substances in the soil *Bull. of normative acts of federal laws of the executive power* **10** 10-38
[9] GN 2.1.7.2042-06 2006 Approximately permissible concentrations (APC) of chemicals in soil. *Bull. of normative acts of federal laws of the executive power* **20** 18-52
[10] GOST R 1.4.3.07-2001 20012001 *Nature protection. Soils. Requirements for sewage sludge use for fertilization* (Moscow: Standards Publishing House) p 7

Use of phased supply of hydrogen additives for improvement of the ecological characteristics of the Wankel rotary engine

E A Fedyanov, Y V Levin[1], E M Itkis and K V Prikhodkov

Volgograd State Technical University, Lenin Avenue, 28, Volgograd, 400005, Russia

[1]E-mail: levin-yv@yandex.ru

Abstract. The Wankel rotary engine, designed according to Wankel's scheme, is more adaptable for running on hydrogen fuels than traditional reciprocating engines, possessing lower pre-ignition and backfire probability. Hydrogen additive to main air-fuel mixture helps to decrease incompleteness of combustion in the vicinity of rear rotor apex. Experimental data for Wankel rotary engine VAZ-311 performance on hydrogen blends with hydrocarbon fuel is presented. Researches are performed on partial loads, where hydrogen mass fraction additions to the fuel mixture didn't exceed 5%. It is shown that for partial loads (20% of full load) and engine speed n= 2000 min^{-1} 5% hydrogen addition by mass yields to 4.2% decrease of brake specific fuel consumption (hydrogen consumption was converted to gasoline consumption computed proportionally to their combustion heats ratio). Amount of unburned hydrocarbons for the conditions mentioned reduces in a factor of 2 and emission of carbon monoxide reduces to 40%.

1. Introduction

The majority of internal combustion engines for transportation vehicles are reciprocating piston engines. Engines of such kinematic scheme are constructible and have the optimum ratio of combustion chamber surface area to its volume. At the same time the presence of rather massive parts executing reciprocating motion is one of disadvantages of such engines due to considerable inertial forces [1]. To resist the loads from inertial forces the parts mentioned demand strengthening and that correspondingly requires higher weight and leads in turn to decrease of engine speed.

The intention to overcome reciprocating engines disadvantages caused many attempts to create various types of internal combustion rotary engines. The majority of schemes developed appeared to be unworkable [2]. Recently functional design of rotary engines with acceptable parameters is based on the Felix Wankel scheme [1–4].

Due to the absence of crank and gas distribution mechanisms balancing of such engines is easy and dynamic characteristic and specific quantity of metal are also good. Wankel rotary engines (WRE) posses high specific capacity and dense of components arrangement. To obtain the capacity series a certain number of same-type sections must be used while the unification level reaches 85-90%. The WRE advantages presented make it possible to consider this engine type to be the serious alternative to two-stroke gasoline engines of small boats and four-stroke engines of light aerial vehicles [5–7]. However the application of WRE in transportation industry is limited by relatively high fuel consumption and unburned hydrocarbons level in exhaust gases. The latter is the result of incomplete fuel burning in the vicinity of rotor rear (counter rotation) apex. It is the bulk air-fuel mixture flow that follows rotor rotation and prevents flame front propagation to the area mentioned. Relatively high square/volume ratio of WRE combustion chamber comparing to reciprocating engine and its oblong shape cause flame hindering in the boundary layer. As the result the considerable air-fuel mixture amount remains unburned to the start of exhaust stroke.

So it's obvious that the reduction of unburnt fuel amount can be obtained by complete flame coverage of the rear apex vicinity volume. In principal there are two ways of this problem solution. In principal there are two ways of this problem solution. First – to decrease the distance of flame front propagation from the ignition source to the rotor rear apex. The second way is to increase flame propagation speed.

Currently the installation of additional spark plug in WRE is considered to be the main method for incomplete burning reduction [8, 9]. It's also determined by engine construction – oblong shape of combustion chamber in the rotor side direction allows installation of two ignition spark plugs, one closer to front rotor apex and the other closer to the rear for the ignition event start.

Unfortunately the installation of additional spark plug displaced to the rear rotor apex doesn't solve the problem completely [9]. For the majority of WRE performance modes the flame front propagating from the additional spark plug does not reach the rear rotor apex till the exhaust stroke starts.

Flame propagation speed depends on air-fuel mixture ratio. But generally it's necessary to take into account the type of fuel. If WRE runs on gasoline the ratio for all speed and loading modes is close to stoichiometric and the normal flame propagation speed reaches its maximum. It's known that the increase of flame propagation speed in air-gasoline mixture can be achieved by gaseous hydrogen addition [10–12]. The increased concentration of hydrogen that posses high reactivity level and combustion heat enables to raise flame propagation speed in the air-fuel mixture towards the rotor rear apex direction[13–15].

Thermophysical properties of hydrogen differ significantly from those of hydrocarbon fuels. It should be noted that hydrogen has the highest diffusion coefficient comparing to all gaseous fuels. They also observed the highest laminar flame propagation speed in air-hydrogen mixtures, and the lowest ignition energy.

However computational study of WRE combustion process showed that elimination of incomplete burning in the rear rotor apex vicinity requires more than 10% hydrogen addition to the basic fuel. [16, 17].

The necessity of such considerable hydrogen amount adds complexity to practical realization of the method presented because devices that can generate the latter onboard the vehicle and providing technical and economical efficiency are unknown [18,19]. To reduce the quantity of gaseous hydrogen that provides the complete air-fuel burning it's preferable to consider the opportunity of stratification of the air-fuel mixture properties varying the value of hydrogen concentration in WRE combustion chamber volume, i.e. charge stratification.

2. Time-phased fuel injection

The directed motion of air-fuel bulk flow in WRE combustion chamber enables to organize the stratification of incoming charge. Thus to reduce the incomplete combustion it's desirable to avoid fuel penetration into the rotor rear apex vicinity volume. Considerable success was reached by firm Curtiss-Wright [20, 21]. The direct injection process in WRE combustion chamber that was organized by application of two nozzles one injecting parallel to the big stator axis in the main flow direction and the other placed near spark plug made it possible to reach efficiency comparable with those of spark ignition piston engines. Representatives of Curtiss-Wright also declared that stratification of air-fuel charge and when near spark plug there's ignition mixture enhance control of WRE performance. However the organization of such direct-injection process demands the application of high pressure fuel system.

The stratification of hydrogen addition in WRE combustion chamber volume can be achieved by phased hydrogen supply during intake stroke. To concentrate the hydrogen addition in the working chamber volume that boarders with rear rotor apex where the incomplete burning takes place, the start of hydrogen injection process must be shifted from the moment of inlet port opening and intake stroke starts.

Evaluation of this hydrogen injection shift from the moment when the intake stroke starts can be based on the analysis of inlet process procedure that is specific for WRE design scheme presented in figure1.

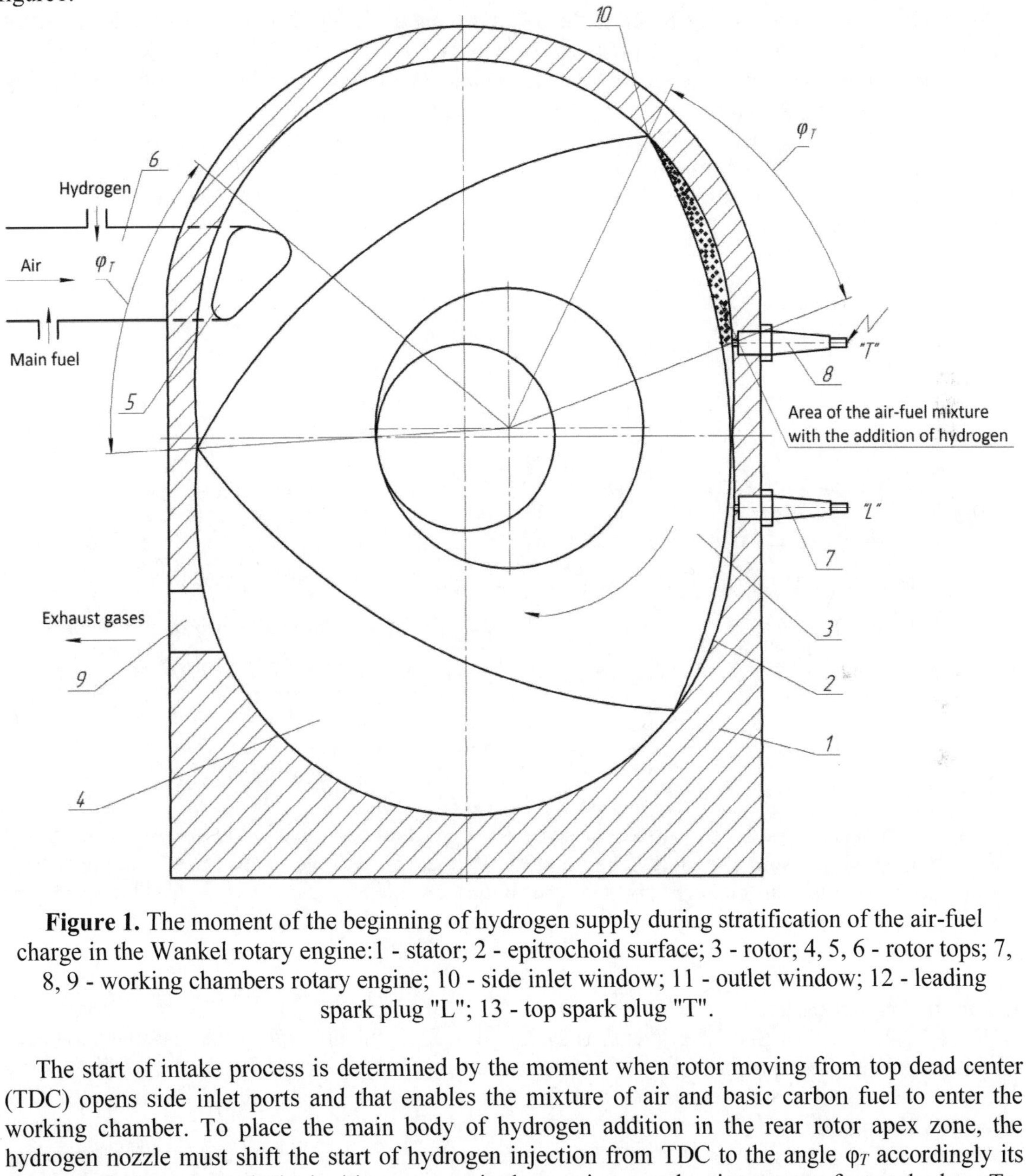

Figure 1. The moment of the beginning of hydrogen supply during stratification of the air-fuel charge in the Wankel rotary engine:1 - stator; 2 - epitrochoid surface; 3 - rotor; 4, 5, 6 - rotor tops; 7, 8, 9 - working chambers rotary engine; 10 - side inlet window; 11 - outlet window; 12 - leading spark plug "L"; 13 - top spark plug "T".

The start of intake process is determined by the moment when rotor moving from top dead center (TDC) opens side inlet ports and that enables the mixture of air and basic carbon fuel to enter the working chamber. To place the main body of hydrogen addition in the rear rotor apex zone, the hydrogen nozzle must shift the start of hydrogen injection from TDC to the angle φ_T accordingly its rear apex that corresponds the ignition moment in the previous combustion process for spark plug «T».

3. Experimental conditions

To determine the optimum moment of hydrogen injection during the intake stroke experiments were carried out on one-sectional WRE VAZ-311.The engine with 649 cm^3 volume of each chamber and compression ratio 9.3 was placed on the test stand equipped with the measuring tools necessary. As the main fuel gasoline was used, feeding by injector. The additional injector for hydrogen supply was installed near the WRE intake manifold. It was connected with hydrogen jet built in the intake manifold by short flexible pipe. The axis of the jet formed the acute angle with axis of gasoline injector. The control system provided the opportunity to vary in a broad range the injection start and duration for both nozzles.

Tests were carried for averaged mode of the urban driving cycle: n=2000 min^{-1}, p_e=0.2 MPa. Spark advance angles for both spark plugs were fixed to provide optimum performance for running on gasoline and were not varied when hydrogen was added. The advance angle for leading spark plug «L» was 26° degrees of eccentric angle(DEA)before TDC, for spark plug «T» 30° DEA BTDC.

The registration of pressure in WRE working chamber was carried out by «Kistler» piezoelectric sensor (type 6118B) installed in spark plug «L». Besides pressure impulses from hydrogen injector and TDC position sensor were transmitted to other analog-to-digital converter channels. The example of record is presented in figure 2.

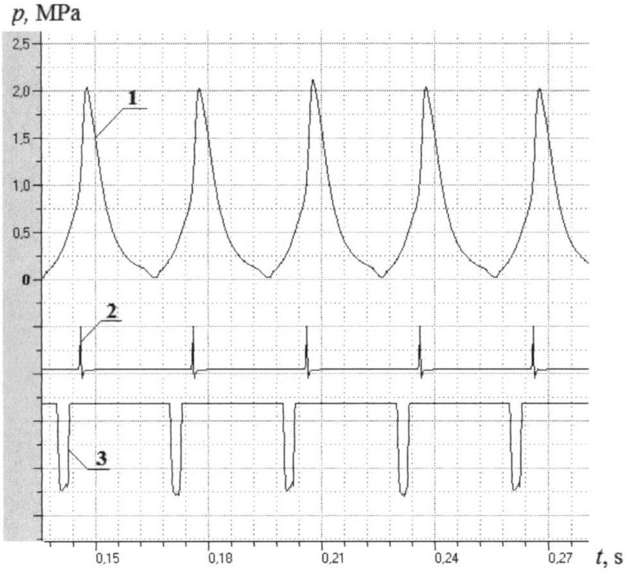

Figure 2. Example of recording signals from sensors of the engine VAZ-311:1 – pressure in the chamber of the rotary engine; 2 – TDC; 3 – hydrogen injector open signal.

The experimental study was carried in two steps. At the first step hydrogen feeding started simultaneously with gasoline feeding. The quantity of injected hydrogen was varied. At the second step quantity of hydrogen remained constant, but hydrogen injection start was varied thus changing relative position of gasoline and hydrogen injection phases. For all tests the air-fuel mixture was kept stoichiometric. Injection timing and injection duration for gasoline remained constant.

4. Results and discussion

In figure 3 the results of first experimental stage are plotted as the unburned hydrocarbons (CH) and carbon monoxide (CO) quantities versus hydrogen addition fraction for averaged urban driving cycle mode. It can be seen that 5% hydrogen addition leads to twice reducing CO emission and to 42% reducing CH emission.

Values of eccentric shaft angles hydrogen corresponding the start of hydrogen injection at the second stage of experiment from specific fuel consumption and standard deviation for peak cycle pressure are presented in table 1. The angle Ψ_1 is the moment of hydrogen injection start relative to the beginning of inlet port opening by front top of the moving rotor, while Ψ_2 is the same relative to TDC of adjacent in the clockwise direction working chamber. The first line of data presented corresponds to the case when hydrogen and gasoline injection start simultaneously. For such conditions volume averaged hydrogen mass fraction is 2%.

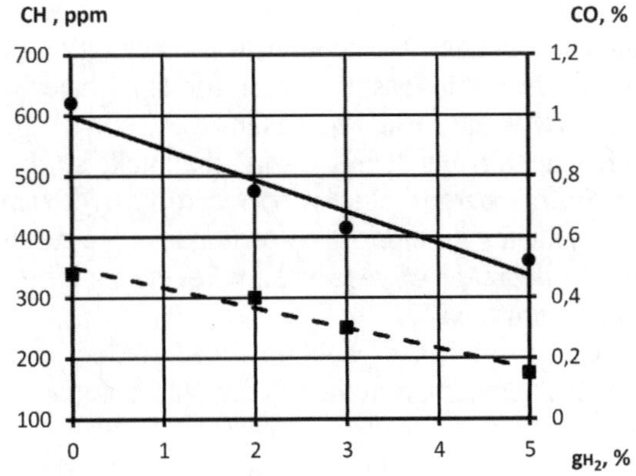

Figure 3. Relationship between the incomplete burning products quantities and the hydrogen addition for the urban driving cycle.

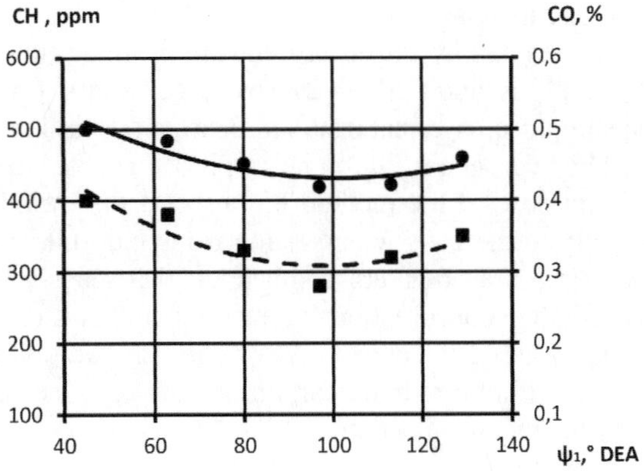

Figure 4. Relationship between the incomplete burning products quantities and the moment of supply of 2% hydrogen for the urban driving cycle.

To obtain quantitative information about the cycle-to-cycle variability (CCV), pressure traces were processed in at least 30 consecutive cycles. As a quantitative measure of the CCV, the value of the standard deviation of the maximum pressure was chosen:

$$\sigma_{P_z} = \left(\frac{\sum_i^n (p_i^{max} - \overline{p^{max}})^2}{n}\right)^{1/2} \quad (1)$$

In equation (1) $p_i^{max}$is the maximum pressure of the cycle, MPa;$\overline{p^{max}}$ is the mean value of maximum pressure of the *n* consecutive cycles, MPa; *n* is number of the consecutive cycles.

The analysis of data presented in table1 enables to get the optimum hydrogen injection shift that corresponds maximum effect of hydrogen addition. It is seen that the lowest specific fuel consumption and standard deviation for peak cycle pressure are reached for Ψ_1 in the interval 90 to 100°DEA. It corresponds 45 – 55° delay of hydrogen injection relative to gasoline injection start.

Table 1. Charge stratification test results.

Ψ_1° DEA	Ψ_2° DEA BTDC	σ_{P_z}, MPa	g_e, g/(kW·h)
45	105	0.060	526.9
63	87	0.057	524.4
80	70	0.056	514.7
97	53	0.051	504.8
113	37	0.057	517.1
129	21	0.068	529.1

In figure 4 variation of unburned hydrocarbons and carbon monoxide emissions are plotted as functions of angle Ψ_1.

For optimum phasing adjustment of hydrogen and gasoline injection fraction of the unburned hydrocarbons in exhaust reduced in 16.2%and for carbon monoxide in 30% comparing to the simultaneous injection start. Optimum phasing adjustment also produces the considerable effect on specific fuel consumption, reducing it in 4.2%.

So for averaged urban driving cycle mode the optimum moment for hydrogen injection start is 97° after intake begins that is 53° DEA BTDC for combustion process in the previous chamber. However it should be mentioned that for better accuracy hydrogen transporting system delay must be taken into consideration that shifts the injection start closer to TDC.

5. Conclusions

Using of small hydrogen additions to the main fuel enables to enhance the ecological characteristics of WRE. Thus addition of 2% hydrogen mass fraction to the main gasoline fuel for simultaneous injection decreases unburnt hydrocarbons emission in factor 1.2 and carbon oxides in 1.3.

The hydrogen injection being phased enables to stratify the air-fuel charge during the intake stroke. The purpose of the method proposed is the transportation of charge volumes enriched by hydrogen directly to the zone where flame front and bulk flow propagates in opposite directions. As the result there is more complete combustion that leads in turn to decrease of required hydrogen additions, specific fuel consumption and incomplete burning products emission.

The phased injection of 2% hydrogen addition to the main fuel mass fraction further reduces the amount of unburnt hydrocarbons in exhaust gases by 16.2% and carbon monoxide by 30% comparing to simultaneous injection.

Moreover hydrogen economy is important result because generation and storage of hydrogen onboard the vehicle in considerable quantities is a complex problem.

Reference

[1] Pulkrabek W W 2004 *J. Eng. Gas Turbines Power* **126** 198
[2] Hu C, Cunyun P, Xiaojun X, Xiang Z and Haijun X 2015 *Conf. on Ind. Electr. and Appl.* (London: Atlantis Press) 180-183
[3] Ozcanli M, Bas O, Akar M A, Yildizhan S and Serin H 2018 *Int. J. of Hydrogen Ener.* **43** 18037-45
[4] Wakayama N, Morimoto K, Kashiwagi A and Saito T 2006 *In 16th World Hydrogen Ener.Conf.* (France: Lyon)
[5] Amrouche F, Erickson P A, Park J W and Varnhagen S 2016 *Int. J. Hydrogen Ener.* **41** 14261-71
[6] Boretti A, Jiang S and Scalzo J 2015 *Advances in Aircraft and Spacecraft Science* **2** 445-467
[7] Ribau J, Silva C, Brito F P and Martins J 2012 *Ener. Conversion and Management* **58** 120-133
[8] Ohkubo M, Tashima S, Shimizu R, Fuse S and Ebino H 2004 *SAE Technical Paper* **01** 1790
[9] Kawahara N, Tomita E, Hayashi K, Tabata M, Iwai K and Kagawa R 2007 *Proceedings of the Combustion Institute* **31** 3033-40
[10] White C M, Steeper R R and Lutz A E 2006 *Int. J. Hydrogen Ener.* **31** 1292-1305
[11] Cipriani G, Di Dio V, Genduso F, La Cascia D, Liga R, Miceli R and Galluzzo G R 2014 *Int. J. Hydrogen Ener.* **39** 8482-94
[12] Jain I P 2009 *Int. J. Hydrogen Ener.* **34** 7368-78
[13] Su T, Ji C, Wang S, Cong X and Shi L 2018 *Int. J. Hydrogen Ener.* **43** 1902-08
[14] Yang J, Ji C, Wang S, Zhang Z, Wang D and Ma Z 2017 *Ener. Conversion and Management* **151** 136-146
[15] Shi C, Ji C, Wang S, Yang J, Ma Z and Ge Y 2019 *Ener. Conversion and Management* **195** 1124-37
[16] Fedyanov E A, Zakharov E A, Prikhodkov K V and Levin Y V 2017 *IOP Conf. Ser.: Materials Science and Engineering* **177** 012076
[17] Itkis E M, Fedyanov E A and Levin Y V 2018 *Int. Conf. on Industrial Engineering* 2079-87
[18] Verhelst S and Wallner T 2009 *Progress in Ener. and Combustion Science* **35** 490-527
[19] Singh A P, Agarwal R A, Agarwal A K, Dhar A and Shukla M K 2018 *Prospects of Alternative transportation fuels* (Singapore: Springer)
[20] Kagawa R, Okazaki S, Somyo N and Akagi Y 1993 *SAE Transactions* 918-926
[21] Loyd R 1976 *Combustion Science and Technology* **12** 47-61

IOP Conf. Series: Materials Science and Engineering 1155 (2021) 012078 doi:10.1088/1757-899X/1155/1/012078

Investigation of the throughput of nonlinear surge arresters using the MatLab software environment

N M Kuznetsov[1], I N Morozov[1,2] and S A Smirnov[2]

[1]Center for Physical and Technical Problems of Energy of the North of the Federal State Budgetary Institution of Science of the Kola Scientific Center of the Russian Academy of Sciences, Apatity, Russia
[2]Federal State Budgetary Educational Institution of Higher Education "Murmansk Arctic State University" Murmansk, Russia

E-mail: moroz.84@mail.ru

Abstract. This work is devoted to the topical issue of researching the throughput of nonlinear surge suppressors. The article discusses various options for test installations and offers simulation models. The MatLab environment, the Simulink simulation application, was chosen as the simulation software environment. The impulse current curve was calculated, after which the tests of nonlinear arresters using rectangular pulses were carried out. The simulation results of installations for operating tests of the surge arrester were compared with the real indicators of laboratory studies for arresters of the OPNp-110 UHL1 brand.

1. Relevance

Operational tests are the main type of tests for surge arresters (SPD), during which they are exposed to both currents and voltages of various types, most often simulating real operating conditions [1-5].

Operating experience shows that the continuous current flowing through the arrester does not significantly damage the protective characteristics. Nevertheless, the question of changing the characteristics after the flow through the varistors of large impulse currents of lightning overvoltages is not closed [7,8,10,11]. It can be assumed that the reason for the appearance of such changes is associated with the application of the operating voltage to the surge arrester, both immediately after the flow of the impulse current, and after seconds or even minutes. In such cases, the characteristics of the arrester become unstable, which can cause damage during further use.

2. Description of simulation models and the results of their work

In most operational tests of surge arresters, the test setups are very similar. They imply the application of an industrial combination of impulse actions and a certain combination of impulse actions that recreate a direct lightning strike, lightning impulses (normalized current pulses 8/20 μs) or switching overvoltages (rectangular pulses of a given current with a duration of 2– 3.2 ms).

The normalized current pulse 8/20 μs is a unipolar current pulse with the conditional rise time Tf = 9 ± 1 μs and the pulse duration Ti = 20 ± 2 μs. The approximate rise time is the time that is 1/4 longer than the current rise time from 10% to 90% of the maximum value of the current impulse. The point of intersection of the time axis and the straight line drawn through the points where the impulse value is equal to 10% and 90% is considered to be the conditional beginning of the impulse (figure 1).

IOP Conf. Series: Materials Science and Engineering 1155 (2021) 012078 doi:10.1088/1757-899X/1155/1/012078

The time interval between the approximate beginning of the impulse and the moment of decay, when the current reaches half of the maximum, is taken as the duration of the impulse.

The calculation of the pulse current curve was performed using the Matlab application [6, 9, 12]. In this case, you can use only those points of the current-voltage characteristic that are within the calculated currents. For a varistor, whose current-voltage characteristic is shown in table 1, a current pulse “8/20” can be obtained in the circuit shown in figure 1.

Table 1. Calculated volt-ampere characteristic of the varistor.

I, A	0.003	35	75	165	1100	2000
U, kV	1.58	3.25	3.46	3.92	4.23	4.72

At the same time, as the analysis shows, the parameters of the circuit should have the following characteristics: C1 = 10 μF; L = 18 μH; R = 2.2 ohms. For these parameters, the duration of the leading edge of the pulse is Tf = 10 μs and the duration of the pulse is Ti = 30 μs (figure 2). With the battery charging voltage Uc = 11000 V, the maximum current impulse value is 2000 A. To obtain results with a different current value, the charging voltage value must be changed. The rectangular current pulse test is a type of test for non-linear surge arresters. Such tests also belong to one of the stages of operational tests of surge arresters with lightning impulses.

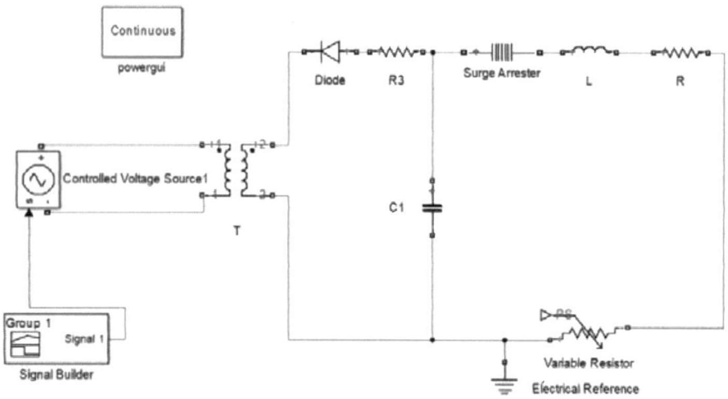

Figure 1. Schematic diagram of an arrester test setup.

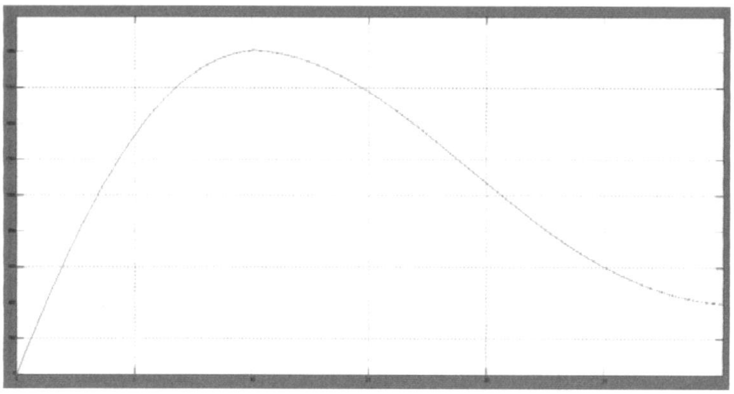

Figure 2. To test the arrester: current impulse 8/20”.

According to the recommendations of the International Electrotechnical Commission, a rectangular impulse wave is a wave with a rapid increase to a maximum value, holding at a constant value for a given period of time, and then a rapid drop to zero.

Parameters such as polarity, maximum value, nominal maximum value duration, and nominal total duration define a square wave pulse. For the conditional duration of the maximum value of the square wave T (figure 3) is taken the time in the duration of which the instantaneous value of the wave exceeds 9/10 of the maximum value. The total conditional duration of the wave Tp is the time during which the instantaneous value of the wave is greater than 1/10 of the maximum value. If there are minor fluctuations at the front of the wave, then an average curve should be drawn to determine when a value of 10% is reached.

When tested in the switching mode of surge arresters with a nominal discharge current of 10 thousand and 20 thousand A, the pulse generator is a model of a power line, and the normalization of the charging voltage of the generators, the line impedance and the relative duration of the pulse amplitude is noticeable. During tests of arresters with a comparatively lower throughput, the parameters of the rectangular current pulse supplied to the samples are normalized, and not the parameters of the simulated power line. These pulses must meet the following requirements:

- the relative duration of the maximum value should be 100% -120% of the normalized value;
- the relative total duration should not be ~ 150% of the conditional duration of the maximum value;
- fluctuations or primary inrush should not exceed 10% of the maximum current value; if there are fluctuations, an average curve must be drawn to find the maximum current value.
- The maximum current value should be 90% -110% of the rated value.

A rectangular current pulse with a normalized conditional duration T = 30 ms can be obtained in the circuit (figure 3) with the number of cells n = 7. For a varistor with a volt-ampere characteristic from table 2, the following circuit parameters are required:

Table 2. Calculated volt-ampere characteristic of the varistor.

I, A	185	285	440	5000
U, kV	8.1	8.4	8.7	11.1

Let us assume that L = 2.2 mH; C = 15 μF; L1 = 4 mH and R = 82 ohms. To obtain a current pulse with a maximum value of 0.2 kA, the charging voltage of the capacitors in the chain must be 18.5 kV. As can be seen from figure 4, the conditional pulse duration is T = 0.03 s and the total duration is T = 0.045 s, which is within the specified range. The resistor provides a drop in current to zero. At a higher value, the current pulse does not fall to zero, and at a lower value, a reverse pulse occurs (after the end of the pulse itself), which has a serious maximum value. Also, the resistor R shortens the leading edge of the pulse and at the same time lengthens its tail, which causes a sharp increase in the total pulse width.

The required parameters of the current pulse through the test sample are determined by the number of links in the chain circuit and the size of its individual components. Calculations show that the conditional pulse duration T and the total duration Tp increase in direct proportion to the inductance of the link L or its capacitance C, as well as the number of links. A change in L1 practically does not affect the relative and total pulse duration, although at small values of L1, strong current fluctuations are observed at the front and the maximum value of the pulse, exceeding the permissible 1/10 of the maximum value. The pulse parameters corresponding to the norms, as a rule, can be obtained only for L1> L (figure 4). However, as the output inductance is increased, there is a large smoothing of the pulse edge, mainly at the level of 90% -100% of the maximum value, and a noticeable current drop within the specified values. This drop can be leveled by reducing the capacity of the output link, as well as by shunting the two inductances closest to the surge arrester with resistors close in resistance to the characteristic impedance of the circuit circuit.

Tests of zinc oxide varistors in surge arresters with 8/20 μs current pulses and square wave currents help evaluate their carrying capacity. However, all information about the protective characteristics and capacity of arresters can only be given by their operational tests. International standards provide for two types of operational tests for metal oxide arresters without spark gaps (arresters):

- working tests by lightning impulses;
- working tests with switching impulses.

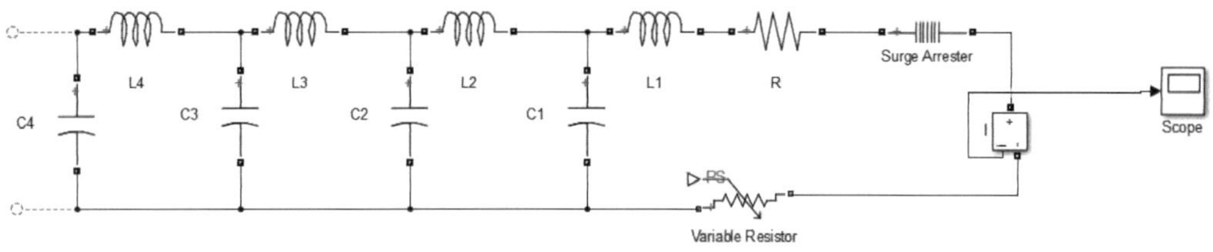

Figure 3. PCG of rectangular pulses: schematic diagram.

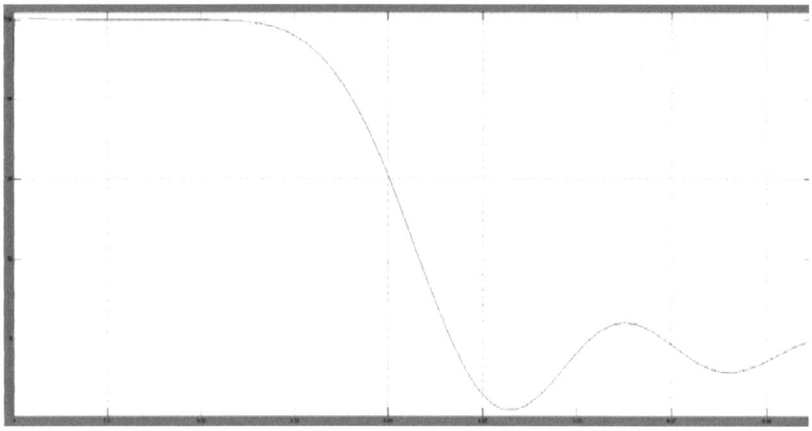

Figure 4. PCG of rectangular pulses: pulse shape.

During operational testing of surge arresters with lightning impulses, the tested models are tested with two high current impulses (from 10,000 to 100,000 A, depending on the class of the arrester) 4/10 μs. Between the first pulse and the second, the tested model is heated to a temperature of 55-60 degrees Celsius. Approximately 100 ms after the last high current impulse, the sample should be subjected to a power frequency voltage equal to the highest operating voltage Unb, which after 10 s should be reduced to the normalized voltage Unnorm. The last voltage is maintained on the sample for half an hour to confirm the thermal stability or thermal instability of the surge arresters. To monitor this stability, one of three sample parameters must be continuously recorded:

- active component of the current;
- temperature of varistors;
- power loss in varistors.

Figure 5 shows a diagram of a test bench during operational tests of surge arresters with lightning impulses as recommended by the International Electrotechnical Commission.

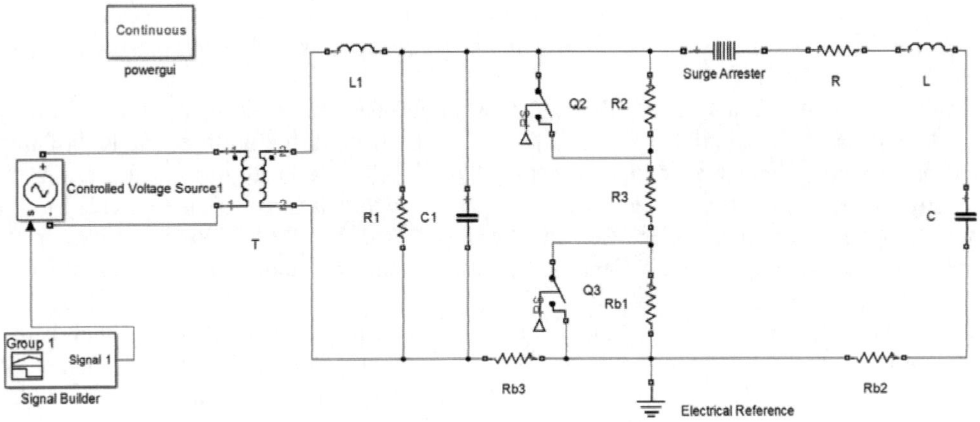

Figure 5. Schematic diagram of an arrester operational test setup.

The installation consists of a pulse current generator and a power frequency voltage source. The generator, in turn, consists of a capacitor bank C, a snubber resistor R, a reactor L and a spark gap F. To the test sample R03 through a transformer T from a shock generator or a powerful power supply network, which are connected to points a - b (see figure 3) power frequency voltage is applied. Reactor L1 and capacitor C1, as well as surge arrester R1 protect the transformer from impulse effects. Switch Q2, which bypasses the nonlinear resistor R2 at the right time, provides a step change in the power frequency voltage across the model under test. To measure currents in the corresponding circuits, low-resistance resistors are installed - shunts Rb1, Rb2 and Rb3. The supply voltage to the power transformer is supplied by the Q1 switch. To prevent overvoltage being applied to the recording device, switch Q3 shunts resistor Rbl for the duration of the pulse current. The operation of the switching devices is provided by a special automatic control device.

The simulation results of the installations for operational tests of the arrester were compared with the real indicators of laboratory research for the arresters of the OPNp-110 UHL1 brand. The comparison results showed the maximum deviation of the throughput values of the order of ± 0.15% from the physically measured values for all tests with different amplitudes.

Thus, the use of the simulation tools of the MatLab software environment lets to conduct laboratory studies of nonlinear surge arresters for throughput without physical tests.

References

[1] Veselov A E, Efimov B V, Kuznetsov N M, Nevretdinov Yu M and Tokareva E A 2008 Limiting arc overvoltages in power supply networks of mining enterprises *Izvestia of higher educational institutions. Mining Journal* **6** 46-51

[2] Vlasko D I, Vostrikov A O, Domonov A P and Nevretdinov Yu M 2011 Problems and prospects of registration of lightning overvoltages in the operating network *Proceedings of the Kola Science Center RAS* **2(5)** 54-65

[3] Efimov B V, Kuznetsov N M, Nevretdinov Yu M, Fastiy G P and Yaroshevich V V 2008 Arc overvoltages during the transition of single-phase earth faults in double *Electrics* **5** 8-11

[4] Efimov B V, Nevretdinov Yu M, Vlasko D I and Vostrikov A O 2012 Registration of lightning overvoltages at a substation *Proceedings of the Kola Scientific Center of the RAS* **5(12)** 29-38

[5] Khalilov F Kh *et al.* 2002 *Protection of 6-35 kV networks from overvoltage* (SPb.: Energoatomizdat) p 272

[6] Morozov I N and Kirillov I E 2019 Modeling the probability of overlapping a string of insulators during lightning strikes into lightning rods *Industrial power engineering* **9** 10-4

[7] Nevretdinov Yu M, Tokareva E A and Vlasko D I 2008 Development of the method of

experimental studies of lightning protection of operating substations *Modeling of transient processes and steady-state modes of a high-voltage network: collection of articles. scientific. tr. Apatity* 96-110
[8] Novikova A N, Shmarago O V, Efimov B V, Danilin A N, Nevretdinov Yu M and Selivanov V N 2011 Issues of lightning protection of overhead lines in the north of the Kola Peninsula: requirements, operating experience of 110-150 kV overhead lines, calculation method *Proceedings of the Kola Science Center of the Russian Academy of Sciences* **2(5)** 9-23
[9] Morozov I N, Kuznetsov N M, Belova L A and Borozdina E D 2020 Investigation of non-stationary modes of a 110 kV electrical substation using simulation in Matlab *Vestnik Chuvash University* 113-22
[10] Wu Z, Tian B, Xu A, Guo X, Cai Y and Wang G 2016 Analysis of overvoltage caused by intermittent earth fault and its suppression method *China International Conference on Electricity Distribution, CICED 2016 China International Conference on Electricity Distribution, CICED 2016 – Proceedings* 7576157
[11] Xiao H, Ding H, Peng T, Li L and Diangong Jishu Xuebao 2009 *Constraining overvoltage existed in the power supply for the pulse magnetic field* **24(1)** 14-7
[12] Morozov I N *et al.* 2020 *IOP Conf. Ser.: Earth Environ. Sci.* 539 012148

Exploring of an open hammer die life with an expanding gutter using information technologies

I N Khaimovich and D A Elistratov[1]

Samara University, Moscow Highway, 34, Samara, 443086, Russia

[1]E-mail: dan.elistratow@gmail.com

Abstract: Currently, open dies with a parallel bridge are widely used. The design of such stamps is carried out according to proven methods. However, the dimensions of the bridges obtained do not always provide a sufficiently high durability of the dies. The decrease in the durability of the dies caused by the overestimated resistance of the bridge is especially noticeable in the manufacture of simple forgings that do not have complex cavities filled by extrusion. Thus, industries need more durable tooling that can provide a large number of stamped parts, as well as save money on the development, design and manufacture of a new stamp. To this end, in this paper, an analysis of the application of an expanding burr groove in a tool is considered. Practical experience of how the tooling is worn out with the number of stampings, the geometry of the tooling is measured, an analysis is carried out, and a mathematical model of the behavior of wear based on the experiment is presented. The same equation was obtained using Excel, which indicates that the mathematical model was obtained correctly. In addition, the paper considers the value of individual parameters of the tool geometry on the overall tool life.

1. Introduction

An important direction in the development of forging and stamping production is to increase the durability of technological equipment for stamping heat-resistant and stainless steels, titanium alloys and other hard-to-form alloys. The durability of open dies is small and amounts to 800-1000 pieces before the first restoration [1].

2. Relevance, scientific importance of the issue

The shape and dimensions of the flash gutter have a great influence on the durability of the open die. Being a pressure regulator in the die cavity, the flash gutter works under very difficult conditions: high normal and shear stresses, high surface temperature. All this leads to significant crushing and abrasion of the surface of the tooling [2].

3. Formulation of the problem

In the course of this work, it is necessary to obtain mathematical one-factor models that give an idea of the influence of the geometric parameters of the flash gutter on the overall wear of the die. A model obtained from the results of a practical experiment, as well as a model obtained using information technology (Excel) [3].

4. Theoretical part

One of the main parameters responsible for tool life is geometry. Consider the value of the geometric dimensions of the flash gutter on the durability of the stamp as a whole.

Flash refers to excess metal usually found at the periphery of the workpiece, which is subsequently trimmed in a separate die. It forms on the flaking surface, usually near the cavity of the stamp engraving. The flash design is critical as it plays two conflicting but important roles during stamping.

1) Acts as a limitation for external metal flow during the first phase of part formation. In this way, deep cavities as well as hard-to-reach and remote corners can be filled.

2) It removes excess metal from the cavity in the second phase after complete filling of the cavity in order to get rid of excess pressure.

Also, it acts as a cushion for hammer blows when shaping at high speeds.

Most finishing dies are designed with a burr groove in them, however, in practice, they are used only in the following cases:

- Fewer forming steps required
- A workpiece of complex shape with large multidirectional deformation should be obtained

Flash gutter thickness has always played an important role: as a lower value will require more energy to bring the forging to the desired size, while a higher value can cause insufficient die filling [4].

5. Practical importance, proposals and implementation results

2000 forgings were stamped in the experimental die.

To study the integral wear, 5 control forgings were selected within the manufactured batch.

To calculate the integral value of the die wear, the cross-sectional areas of the forgings were measured graphically.

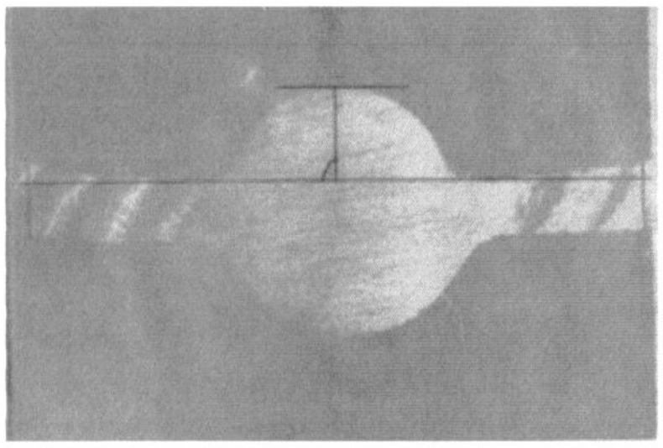

Figure 1. Cross-section of the control forging after the manufacture of 100 parts, sample No.1.

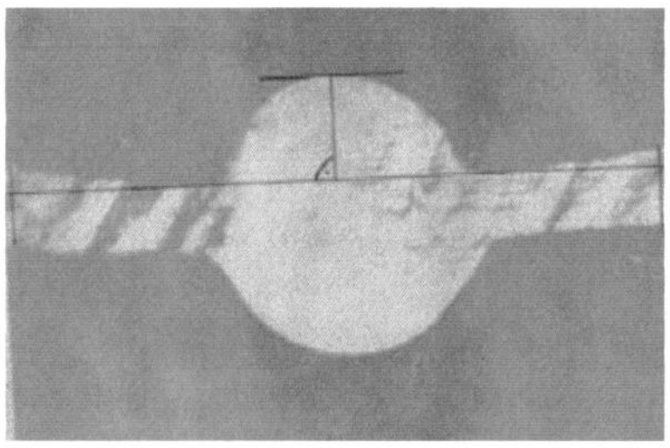

Figure 2. Cross-section of the control forging after the manufacture of 500 parts, sample No.5.

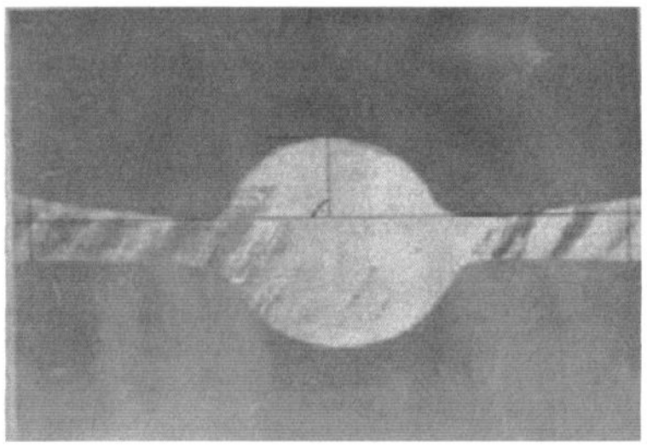

Figure 3. Cross-section of the control forging after the manufacture of 1100 parts, sample No.1`.

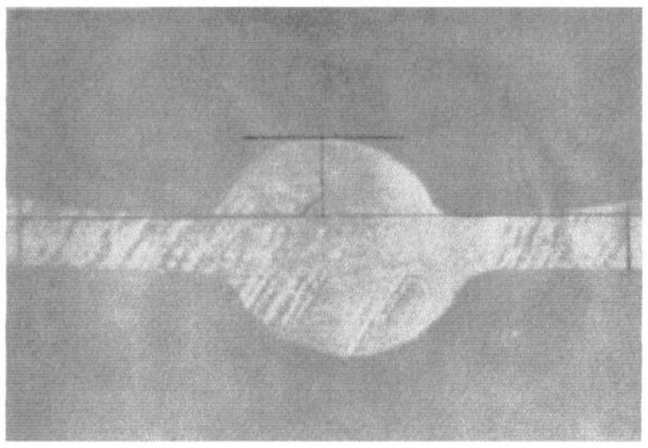

Figure 4. Cross-section of the control forging after the manufacture of 1600 parts, sample No.6`.

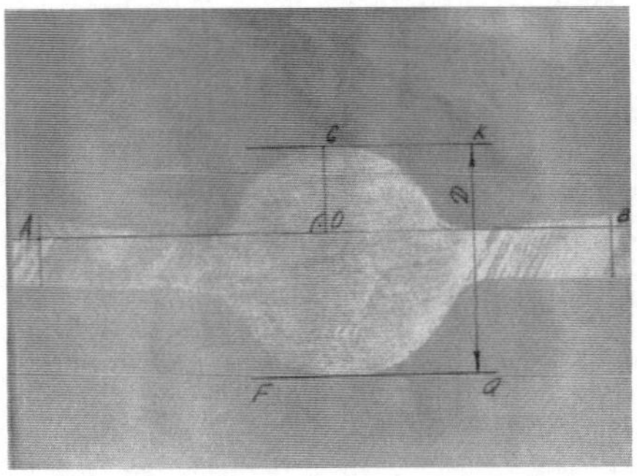

Figure 5. Cross-section of the control forging after the manufacture of 2000 parts, sample No.10`.

Table 1 shows measurements of the groove thickness:

Table 1. Measurement results for minimum burr thickness.

Part No.	D, (mm)	ΔD, (mm)	h_1, (mm)	h_2, (mm)	h_{cp}, (mm)	h_i, (mm)	S, (mm)	$S \cdot h_i$
1	7.275	0.275	1.59	1.605	1.597	1.322	100	132.2
2	7.215	0.215	1.58	1.515	1.547	1.332	200	266.4
4	7.15	0.15	1.41	1.505	1.457	1.307	400	522.8
5	7.215	0.215	1.63	1.54	1.58	1.365	500	682.5
6	7.18	0.18	1.53	1.515	1.522	1.342	600	805.2
7	7.18	0.18	1.575	1.56	1.567	1.387	700	970.9
9	7.18	0.18	1.55	1.54	1.545	1.365	900	1228.5
10	7.29	0.29	1.595	1.55	1.572	1.282	1000	1282
1′	7.0	0	1.39	1.405	1.397	1.397	1100	1536.7
2′	7.08	0.08	1.62	1.33	1.47	1.39	1200	1668
4′	7.21	0.21	1.53	1.54	1.535	1.325	1400	1855
6′	7.20	0.20	1.57	1.63	1.6	1.4	1600	2440
7′	7.10	0.10	1.64	1.54	1.59	1.49	1700	2533
9′	7.18	0.18	1.645	1.645	1.645	1.465	1900	2783.5
10′	7.16	0.16	1.68	1.605	1.642	1.482	2000	2964

The best approximation to the experiment will be such a functional dependence of the sum of the squared distances, from the point to which it will be minimal. Let us assume that this functional dependence is the equation of the straight line (1):

$$y = ax + b. \tag{1}$$

Thus, applying the least squares method, we get:

$$S = \sum_{i=1}^{n}(ax_i + b - y_i)^2 \rightarrow \min \tag{2}$$

Thus, applying the least squares method, we get:

$$S = \sum_{i=1}^{n}(ax_i + b - y_i)^2 \rightarrow \min \tag{3}$$

where $x_i y_i$-coordinates of experimental points; a, b -are unknown coefficients.
Let's find the partial derivatives:

$$\frac{\partial s}{\partial a} = 2\left[\sum_{i=1}^{n}(ax_i + b - y_i)x_i\right] = 0; \tag{4}$$

$$a\sum_{i=1}^{n}x_i^2 + b\sum_{i=1}^{n}x_i - \sum_{i=1}^{n}x_iy_i = 0; \tag{5}$$

$$\frac{\partial s}{\partial b} = 2(\sum(ax_i + b - y_i)] = 0; \tag{6}$$

$$a\sum_{i=1}^{n}x_i + nb - \sum_{i=1}^{n}y_i = 0; \tag{7}$$

Can write a system of equations:

$$\begin{cases} a\sum_{i=1}^{n}x_i + nb - \sum_{i=1}^{n}y_i = 0 \\ a\sum_{i=1}^{n}x_i^2 + b\sum_{i=1}^{n}x_i - \sum_{i=1}^{n}x_iy_i = 0 \end{cases} \tag{8}$$

$$b = \frac{\sum_{i=1}^{15}y_i - a\sum_{i=1}^{15}y_i=0}{15}; \tag{9}$$

$$a = \frac{-\frac{\sum_{i=1}^{15}y_i}{15}\cdot\sum_{i=1}^{15}x_iy_i + \sum_{i=1}^{15}x_iy_i}{\sum_{i=1}^{15}x_i^2 - \frac{\left(\sum_{i=1}^{15}x_i\right)^2}{15}}; \tag{10}$$

When calculating the coefficients a and b, it is necessary to take into account that

$$x_i = \varphi_i, ay_i = h_i$$

$$\left(\sum_{i=1}^{15}x_i\right)^2 = 234090000;$$

$$\sum_{i=1}^{15}x_i^2 = 20790000.$$

We get:

$$a = 7{,}84\cdot 10^{-5};$$

$$b = 1{,}297.$$

Accordingly, the results of the experiment can be described by the following straight line (11):

$$h_i = 7{,}84\cdot 10^{-2}\cdot S + 1{,}297 \tag{11}$$

where h_i- is the minimum thickness of the expanding bridge; S — is the number of stamped parts.

A similar study can be carried out using the Excel package, only for different types of approximation.

Let us consider the influence of the number of stamped products on the groove thickness with a quadratic dependence, shown in figure 6.

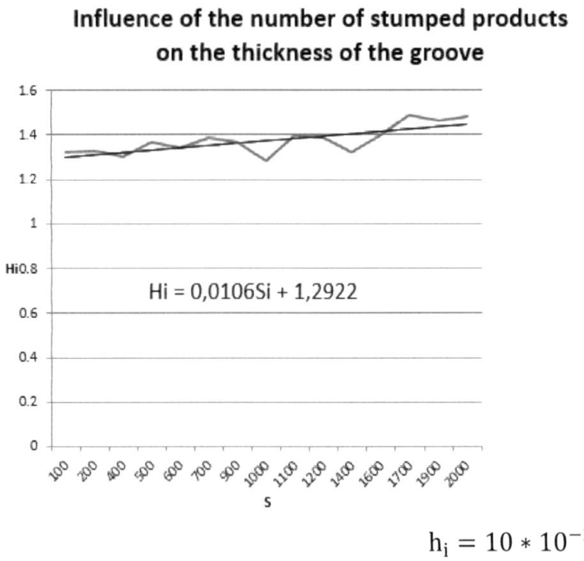

Figure 6. Influence of the number of stamped products on the minimum groove thickness (quadratic dependence).

$$h_i = 10 * 10^{-2} * S_i + 1{,}2922 \tag{12}$$

As a result, a dependence in Excel is created, we got an equation close to the equation, obtained mathematically. They are identical. The Excel equation can be used with the score the degree of confidence which, the equality $R^2 = 0.5595$.

6. Conclusion

Flash land is not directly related to the tool life, since the coefficients for the minimum thickness of the expanding flash land, the number of stamped parts, and the diameter are equal to zero, and at heights they are insignificantly small.

The durability of the punch depends most of all on the deviations of the diameter of the workpiece during stamping, depends on the input and output thicknesses of the burr, as well as on the combined effect of durability and the thickness of the flash, i.e. tool life depends on the geometry of the flash gutter.

With an increase in the difference in diameter from the standard, the resistance will increase. With an increase in the thickness of the entranceflash gutter at the entrance and exit, the tool life decreases. The combined effect of durability and thickness increases die durability.

References

[1] Hill R 1971 The Mathematical *Theory of Plasticity* (Gos.tech.isdat, Moscow).

[2] Perlin I L and Ermanok M Z 1971 *Drawing theory* (Metallurgy, Moscow.) ‘Reference (no number)’.

[3] Khaimovich I N 2014 Computer-aided design of blank forging production facilities for aircraft engine compressor blades *Russian Aeronautics* 169-74

[4] Tirosh J and Iddan D 1991 On the Limit Analysis of High Speed Forming Processes in Cold or Hot Conditions *Advances in Continuum Mechanics* 371-86

[5] Demyanenko E G 2014 Research on the method of shaping barrel-shaped parts (News of Higher Educational Institutions *Aircraft Engineering* 72-6

[6] Khaimovich I N 2014 Computer-aided design for objects of blank and stamp production of compressor blades of aircraft engines *News of Higher Educational Institutions Aircraft Engineering* 44-8

[7] Levanov A N, Kolmogorov V L and Burkin S P 1976 *Contact Friction in Metal Forming Processes* (Metallurgy, Moscow)

[8] Aronov B M, Bibikov V V, Kovalkova I N, Nenashev V Y, Ivashchenko V I and Pantazieva I N 2000 *Computer-aided design system for blank blades and die forgings*, patent no 2000610439, application priority: 2000-04-03

[9] Kachanov L M 1969 *Fundamentals of the Theory of Plasticity* (Nauka, Moscow.)

[10] Surkov O S, Khaimovich A I and Khaimovich I N 2016 Practice in the application of specialized technologic process patterns of 5-axis machining operation of blade rings of blisks *Russian Aeronautics* 112-7

[11] Mikheev V A and Khaimovich A I 2011 Mathematical modeling of the processes of dynamic recrystallization of polycrystalline materials under conditions of intense plastic deformation *Forging and stamping production* 37-44

[12] Frolov M A and Haimovich I N *2016* Improvement of Technological Process of Multiproduct Production on the Bases of Simulation Modeling of Production Unit *Key engineering materials* 487-507

[13] Grechnikov F V, Nenashev V Y and Khaimovich I N 2008 Control over process engineering of compressor blades based on the integration of computer-aided design and engineering analysis *Forging and stamping production* 42-6

Air heating in an air heat pump installation in the expander - generator set

A A Mukolyants[1], M D Buranov, I V Sotnikova and A A Taubaldiev

Tashkent State Technical University named after Islam Karimov, 100095 Uzbekistan,

[1] E-mail: arsen5675@mail.ru

Abstract. The article touches upon the actual problem of the use of secondary energy resources in the transport and distribution of natural gas. Excess gas pressure as the main component of secondary energy resources for technological processes is practically not used at gas distribution stations, where throttling devices are used. The proposal to use excess gas pressure in turboexpanders both at gas distribution stations and at compressor stations of main gas pipelines without preheating has not yet found widespread use, and therefore the replacement of throttling devices with turboexpander units will be determined by energy and economic efficiency. Expander-generator technology is one of the effective technologies that allows to reduce the consumption of fuel and energy resources. In combination with heat pump units, expander-generator units make it possible to create highly efficient power generating complexes that are capable of generating electricity without burning fuel. The possibility of using heat pumps is considered, in which air is used as a working medium, the purpose of which is to increase the temperature potential of heat to ensure the possibility of heat exchange when transferring this heat to the heated medium. An analytical dependence of the electric power of the heat pump installation on the difference between the total power consumption of the compressor and the power of the air turbine is obtained.

1. Introduction

The electric power industry of Uzbekistan is the basic branch of the national economy of the republic and, having significant production and scientific and technical potential, has a significant impact on the development of the entire national economic complex.

The unified electric power system of the republic fully meets the needs of economic objects and the population in electric energy. Uzbekistan is the largest energy power in the Central Asian region.

The technological requirements for the operation of gas pipelines force the gas to be heated upstream of the EGS expander to a temperature of 80-100 ^{0}C. Usually, this is done using the energy of the high temperature potential obtained from the combustion of fuel. However, there are technical solutions that make it possible to operate the EGS even without fuel combustion (figure 1).

IOP Conf. Series: Materials Science and Engineering 1155 (2021) 012080 doi:10.1088/1757-899X/1155/1/012080

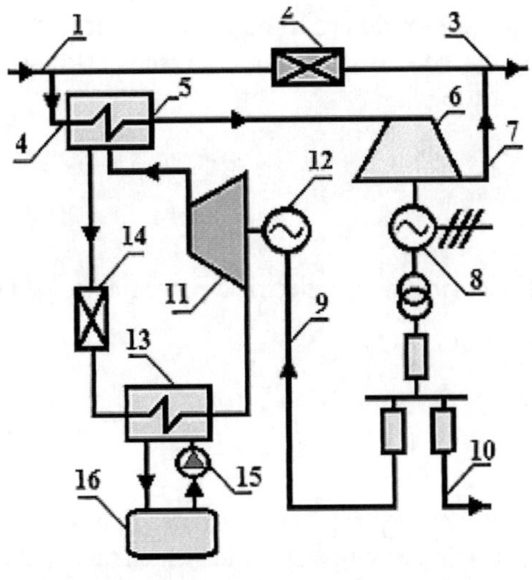

Figure 1. Process flow diagram of a traditional plant with an expander-generator set and a heat pump for heating the gas before the expander. 1 - high pressure gas pipeline; 2 - throttling device; 3 - low pressure gas pipeline; 4 - high pressure gas supply to the expander; 5 - heat exchanger for heating gas before the expander; 6 - turbo expander; 7 - low pressure gas pipeline at the expander outlet; 8 - electric generator; 9 - line of supply of electricity to the electric motor drive the compressor; 10-line for supplying electricity to the power grid; 11 - compressor; 12 - electric motor - compressor drive; 13 - evaporator; 14 - throttling device; 15 - pump for supplying heat carrier of heat of low temperature potential; 16 - heat source of low temperature potential.

2. Materials and methods

The efficiency of the EGS depends on the method of heating the gas before the expander [1-4].

In this connection, recently, a certain interest has begun to appear in air heat pump installations (AHPI) - heat pumps in which air is used as a working medium [5-7]. The purpose of AHPI is to increase the temperature potential of heat to ensure the possibility of heat exchange when transferring this heat to the heated medium. Figure 2 shows a schematic diagram of the AHPI.

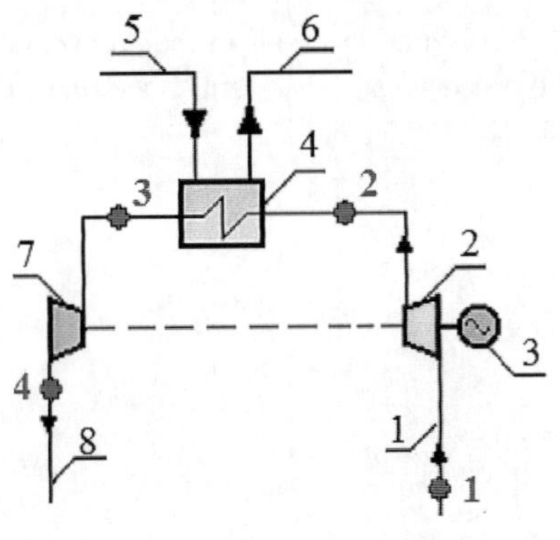

Figure 2. Schematic diagram of an air heat pump installation. 1 - air inlet; 2 - compressor; 3 - electric motor; 4 - heat exchanger; 5, 6 - inlet and outlet of the heated medium; 7– turbine; 8 - air outlet.

The principle of operation of the installation is as follows. Atmospheric air, which is a working fluid in this installation, enters through line 1 into compressor 2, driven by an electric motor 3. In the compressor, the pressure and, accordingly, the air temperature increase. The pressure to which it is necessary to compress the air in the compressor must be such that the corresponding air temperature at the outlet of the compressor is higher than the temperature to which it is necessary to heat the coolant in the heat exchanger 4. In figure 1, for the sake of clarity, a counter-flow heat exchanger scheme is selected, therefore, the positions 5 and 6 in figure 2 - inlet and outlet of the heated medium, respectively. The air pressure leaving the heat exchanger remains higher than atmospheric pressure. In the air turbine 7, where the air enters after the heat exchanger, it expands, the energy of the air flow is converted into mechanical work, while the air is cooled. After the turbine, the air is directed to the atmosphere. The mechanical work obtained in the air turbine is used to drive the compressor, thereby reducing the required energy supplied to the installation from the outside.

The heat conversion coefficient is usually used as a criterion for determining the efficiency of heat pump installations V , which is the ratio of the heat Q_1 of high potential transferred to the consumer to the consumed electric power [8-10].

$$V = \frac{Q_1}{N_Э}, \tag{1}$$

where Q_1 - the heat transferred to the air to the gas in the heat exchanger, N_e - electrical power consumed by the compressor.

The amount of heat is determined by the enthalpy difference:

$$Q_1 = G_a(h_2 - h_3), \tag{2}$$

where Ga is the air flow rate.

This criterion characterizes the efficiency of the installation and depends, among other things, on the temperatures of low heat and high heat potential. Due to the fact that the temperature T_1, which the air at the outlet from the compressor should have, is determined by the operating conditions and cannot be arbitrarily changed, it is of certain interest to study the effect of the air temperature at the inlet to the compressor on the conversion coefficient of the AHPI. Figure 3 shows the hs-diagram of the processes in the AHPI with an increase in the air temperature at the compressor inlet. In this paper, an increase in temperature is considered as a result of preliminary heating of air with low-grade heat from secondary energy resources (SER). The process 1_0-2_0-3_0-4_0 is considered as the initial one. Then the temperature T_{atm}, corresponding to point 1_0, rises to the temperature T_{inlet}, corresponding to point 1_1. In this case, constant operating conditions of the heat exchanger must be ensured.

To fulfill these conditions, as can be seen from the diagram shown in figure 3, the air in the compressor must be compressed to a lower pressure than the initial pressure p_1 (the process 1_1-2_1-3_1-4_1 must be organized in the AHPI).

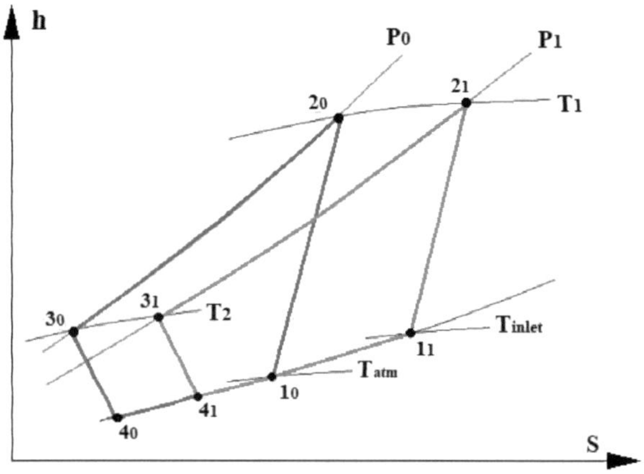

Figure 3. Processes in APHI in hs-diagram with increasing air temperature at the compressor inlet.

Under the condition of a slight change in the air flow rate ($G_{A0} = G_{A1} = G_A$) - such an assumption is quite possible in the range of real temperatures and pressures of air in the high-voltage pump unit - the energy that must be supplied to the compressor decreases

$$(h_{2_1} - h_{1_1}) < (h_{2_0} - h_{1_0}). \tag{3}$$

This also reduces the energy supplied to the compressor from the air turbine.

$$(h_{3_1} - h_{4_1}) < (h_{3_0} - h_{4_0}). \quad (4)$$

The increase in the heat conversion coefficient with an increase in the air temperature at the inlet to the compressor or decreases, depends on how intensively the total required power of the compressor and the power of the air turbine change. This is determined by the fact that the electrical power Ne required to ensure the operation of the AHPI is the difference between the total required power N_C of the compressor and the power of the N_{AT} of the air turbine

$$N_E = N_C - N_{AT}. \quad (5)$$

Transforming formula (5), in accordance with the designations in figure 3, it can be reduced to the form

$$N_E = G_a\left[(h_{2_1} - h_{1_1}) - (h_{3_1} - h_{4_1})\right]. \quad (6)$$

3. Results and discussion

Calculations were carried out using formulas (1), (2) and (6) with the following initial data:

$$T_1 = 100^0C;\ T_2 = 25^0C;\ p_{atm} = 0{,}15\ \text{MPa};$$

$$\eta_{oiC} = 0{,}82;\ \eta_{oiT} = 0{,}80.$$

The air temperature T_{inlet} at the compressor inlet varied from 5 to 25 ^{0}C.
The calculation results are presented graphically in figure 4.

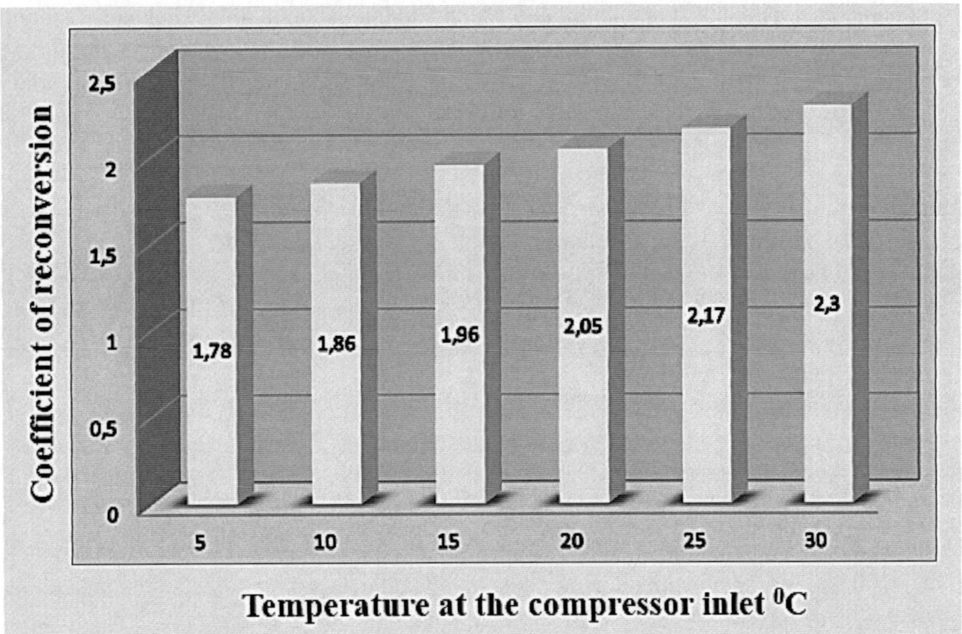

Figure 4. Dependence of the conversion factor of an air heat pump on the air temperature at the compressor inlet.

4. Conclusion

The calculations showed that with an increase in the air temperature at the compressor inlet, the conversion coefficient increases.

Thus, under the accepted conditions, preheating the air in front of the compressor makes it possible to increase the efficiency of the AHPI operation.

References

[1] Urvanov S V, Kondrashova Yu N, Gazizova O V and Skvortsov D S 2017 Development and research of existing possibilities of using an expander-generator unit for a gas distribution station using a heat pump unit as a gas heating system *Bulletin of YuUrGU. "Energy" series* **2** 5-13

[2] Buranov M, Mukolyants A and Sotnikova I 2019 Generation of electrical energy at gas pipelines using a transported natural gas technological pressure drop *J. Phys.: Conf. Ser.* **1399** 055038 https: // doi: 10.1088 / 1742-6596 / 1399/5/055089

[3] Agababov B C 2011 Analysis of the influence of operation parameters of a fuel-free energy generating unit on its efficiency *Energy Saving and Water Treatment* **6(74)** 26-8

[4] Mukolyants A A, Buranov M D, Sotnikova I V and Makhmudov H F 2020 Operation analysis of expander-generator unit at a gas distribution station *Journal of Physics: Conference Series* **1515** 022053 https: // doi: 10.1088 / 1742-6596 / 1515/2/022053

[5] Taslimov A D, Berdishev A S, Melikuzuev M V and Rakhimov FM 2019 E3S Web of Conferences B 139 010 https://doi.org/10.1051/e3sconf/201913901082Federal norms and regulations in the field of industrial safety *"Instruction for the calculation and use of anchor support in coal mines"* (approved by the Order of Rostekhnadzor No. 610 from December 17, 2013)

[6] Aleksandrov A A, Agababov V S, Dzhuraeva E V, Koryagin A V and Utenkov V F 2004 Analysis of the joint work of the expander-generator unit and the heat pump *Problems of power engineering* **7-8** 50-60

[7] Mukolyants A A, Buranov M D, Sotnikova I V and Azimova M M 2020 Analysis of the influence of the input parameters on the efficiency of the operation of the detander generator unit in the gas network system of the republic of Uzbekistan *E3S Web of Conferences* **216** 01130 https://doi.org/10.1051/e3sconf/20202/601130

[8] Belousov A J (2016) Oil and Gas Technologies B 3 51-55 *Website of Public company "EVRAZ"* Retrieved from: http://www.evraz.com/ru/

[9] Agababov V S, Dzhuraeva E V and Arkharova A Yu 2007 Installation for electricity generation based on DGA, air turbine and compressor *MPEI Bulletin* **2** 48-53

[10] Rakhmonov I and Reymov K 2019 *j. Energetika* B **62(6)** 528-3 https://doi.org/10.21122/1029-7448-2019-62-6-528-535

Ecological engineering as a way to reduce industrial waste

O B Sheveleva[1,3], E V Slesarenko[1], O V Zonova[1], T F Mamzina[2] and N D Vagina [1]

[1] T.F. Gorbachev Kuzbass State Technical University, Department of Finance and Credit, 28 Vesennya st., Kemerovo, 650000, Russian Federation
[2] T.F. Gorbachev Kuzbass State Technical University, Department of Theory and Technology of Management, 28 Vesennya st., Kemerovo, 650000, Russian Federation

[3] E-mail: shob.fk@kuzstu.ru

Abstract. The annually increasing volume of human consumption entails a constant increase in the volume of consumption waste and industrial waste as a result of the need to meet the growing demand. The commodity dependence of the Russian economy only aggravates this problem, since the mining industry is the main source of industrial waste generation. The share of recycled waste tends to decrease, and the share of waste disposed in the territories owned by enterprises is growing. One of the modern methods for solving environmental problems is ecological engineering, which promotes environmentally friendly decision-making and is accompanied by the development of environmental programs as part of industrial enterprise engineering, which, along with increasing the production efficiency and performance, reduce the generation of harmful and hazardous waste. Therefore, in order to achieve production goals, it is necessary to identify the needs of target markets and ensure that demand for them is met through the use of more efficient, cost efficient and environmentally friendly means of production.

One of the most important global environmental problems at the present time is waste management [1-4].

The level of human consumption increases annually, and this entails both a direct increase in the volume of the population waste [5-6], and an increase in industrial waste resulting from increasing production volumes to meet growing demand.

In the Russian Federation, this problem is aggravated by the commodity dependence of the economy, which leaves a significantly influences the industrial waste generation in the country, since it is known that the extractive industry is the industry that generates the largest volume of waste [7-8]. In this regard, issues related to the reduction of industrial and consumption waste, their environmentally safe disposal are extremely urgent for the Russian Federation. This is evidenced by a significant number of legislative and regulatory acts in the field of waste management that are currently in force in the country.

National expenditures on environmental protection in the area of "Waste management" in the Russian Federation are the third after such areas as "Waste water collection and treatment" and "Air protection and climate change prevention".

There is an increase in national expenditures in the area of waste management, and both the amount of expenditures increases (from 51.611 million rubles in 2013 to 79.517 million rubles in 2017, i.e. by 54%), and the share of these expenditures in the total value of national expenditures on environmental protection (from 10.8% in 2013 to 12.1% in 2017).

Business sector accounts for 71.5% of national spending in this area of environmental activities, and the specialized providers of environmental services account for the remaining 28.5%.

Over the period 2013-2017, the final consumption of environmental services increased by 31.8 times, although the total amount of household expenditures for environmental services compared to national expenditures is extremely insignificant, amounting to 1.621 million rubles in 2017.

Despite all this, according to the Federal State Statistics Service, the generation of industrial and consumption waste in the Russian Federation has been steadily increasing over the past five years (2015-2019) (table 1).

Table 1. Dynamics of the volumes of generation, utilization and disposal of industrial and consumption waste in the Russian Federation for 2005-2019.

Years	Waste generation, million tons	Share of hazardous waste in the total waste volume, %	The volume of utilized and neutralized waste, million tons	The share of recycled and neutralized waste in the total volume of waste generated, %
2005	3035.5	4.69	1265.7	41.70
2015	5060.2	2.18	2685.1	53.06
2016	5441.3	1.81	3243.7	59.61
2017	6220.6	1.72	3264.6	52.48
2018	7266.1	1.35	3818.4	52.55
2019	7750.9	1.30	3881.9	50.08

Thus, the volume of waste increased from 5.06 billion tons to 7.75 billion tons, i.e. by 53.2%, and over 15 years - by 2.55 times.

Along with an increase in the generation of industrial and consumption waste, there is an increase in the volume of utilized and neutralized waste - from 1.27 billion tons in 2005 to 2.69 billion tons in 2015, and 3.88 billion tons in 2019. However, the volume of waste generated is increasing faster than the utilized waste volume - in 2015-2019, the rate of increase in the volume of industrial waste amounted to 53%, with an increase in the utilized waste volume by 45%. As a result, the share of recycled waste in the total amount of waste generated has decreased from 53.1% in 2015 to 50.1% in 2019.

At the same time, there is an increase in the volume of waste disposed at facilities owned by enterprises - from 2.08 billion tons in 2005 to 2.33 billion tons in 2015 and 3.8 billion tons in 2019, i.e. by 62.9% in five years and by 83% in 15 years (table 2).

Table 2. Dynamics of disposed production and consumption waste at facilities owned by enterprises for 2005-2019.

Years	Volume of production and consumption waste disposed at facilities owned by enterprises, million tons	Share of disposed waste in storage areas,%	Share of disposed waste in landfills, %
2005	2077.3	80.4	19.6
2015	2333.1	84.8	15.2
2016	2620.8	80.3	19.2
2017	3204.5	74.2	25.8
2018	3575.4	71.2	28.8
2019	3800.8	69.0	31.0

The share of industrial and consumption waste disposed at facilities owned by enterprises increased from 46.1% to 49%. At the same time, 69% of the waste disposed by enterprises in 2019 is located in storage facilities, and only 31% are buried.

Positive dynamics can be traced in the share of hazardous industrial and consumption waste - it decreased from 4.7% in 2005, to 2.2% in 2015, and to 1.3% in 2019, i.e. over 15 years by 3.4%. The largest share of industrial and consumption waste falls on the V class of hazard, which is classified in accordance with legally established criteria as virtually non-hazardous waste.

Analysis of economic activities in the Russian Federation showed that the largest share of industrial and consumption waste generation falls on such an activity as mining - and the share of this activity tends to increase - from 92.0% in 2015 to 93.6% in 2019. At the same time, the share of coal and metal ore mining in 2019 accounts for 67.1% and 22.5%, respectively, of the total amount of industrial and consumption waste generated.

As for the processing industries, here the main activities that generate the largest volume of waste are metallurgy, chemical production, and production of textiles - their shares in 2019 accounted for 52.4%, 14.3%, and 13.9% of the total amount of processing industries waste.

All of the above reflects the relevance of issues related to the need to reduce the volume of industrial and consumption waste, as well as their safe disposal.

One of the modern trends in solving environmental problems is ecological engineering [9-11]. It is a fairly young branch of ecology, although it is part of engineering in itself, since it includes a sequence of steps and procedures similar to those of engineering science, starting with defining goals and objectives, identifying problems, searching for solutions to problems, and ending with choosing the final solution.

The results of the implementation of ecological engineering mechanisms is reducing the impact of adverse environmental factors on the population, improving the quality of the natural environment. Ecological engineering, including a set of engineering techniques and methods, can serve as the basis for solving environmental problems, reducing their severity. Moreover, certain developments of environmental engineering can find application both at the stage of waste generation and at the stage of their disposal.

Eco-engineering is a set of procedures and actions, the result of which is the minimization of damage to the environment, rationalization of the use of natural resources. It is characterized, on the one hand, by the participation of the state, and, on the other hand, by the interaction of entities - "polluters" of the environment - with engineering companies developing projects that help to minimize the harmful effects of human waste on nature.

Ecological engineering is an area closely related to environmental monitoring, controlling, auditing and marketing.

Together, these areas of environmental science make it possible not only to assess the state of the environment (including on the basis of information from measuring instruments), but also contribute to the identification of problems in the field of environmental safety at an early stage and the timely development of ways to solve them.

In the chain of environmental sciences, at the first stage, there are environmental auditing and environmental monitoring, which contribute to the identification of environmental problems at industrial enterprises [12].

At the second stage, environmental marketing is put into action, on the basis of research of which a set of options for solving the identified environmental problems is formed.

And at the final, third, stage, ecological engineering is used, contributing to the development of a specific management decision in the field of environmental safety.

All three stages are accompanied by the implementation of environmental controlling procedures as part of the environmental management of an enterprise.

Thus, eco-engineering is a mechanism for making environmentally sensitive decisions, accompanied by the development of environmental programs as part of industrial enterprise engineering, which, along with an increase in the production efficiency and performance, contribute to the reduction of the generation of harmful and hazardous waste.

The need for environmental studies of production systems is associated with determining the possibility of:

- modernization of technological systems and units by improving their environmental performance;
- development of more advanced production methods and processes through the relationship of indicators "efficiency" and "environmental friendliness";
- identification of more efficient methods for recycling industrial waste.

All other things being equal, products that are more affordable and comparable in quality and the level of environmental safety for humans are more in demand than competitive products. Therefore, in order to achieve production goals, it is necessary to identify the needs of target markets and ensure that demand for them is met through the use of more efficient, cast-efficient and environmentally friendly means of production.

References

[1] Yacout D, Abd El-Kawi M A and Hassouna S 2014 Industrial waste management implementation *The 15th Arab international conference of materials science «Materials for Water Applications»* 180-6
[2] Rasel Md, Zerin I and Hasanul Hoque K M 2019 Industrial Waste Management by Sustainable Way *European Journal of Engineering Research and Science* **4(4)** 111-4
[3] Gurjanov A V, Shukalov A V and Zharinov I O 2020 Solid industrial waste management cyber-administration *IOP Conf. Series: Earth and Environmental Science* **548** 052057
[4] Perfilov V A, Volskaya O N 2016 Industrial waste management to improve environmental security *South of Russia ecology development* **11(2)** 205-12
[5] Supangkat S and Herdiansyah H 2020 Analysis Correlation of Municipal Solid Waste Generation and Population: Environmental Perspective *IOP Conf. Series: Earth and Environmental Science* **519** 012056
[6] Read Q D and Muth M K 2021 Cost-effectiveness of four food waste interventions: Is food waste reduction a «win–win?» *Resources Conservation and Recycling* **168(5)** 105448
[7] Abiahu C M G, Fagorite V I, Oli I C and Okeke O C 2019 Mine Waste: sources, problems and mitigations *International Journal of Applied Science* **5-7** 1-12
[8] Hlavata M, Cablík V and Svoboda M 2020 Utilisation of coal mining waste *GeoScience Engineering* Vol. **66-1** 53-9
[9] Awasthi A K, Sankar Cheela V R, D'Adamo I and Li J 2021 Zero waste approach towards a sustainable waste management *Resources, Environment and Sustainability* **3** 100014
[10] Alzeyadi A T and Al-Ansari N 2020 Environmental Engineering and Management *Environmental Engineering and Management Journal* **19(4)** 643-53
[11] Mukaba Mukaba J-L, Emmanuel A and Petrik L F 2020 Environmental Engineering and Management *Environmental Engineering and Management Journal* **19(3)** 475-83
[12] Sheveleva O B, Slesarenko E V, Vagina N D, Kudrevatykh N V and Mamzina T F 2019 Environmental monitoring as an information system to ensure environmental safety of a region (as exemlified in kemerovo region) *IOP Conf. Series: Materials Science and Engineering* 012034

Investigation of composition and properties of EAF dust for metal reduction

D A Kalganov[1,2], D A Pavlov[1], A P Anzulevich[1], L N Butko[1], V A Tolkachev[1] and Z Peng[3]

[1] Chelyabinsk State University, 129 Br. Kashirinykh Str., Chelyabinsk, 454021, Russia
[2] ITMO University, 49 Kronverkskiy pr., St. Petersburg, 197101, Russia
[3] Central South University, Changsha, Hunan, 410083, China

E-mail: dezwark@gmail.com

Abstract. In this work, we investigated the composition and structure of the dust of electric arc steelmaking furnaces. We considered the process of direct reduction of metals (mainly Zn and Fe) with biochar from EAF dust. For better control over chemical transformations, it is proposed to use microwave heating of a mixture of EAF dust and biochar. Possible chemical and phase transformations for complex oxides in the composition of EAF dust have been determined.

1. Introduction

The tasks of recycling of waste from the metallurgical and pulp and paper industries are acutely faced by the world industrial community.

The possibility of using such raw materials as secondary raw materials for obtaining useful materials is an important applied problem of related scientific research [1]. Good opportunities for these technologies are provided by the development of new metallurgical technologies using microwave energy [2].

During the production and processing of various types of steel in Russia, up to 2000 tons of electric arc furnace (EAF) dust are generated daily. Dust and gas cleaning systems effectively capture about 99% of this secondary raw material, but its further processing is complicated by both the relatively low content of useful metals and the presence of thermally highly stable phases [3,4]. The need to reduce dust formation and develop a technology for utilization of dust captured by filters is due to its negative impact on the environment and human health.

Also, in recent years, technologies of iron reduction from iron-containing compounds using pyrolyzed carbon-containing biomass - biochar - have attracted much attention [2,5]. The main advantages of their use are a decrease in the reaction temperature, a decrease in slag formation and the absence of harmful substances in the reducing agent. The use of microwave energy makes it possible to carry out all stages of this process, including endothermic ones, and to flexibly regulate the course of the process by changing the power of microwave radiation [6].

2. Methods and materials

X-ray phase analysis (XRD) of the samples was performed using a Bruker Advance D8 diffractometer (Cukα radiation, standard powder technique).

For the process of metal reduction from oxides, biochar was added to the initial samples of EAF dust in a ratio of 4:1.

To determine possible chemical and phase transformations, we used a NETZSCH Jupiter F5 differential scanning calorimetry (DSC) complex heated to 1473K in an argon atmosphere at a constant rate of 10°/min.

The sizes of EHP dust microparticles and the phase composition uniformity were determined from scanning electron microscopy images obtained using a Jeol JSM 6510la with preliminary deposition of a thin conductive Pt layer on the samples.

3. Results

According to the data of electron microscopy (figure 1), it can be concluded that the sizes of dust particles are in the range of up to 500 nm. The samples under study represent a statistically homogeneous mixture of phases:

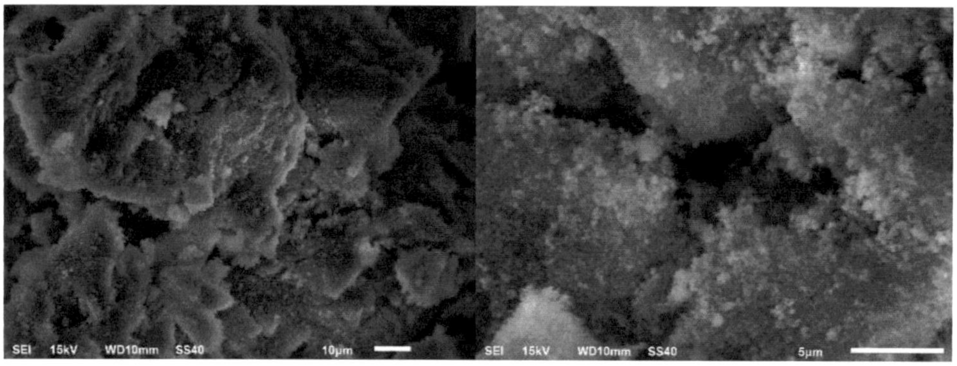

Figure 1. Microimages of EAF dust.

The composition of the samples was determined using the PDF2 database file (figure 2).

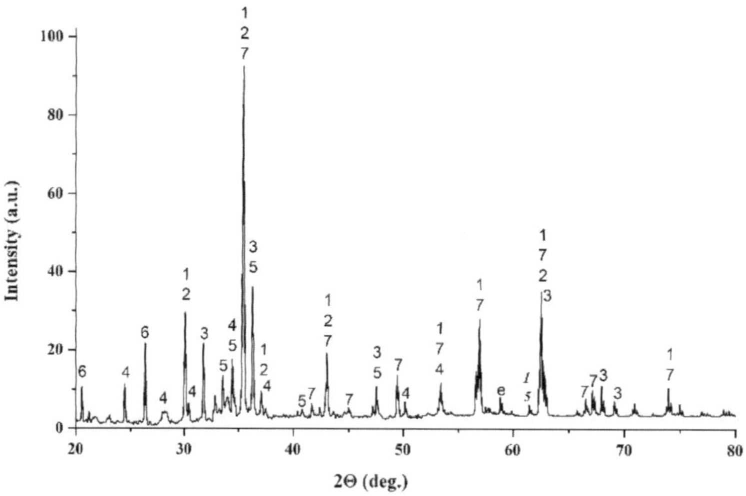

Figure 2. X-ray diffraction plot of EAF dust samples.

Based on the XRD data, the following phase and elemental composition of the EAF dust sample was obtained:

Table 1. Phase and elemental composition of EAF dust.

№	Phase composition	Content, % vol.	Equivalent elemental composition: Chemical element	Content, % vol.
1	$Fe+2Fe_2+3O_4$	3.8	O	37.6
2	$Mg_{1,55}Fe_{1,6}O_4$	36.2	Mg	9.7
3	ZnO	10.0	Si	15.1
4	$CaMgSiO_4$	15.0	Ca	7.0
5	$Ca_{14}Mg_2(SiO_4)_8$	7.6	Fe	22.6
6	SiO_2	23.7	Zn	8.0
7	$(Fe_{0,914}Si_{0,086})(Fe_{0,998}Si_{0,002})2O_4$	3.8		

Common analysis and approximation of DSC and DTG curves by Gaussian curves showed the presence of 7 stages of thermolysis (figure 3):

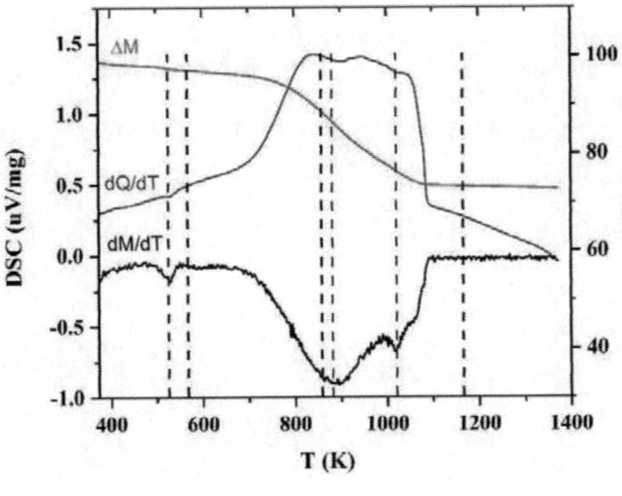

Figure 3. DSC samples of the mixture: EAF dust - biochar in a stoichiometric ratio of 4:1.

The main chemical processes occurring during the direct reduction of iron with solid carbon C depend on the initial composition of the EAF dust, but generally correspond to redox reactions:

$$6Fe_2O_3 + C \xrightarrow{750-1090K} 4Fe_3O_4 + CO_2 \quad (1)$$

$$Fe_3O_4 + C \xrightarrow{1090-1270K} 3FeO + CO \quad (2)$$

$$FeO + C \xrightarrow{1270-1455K} Fe + CO \quad (3)$$

$$ZnFe_2O_4 + 3CO \xrightarrow{1170K} ZnO + 2Fe + 3CO_2 \quad (4)$$

$$ZnO + C \xrightarrow{1210K} Zn + CO \quad (5)$$

4. Conclusion

As a result of the comparison shown in figure 3 of the stages of thermolysis with possible chemical transformations (1) - (5), we can say that the peaks at temperatures of 863K and 870K are most likely due to reaction (1) and the concomitant reaction with carbon monoxide, the peaks 1010K and 1020K correspond to reaction (2) in a similar way, and the peak at 1176 K corresponds to the Gibbs energy for reaction (5).

The recovery of EAF dust with the addition of biochar in the stoichiometric ratio was carried out in a multimode microwave resonator [4]. The resulting mixture consisted of a ferromagnetic fine powder of iron and silicates.

Acknowledgments

The reported study was funded by RFBR and NSFC, project number 21-58-53044, and was supported by the National Natural Science Foundation of China under Grant 52111530046.

References

[1] Yakornov S A, Pan’shin A M, Kozlov P A, Ivakin D A and Golubeva E V 2018 *Metallurgist* **61(9)** 891-8
[2] Ye Q, Li G, Peng Z, Lee J, Lin X, Rao M, Zhang Y and Jiang T 2019 *Powder Technology* **342** 224-32
[3] Simonyan L M, Alpatova A A and Demidova N V 2019 *Journal of Materials Research and Technology* **8(2)** 1601-7
[4] Butorina I V and Fedotova E S 2016 *Steel in Translation* **46(6)** 447-52
[5] Anzulevich A P, Kalganov D A, Anzulevich S N, Bychkov I V and Peng Z 2020 *Journal of Physics: Conference Series* **1461(1)** 012007
[6] Wang L, Peng Z, Lin X, Ye Q, Ye L, Zhang J, Liu Y, Liu M, Rao M, Li G and Jiang T 2021 *Powder Technology* **383** 509-21

Comparative analysis of CAD software packages for engineering design

I A Magomedov[1,4], A M Bagov[2] and A A Elmurzaev[3]

[1] Faculty of information technology, Chechen State University, 32 Sheripov Street, Grozny, 364024, Russia
[2] Scientific Research Department, Kabardino-Balkarian State University, 173 Chernyshevsky Street, Nalchik, Russia
[3] Department of Technological Machines and Equipment, Grozny State Oil Technical University named after academician M. D. Millionshchikov, 100 H A Isaeva pr, Grozny, 364051, Russia

[4] E-mail: ismwork@mail.ru

Abstract. This work is focused on the comparison of two software packages Solidworks and Inventor. The following two software packages are well known for their ability to design and they share a leading position in the market. A comparison of the functionality of the two software packages will be made. The difference will be outlined on the divorce tools they have. The market share of CAD software packages will also be briefly outlined. Both software packages will then be compared for how easy they are to use for customers. Lastly, the availabilities of the following software will be discussed.

1. Introduction

In the present day, we are granted to use software packages to fulfill the required needs of daily life whether it's communicational, computational, or others. One can easily say that today's software packages are more efficient and intelligent. Meaning that they weigh less space and the functionality is way wider. The usage of software in modern times is immense. It is integrated with every aspect of human life. These software packages can be as simple as an alarm clock or complex as combinations of software packages that are used to analyses huge data such as 2D/3D modeling and analyzing tools. Here and hence the usage of the last-mentioned software packages will be covered. These tools are widely used in almost every industry and they can be well adapted for the specifications of the fields. Meaning that they have a diverse range of tools to work with.

These software packages are named as CAD, which means Computer-Aided Design and as well as known in the past Computed Aided Draughting. The CAD usage migrated from aircraft and then permitted into the following fields: automotive, architecture, animation, artificial intelligence, etc. As in any case these software packages are invented to result in lessening time waste (or reduction of the processing time), labor cost reduction (labor reduction), analyze complex problems, that are hard to achieve in real-time. For further increasing the production flow standardized parts files are included in the modern software packages. Today almost every CAD related package have so-called ready files to use as templates or standardised parts files to quicken the process. These standardizers can vary from different types of doors, windows, stairs, and other similar types of objects. However, it was not that simple in the past time. Companies of manufacturing, architects' offices, engineering firms had immense premises devoted to labors working on drawings of the products (parts of buildings, bridges, vehicles etc.) [1]. One can argue that some skills were left out with the introduction of software packages and that led to massive labor reduction, which is good for one side and similarly bad for the other side. Nonetheless, it pushed the draughting to a new level.

The following software packages can be divided into two main groups: media and engineering. Where in the first the usage of the software is mainly considered to create the objects and simulate them as to how they would occur in the real world. Hence the objects can have unprecise parameters and no real-life representation. However, in the second group, the preciseness is very essential. Therefore, the correct outcome from the software packages in engineering is the main characteristic when choosing software.

The following group of software packages can be classified into the first group: Blender, 3Ds max, Unity, Cinema 4D, etc. These software packages offer a wide range of tools to work with and are mainly utilized in media and entertainment. Surely, they can be used in other fields such as architecture, mechanical engineering, and similar areas, but due to the specifications, they are commonly used as tools when creating animation, movies, advertisements, or generally in entertainment. Similarly, software packages from the second group can perform some of the function that can be done by using the first mentioned software packages. However, the limitation will be met when performing some of the functions, thus these software packages are mainly used to build 3D models and if necessary, to analyses them under some different conditions to see how they would perform. Into these packages can go the following software: SOLIDWORKS, Inventor, Onshape, Abaqus, Fusion 360, Ansys, Rhinoceros, Solid Edge and etc.

2. Solidworks and Inventor comparison

The following two software comes from different companies, but both contribute to the same field. The main reasons that these two software packages are popular in the market are the vast number of tools to be used for modeling the given objects, the long-run reputation of the companies, numerical analysis capability, and mostly the user-friendly interface. In addition, these software packages look the same and have almost identical tools. However, there are some major differences and thus the work will look into and outline them.

The following paragraph will be devoted to the comparison of the two software packages. The comparison will be done using two popular software packages Solidworks and Inventor The main aspect is how good the software is and it can be determined by different features. As was described in this work there are a lot of similar features in these software packages. Therefore, the advantages and disadvantages of both software will be described.

2.1 Functionality

As mentioned before both are similar to each other and both focus on 3D rendering and machine part design. Both come from well-established companies, which have been in the market for quite a while and earned the trust of the users [2]. But by diving, a bit deep one can find out that each of them can provide some unique function the other one does not have.

2.1.1 Solidworks. It can be said that the simulation part of the software can give the necessary requirements for today's times. It can work on stress tests, temperature, electronics and so on and its giving to user the opportunity to work with a wide range of parameters. Solidworks has a sustainability tool to determine the environmental impact on the built part or parts. Its worth mentioning that Solidwork wast the first to recognize the benefits of the usage of AR/VR that pushed the software to the next level. And the most important one of all the large design review. It offers the power to design, test, and model pieces with millions of components without burning out a computer's CPU. Lastly, due to the fact that Solidworks has been around for a long time, it can offer a large library of tutorials to learn from [3, 4].

2.1.2 Inventor. As an alternative to parametric design's prediction-based process, Inventor offers direct-edit and free-form modeling. The software has its way to automate the math behind scenes for complex moving parts, like kinetic blades or support wires. This allows one to focus on the big picture of a design without getting bogged down by small details. Also, the software has a quick loading time due to its being able to distinguish and load the graphic parts of a design separately from the material and geometric data. Inventor has its own ways to test 3D designs in real-world situations. Its Dynamic Simulation module, in particular, applies specific types of pressure to key points, like torque to joints. A favorite among many is the burst weldment tool. This lets one undo a weld to simulate what would happen if it gave way in an emergency situation [5].

2.2 Market share

The main players in the market are Solidworks, and Autodesk's Fusion 360, Autocad and Inventor. These software packages hold more than half of the overall CAD market share. The survey done by the CNC Cookbook's on the dates of 2016 and 2017 of the CAD market shares illustrates the four main leaders in the market (figure 1) [6,7]. Over the years the following software packages shifted in the direction of losing or gaining the market share. It is interesting to note that by taking these software packages individually Solidworks around 2015 is still holding first place, but taking Autodesk's software together it shifts to second place. However, in the following years, the tendency to the leading place takes Fusion 360. If the two of the software packages Solidworks and Inventor to be compared to each other Solidworks stays in the leading place throughout the years.

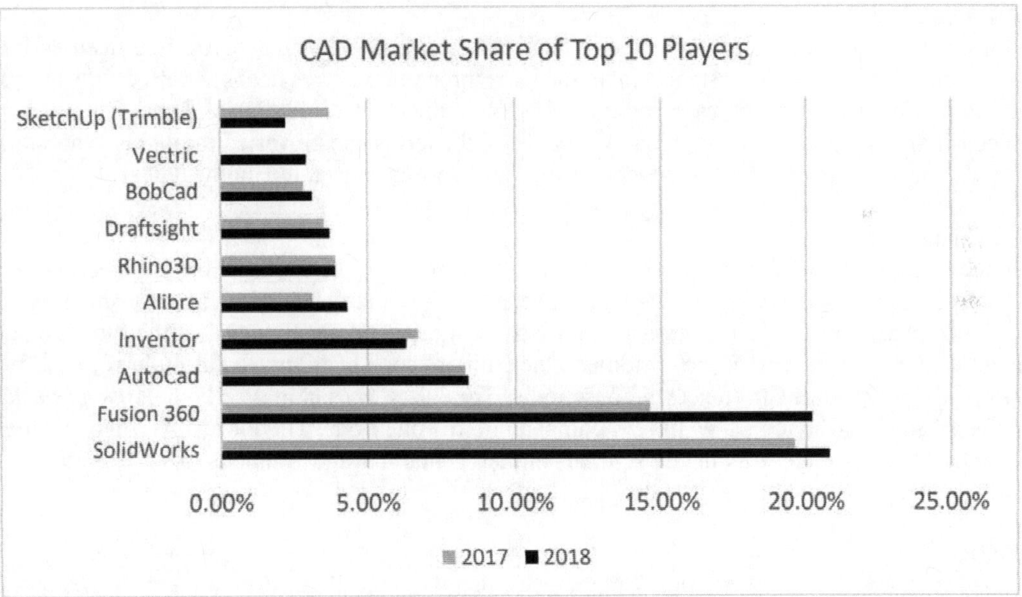

Figure 1. CAD market share.

2.3 User-friendly

Both of the software is extremely easy to use. Both have a similar interface with the depictive tools to guide the user. Sketching parts of both tools have similar functionality with a bit discrepancy, but these unmatched parts have few influences on the total view. The similarities can be seen in the survey done by the G2 [8, 9]. The survey illustrated that both of the software packages received 8.3 out of 10 marks when the respondents were asked on the terms of easy use (figure 2) [7]. The same thing can be said about the other parameters they are almost indistinguishable.

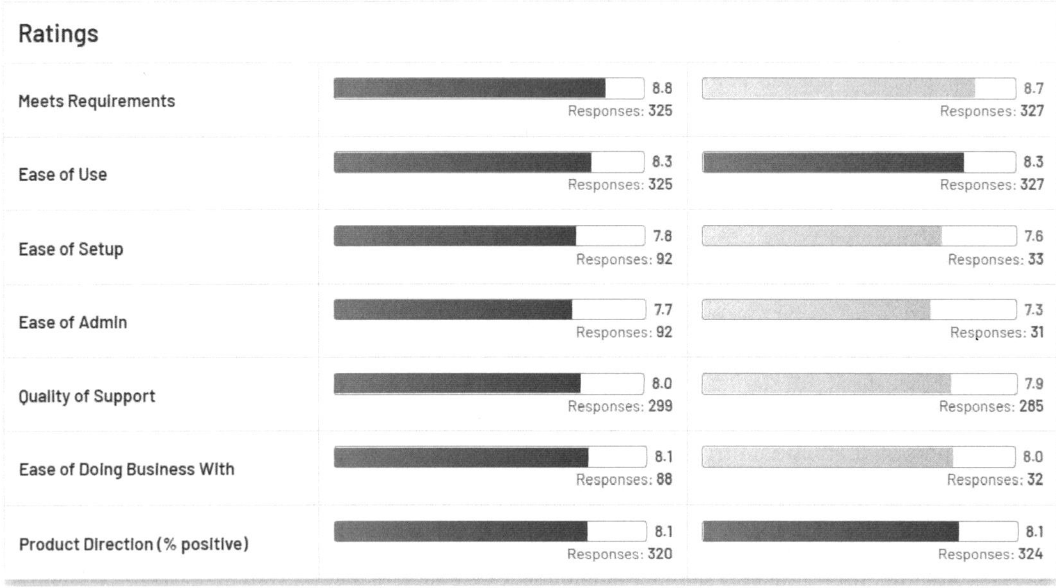

Figure 2. Rating of two software packages (Inventor-left and Solidworks-right).

2.4 Availability

Bot software packages are not free and have different prices. However, there are free licenses for the narrowed circle of people. For instance, Inventor gives three years of free licenses for the students, tutors generally for educational purposes. Similarly, the same thing can be said for Solidworks, but instead of free licensing, it provides cost reduction. Nonetheless, the terms can be varied throughout the years. In addition, the availability of the Solidworks for the Apple products is at the moment absent.

3. Conclusion

To conclude, the following work was done on the comparison of the tools for the engineering design. The two software packages were selected on the behalf of their market share and due to similarity of the two. The comparison showed that both are similarly designed and perform almost the same functions. besides, according to survey of the customers, the similar scores were given for both of the software. However, there are some differences in these tools. The differences come for both software packages from the unique build-in tools they have if compared to the other one. Therefore, one can conclude that the compared tools have for the most time similar functions and methods only differing in the uniqueness.

References

[1] Allen J and Kouppas P 2012 Computer Aided Design: Past, Present, Future. *Design and Designing: A Critical Introduction,* 97-111
[2] Zolkin A L, Tychkov A S and Kalyakulin A N 2010 *Samara: SamGUPS* 63-6
[3] Magomedov I A and Sebaeva Z S 2020 *Journal of Physics: Conference Series* **1515** 032073
[4] Matt L 2018 *Mastering SolidWorks.* John Wiley & Sons, p 13-57
[5] Kishore T 2017 Getting Started with Autodesk Inventor 2018 (Learn Autodesk Inventor 2018 Basics), 1-14
[6] *CNCCookbook 2016 CAD Survey Results, Part 1: Market Share, CNC Cookbook* Retrieved from https://www.cnccookbook.com/cnccookbook-2016-cad-survey-results-part-1- market-share/
[7] Bob Warfield 2017 *CNCCookbook 2017 CAD Survey Results [+Customer Satisfaction Awards], CNC Cookbook* Retrieved from https://www.cnccookbook.com/cnccookbook-2017-cad-surveyresults/G2

[8] *Compare Inventor and SOLIDWORKS, CAD Survey Results, Website of Public company G2* Retrieved from https://www.g2.com/compare/inventor-vs-solidworks
[9] *Hitechcaddservices 2020 SolidWorks Vs Inventor: Feature Based Comparison of Two Major 3D CAD Platforms* Retrieved from https://www.hitechcaddservices.com/news/solidworks-vs-inventor-feature-based-comparison-of-two-major-3d-cad-platforms

Innovative technologies in the production of curd

Y A Alekseeva[1], D Ts Garmaev[2], T A Khoroshailo[3] and I V Serdyuchenko[3]

[1]Irkutsk State Agricultural University named after A.A. Ezhevsky, Irkutsk, 664038, Russian Federation
[2]Buryat State Agricultural Academy named after V.R. Filippova, Ulan-Ude, 670024, Russian Federation
[3]Kuban State Agrarian University named after I.T. Trubilina, Krasnodar, 350044, Russian Federation

E-mail: yulia_a72@mail.ru, dylgyr56@mail.ru, tatyana_zabai@mail.ru, 79184370664@yandex.ru

_**Abstract.** At the moment, the Russian Federation has mastered various methods of making dairy products. Cottage cheese is considered one of the most beloved goods of the population. This is explained not only by its taste, but also by its nutritional and bioavailability. Future progress in the production of cottage cheese must be achieved both through the introduction of comprehensive mechanization and automation of production processes, and through the use of investments for the technical reequipment of equipment and the installation of new lines. During the reconstruction and modernization of production facilities, mainly automated lines of manufacturing plants from Germany, Denmark, Poland, Bulgaria are installed. At some factories, work continues on the lines for the production of acid curd Ya9-OPT (Russia). In this regard, it is important to improve the existing and create new equipment for the production of curd and curd products. Along with the development of new methods for the production of curd and curd products, in order to increase their competitiveness, it is quite reasonable to use technologies with a closed method of conducting the process, i.e. it is necessary to exclude the influence of the human factor on a complex technological process. For the successful sale of products, it is necessary to expand the range of cottage cheese products, improve their quality and price competitiveness, develop production with a high proportion of prepackaged and packaged products with a design that meets consumer demand. It is necessary to increase the degree of factory readiness of products for direct consumption.

1. Introduction

In Russia, there is a constant increased demand for the production and consumption of cottage cheese, which is due to both the consumer's habit and the attractive taste of the product. In addition, cottage cheese and products made from it have a high nutritional value and are affordable [1,2,3].

Fatty cottage cheese is best used in its natural form, low-fat - for making curd products (cheesecakes, puddings, etc.). From a mixture of buttermilk and skim milk, table dietary cottage cheese (3% fat) is produced. Soft dietary cottage cheese, low-fat (5 or 11%), as well as sweet, has a delicate consistency. They produce dietary fresh low-fat cottage cheese, its acidity is 2-3 times less than that of other species.

Cottage cheese has a significant content of well-balanced protein (14-16%). Cottage cheese proteins are digested better than meat, fish and even milk proteins: 300 g of cottage cheese almost completely satisfies a person's daily protein requirement. It should be noted that during the production of cottage cheese, almost all the valuable components of milk pass into the finished product. Also, the high nutritional value of cottage cheese is due to the increased content of amino acids important for the human body - methionine, lysine and choline. Methionine promotes the elimination of cholesterol from the body, choline is necessary for the normal functioning of the nervous system. Methionine and choline also normalize liver function, preventing obesity. Curd has a lipotropic effect and is widely used in diseases of the liver, cardiovascular system, obesity, diabetes, after burns and bone fractures, and many other diseases [4].

2. The results and discussion

Features of technological equipment for the production of cottage cheese are determined by the methods of its production. In one of the farms of the Irkutsk region, cottage cheese is produced in the traditional way, the scheme of which is shown in figure 1.

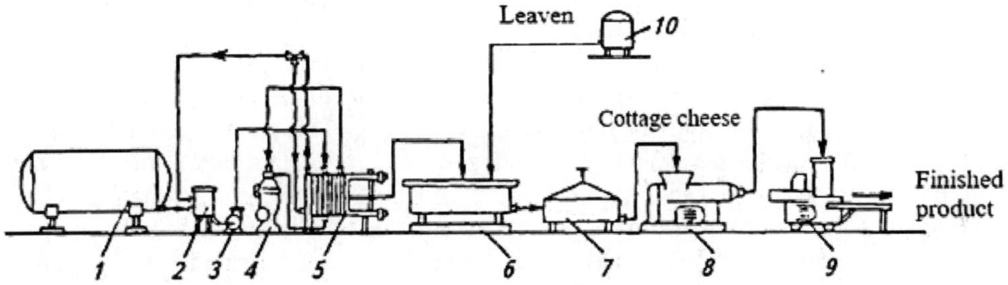

Figure 1. Scheme of the technological line for the production of curd in the traditional way: 1 – milk container; 2 – balancing tank; 3 – pump; 4 – separator-cleaner; 5 – plate pasteurization and cooling unit; 6 – curd bath; 7 – press cart; 8 – cooler for curd; 9 – automatic cheese filling machine; 10 – fermentor.

In the production of curd in the traditional way, for fermentation of normalized milk, special tanks are used to separate whey from the curd mass, presses and self-pressing baths. All equipment for the production of curd can be divided into several groups: equipment for obtaining and processing a curd, cooling and packaging of curd [5,6].

Milk normalization is the regulation of the composition of raw materials to obtain a finished product that meets the requirements of the standard. Milk standardization can be carried out by mixing the main milk constituents such as cream or skim milk. Mixing normalization is carried out in storage tanks and baths equipped with agitators. To reduce the mass fraction of fat in whole milk, it is mixed with skim milk, and to increase it – with cream. In the flow, milk is standardized in separators-cream separators equipped with special devices for standardizing [4,5,6].

At enterprises, normalization is carried out in three ways:

- in the presence of the required amount of cream and skim milk, they are added to whole milk, mixed and at the same time the mass fraction of fat is regulated in it;
- part of the whole milk supplied for processing is separated, cream and skim milk are obtained, and then the rest of the non-skimmed whole milk is mixed with skim milk and cream, while regulating the mass fraction of fat;
- all milk supplied for processing is normalized, and the part of the cream and skim milk remaining from the normalization is taken away for further processing.

Milk cleaning is carried out to remove mechanical impurities. Cleaning is carried out in a milk separator with centrifugal sludge discharge. The cleaning process is best done at temperatures between 40 and 45°C.

The normalized pasteurized mixture is fermented with a sourdough prepared on pure cultures of lactococci at a temperature of 30°C in the cold season and 28°C in the warm season. With the accelerated fermentation method, a symbiotic starter culture is used, prepared on the cultures of lactococci and thermophilic lactic acid streptococci. In this case, the mixture is fermented at a temperature of 32°C [7,8,9].

The dose of the starter culture, depending on its activity and the required duration of fermentation, is from 30 to 50 kg per 1000 kg of the fermented mixture; in order to accelerate the fermentation process, up to 100 kg of sourdough is added to the mixture per 1000 kg of the fermented mixture.

After adding the starter culture to the mixture, calcium chloride is added at the rate of 400 g of anhydrous calcium chloride per 1000 kg of the mixture to be fermented. Calcium chloride is introduced in the form of an aqueous solution with a mass fraction of calcium chloride from 30% to 40%. A solution of calcium chloride is prepared according to the instructions for techno chemical control at dairy enterprises, approved in accordance with the established procedure [8,10].

After adding a solution of calcium chloride, rennet powder, food pepsin or an enzyme preparation in the form of a solution with an enzyme mass fraction of not more than 1% are added to the mixture. The dose of the enzyme with the activity of 100.000 IU per 1000 kg of the fermented mixture is equal to 1 g.

Stirring the mixture after fermentation is continued from 10 to 15 minutes, then the mixture is left alone until a clot of the required acidity is formed: 60±2°T for cottage cheese with a mass fraction of fat 18.0; 9.0%; 73±2°T for cottage cheese with a fat mass fraction of 5.0; 4.0%; 75±5°T for fat-free cottage cheese (fat mass fraction 1.8%). The duration of fermentation of the mixture with an active bacterial starter culture at the temperatures indicated above is from 6 to 10 hours from the moment of adding the starter culture, with an accelerated method – from 4 to 6 hours [1].

To accelerate the separation of whey, the finished curd is heated for 30 to 60 minutes by introducing steam or hot water into the interstitial space of the bath. For uniform heating of the clot, the upper layers are gently mixed from one wall of the bath to the other, due to which the lower heated layers gradually rise up, and the upper layers go down. The curd, heated to the above temperature, is kept for 20 to 40 minutes, then cooled by at least 10°C by running cold or ice water into the interlink space [8,9,10].

The released whey is discharged from the bath and collected in a separate container. The duration of pressing the curd in the press cart lasts at least 1 hour. Then a metal plate is placed on the serpyanka, on which pressure from the press screw is transmitted through a special frame. Pressing continues until the curd reaches the mass fraction of moisture, due to the current technical documentation for this product, but no more than 4 hours. It is allowed to press cottage cheese in a cold chamber for no more than 10 hours.

For packing cottage cheese, parchment, cups, boxes made of polystyrene tape, from combined materials are used with the application of the appropriate requisites. Marking of consumer containers, cottage cheese must comply with STB 1100. Each packaging unit of the product must be coated with an indelible, odorless paint approved for use by the Ministry of Health of the Russian Federation and in other ways the following information: product name; name and location (legal address, including country) of the manufacturer; manufacturer's trademark (if any); net weight, g (kg); the composition of the product; the nutritional value; the number of lactic acid microorganisms; thermal state (for frozen product); storage conditions; date of manufacture; shelf life; designation of this standard; designation of a technological document in the presence of expiration dates other than those established by this standard; information on confirmation of conformity (if any); barcode identification.

The shelf life of frozen cottage cheese is calculated from the moment of their release from the manufacturer. In the absence of cold at the trading enterprise, the shelf life of cottage cheese is reduced to 12 hours, and the sale of other products is not allowed. Long-term storage of frozen cottage cheese is possible only in refrigerators or wholesalers [8,9,10].

At a dairy plant in the curd shop, work begins with a milk pasteurization tank, where milk has already arrived from the equipment shop. The tank for pasteurized milk has a volume of 9.500 tons, there are two of them in production. Milk is brought to a certain temperature in the warm season up to 30°C, in the cold up to 32°C.

Then the milk goes into curd baths of VTN-3.5P type. The bath has a working capacity of 3500 liters. The design of the bath for the production of curd is a stainless steel tank mounted on supports. The tank is equipped with a heat exchange jacket, in which a constant temperature is maintained by supplying water or steam. The device provides a branch pipe for supplying water of different temperatures, overflow and drainage systems. The curd bath is shown in figure 2.

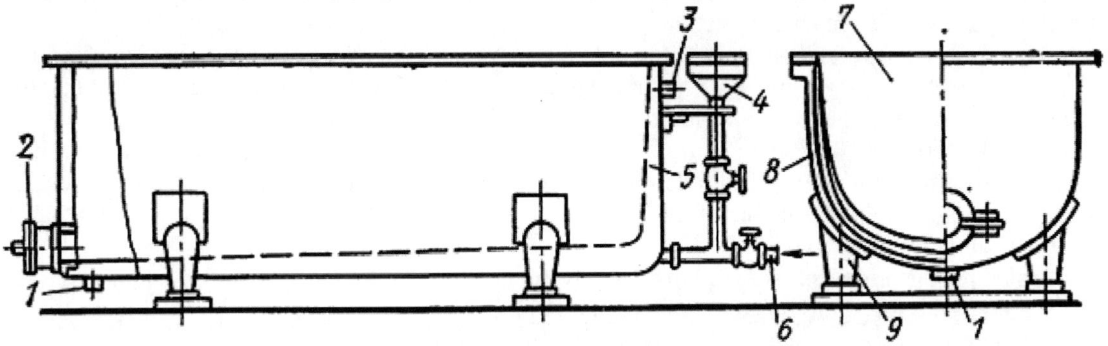

Figure 2. Curd bath:
1 – water drain pipe; 2 – valve for draining whey and curd; 3 – overflow pipe; 4 – funnel;
5 – park shirt; 6 – steam pipe; 7 – inner perforated half-bath; 8 – external bath; 9 – supports

The bath is filled with milk mixture at the fermentation temperature. Then the water jacket is filled with water to the overflow level. Steam is supplied to the water and the boiling process begins.

At the end of the boiling process, ice water is fed into the space between the walls to the overflow level. The cooling process takes place. The bath is equipped with a coil for more efficient cooling. The coil allows you to supply ice water under a pressure of 0.15 MPa, which significantly improves the cooling process [8,9].

After the end of cooling, it is necessary to separate the whey from the curd grain. For this, a disc drain valve is provided in the unloading hatch. Part of the formed whey is drained through the tap. The main process of separating the whey is carried out using a perforated press. The press is driven by two pneumatic cylinders. When the press is lowered to the lower position, the curd is pressed against the walls of the bath, and the whey is separated through the openings of the press. In the lower position of the press, the whey is pumped out through the upper open part of the bath using a food hose and pump. To achieve the required consistency, it can be carried out several times.

Next, the curd is transported by trolleys to the screw bath for cooling. Cooler of curd with two screws, model OT-2. Designed for continuous cooling of the curd and obtaining a smearing consistency of curd at the exit. The curd is fed by trolleys to the cooler hopper. The augers, rotating towards each other, move the cottage cheese from the side walls of the hopper to the center, capturing it in turns and move it inside the cooling cylinders. Ice water circulates inside the cylinders and augers, with the help of which the curd is cooled. From the outlet covers of the cylinders, the curd goes into the curd cart.

The temperature of the curd at the inlet to the cooler is from 28 to 30°C. The temperature of the curd at the outlet from the cooler is from 8 to 10°C. After the curd has cooled, it is moved to a cart and transported to the filling machine [6,7].

Automatic filling machine for curd M6-AR2T manufactured in the city of Kirov at the Plant named after V.A. Degtyareva. It is intended for packing and packaging cottage cheese, sweet curd mass in briquettes of 125 g and 250 g. Products are packaged in parchment or aluminum foil with a pre-printed label. The technological cycle of the machine consists of the following operations: loading the bunker with the product; filing of packaging material and date stamping; pieces of tape of packaging material; transfer of the scan on the forming die; forming the box and placing it in the nest of the forming table; dosage of a certain portion of the product in the box; folds of the edges of the briquette; the end of pressing, ensuring the density and final embedding of the briquette; shooting of finished briquettes from the forming table to the inverter, which overturns the briquettes by 180° and feeds them to the conveyor belt. All operations take place simultaneously in all slots of the forming table with its sequential periodic movement.

The production of a product on such equipment has a number of significant drawbacks: an open method, a large proportion of manual operations, and the inexpediency of automation in this hardware design. At the same time, cottage cheese can be obtained with a different mass fraction of fat by methods of both acidic and acid-rennet coagulation.

Taking into account the modern requirements for the production of cottage cheese and the features of traditional technology, a number of equipment has been developed and a complete automated line for the production of cottage cheese in a closed way A-TL-6 has been formed.

The open method of producing curd is not very hygienic. During the manufacturing process, it is possible that harmful microorganisms and foreign mechanical impurities can enter, so that this does not happen, an automated installation of a closed production line A-TL-6 is made. Basic automated operations of the A-TL-6 line:

1. Obtaining and processing of curd in a curd manufacturer;
2. Filling and dosing curd by means of a rotary pump;
3. Self-pressing, pressing and cooling of the curd in the installation;
4. Sanitization of basic equipment and pipelines.

Two people are required to service this installation, the daily output of finished products is 1800 kilograms. The advantage of A-TL over the existing equipment at the dairy is that it requires less human influence to produce curd. The basic configuration of this equipment includes: a curd manufacturer and a curd pressing machine. The A-TL-6 unit is shown in figure 3.

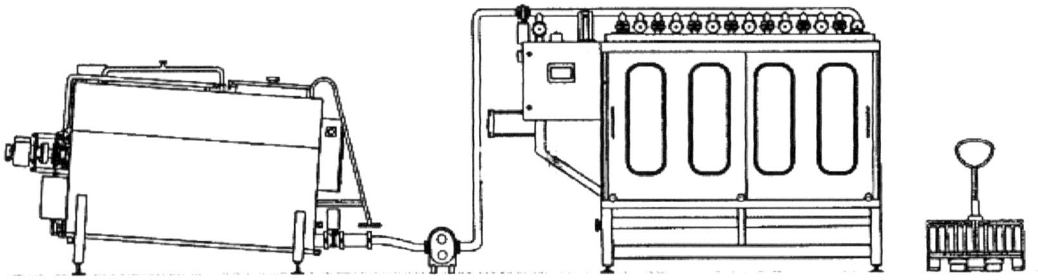

Figure 3. Installation A-TL-6.

The organization of the technological process is as follows:

- 1 – the curd from the manufacturer of curd is fed by a volumetric pump to the pressing installation and, with the help of a dispenser, is distributed over the filter elements (bags) in several cycles;
- 2 – the number of cycles is set by the program based on the quality and volume of the curd.
- 3 – after filling the filter elements with the entire volume, self-pressing of the curd takes place with spontaneous separation of whey;
- 4 – further bringing the moisture content of the product to the required parameters is carried out by cyclic pressing with a holding time set by the program;
- 5 – cooling is carried out simultaneously with the self-pressing and pressing processes.
- 6 – after determining the readiness of the curd, the filter elements with the finished product are automatically discarded;

- 7 – an operator servicing the pressing installation transfers the finished product from the filter elements into polymer boxes with a polyethylene liner, located on a euro pallet. Filled boxes are moved to the storage room or for packing;
- 8 – equipment sanitization is carried out by sequential mechanized supply of washing solutions along the circuits: cottage cheese manufacturer, UTS installation, process pipelines;
- 9 – sanitization is carried out in accordance with the instructions for sanitization of equipment, inventory and containers at dairy enterprises.

3. Conclusion

An innovative solution is a more convenient installation, since the closed type of curd production is protected from external influences. The compactness of the line allows you to significantly reduce the occupied production area, which ultimately is reflected in a decrease in production costs. Possibility of producing cottage cheese in cottage cheese makers using the acid-rennet method. A closed process, the design of each individual piece of equipment allows it to be washed in an automatic mode.

The entire line is controlled by a programmable controller. In the production of cottage cheese, one of the difficult tasks is also the creation of a product, the quality characteristics of which remain constant from batch to batch.

Manual production is the cause of the substandard end product. And most importantly, the traditional structure of the product has been preserved, which is important for the end consumer.

References

[1] Kozub Y A 2018 Development of the dairy industry in the Irkutsk region. In the collection: Problems in animal husbandry *Materials of the international scientific and practical conference* pp 30-6

[2] Kozub Y A 2013 Influence of carbohydrate-vitamin-mineral concentrate (UVMK) on milk quality *Vestnik IrGSKhA* **59** 92-6

[3] Vlasov B, Karelina L and Kozub Yu 2012 Metabolic aspects of cows' productivity when feeding "Felucene" *Dairy and beef cattle breeding* **5** 19-20

[4] Kozub Y A 2017 Improving the efficiency of milk production *Bulletin of the Irkutsk State Agricultural Academy* **81(2)** 50-4

[5] Komlatsky V I, Podoinitsyna T A, Verkhoturov V V and Kozub Y A 2019 Automation technologies for fish processing and production of fish products *J. Phys.: Conf. Ser.* **1399** 044050

[6] Kozub Y A, Komlatsky V I and Khoroshailo T A 2020 About some automated processes in the production of dairy products *IOP Conf. Ser.: Mater. Sci. Eng.* **862** 032021

[7] Kozub Y A 2019 *Improving the quality and improving the use of livestock products State and prospects for the development of animal husbandry and veterinary medicine in Siberia and the Far East* (Federal State Budgetary Educational Institution of Higher Education "Buryat State Agricultural Academy VR Filippov", Ulan-Ude) pp 131-5

[8] Komlatsky V I, Takho-Godi A Z and Podoinitsyna T A 2017 To the problem of automation of technological processes of milk processing and production of dairy products *Proceedings of the Kuban State Agrarian University* **69** 266-42

[9] Komlatsky V I *et al.* 2020 Technological process intensification trends in livestock *J. Phys.: Conf. Ser.* **1515** 022009

[10] Khoroshailo T A and Kozub Y A 2020 Robotization in the production of dairy, meat and fish products *J. Phys.: Conf. Ser.* **1515** 022007

IOP Conf. Series: Materials Science and Engineering 1155 (2021) 012085 doi:10.1088/1757-899X/1155/1/012085

Development of an automated information system to support computational procedures for scheduling

O N Lopateeva

Reshetnev Siberian State University of Science and Technology, Krasnoyarsk, prospect “Krasnoyarsky rabochy “31, Russia

E-mail: olga-lopateeva@rambler.ru

Abstract. One of the most important problems of the high-quality organization of the educational process in a higher educational institution is the task of forming a high-quality educational timetable. This task is the main one in the activities of the dispatching service of the university. A well-designed timetable should ensure an even workload of student groups and faculty in the next semester. Therefore, much attention is paid to researching and analyzing the operation of existing alternative means, formalizing and solving problems of optimal scheduling. In this paper, it is assumed that the problem of drawing up an effective university schedule can be effectively solved within the framework of an automated information system. The tasks of optimizing the curriculum at the University are being solved, which will help to take into account the current sanitary and epidemiological rules and norms through the introduction of the information system "University Schedule". The paper presents diagrams of use cases and a diagram of the databases of the future system. The goal is to improve the level and quality of the preparation of the curriculum through the introduction of the system "University Schedule". The software tool created as a result of the project will be used directly by the teachers who are the people responsible for scheduling.

1. Introduction

Drawing up and promptly adjusting the class schedule is one of the difficult problems of managing the educational process of a large university. Significant difficulties arise in scheduling. First of all, this is due to the large dimension of the problem, which is characterized by the number of disciplines, rooms for classes, teachers and study groups, it is for these reasons that it is very difficult to draw up a schedule without using a computer.

The main requirements for scheduling include the following: conducting classes in accordance with the structural - logical scheme of disciplines; lack of "windows"; scheduling lectures for the first hours of classes (first, second and third lessons); alternation of disciplines from different cycles; taking into account the requirements of teachers for the rational use of working time.

The procedure for scheduling should not be considered only as a kind of program that implements the function of enumerating the distribution of classes before the start of the semester and its current adjustment, taking into account the requirements of the educational process. It is necessary to identify the normative models of the activities of students, teachers and staff, assess the preferences of departments and individual teachers, resolve contradictions, accumulate knowledge in the form of effective and unsuccessful schedules, keep records and analysis of mistakes, etc.

In other words, the task of drawing up an optimal timetable is only part of a complex set of tasks for managing the educational process.

At the moment, there are a number of similar automated information systems for creating an optimal schedule.

For example, the automated system "1C: Automated scheduling" [1]. This system is designed to solve the problems of automated scheduling of educational schedules and operational management of premises in universities. With its help, you can create a schedule in automatic, manual and mixed modes, taking into account many restrictions and conditions. At the same time, you can build both a valid schedule and an optimized one, in which the number of windows or the number of rooms used is reduced.

Also on the market for these products there is such a system as "Galaxy" [2].

The system "Galaxy" offers a new look at the optimization of the resources of an educational institution and has a number of competitive advantages. Among the features that were absent in the analogue programs at the time of the appearance on the market of the system "Galaxy" is the automatic generation of the reference schedule, a high-tech user interface. The system "Galaxy" is addressed to universities and institutions of secondary vocational education.

Most of these automated systems have a large number of disadvantages:

- cost;
- implementation time is an important criterion, since project implementation time limited;
- hardware requirements - which is responsible for the conformity of the characteristics of the PC so that the embedded software can be used on it;
- difficulty in mastering the program - a criterion that should comply with the minimum efforts of users to prepare the initial information, use the software product and evaluate the results;

The main purpose of this work is the development of an automated information system for the formation of the educational schedule.

This project will solve the problem of optimizing the educational schedule of an educational institution based on the use of information received from the teaching staff.

Thus, the automation of the process of scheduling classes and the flexibility of the developed automated system gives advantages when using it in the education system, while improving the activities of personnel, and at the same time increasing the quality of the education provided by the university.

2. Methods

An automated information system (AIS) is a set of software and hardware intended for implementation at enterprises in order to automate activities related to the storage, transmission and processing of information [3].

The process of drawing up and tracking the school schedule lasts continuously throughout the school year. This process involves: the educational and methodological department, dean's offices, departments, the dispatch service of the university. The main responsibility for the formation of the curriculum lies with the workers of the dispatch service, which is part of the educational and methodological department.

Software products make it possible to quite successfully form a class schedule in an automatic mode, but at the same time, experience and a stable habitual schedule structure obtained during manual compilation are lost.

The order of formation of the training schedule is as follows.

On the basis of “State educational standards” approved by the Ministry of Education and Science of the Russian Federation, “Working curricula” are formed in areas and specialties. On the basis of "Working curricula" in the specialties in the departments of the faculties, before the beginning of each semester, "Teaching Reviews" are drawn up. Each line of the “Teaching Survey” is assigned to a department.

From the "Teaching Reviews" of all faculties, lists of disciplines are compiled for each department, which will be conducted by the teachers of this department. Such lists are usually called "extracts" for departments.

After solving these problems at the departments, "Applications for inclusion in the schedule" are drawn up and sent to the dispatching service. The compilation and operational monitoring of the status of these three journals is the essence of "manual" scheduling.

A diagram of the process of forming the curriculum at the university is shown in figure 1.

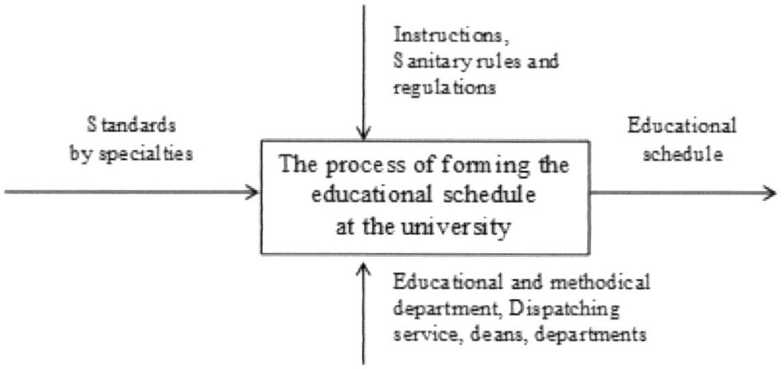

Figure 1. Scheme of the process "Formation of the curriculum at the university".

As can be seen from the diagram in figure 2, the ratio of the volume of input and output data is so great that there is an unreasonable loss of time arising from the preparation of a significant amount of input data only for scheduling. The possibility of using the database for solving other problems is absent.

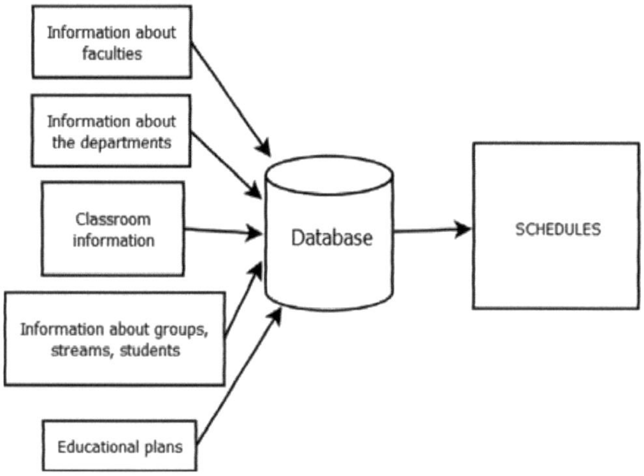

Figure 2. Data flows in a local approach.

To solve this problem, it is necessary to charge the input of each block of information to the appropriate services directly responsible for this data, which will control, maintain relevance and be responsible for the information entered. Information about teachers - HR department; curricula and disciplines - educational and methodological department; groups and students - dean's offices of faculties. In addition, it is necessary to interest these services in entering data into the information system, for example, the ability to automate their business processes, establish the search and elimination of critical errors, as well as the ability to form various samples using consolidated data from various blocks of information.

Thus, the data flow diagram will take the form shown in figure 3.

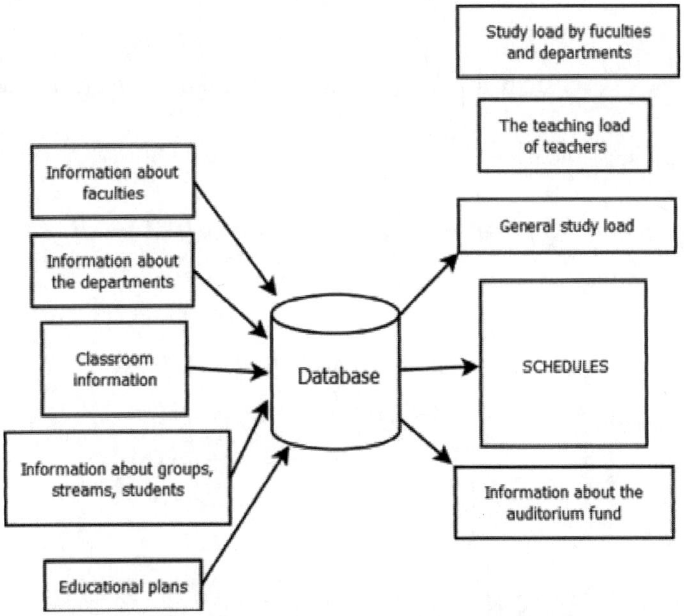

Figure 3. Data flows in a systems approach.

This is the schematic diagram of a systematic approach to automating the scheduling of a university, i.e. automation of all business processes related to the subject area of the schedule. It is this that will automate the process, and leave the final choice when making a decision to the user [4].

When forming the calendar curriculum, it is necessary to take into account the current sanitary and epidemiological rules and regulations [5], the division of student groups into subgroups, the presence of part-time teachers. When drawing up the schedule, the dispatcher must take into account the didactic requirements for the organization of the educational process, the instructions of the administration on the distribution of the classroom fund, as well as the personal wishes of teachers and student groups for drawing up a calendar training schedule.

Let us introduce the following notation:

P - many teachers; $\{p\}$- many indexes to identify teachers, $p = 1,2,\cdots,\overline{P}$;

r_p - teacher rank *p*, according to the unified tariff scale;

G- many groups; $\{g\}$ - set of indices to identify groups, $g = 1,2,\cdots,\overline{G}$; s_g - number of students in a group *g;*

A- many audiences; $\{a\}$ - many indexes for audience identification, $a = 1,2,\cdots,\overline{A}$;

$\{k\}$ - many indexes for identifying enclosures, $k = 1,2,\cdots,\overline{K}$;

T - many working hours;

D – many days of the week; $\{d\}$ - many indices to identify the days of the week, $d = 1,2,\cdots,\overline{D}$

L – many lesson; $\{l\}$ - multiple indices to identify the lesson, $l = 1,2,\cdots,\overline{L}$;

Z- many academic disciplines; $\{z\}$ - many indexes for identifying lessons ("Teaching Reviews" strings), $z = 1,2,\cdots,\overline{Z}$;

c_z - indicators of the complexity of mastering the discipline *z*,; $z = 1,2,\cdots,\overline{Z}$;

$\{v\}$ - many indices to identify occupations, $v = 1,2,\cdots,\overline{V}$;

b_{vz} - number of hours for each type v lessons of each subject z "Teaching Review"; $v = 1,2,\cdots,\overline{V}$; $z = 1,2,\cdots,\overline{Z}$;

η_d - the share of the weekly study load of the student group falling on the d-th day of the week, and $\sum_{d=1}^{\ddot{D}} \eta_d = 1$.

Let's introduce two groups of boolean numbers: x_{pdl} and x_{gdl} (p=1,2,…,P; g=1,2,…G; d=1,2,…,D; l=1,2,…,L)

$$x_{pdl} = \begin{cases} 1, & \text{if teacher } p \text{ on day } d \\ & \text{conducts classes } l-th \text{ class lesson,} \\ 0, & \text{otherwise.} \end{cases} \tag{1}$$

$$x_{gdl} = \begin{cases} 1, & \text{if group } g \text{ on day } d \text{ is} \\ & \text{in class } l-th \text{ lesson,} \\ 0, & \text{otherwise.} \end{cases} \tag{2}$$

and a group of variables $x_{gd} = \sum_{l=1}^{\overline{L}} x_{gdl} \quad (g = 1,2,\cdots,\overline{G}, \quad d = 1,2,\cdots,\overline{D})$.

Let us introduce six groups of functions: $\psi(x_{gd})$, $f_{kp}(x_{pdl})$, $f_{kg}(x_{gdl})$, $f_{vz}(x_{pdl})$, φ_{pdl}, φ_{gdl} $(p = 1,2,\cdots,\overline{P}, g = 1,2,\cdots,\overline{G}, d = 1,2,\cdots,\overline{D}, l = 1,2,\cdots,\overline{L}, k = 1,2,\cdots,\overline{K})$.

The function $\psi(x_{gd})$ - is a complex dependence, with the help of which the total difficulty of the educational load of the g–th group on the в d-th day is calculated.

The function $f_{kp}(x_{pdl})$ has a complex dependence that allows x_{pdl} using a pre-formed database to determine the number of the corpus in which teacher p on day d conducts a lesson with the l–th lesson.

The function $f_{kg}(x_{gdl})$ is a complex dependence, which allows for x_{gdl} using a pre-formed database to determine the number of the corpus in which the group g on day d is engaged in the l–th lesson.

The function $f_{vz}(x_{pdl})$ is a complex dependence that allows for x_{pdl} using a pre-formed database to determine the subject and type of classes (for example: physics, laboratory), which teacher p on day d conducts on the l–th lesson;

The arguments to the function φ_{pdl} are the values of the function $f_k(x_{pdl})$. The function φ_{pdl} can only take two values: 0 or 1.

$$\varphi_{pdl} = \varphi(x_{pdl} - x_{pd(l-1)}) = \begin{cases} 1, & \text{if } f_k(x_{pdl}) \neq f_k(x_{pd(l-1)}) \\ 0, & \text{if } f_k(x_{pdl}) = f_k(x_{pd(l-1)}) \end{cases} \tag{3}$$

If the function φ_{pdl} takes on the value 0, it means that the teacher p on the day d l–th and $l+1$–th lesson spends in the same building (no teacher's transition from one building to another).

If the function φ_{pdl} takes the value 1, it means that the teacher p on the day d l–th and $l+1$–th lesson spends in different corps (the presence of the teacher's transition from corps to corps).

The arguments to the function φ_{gdl} are the values of the function $f_k(x_{gdl})$. The function φ_{gdl} can only take two values: 0 or 1.

$$\varphi_{gdl} = \varphi(x_{gdl} - x_{gd(l-1)}) = \begin{cases} 1, & if \quad f_k(x_{gdv}) \neq f_k(x_{gd(l-1)}) \\ 0, & if \quad f_k(x_{gdl}) = f_k(x_{gd(l-1)}) \end{cases} \tag{4}$$

If the function φ_{gdl} takes the value 0, it means that a group of students g on the day d l–th и $l+1$–th lesson is engaged in the same building (no transition of a group of students from building to building).

If the function φ_{pdl} takes the value 1, it means that a group of students g on the day d l–th и $l+1$–th lesson is engaged in different buildings (there is a transition of a group of students from building to building).

The criterion - the minimum of "windows" for teachers, can be expressed by function (5).

$$\sum_{p=1}^{P}\sum_{d=1}^{D}\sum_{l=2}^{L}\left|x_{pdl} - x_{pd(l-1)}\right| \Rightarrow \min \tag{5}$$

The criterion - the minimum of "windows" for groups, can be expressed by function (2.6).

$$\sum_{g=1}^{G}\sum_{d=1}^{D}\sum_{l=2}^{L}\left|x_{gdl} - x_{gd(l-1)}\right| \Rightarrow \min \tag{6}$$

The criterion - the minimum losses from the transitions of teachers from corps to corps, can be expressed by function (7).

$$\sum_{p=1}^{P} r_p \sum_{d=1}^{D}\sum_{l=2}^{L}\varphi_{pdl} \Rightarrow \min \tag{7}$$

The criterion - the minimum losses from the transitions of students from corps to corps, can be expressed by function (2.8).

$$\sum_{g=1}^{G} s_g \sum_{d=1}^{D}\sum_{l=2}^{L}\varphi_{gdl} \Rightarrow \min \tag{8}$$

According to the recommendations of the Sanitary and Epidemiological Services (SES), the complexity of the study load should be minimal on Monday, increase by Wednesday, and decrease by the end of the school week.

In this case, the function $\beta\,(x_{gd})$ for the criterion for taking into account the load balance in the schedule of student groups is as follows.

$$\beta\,(x_{gd}) = \sum_{g=1}^{\ddot{G}}\sum_{d=1}^{\ddot{D}}\left|\psi(x_{gd}) - \eta_d \sum_{d=1}^{\ddot{D}}\psi(x_{gd})\right| \Rightarrow \min \tag{9}$$

Constraints (2.10) require that the hours of all disciplines and all types of classes planned by "Teaching Reviews" should be included in the class schedule [6].

$$\sum_{p=1}^{P}\sum_{d=1}^{D}\sum_{l=2}^{L}\left(2 * x_{pdl}\right)=\sum_{g=1}^{G}\sum_{d=1}^{D}\sum_{l=2}^{L}\left(2 * x_{gdl}\right)=b_{vz}\ ;\ v=1,2,\ldots,V;\ z=1,2,\ldots,Z. \quad (10)$$

The formation of the optimal training schedule is formulated as follows.

Find an optimal partition of the set of works planned by "Teaching Reviews" (by discipline and type of occupation) into $p{\times}d{\times}l+g{\times}d{\times}l+a{\times}d{\times}l$ disjoint subsets, which would ensure that the minimum of function (5), the minimum function (6), minimum of function (7), minimum of function (8), minimum of function (9), under constraints (10). In this case, the required boolean variables x_{pdl} and x_{gdl} ($p=1,2,\ldots,P$; $g=1,2,\ldots G$; $d=1,2,\ldots,D$; $l=1,2,\ldots,L$) are defined by relations (1) and (2), and the components of formulas (7) and (8) of the function are determined by relations (3) and (4), respectively.

Thus, the formalization and solution of this problem of forming the optimal timetable of the University will improve the level and quality of drawing up the educational timetable through the introduction of the system " University Schedule".

3. Results

In the course of the work performed, the UML diagram (figure 4) and the ER-diagram of the database (figure 5) of the future automated information system for the formation of the university curriculum [7] [8] were designed.

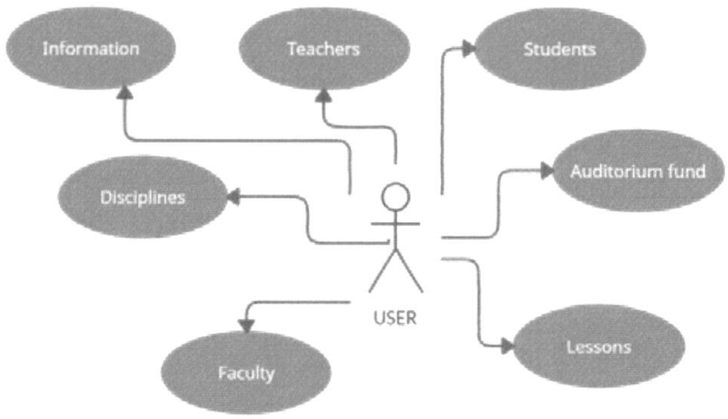

Figure 4. UML diagram of use cases.

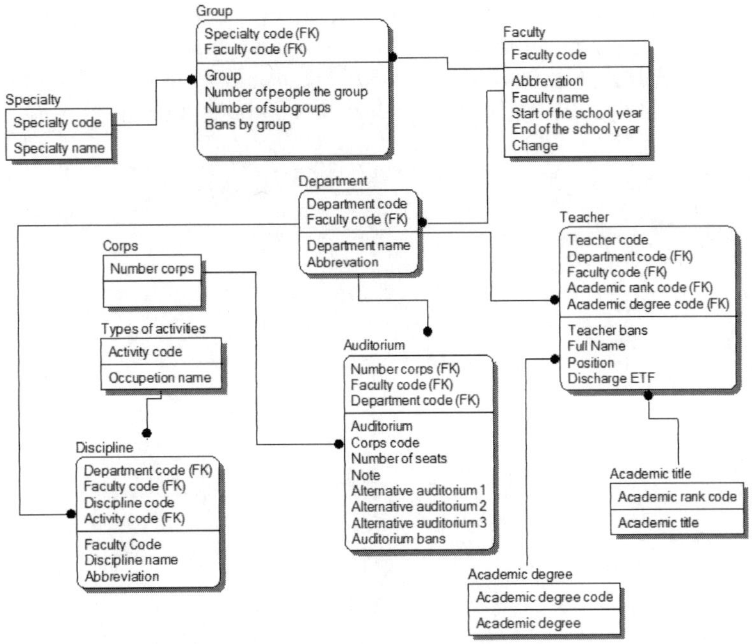

Figure 5. ER-diagram of the reference book of the "University Schedule" system.

The result of the design of the AIS "University Schedule" indicates that the prototype of such an information resource, implemented at the next stage, will automate the process of forming the class schedule.

4. Conclusion

The analysis of the existing procedure for the formation of the curriculum at the university is carried out. It has been established that the educational and methodological department, dean's offices, departments, the dispatch service of the university are involved in the process of drawing up and tracking the curriculum. The main responsibility for the formation of the curriculum lies with the workers of the dispatch service, which is part of the educational and methodological department.

For the purpose of the automated formation of the university curriculum, it is proposed to create an information system "University Schedule", which is one of the subsystems of the automated information system for organizing the educational process at the university.

A methodology for designing the information system "University Schedule" is proposed, the basis of which is a systematic approach, according to which any system is a set of interrelated objects (elements) that function together to achieve a common goal.

In the future, it is planned to create an application to solve this problem.

References

[1] *1C: Automated scheduling. University* Retrieved from: https://clck.ru/Th4Ut

[2] *Galaxy. Schedule of training sessions* Retrieved from: https://galaktika-it.ru/spb/ruz

[3] *Definition and description of the automated information system* Retrieved from: https://dic.academic.ru/dic.nsf/ruwiki/334809

[4] Farkhetdinov I R 2013 Improving the compilation process electronic timetable at USPTU *Oil and Gas Business* **1**

[5] SaNPiN 2.4.2.1178-02 *Hygienic requirements for learning conditions in educational institutions* Retrieved from: http://docs.cntd.ru/document/901835064

[6] Vorobovich N P, O N Lopateeva 2006 On the NP-completeness of the problems of forming the schedule in the university *Vestnik KrasGAU* **11**
[7] Gorbachev I V 2005 *CASE-technology for modeling processes using BPWin and ERWin* p 156
[8] Glushakov S V 2002 *Databases* (St. Petersburg: AST Publishing House) p 504

Research of the process of crushing liquid in a swirl gas flow

V G Afanasenko, A V Rubtsov, P A Kulakov and E V Boev

FSBEI of Higher Education "Ufa State Petroleum Technological University", 1, Kosmonavtov St., Ufa, 450064, Russia

E-mail: afanasenko.v.g@yandex.ru

Abstract. Liquid spraying is a common process in the chemical, oil and gas industries. For its implementation, nozzles of various designs are used. The outflow of liquid from the nozzle leads to the fragmentation of the continuous jet into separate drops. This process depends on many factors, including the movement of the continuous phase in which spraying is carried out. This article discusses the relative motion of a liquid when it flows from a nozzle into a rotating gas flow and the conditions for re-crushing of liquid particles.

1. Introduction

Liquid atomization is a complex process that depends on many factors. The jet at the exit from the atomizer under the action of external and internal forces breaks up into separate drops, in the future it is possible to re-crush large drops into smaller ones.

External - aerodynamic forces include the interaction of the sprayed component with the medium into which it is injected. Their value depends on the density of the environment, the speed of the jet and the size of the liquid droplets. External forces also include the forces of interaction when the jets intersect or when the jet meets a solid wall.

Internal forces include molecular forces and turbulence. Turbulent pulsations arise in the flowing liquid stream, the intensity of which depends on the density, viscosity, pressure drop, and also on the design of the atomizer. An increase in the outflow rate contributes to an increase in the intensity of turbulent pulsations, which in turn improves the quality of liquid dispersion [1-6].

2. The physical basis of the process of crushing liquid droplets

Liquid crushing and the formation of droplets occurs as follows. During injection, a sheet is formed, which, under the action of external forces and turbulent pulsations, breaks down into particles of various shapes and sizes. Small particles, under the action of surface tension, take the shape of a ball and form drops, while large ones continue to disintegrate.

The scheme of the primary crushing of the jet into droplets due to the turbulent action can only serve as a simplified model of the liquid spraying process. And they propose a decay scheme based on the assumption that cavitation processes are the cause of the destruction of a single fluid flow into droplets.

The appearance of cavitation recesses in the liquid occurs when the pressures in the considered volume are equal and the saturated vapor of the liquid under the given conditions. In the injector nozzle, cavitation occurs due to an increase in the hydrodynamic head of the liquid, at which the hydrostatic head and, accordingly, the pressure decrease. An increase in the flow rate, which increases the hydrodynamic head, is achieved by a decrease in the flow area of the channel [7-11].

Caverns formed in the injector nozzle at the nozzle exit, where the pressure is usually close to atmospheric, disappear, destroying the integrity of the jet. The formation of cavitation recesses is periodic, with a frequency that depends on the flow rate. With an increase in the flow rate, cavitation bubbles are formed not only on the surface, but also inside the jet, and a vapor-liquid mixture comes out of the nozzle. The swirling movement of the flow also contributes to the intensification of cavitation in the entire section of the jet.

The maximum drop size that can exist in a plume is determined from the condition of equality of the surface tension force [4, 12-15]:

$$F_{sur} = 2\pi \cdot r_d \cdot \sigma_l , \tag{1}$$

and the forces of aerodynamic pressure:

$$F_{aer} = \zeta \cdot \frac{\pi \cdot r_d^2 \cdot \rho_l \cdot u^2}{2} . \tag{2}$$

Where is the maximum drop diameter:

$$d_d = \frac{2 \cdot \sigma_l}{\zeta \cdot \rho_l \cdot u^2} . \tag{3}$$

The mode of crushing a liquid droplet in a gas flow depends on the physical properties of the contacting media (surface tension of liquid - σ_l, medium resistance coefficient - ξ), the droplet diameter and the velocity of the gas blowing over it. It is difficult, and sometimes impossible, to influence the physical properties of gas and liquid under given technological conditions. The maximum droplet size is limited by the outflow rate, since large droplets formed at the initial stage, even with a small impact, are deformed and destroyed, resulting in smaller ones. Therefore, the main criterion that determines the mode of droplet crushing and dispersion of the liquid at the outlet from the nozzle as a whole is the difference in the velocities of the liquid droplet and the surrounding environment.

3. Liquid phase injection unit in vortex apparatuses

In industrial devices, liquid is injected into a moving medium, which makes it necessary to take this fact into account when designing and selecting nozzles. So, for example, in spray absorbers, when entering the apparatus, the liquid interacts with a swirling gas stream that enters through a tangential pipe located at the periphery of the mixing chamber. To increase the difference between the velocities of the liquid and the gas blowing over it, it is possible, using a swirling device located in the nozzle body, to impart a tangential component to the liquid velocity vector, and the swirling of the flow should be carried out in the direction opposite to the gas (figure 1).

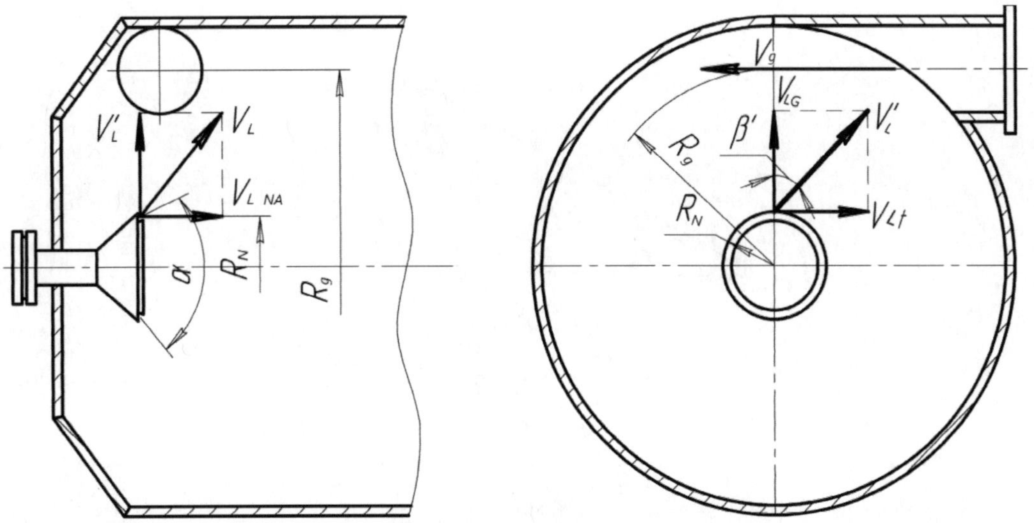

Figure 1. Swirl-chamber of the absorber.

4. Change in the velocity difference during the movement of the dispersed phase in a swirling flow

Let us consider the change in the difference between the velocity vectors of a single drop and the swirling gas flow blowing over it when the liquid outflows at an angle to the radial-axial direction. To do this, consider the projections of the corresponding velocity vectors on a plane perpendicular to the axis of rotation [16].

The angular velocity of swirling of a continuous medium will depend on the design features of the gas supply device, its flow rate and other factors. We denote it by ωg, then the gas velocity at an arbitrary point:

$$V_{qi} = \omega_q \cdot R_i \,. \tag{4}$$

Without taking into account the friction forces, the liquid flowing out of the injector nozzle with an initial velocity Vl will retain the magnitude and direction of the velocity vector when moving to the periphery of the mixing chamber.

We divide the initial speed of movement of the liquid droplets leaving the nozzle into axial, radial and tangential components. The magnitude of the velocity in the axial and radial directions is most influenced by the angle of deviation of the velocity vector from the axis of rotation; with an increase in this angle, the radial component increases and the axial component of the velocity vector decreases and vice versa. The tangential component appears when using various swirling devices in the nozzle body or nozzle. Swirling the flow allows you to increase the spray angle due to the appearance of centrifugal forces, as well as increase the difference in the velocity vectors when injecting a rotating gas into the flow (see figure 2).

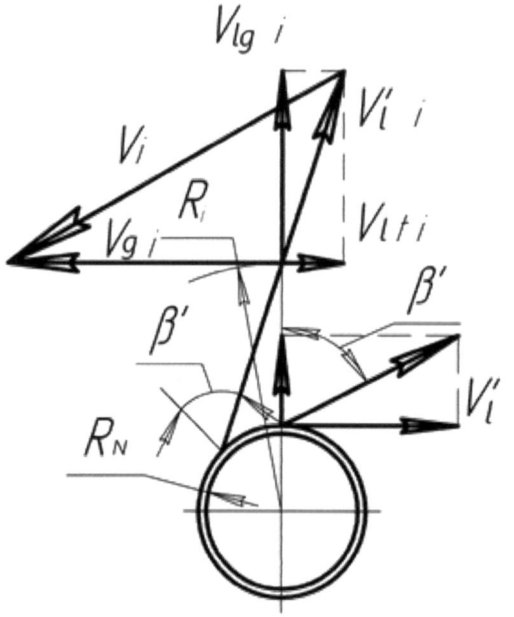

Figure 2. Projections of the vectors of gas and liquid velocities on a plane perpendicular to the axis of rotation.

When changing the distance to the axis of the nozzle.

The axial component of the fluid velocity vector will be constant and equal to:

$$V_{lNA} = V_l \cdot \cos\beta \cdot \cos\frac{\alpha}{2}. \tag{5}$$

The tangential component of the fluid velocity vector, at the point the distance, to the axis of rotation R_i:

$$V_{lti} = \frac{R_n \cdot \sin\beta'}{R_i} \cdot V'_l. \tag{6}$$

The radial component of the fluid velocity vector, at the point under consideration:

$$V_{lri} = V_l \cdot \sqrt{\left(1 - \cos^2\beta \cdot \cos^2\frac{\alpha}{2}\right)\frac{R_i^2 - R_n^2 \cdot \sin^2\beta'}{R_i^2}}. \tag{7}$$

The vector of the difference between the velocities of a gas and a liquid, as well as the velocity vector of a single drop, can be divided into the same mutually perpendicular components axial, radial and tangential. Moreover, the first two of them will be numerically equal to the corresponding projections of the velocity vector of the liquid drop, and the tangential component is determined by the formula:

$$u_{ti} = V_{lti} + V_{gi} = V_l \cdot \frac{R_n \cdot \sin\beta'}{R_i} \cdot \sqrt{1 - \cos^2\beta \cdot \cos^2\frac{\alpha}{2}} + \omega_z R_i. \tag{8}$$

The total value of the velocity difference vector is:

$$u = \sqrt{u_{nai}^2 + u_{ri}^2 + u_{ti}^2}\,. \tag{9}$$

After substitution of (5,7,8) in (9) and transformation, the dependence of the value of the vector of the difference in velocities at the outflow of liquid from the injector nozzle at an angle to the radiaxial direction on the distance to the axis of rotation was obtained:

$$u = \sqrt{V_l^2 + \omega_q^2 \cdot R_i^2 + 2V_l \cdot \sqrt{1 - \cos^2\beta \cdot \cos^2\frac{\alpha}{2}} \cdot R_l \cdot \sin\beta' \cdot \omega_q}\,. \tag{10}$$

In real conditions, the movement of the contacting phases is significantly influenced by resistance forces, under the influence of which the direction and magnitude of the velocity vectors change. In this article, we examined the radiaxial outflow of liquid from the injector nozzle, provided that the momentum of the gas is much greater than the momentum of the liquid, that is, provided that the movement of the gas changes insignificantly.

5. Conclusion

Based on the studies carried out, a method has been developed for calculating the process of crushing a dropping liquid in a swirling gas flow, which includes the following main stages:

1) Determination of the difference between the velocities of the liquid and the gas blowing over it at the outlet of the nozzle (u_{init}).

2) Depending on the physicochemical properties of the components and the initial speed difference, finding the maximum diameter of the resulting liquid droplets (d_{maxN}).

3) For the largest particles, determining the change in time and the largest value of the vector of the difference between the velocities of the dropping liquid and the gas flow blowing over it ($u=f(t)$; u_{max}).

4) Re-finding the maximum size of the particle diameter of the dispersed phase depending on the largest difference in velocities (d_{max}).

After the calculations, it is possible to determine the specific surface area of the liquid phase and, accordingly, the contact area of the “liquid-gas” system to calculate the efficiency of mass exchanging processes.

References

[1] Pagi D G, Koryagin A A and Lamm E L 1975 *Spraying devices in the chemical industry* (Moscow: Chemistry)

[2] Dytnersky Yu I 1995 *Processes and devices of chemical technology: Hydromechanical and thermal processes and devices* (M.: Chemistry)

[3] Kulagin L V and Okhotnikov S S 1967 *Combustion of heavy liquid fuels* (M., Nedra)

[4] Vitman L A, Katsnelson B D and Paleev I I 1962 *Spraying liquid with nozzles* (M.: State Energy Publishing House)

[5] Golovachevsky Yu A 1967 *Sprinklers and nozzles of scrubbers of the chemical industry* (M.: Mashinostroenie)

[6] Afanasenko V G, Kulakov P A, Boev E V, Imaeva E Sh, Davletov O B and Mazidullin D N 2020 Optimization of ultrasound emulsion under mechanical mixing *Bulletin of the Tomsk Polytechnic University, Geo Assets Engineering* **331(4)** 148-55

[7] Xiong X, Li L and Zhou X 2017 Numerical Analysis and Optimization Research on Backflow Effect of Cooling Tower *Procedia Engineering* **205** 2003-10

[8] Li S, Moradi A, Vickers B and Flynn M R 2018 Cooling tower plume abatement using a coaxial plume structureInternational *Journal of Heat and Mass Transfer* **120** 178-93

[9] Khafizov F Sh, Afanasenko V G, Khafizov I F, Khaibrakhmanov A Sh and Boev E V 2008 Use of vortex apparatuses in gas cleaning process *Chemical and Petroleum Engineering* **44 (7-8)** 425–8

[10] Nasyrova M I and Kulakov P A 2019 Influence of the shape of quartz sand particles factor on

single particle erosion damage *Journal of Physics: Conference Series* **1384(1)** 012032
[11] Kulakov P A, Rubtsov A V, Afanasenko V G, Zubkova O E, Sharipova R R and Gudnikova A A 5 December 2019 Influence of technical condition parameters on the residual resource of capacitive equipment *Journal of Physics: Conference* Series **1399** 055052
[12] Kuleshov V S, Moiseev K V, Khizbullina S F, Mikhaylenko K I and Urmancheev S F 2018 Convective Flows of Anomalous Thermoviscous Fluid *Mathematical Models and Computer Simulations* **10(4)** 529–37
[13] Kuleshov V S, Moiseev K V and Urmancheev S F 2019 Isolated Convection Modes for the Anomalous Thermoviscous Liquid in a Plane Cell Fluid *Dynamics* **54(7)** 983–90
[14] Dmitriev A V, Zinurov V E and Dmitrieva O S 2018 Influence of elements thickness of separation devices on the finely dispersed particles collection efficiency *MATEC Web of Conferences* **224** 02073
[15] Boev E V 2020 Aerodynamic and hydroaerothermic tests of cooling tower sprinkler *Journal of Physics: Conference Series* **1515(2)** 022066
[16] Naik B, Choudhary V, Muthukumar P and Somayaji C 2017 Performance Assessment of a Counter Flow Cooling Tower *Energy Procedia* **109** 243-52

A roller for soil cultivation and its optimal operating parameters in vibration mode

Gennady Maslov, Valery Tsybulevsky, Bulat Tazmeev and Sergey Rudnev

Kuban State Agrarian University named after I.T. Trubilin, Krasnodar, Russian Federation

E-mail: tazmeevb@mail.ru

Abstract. The optimal parameters and operating modes of vibratory roller for soil compaction after sowing winter wheat were substantiated. For objective function - the optimum coefficient of variation of density of soil during operation of vibratory roller for compacting winter wheat regression equation was obtained in the planning according to the B_K plan of the experiment. The adequacy of model according to Fisher criterion, the significance and reliability of coefficients of regression equation was established. The response surface obtained by the shape of hyperboloid rotation was studied, two-dimensional sections of three factors were constructed on optimization criterion. With its minimum value, the optimal parameters of vibratory roller were established: the mass of roller is 293 kg, the spring stiffness is 11.95 kN/m, the working speed of roller is 9.6 km/h. The minimum value of optimization criterion is 10.43%.

1. Introduction

A further increase efficiency of agricultural production is impossible without improving machine technologies, as well as optimizing parameters, rational operation modes of machines and their use [1, 2, 3]. The competitiveness of manufactured products is determined by their cost, productivity of machines and quality of their work. Optimization of parameters and operating modes of machines in this area [4, 5] is an urgent task, because production efficiency depends on technology. Scientific approaches [6, 7] to the substantiation parameters of technology determine results of research. The purpose of article submission - based methods of planning 3 factorial experiment by the B_K plan optimal parameters and operation of vibrating roller for optimal compaction in winter wheat.

2. Research materials

The influence of following three factors established as a result of previously conducted one-factor experiments, as well as their fixed values at optimal levels by the uniformity of soil density in treated area, was studied: operating speed, km/h; roller mass, kg; spring stiffness, kN/m.

Other parameters of developed vibratory roller are impractical to use as controlled factors, since they are not decisive for reducing the unevenness of soil density in the packed area.

The symmetric compositional plan of the B_K, type was chosen to build plan for 3-factor experiment, whose star points are equal ± 1. Factors, intervals of variation and their levels for the experiment are shown in table 1.

Table 1. Factors, intervals and their levels of variation.

Factors	Coded representation	Variation interval	Factor variation levels		
			-1	0	+1
Operating speed (υ), km/h	x_1	4	4	8	12
Roller mass (G), kg	x_2	100	150	250	350
Spring stiffness (N), kN/m	x_3	6	6	12	18

The operating speed υ of movement was set by shifting gears in the gearbox and supplying fuel. The stopwatch measured time of movement of the unit on a certain section of path in field marked by pegs.

The mass of the roller was weighed on platform scales. The stiffness of the springs was selected depending on diameter of wire, cross-sectional area of spring and material of the spring.

The three-factor experiment planning matrix is shown in table 2.

In the experiment, a second-order polynomial with three variables was used as a response function. After processing the obtained data, was obtained a regression equation with imaginary coefficients.

Table 2. Experiment planning matrix.

Coded value variables			Natural value variables			Coefficient of soil density variation ν, %
x_1	x_2	x_3	Operating speed (υ), km/h	Roller mass (G), kg	Spring stiffness (H), kN/m	
+1	+1	+1	12	350	18	21.1
-1	+1	+1	4	350	18	16.2
+1	-1	+1	12	150	18	29.3
-1	-1	+1	4	150	18	28.9
+1	+1	-1	12	350	6	21.2
-1	+1	-1	4	350	6	19.9
+1	-1	-1	12	150	6	22.8
-1	-1	-1	4	150	6	26.1

+1	0	0	12	250	12	10.3
-1	0	0	4	250	12	8.9
0	+1	0	8	350	12	10.8
0	-1	0	48	350	12	18.1
0	0	+1	8	250	18	20.9
0	0	-1	8	250	6	21.2

Criteria for optimizing problem (response) is the coefficient of variation ν of soil density series in corresponding series of experimental plan. The minimum value of the coefficient ν is corresponds to the best result of uniformity of compaction and determines the optimal design and regime parameters of the vibratory roller: its mass G, spring stiffness H and operating speed υ of the unit.

The regression equation (1) of the response function has the form:

$$Y = 11.108 + 0{,}469x_1 - 3.551x_2 + 0.523x_3 + 1.377x_1x_2 + 0.912x_1x_3 - \\ -1.638x_2x_3 - 1.357x_1^2 + 3.46x_2^2 + 10.094x_3^2 \quad (1)$$

where Y – coefficient of soil density variation;
x_1 – coded value of the unit speed;
x_2 – coded value of the roller mass;
x_3 – coded value of stiffness of the spring.

Below is the equation for translating coded values of factors into valid ones:

$$X_1 = 8x_1 + 4; X_2 = 250x_2 + 100; X_3 = 12x_3 + 6 \quad (2)$$

where X_1 – valid value of unit speed, km/h;
X_2 – valid value of roller mass, kg;
X_3 – valid value of spring stiffness, kN/m.

The coefficients of regression equation (1) obtained as a result of calculation were checked for significance by tentative calculating the confidence interval using Student's criterion. As a result, all coefficients of obtained equation turned out to be significant.

$$db_0 = 0.41; db_i = 0.073; db_{ij} = 0.066; db_{ii} = 0.412; \text{at } t = 2.57.$$

To verify adequacy of the model according to the Fisher criterion, used the obtained experiments in the center of the plan to find variance of experience.

The natural values of variables of the plan in center of the plan are presented in table 3.

Table 3. Natural values of variables in center of the plan.

Coded value variables			Natural value variables			Coefficient of soil density variation ν, %
x_1	x_2	x_3	Operating speed (υ), km/h	Roller mass (G), kg	Spring stiffness (H), kN/m	
0	0	0	8	250	12	10.29

0	0	0	8	250	12	10.88
0	0	0	8	250	12	10.51
0	0	0	8	250	12	10.56
0	0	0	8	250	12	10.32

As a result of calculations variance of experience is obtained
$S_y^2 = 0.255$; $F_{\text{tabular}} = 4.74$; $F_{\text{calculated}} = 0.78$
Equation is adequate $F_{\text{tabular}} \geq F_{\text{calculated}}$.
System of linear equations is obtained by differentiating equation for each variable and equating it to zero:

$$\begin{cases} dy/dx_1 = 0{,}469 - 2.714x_1 + 1.376x_2 + 0.912x_3 \\ dy/dx_2 = -3.55 + 1.376x_1 + 6.92x_2 - 1.638x_3 \\ dy/dx_3 = 0.523 + 0.912x_1 + 1.638x_2 + 20.188x_3 \end{cases} \quad (3)$$

Response surface center coordinates can be found by solving system of linear equations: $x_1 = 0.39$; $x_2 = 0.433$; $x_3 = -0.0083$.

Substituting into original regression equation (1) values of x_1, x_2, x_3, it can be received value of optimization parameter v in center of response surface. This is the response value at new origin (free term of the canonical equation) $Y_s = 10.428$.

Perform to canonical conversion. For this, equation (1) was transferred to a new coordinate system and the coordinate axes were rotated. In this case, equation (1) is simplified to the form:

$$Y - Y_s = B_{ii} \cdot x_i^2 + B_{jj} \cdot x_j^2 + B_{kk} \cdot x_k^2 \quad (4)$$

where B_{ii},$B_{jj,}$ B_{kk} – canonical equation coefficients.

The angle α rotation of original coordinate axes of response surface is determined unto align with main axes of figure $\alpha = 6.93$ deg.

After transformations, equation (1) in canonical form has following form:

$$Y - 10.428 = -1.357X_1^2 + 3.36X_2^2 + 10.194X_3^2 \quad (5)$$

To further study response surface (4) it's two-dimensional cross-sections were constructed. Initially, the cross-section of response surface with X_1SX_2 plane is considered. To do this, substitute $x_3 = -0.083$ in equation (1) and get:

$$Y_{12} = 11.105 + 0.461x_1 - 3.537x_2 + 1.376x_1x_2 - 1.357x_i^2 + 3.468x_2^2 \quad (6)$$

where Y_{12} – coefficient of variation in the density (6) of the soil in the interaction of 1 and 2 factors, when the 3rd factor is in plan center of experimental design.

Performing canonical transformation and solving system of linear equations is found response surface central coordinates $Y_s = 10.428$ when $x_3 = -0.0083$.

3. Results and discussion

The angle α rotation of coordinate axes is determined. It is equal – 7,97 deg., And coefficients at unknowns in canonical form are equal: at $B_{11} = -1.453$; at $B_{22} = 3.556$. The response surface equation in canonical form is obtained:

$$Y_{12} - 10.428 = -1.453X_1^2 + 3.556X_2^2 \quad (7)$$

The resulting response surface (fig. 1) is hyperboloid rotation, which is indicated by the coefficients of regression equation (7) in canonical form, since they have different signs. The optimal value of response function is at point with coordinates: $X_1 = 0.39, X_2 = 0.433$.

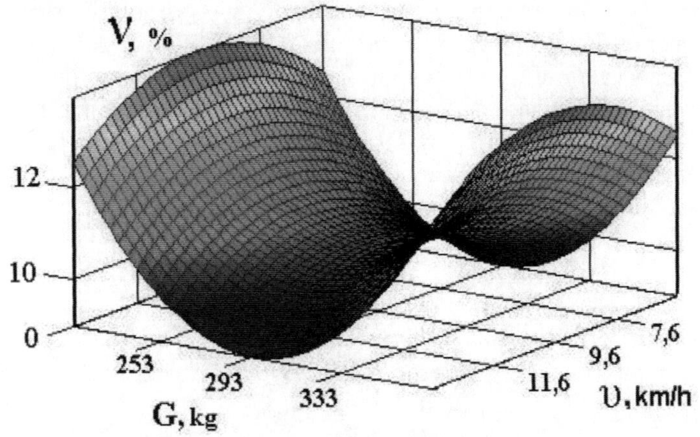

Figure 1. Response surface for depending on unit speed υ and roller mass G.

Given values of variables and substituting them in equation (7), a family of conjugate isolines is obtained (figure 2).

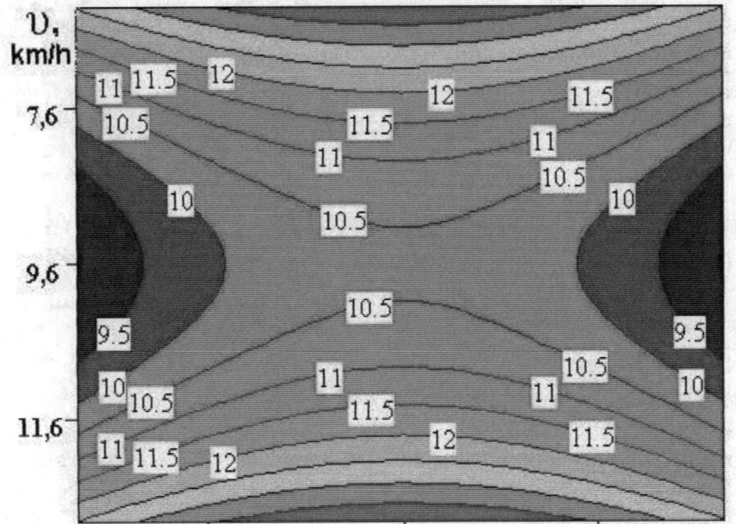

Figure 2. Two-dimensional cross-section of response surface depending on machine speed υ and roller mass G.

The contours obtained as a result of cross section of response surface are elongated along axis of corresponding operating speed of unit. Therefore, this factor 7.3 times less affects uniformity of soil density during compaction with vibratory roller than roller mass.

When $x_3 = -0.0083$ in coded form, i.e. when coefficient of soil density variation in the center of the experiment plan and equal to 10.428%, the working speed is $\upsilon = 9.6$ km/h, and roller mass G is 293 kg.

Consider the cross section of response surface with density X_1SX_3. For this, $x_2 = 0.434$ in the regression equation (1) is substituted. The regression equation took form:

$$Y_{13} = 10.22 + 1.065x_1 - 0.188x_3 + 0.913x_1x_3 - 1.354x_1^2 + 10.094x_3^2 \quad (8)$$

Having performing canonical transformation and solving system of linear equations, was founding coordinates of center response surface: $x_1 = 0.39$, $x_3 = -0.0083$.

Substituting found value of x_1 and x_3 into equation (1) was determining value of parameter of optimization in center of response surface at $x_2 = 0.433$. The value of parameter of optimization – coefficient of soil density variation $Y_{13} = 10.428$. The rotation angle of coordinate axes α is – 2.28 degrees, and regression coefficients in canonical form are equal: at X_1 $B_{11} = -1.375$, at X_3 $B_{33} = 10.113$. Now response surface equation in canonical form takes the form:

$$Y_{13} = 10.428 - 1.375X_1^2 + 10.113X_3^2 \quad (9)$$

The response surface has form of hyperboloid rotation (figure 3), because coefficients of regression equation are B_{11} and B_{33} with different signs.

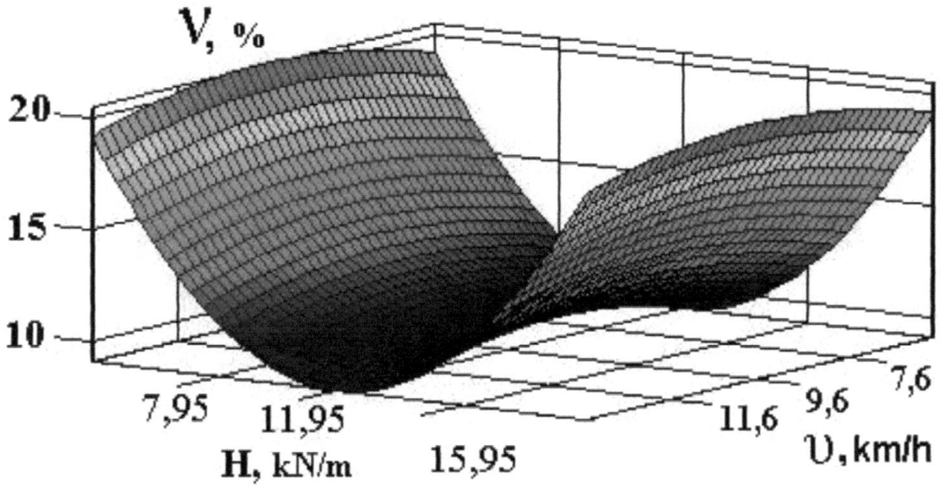

Figure 3. Response surface depending on operating speed of unit and spring stiffness.

The optimal value of response function is at the point with coordinates: $x_1 = 9.6\kappa m/h$, $x_3 = 11/95\kappa N/m$.

Consider two-dimensional cross section of response surface of X_1SX_3 plane (fig. 4), when $x_2 = 0.434$, i.e. at its optimum value. The cross section of response surface in experiment area gives isolines (fig. 4), from which it follows that change in operating speed of unit affects less than spring stiffness, because the elongation of isolines by unit speed is higher than along x_3 axis, and coefficient of regression equation is $|B_{11}| < |B_{33}|$.

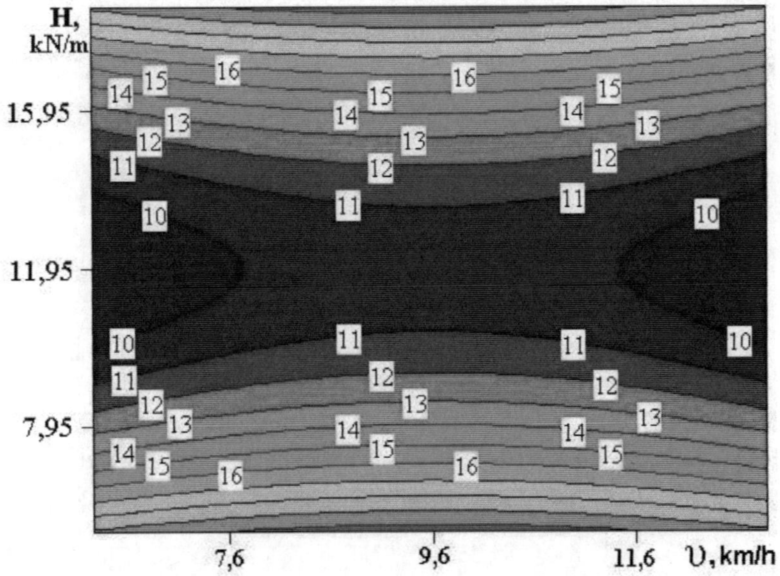

Figure 4. Two-dimensional cross-section of response surface depending on operating speed of unit and spring stiffness.

Consider cross section of response surface by X_2SX_3 plane. To do this, substitute value $x_1 = 0.39$in equation (1) and obtain following expression

$$Y_{23} = 11.085 - 3.14x_2 + 0.878x_3 - 1.638x_2x_3 + 3.468x_2^2 + 10.094x_3^2 \quad (10)$$

Having performing canonical transformation and solving system of linear equations, were founding coefficients of center response surface: $x_2 = 0.434$ and$x_3 = -0.0083$. Substituting found value in equation (1), was determined the value of parameter of optimization in center of response surface at optimal value $x_1 = 0.39$.

Equation (10) has pair interaction (x_2x_3), and rotation angle of coordinate axes α is 2.28 degrees. The motion of regression coefficients in canonical form are equal: at X_2 $B_{22} = 3.36$, at X_3 $B_{33} = 10.194$.

The regression equation of response surface in canonical form is:

$$Y_{23} - 10.128 = 3.36X_2^2 + 10.194X_3^2 \quad (11)$$

The response surface (fig 5) is paraboloid of rotation, and optimum value of response function is at point with coordinates: $x_2 = 0.434$ and $x_3 = -0.0083$ with positive values. The isolines of two-dimensional sections (Fig. 6) were obtained in form of an ellipse, and its long axis indicates a lesser effect on uniformity of soil density by roller mass compared to spring stiffness: $|B_{22}| < |B_{33}|$ at $B_{22} = 3.36$ and $B_{33} = 10.194$.

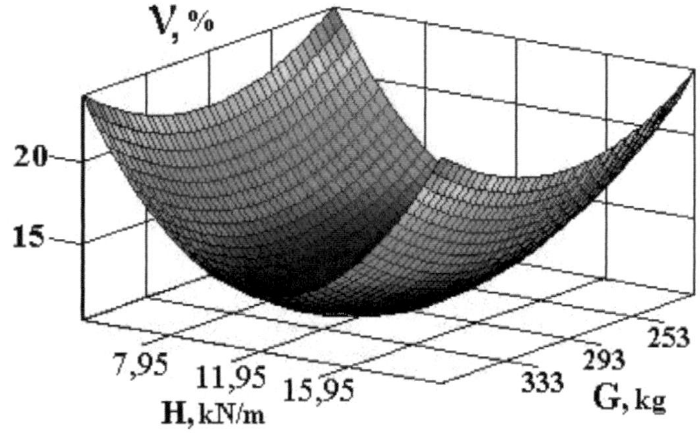

Figure 5. Response surface depending on roller mass and spring stiffness.

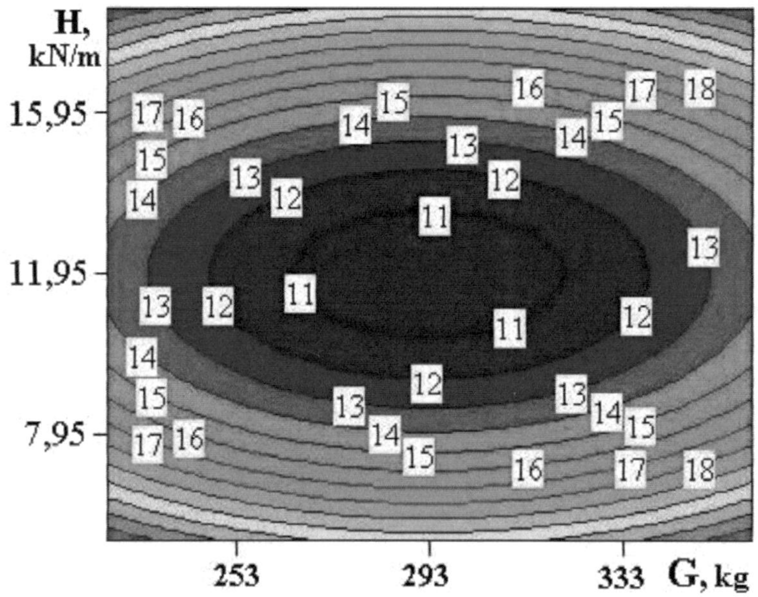

Figure 6. Two-dimensional cross-section of response surface depending on roller mass and spring stiffness.

Thus, in proposed vibratory roller the largest influence on quality of soil compaction is affected by spring stiffness, which creates vibration of the compacting drum of roller.

4. Conclusion

The optimal parameters of vibration roller and its operating modes were established by planning the experiment according to plan B_K using a 3-factor experiment, taking into account the fulfillment of agricultural requirements for the uniformity of soil compaction by rollers.

According to obtained regression equation, according to criterion of minimum value of coefficient of soil density variation after the passage of vibratory roller, the optimal operating speed of the unit is υ = 9.6 km/h, the roller mass G = 293 kg, the spring stiffness is H = 11.95 kN/m. At a minimum value of optimization criterion v = 10.43 %.

References

[1] Maslov G, Borisova S, Yudina E, Tsybulevsky V and Nda M N 2019. Development Parameters of Pneumatic Slotted Sprayers for Treatment Potato Tubers *Int. J. Innov. Technol. Explor. Eng.* **9** 1200-3 https://doi.org/10.35940/ijitee.b7848.129219

[2] Maslov G, Palapin A, Tsybulevsky V, Lavrentiev V and Yudina E 2019. Development of Parameters of a Gear Spring Harrow by Planning an Experiment *Int. J. Recent Technol. Eng.* **8** 8030-4 https://doi.org/10.35940/ijrte.d4513.118419

[3] Maslov G, Lavrentiev V, Tsybulevsky V, Yudina E and Tkachenko V 2019. Optimization of parameters of a multifunctional unit based on a spring harrow *Int. J. Eng. Adv. Technol.* **9** 1915-8 https://doi.org/10.35940/ijeat.A1054.109119

[4] Tarasenko B F, Gorovoy S A and Tsybulevskiy V V 2010 Universal plow for moldboard-free tillage with cylindrical chisels and swiveling paws and optimization of its parameters during deep loosening *Polythematic online scientific journal of Kuban State Agrarian University* **60** 134-46

[5] Drobot V A and Tsybulevskiy V V 2010 Optimization of the parameters of the process of surface tillage by horizontally located disc working bodies *Trudy Kubanskogo gosudarstvennogo agrarnogo universiteta* **23** 181-6

[6] Adler V A 1999 *Vvedeniye v planirovaniye eksperimenta* (Moskva: Metallurgiya)

[7] Novik F S and Arsov J B 1980 *Optimizacija processov tehnologii metallov metodami planirovanija jeksperimentov* (Moskva: Mashinostroenie)

Analysis of the state and prospects of application of the plasma sputtering method for obtaining wear-resistant coatings

G I Bondareva[1,3], S K Toigambayev [2], N M Tambovsky [2]

[1]All-Russia Research Institute of Hydraulic Engineering and Land Reclamation of A. N. Kostyakov, Moscow, 127434, Russia
[2] Russian State Agrarian University-Moscow Agricultural Academy named after K. A. Timiryazev, Moscow, 127550 Russia

[3] E-mail: boss2569@yandex.ru

Abstract. The analysis of the properties of structural materials according to the wear resistance criterion is carried out. The prospects for the application of the plasma spraying method for obtaining wear-resistant coatings are given. Practice shows that methods of increasing the wear resistance of parts by applying on them thin coatings of ceramic, cermet and other highly wear-resistant materials by surfacing, vapor deposition, detonation and plasma spraying are very effective.

1. Introduction

The most important performance indicator of the quality of a material is its wear resistance. If we analyze the properties of known structural materials, it turns out that the level of wear resistance of most of them is low. And as you know, the durability, reliability and efficiency of friction units are largely determined by their ability to resist wear. This explains the relevance of research in the direction of developing technologies that would improve the wear resistance of the working surfaces of parts that have small tolerances for final processing and are made of easily processed and cheap materials.

The literature gives many examples indicating the high-performance properties of coatings obtained by these methods. However, there is no information on a specific technology for obtaining such coatings either in the literature on this topic known to us, or in the practice of manufacturing friction units. If we consider wear resistance as the resistance of a material to separation under dry friction in a system of two solids, then the whole variety of friction patterns can ultimately be represented by five types of frictional interaction: elastic displacement, plastic deformation, cutting, adhesive interaction of the film surface, seizure accompanied by deep pulling out of the material. [1,2]. Under certain conditions, these types of interaction can change the value of the contribution to the total friction force, passing from the predominant type to the accompanying one.

2. Research methods and results

Ordinary functioning of friction units is possible if the first, second and fourth types of interaction make a contribution to the magnitude of the frictional contact. The third and fifth types of interaction are manifested as the main ones only in emergency situations and therefore, when designing a unit, friction should be minimized [3].

Table 1 shows the mechanical properties and wear resistance of some structural materials, as well as the properties of plasma-sprayed materials according to various literature sources. It is stimulating to note that despite the huge number of structural plasmatrons, feeders, power sources and types of powders, the properties of materials sprayed in the optimal mode are approximately at the same level, an order of magnitude or more inferior to the properties of compact materials, even substantiated the fundamental impossibility of obtaining other properties associated with ultimately with a discrete nature of the solidification of particles during spraying, which is a distinctive feature of the method.

Table 1. Properties of some structural and plasma-sprayed materials.

Structural materials							
Material	ϱ, kg/m^3	$\sigma_в$ kg/mm^2	$H_V, kg/mm^2$	$E\cdot10^{-6}$ kg/mm^2	$\alpha\cdot10^{-6}$, 1/deg	Relative wear resistance	Conditions of receiving
Beryllium	1850	40	320	30	12	44	Sintering
Tungsten	19300	100	250-400	36	7	60	-
Molybdenum	10200	50	250	32	5,5	40	-
SNGN-60	7700-7900	40	860	13,5	5,5	40	Reflow at 1100 ° C
Relit	16500	40	2500	72	4-5	330	Casting
NK-8	14600	150	1200	55	4-5	70	-
Microlite	3960	9	2100	37,6	8,5	55	-
Plasma-sprayed materials							
Material/ properties	Be	W	Mo	Relit	WC+Co	Al_2O_3	ZrO_2
Density, kg/m^3	1840	19250	10300	13000-15000	1400	2800-3200	5700
Porosity, P	90%	80-90%	87-90%	7-11%	1-4%	8-25%	16-28%
Hardness, kg/mm^2		330-390	320-370	100-250	700	1000	640-800
Strength, kg/m^2	2	4.4	4.3	2-10	9	4-11	0.4-0.8
$\alpha, 10^{-6}$, 1/deg	-	2.7-3	3-3.5	-	-	8-8.5	-

Within the framework of the accepted assumptions, this conclusion is correct, which is confirmed by the overwhelming majority of the results of determining the properties of sprayed materials, some of which are presented in table 1. Taking into account the established relationships between the strength characteristics of materials and their wear resistance, it can be argued that the possibilities for using sprayed coatings obtained within of this method, as frictional or wear-resistant, are minimal, due to the low mechanical properties of the sprayed layers, and attempts to solve this problem by selecting any combination of initial powders and finding the traditional optium in terms of current, distance, granulation, gas consumption, relative movement rates, etc. to fail.

This is the first important conclusion drawn from the analysis of the relationship between the strength characteristics of materials and their wear resistance. It is also confirmed by the fact that there is no information in the literature on the successful use of plasma-sprayed materials in friction units obtained by one-step spraying technology in air or in a controlled atmosphere; the field of application of sprayed materials having serious practical value, today, in essence, is limited only to heat-resistant, heat-protective and corrosion-resistant coatings based on *W, Mo, Ni,* Al_2O_3, ZrO_2, respectively; tests of molybdenum coatings sprayed by conventional technology as friction disc brakes and coatings based on Al_2O_3 as wear-resistant coatings did not give positive results.

Successful use of nickel-cobalt alloy steels as wear-resistant sprayed coatings is due solely to the operation of subsequent reflow. The properties of these coatings immediately after spraying do not fundamentally differ from the properties of other sprayed materials and, therefore, this case goes beyond the scope of processes that limit the method of high-temperature spraying [4].

Consequently, without a qualitative improvement in the properties of plasma-sprayed coatings, their use to increase wear resistance in the overwhelming majority of practically important cases (under conditions of real specific loads and sliding velocities of a rubbing pair) is fundamentally impossible. What are the possible ways of such a qualitative improvement in the strength properties of coatings? In our opinion, today, there are 2 promising areas, effective to varying degrees:

1. Application of multi-operational technologies, including operations for the following high-temperature treatment: spraying and subsequent reflow; spraying and heat treatment.
2. More complete utilization of the heat contained in the applied material particles by increasing the compactness of the metallization jet, provided that inert atmospheres are used where necessary.

The prospects of the first of these directions are confirmed by the successes achieved in obtaining wear-resistant coatings by plasma spraying of alloys based on *Ni-Cr-B-Si* [5].

There are, however, significant difficulties that limit the possibilities of improving the quality of sprayed coatings by this group of techniques. These difficulties are as follows: the obligatory combination of a more refractory substrate with a less refractory coating; the need for uniformity of mechanophysical properties (thermal expansion coefficient, elastic modulus), which makes it possible to jointly heat up to temperatures close to the melting temperatures of the coating, the coated part, while not all materials allow prolonged heating to such high temperatures, due to irreversible changes in the chemical and phase composition, leading to a sharp deterioration in their properties.

In our opinion, the success in the creation of technologies for obtaining wear-resistant coatings by the plasma method will largely depend on the success in overcoming the listed difficulties.

The essence of the second direction of qualitative improvement of the properties of sprayed coatings is described in detail in [6]. Here we would like to dwell only briefly on some features of the method, the results obtained and the prospects for further research in this direction in relation to the problem of obtaining wear-resistant coatings.

The main difference of this method from the classical method of plasma spraying is that when spraying in a compact jet mode, it is possible to use the cumulative thermal effect of the deposited particles in order to create conditions for favorable thermodynamic and metallurgical reactions. Under certain conditions, the high-temperature region turns out to be localized in a thin surface layer of the coating, which makes it possible to almost completely eliminate the heating of the substrate and thereby make it possible to obtain high-strength coatings on materials with a significantly lower melting point and strongly differing in their physical and mechanical properties from similar properties of the applied cover. For example, it was possible to obtain Al oxide coatings containing from 80 to 90% of the α-phase on sintered aluminum alloy substrates. The coatings were tested for wear in a pair with sintered tungsten carbide and showed characteristics close to the wear resistance of a microlite.

Structures of coatings made of Al_2O_3 with additives of magnesium and chromium oxides. zirconium samples obtained in the mode of a non-compact and compact jet show that in the latter case, the structure of the material is characterized by the absence of layering, typical for plasma coatings, with a small number of dead-end pores. By comparing some of the obtained etched structures with the atlas of microlite microstructures, it was possible to establish their identity. Other properties of alumina coatings and wear test results when paired with other materials are shown in table 2.

Table 2. Properties of plasma sputtered alumina coatings in compact jet mode.

Original structure	Microhardness coverings	Phase structure	Porosity, %	Diameter blocks, micron	Surface cleanliness class	-
$Al_2C_3$99.99%	2200	$\alpha - Al_2O_3$-95 -97%	9.4	11.4	11-12	-
Al_2O_3+1.5+2% Cr_2O_3	2400	$\alpha - Al_2O_3$-96 -98%	7.65	14.8	11-12	-

Al_2O_3+1%MgO	2100	$\alpha - Al_2O_3$-93-96%	6.25	7.2	11-12	-
Al_2O_3+1%MgO +3% ZrO_2	2100	$AlMgO_4$-3-5%	4.4	5	10-12	-
CM -322	2182	$\alpha - Al_2O_3$-98%	4.7	1-3	10-11	-
Frictional properties						
Sample	Counterbody	Friction coefficient	Pressure, kg/	Speed, m/s	Phase composition	$E=\frac{\Delta H}{l}$
Al_2O_3	Al_2O_3	0.127-0.133	2.7	0.5-0.8	$\gamma - Al_2O_3$	10^{-2}
Al_2O_3	Steel 2X13	0.24-0.44	5-14	0.5-0.8	70-100% $\gamma - Al_2O_3$	10^{-3}
$Al_2O_3 + 0.5Cr_2O_3 + 0.7MgO + 0.5BeO$	Cast iron ZhCHKh	0.33-0.39	32	0.5-0.8	90% $\alpha - Al_2O_3$	10^{-7}

Analysis of the results obtained allows a more optimistic assessment of the capabilities of the plasma spraying method in order to obtain wear-resistant coatings. The creation of the required temperature regime in the macro-volumes of the coating formation region (determined experimentally so far) allows bringing the mechanical characteristics of the sprayed layers to a level close to the corresponding characteristics of compact materials [7].

The data available on the structure, strength, and wear resistance of ceramics sprayed in a compact jet mode indicate the identity of the listed properties and properties of sintered ceramics. Thus, the possibility of obtaining plasma-sprayed materials with the properties of known structural materials, including the property of wear resistance, has been proved in principle. At the same time, all the advantages of the economic nature of the plasma method, considered at the beginning of the article, are preserved.

Further studies related to the second of the listed areas of qualitative improvement of the properties of sprayed coatings should be carried out in order to study the conditions for obtaining wear-resistant coatings on a Monolit-type installation from other materials, both homogeneous (metal, ceramic) and composite (cermet). The main task of these studies is to find a correct quantitative description of the relationship between the deposition conditions and the values of nonstationary temperature fields in the forming coating and substrate [8].

3. Conclusion

The possibilities of using plasma-sprayed coatings obtained within the framework of the traditional "classical" method of high-temperature spraying as wear-resistant are minimal, due to the low strength properties of these coatings.

Progress in this matter can be achieved in 2 ways: 1) through the use of multi-operational technologies for obtaining sprayed coatings, which necessarily include the operations of subsequent high-temperature processing; 2) by improving the existing plasma spraying process by making the most complete use of the heat contained in the deposited material particles during the formation of coatings on the substrate.

References

[1] Bondareva G I 2011 Redistribution of residual stress in plasma-sprayed coatings *Russian Engineering Research* **9** 853-5
[2] Bondareva G I, Evgrafov A V and Ermolaeva A Yu 2020 Metrological support characteristics for life cycle of agricultural machines *Journal of Physics Conference Series* **1515** 042087
[3] Karymsakova I B, Denissova N F and Krak I V 2017 Criteria for implants classification for coating implants using plasma spraying by robotic complex *Eurasian Journal of Mathematical*

and Computer Applications **3** 44-52
[4] Kravchenko I N, Kuznetsov Yu A, Goncharenko V V and Kalashnikova L V 2020 Investigation of influence of plasma spraying process parameters on the level of residual stresses *Key Engineering Materials* **864** 198-203
[5] Zhao X, Zhuo Y, Liu S, Zhou Y, Zhao C, Wang C and Yang Q 2016 Investigation on WC/TIC interface relationship in wear-resistant coating by first-principles *Surface and Coatings Technology* **305** 200-7
[6] Karpuzov V, Golinitskiy P, Cherkasova E, Antonova U and Toygambaev S 2020 Development of knowledge management process at the enterprise of technical service of the agro-industrial complex *Journal of Physics: Conference Series* **1691** 12031
[7] Sorokin G M and Malyshev V N 2013 Materials science and corrosion protection method for ranking steels based on wear resistance and strength characteristics *Chemical and Petroleum Engineering* **48** 583-8
[8] Stukhlyak D P, Mytnyk M M, Holotenko S, Dobrotvor G, Skorokhod Z and Marukha V 2019 Modeling of the wear resistance of epoxy composites according to changes in their mechanical characteristics *Materials Science* **54** 697-704

IOP Conf. Series: Materials Science and Engineering 1155 (2021) 012089 doi:10.1088/1757-899X/1155/1/012089

A planetary mill modelling in chaotic mode

V S Kalnitsky[1,2,3], O V Rusakov[1] and V E Sergeenkov[2]

[1] Saint-Petersburg state university, 7/9 Universitetskaya Emb., St Petersburg, 199034, Russia
[2] Military Academy of Logistics, 8 Makarov Emb., St Petersburg,199034, Russia

[3] E-mail: st006987@spbu.ru

Abstract. Consider a random walk in a regular m-polygon inscribed in the unit circle. The movement of the point to a randomly selected vertex determines the new position of the point. The point passes a prescribed distance and stops. For such a process, there is a limit distribution of the probabilities of the position of the point. This algorithm was used for modelling different processes. Now we apply it for high-energy ball milling. The impact frequency spatial distribution calculated with the stochastic model simulation has coincided with the distribution predicted by the distinct element method. The real granulometric analysis also can be explained by the model.

1. Introduction

The planetary ball mill processing has been modelled within decades. We can point two main branches of investigations: analytical and numerical modelling. Both are aimed to study the dependence between numerous grinding parameters. In order to fully understand the grinding process, the following issues should be considered:

- the nature and quantification of the interactions between the ball and powder, including a complete description of the kinetics of the ball during grinding;
- deformation phenomena arising from the interaction of one powder particle with a ball;
- physical phenomena occurring inside the powders.

The analytical approach gives the most reliable results but small quantity of parameters can be included. The numerical methods are the power instrument of investigation of relations between parameters, but does not give clear understanding of processes. To illustrate, a numerical dynamic-mechanical modelling indicates just a correlation between the grinding parameters and the resulting microstructure of the ground material.

In work [1], supplementing the results of the EURAM project implemented by P. Le Brun [2], the kinetics of the balls was investigated. The calculated trajectories were compared with the trajectories observed with the high-speed camera. The authors introduced into consideration the angular velocity ratio $R = \omega/\Omega$ of the vials a. v. ω and the sun disc a. v. Ω, which are rotating in opposite direction. It turned out that the introduced parameter determines the types of the motion of the balls in the mill. The authors identified three different modes and defined the parameter value limits for them. For $R < R_{limit}$ the mode was termed "chaotic", for $R_{limit} < R < R_{critical}$ – "impact-friction" mode, and for $R > R_{limit}$

– "friction" one. It is for the chaotic mode that the most effective grinding process is observed. However, numerical analysis becomes impracticable. Moreover, the authors noted a strong discrepancy between the results of numerical analysis of trajectories and observed ones.

In [3] a numerical dynamic-mechanical model of a planetary ball mill was developed to study the dependence of the process efficiency on grinding parameters, such as the size and number of the ball, the jar geometry and the velocity of rotating parts. Modeling indicates a correlation between the grinding parameters and the resulting microstructure of the ground material. Analogously, the maximum efficiency of the grinding process is observed with the most disordered movement of the ball, which is achieved in a well-defined range of the parameter *R*.

In the paper of Gy. Kakuk et al. [4] also the relationship between the ratio of the angular velocity of the sun disk and the containers and the geometrical parameters of the mill were considered. Having studied the relationship between the model created for the grinding process that occurs in a planetary ball mill and the grinding parameters that depend on the mill, and using the calculations performed, the authors obtained more reliable than the previous data on the energy transmitted to the mill.

In the article of Abdellaoui M. and Gaffet E. [5], on the basis of kinematic modeling of a planetary ball mill, kinematic equations are given that determine the velocity and acceleration of the ball. A comparison of those calculated with some experimental results documented in the literature shows that neither the impact energy nor the frequency of the impact, taken separately into account, do not affect the grinding result, but their product has the determining value.

2. Simulation

In the study of the authors [6], a new stochastic model of a planetary mill based on the concept of random walk on a polygon was proposed. This model was designed for chaotic mode of grinding.

Consider a regular *m*-polygon inscribed in the unit circle. Denote by z_0, z_1, ... the wandering points. The position of the point at each subsequent moment of time is determined by the result of the point moving to a randomly selected vertex (the probabilities of selecting vertices are the same), so that the point passes the fraction α of the distance to the selected vertex and stops, etc. This recurrent algorithm is written in the form of an autoregression equation of the 1st order.

$$z_n = (1-\alpha)z_{n-1} + \alpha\varepsilon_n, \qquad n = 0,1,\ldots$$

For such a process, subject to the condition $|1-\alpha|<1$, there exists a limiting probability distribution of the position of a point, regardless of the initial position. For example, if we consider a regular triangle and $\alpha = 1/2$, then the limit set will be the Sierpinski triangle. In any case, the limit set is a fractal figure. Note that the recurrence algorithm itself allows modifications that allow it to be used for modeling many real processes, for example, in financial mathematics [7].

The described stochastic process is applicable to the description of the distribution density of grinding element impacts in a planetary mill with continuous feed.

Here are the basic assumptions:

- the movement of one grinding element is considered in projection on the base of the cylindrical body of the mill;
- we consider the movement of the element to be straightforward before the impact (stop);
- the direction after the impact will be considered a normally distributed vector in a regular polygon with a large number of vertices (in our numerical experiments, it is 1000);
- for the primary model, we assume that the path length μ for a particle has a fixed length in fractions of the drum radius;
- the individual trajectory of the wandering of the particle will be considered common to all homogeneous grinding elements;
- the density of distribution in the drum of the number of particle stops is taken as the density of collisions of a collective of particles.

The first assumption means that we use two-dimensional model of the process, which is a typical assumption in the majority of models. The third assumption reflects the chaotic nature of the process.

Note that the path length, of course, depends on many factors – the density of the material being dispersed, the mass of grinding elements, the size of the mill, the speed of rotation of the mill, etc. After constructing the primary density distribution model, an adjustment should be made to the law that determines the path length of the particle.

3. Result and analysis

The simulation of the stochastic process described above was carried out using the "FractalDemo" program created by O.V. Rusakov.

We set the following parameters: m = 1000, the number of steps for calculating densities is from 8 to 30 million, depending on the path length μ. The duration of the numerical experiment did not exceed 1 minute. Below are the results of a numerical experiment for calculating the particle impact frequency distribution in the cross section of a mill drum (figure 1). The color scale is from low (green) to large (red) frequency.

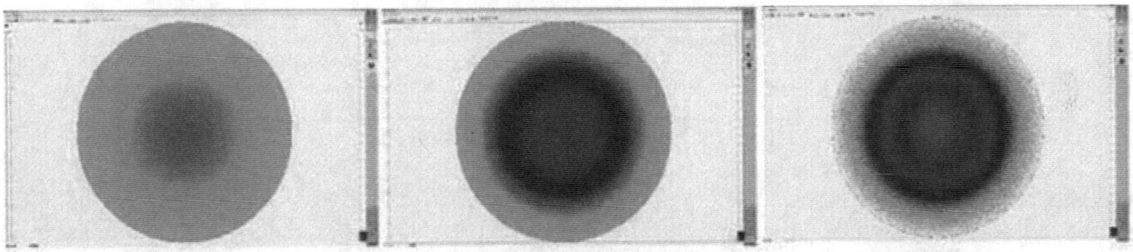

μ=0.4 μ=0.3 μ=0.2

Figure 1. Limit impact frequency distribution of stochastic walk.

As you can see, even with a path length of 0.4 radii and above, a uniform distribution of particle collisions over the sectional area is observed. Apparently, this is the threshold beyond which the model describes the behavior of the grinding elements when they are low in the substance being dispersed.

For small runs, a fundamentally different picture of the density distribution is observed: symmetrically, with increasing density towards the center of the drum. With a path length less than 0.1 radius, the simulation gives almost zero density at the edge of the drum, which corresponds to the friction-mode.

If we introduce into the model the dependence of the path length on the density of the grinding elements in the cross section of the drum, then this effect of uneven frequency distribution will increase significantly, because the length of the run (before impact) outside the red zone will be longer than in the red one. Thus, it can be assumed that due to the limited residence time of the dispersible substance in the attritor, a part of the substance will experience less impact from the grinding elements. The particle size analysis of the material obtained as a result of grinding shows the presence of two pronounced peaks in the histogram of the distribution of particle sizes [6]. What can be explained by the model we built. To overcome this effect, an interesting innovative jar design has been proposed in work [8]. A flat wall portion was inserted halfway between the original curved wall and the axis of the jar, resulting in an half-moon (HM) cross-section. HM jar produces an homogeneous size distribution powder for a wider range of parameter *R*.

The DEM (distinct element method) modelling has been used in [9] to analyze the planetary ball mill. The fact that DEM modelling is suited to the analysis was first realized by Mishra et al [10] and was used in horizontal tumbling mills (see also [11]). The computer simulation incorporated a modified Kelvin viscoelastic spring/damper system. The collision spatial distribution was analyzed. The modeling was conducted for different values of the filling fraction

$$n_v = \frac{N_b}{N_{b,max}},$$

where N_b is the number of balls included in the vial and $N_{b,max}$ is the maximum number of balls, that could be included in the vial. Figure 2 shows the spatial distribution of impact frequency for a range of vial filling fractions ($n_v = 0.26, 0.42, 0.63, 0.84$). The computational simulation in [9] were correlated with experimental measurements. Comparing the results of numerical analysis and the impact frequency distribution obtained by us, we can conclude that stochastic analysis effectively simulates the chaotic regime of a planetary mill.

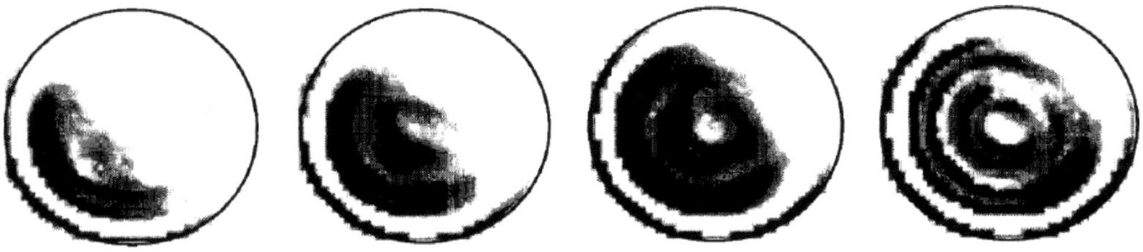

$n_v = 0.26$ $n_v = 0.42$ $n_v = 0.63$ $n_v = 0.84$

Figure 2. Predicted impact frequency spatial distribution for a range of vial filling fractions [9].

The apparent discrepancy in the structure of the collision density distribution is explained by the initial choice of a fixed impact-free path length μ. The next step in the computational procedure should be the adjustment of this parameter μ depending on the obtained distribution. This process converges and is likely to produce a result similar to [9].

4. Conclusions

The analytical solution to the problem of a collision of a large number of particles is difficult. The DEM method requires huge resources for a relatively small number of particles. Real 3D simulation of collisions becomes quite difficult if the number of particles exceeds a thousand. In our proposed stochastic model, we completely remove restrictions of this kind, replacing the collective behavior of particles with the chaotic motion of one particle, setting only the parameters of the random walk.

One of the difficult technical problems, the solution of which has not been found up to the present, is the scaling of existing samples of attritors. The production of scale models is not justified in terms of costs. Mathematical modeling actually remains the only available method for solving this problem. One of the characteristics for a continuous type mill is the ratio of the length of the drum and its radius. If the effect of the separation of a dispersible substance found by us is stable along the drum, then this problem should be solved by the method of separation rather than by changing this parameter. The question of stability, in turn, requires the construction of an additional model in projection onto the generator cylinder.

We list only a few large unsolved problems: controllability of grinding parameters and the problem of design scaling. It is assumed that both problems can be solved if an adequate model of the grinding process as a whole is created. This task is far from being solved today. Significant results were achieved neither on the path of analytical, nor on the path of computer modeling. A fundamental obstacle to the practical verification of many of the obtained theoretical extrapolating results is the high cost of the industrial design.

Our approach may perhaps complement existing models. We obtained results that are in good agreement with the numerical analysis in other models. By varying the path of particles between impacts, we can also simulate different modes of operation of the mill. The integration with other numerical modelling procedures may be essential to achieve a more complete understanding of the processes.

References

[1] Le Brun P, Froyen L and Delaey L 1993 The modelling of the mechanical alloying process in a planetary ball mill: comparison between theory and in-situ observations *Materials Science and Engineering* **A161** 75-82
[2] Le Brun P 1989 *Internal Rep.* (Katholieke Universiteit Leuven)
[3] Broseghini M, Gelisio L, D'Incau M, Azanza Ricardo C L, Pugno N M and Scardi P 2016 Modeling of the planetary ball-milling process: The case study of ceramic powders *Journal of the European Ceramic Society* **36(9)** 2205-12
[4] Kakuk Gy, Zsoldos I, Csanady A and Oldal I 2009 Contributions to the modelling of the milling process in a planetary ball mill *Rev. Adv. Mater. Sci.* **22** 21-38
[5] Abdellaoui M and Gaffet E 1995 The physics of mechanical alloying in a planetary ball mill: mathematical treatment *Acta metall. mater* **43(3)** 1087-98
[6] Kalnitsky V S and Rusakov O V 2018 Random walk modelling in the high-energy ball milling *J. Pharm. Sci. Emerg. Drugs* **6** 77
[7] Rusakov O and Laskin M 2017 A stochastic model for stationary dynamics of prices in real estate markets *AIP Conference Proceedings* **1836** 020087
[8] Broseghini M, D'Incau M, Gelisio L, Pugno M N and Scardi P 2016 Effect of jar shape on high-energy planetary ball milling efficiency: Simulations and experiments *Materials & Design* **110** 365-74
[9] Dallimore M P and McCormick P G 1997 Distinct element modelling of mechanical alloying in a planetary ball mill *Materials Science Forum* **235-8** 5-14
[10] Mishra B K and Rajamani R K 1990 Motion analysis in tumbling mills by the discrete element method *KONA Powder and Particle* **8** 92-8
[11] Hiroshi M, Junya K and Fumio S 2004 Scale-up method of planetary ball mill *Chemical Engineering Science* **59(24)** 5909-16

IOP Conf. Series: Materials Science and Engineering 1155 (2021) 012090 doi:10.1088/1757-899X/1155/1/012090

Designing a mobile medical expert system using the Yandex.Alice voice assistant

E M Markushin, K E Ognegin[1], P S Polskaya, A Shinkaruk and S P Yakimov

Reshetnev Siberian State University of Science and Technology, 31 Krasnoyarsky rabochy avenue, Krasnoyarsk, Russia

[1]E-mail: ogneginkirill@gmail.com

Abstract. Solving the problem of making a primary diagnosis in the absence or shortage of qualified doctors. The main task is to reduce the risk of incorrect assessment of the health status of people who do not have the opportunity to receive prompt qualified medical advice and assistance. We offer a project solution – Alice's skill (voice assistant from Yandex) "Pocket Doctor" with a voice interface, which is a diagnostic expert system containing rules for logical inference based on symptoms, the assessment of which does not require special laboratory tests and can be performed by people who do not have a medical education. Based on the collected information from the user, the primary diagnosis and recommendations for emergency medical care are determined.

1. Introduction

If a person feels seek, it is necessary to go to a doctor. Unfortunately, this is not always possible – there are lots of reasons for this, which can be both objective and subjective. How to be a hunter in the taiga, a tourist on a long hike, a shift worker, a sailor on a fishing seiner? Even in the city, it is often quite difficult to get primary medical care even it the city because of long queues in local clinics, which do not disappear despite the best efforts of medical staff.

It would seem that the Internet at least partially solves this problem, but in practice, the availability of poorly structured information often leads to sad consequences. People at the first signs of illness, try to independently diagnose themselves and self-medicate, without having medical education and often the consequences of such actions can lead to serious consequences, as well as fatal outcomes.

A striking example was the global problem with the COVID-19 coronavirus infection. People who falsely assume that they have experienced a common cold do not receive the medical care they need, which leads to serious consequences, as well as fatal outcomes. On the other hand, the workload of medical staff forces doctors to recommend that patients limit their visits to medical institutions as much as possible in order to reduce the risk of infection.

Thus, it becomes urgent and necessary to create an expert system that is constantly in the zone of accessibility and allows you to make a primary diagnosis and give recommendations for providing first aid to the patient, and reduce the risks that arise. The technical side of the issue is not fundamentally complicated. The life of a modern person is closely connected with mobile devices. Today, people can't imagine their life without a smartphone that allows them to make online purchases, make an appointment with doctors, and search for the necessary information in a matter of minutes.

The complexity of the task is due to the need to process a large amount of information using logical output and the organization of an intuitive data input interface. To solve this problem, it seems appropriate to use the achievements in the field of creating a voice interface [1].

Voice interfaces have become commonplace, there are General-purpose voice assistants (Alexa, Siri, Cortana, Google Assistant, Alice), as well as specialized software solutions for narrow specialists or certain areas of business and social sphere.

The voice interface is convenient in cases when the usual ways of interacting with the information system may be difficult due to the user's type of activity, or when the user has reduced mobility or other physical disorders.

The main goal of the project is to create the Alice skill using the voice interface to determine the initial diagnosis and assign the appropriate specialist (Created on the "Yandex platform.Dialogs").

2. Methods

An important criterion for a good voice interface is its naturalness, that is, interaction with such an interface should create a sense of communication with a real person. Such a voice interface should work correctly, have a huge vocabulary, conduct a "dialogue" with the user, ask leading questions and give useful hints.

The most popular voice assistant in our country is Yandex Alice. Alice is a voice assistant that can recognize and synthesize speech in Russian, give answers to user questions, and solve applied tasks using programmed skills. Alice is based on a neural network that allows you to maintain conversations with the user without pre-defined scenarios.

For third-party developers, Yandex has launched the "Yandex platform.Dialogs" is intended for publishing various automated skills or chats with operators, followed by connecting to Alice [2].

Alice's skill is a specialized voice assistant mode that allows the user to call automated processes of a certain kind and get the necessary information. The skill is placed in Alice's public skill catalog. This means that It is available to a wide range of users who can call this skill by keywords or its name.

Advantages of creating a skill for Alice:

- Ability to access more than 35 million users.
- The skill appears in Yandex apps, as well as on various devices, such as phones, watches, smart speakers, and others.
- No need to install a third-party app.
- Reliability and accuracy of voice recognition provided by Alice.
- Continuous improvement of Alice's functions.

The functionality of the designed information system assumes two user roles:

- Administrator responsible for its operation and the correct operation of individual components (figure 1).
- User who needs to be diagnosed (figure 2).

The administrator can maintain a dictionary of keywords and nodes in the inference tree (add, edit, and delete). When editing the corresponding directories, the same interface solutions should be used for adding and changing entries. The logical output tree in the designed information system is binary and, accordingly, has two types of nodes:

- Symptom - is a branch of the tree that has a logical connection with subsequent nodes: the "Yes" link-confirmation of the presence of the symptom and the "No" link, which determines the conclusion if it is absent.
- Diagnosis - has no logical continuation and is a diagnosis and recommendations for further actions.

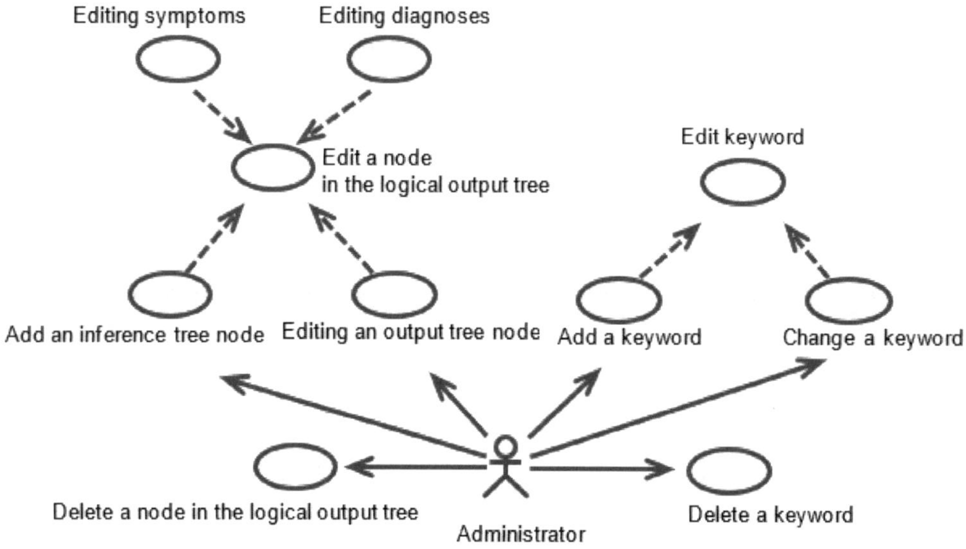

Figure 1. Use case diagram for the administrator.

The user's application must have access to the corresponding database on the server via the Internet in order to function. The user is consistently given questions about the presence of a particular symptom from the output tree until the node corresponding to the diagnosis is reached. The answer is selected using the Alice voice interface.

Scenario of interaction of the voice interface with the user:

- A greeting, and a quick hint on how to use the skill.
- Getting symptoms from the user (as a set of replicas).
- Answer in the form of clarifying questions and go to step 2, otherwise-issue a list of possible diagnoses, if the information received from the user is sufficient.

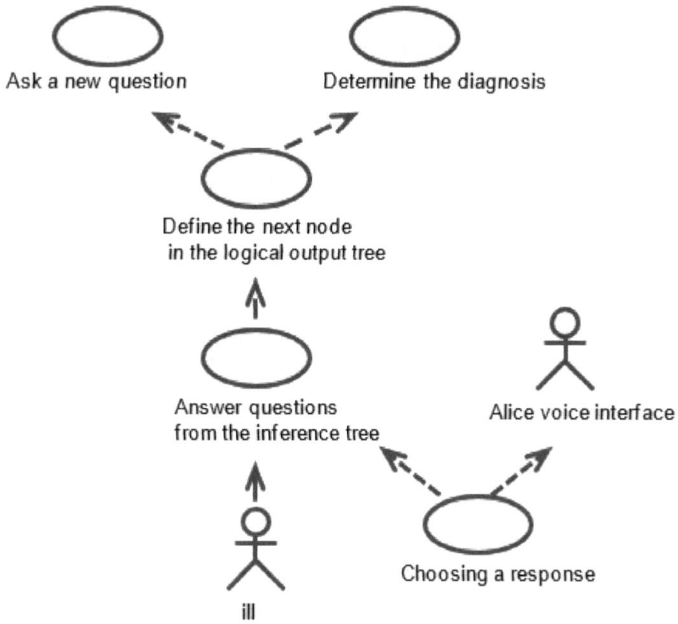

Figure 2. Use case diagram for the user.

To develop the functional model of the application, the unified modeling language UML was used in the Rational Rose design environment [3], which provides users with the ability to create, analyze, modify, and manage models using a single object-oriented approach and a single modeling language.

In order for the developed application to be user – friendly and have the most complete functionality, it is necessary to perform an important stage of the software lifecycle - design, according to the ISO/IEC/IEEE 12207:2017 standard [4].

Database design is an important stage in the development of an information system (IS). Data storage and processing is one of the main tasks of any information system. According to GOST 20886-85, a database is a collection of data organized according to certain rules that provide for General principles of data description, storage, and manipulation [5].

The developed IS will include a large amount of information: text, graphics, links, and other types. The database structure is implemented in a classical way using a relational model, which is represented as two-dimensional tables [6]. The database is designed using MySQL Workbench 8.0.19.

The administration application will be implemented using the classic client-server architecture. The server part will be written in C# using the framework ASP.NET, the client part will be written using one of the popular JavaScript frameworks: Angular, React, or Vue. HTTP entry point (webhook) for interaction with the "Yandex platform.Dialogs" will also be implemented on ASP.NET.

3. Results

Created based on UML use case diagrams, the entity-relationship diagram contains tables (figure 3):

- Symptoms (this table stores questions indicating the presence of the symptom, and foreign keys to go to either the next symptoms or the diagnosis).
- Diagnoses (this table stores the names of diagnoses and their descriptions).
- Node (this table is the link between the symptoms and diagnoses tables).
- Node type (this table contains information about the node type).

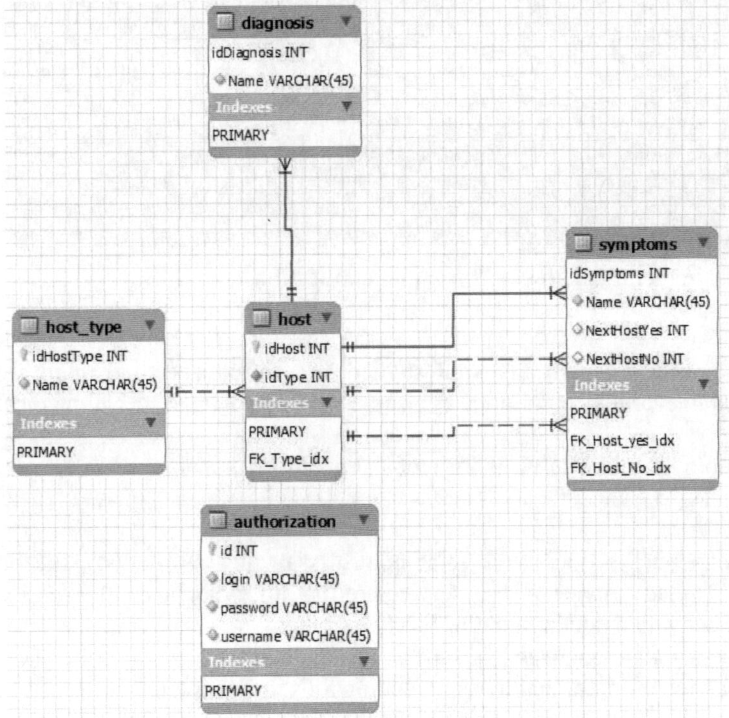

Figure 3. entity-relationship diagram.

The interface part of the application for the administrator requires a lot of attention. In order for the app to be user-friendly, when designing the interface, you must take into account:

- Intuitive layout of controls for faster operation.
- Tips, informational messages to reduce the number of human errors.
- Adjust the color scheme to reduce eye fatigue when working long hours.

Based on these conditions, an application was developed for database administration and setting up links and branches between issues.

Before using the program, the administrator must log in to their account. After that, the main menu window of the program will open, where the user can search for existing symptoms or diagnoses, add a new symptom or diagnosis, create a new branch, and test the operation of the primary diagnosis, as shown in figure 4.

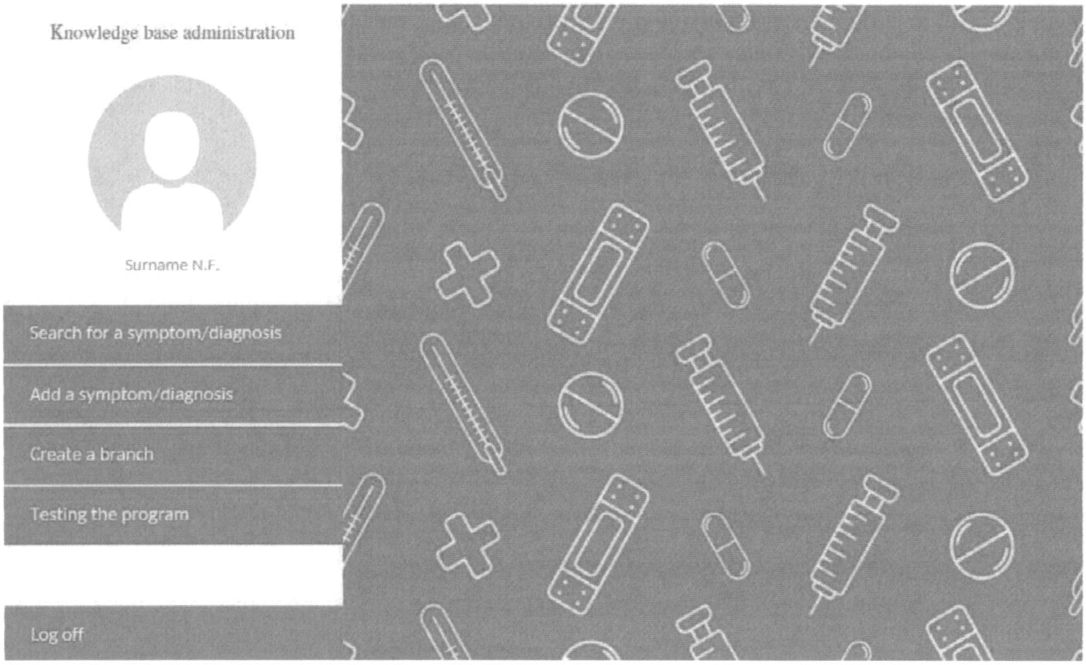

Figure 4. Main menu window of the admin application.

4. Discussion

The procedures for making a diagnosis are currently well studied and formalized, and there are descriptions of them adapted for use in everyday conditions by people who do not have special medical education [7].

Currently, there are a large number of online services for medical diagnostics that meet the subject area:

- Symptomate - medical diagnostics service with voice control via voice assistants: Amazon Alexa, Microsoft Cortana. The advantages of this service are that after diagnosis, the service offers to make an appointment with a doctor, and the disadvantages of the system are that voice assistants do not support the Russian language [8].
- WebMD - a website for making an initial diagnosis, including a list of symptoms, information about medications and diseases. The website lacks a voice interface and support for the Russian language [9].

- Therapist - a mobile app that allows you to make a primary diagnosis. The complexity of the app is that the user can't start diagnostics until they select a body part and a disease, which in turn can't always make a correct diagnosis. Also, there is no voice interface [10].

Advantages of the Alice skill being developed include the following:

- The presence of a voice interface.
- Usability.
- Integration into the Yandex ecosystem.
- The ability to use the app anywhere in the world if you have Internet access.

5. Conclusion

Thus, work was done on the development of an information system for database administration and on the design of Alice's skill.

Currently, the expert system is being actively tested under the supervision of qualified doctors in the city of Krasnoyarsk, and work is underway to accumulate a database that will make the application the most useful and provide information to the user in full.

In the future, it is planned to improve Alice's skill, which will allow you to offer an appointment to the appropriate specialist after determining the initial diagnosis. This design solution will reduce the risk of incorrect initial diagnosis and avoid various complications, which will undoubtedly have a positive impact on people's health.

References

[1] *Voice interfaces: how a computer recognizes speech and learns to speak.* Retrieved from: http://te-st.ru/2017/11/08/voice-interface/
[2] *Yandex. Dialogues.* Retrieved from: https://dialogs.yandex.ru/
[3] Leonenkov A V 2006 *Object-oriented analysis and design using UML and IBM Rational Rose* (Moscow: Binomial. Knowledge lab)
[4] *ISO/IEC/IEEE 12207:2017 Systems and software engineering – software life cycle processes.* Retrieved from: https://www.iso.org/ru/standard/63712.html
[5] *GOST 20886-85 data Organization in data processing systems*. Retrieved from: https://internet-law.ru/gosts/gost/12548/
[6] Markov A S and Lisovskiy K Yu 2006 *Database. Introduction to theory and methodology* (Moscow: Finance and statisctics) p 51-77
[7] Vorobiev A I and Shklovskiy N E 1992 *Your family doctor. Home adviser* (Moscow: World)
[8] *Symptomate.* Retrieved from: https://symptomate.com
[9] *WebMD.* Retrieved from: https://webmd.com
[10] *Therapist.* Retrieved from: https://clck.ru/N4nWM

Development of a thermal comfort management system in heat-protective clothing for people with disabilities

L G Babenko[1] and N Yu Saveliefa

Institute of service and business (branch) of Don state technical University, in Shakhty, Rostov Region, 346500, Russian Federation

[1]E-mail: liana-babenko@mail.ru

Abstract. The article deals with the development of a thermal comfort control system in thermal protective clothing for people with limited motor capabilities. In particular, the stages of forming the thermal comfort control system are described, and testing of the proposed system in real environmental conditions is carried out. The aim of the work is to develop and scientifically substantiate a thermal comfort control system used in thermal protective clothing for people with limited motor capabilities. The relevance of the paper is conditioned by the increased need for special heat-protective products with the function of control and monitoring of thermal comfort for people with limited motor abilities, who use a wheelchair as a means of transportation in everyday life. The work is based on the use of a systematic approach to the design of heat-protective adaptive clothing for people with limited motor capabilities. To solve a set of tasks the methods of mathematical modeling and experimental research were applied. The methods of system analysis and the basics of digital circuitry were used in theoretical research. The algorithm of the thermal comfort control system operation was substantiated and software for a smart phone was developed to control and monitor the temperature of under clothing space in a heat protective product for people with limited motor capabilities and findings are presented in the research work.

1. Introduction

The problem of universal integration of people with disabilities into social life is a strategic task of the Government of the Russian Federation. At the same time, people with limited motor capabilities (PLMC) are not provided with comfortable clothing that meets their physiological needs [1].

So, the important factors are to be taken into account when designing heat-protective adaptive products for PLMC are thermoregulation disorders, the absence of an objective reaction of the hypo dynamic system of the lower extremities of PLMC to external stimuli (negative ambient temperatures). These factors increase the risk of freezing and uncontrolled frostbite of the lower extremities.

At present, the theoretical and regulatory framework for designing adaptive clothing for PLMC is poorly developed, including the temperature changes on the skin surface of PLMC as compared to the group of healthy people, there is no general understanding of thermoregulation processes and circulatory disorder mechanism, there are no criteria for determining a comfortable physiological state of a person with limited motor capabilities (LMC) in connection with a huge number of causal relationships leading to a person's disability [2].

In the Russian Federation, the leading companies for the production of functional thermal protective clothing for PLMC are Bryansk Prosthetic and Orthopedic Enterprise [3], LLC Center of Designing

Special Purpose Shoes ORTOMODA (LLC CDSPS ORTOMODA, Moscow) [4], LLC «Cryptomed» [5] (Moscow) and the Federal State Budgetary Institution «St. Petersburg Scientific and Practical Center for Medical and Social Expertise, Prosthetics and Rehabilitation of Disabled People named after G.A. Albrecht of the Ministry of Labor and Social Protection of the Russian Federation» (St. Petersburg) [6]. The leaders in the production and sale of special (adaptive) clothing for PLMC among foreign manufacturers are the brands Buck & Buck [7], ABL Denim [8], Adaption by Adrian [9], Easy Access Clothing [10] (USA), Izadaptive [11], Silvert's [12] (Canada), Able2Wear [13] (Great Britain). At the same time, as a detailed consideration of the issue shows that the problem of designing special heat-protective clothing for PLMC, with a thermal control device, remains insufficiently studied [14].

The solution of the problem of maintaining a comfortable thermal state of PLMC in negative ambient temperatures is the use of smart clothing elements in the design of the product, in particular, a local heat system (LHS) and a thermal comfort control system (TCCS) based on the ESP-WROOM-32 microcontroller.

One of the most important tasks in the design of a heat-protective product with LHS is the choice of adequate heat temperature for various climatic conditions. Taking into account the complexity of an objective assessment of PLMC based on thermal sensations and preferences, the authors experimentally determined the permissible temperature conditions for PLMC in heat-protective clothing with a local heating system at negative ambient temperatures.

For the first time, the average values of the skin surface temperature in the areas of the hypo dynamic system of the lower extremities of the PLMC in the comfort zone, and its variability limits were determined as a theoretical justification for the technical task of designing heat-protective products for the PLMC. It was found that the average surface temperature of the skin of the hypo dynamic system of the lower extremities of healthy people exceeds the average value of the skin surface temperature of PLMC by 0.8°C with a variation in individual areas from 0.5 to 4.4°C. The areas of the greatest and the least deviation of the mean values of the skin temperature of the lower extremities of the group of healthy people and PLMC were determined.

The results of the previously obtained experimental data formed the basis for the development of an algorithm of the TCCS operation with the function of controlling and monitoring by means of the smart phone software. The authors consider that it possible to present the results of the experimental and theoretical research in the framework of achieving the goal – the development of a thermal comfort management system in a heat-protective product like a bag for feet for PLMC.

2. The development of a thermal comfort management system in a thermal protective product for PLMC

One of the important factors in equipping a heat-protective product with heating elements for PLMC is the choice of a temperature regime range under various situational environmental conditions, i.e. heat in the product should be controlled automatically in response to changes in preset conditions. In other words, the heat system must be safe during operation and take into account main person's physiological characteristics with physical inactivity of the lower extremities, first of all, such as impaired skin sensitivity – reactions to external stimuli. Since a person with LMC, when exposed to heat on body parts, without an adequate response to a stimulus (for example, with a violation of body functions as paraplegia) can get a thermal burn.

This increases the need to control temperature changes due to additional heating sources. To solve this problem, an operation algorithm of the thermal comfort management system has been developed, taking into account the use of a local heat system. The proposed algorithm is shown in figure 1.

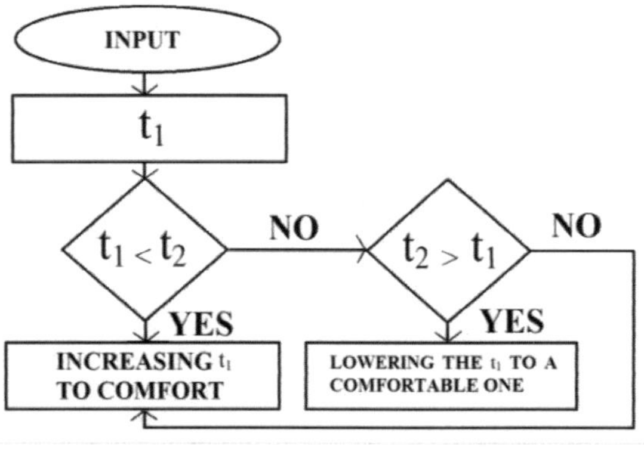

Figure 1. Algorithm of the thermal comfort control system, where t_1 – is the temperature of the heat element; t_2 – is comfortable temperature.

The presented algorithm of the TCCS operation has a temperature hysteresis equal to 2°C, embedded into the logic of the thermostat. Suppose that the user has set the temperature of the heating elements to 36°C, while the temperature hysteresis is 2°C. If the temperature drops to 34°C, the thermostat turns on and when the temperature inside of the product reaches 36°C, the thermostat goes off. For the safe operation of the product with heat elements the use of an external mobile battery («Power bank») is proposed.

Figure 2 shows the TCCS diagram. The principle of operation of the TCCS is that the product is equipped with a control microcontroller (ESP-WROOM-32) (1), with a power management system with a built-in Bluetooth bridge (2), a system that reads the temperature of the heating elements, equipped with a thermal sensor (3), which in its turn measures the temperature inside of the under-clothing space, as well as a system of switching on and off the heat elements (4).

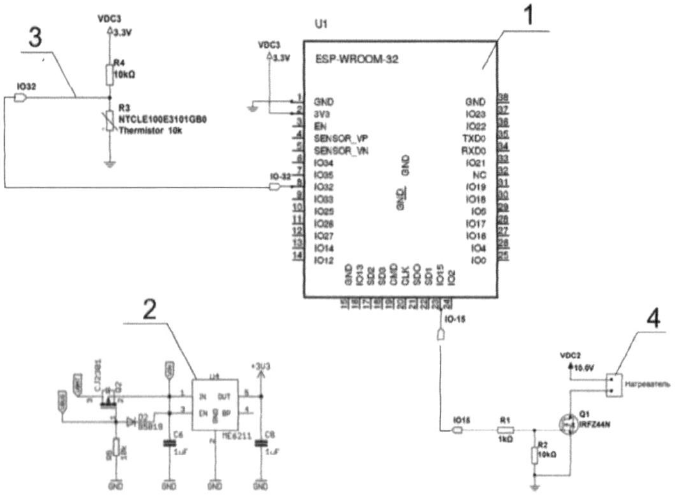

Figure 2. Diagram of the thermal comfort control system.

To control and monitor the desired temperature inside of the product, smart phone software has been developed

The algorithm of TCCS operation is recorded on the control microcontroller. Using Bluetooth on the smart phone, the user connects to the heat-protective product, and sets the desired temperature inside of the product by means of the application. Figure 3 shows the interface of a prototype application for a smart phone with the working title «THERMOBAG»

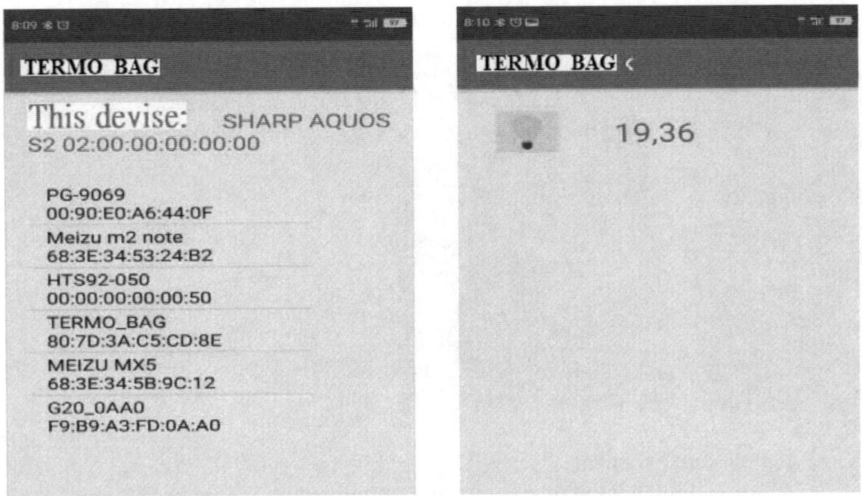

Figure 3. The interface of the «THERMOBAG» application.

To the right on figure 3, the temperature values are shown separated by commas: the first value is the current temperature inside of the product; the second value is the preset heat temperature.

The above pointed facts let us conclude that the proposed thermal comfort management system allows not only to create and maintain the thermal comfort of the TCCS in conditions of negative temperatures, but also meets the safety requirements during operation.

3. Approbation of a heat-protective product for PLMC with a thermal comfort control system

The approbation of a prototype of a heat-protective product with TCCS was carried out according to a method similar to that given in [15]. At first, a heat-protective product was selected, consisting of the main material – membrane fabric, a gasket insulation material-Tinsulate® and a lining material-fleece Mesh neoprene was used to make pockets for the local system and TCCS elements. The elements of the additional insulation layer are made of foiled material penofol

The tests were carried out in laboratory conditions with artificially set parameters, in the initial stage of the test the subjects were sitting (when simulating the air temperature of minus 15°C as the extreme temperature of the declared range), relative air humidity (60-65%) and air mobility no more than 0, 2-0.4 m/s), the participants were six men aged 45-60 who can move by means of wheelchairs, being the second group of limitation of motor activities according to the PLMC classification (with spinal lesions brain at the level of the lumbar spine) (with a confidence level of 0.95 and an acceptable error of 5%). The total disability experience of the subjects was more than 10 years. All the subjects were dressed in household clothing: underwear, thermal underwear, an over shirt, warm trousers, a sweater, warm socks, an insulated jacket, a hat and gloves. All studies were carried out under the supervision and direct participation of workers of medical institutions, as evidenced by the act of conducting joint research on the basis of state budgetary institution of social services of the population of Rostov region Shakhty boarding house for the elderly and disabled (Shakhty, Rostov region). Figure 4 shows photographs before the experiment (10 minutes of rest before the experiment).

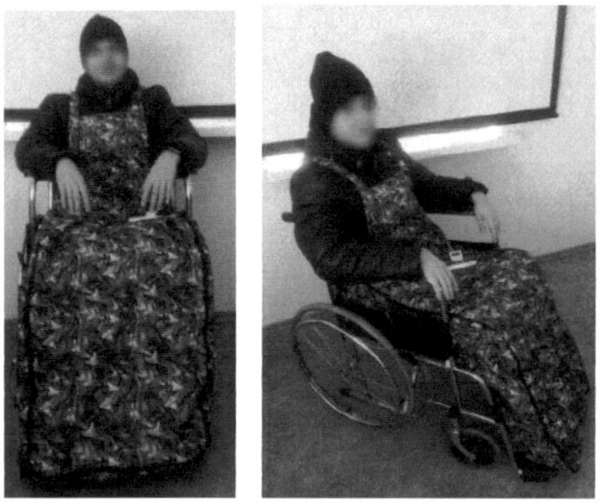

Figure 4. Carrying out experimental studies of a prototype of a heat-protective product with TCCS.

The results of experimental wearing for five days (taking into account the duration of the walk no more than 1 hour) showed a high degree of statodynamic conformity of the product and thermal comfort. Additionally, an improvement in the emotional state of the subjects during and after the walk was recorded. At the same time, the subjects themselves noted that the complex heat-protective product is convenient for self-putting on and taking off. Increasing the degree of self-service, comfort and safety of the product for all subjects became fundamental in determining the «feet bag» as a priority and desirable for everyday wear in winter.

4. Conclusion

The use of modern technologies and various technical devices leads to an increase in the quality of life of people with disabilities. Thus, in the course of the research work, the following results were obtained:

- for the first time, the use of the thermal comfort management system of the hypo dynamic system of the lower extremities in the operation of heat-protective products for PLMC in conditions of negative ambient temperatures is proposed, which allows creating and controlling a safe and comfortable temperature of the under-clothing space;
- the software for the smart phone is developed to control and monitor temperature of the under-clothing space;
- the functional and constructive device of the thermal protection product for PLMC has been improved, taking into account the use of a local heat system and STCS. A method for designing the feet bag in the conditions of functioning of computer-aided design system (CADS) «Grace» is proposed.

As part of the further development of research on the problem considered in this article, it is planned to improve the methodological apparatus for analyzing thermo physical processes and the choice of structural elements of heat-protective clothing.

References

[1] Babenko L G, Molev M D, Mikhailov A B and Savelyeva N Yu 2019 *IOP Conf. Ser.: Mater. Sci. Eng.* **680** 12-31
[2] Babenko L G, Molev M D, Savelyeva N Yu, Mikhailov K A and Mikhailova I D 2020 *E3S Web of Conf.: Inter. Sci.and Pract. Conf.on St. and Pr. for the Develop. of Agr.* **175**
[3] Department of rehabilitation treatment "Phoenix" *"Bryansky" branch of FSUE «Moskovskoe PrOP» of the Ministry of Labor of Russia)* Electronic resource: Access mode http://bpop32.ru/pages/podrazdeleniya-predpriyatiya/proizvodstvo
[4] *LLC "TsPOSN" "ORTHOMODA"* Electronic resource: Access mode: http://www.orthomoda.ru/
[5] *Online shop of adaptive clothing "Cryptomed"* Electronic resource: Access mode: http://

http://www.kriptomed.com//
[6] *St. Petersburg Scientific and Practical Center for Medical and Social Expertise, Prosthetics and Rehabilitation of Disabled Persons named after G.A. Albrecht of the Ministry of Labor and Social Protection of the Russian Federation (St. Petersburg)* Electronic resource: Access mode: http://www.center-albreht.ru/
[7] *Online shop of adaptive clothes "Buck & Buck"* Electronic resource: Access mode: http://www.buckandbuck.com/
[8] *Online shop of adaptive clothing "ABL Denim"* Electronic resource: Access mode: http://abldenim.com/
[9] *Online shop of adaptive clothing "Adaptations by Adrian"* Electronic resource: Access mode: http:// www.adaptationsbyadrian.com//
[10] *Online shop of adaptive clothing "Easy Access Clothing"* Electronic resource: Access mode: http://www.easyaccessclothing.com/
[11] *Online shop of adaptive clothing "Izadaptive"* Electronic resource: Access mode: http://izcollection.com
[12] *Online store of adaptive clothing Silvert's* Electronic resource: Access mode: http://www.silverts.com/
[13] *Online store of adaptive clothing "Able2wear"* Electronic resource: Access mode: http:// www.able2wear.co.uk//
[14] Babenko L G, Molev M D and Savelyeva N Yu 2020 *IOP Conf. Ser.: Mater. Sci. Eng.* **1079** 072-35
[15] Babenko L G, Savelieva N Yu and Volkova G Yu 2018 *Design and Technology* **67** 44-8

IOP Conf. Series: Materials Science and Engineering 1155 (2021) 012092 doi:10.1088/1757-899X/1155/1/012092

Model predictive control for path planning of UAV group

D L Vinokursky[1], O S Mezentseva[1], Ph V Samoylov[1] K Y Ganshin[1] and O A Baklanova[2]

[1] Institute of Information Technologies and Telecommunications, North-Caucasus Federal University, Stavropol, Russia
[2] Department of Philosophy, North-Caucasus Federal University, Stavropol, Russia

E-mail: fsamoilov@ncfu.ru

Abstract. This paper addresses a predictive control strategy for Unmanned Aerial Vehicles. The goal is to guarantee tracking capabilities with respect to a reference trajectory which is predefined using Gradient descent method. The proposed method provides effective performance validated through flight experiment using micro-UAVs.

1. Introduction

Model Predictive Control (MPC) is an optimization-based control technique, where at every sampling instant a finite-horizon optimal control problem (OCP) is solved. Historically, MPC has its roots in the process industry, where it has been applied successfully for a long time and is considered a mature technology [11]. The method has been developed under many different names throughout the years, and is also known under other names such as Receding Horizon Control (RHC), Dynamic Matrix Control (DMC) [12], or Generalized Predictive Control (GPC) [13]. What unifies these methods is their use of models of the system dynamics to obtain input signals by minimizing an objective function, and that the optimization is done according to the receding horizon principle, as is illustrated in figure 1. Though initially applied primarily to relatively slow processes, recent algorithmic advances and modern hardware have enabled successful implementations of MPC in fast systems that have sampling times of fractions of seconds, for example drones [14] and trucks [15]. Currently, there is a growing interest in group control of multi-agent systems. It is necessary to have a control system that can support the formation of groups, prevent collisions of agents and avoid obstacles by the whole group without losing the integrity of the entire system.

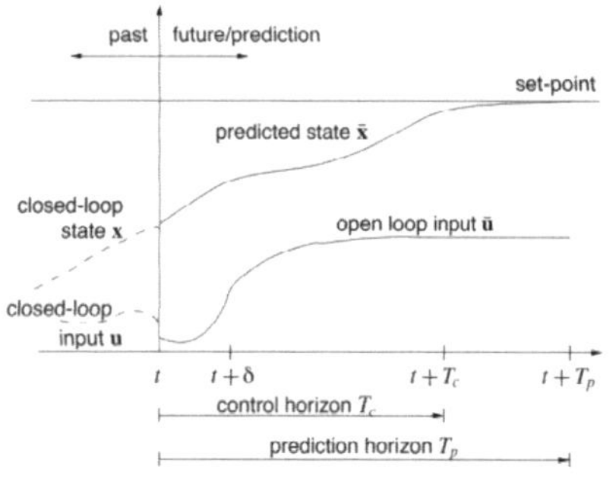

Figure 1. MPC structure.

2. Methods and materials

The use of a single MPC controller to control the entire group is difficult from the view of the fact that the used mathematical model must describe the dynamics of the entire group at once, which also leads to a manifold increase in the number of manipulating, manipulated and output signals. Considering the fact that this controller makes a prediction of system states, then this process can be computationally complex, and the optimizer will not be able to provide the required output states in real time. It should also be considered that in the case of swarm control, each individual agent must independently perform control operations on its own platform moving in space according to general swarm strategy [1].

To solve the problem of group and swarm control, a decentralized MPC controller (DMPC) is most often used [2 - 4]. In work [2], the authors describe a decentralized control system for multiple aircraft based on the Delaunay triangulation algorithm, due to which the agents line up into a regular network-like structure that can be rebuilt in the process of avoiding obstacles, and then returning to their original form. The method of virtual potentials is used to determine the required azimuth and maintain the required speed to keep the grid structure of the group. In this article, the MPC is used for trajectory tracking and obstacle avoidance maneuvers.

Figure 2 describes the architecture of the trajectory control and collision avoidance system. This system works on each UAV and consists of four blocks:

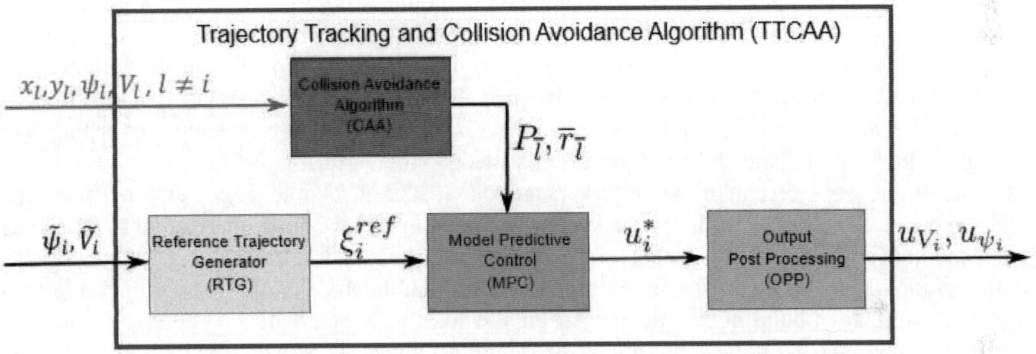

Figure 2. Architecture of the trajectory control and collision avoidance subsystem. RTG - reference trajectory generator; CAA - a collision avoidance algorithm that limits the MPC when an obstacle is detected; MPC - a predictive controller that calculates the acceleration vector required to maintain a reference path and avoid obstacles; OPP - is a post-processing unit that converts the acceleration vector into a control signal for the aircraft control system, consisting of acceleration and turn rate.

As can be seen from the presented block diagram of the algorithm, at each time moment k, the optimal control action $u_i^*(k|k)$ is determined, obtained by solving the optimal control problem on the finite horizon of predictions $[k, k + n_p]$. To generate a reference trajectory, states about the azimuth and velocity of each individual agent in the group are sent to a specific MPC. To obtain the optimal control action, it is necessary to find the optimal sequence $U_i^*(k)$. The numerical solution of this problem is carried out using quadratic programming techniques. In order to prevent collisions, only the closest obstacle is taken into account, due to which a finite set of restrictions is formed.

The performance system checks [2] were carried out under the conditions of different number of aircrafts in the simulation environment. To conclude of the simulation, it was noted that UAVs are formed into a network, and when flying around large obstacles, they are divided into several small groups supporting a mesh structure of triangles, which subsequently tend to merge into a common formation, if other obstacles are not avoided at this time. The most important indicator of simulations is the computational execution time: in the case of the technical equipment of the authors, a maximum of 100 ms is required to perform computations in all MPC simulation blocks, as shown in figure 2.

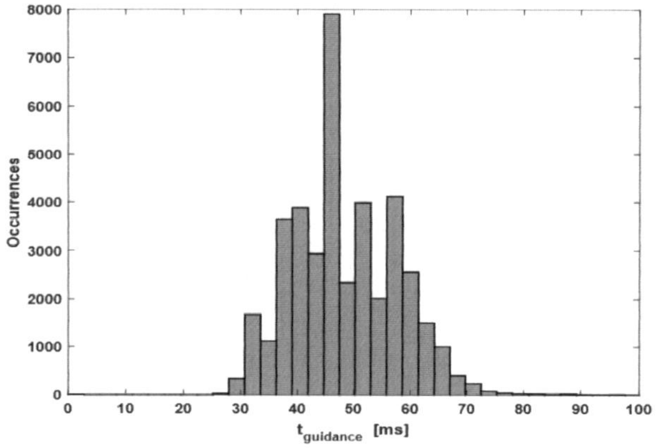

Figure 3. The number of MPC iterations for different periods of time.

Thus, the presented approach based on decentralized MPCs can be used on modern low-power embedded computers to implement a real swarm UAVs. The work [3] proposes the use of DMPC for the implementation of group control and obstacle avoidance of many multi-rotor UAVs. This study indicates that each individual agent performs a separate subtask which was obtained by the swarm, and at the same time must implement collision avoidance without disrupting the execution of both its own and others' subtasks. To implement the control system, a simplified mathematical model of a quadcopter is set, according to which each UAV flies slowly enough not to take into account the influence of the environment and changes in air flows from the propellers, the engines respond instantly to control signals, and the yaw angle is not taken into account. Thus, the model is represented by a fourth-order system in the horizontal plane and a second-order system in the vertical axis.

For the simulation, a system of two followers and one leader quadcopter is presented. Three types of simulation are used: using algorithms for specifying the formation based on consensus; determining whether the multi-agent system is capable of moving around obstacles; calculating the execution time of optimizations. As a result of the first simulation, the maintenance of the formation was revealed, however, at the stage of building into the formation, the possibility of a collision is noted due to the non-compliance of the UAV with the condition of the minimum distances between each other. In the second case, a formation with successful collision avoidance is noted, while the third simulation shows results similar to the results of the second, but with smoother and more optimal trajectories. At the same time, the execution time of the optimization stage for the first and second types of simulations turns out to be approximately equal (0.028 and 0.029 seconds, respectively) and differs from the execution time of the third type by 2.3 times. The authors conclude that the resulting architecture can be extended by adding constraints on the vectors of velocities and Euler angles. In [4], the authors propose to combine MPC with feedback linearization. In this work, the emphasis is placed on the formation of a group of quadcopters of the "leader - follower" type. It is noted that in many cases the control of groups is carried out on the basis of linear controllers, in particular, the maintenance of the group of the UAV formation in work [5] is carried out using a set of PI-controllers. Linear controllers turn out to be easy to understand and easy to synthesize; however, they cannot be used directly to work with constraints, and their performance is not guaranteed in a wide range of values, because they are configured to operate within narrow boundaries around the target point. The use of MPC in this work is due to the following advantages of controlling a group of UAVs:

- Limitations of a dynamic system with multiple signal inputs and multiple outputs of the quadcopter are introduced directly into the mathematical model in the controller cycle;
- Kinematics and dynamics of a quadcopter are an integral system, which makes it possible to represent their group as an integrated flight structure;
- Functions for constructing local trajectories are synthesized by combining the prediction of future states with information about the environment in which obstacles may be located.

Obstacle avoidance is based on mixed integer linear programming and combined with MPC [6]. It is noted that collision avoidance restrictions can lead to a collision due to the fact that the width of the obstacle is less than the predicted distance which can be traveled by the agent at the next iteration of the algorithm. MPC introduces the following limitation to prevent collisions of this kind:

$$norm(\hat{x}_{k+1} - x_k) \leq H_i^* \tag{1}$$

where $\hat{x}_{k+1}$and x_k - next predicted and current positions, respectively.

The operation of a controller with a prediction horizon of 10 steps takes on average 0.14 seconds when using a personal computer based on Intel Core Duo. When a single MPC controller is used for trajectory control, the computation time increased to 1.75 seconds.

MPC control can be used as a local path planner [7]. In this case, the MPC control will generate position and speed commands so that the UAV group follows the general desired trajectory that meets these restrictions [8]. The advantage of this approach is the calculation of coordinates and velocities at each time step, depending on the environment and the drone.

Consider the simplest linear model for determining coordinates and velocities:

$$\vec{x} = [x, y, z, \dot{x}, \dot{y}, \dot{z}]^T \tag{2}$$

Control vector:

$$\vec{u} = \left[f_x, f_y, f_z, \dot{\psi}_\tau\right]^T \tag{3}$$

Obstacle flight condition:

$$\left|x_j - x_k\right|^2 \geq R^2 \qquad j, k = 1,2,3 \tag{4}$$

As a result, we get the following formulation of the MPC control problem:

$$\begin{gathered} \min_{\Delta\vec{u}} x_k(k+1|k) = AX(k-1|k) + B\Delta\vec{u}(k-1|k) \\ \Delta u_{min} \leq \Delta u(k+1|k) \leq \Delta u_{max} \\ V_{min} \leq v_{min}(k+i|k) \leq V_{max} \end{gathered} \tag{5}$$

Solving this problem using the gradient descent method, we find new coordinates and speeds of the UAV group.

3. Experiments

For the experiment, the UAV Crazyflie 2.1 was used [9]. Communication between the ground station and the UAV was carried out using the Crazyradio PA module, operating in the 2.4 GHz ISM band with a 2 MHz channel width. The UAVs are equipped with an inertial measuring unit, consisting of a three-axis gyroscope, an accelerometer, and a barometer. The Loco Positioning System (LPS) is used which allows to define position of UAV in the space. LPS consists of 8 radio beacons and receiver modules installed on each UAV. The initial positions of the three UAVs are presented in figure 4.

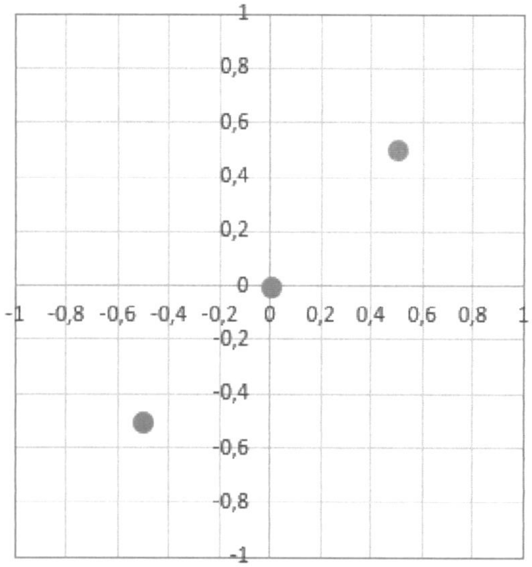

Figure 4. Initial coordinates of three UAVs.

The following graphs describe current and desired trajectory of the leading UAV:

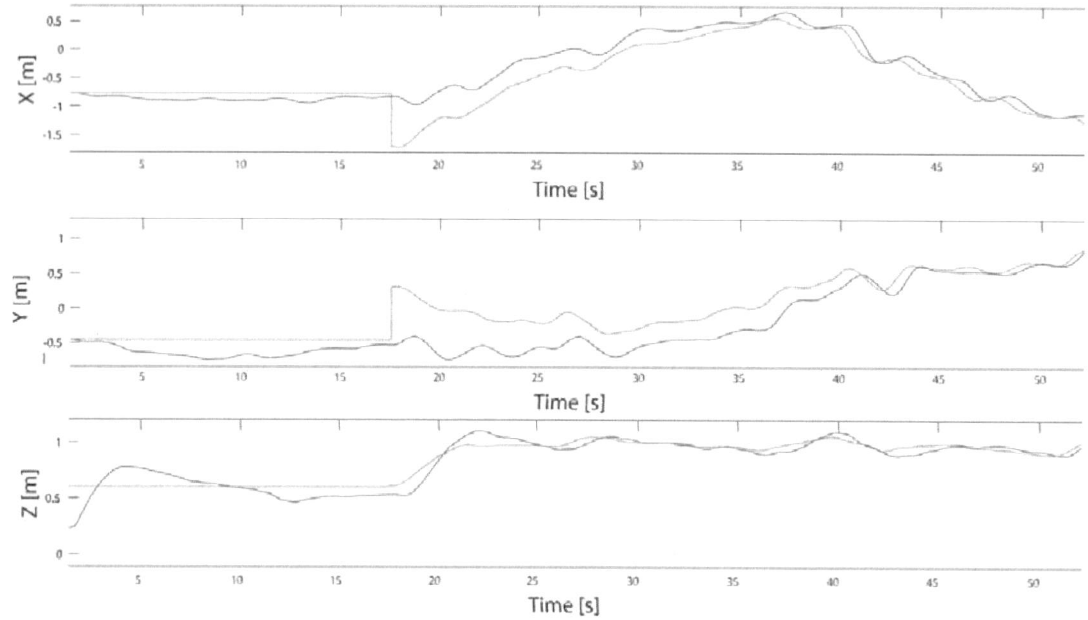

Figure 5. Graphs of changes in the leader's position over time. Orange lines – desired coordinate, blue – current coordinate.

4. Conclusion

In the future, for minimization, the variable gradient method will be used, as specially developed for solving nonlinear optimization problems.

Reference

[1] Beni G and Wang J 1993 Swarm Intelligence in Cellular Robotic Systems *Springer* **102** 703-12
[2] Bassolillo S R 2020 Decentralized mesh-based model predictive control for swarms of UAVs *Sensors* **20** 4324

[3] Kuriki Y and Namerikawa T 2015 Formation control with collision avoidance for a multi-UAV system using decentralized MPC and consensus-based control *SICE Journal of Control Measurement and System Integration* **8** 285-94
[4] Zhao W and Go T H 2014 Quadcopter formation flight control combining MPC and robust feedback linearization *Journal of the Franklin Institute* **351** 1335-55
[5] Proud A, Pachter M and D'Azzo J 2001 Tight formation flight control *Journal of guidance, control and dynamics* **24** 246-52
[6] Eren U 2017 Model predictive control in aerospace systems *Journal of Guidance, Control and Dynamics* **40** 1541-66
[7] D'Amato E, Mattei M and Notaro I Bi-level flight path planning of UAV formations with collision avoidance *Journal of Intelligent & Robotic Systems* **93** 193-211
[8] Viana I B, dos Santos D A and Góes L C S 2018 Formation control of multirotor aerial vehicles using decentralized MPC *Journal of the Brazilian Society of Mechanical Sciences and Engineering* **40** 1-12
[9] Vieira Lima G, Monteiro Jorge Alves de Souza R, Silva de Morais A, Oliveira-Lopes L C and Mara Vieira Ladeira G 2019 Stabilization and Path Tracking of a Mini Quadrotor Helicopter *IEEE Latin America Transactions* **17** 485-92
[10] Mansouri S S Remaining useful battery life prediction for UAVs based on machine learning *IFAC-PapersOnLine* **50** 4727-32
[11] Qin S J and Badgwell T A 2003 A survey of industrial model predictive control technology *Control engineering practice* **11** 733-64
[12] Cutler C R and Ramaker B L 1980 Dynamic matrix control a computer control algorithm *Joint automatic control conference 1* (San Francisco, California), 72
[13] Clarke D W, Mohtadi C and Tuffs P S 1987 Generalized predictive control - **I**. The basic algorithm *Automatica* **23** 137-48
[14] Neunert M 2016 Fast nonlinear model predictive control for unified trajectory optimization and tracking *IEEE international conference on robotics and automation (ICRA)* (Stockholm, Sweden), 1398-404
[15] Lima P F 2017 Spatial model predictive control for smooth and accurate steering of an autonomous truck *EEE Transactions on Intelligent Vehicles* **2** 238-50

IOP Conf. Series: Materials Science and Engineering 1155 (2021) 012093 doi:10.1088/1757-899X/1155/1/012093

Influence of plasma power on the size distribution of deposited zinc oxide nanorods

T S Sazanova[1], L A Mochalov[1], A A Logunov[1], D G Fukina[2] and I V Vorotyntsev[1,3]

[1]Nizhny Novgorod State Technical University n.a. R.E. Alekseev, Nizhny Novgorod, Russia
[2]Lobachevsky State University of Nizhny Novgorod, Nizhny Novgorod, Russia
[3]Mendeleev University of Chemical Technology of Russia, Moscow, Russia

E-mail: yarymova.tatyana@yandex.ru

Abstract. Zinc oxide nanopowders were obtained by plasma-enhanced chemical vapor deposition (PECVD) with optical emission spectrometry control at different plasma discharge powers (30, 50, and 70 W). Dimensional and morphological characteristics of the zinc oxide nanoparticles were studied by scanning electron microscopy (SEM). The received SEM data were statistically processed. The zinc nanoparticles had a rod-like shape with different transverse diameters depending on the plasma power. With an increase in the plasma power, the diameter of the zinc oxide rods decreased with a simultaneous narrowing of the variability for their size range. The transverse size of the zinc oxide rods decreased by 2 and more than 15 times with an increase in the plasma discharge power to 50 and 70 W, respectively.

1. Introduction

Zinc oxide is one of the most important materials of our time. Zinc oxide is widely used in the field of electronics as a material for solar cells [1–3], fluorescent devices [4], backup power supplies [5], fuel cells [6,7], components of artificial intelligence systems [8], as well as various types of sensors [9–15]. Also, zinc oxide is a promising material for use in membrane gas separation technologies as filler for polymer membranes with mixed matrix to remove carbon dioxide from natural and furnace gases [16].

Currently, nanosized zinc oxide is produced in the form of rods (threads), ridges, honeycombs, rings, ribbons, springs (spirals), cells, tetrapods, as well as thin films and coatings. Such a variety of zinc oxide nanostructures, both in morphology and in properties, distinguishes it from other materials; it should be noted that the initial reagents are very cheap and the synthesis of such structures is relatively simple. For the synthesis of all these numerous modifications, a large number of preparation methods have been proposed, for example, chemical vapor deposition (CVD) [17] and deposition from solutions [18] in their various modifications. However, the existing methods for the synthesis of nanomaterials based on zinc oxide, as a rule, are complex and multi-stage, require additional solvents or organometallic precursors, do not provide the chemical purity of the final material, are not technologically advanced due to the complexity of the hardware design and are not scalable, and do not allow one to obtain all types of nanostructures by one method. The method of plasma-enhanced chemical vapor deposition (PECVD) can help to avoid these limitations, with the help of which it is possible to achieve versatility, controllability, one-stage, high purity of the final material, cost efficiency, and scalability [19-21].

Nevertheless, for the widespread use of the PECVD method to obtain nanosized zinc oxide with variable morphology, it is necessary to carry out fundamental research on the effect of process technological parameters on the final product. Given this, the current work is devoted to studying the effect of plasma power on the size and morphological characteristics of the resulting zinc oxide nanoparticles.

2. Experimental section

2.1. Materials

As components of the plasma-forming mixture, zinc of 5N purity (Changsha Rich Nonferrous Metals Co Ltd, China), high purity hydrogen (99.9999%), and high purity oxygen 6.0 (99.9999%) (Horst Technologies Ltd, Dzerzhinsk, Russia) were used. Hydrogen also were used as a carrier gas.

2.2. Plasma-chemical synthesis

A scheme of the plasma-chemical installation is shown in figure 1. The installation consists of a gas supply system, a pumping system, a tank with initial zinc, and a pear-shaped plasma-chemical reactor made of high-purity silica glass with a total volume of about 1200 cm^3, equipped with an external heater. An inductor located on the outer reactor surface is used to excite an inductively coupled or mixed nonequilibrium plasma discharge. To organize a capacitive discharge, external ring electrodes are used, as well as an RF generator (with an operating frequency of 40.68 MHz and a maximum power of 500 W), and a universal matching device. Also, the facility is equipped with an HR4000CJ-UV-NIR high-resolution optical emission spectrometer (Avantes, Holland) operating in the range of 180 ÷ 1100 nm to control an excited intermediate gas phase during the plasma-chemical process.

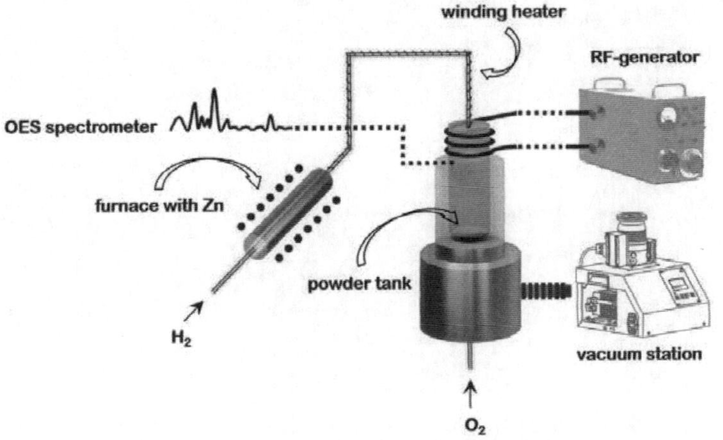

Figure 1. The scheme of a plasma-chemical installation with a pear-shaped reactor.

The loading tank with metallic zinc, made in the form of a boat, was placed directly into the reactor, in that part of it, which is equipped with the external resistive heating element. The collecting powder tank was placed in the reactor through a stainless steel vacuum loading flange and was installed perpendicular to the carrier gas flow on a special movable substrate holder. The loading tank, the collecting powder tank, and the movable substrate holder are made of high-purity quartz glass.

Before the start of the experiment, the installation was evacuated to a pressure of 1×10^{-3} Pa for several hours to remove traces of nitrogen and water from the walls of the reactor. After that, the source with zinc was heated to 470 °C. The temperature of the collecting powder tank was maintained at a level of 250 °C. Until the system entered the operating mode, the powder tank was closed with a magnetic diaphragm of a special design. The total gas flow through the plasma-chemical reactor was set equal to

30 ml min^{-1} at a total pressure in the system of 0.1 Pa. The power of the plasma discharge varied from experiment to experiment and was maintained at levels of 30, 50, and 70 W.

During the synthesis, zinc vapors were entrained by the carrier gas flow and delivered to the RF discharge zone. Oxygen was supplied from the bottom towards the main one of the carrier gas with zinc vapor directly in the plasma zone. In fact, the formation of zinc oxide particles occurs in the gas phase in front of the collecting powder tank with their subsequent deposition. Hydrogen was applied as a terminator and growth regulator. The duration of each of the three experiments (with different plasma discharge power) was one hour.

2.3. Scanning electron microscopy

Dimensional and morphological characteristics of zinc oxide nanoparticles were studied by scanning electron microscopy (SEM) using an electron microscope JSM-IT300LV (JEOL, USA) with the electron probe diameter of about 5 nm and probe current below 0.5 nA (operating voltage 20 kV). SEM scanning was performed using low-energy secondary electrons and backscattered electrons under a low vacuum to eliminate the charge.

3. Results and discussions

Figure 2 shows the optical emission spectra of zinc-hydrogen-oxygen plasma for different plasma discharge powers (30, 50, and 70 W). An increase in the plasma discharge power caused a significant increase in the intensity of the O (I) lines at 777.24 and 844.69 nm with a simultaneous moderate decrease in the intensity of the Zn (I), H (I), and O (II) lines at 307.59, 486.16, 546.10 and 656.27 nm. The emission Zn (II) lines were not observed.

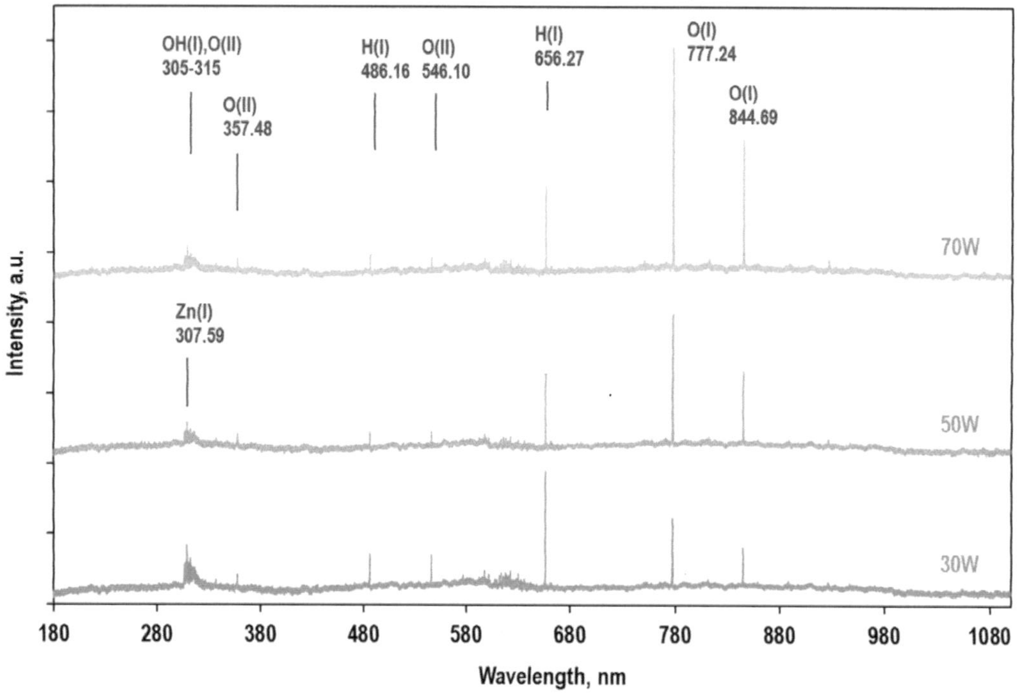

Figure 2. The optical emission spectra of zinc-hydrogen-oxygen plasma (Zn: H_2: O_2 = 2: 1: 1) at various plasma discharge power.

As a result of the plasma-chemical synthesis, rod-shaped zinc oxide nanoparticles with different transverse diameters were obtained. According to the SEM images of the deposited powders (figure 3), the diameter of these rods depends on the plasma discharge power, which was 30, 50, and 70 W. According to statistical processing (the sample size was 50) of the obtained SEM data, it was found that the transverse diameter of the zinc oxide rods decreases by 2 times with an increase in the plasma discharge power from 30 to 50 W, and more than 15 times when going from 50 to 70 W (figure 4).

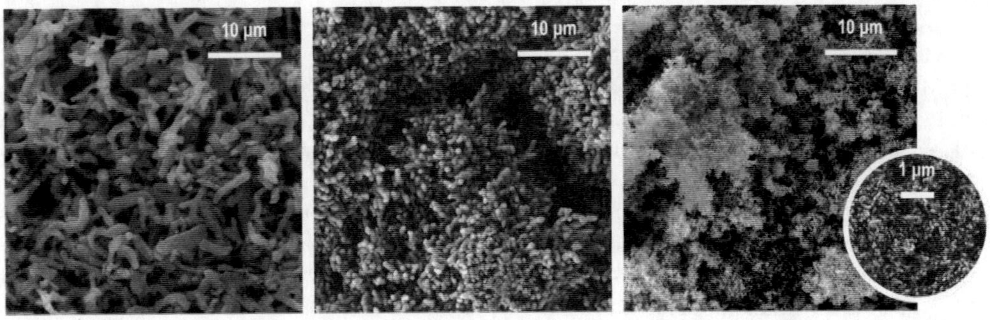

(a) (b) (c)

Figure 3. SEM images of zinc oxide nanopowder obtained by the PECVD method at various plasma discharge power: (a) 30 W, (b) 50 W, (c) 70 W.

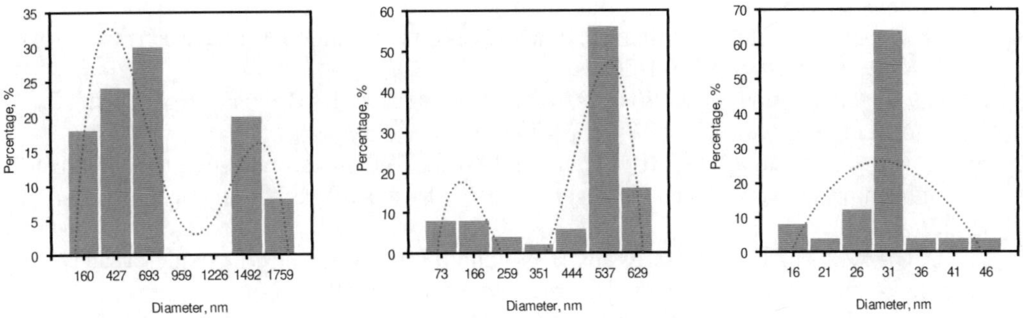

(a) (b) (c)

Figure 4. Distribution of ZnO rod-shaped structures by the size of their transverse diameter.

In addition to a noticeable decrease in the size of the zinc oxide rods, the behavior of their statistical distribution also changes. As the discharge power increases, the size distribution series narrows, i.e. reducing its variability. At 30 W of plasma power, the distribution is so wide that the coefficient of variation is close to 100 %, which in fact emphasizes the random behavior of zinc oxide rods' formation during deposition. However, an increase in the plasma power to 50 and 70 W leads to a unification of the size range for the zinc oxide nanorods and a decrease in the coefficient of variation to 32 and 23 %, respectively. Also, there is also a transition from a bimodal ZnO rods size distribution to a unimodal one with an increase in the plasma power, which may indicate an increase in the controllability of the ongoing process.

4. Conclusions

Zinc oxide nanopowders were obtained by plasma-enhanced chemical vapor deposition with optical emission spectrometry control at different plasma discharge powers (30, 50, and 70 W). According to the data of scanning electron microscopy, the zinc oxide nanoparticles have a rod-like shape with different transverse diameters depending on plasma discharge power. Using a statistical analysis of the results of measuring the diameters for the zinc oxide rods, it was found that the formation of the powder particles is random at a low plasma power (20 W), which was confirmed by a very wide particle size distribution. With an increase in the plasma discharge power, the diameter of the zinc oxide rods decreased with a simultaneous narrowing of the variability for their size range. The transverse diameter of the zinc oxide rods decreased by 2 times with an increase in the plasma discharge power from 30 to 50 W, and more than 15 times when going from 50 to 70 W.

Acknowledgements
The SEM study was performed by the Collective Usage Center "New Materials and Resource-saving Technologies" (Lobachevsky State University of Nizhny Novgorod).
This work was supported by the Russian Science Foundation (grant no. 20-79-00138).

References
[1] Soultati A, Verykios A, Panagiotakis S, Armadorou K-K, Haider M I, Kaltzoglou A, Drivas C, Fakharuddin A, Bao X, Yang C, Yusoff A R bin M, Evangelou E K, Petsalakis I, Kennou S, Falaras P, Yannakopoulou K, Pistolis G, Argitis P and Vasilopoulou M 2020 Suppressing the photocatalytic activity of zinc oxide electron-transport layer in nonfullerene organic solar cells with a pyrene-bodipy interlayer *ACS Appl. Mater. Interfaces* **12** 21961-73
[2] Mohammed M K A, Dehghanipour M, Younis U, Shalan A E, Sakthivel P, Ravi G, Bhoite P H and Pospisil J 2020 Improvement of the interfacial contact between zinc oxide and a mixed cation perovskite using carbon nanotubes for ambient-air-processed perovskite solar cells *New J. Chem.* **44** 19802-11
[3] Geleta T A and Imae T 2020 Influence of additives on zinc oxide-based dye sensitized solar cells *Bull. Chem. Soc. Jpn.* **93** 611-20
[4] Shukla S and Sharma D K 2021 A review on rare earth (Ce and Er)-doped zinc oxide nanostructures *Mater. Today Proc.* **34** 793-801
[5] Subramani K and Sathish M 2019 Facile synthesis of ZnO nanoflowers/reduced graphene oxide nanocomposite using zinc hexacyanoferrate for supercapacitor applications *Mater. Lett.* **236** 424-7
[6] Xia C, Qiao Z, Feng C, Kim J-S, Wang B and Zhu B 2017 Study on zinc oxide-based electrolytes in low-temperature solid oxide fuel cells *Materials (Basel).* **11** 40
[7] Xing Y, Hu E, Wang F, Muhammad N, Wang B, Wang J, Maryam A, Rasheed M N, Asghar M, Xia C, Yun S and Zhu B 2020 Cubic silicon carbide/zinc oxide heterostructure fuel cells *Appl. Phys. Lett.* **117** 162105
[8] Effendy N, Ab Aziz S H, Mohamed Kamari H, Mohd Zaid M H and Abdul Wahab S A 2020 Ultrasonic and artificial intelligence approach: Elastic behavior on the influences of ZnO in tellurite glass systems *J. Alloys Compd.* **835** 155350
[9] Panth M, Cook B, Zhang Y, Ewing D, Tramble A, Wilson A and Wu J 2020 High-performance strain sensors based on vertically aligned piezoelectric zinc oxide nanowire array/graphene nanohybrids *ACS Appl. Nano Mater.* **3** 6711-8
[10] Mai L, Mitschker F, Bock C, Niesen A, Ciftyurek E, Rogalla D, Mickler J, Erig M, Li Z, Awakowicz P, Schierbaum K and Devi A 2020 From precursor chemistry to gas sensors: Plasma-enhanced atomic layer deposition process engineering for zinc oxide layers from a nonpyrophoric zinc precursor for gas barrier and sensor applications *Small* **16** 1907506
[11] Cherumannil Karumuthil S, Singh K, Valiyaneerilakkal U, Akhtar J and Varghese S 2020 Fabrication of poly (vinylidene fluoride-trifluoroethylene) – Zinc oxide based piezoelectric pressure sensor *Sensors Actuators A Phys.* **303** 111677
[12] Wang D, Chi M, Zhang D and Wu D 2020 Ammonia sensing properties of metal–organic frameworks-derived zinc oxide/reduced graphene oxide nanocomposite *J. Mater. Sci. Mater. Electron.* **31** 4463-72
[13] Lin C, Kao C, Lin C, Chen K and Lin Y 2020 NH3 plasma-treated magnesium doped zinc oxide in biomedical sensors with electrolyte–insulator–semiconductor (EIS) structure for urea and

glucose applications *Nanomaterials* **10** 583

[14] Ren Q, Cao Y-Q, Arulraj D, Liu C, Wu D, Li W-M and Li A-D 2020 Review — Resistive-type hydrogen sensors based on zinc oxide nanostructures *J. Electrochem. Soc.* **167** 067528

[15] Kim D W, Park K H, Lee S-H, Fàbrega C, Prades J D and Jang J-W 2021 Plasmon expedited response time and enhanced response in gold nanoparticles-decorated zinc oxide nanowire-based nitrogen dioxide gas sensor at room temperature *J. Colloid Interface Sci.* **582** 658-68

[16] Saeedi Dehaghani A H and Pirouzfar V 2017 Preparation of high-performance membranes derived from poly(4-methyl-1-pentene)/zinc oxide particles *Chem. Eng. Technol.* **40** 1693-701

[17] Manzoor Z, Saravade V, Corda A M, Ferguson I and Lu L 2020 Optical and structural properties of nickel doped zinc oxide grown by metal organic chemical vapor deposition (MOCVD) at different reaction chamber conditions *ES Mater. Manuf.* **8** 31-5

[18] Abdullin K A, Zhumagulov S K, Ismailova G A, Kalkozova Z K, Kudryashov V V. and Serikkanov A S 2020 Synthesis of heterogeneous ZnO/Co3O4 nanostructures by chemical deposition from solutions *Tech. Phys.* **65** 1139-43

[19] Mochalov L, Logunov A and Vorotyntsev V 2019 Structural and optical properties of As-Se-Te chalcogenide films prepared by plasma-enhanced chemical vapor deposition *Mater. Res. Express* **6** 056407

[20] Mochalov L, Logunov A, Markin A, Kitnis A and Vorotyntsev V 2020 Characteristics of the Te-based chalcogenide films dependently on the parameters of the PECVD process *Opt. Quantum Electron.* **52** 197

[21] Mochalov L, Logunov A, Prokhorov I, Sazanova T, Kudrin A, Yunin P, Zelentsov S, Letnianchik A, Starostin N, Boreman G and Vorotyntsev V 2021 Plasma-chemical synthesis of lead sulphide thin films for near-IR photodetectors *Plasma Chem. Plasma Process.* **41** 493-506

IOP Conf. Series: Materials Science and Engineering 1155 (2021) 012094 doi:10.1088/1757-899X/1155/1/012094

Switching strategies for a multi-level inverter

S S Moldakhmetov, V P Ivel, Y V Gerasimova, P A Petrov and N B Kaliaskarov

M. Kozybayev North-Kazakhstan University, 86 Pushkin Street, Petropavlovsk, 150000, Kazakhstan

E-mail: sayatmoldakhmetov@gmail.com

Abstract. The article examines the methods of switching the stages of a multi-level inverter in order to reduce the harmonic factor of the output voltage of the inverter. These methods use pulse-width modulation for switching, which is obtained by comparing a reference sinusoidal signal and a sequence of triangular pulses. The article describes the algorithms for obtaining the control sequence of pulses and simulates each of the strategies. The dependence of the nonlinear distortion factor on the frequency of the carrier wave for different switching strategies is obtained based on the simulation results. The influence of the LPF on the quality of the output voltage of the inverter is also investigated. In addition, an experimental study of a six-level inverter was carried out. The waveforms and spectrograms of the inverter were obtained using the Fluke 434-II electrical energy analyzer. The harmonics of the output voltage, which are distinguished by power from the entire spectrum, were identified.

1. Introduction

The topology of a multi-level inverter allows switching voltage levels. At the same time, in most cases, a particular topology differs only in the number of switching keys used, power supplies and other performance elements and does not allow obtaining a significant advantage in reducing the harmonic components. To reduce them, a certain strategy to select the switching time of each of the stages of a multi-level inverter was used. At the moment, there are various strategies used to implement a multi-level inverter [1-3]. Most of them are based on the use of pulse-width modulation [4-6].

The use of PWM is widely used to obtain a sinusoidal voltage at the output of a single-level inverter, where the pulse width of the carrier wave varies according to a sinusoidal law. But PWM can also be used for switching in a multi-level inverter [7-10] with the only difference that the carrier waves form each of the stages of the inverter separately.

It is proposed to investigate various strategies for reducing the harmonic factor basing on the topology of constructing a multi-level inverter presented in [11].

2. Switching Strategies

The PWM-based switching strategies discussed in this article are as follows:

- co-phased PWM strategy;
- antiphase PWM strategy;
- alternating antiphase PWM strategy;
- PWM strategy with superimposed carrier waves;
- PWM strategy with different frequency.

Let us consider a co-phased PWM stage switching strategy. U_sA sinusoid with an amplitude of 312 V and a frequency of 50 Hz was set as a modulating signal according to the law

$$U_s = 312 \sin 100\pi t.$$

A triangular signal of a symmetrical shape shall act as a carrier wave, i.e. the growth duration of such a signal is equal to the duration of its downfall. The frequency of the carrier triangular signal will be constant but it must be several times greater than the frequency of the modulating sinusoidal signal. The amplitude of the triangular signal is selected depending on the amplitude of the multi-level inverter stage.

Let us introduce the concept of a frequency factork_f, which is equal to the ratio of the carrier wave frequency $f_c$to the modulating signal frequencyf_s.

$$k_f = \frac{f_c}{f_s}$$

Thus, it is possible to study the influence of this factor on the nonlinear distortion factor. Similarly, let us introduce the concept of the k_A amplitude factor, which will be calculated by the following formula

$$k_A = \frac{n \cdot A_c}{f_s},$$

where n is the number of inverter stages, A_c is the amplitude of the carrier wave, and A_s is the amplitude of the modulating signal.

Let us consider strategies based on a three-level inverter. Let the level of the stages be the same and conditionally equal to 1. In this case, the amplitude of the modulating signal is 3.

Figure 1a shows a diagram of the switching time selection according to the co-phased PWM strategy. For this case, the value of the frequency factor k_f is 20, i.e. the frequency of the carrier triangular signals is 20 times higher than the frequency of the sinusoidal wave. The amplitude factor k_A equals 1.

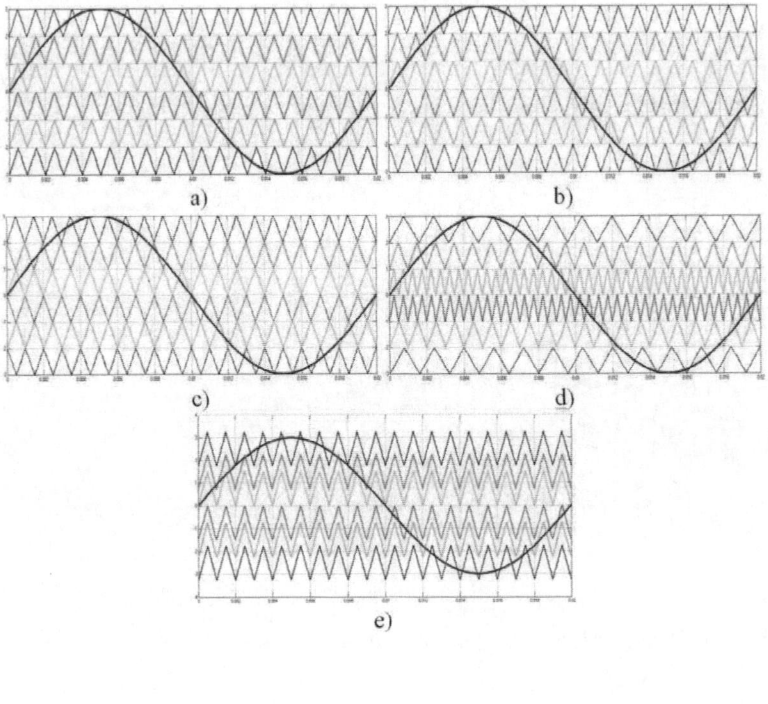

a) b) c) d) e)

Figure 1. Carrier Wave Position Diagrams for Different Strategies.

As can be seen from the figure, the carrier waves for each stage are in the same phase. The number of carriers required to obtain the switching time at each level is equal to the number of inverter stages multiplied by two. It should be noted that all carriers are of the same amplitude. According to this strategy, if the value of the carrier triangular wave for a given time is higher than the value of the sinusoidal signal, then a high-level value is formed, otherwise a low-level value is formed. Thus, the commuting pulse is generated whenever the triangular pulse of the stage is greater than the sinusoid.

According to the principle of forming control pulses according to the law of antiphase PWM strategy, all the carrier waves of the positive half-wave are in the same phase relative to each other and all the carrier waves of the negative half-wave are also in the same phase, but in the opposite phase to the carrier waves of the positive half-wave (figure 1b).

For an alternating antiphase PWM strategy, the carrier waves of the same amplitude are in antiphase relative to the neighbouring ones (figure 1c). In fact, this and the previous strategies are convenient from the point of view of application, since a phase shift of 180° shifts the carriers so that the positive and negative half-waves are in the same position relative to the carriers. Thus, the pulses are formed for the positive and negative half-waves equally, which is very convenient.

Since the sinusoid has a characteristic steepness, the number of switches for the lower levels of the inverter is less than the number of switches for the upper levels. Different frequencies for the carrier waves can be used in order to equalize the number of switches for all levels. This is implemented in the PWM strategy with frequency. The carrier position diagram for this strategy is shown in figure 1d (k_{f1} = 40, k_{f2} = 20, k_{f3} = 12.5, k_A = 1).

Using PWM with superimposed carrier waves allows increasing or decreasing the duration of switching pulses and eliminate short-term bursts. This can be achieved by increasing the amplitude of the carrier waves, as the triangular signals are cut off and become trapezoidal. The carrier wave position diagram for the PWM strategy with carrier wave superposition is shown in figure 1e (k_f = 20, k_A = 1.5).

3. Simulation

In order to study the effect of switching strategies on the harmonic factor, let us build a special model in the Simulink visual simulation environment.

Figure 2 shows a model of a 6-level inverter for the study of switching methods [12].

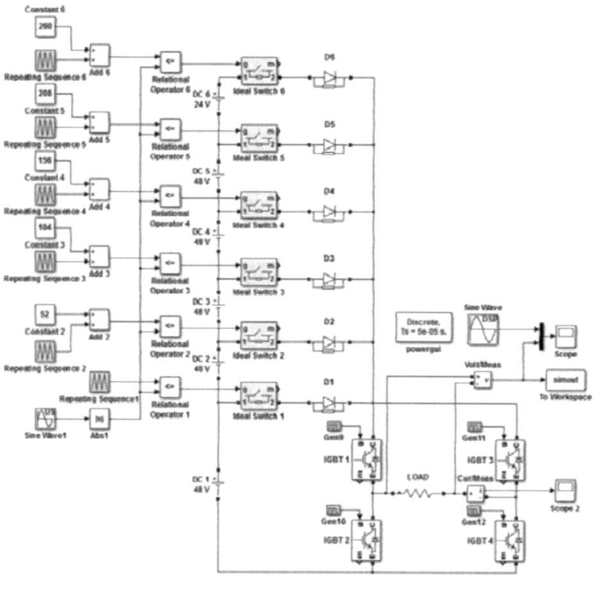

Figure 2. Model of a 6-Level Inverter for the Study of Switching Methods.

The output voltage obtained from the simulation differs from the usual step voltage by the presence of pulse-width modulation (figure 3). It should be noted that the closer the pulses are to the border of the transition to a high level, the wider the pulses and vice versa.

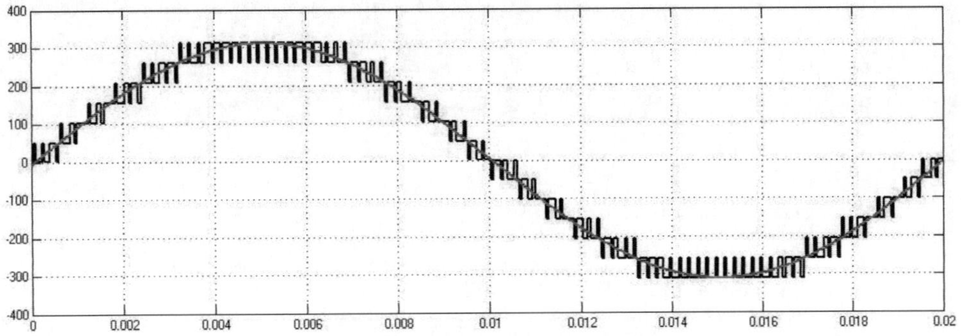

Figure 3. Output Voltage Waveform.

To compare strategies, there can be used the nonlinear distortion factor, which is automatically calculated using the powergui block by means of a fast Fourier transform. The data for the strategies are shown in table 1. Conspicuously, the nonlinear distortion factors are quite big due to high frequencies in the output voltage of the inverter. Therefore, these strategies are applied with a low-pass filter. In order not to distort the low frequencies, it is necessary to choose a filter with a smooth amplitude-frequency response at the passband frequencies [13,14]. A typical third-order Butterworth filter with a cutoff frequency of 1000 Hz was used.

Table 1. Nonlinear Distortions Factors of Various Methods.

Period, sec	Frequency, Hz	AP PWM		AAP PWM		PWM with SC		PWM with DF	
		without the LPF	after the LPF	without the LPF	after the LPF	without the LPF	after the LPF	without the LPF	after the LPF
0.0013	769.2	9.3	7.04	10.05	7.51	14.24	11.49	9.27	6.98
0.0012	833.3	9.75	6.95	9.6	6.42	15.65	11.8	9.54	6.93
0.0011	909.1	8.78	6.08	7.91	4.44	13.8	9.69	9.7	7.8
0.001	1000	8.38	5.85	10.08	7.1	15.12	9.6	9.46	7.23
0.0009	1111.1	8.99	4.88	9.15	5.17	14.73	7.98	7.86	4.44
0.0008	1250	8.98	4.33	9	5.26	14.39	6.37	8	4.38
0.0007	1428.6	8.94	3.21	9.22	4.59	14.3	4.84	8.92	3.75
0.0006	1666.7	9.14	2.63	9.08	3.79	14.4	3.19	9.12	3.05
0.0005	2000	8.27	2.28	8.98	2.31	15.05	2.11	8.19	2.54
0.0004	2500	8.71	1.65	9.01	1.43	14.29	2.35	8.65	2
0.0003	3333.3	9.01	1.4	9.17	1.67	14.8	2	8.96	1.38
0.0002	5000	9.86	1.74	9.64	1.7	14.6	2.47	9.22	1.37
0.0001	10000	13.1	4.18	13.3	4.09	21.11	3.63	11.42	1.85

A graph of the dependence of the nonlinear distortion factor on the frequency of the carrier wave is constructed based on the simulation results. In this graph, the dashed lines show the results of simulation strategies without filtering, and the solid lines show the simulation with the LPF.

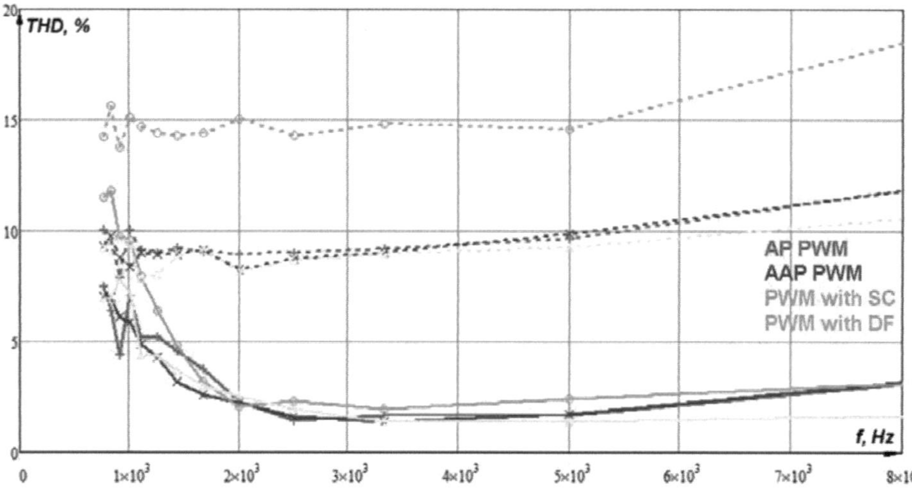

Figure 4. Graph of the Dependence of the Nonlinear Distortion Factor on the Carrier Wave Frequency.

As can be seen from figure 4, k_fa high value of the nonlinear distortion coefficient can be observed for all strategies with low values of the frequency factor, and with an increase in the carrier frequency, this factor decreases to a certain minimum and then increases slightly. At the same time, the lowest factor of nonlinear distortion is observed when using a switching strategy based on PWM with DF. When implementing a power inverter, however, a higher frequency should be chosen, since the frequency increases the energy loss for transients. The fact that switching is carried out by power switches, such as IGBT or MOSFET, transients in which at a high switching frequency can disable them [15,16], should be also taken into account.

4. Experiment

An experimental setup was assembled in the MATLAB environment after checking the correctness of the diagram. Figure 5a shows the block diagram of the inverter, and figure b shows the experimental setup. The inverter consists of a 6-channel switch, an H-bridge and a control system built on an ATmega32A microcontroller.

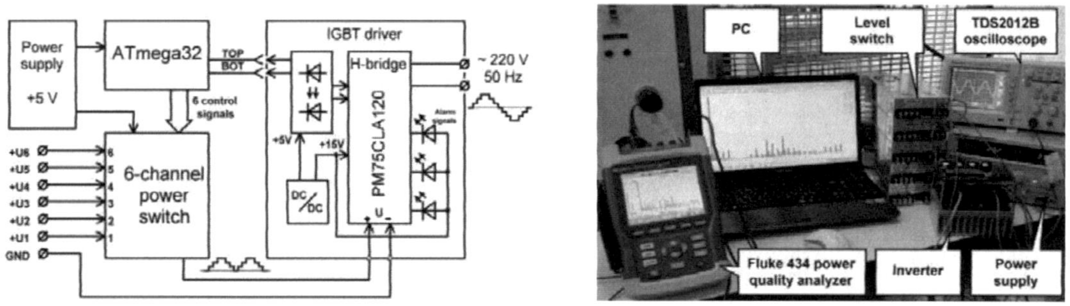

a) b)

Figure 5. 13-Level 5 kW Voltage Inverter.

Waveforms and spectrograms of the output voltage were obtained as a result of the experimental study. Figure 6a shows the waveform, and figure 6b shows the output voltage spectrogram for a six-level PWM inverter.

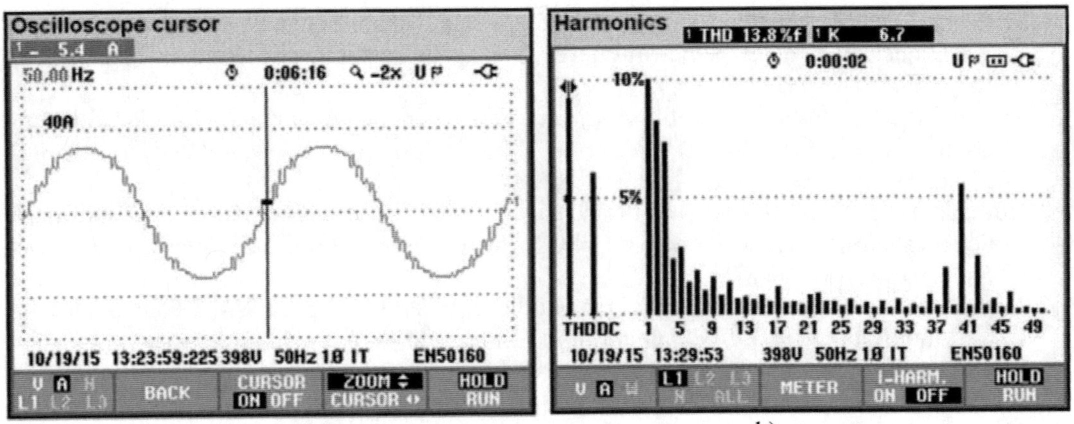

a) b)

Figure 6. Waveform and Spectrogram of the Output Voltage of a six-level PWM-Strategy Inverter.

As can be seen from figure 6a, the shape of the output voltage of the inverter has spikes at each of the stages, but it is visually close to sinusoidal. Of particular interest are the spectrograms of the output voltage of inverters. It can be noted that the 37th, 39th, 41st and 43rd harmonics, whose frequencies are 1850, 1950, 2050 and 2150 Hz, respectively, are expected to stand out in power from the entire spectrum. This is naturally related to the pulse frequency of the pulse width modulation stages. Harmonics at this frequency are very convenient to filter since the filter is set to high harmonics, i.e. with a higher cutoff frequency. Such a filter is lighter and more space-saving than a conventional LPF and is very simple in design.

5. Conclusion

The dependences of the nonlinear distortion factor on the frequency for various strategies without filtering the output voltage of the inverter, and with the low-frequency filter are obtained based on the results of the Simulink simulation. It was found that the lowest factor of nonlinear distortion can be obtained by using a switching strategy based on PWM with different frequencies for the stages. The results of the experimental study confirmed the presence of harmonics at the carrier frequency, which can be eliminated using a low-pass filter.

References

[1] Rodríguez J, Lai J S and Peng F Z 2002 Multilevel inverters: A survey of topologies, controls and applications *IEEE Trans. Ind. Electron.* **49(4)** 724-38

[2] Singh B, Mittal N, Verma K S, Singh D, Singh S P, Dixit R, Singh M and Baranwal A 2012 Multi-level inverter: a literature survey on topologies and control strategies *Int. J. of Reviews in Computing* **10** 1-16

[3] Sandhu M and Thakur T 2014 Multilevel Inverters: Literature Survey – Topologies, Control Techniques and Applications of Renewable Energy Sources - Grid Integration *J. of Engineering Research and Applications* **4(3)** 644-52

[4] Govindaraju C and Baskaran K 2009 Optimized Hybrid Phase Disposition PWM Control Method for Multilevel Inverter *Int. J. of Recent Trends in Engineering* **1(3)** 129-34

[5] Arun V, Shanthi B and Natarajan S P 2013 Performance Analysis of Multicarrier SPWM Strategies for Three Phase Z - Source Seven Level Cascaded Inverter *Int. J. of Modern Engineering Research* **3(1)** 204-11

[6] Johnson U R, Natarajan S P and Padmathilagam V 2012 A New Three Phase Seven Level Asymmetrical Inverter with Hybrid Carrier and Third Harmonic Reference *Int. J. of Modern Engineering Research* **2(4)** 1814-18

[7] Seyezhai R and Mathur B L 2010 Implementation and Control of Variable Frequency ISPWM Technique for an Asymmetric Multilevel Inverter *European J. of Scientific Research* **39(4)** 558-68
[8] Seyezhai R 2011 Investigation of Performance Parameters For Asymmetric Multilevel Inverter Using Hybrid Modulation Technique *Int. J. of Engineering Science and Technology* **3(12)** 8430-43
[9] Jeevananthan S, Nandhakumar R and Dananjayan P 2007 Inverted Sine Carrier for Fundamental Fortification in PWM Inverters and FPGA Based Implementations *Serbian J. of Electrical Engineering* **4(2)** 171-87
[10] Seyezhai R and Mathur B L 2010 Hybrid Multilevel Inverter using ISPWM Technique for Fuel Cell Applications *Int. J. of Computer Applications* **9(1)** 41-7
[11] Moldakhmetov S, Issembergenov N and Kasymov A 2015 Multilevel inverter based on level switch and H-bridge *ARPN J. of Engineering and Applied Sciences* **10(16)** 6884-7
[12] Chakravarty S 2012 Technology and Engineering Applications of Simulink *InTech* 266
[13] Gupta A and Tripathi G C 2010 Design and Analysis of Low Pass FIR & IIR Filter and Find Optimum Result Using Neural Network *Global J. of Researches in Engineering* **10(7)** 61-4
[14] Burlaka V V, Gulakov S V, Bublik S K and Dyachenko M D 2009 Shunt Active Filter With Improved Suppression Of High-Order Current Harmonics *Visnik Priazovskogo Derzhavnogo Technichnogo Univercitetu* **19** 237-41
[15] Kulkarni V V and Swami L B 2014 MOSFET Based High Frequency Inverter for Induction Heating Equipment Using MATLAB/Simulink Environment *Int. J. Of Innovations In Engineering Research And Technology* **1(1)** 1-7
[16] Rajesh J, Deepak S V, Ramjee S V , Sai kiran M, Durga Prasad T N V 2014 High-Frequency Resonant Matrix Converter using IGBT-Based Bidirectional Switches for Induction Heating *Int. J. of Engineering Research and Applications* **4(2)** 451-7

IOP Conf. Series: Materials Science and Engineering 1155 (2021) 012095 doi:10.1088/1757-899X/1155/1/012095

Digital multi-agent system for simulating natural fire fighting processes

G A Dorrer[1], S V Kobijzhakova[2,3] and S V Yarovoy[1]

[1] Reshetnev Siberian State University of Science and Technologies, Krasnoyarskiy Rabochiy street, 31, Krasnoyarsk, 660037, Russia
[2] FSBOU VO Siberian Fire and Rescue Academy of the Ministry of Emergency Situations of Russia, North Str., 1, Zheleznogorsk, 662972, Russia

[3] E- mail: svetkob84@mail.ru

Abstract. A system for modeling and supporting decision-making on the use of forces and means to eliminate wildfires, based on a digital model, implemented in the software "Taiga-Analyst", is proposed. The system works on the basis of agent-based simulation and allows you to select forces and means to fight a fire, simulate its propagation and elimination with reference to a digital terrain map using a geographic information system. The possibilities and scope of the "Taiga-Analyst" system are described. A decision-making scheme for fighting natural fires is shown, and the use of the system in the process of training fire safety specialists is described.

1. Introduction

The Fourth Industrial Revolution, which is currently taking place, and Industry 4.0, which is being created at the same time, presuppose the mass introduction of cyber-physical CFS systems into the production and service of human needs, including life, work, security and leisure. It is expected that the changes caused by this revolution will cover a variety of aspects of life: the labor market, the living environment, political systems, the technological way of life, human identity, and others. Human interaction with the environment will not be left aside, which should move to a new level, ensuring the harmonization of human needs and the preservation of the natural environment.

The cyberphysical system assumes the presence of a digital model of a physical object and the possibility of influencing it (feedback). Within the framework of the development of Industry 4.0, depending on the degree of automation of the data exchange process, the following systems are distinguished [1, 2]: digital model, digital shadow, digital double. A review of publications from 2015 to 2020 [2] shows that the most common areas of application of FSCs are the following: smart homes, smart cities, manufacturing, and healthcare.

Cyber-physical systems can play an important role in the environmental sphere. However, in the literature, this direction is not given due attention.

In this paper, we consider a digital model of the processes of spreading wildfires and fighting them, implemented in the form of a multi-agent system using a digital terrain model

2. The concept of agents and multi-agent systems

Let's look at the basic terms related to agent systems.

A digital agent is a software system that has specific properties: the ability to show their own activity, to act independently and proactively to achieve the set goal, to interact with other agents, as well as with the external digital environment, to react and influence its state. Dynamic agents have the property of simulating movement in the external environment. Many interacting agents form a multi-agent system. The property of such systems is considered in [3 – 6].

The multi-agent system "Taiga-analyst" developed by the authors is designed for digital modeling of the processes of propagation and localization of wildfires [7, 8].

Let's list the main functions of this system.

- The system contains a digital model of the environment in which a natural fire occurs. The environment model contains all the necessary data for calculating the processes of fire propagation and elimination, and also displays the dynamics of these processes on the map of the real area of the required scale.
- The system simulates the processes of gorenje and the spread of natural fire in a given environment, and this process is displayed on the map of the area.
- The system simulates the process of interaction of a fire with fire-fighting forces and means with various methods and tactics of fire fighting.
- The system is implemented as an independent software product "Taiga-Analyst" with the possibility of integration or joint use with both existing remote monitoring systems and e-learning systems.

Currently, the system contains the following software components: a digital model of the environment; agents that simulate the dynamics of the fire front (A-agents), agents that simulate the action of fire-fighting forces and means (B-agents), agents that affect the digital model of the environment (D agents); agents-managers that coordinate the actions of other agents (M-agents). If you need to refine or change the model, new types of agents can be introduced. The operating components of the system are described below.

The operation of the digital system is controlled by the user-operator, who determines the goals and objectives of modeling processes in the system and evaluates their results.

3. Digital environment model

A digital model of the environment that determines the conditions for the spread of fire is created using a geographic information system. The map of the area and its associated database contain information about the topography and infrastructure of the area, about the properties of plant fuels and weather conditions, as well as about the available means of fire control and their characteristics.

The processes of fire propagation and control, calculated using the above-mentioned agents at discrete time points (simulation steps), are displayed on the map.

Agents that simulate the dynamics of the fire front (A-agents). Each A-agent models the behavior of a fire edge fragment. The set of A-agents forms a digital fire contour (figure 1). The spatial coordinates of this type of agents are calculated at each step of the simulation based on the solution of the Hamilton-Jacobi equations and using the mobile grid algorithm, which are described in detail in [6-9]. In addition, in accordance with the mentioned algorithm, in the process of modeling, under certain conditions, agents can "go out" or, on the contrary, new ones can arise. The Gorenje properties of A-agents are determined by the parameters of the combustion of vegetable fuel, the speed of the fire front movement and the possible directions of its propagation.

Type A agents generate a heat flow and a smoke jet, which are calculated using a separate model. Gorenje heat flow depends on the combustion conditions and can be reduced to zero under the influence of agents of type B, while the A-agent "goes out" and no longer enters the digital fire circuit.

Agents that simulate the action of fire-fighting forces (B – agents). Type B agents act on type A agents, thereby simulating the action of fire-fighting teams. Their goal is to "extinguish" all the gorenje nodes. To do this, each B-agent moves through the digital simulation environment to the nearest A-

IOP Conf. Series: Materials Science and Engineering 1155 (2021) 012095 doi:10.1088/1757-899X/1155/1/012095

agent and, approaching it, at each step of the simulation, reduces the intensity of its heat release by a certain amount – until agent A "goes out". Agent B then moves to the nearest active agent A and starts interacting with it in a similar way.

Agent B is characterized by the following parameters:

- current spatial coordinates;
- the number of the nearest agent of type A by distance;
- the direction and speed of movement in the area;
- the direction of the fire contour bypass-clockwise or counterclockwise;
- the intensity of fire extinguishing, which is determined by the means used to combat wildfires [10,11].

Theinitial location, number and characteristics of A-and B-agents are set by the user at the beginning of the modeling process.

Agents acting on a digital terrain model (D-agents). D-type agents simulate the forces and means used in an indirect method of fighting a fire by changing the digital terrain model, which leads to restrictions for a spreading fire. These agents can simulate the operation of various types of fire-fighting equipment, such as tractors, air tankers, and others, while they do not directly interact with the A-agents, but obey the commands of the agent-manager M and the system user.

For example, the D-agent in modeling the process of laying a fire strip is characterized by the following parameters:

- current status: traffic, lane laying, parking.
- current spatial coordinates;
- speed and direction of movement on the terrain;
- speed of laying fire strips.

When executing these agents, they use direct user commands that determine the type and necessary parameters of the work. For example, after receiving a command to lay a barrier, the D-agent adds special polygons of a given width to the map when moving. As a result, there is a certain strip on the map, which is perceived by A-agents as an impassable section.

The speed of movement, duration of operation and other characteristics of D-agents are set according to the specifications of the specific equipment and reference books.

The number and types of available equipment (D-agents), as well as their characteristics, are set in advance, based on the modeling goals, and are stored in a database. The user in the process of modeling has the opportunity to use the available technology, issuing tasks using a special interface.

Agent-manager who coordinates the actions of other agents (M-agent). An agent of type M in the system under consideration is present in the singular. It is designed to interact with the user and independently control the actions of other agents in the simulation process. In particular, it synchronizes the movement of type A and B agents on a digital terrain model, analyzes the fire contour, and adds and removes type A agents in accordance with the mobile grid algorithm [9].

4. Implementation of a multi-agent system

Based on the multi-agent model described above, the Taiga-Analyst software package was developed, which has a client-server architecture and a web interface [8]. The server stores a database that contains data about the digital terrain model. The program code of the server part is implemented in PHP. The client side of the system is implemented in JavaScript (JS) using the open source library OpenLayers. The library is designed for developing maps based on a software interface (API) and allows you to create a web interface for displaying map materials, working with data from map services such as OpenStreetMap, Google, Yandex, etc.

During the simulation process, a protocol of all user and system actions is compiled. The success and time of fire localization, the area covered by the fire, the cost of fire prevention measures, and other indicators can be evaluated. The following describes how to use the system and provides an example of a single simulation session.

5. Using the developed system

The digital multi-agent system "Taiga-analyst" is currently used in the Siberian Fire and Rescue Academy of the Ministry of Emergency Situations of Russia as a training simulator for cadets when preparing them in the specialty 20.05.01 "Fire Safety" and the direction 20.03.01 "Technosphere safety" in the course "Physical and chemical foundations of fire development and extinguishing", when studying the topic "Natural fires" in the 6th semester of training (3rd year). Working with the system allows you to increase the competence of students in the use of distance learning technologies and management. In addition, when modeling the process of spreading a fire, its elimination, students develop decision-making skills on the use of forces and means to extinguish a fire, with an assessment and understanding of the consequences of the decisions made. For one course of training in the "Taiga – analyst" system, an average of up to 500 protocols are compiled to solve these problems. It is also possible to use the system for conducting competitions for the best solution to the task of fighting a fire.

At the same time, the experience of practical application of the "Taiga - analyst" system as a disadvantage indicates a small choice of fire fighting equipment (type B agents), in particular, the lack of aviation extinguishing equipment in the system and restrictions on the number of foot firefighters, which does not comply with regulatory documents.

Example of modeling the processes of spreading and fighting a forest fire

The digital terrain model is based on the use of GIS, which contains a map of the forest area where the fire occurs. Natural and weather conditions, as follows:

- the forest area with a predominant spruce stand is located on a slope, as indicated by the level lines on the map of the area;
- the type of predominant plant fuel is lichen;
- fire hazard class-2;
- wind speed under the forest canopy – 1 m/s, wind direction-southwest;
- the area of the fire center when detected is 4 hectares.

These data indicate that the expected intensity of the simulated fire is low, so it is acceptable to use manual means to influence the fire around the entire perimeter. It was decided to involve three fire-fighting teams of 6 people each, which are placed around the edge of the fire and were assigned to bypass the fire counterclockwise. The performance of the selected teams to extinguish the edge of the fire was determined by reference books [10,11]. Figure 1A shows the initial configuration of the fire and the initial position of the fire brigades.

A B

Figure 1. An example of modeling the process of fire localization. The light points on the fire contour are the agents of type A, the pictograms of the firemen are the agents of type B. A is the initial configuration of the fire, B is the configuration of the fire through 4 operations of the fire brigades.

The simulation was carried out with a time step of 5 minutes. Figure 1B shows the position of the commands and the outline of the fire after 48 steps of the simulation process, i.e. 4 hours after the start of extinguishing. The figure shows that the fire continues to develop, the length of the burning edge has not decreased, and the area covered by the fire has more than doubled. All this indicates an erroneous choice of tactics and means of fighting the fire, that three groups of firefighters of 6 people are not able to cope with this fire.

The given simple example of a negative result when planning a fire-fighting event speaks, in our opinion, about the realism and usefulness of the process of digital agent modeling in comparison with volitional decision-making, since the reference books give rather rough estimates that do not allow us to take into account the specifics of the current situation and can lead to erroneous decisions.

6. Conclusion

The digital multi-agent system for modeling the processes of fighting wildfires presented in this paper allows us to solve a number of tasks: modeling the dynamics of wildfires, making decisions on tactics and methods of fighting fires, and attracting the necessary forces and resources for this purpose. Further development of the system is associated with the expansion of its functionality by increasing the number of active agents, in particular fire planes and helicopters and other means, with the creation of knowledge bases for decision-making and with the gradual transformation of the system into a digital twin of the process of fighting natural fires.6 It is planned to gradually introduce the system into the work of the operational headquarters of the Ministry of Emergency Situations to support decision-making in the fight against wildfires. In addition, as the existing experience has shown, the system can serve for training personnel in the areas of fire safety.

References

[1] Madni A 2019 Leveraging Digital Twin Technology in Model-Based Systems Engineering *Systems* **7(1)** 7
[2] Fuller A 2020 Digital Twin: Enabling Technologies, Challenges and Open Research *IEEE Access* **8** 108952-71
[3] Tarasov V B 1998 Agents, multi-agent systems, virtual communities: strategic direction in the computer science and artificial *Intelligence news artificial intelligence* **2** 5-63
[4] Fisher K Intelligent Agents in Virtual Enterprises *Proc. Of the First International Conference on the Practical Applications of Intelligent Agents and Multi-Agent Technology (London, UK)* 205-24
[5] Dorrer G A 2018 Multiagent system for modeling dynamic processes on the Earth's surface *Control systems and Information technologies* **2(72)** 42-7
[6] Dorrer G A 2018 Modeling of dynamic processes on the Earth's surface *ITNOU: Information technologies in Science Education and Management* **4(8)** 53-7
[7] Dorrer G 2018 Use of Agent-Based Modeling for Wildfire Situations *Conf. Ser :3rd Russian-Pacific Conference on Computer Technology and Applications*
[8] Spring S V 2016 Agent-based approach to modeling forest fire situations *Software products and systems* **29(3)** 101-8
[9] Dorrer G A 2008 *The Dynamics of forest fires* (Novosibirsk: publishing house of SB RAS) p 404
[10] Ivanov V A 2011 *Handbook on extinguishing natural fires* (Krasnoyarsk: UNDP/MKI) p 130
[11] Shchetinsky E A 2002 *Extinguishing forest fires: a manual for forest firefighters* 104

Design, fabrication and testing of the package-on-package microcircuit

K R Marcinkevich[1], P V Lunkov[1], D A Kirienko[1], V V Putrolaynen[1], M A Belyaev[1] and A V Yartsev[2]

[1]Institute of Physics and Technology, Petrozavodsk State University, 33 Lenin str., 185910, Petrozavodsk, Russia
[2]GS Nanotech, 11 Industrialnaya str., Gusev, 238050, Russia

E-mail: marcink-kristina@yandex.ru

Abstract. The paper presents the results of the development of a package-on-package multichip microcircuit with three-dimensional integration of processor and memory dies. The stages of fabrication the upper part of the microcircuit with several memory dies, which are mounted by the wire bonding method, and the lower part of the microcircuit, on which the processor die is attached by the flip-chip method, are described. After the fabrication of the microcircuit by combining the upper and lower substrates, a multi-stage functional testing of the processor and memory dies is performed using a test board and software loaded into the microcircuit memory.

1. Introduction
In recent decades, the trend towards miniaturization of electronic devices pushes manufacturers to search for solutions to reduce the size of printed circuit boards and component base [1]. At the same time, there is a demand to reduce the power consumption of devices along with increasing their performance [2]. One of the options for solving these problems is to reduce the topological norms in the processes of manufacturing silicon dies. However, by now the microelectronic industry has come close to the fundamental limitations of miniaturization of field-effect transistors [3] - the main elements of digital logic and memory. Therefore, planar technologies for manufacturing silicon dies are being replaced by three-dimensional ones, in which a plurality of planar layers connected by vertical channels are made on a silicon wafer [4]. Such technologies, called monolithic three-dimensional integration, are currently successfully used in the manufacture of DRAM [5] and NAND [6] memory dies, but are not used in the manufacture of three-dimensional logic integrated circuits due to the high heat emission and the large number of input/output (I/O) lines.

In the context of the development of electronic devices, three-dimensional integration of logic and memory integrated circuits is of great interest, since most modern devices include these components. However, there are currently no successful examples of monolithic 3D integration of such devices. Instead, three-dimensional integration at the package-level is used [4], in which several dies are connected at the stage of packaging. Dies can be connected using vertical channels formed by deep silicon etching [6], or using intermediate inserts or substrates [7, 8]. The second option has more flexibility both in layout options and in the selection of dies manufacturers, allowing to change the number and model of processor, DRAM, NAND and NOR memory dies in the microcircuit. The three-dimensional integration method called "package-on-package" [7,9] is very popular among such packaging methods.

A "package-on-package" microcircuit consists of two substrates that are connected to each other using solder balls. The upper substrate contains the memory dies, and the lower one contains the processor die. There are solder balls on the underside of the bottom substrate for attaching to the motherboard. The location of the processor die at the bottom is caused by the requirements for a large number of I/O lines for communication with peripheral devices located on the motherboard. The solder balls connecting the substrates together are used to establish communication between the processor and memory. One of the main advantages of such vertical integration is the reduction in the size of the motherboard (and the final device) because of the fact that memory microcircuits do not occupy separate places on the motherboard [10]. The length of the conductors between the processor and the memory is also reduced, which makes it possible to increase the speed of data exchange between them [10], while reducing energy losses from resistance heating. In addition, the use of "package-on-package" microcircuits allows to reduce the number of layers on the motherboard, thanks to fewer intersecting lines compared to direct mounting of processor and memory microcircuits on the motherboard.

This paper describes the development, fabrication and testing of "package-on-package" microcircuits using processor, NOR, NAND and DRAM memory dies. The paper describes the main stages of developing a microcircuit using a computer-aided design system and the stages of its fabrication with a description of the sequence of applied technological processes. In addition, the paper presents a procedure for testing a microcircuit after its fabrication, including testing the basic functions of the processor core, NOR, NAND and DRAM memory, as well as interfaces for interaction with peripheral devices.

2. Microcircuit development

The development of the microcircuit was carried out taking into account its purpose as a control computing module as part of a modular system for collecting and analyzing information. This module must interact with external peripheral devices for collecting and aggregating data using network protocols of the Ethernet standard. For processing and analyzing data, including using artificial neural networks, hardware acceleration is used with the help of a tensor computing module connected via the USART interface. The storage of primary and processed data is carried out by data storage modules, interaction with which is carried out via the SATA interface. Also, a 9-axis inertial sensor is connected directly to the control computational module using the SPI interface.

To implement the presented tasks, a sufficiently powerful processor with hardware implementation of the specified interfaces is required. In addition, the processor should have reduced power consumption, since due to the packaging method, the heat generated from the processor can be efficiently dissipated only through the motherboard on which the microcircuit is mounted. Heat sink through the upper radiator is difficult due to the presence of the upper substrate with memory microcircuits and, as a consequence, significant thermal resistance between the processor die and the upper side of the microcircuit. Taking into account the above, when developing the microcircuit, a 64-bit single-core processor with MIPS architecture was used, which has a maximum power consumption of 2 W when the core is running at 400 MHz. frequency. The Linux operating system was chosen to control the modular system for collecting and analyzing information.

Based on the results of the preliminary assessment of memory requirements, it was found that to effectively solve the assigned tasks it is necessary to have: 1 GB of random-access memory, 512 MB of non-volatile memory for storing operating system files and program modules, and 32 MB of non-volatile memory for storing boot software and service information required for the processor work. To do this, it was proposed to place on the upper substrate 2 dies of DDR3L random-access memory with a volume of 512 MB each, one die of NAND memory with a volume of 512 MB and two dies of NOR memory with a volume of 16 MB each.

When developing a package-on-package microcircuit, one of the main difficulties is the design of substrates for attaching dies. During this design it is necessary to solve a number of problems. The first problem is related to the density of the memory dies on the upper substrate. Memory dies have a significant area (DDR - 10x8 mm, NAND - 9x7 mm, NOR - 3x3 mm) and are mounted using the wire bonding method, in which part of the substrate area is occupied by contact pads, while the substrate size is only 21x21 mm. The second problem is associated with a large number of I/O lines of the processor, which is mounted on a substrate using the flip-chip method, and the need to organize heat sink through the substrate to the motherboard. The third challenge relates to the need to use the bottom substrate to make electrical connections to both the top substrate and the motherboard. An effective solution to these problems was achieved through the use of the Cadence SIP Layout software package, which allows to develop substrates for multichip microcircuits, taking into account various dies sizes, the location of contact pads, the mounting method (wire bonding, flip-chip), and analyze the result on a three-dimensional model of the microcircuit. Figure 1a shows an image of the upper and lower substrate of the developed microcircuit. During the design, it was found that to solve the set tasks, the upper substrate should have 6 copper layers, and the lower one – 8 layers. Using the Cadence Sigrity power circuit analysis package, "hot spots" were identified and eliminated; the current in them exceeded the recommended one and caused a drop voltage more than by 1%. The traces and vias have been optimized. Figure 1b shows the simulation results in the upper copper layer of the lower substrate.

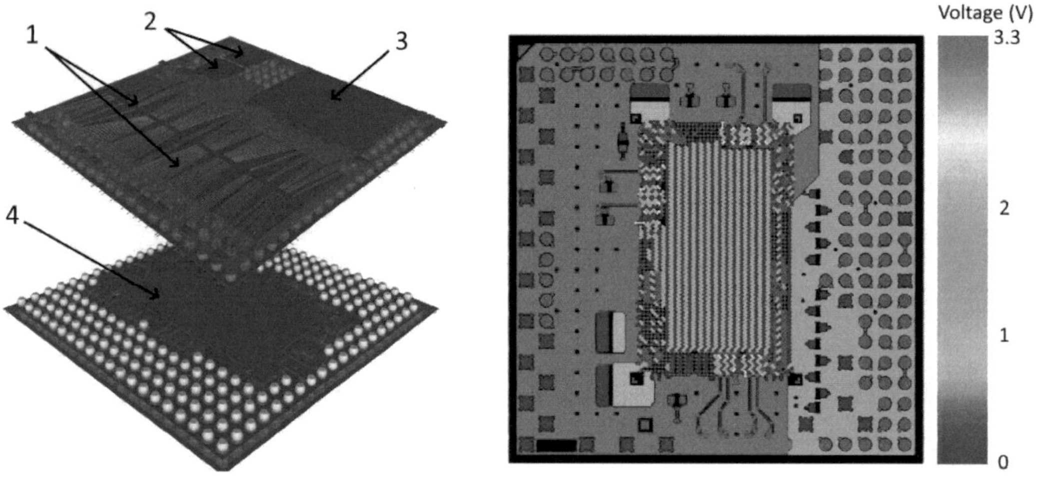

(a) (b)

Figure 1. Image of the upper and lower substrates (a), developed in Cadence SIP Layout, with DDR (1), NOR (2), NAND (3) memory and processor (4) dies attached, and the results of voltage simulation in the upper copper layer with using Cadence Sigrity (b).

3. Microcircuit fabrication

Package-on-package microcircuits fabrication is a multi-stage process: thinning and dicing silicon wafers into separate dies, attaching of dies on a substrate, creating electrical connections between dies and a substrate, sealing with a mold compound, mounting solder balls, dicing strips into separate substrates, connection of the upper and lower substrate into a single microcircuit.

Silicon wafers were thinned using a grinding and polishing machine DGP 8761. This operation was performed to reduce the thickness of the dies from 900 μm to 180-250 μm, which made it possible to reduce the final thickness of the microcircuit. Before the thinning operation, the face side of the wafer (figure 2a) was laminated with a protective film to avoid damage to the functional layer. The thinning process was monitored using built-in thickness gauges and included the stages of coarse and fine grinding and dry polishing. After thinning, the plates were mounted on a frame with a film for further dicing into individual dies (figure 2b). The film on the frame, on which the DDR and NAND memory dies were mounted, had an additional adhesive layer, which was used later when mounting the dies on the substrate. This type of film is called DAF (die attach film). After mounting of the wafer on the frame, the protective film, which was glued before thinning, was removed from the front side of the wafer. To do this, it was exposed to ultraviolet radiation, which led to the destruction of the photosensitive adhesive layer.

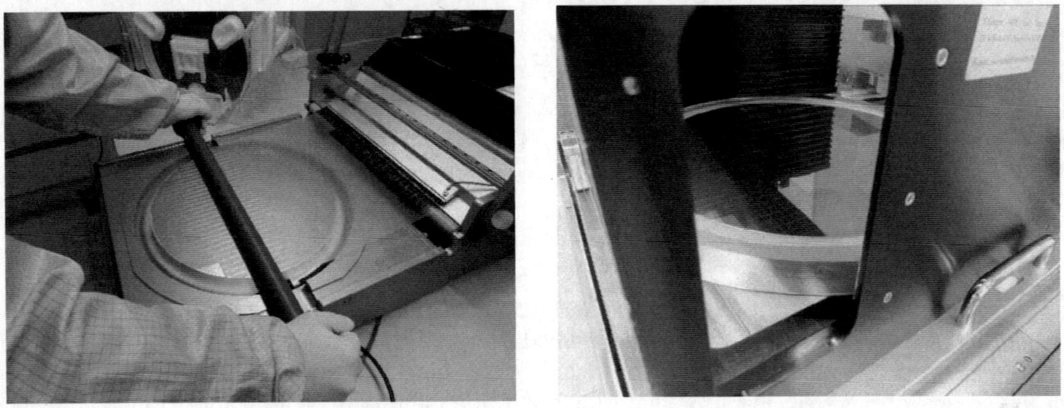

(a) (b)

Figure 2. Lamination of the front side of a silicon memory wafer before thinning (a) and a thinned wafer mounted on a frame with the film before dicing (b).

All silicon wafers were diced using diamond circular saws by two-stage method (step cut). First, most of the wafer thickness was diced with a wider saw, and then the remainder was diced with a narrower saw. This method makes it possible to reduce the size of the chips of the silicon wafer. If the film on the frame had an additional adhesive layer (DAF film), then it was also cut at the second stage. The dicing of silicon memory wafers was carried out on the DFD 6361 HC machine, and processor wafers were diced on the DFD 6560 machine. Figure 3 shows the cutting lines of the silicon memory and processor wafers. It should be noted that before dicing the processor wafers, the grooving of the surface polyamide layers in the area of the planned cutting line was used with the help of a laser beam on the DFL 7160 machine. If you do not carry out the grooving procedure and immediately start dicing with diamond saws, then the size of the resulting chips could exceed acceptable values, up to the formation of cracks in the wafer. This could also lead to the appearance of hidden defects in the form of delamination of the wafer layers. After dicing, the film on the frame was exposed to ultraviolet radiation to facilitate the separation of the die from the film by reducing adhesion.

At the next stage, the dies were attached on the substrates and electrical connections were created between the contact pads of the dies and the substrate. The size of the substrates was only 21x21 mm, therefore, to optimize production processes, several substrates were combined into one, the so-called strip, which in our case consisted of 18 single substrates (2 rows of 9 pieces). It should be noted that not all dies on the wafer were functional and some of them had internal defects that were identified during testing at the manufacturing plant. Dies that were working correctly were marked as known good die on the map used on the die mounting machine. Datacon 2200 EVO was used to mount dies and passive components (resistors and capacitors) on substrates. It had an optical positioning system, a set of accessories for separating dies from a film and placing them on a substrate, and a set of dispensers for applying liquid glue, flux, or solder paste. The memory and processor dies were attached using different mounting methods, so the further description of technological processes for them varies.

(a) (b)

Figure 3. Image of cutting lines of a memory (a) and a processor (b) wafers.

The memory dies were attached on the upper strip in three stages: at the first stage, passive components (resistors and capacitors) were attached using solder paste by the surface mount method. These components were then soldered by reflowing of the solder paste in an oven. At the second stage, two NOR memory dies were attached. Since the film on the frame with NOR memory dies did not have an additional adhesive layer, liquid glue was used to mount the dies, which was applied to the strip using a dispenser. To separate the dies from the film, a needle ejector was used - a special tool that presses on the die from under the film during the rise of the die using a vacuum gripper. It is used to stretch the film and facilitate the separation of the die from the film, which could otherwise lead to damage (cracking, chipping, etc.). After separation of the die, it was transferred to the attachment site and glued to the strip. After attaching all the NOR memory dies, the strip was placed in an oven to polymerize the glue. At the third stage, NAND and DDR memory dies were attached. At this stage, no liquid glue was applied to the strip, since there was a film adhesive layer on the bottom of the dies thanks to the use of a DAF film. After separation of the dies from the film (the process is similar to the separation of NOR memory dies), they were attached on the strip with a force sufficient for the adhesive film to glue evenly to the surface of the strip without the formation of air bubbles. Thereafter, the strip was placed in an oven to polymerize the adhesive film layer. Figure 4 shows the images of a single substrate and a strip after attaching all the memory dies.

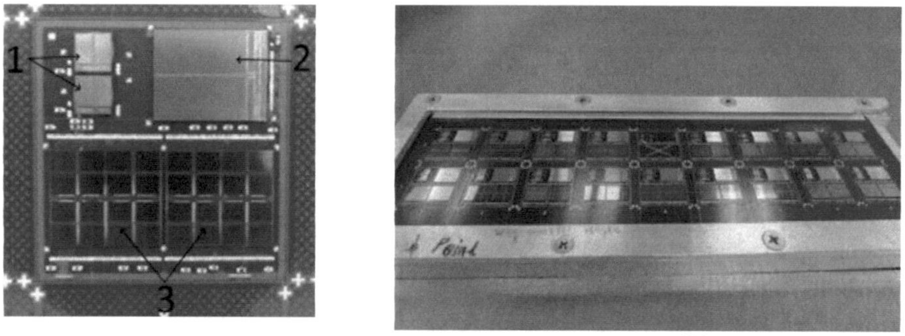

(a) (b)

Figure 4. Image of a single substrate (a) and a strip (b) with attached NOR (1), NAND (2) and DDR (3) memory dies.

After attaching the memory dies between the contact pads of the dies and the strip, wire connections were created using the technology of thermosonic wire bonding. Immediately before the wire bonding, the strips with dies were subjected to plasma treatment to remove contamination from the surface of the contact pads and to activate the surface to improve the wire bonding process. The strips were placed in the ASM Eagle Express GoGu machine where they were preheated to reduce the ultrasonic power during bonding. At the first stage of wire bonding, a ball was formed at the free end of the wire because of melting under the action of an electric discharge. Then, the formed ball was pressed against the contact of the die by a bonding capillary with the supply of ultrasound, forming a welded joint. Then the capillary rose and formed a loop of wire, which ended at the second contact on the contact pad of the strip. Here the capillary pressed the wire again and, under the influence of ultrasound energy, formed a second welded joint. After the second connection was made, the wire was broken so that a new wire bonding cycle could be started. To create connections between the required contact pads, a "die-strip" connection map was loaded into the software of the wire bonding machine. The search for the required contact pads was carried out using the optical system of the wire bonding machine. Figure 5 shows an image of NOR, NAND and DDR memory dies after the thermosonic wire bonding procedure.

At the next stage, the strips were sealed using an epoxy mold compound to create mechanical and chemical protection of dies and welded joints from external influences. To improve the adhesion of the liquid mold compound to the strip surface, it was cleaned in an oxygen-hydrogen plasma. The Fico AMS-i306 machine was used to seal the strips. Immediately before placing the strips into the mold, they were heated to minimize the temperature difference with the molten mold compound and to avoid sudden temperature changes that can lead to large warping of the strips. After heating, the strips were placed into the mold in two pieces and were poured under pressure with molten mold compound. Then, heat treatment was carried out to complete the process of polymerization of the mold compound, which contributed to the partial relaxation of mechanical stresses in the strip caused by the difference in the coefficients of thermal expansion of the materials used (silicon, strip material, mold compound). Figure 6 shows the image of the strips before and after sealing with a mold compound.

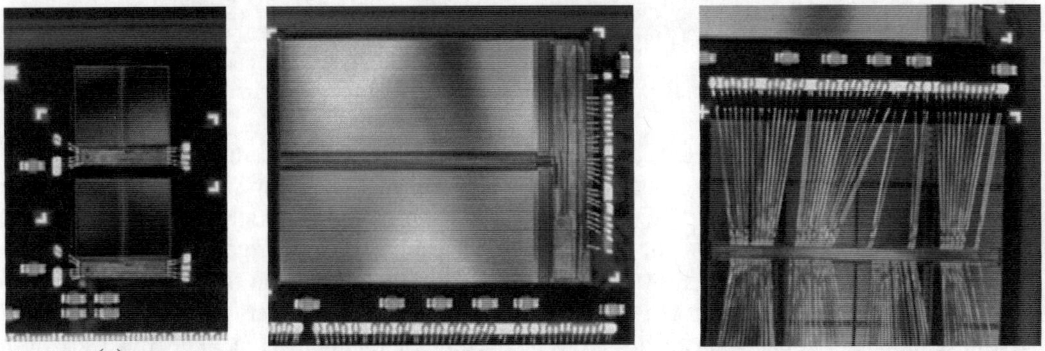

(a) (b) (c)

Figure 5. Image of NOR (a), NAND (b), DDR (c) dies together with wire electrical connections after the wire bonding process.

(a) (b)

Figure 6. Strips before (a) and after (b) sealing with mold compound.

The processor was mounted on the lower strip by the flip-chip method, in which the die was soldered to the strips using solder balls mounted by the manufacturer at the stage of fabricating silicon wafers. Before attaching the dies, the strips were treated with an oxygen-argon plasma to clean the contact pads and improve the quality of soldering. The attachment of the processor die and passive components (resistors and capacitors) was carried out using the Datacom 2200 EVO machine. First, using dispensers, solder paste was applied to the strip for mounting passive components and flux for soldering the die. Then, using vacuum grippers, the passive components and the die were attached on the substrate. After that, the strips were placed into an oven to reflow the solder paste and solder balls. Then the space between the die and the strip was filled with a special glue (underfill), which reduces the likelihood of mechanical destruction of the solder balls due to differences in the coefficient of thermal expansion between silicon and the substrate material during the heating/cooling cycles of the microcircuit during its operation. To polymerize the underfill, the strip was again placed into the oven. Since on the upper side of the lower strip there are contact pads for connecting the lower and upper parts of the "package-on-package" microcircuit using solder balls, the lower strip is not sealed with a mold compound. Figure 7 shows the image of the strip after mounting the processor dies.

Figure 7. Image of a strip with attached processor dies.

The following technological processes are the same for the bottom and top strips. Firstly, the solder balls were mounted on the underside of the strips. The balls on the upper strip will be used to connect the substrates to each other, and on the lower one - to connect the fabricated microcircuit to the motherboard. Before placing the balls, the strips were treated in an oxygen-argon plasma to clean the contact pads. Then, using a Shibuya SBM 351 machine, flux was applied on the contact pads of the strips and solder balls were mounted. Then the strips were placed into a Rehm V8 Nitro oven to melt the balls and solder them to the contact pads. After that, using the dicing machine the strips were diced into separate substrates. Figure 8 shows the image of the resulting lower and upper substrates.

(a) (b)

Figure 8. Upper (a) and lower (b) substrates of a “package-on-package” microcircuit.

At the last stage of fabrication the microcircuit, the lower and upper substrates were soldered to each other. Before this, the lower substrate was treated in an oxygen-argon plasma to clean the contact pads before soldering. Using a Datacom 2200 EVO machine, flux was applied to the contact pads of the lower substrate, and the upper substrate was mounted on the lower one. The assembly was then placed into an oven to reflow the solder balls and form a soldered joint. Figure 9 shows the image of the microcircuit after soldering, side view. Then the space between the microcircuits was filled with glue to mechanically protect the bottom die and improve thermal contact between the two substrates. Polymerization of the glue was carried out in the oven.

Figure 9. Image of the "package-on-package" microcircuit after soldering, side view.

4. Microcircuit testing

To test the fabricated microcircuit, a test board was used, which had ports for connecting peripheral devices with USB, SATA, PCI Express, Ethernet, I2C interfaces. The testing process was controlled using a workstation that was connected to the test board using USB and Ethernet interfaces. The boot software was installed into the NOR processor memory, with the help of which the test software was loaded into the microcircuit memory and the commands were transmitted from the workstation.

The first group of tests checked the operation of the cache memory, the translation lookaside buffer, the control and arithmetic coprocessor. The procedures of writing and reading a predefined sequence of data were used to check the data cache of the 1st and 2nd levels. The 1st level instruction cache was checked by filling in branch instructions with writing and reading from memory. The operation of the translation lookaside buffer was checked both through the operations of writing and reading data into the buffer, and through checking the mechanism of translating virtual addresses. To check the control coprocessor, the command to read the processor identifier was executed. The arithmetic coprocessor was tested using the matrix addition operation. After that, the operation of the processor timers were tested using the following procedures: writing of the counting interval into the boot registers, starting and stopping counting, checking with enabled and disabled interrupts, checking in single and cyclic counting modes, checking the generation of a non-maskable interrupt, checking the timer outputs on the block port lines of discrete signals, interrupt priority check.

The next step was functional testing of DDR, NOR and NAND. Firstly, we checked the processor's operation with DDR RAM. After measuring the amount of available memory, the correct operation of the data bus and the address bus was checked using sequential data writing and reading according to the "walking ones" algorithm. Then the complex testing of DDR memory was carried out by means of writing and reading data according to the algorithms of "walking ones and zeros", "moving inversions" and "modulo-X". At the last stage, the mode of direct memory access was tested with the transfer of data blocks of different sizes and with a change in the offset of the address of receiving and sending data.

Then the NOR memory and the operation of the SPI interface were tested using the following procedures: reading the memory device identifier, reading and checking the status register, multiple data readings with the calculation of the checksum in standard and fast reading modes, checking the checksum written into the memory of the boot program, writing and erasing data in the free memory area, writing into the status register, checking the operation of interrupts at the end of the cycle on the SPI interface.

When checking NAND memory, the identifier of the memory device was read and checked, the number of bad blocks of memory was requested, which was compared with the threshold value allowed for the correct operation of the microcircuit. Then, multiple tests were performed with reading data and calculating a checksum, as well as sequential writings into the memory of predefined constants, followed by verification using the read data.

After testing the basic functionality of the processor and memory, we checked the operability of the interfaces for interacting with peripheral devices: USB, SATA, Ethernet, PCI Express, I2C. To test the USB interface, a USB flash drive connected to the test board and the values of the block registers were read, the device identifier was checked, and a connection to the USB flash drive was established.

To check the SATA interface, SATA disks were connected to the test station, after which test procedures were performed: checking the connection of devices, performing a self-diagnosis procedure, reading device parameters (model, serial number, amount of memory, logical sector size, etc.), writing data into various sectors of the device with subsequent verification using the sector reading procedure, reading sectors in multi-sector mode, verification of reading and writing operations in direct memory access mode.

When testing the Ethernet interface, we used three transmission speed modes (10, 100 and 1000 Mbit s) with forced and automatic speed settings. During testing, the following checks were performed: checking the device identifier, transmitting and receiving the tuning data packet, changing the size of the transmitted data, changing the number of descriptors for receiving and transmitting, checking the data buffer padding, checking interrupts for receiving data packets, checking the JUMBO FRAME mode with transmission data up to 8192 bytes.

To test the PCI Express interface, the expansion board connected to the test board was used. During testing the following procedures were carried out: initializing the device on the PCI Express bus, checking the connection with the device, checking the set operating mode and baud rate, searching for connected devices and reading their identifiers.

To test the I2C protocol, we used the EEPROM memory installed on the test board. In the course of testing, the connections with the device were checked, indicating the address and type of the device, the amount of available memory was read, and the procedure for writing the memory with test data was performed, followed by verification by reading the data.

After successfully passing all the tests, the barebox boot loader was installed in the NOR memory of the microcircuit, with the help of which the Linux operating system was installed in the NAND memory of the microcircuit.

Acknowledgement
This research is implemented in Petrozavodsk State University with financial support by the Ministry of Science and Higher Education of Russia within Agreement no. 075-11-2019-088 of 20.12.2019 on the topic "Creating the high-tech production of mobile microprocessor computing modules based on SiP and PoP technology for smart data collection, mining, and interaction with surrounding sources".

References

[1] Taylor R, Baron D and Schmidt D 2015 The world in 2025 - Predictions for the next ten years

2015 10th International Microsystems, Packaging, Assembly and Circuits Technology Conference, IMPACT 2015 - Proceedings pp 192-5
[2] Mittal S 2014 A survey of techniques for improving energy efficiency in embedded computing systems *Int. J. Comput. Aided Eng. Technol.* **6** 440-59
[3] Mamaluy D and Gao X 2015 The fundamental downscaling limit of field effect transistors *Appl. Phys. Lett.* **106** 193503
[4] John H and Lau P D 2016 *3D IC Integration and Packaging* (New York: McGraw-Hill Education)
[5] Belmonte A, Oh H, Rassoul N, Donadio G L, Mitard J *et al.* 2021 Capacitor-less, Long-Retention (>400s) DRAM Cell Paving the Way towards Low-Power and High-Density Monolithic 3D DRAM pp 28.2.1-28.2.4
[6] Kim C, Kim D H, Jeong W, Kim H J, *et al.* 2018 A 512-Gb 3-b/Cell 64-Stacked WL 3-D-NAND Flash Memory *IEEE J. Solid-State Circuits* **53** 124-33
[7] Chen M F, Chen F C, Chiou W C and Yu D C H 2019 System on integrated chips (SoIC(TM) for 3D heterogeneous integration *Proceedings - Electronic Components and Technology Conference* pp 594-9
[8] Hsu F C, Lin J, Chen S M, Lin P Y, Fang J, Wang J H and Jeng S P 2018 3D Heterogeneous Integration with Multiple Stacking Fan-Out Package *Proceedings - Electronic Components and Technology Conference* pp 337-42
[9] Yoshida A, Taniguchi J, Murata K, Kada M, Yamamoto Y, Takagi Y, Notomi T and Fujita A 2006 A study on package stacking process for Package-on-Package (PoP) *Proceedings - Electronic Components and Technology Conference* pp 825-30
[10] Menon A R, Karajgikar S and Agonafer D 2009 Thermal design optimization of a package on package *Annual IEEE Semiconductor Thermal Measurement and Management Symposium* pp 329-35

Capacitive sensors of mechanical strain

V S Ignakhin, I V Sekirin and N A Shilovskiy

Institute of Physics and Technology, Petrozavodsk State University, 33 Lenin Avenue, 185910, Petrozavodsk, Russia

E-mail: art101@petrsu.ru

Abstract. The paper reviews recent research and applications of both laboratory prototypes and commercially fabricated capacitance gauges of mechanical strain. Basic operational principles as well as advantageous points of each type of sensors are considered. The prospective areas of applications of the capacitive gauges are briefly considered. The properties of other strain gauges are briefly discussed and compared with those of the capacitive sensors.

1. Introduction

Measurement of deformations and mechanical stresses is widely used in the study and monitoring of building constructions and structures, as well as rocks, machine parts and mechanisms, measuring the mechanical properties of materials and performing strength tests. Strain measurement is used in technical diagnostics, as well as in the measurement of related physical quantities – forces, moments, pressure, flow velocity, etc., which are converted into strain of the sensitive element [1, 2]. The use of sensors for smart health monitoring and biomechanics [3], [4], wearable electronics and "electronic skin" [5], industrial Internet of things (IIoT) [6] is becoming more and more relevant. In particular, the IIoT monitoring system would benefit from real-time measurements of the deformation state of the observed structure [7].

Strain measurement can currently be performed using various physical sensors: resistive (piezoresistive), piezoelectric, capacitive, optical, etc. Similarly, for mechanical stress measurement – in most methods of such a measurement the sensor perceives the absolute or relative value of the strain, since the natural input value of the transducers used is displacement. Sensors of any type have advantages and disadvantages, determined by the physical principles underlying their functioning, which determines the specifics of choosing a device for a specific task.

The capacitive sensors are attractive because of their effectiveness for wireless applications and wearable electronics due to low power consumption, ability to register steady strains, stretchable applications such as "electronic skin" with its high strain limits etc. A series of cheap commercial solutions for signal processing is available in the wide market, also. In this review, capacitive strain sensors will be considered. Other types are excluded, but their properties are briefly discussed and compared with those of capacitive sensors.

2. Strain characterization and sensors parameters

The elementary types of strains are linear tensile and compressive strains, which can be expressed in terms of absolute and relative elongation (normal strain):

$$\Delta l = l - l_0;\ \varepsilon = \frac{\Delta l}{l_0} \quad (1)$$

where Δl – absolute elongation, l_0 and l – original and resulting length, respectively. When a compressive strain is applied both values are negative [8]. Except for linear strain and compression, different types of more complex deformation can be applied to a rigid body such as bending, twisting, and shear.

The tensile and compressive strain correspond to normal stresses which are perpendicular to the surface and given by the ratio of applied force F and the cross-sectional area A:

$$\sigma = \frac{F}{A} \quad (2)$$

Shear stresses correspond to off-diagonal components of strain tensor and usually correspond to shifts parallel to the surface of the deformed body. Practically shear stresses always occur together with normal ones. Mechanical stresses are important to be monitored for study and design of building constructions and machinery, fundamental science such as geology, but at the same time they are not accessible for direct measurement. Stresses can be either calculated according to the theory or from strain measurements [8].

Typically, key demands for strain gauges are the ability to detect small strains and high accuracy, but sometimes key demand can be another. For instance, this could be contactless measurement requirement or large strain working limits. In the branch standards (e.g. [9]), one introduces the notions of maximum capacity, sensitivity of a measuring system and accuracy class for characterization of strain gauges. However, the analysis of these notions shows that maximum capacity can be improved by adding springs, sensitivities of a measuring system for different types of sensors have different dimensions and are not comparable, and the accuracy class can be determined by an external ADC. Practically, the sensitivity threshold as the minimal value of mechanical stress or relative strain the sensor and the accuracy in terms of measuring of strains (or loads) are more important. The first one is determined as the minimal deformation $\Delta l / l_0$ which can be confidently detected. To describe the accuracy of the measurements with the use of strain gauges, the following quantities are introduced in [9]: hysteresis, creep, repeatability, durability; the influence of extraneous factors is introduced separately. The term "expanded uncertainty" is used in [9] for accuracy in the usual sense – the width of the interval around the measurement result into which the true value is likely to fall. Therefore, when describing specific strain gauges, it is desirable to have their sensitivity threshold and expanded uncertainty values. Unfortunately, the authors of academic works often do not provide such data. Sometimes these data can be estimated from the figures etc.

All the sensors could be either self generating (active transducers) or modulating (passive) [10]. A useful tool for comparison parameters of the latter type is gauge factor (*GF*):

$$GF = \frac{\Delta Y / Y_0}{\Delta X / X_0}, \quad (3)$$

where Y_0 and X_0 are initial values of output and input signals, ΔY and ΔX are their increments, respectively. This can be applied to input and output signals of different nature. In case of strain $\Delta X / X_0$ is the normal strain $\varepsilon = \frac{\Delta l}{l_0}$.

3. Sensors review

The operation principle of capacitive sensors is based on the transformation of the controlled parameter into a change in the electrical capacitance of the capacitor, which is measured directly [11]. A series of techniques for capacitance measurements have been developed such as charge-discharge time, resonant frequency shift of LC circuit, capacitance-to-digital $\Sigma - \Delta$ conversion. Specialized IC chips existing on the commercial market are able to measure small quantities of electrical capacitances with digital resolution and accuracy of several fF, at the same time they have low power consumption and low cost [12]. The examples of such IC's are products of Analog Devices (AD7745-AD7747), Texas Instruments (FDC1004, FDC22XX), etc.

3.1. Stretchable elastomer based sensors

Recently, flexible capacitive sensors have been actively developed for use as wearable devices on the body, "electronic skin", and also for monitoring specific technological processes. These applications impose strict requirements for weight, convenience, flexibility, mechanical strength and sensor's operation life. Compact and flexible capacitive sensors based on elastomers (polymers with high elasticity) are being developed [13-17]. Commercial sensors are available, also [18,19]. These transducers are able to track both small and large strain levels, significantly exceeding the limit of metal (resistive) strain gauges.

The use of elastomers as a deformable dielectric material is one of the approaches to expanding the limits of measured strains. An uniaxial strain ε causes increase of the initial length l_0 to $(1+\varepsilon)l_0$, whereas due to the Poisson effect the width of the plates w_0 and the interelectrode distance d_0 decrease to $(1-\mu\varepsilon)w_0$ and $(1-\mu\varepsilon)d_0$, respectively [15, 20]. Here, μ is the Poisson ratio. Thus, in the case of a plane capacitor the capacitance C under applied strain and GF are as follows:

$$C = (1+\varepsilon)C_0, \; GF = \frac{\Delta C / C_0}{\varepsilon} = 1, \tag{4}$$

A series of academic research papers confirms this result [13,14,16,17].

In [13], an elastomer-based sensor obtained using laser techniques is described. Manufacturing of the sensor consists of the following stages: obtaining a silicone elastomer substrate, fabricating the surface as a set of grooves using a DPSS laser, pre-stretching the elastomer up to ~80%, applying layers of metal that serve as electrodes, and cutting sensors of a given size from the workpiece. The dependence of the capacitance change on the applied strain is perfectly approximated by the linear dependence (R^2=0.998) and has an extremely small hysteresis. The gauge factor $GF = \frac{\Delta C / C_0}{\varepsilon}$ was found to be ~0.9. The noise level measured at the 50 Hz bandwidth allows absolute strains of ~0.5 mm to be resolved with the gauge length of 40 mm. In a series of loading cycles, a sharp drop in the relative change in capacity is observed for strains $\varepsilon > 0.8$ during the first cycles, it is caused by the elastic properties of the elastomer, and then this value is stabilized. In case of $\varepsilon = 1$, the capacity continues to fall with an increase in the number of loading cycles, which is associated with the successive crack formation in the metalized layers. The sensor delay time was found 0.146 s, which makes it possible to measure strains with frequencies less than 10 Hz. [13] demonstrates the potential use of the sensor as a sensor of movement and position of human body parts.

In [16, 21] carbon nanotubes coatings embedded in a polydimethylsiloxane (PDMS) matrix are used as top and bottom electrodes of the capacitor separated by an elastomer layer. The sensor [16] is transparent and can detect deformations both when pressed or stretched. The sensor [21] is characterized by both a change in the resistance of the electrodes ("piezoresistive" response) and a change in the capacitance during deformation. The dependence of the resistance change on the strain shows a significant hysteresis, which is expressed in a change in the resistance in the unloaded state at a level of more than 100%. The change in the system capacity, however, shows excellent linearity and no visible hysteresis (depending on the prehistory of deformations). The sensor [21] shows a capacity change of less than 3% at 3000 load cycles of 100% strain. The gauge factor value is close to the theoretical limit and is 0.99 with a temperature error of 0.1%/°C. In [22], similar results were obtained for a sensor based on silicone rubber and nanotube plates. The sensor can withstand 1800, 3800 and 10000 cycles with strains of 200%, 150% and 100%, respectively, though the potentially detected strains are of 1 % to 300 %.

Graphene [23-24] is also considered as a material for the plates of capacitive sensors due to its elasticity and optical transparency. The latter is especially important when building optical touchpad displays. A capacitive sensor based on graphene nanoplastics and PDMS was proposed in [24]. The sensor demonstrates linearity in the range of applied deformations up to 0.2%, the GF is about 3.

Elastomer based sensors are available in the market for use in wearable electronics [18]. The SSD18 sensor [18] can withstand stretching up to 80 % and has an elongation sensitivity of ~5 pF/mm, which gives a gauge factor of $GF = \frac{\Delta C / C_0}{\varepsilon} \sim 0.8$ for the total length of the sensitive area $l_0 = 70$ mm and an initial capacity of $C_0 \approx 450$ pF. Flexible capacitive bending strain sensors based on silicone elastomer are commercially available by Bend Labs [19]. The temperature coefficients of expansion of the materials are matched together, which is especially important taking into account the large length of the sensor. The company has developed single-axis sensors that respond only to strains along one of the axes, and two-axis sensors are offered to control two-dimensional deformations. The internal capacitive sensing elements are connected according to a differential scheme, which allows one to exclude the influence of errors due to temperature, tensile strain, etc. The bending sensitivity of the sensors is 0.274 pF/°, the power consumption of the single-axis sensor in active mode 200 μa at 3.3 V at the acquisition frequency of 100 Hz.

A number of ways have been proposed to improve elastomer based strain sensors. An elastomer-based sensor described in [14] allows one to enhance the sensitivity for potential miniaturization by increasing relative permittivity ε_r. The increase in permeability is achieved by introducing a small amount of a conductive polymer into the elastomer matrix of the SEBS-g-MA with Young's modulus E_Y=4.5 MPa and the permittivity ε_r=2.0. The increase in permittivity is achieved due to hyperelectronic polarization – the ability of mobile charges to shift significantly, forming a macromolecular dipole. At a volume concentration of 2.3 % of the doping substance, ε_r increases up to 122 at a frequency of 100 Hz. The change in capacity is linear and is ~10 % at the concentration of the doping impurity 2 and 2.3 % by volume and 10% of applied strain. Increasing the dopant concentration enhances both the aggregate capacitance and its increment at an applied strain. The deviation of the capacitance $\Delta C / C_0$ (estimated noise) at 20% strain during 10 s was about 10^{-4}, which corresponds to the absolute strain of ~10 microns. At the same time, the level of fluctuations in long-term exposure (3600 s) is significantly higher.

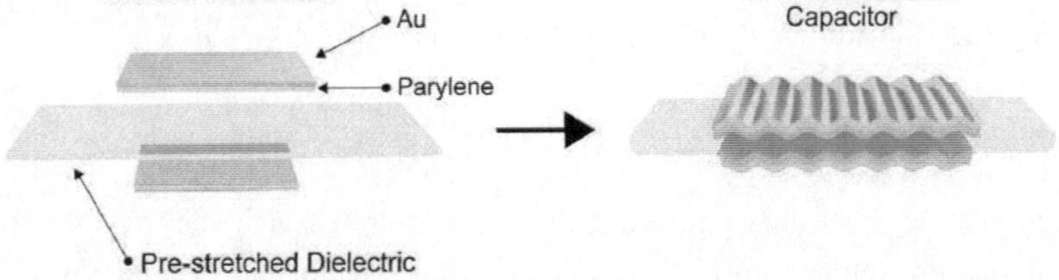

Figure 1. Manufacturing scheme and design of a film capacitive load cell based on a "wrinkled" Au film [13].

One more approach to enhance the sensitivity of sensors on elastomers and their working range is possible increase of the gauge factor. Thus, [13] describes a capacitive sensor based on wrinkled structure with gauge factor reaching 3 at the maximum value of the applied strain of $\frac{\Delta l}{l_0}$=140% (1.4 ε). The higher GF is obtained by introducing an additional degree of freedom to strain the film in the transverse direction when stretched. This is achieved by the fact that the film in the unstrained state has a structure full of folds (figure 1) and its following transformation into a completely flat capacitor, which allows one to change the effective area of the capacitor by about three times.

While width contraction is cancelled, the strained capacitance can be found as

$$C = \frac{(1+\varepsilon)C_0}{(1-\mu\varepsilon)}. \quad (5)$$

The folds are formed by sputtering a gold film on a prestressed (stretched) layer of a dielectric elastomer with an intermediate layer of Poly-para-xylylene (PPX, parylene), which is necessary to improve the adhesion of the gold layer and prevent its destruction under large strains. After deposition, the dielectric film is unloaded and folds appear. The optimal thickness of the PPX layer found in the study was from 500 nm to 2.5 μm. The sensor shows a fairly high linearity $R^2 = 0.98$ and minimal hysteresis, no sensitivity changes after 1000 load cycles with the change of capacitance of 5.4 % when heated to 50 °C. The paper demonstrates the possibilities of using the sensor for applications in flexible wearable electronics.

Another noticeable type of the elastomer based sensors is that with flat printed strip conductors [12, 25-27], the so-called interdigital capacitors being used as plates. A similar rubber-based sensor for diagnosing vehicle tire pressure is proposed in [25-26]. The sensor consist of a rubber substrate with printed by photolithography gold interdigital structure upon (figure 4). The pressure applied in the normal direction to the plane of the conductors causes a change in the distance between them and, consequently, in the capacity. The configuration of the capacitor plates (outline drawning) is shown in figure 2: a – the length of the overlapping part of the strip, b - the width of the conductor, c - the total length of the plate, g – the gap, l - the total length of the capacitor.

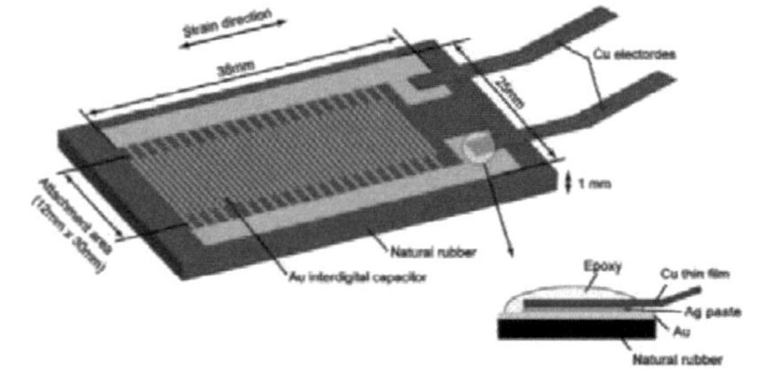

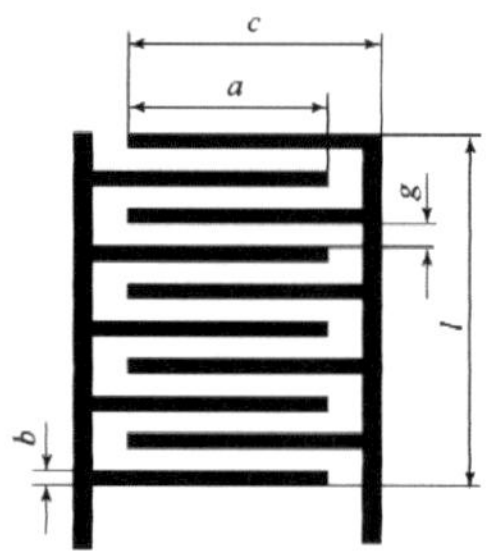

Figure 2. Rubber-based capacitive sensor (left) and the outline drawing (right) [25-26].

When applying a tensile strain ε the change in the capacity of the system is:

$$C = C_S + C_P = (2n-1)\varepsilon_r\varepsilon_0 \frac{a(1-\mu\varepsilon)h}{g(1+\varepsilon)} + C_P \quad (6)$$

where n is the number of electrodes pairs, μ is the Poisson's ratio, h is the thickness of the dielectric (rubber), ε_r is the permittivity, $C_P = \alpha C_{S0}$ is the parasitic capacitance determined by the connecting conductors, C_{S0} is the initial capacitance of the interdigital structure and α is the coefficient. The GF of the sensor is as follows: $GF = \frac{\Delta C / C_0}{\varepsilon} = -\frac{1+\mu}{(1+\varepsilon)(1+\alpha)}$. From this ratio, the gauge factor does not depend on the sensor's design, but only on the coefficient that determines the ratio between the parasitic and net capacity. The sensor reveals a small hysteresis of fairly linear dependence on the strain with the maximum deviation from linearity equal to 8 %. The sensor lifetime reaches 10^5 load-discharge cycles, which is due to degradation of the electrodes, and when applying an additional protective coating based on silicone rubber, it increases the life to 10^6 cycles. The maximum measured strain reaches more than 14 %.

[12] describes a interdigital sensor based on a flexible substrate. The substrate material is polyamide, a liquid crystal polymer. The gold conductors are built by photolithography and PVD, and sealed with an acrylic resin. The process limitations allowed one to fabricate an electrode of width b = 45 μm, height h = 300 nm, and gap g = 15 μm. The sensors are connected according to the half-bridge, which eliminates the influence of common mode errors acting on both capacitors simultaneously (temperature, humidity etc.). Both single sensors on polyamide (PI-45/15) and LCD (LCP-45/15) substrates show excellent linearity at low strain levels of less than 1 %, while the change in capacitance lies in the sub-pF region at the initial value of ~48 pF in a dry atmosphere. The power consumption of the sensor system based on a half-bridge sensor together with the measuring electronics is about 0.6 mW, which makes it attractive for use in wireless systems.

3.2. Elastic sensors with liquid electrodes

Measuring large deformations using the capacitive effect imposes strict requirements on the mechanical and electrical properties of the electrode material. Despite certain advances in the design of capacitive sensors based on solid-state metal conductors, the lifetime of such sensors under periodic large strains can be significantly limited. In this regard, solutions based on liquid conductors that can withstand large strains with no disturbing electrical conductivity are relevant [28-31]. Recently liquid metals such as eutectic based on Ga and In (EGaIn), gallium–indium–tin (Galinstan) with low melting points below room temperature.

[28] and [29] demonstrate a possibility comb capacitor (interdigital structure) in which EGaIn or Galinstan is embedded in a soft silicone elastomer. The sensor can distinguish the stretch in directions parallel and perpendicular to the electrodes. The predicted values of the relative changes of the capacitance are $\Delta C / C_0 = \varepsilon$ and $\Delta C / C_0 = 1/\sqrt{\varepsilon+1} - 1$ for the parallel and perpendicular axes, respectively, and the corresponding gauge factors are $GF = (\Delta C / C_0)/\varepsilon = 1$ and $GF = (\Delta C / C_0)/\varepsilon = \dfrac{1-\sqrt{\varepsilon+1}}{\varepsilon\sqrt{\varepsilon+1}}$. These predicted relations are confirmed by the experimental dependences. A spiral geometry of capacitor is fabricated and discussed in [28], also.

In [30], a stretch strain textile-mounted sensor based on a silicone elastomer and a conducting ionic liquid is proposed, a method for manufacturing is given, also. The liquid serves as the plates of the capacitor, which is built according to the coaxial scheme (figure 3).

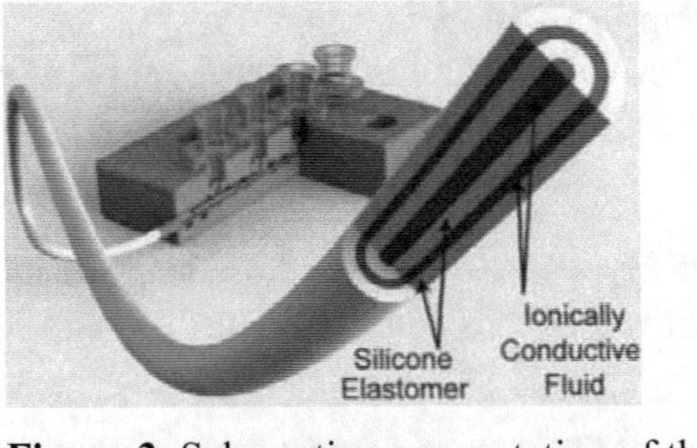

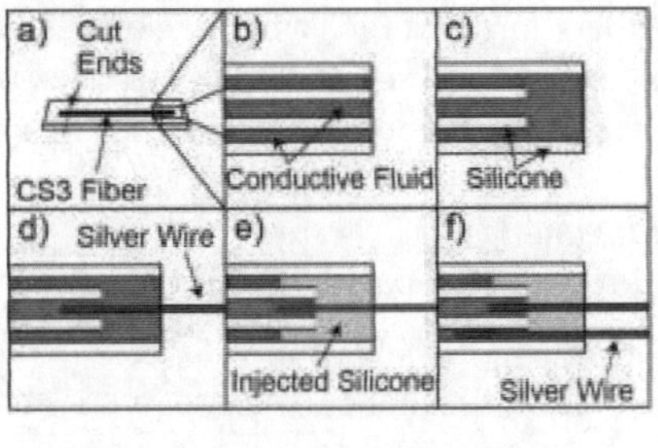

Figure 3. Schematic representation of the design of a capacitive sensor based on an elastomer and a conducting liquid [30].

The core-shell coaxial threads were made using 3D printing, after which they were filled with a ionically conductive liquid (figure 3). Electrical contacts were created by inserting a silver wire into a liquid conductor. When the sensor is stretched, the capacitance of this coaxial system changes proportionally to the strain:

$$C(\varepsilon) = \frac{2\pi\varepsilon_r\varepsilon_0(\varepsilon+1)l_0}{\ln(\frac{r_{20}}{r_{10}})}, \tag{7}$$

where r_{10}, r_{20} are the outer radii of the liquid core and the inner dielectric layer, respectively. At the same time, the resistance of the conductor of the plates also changes due to the strain effect, but in proportion to the square of the deformation. The sensor can withstand deformations up to 250 % with no hysteresis under static and dynamic loads. For small deformations, the dependencies deviate from the theoretical ones due to the inadequate tension in the dielectric layer to ensure a straight inner tube.

A capacitive sensor of torsion or tension strains based on a liquid-metal conductor placed inside an elastomer fiber is proposed [31]. Conductors based on hollow tubes with a liquid metal (LM) conductor are known to be able to maintain metal conductivity under deformations up to ≈ 800% [32]. The fiber is quite compact and can be embedded in a textile. Eutectic based on Ga and In (EGaIn, Ga 75% and In 25%) has been used as a liquid conductor. Its advantages are the absence of toxicity, negligible saturated steam pressure at room temperature, and low viscosity. The sensor consists of two hollow tubes of triangular cross-section elastomer filled with a liquid metal conductor and twisted into a double helix. The triangular shape of the cross-section allows minimizing the non-linearity and hysteresis of the liquid-metal channels [33]. The sensor allows detecting stretching and torsion deformations, the possible level of the latter reaches Θ=10800 rad/m, which, according to the authors, is two orders of magnitude greater than the existing torsional strain sensors can measure. When maintaining a constant distance between the two ends of the fibers and twisting them, the main change in geometry is associated with an increase in the total length ζ, and this causes a change in the total capacity of the system. The sensitivity of the system increases with the increase in the diameter of the LM-filled fiber, while the limit of detected strain decreases. This is because larger diameter fibers must experience greater tensile strain at an equal torsional level and achieve destructive tensile stress at lower relative torsion angle. The dependencies $C(\Theta)$ for all the selected diameters are linear with $R^2 = 0.99$. However, for large and small torsion, there is a deviation of the actual capacity from the linearity toward smaller values. The maximum values of the relative twisting angle are 10887, 8378, 5585 rad/m for sensors with fiber diameters of 235, 350 and 850 microns, respectively. When applying a cyclic load in the area of small torsional strains, gradually progressive deviations are observed, which is associated with the formation and increase of the gap between the fibers. At the same time, at loads of more than 200 rad/m, the repeatability is significantly higher and does not depend on the number of cycles of the applied strain. As the temperature changes, the capacity per length decreases by 0.0003 pF/cm for each degree when the temperature drops from 25 °C to -10 °C, which is 0.27 %/°C of the measurement limit. Stretching causes the fibers to converge, which also leads to an increase in capacity. Thus, the sensor can also detect tensile strains and, in principle, complex tensile-torsion strain. The paper [31] also provides an example of using a sensor as a tactile sensor in a touchpad type device.

3.3. Thick-film and MEMS sensors

The thick film type of capacitive strain sensors is the type of gauges based on thick films or layers of dielectrics [34-35]. High values of relative permittivity ε_r of some dielectric films allows one to increase the initial capacitance and provide a change of ε_r when compressed. A sensor with a thick-film dielectric based on lead-zirconate–titanate (PZT) (lead–zirconate-titanate, $\varepsilon_r \cong 600$) and polyvinylidene fluoride (PVDF, $\varepsilon_r \cong 13$) is described in [34]. The dielectric layer was fabricated by applying a specially made paste layer 150 μm thick on a substrate of aluminum oxide. The plates were created by applying a layer of a conductive paste. When applying a tensile strain, the film is deformed both longitudinally and transversely, and its thickness decreases. The theoretically predicted *GF* is obtained as $GF = \frac{\Delta C / C_0}{\varepsilon} = \frac{\Delta\varepsilon_r / \varepsilon_r}{\varepsilon} - 1$ and contains two terms: caused by a change in the permittivity

$\frac{\Delta\varepsilon_r/\varepsilon_r}{\varepsilon}$ (the "piezo-capacitive" component) and due to a change in the geometry "−1". When applying a strain of up to 300 microns/m and then removing it, the sensor based on PZT cermet shows a *GF* of 6, and *GF* of that based on PVDF polymer yields *GF*=3.5. The obvious disadvantage of the thick film sensors in comparison with the elastomer based ones is low limiting strains.

Capacitive [36-41], resistive [42, 43-45], piezoelectric [42, 46], and resonant strain sensors [47-52] based on microelectromechanical systems (MEMS) technology are being currently developed. MEMS technology allows one to reach significant miniaturization of the sensor, provide precision manufacturing, reduce power consumption, and increase sensitivity. MEMS capacitive strain sensors are attractive due to high sensitivity, low noise, large dynamic range, and potential for integration with low power CMOS electronics [36-41].

The paper [36] describes a differential capacitive MEMS sensor of tensile and compressive deformations manufactured using a single-layer technology. The sensor's structure is shown in figure 4. The support ribs are supported above the surface by the bent beam suspensions. The edges hold the interdigitated tines, which play the role of capacitor plates. The plates are placed in such a way that when the strain is applied, the capacity increases on one part and decreases on the other. This provides a differential response of the structure. In this case, the ribs A move in the opposite direction, compared to the ribs B and C.

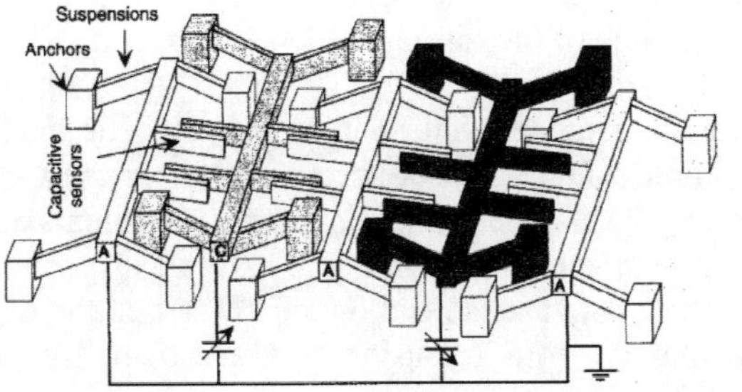

Figure 4. Structure of the differential capacitive MEMS strain sensor [36].

The sensor prototypes were fabricated of polycrystalline silicon for compressive deformations and electroplated Ni for tensile strains. The polysilicon Sensors were manufactured by CVD method on an oxidized silicon substrate. Silicon lithography was performed by the reactive ion etching (RIE), the sacrificial layer was etched in BHF. The MEMS capacitors were manufactured on a silicon substrate with a 2 μm layer of oxide coated with a 500 nm layer of nitride. After that, a 2 μm Ti layer was sputtered, which was covered by a layer of 100/100 Å Cr/Ni. The change of capacitance in the case of the polysilicon sensor amounted to several femtofarads when applying compressive stress of -40÷-42 MPa and tensile stress of 40÷55 MPa.

In [37,38], a capacitive MEMS bending and axial strain interdigital sensor is proposed. Each plane system can be displaced independently of the other when applying a tensile or bending strain. During manufacturing, they both are rigidly bound with ribs, and the ribs are destroyed during installation. Changing the distance between the plates causes a change in capacity, it is the basic operation mode of the device when applying tensile strain. However, when bending, the overlap area of the interdigitated plates also changes. When applying a tensile strain of 1000 με, the change in the system capacity is about 3 pF. The nonlinearity of output characteristic of the sensor is a disadvantage of the capacitive sensors based on changes in the distance between the plates. At the same time, the advantage is high sensitivity.

Another approach for the interdigital MEMS sensors is the use of the overlapping area change which provides a more linear response. The sensor [39-41] consists of three amplifying bent-beam suspensions with interdigitated comb fingers positioned at the structural center. As it is expected, the output characteristic of the sensors [39-41] is close to linear (1.9% full-scale non-linearity), with the sensitivity equal to 265 aF/(μm/m). The temperature change of the capacity in the temperature range from room to

150 ° C was 10 fF with a nominal sensor capacity of 440 fF. The authors developed an original low-noise readout, the measured noise level was calculated as 0.24 aF/√Hz (0.24 aF/√Hz) or 0.9 nm/m/√Hz (nε/√Hz).

4. Comparison with other transducers

Depending on the specific task, sensors can be selected that provide the desired resolution, measurement ranges, and also specific requirements. The obvious advantage of the capacitive sensors is possibility to measure not only tension-compression deformations, but also torsion and bend.

The capacitive sensors are more attractive for high stretch applications than the resistive strain sensors which are generally suitable for measuring moderate constant and time-varying strains up to ~ 10^3 με. The power consumption of the the capacitive trancducers is much less. The read-out for a typical resistive strain system is easier, but measurement of the capacitance changes leads to a slightly greater complexity of processing methods and circuit solutions in comparison with those for resistive and piezoelectric sensors. However, there is currently a wide range of relatively inexpensive specialized commercial IC's and solutions for capacitive sensors. The piezoelectric sensors are self generating (active transducers), they directly produce output voltage with no power supply, but their lack is impossibility to measure static strains.

The elastomer-based sensors, both resistive or capacitive, can be used to measure larger strains up to ~1 ε. To build a "smart" sensors and sensing IC's with high degree of integration, one requires miniaturization of the size of the sensor, extremely low power consumption etc. The MEMS sensors are the most promising for such tasks. They are capable of measuring moderate strains up to 10^3 με with a resolution of less than 1 με.

The magnetic sensors are sensitive to external magnetic fields and electromagnetic interference, but their unique advantage is the possibility of non-contact measurements with no power supply of the sensing element. The FBG (fiber Bragg grating) sensors provide remote measurements and dielectric strength when measuring objects potentially at high voltage, the ability to multiplex signals and measure strain at multiple points using a single line. Distributed sensors allow one to obtain the distribution of deformations along the fiber with a certain spatial resolution, using only one fiber. Nevertheless, the price of a FBG and a read-out is too high.

Table 1 provides a summary of the sensors reviewed.

Table 1. Various types of strain sensors.

Sensor type	Resolution, με	Strain measurement range, με	Advantages	Ref
Capacitive				
- elastomer-based	Up to ~ 1	Up to ~10^6	Large limit deformations, two-axis measurements, torsional strain measurement	[18, 13-15, 19]
- with liquid conductors	~ 10^4	Up to ~10^6	Large limit deformations, torsional and complex strain measurement	[30-31]
- with a thick dielectric	~ 0.1-1	Up to ~10^2	Larger GF	
Resistive (piezoresistive)				
- metal	~0.1	Up to ~10^4	Simple measurement and low price	[1-2]
- semiconductor	~10^{-3}÷10^{-2}	Up to ~10^3	Dynamic and static deformations	[1]
Piezoelectric	Up to 10^{-4}	Up to ~10^3	Direct conversion of strain to electrical voltage	[42, 2, 53-54]
MEMS				
- capacitive	1-10	Up to ~10^2		[36-52]
- piezoresistive	~ 0.1-1	~10^3		

- piezoelectric - resonance	~ 0.1-1 Up to 10^{-4}	10^2-10^3 Up to 10^2	Miniaturization, integration in a single smart chip sensor, low consumption	
Magnetic - magnetoelastic effects in amorphous and nanocristalline ribbons and wires - Barkhausen jumps based (microwires)	 ~ 10 ~ 1-10	 ~ 10^3 ~ 10^3	Non-contact measurements, strains in small spatial areas	[55-57]
FBG sensors	~ 0.1÷1	Up to 10000	The possibility of multiplexing sensors	[58-63]
Fiber-optic distributed sensors - Distributed sensor BTDR - Distributed sensor OFDR - Rayleigh (φ)OTDR	 ~ 1 ~ 1 ~0.001÷0.01	 - - -	Measurement of the strain field along the fiber with spatial resolution	[64-66]

5. Conclusion

In this paper, we reviewed capacitive strain sensors and provided a comparison with some other types. The power consumption of the the capacitive trancducers is much less than that of the resistive ones. The read-out for a typical resistive strain system is easier, but measurement of the capacitance changes leads to a slightly greater complexity of processing methods and circuit solutions in comparison with those for resistive and piezoelectric sensors. However, there is currently a wide range of relatively inexpensive specialized commercial IC's and solutions for capacitive sensors. These solutions as well as the sensors themselves are not so expensive as FBG and their optical read-outs. In comparison with the magnetic transducers the capacitance sensors are not sensitive to external interferences, especially low-frequency interferences.

The prospective areas of applications are wearable electronics, health care and high strain measurement (elastomer based sensors) and MEMS sensors. The elastomer based sensors provide a high sensitivity and large stretchability providing at the same time reliable electrical contact under applied strain and mechanical endurance under repetitive load. The existing commercial sensors show reproducibility in fabrication. The problems of such sensors are mechanical hysteresis, large relaxation time, narrow temperature range [20].

The MEMS sensors are the most promising to build a "smart" sensors and sensing IC's with high degree of integration. They provide required miniaturization of the sensor's size, extremely low power consumption etc. The MEMS sensors are capable of measuring moderate strains up to 10^3 με with a resolution of less than 1 με.

Acknowledgement

This research is implemented in Petrozavodsk State University with financial support by the Ministry of Science and Higher Education of Russia within Agreement no. 075-11-2019-088 of 20.12.2019 on the topic "Creating the high-tech production of mobile microprocessor computing modules based on SiP and PoP technology for smart data collection, mining, and interaction with surrounding sources".

References

[1] Kester W 1999 Section 4: Strain, Force, Pressure, and Flow Measurements *Practical Design Techniques for Sensor Signal Conditioning* (Analog Devices, USA)
[2] Du W Y 2014 *Resistive, capacitive, inductive, and magnetic sensor technologies* (CRC Press)
[3] Laflamme S, Ubertini F and Li J 2019 *Smart Sensors for Structural Health Monitoring* (MDPI)
[4] Miles A W and Kathleen E T 1992 *Strain measurement in biomechanics* (Chapman & Hall)
[5] Mukhopadhyay S C 2015 *Wearable electronics sensors: For safe and healthy living* (*Smart Sensors, Measurement and Instrumentation* **15**) (Springer)

[6] Chen B, Wan J, Shu L, Li P, Mukherjee M and Yin B 2017 Smart factory of industry 4.0: Key technologies, application case, and challenges *IEEE Access* **6** 6505-19
[7] Korzun D, Balandina E, Kashevnik A, Balandin S and Viola F 2019 *Ambient Intelligence Services in IoT Environments: Emerging Research and Opportunities: Emerging Research and Opportunities* (IGI Global)
[8] Hoffmann K 2012 *An introduction to stress analysis and transducer design using strain gauges* (HBM)
[9] OIML TC 9: Instruments for Measuring Mass and Density. 4 Committee Draft OIML/4CD Date: 19 November, 2015. Reference number: OIML R60-1 4CD. Metrological Regulation for Load Cells
[10] Usher M J and Keating D A 1996 Sensors and transducers: characteristics, applications, instrumentation, interfacing *Macmillan International Higher Education*
[11] Baxter L K 1997 Capacitive sensors. Design and Applications *IEEE Press, Piscataway*
[12] Zeiser R, Fellner T and Wilde J 2014 Capacitive strain gauges on flexible polymer substrates for wireless, intelligent systems *Journal of Sensors and Sensor Systems* **3** 77-86
[13] Atalay O, Atalay A, Gafford J, Wang H, Wood R and Walsh C 2017 A Highly Stretchable Capacitive Based Strain Sensor Based on Metal Deposition and Laser Rastering *Advanced Materials Technologies* **2** 1700081
[14] Kollosche M, Stoyanov H, Laflamme S and Kofod G 2011 Strongly enhanced sensitivity in elastic capacitive strain sensors *Journal of Materials Chemistry* **21** 8292-4
[15] Nur R, Matsuhisa N, Jiang Z, Nayeem M O G, Yokota T and Someya T 2018 A highly sensitive capacitive-type strain sensor using wrinkled ultrathin gold films *Nano letters* **18** 5610-7
[16] Lipomi D J, Vosgueritchian M, Tee B C-K, Hellstrom S L, Lee J A, Fox C H and Bao Z 2011 Skin-like pressure and strain sensors based on transparent elastic films of carbon nanotubes *Nature nanotechnology* **6(12)** 788-92
[17] Xu F and Zhu Y 2012 Highly conductive and stretchable silver nanowire conductors *Advanced materials* **24(37)** 5117-22
[18] *StretchSense Limited* Retrieved from:http://www.stretchsense.com
[19] Bend Labs Inc. Retrieved from: http://www.bendlabs.com
[20] Park J, You I, Shin S and Jeong U 2015 Material approaches to stretchable strain sensors *ChemPhysChem* **16(6)** 1155-63
[21] Cohen D J, Mitra D, Peterson K and Maharbiz M M 2012 A highly elastic, capacitive strain gauge based on percolating nanotube networks *Nano letters* **12(4)** 1821-25
[22] Cai L, *et al.* 2013 Super-stretchable, transparent carbon nanotube-based capacitive strain sensors for human motion detection *Scientific reports* **3(1)** 1-9
[23] Kang M, Kim J, Jang B, Chae Y, Kim J H and Ahn J H 2017 Graphene-based three-dimensional capacitive touch sensor for wearable electronics *ACS nano* **11(8)** 7950-7
[24] Filippidou M K, Tegou E, Tsouti V and Chatzandroulis S 2015 A flexible strain sensor made of graphene nanoplatelets/polydimethylsiloxane nanocomposite *Microelectronic Engineering* **142** 7-11
[25] Matsuzaki R, Keating T, Todoroki A and Hiraoka N 2008 Rubber-based strain sensor fabricated using photolithography for intelligent tires *Sensors and Actuators A: Physical* **148(1)** 1-9
[26] Matsuzaki R, Keating T, Todoroki A and Hiraoka N 2009 Rubber-based capacitive strain sensor fabricated using photolithography for intelligent tires *Transactions of the Japan Society of Mechanical Engineers, Part A* **75(750)** 235-42
[27] Ueno T, Mori K and Yoshida T 2003 *Capacitive strain sensor and method for using the same* U.S. Patent No. 6532824
[28] Fassler A and Majidi C 2013 Soft-matter capacitors and inductors for hyperelastic strain sensing and stretchable electronics *Smart Materials and Structures* **22(5)** 055023
[29] Tabatabai A, Fassler A, Usiak C and Majidi C 2013 Liquid-phase gallium–indium alloy electronics with microcontact printing *Langmuir* **29(20)** 6194-200

[30] Frutiger A, Muth J T, Vogt D M, Mengüç Y, Campo A, Valentine A D, Walsh C J and Lewis J A 2015 Capacitive soft strain sensors via multicore–shell fiber printing *Advanced Materials* **27** 2440-6

[31] Cooper C B, Arutselvan K, Liu Y, Armstrong D, Lin Y, Khan M R, Genzer J and Dickey M D 2017 Stretchable capacitive sensors of torsion, strain, and touch using double helix liquid metal fibers *Adv. Funct. Mater.* **27** 1605630

[32] Zhu S, So J H, Mays R, Desai S, Barnes W R, Pourdeyhimi B and Dickey M D 2013 Ultrastretchable fibers with metallic conductivity using a liquid metal alloy core *Adv. Funct. Mater.* **23** 2308-14

[33] Park Y L, Tepayotl-Ramirez D, Wood R J and Majidi C 2012 Influence of cross-sectional geometry on the sensitivity and hysteresis of liquid-phase electronic pressure sensors *Appl. Phys. Lett.* **101** 191904

[34] Arshak K I, McDonagh D and Durcan M A 2000 Development of new capacitive strain sensors based on thick film polymer and cermet technologies *Sensors and Actuators A: Physical* **79(2)** 102-14

[35] Arshak K I, Collins D and Ansari F 1994 New high gauge-factor thick-film transducer based on a capacitor configuration *International Journal of Electronics* **77(3)** 387-99

[36] Que L, Li M H, Chu L L and Gianchandani Y B 1999 A micromachined strain sensor with differential capacitive readout *Proc. 12th IEEE International Conference on Micro Electro Mechanical Systems* 552-7

[37] Aebersold J W, Walsh K, Crain M and Voor M 2009 *MEMS capacitive bending and axial strain sensor* U.S. Patent No. 7509870

[38] Aebersold J W, Walsh K, Crain M and Voor M 2010 *MEMS capacitive bending and axial strain sensor* U.S. Patent No. 7854174

[39] Suster M, Guo J, Chaimanonart N, Ko W H and Young D J 2006 A high-performance MEMS capacitive strain sensing system *Journal of Microelectromechanical Systems* **15(5)** 1069-77

[40] Ko W H, Young D J, Guo J, Suster M, Kuo H I and Chaimanonart N 2007 A high-performance MEMS capacitive strain sensing system *Sensors and Actuators A: Physical* **133(2)** 272-7

[41] Guo J, Kuo H I, Young D J, and Ko W H 2004 Buckled beam linear output capacitive strain sensor *Solid-State Sensor, Actuator and Microsystems Workshop* pp 344-7

[42] Kon S, Oldham K, Horowitz R 2007 Piezoresistive and piezoelectric MEMS strain sensors for vibration detection *Proc. SPIE* **6529** 65292V

[43] Cao L, Kim T S, Mantell S C and Polla D L 2000 Simulation and fabrication of piezoresistive membrane type MEMS strain sensors *Sensors and Actuators A: Physical* **80(3)** 273-9

[44] Mohammed A A, Moussa W A and Lou E 2008 High sensitivity MEMS strain sensor: design and simulation *Sensors* **8(4)** 2642-61

[45] Mohammed A A, Moussa W A and Lou E 2011 High-performance piezoresistive MEMS strain sensor with low thermal sensitivity *Sensors* **11(2)** 1819-46

[46] Yamashita T, Takamatsu S, Okada H, Itoh T and Kobayashi T 2016 Ultra-thin piezoelectric strain sensor array integrated on a flexible printed circuit involving transfer printing methods *IEEE Sensors Journal* **16(24)** 8840-6

[47] Wojciechowski K E, Boser B E and Pisano A P 2005 A MEMS resonant strain sensor with 33 nano-strain resolution in a 10 kHz bandwidth. *Proc. IEEE SENSORS* p 4

[48] Wojciechowski K E, Boser B E and Pisano A P 2004 A MEMS resonant strain sensor operated in air *Proc. 17th IEEE International Conference on Micro Electro Mechanical Systems* 841-5

[49] Azevedo R G, *et al.* 2007 Silicon carbide coated MEMS strain sensor for harsh environment applications. *Proc. 20th International Conference on Micro Electro Mechanical Systems* 643-6

[50] Guo Z, Li B, Gao Y, Cheng F and Cao L 2016 Theory and experimental research for the double ended tuning fork in MEMS *Sensor Review* **36** 217-24

[51] Shi X, Lu Y, Xie B, Xiang C, Wang J, Chen D and Chen J 2018 A Double-Ended Tuning Fork

Based Resonant Pressure Micro-Sensor Relying on Electrostatic Excitation and Piezoresistive Detection *Sensors* **18(8)** 2494
[52] Belsito L, Ferria M, Mancarella F, Masini L, Yan J, Seshia A A, Soga K and Roncaglia A 2016 Fabrication of high-resolution strain sensors based on wafer-level vacuum packaged MEMS resonators *Sensors and Actuators A: Physical* **239** 90-101
[53] *Kistler Instrument Corp* Retrieved from: https://www.kistler.com
[54] *Piezo.com* Retrieved from: https://piezo.com
[55] Praslicka D, Blazek J, Smelko M, Hudak J, Cverha A, Mikita I, Varga R and Zhuko A 2012 Possibilities of measuring stress and health monitoring in materials using contact-less sensor based on magnetic microwires *IEEE Transactions on Magnetics* **49(1)** 128-31
[56] Ignakhin V S, Severikov V S and Grishin A M 2019 Tensile and torsional strain gauge based on $Fe_{48}Co_{32}P_{14}B_6$ metallic glass *Journal of Magnetism and Magnetic Materials* **376** 382-6
[57] Sabol R, Rovnak M, Bajzecerova V, Vojtanik P and Varga R 2015 Application of magnetic microwires for sensing stresses in structures *Journal of Electrical Engineering* **66(7)** 164-7
[58] Campanella C E, Cuccovillo A, Campanella C, Yurt A and Passaro V 2018 Fibre Bragg grating based strain sensors: review of technology and applications *Sensors* **18(9)** 3115
[59] Shivendra and Garima S 2015 Strain sensors for strain measurement: A Review *Int. Journal of Electrical & Electronics Eng.* **2** 144-6
[60] Yuan W, Stefani A and Bang O 2011 Tunable polymer fiber Bragg grating (FBG) inscription: fabrication of dual-FBG temperature compensated polymer optical fiber strain sensors *IEEE Photonics Technology Letters* **24(5)** 401-3
[61] *Hottinger Baldwin Messtechnik GmbH* Retrieved from: www.hbm.com
[62] *Smartec* Retrieved from: http://www.smartec.ch/
[63] *Inversiya-Sensor* Retrieved from: http://i-sensor.ru/
[64] Barrias A, Casas J R and Villalba S 2016 A review of distributed optical fiber sensors for civil engineering applications *Sensors* **16(5)** 748
[65] Li J, Gan J, Zhang Z, Heng X, Yang C, Qian Q, Xu S and Yang Z 2017 High spatial resolution distributed fiber strain sensor based on phase-OFDR *Optics Express* **25(22)** 27913-22
[66] Costa L, Martins H F, Martín-López S, Fernández-Ruiz M R and González-Herráez M 2019 Fully Distributed Optical Fiber Strain Sensor With 10− 12 $\epsilon/\sqrt{}$ Hz Sensitivity *Journal of Lightwave Technology* **37(18)** 4487-95

Adaptive microwave heating control for oily waste processing

R R Kadyrov

Ufa State Petroleum Technological University, Branch in the Sterlitamak,
Pr. Oktyabrya, 2, Sterlitamak, 453118, Bashkortostan, Russia

E-mail: r_kadyrov@mail.ru

Abstract. Processing of petrochemical industry wastes involves the development of specific technologies and devices that require a significant investment of resources and time. To solve these problems, it is possible to develop universal technological, constructive, and managerial solutions capable of ensuring the adaptation of equipment to different raw materials, mode of operation, and yield of target products. The proposed reactor with a microwave effect on the catalyst and oil-containing raw materials with adaptive power control of the magnetron. Adaptive management is necessary because of the instability of the physicochemical characteristics of the raw materials. Adaptive control system takes into account the real state of the system and provides the necessary control. The advantage of the proposed method is that the entire algorithm is implemented in software, that is, for the introduction of this method of adaptive control does not require a change in the design of technological equipment. The implementation of this method comes down to the development of a special adaptive control driver. This approach greatly simplifies the adjustment of adaptive control systems for the considered processes of catalytic cracking, hydrocracking, visbreaking, coking.

1. Introduction

Modern solutions involve the development of specific technologies, apparatus designs for recycling a wastes of petrochemical production industries that require significant resources and time, which, in the context of market demand for products and changes priorities, requires more cost-effective solutions. Important role here plays an ensuring a quick changeover of equipment with preservation of high productivity and ecological cleanliness of production.

One of the acceptable approaches would be the development of new, more versatile technological, constructive, managerial solutions capable of ensuring the adaptation of equipment to different raw materials, mode of operation, yield of target products.

The proposed technology is based on the electrophysical method of influencing the technological environment. Unlike traditional heating of the reaction mass during catalytic cracking, hydrocracking, coking, visbreaking, the use of microwave energy has significant advantages:

- the absence of the traditional coolant provides low-inertia regulation, the heating rate is determined only by the speed of propagation of an electromagnetic wave in the medium, the dielectric properties of the medium and the radiation power;
- provides volume heating at once all material.

Features of exposure using microwave radiation allow us to offer a batch reactor for conducting the above-mentioned processes of universal construction, since the smoothness of controlling the radiation power of a magnetron from small values to large allows consistently heating the reaction mass and gradually recycle multi-component raw materials, selectively obtaining the target products (see figure 1) [1].

The operation of a batch reactor occurs as follows:

- the oil-containing raw material is loaded into the reactor after opening the cover interlocked with the magnetron switch;
- the feed oil-containing raw to the reactor continues until it reaches a required level;
- the required amount of catalyst is loaded into the reactor; after the loading process is over, the top cover closes and the magnetron is turned on for a certain power;
- microwave radiation begins to affect the solid - a catalyst that heats up and transfers heat to the raw materials;
- by the desired temperature is reached, the presence of a catalyst begins to significantly accelerate the reaction, it leads to the formation of light hydrocarbons, which are removed from the reactor;
- discretely increasing the power of microwave radiation, it is possible to consistently achieve the involvement in the reaction of the maximum amount of oil-containing materials present, and the resulting light hydrocarbons are selectively removed from the reactor by product type based on the analysis of the composition of the medium with a gas analyzer.

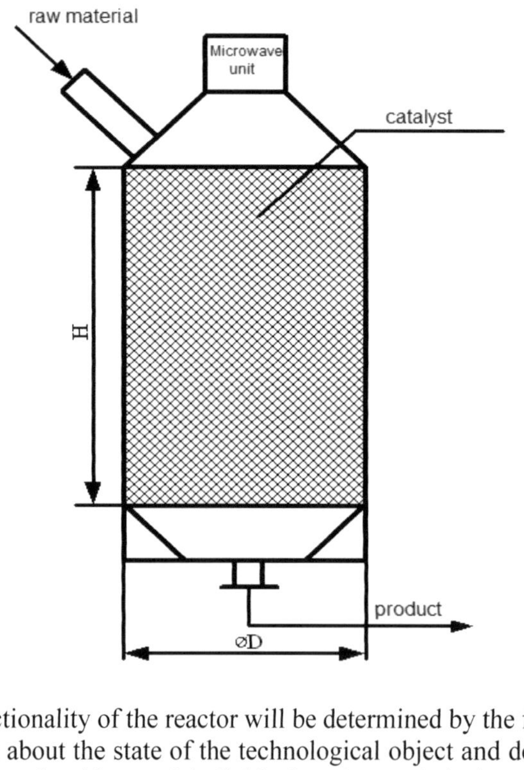

Figure 1. Diagram of the electrodynamic microwave reactor.

The functionality of the reactor will be determined by the features of the control system that collects information about the state of the technological object and determines the operating modes, as well as the behavior algorithms of individual nodes and devices. The main difficulty in organizing the management of the above processes is the instability of the physicochemical characteristics of the raw materials, which makes it difficult both to build a mathematical model and to define empirical dependencies. The way out of this situation would be to use an adaptive control system that takes into account the real state of the system and provides the necessary control actions.

2. Adaptive control block-diagram by process parameter errors

A common drawback of many adaptive systems is the need for a special sensor that needs to be embedded in a block-diagram (see figure 2) [2-8].

In figure 2 the following conventions are used:

- R - regulator;
- CO - control object;
- OS - feedback sensor;
- ACU - Adaptive Control Unit;
- ASFS - Adaptation Signal Feedback Sensor.

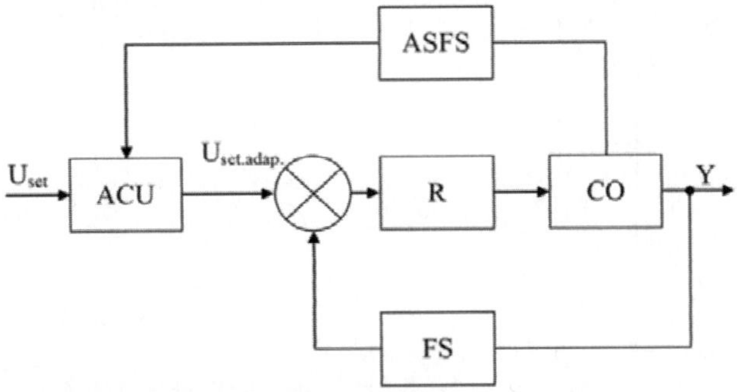

Figure 2. Traditional adaptive control scheme.

3. The logical scheme of adaptive control

The proposed software implementation of the adaptive errors control will allow to organize adaptive control without an additional feedback sensor of the adaptation signal. This is ensured by the fact that the error value in the tracking system is selected as the adaptive control signal for a certain technological parameter S. To do this, the permissible error value ε is set, with which the current error value of the technological parameter is compared. If the current value of $\varepsilon > \varepsilon_{set}$, then the value specified in the program S will decrease by a given discrete Δ. If $\varepsilon < \varepsilon_{set}$, then on the contrary a given discrete Δ is added to S. Thanks to two indicated contradictory trends, the adaptive control system maintains the current value of the technological parameter at the S level specified [9].

The logical scheme of this algorithm is presented in figure 3.

The advantage of this method is that the entire algorithm is implemented in software, that is, for the introduction of this method of adaptive control does not require a change in the design of the process equipment. The implementation of this method is reduced to the development of a special adaptive control driver, which is included in the tasks of the microprocessor and processed in each period interrupt of the timer.

The adaptive error control algorithm can be represented as a unified module of the adaptation unit. The scheme of such a module has the form shown in figure 4.

The following signals are input to the adaptation unit: the specified value of the controlled variable φ_s, the specified error value ε_s, the change discrete of the controlled variable Δ, the current error value ε.

The adaptation signal φ_a is removed from the output of the adaptation unit, then this signal is compared with the actual value of the controlled variable, the difference between the signals $\varphi_a - \varphi_{real} = \varepsilon$ is fed to the input of the regulator.

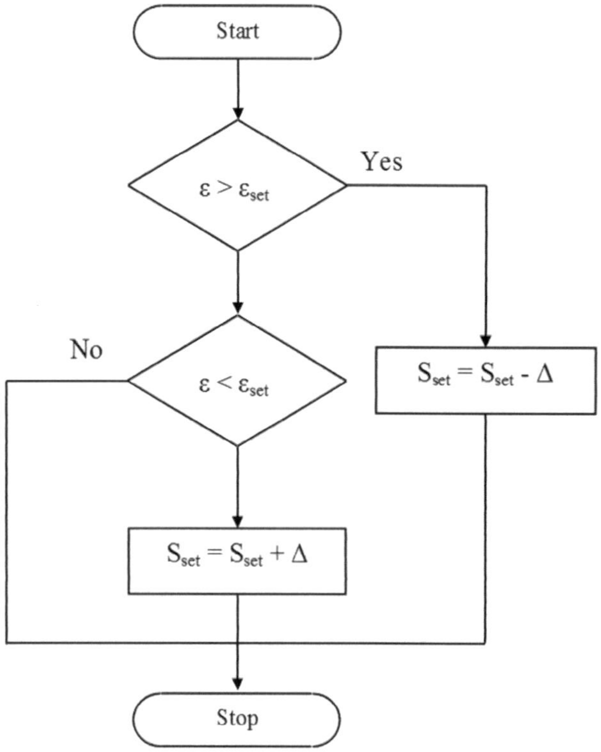

Figure 3. The logical scheme of adaptive control.

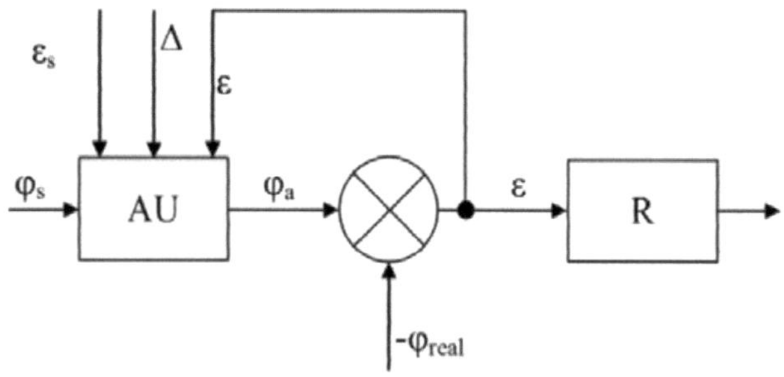

Figure 4. Diagram of the adaptation module.

4. The adaptive control method
The adaptive control method is as follows (see figure 5).

In figure 5 the following conventions are used:

- T_S - temperature sensor;
- V_S - voltage sensor;
- T_R - temperature regulator;
- V_R - voltage regulator;
- ε_T - the temperature error signal;
- ε_H - the voltage error signal;
- T_s - set temperature;
- ΔT – the discrete change in temperature;
- ε_{sT} - the set value of the temperature error;
- T_{ad} - adapted temperature value;
- ΔU - discrete voltage change;

- ε_{sU} - the set value of the voltage error;
- TAU - temperature adaptation unit;
- VAU - voltage adaptation unit.

When the temperature of the catalyst reaches the so-called zero point, the microcontroller (IPC) controls the magnitude of the actual error in the tracking loop at controlling the heating of the raw material. If this error exceeds a predetermined value, then there is a slow change (depending on the error sign) of the microwave field power before the signal "Starting point" arrival from the feedback sensor by temperature in the IPC, after which an instant termination of change of the microwave field power. When the temperature of the catalyst reaches the value required for the reaction (zero point), the actual error in the tracking system is less than the specified value of the constant, which is set programmatically. If this excess becomes greater than the value of the above-mentioned constant, the IPC records the error value and generates a command to change the power of the microwave field before the IPC receives the "Starting Point" signal, and reaches the zero point [10].

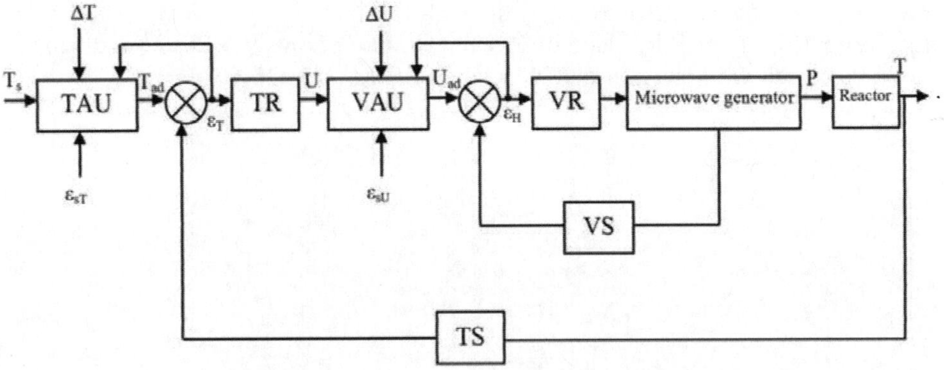

Figure 5. Adaptive control block-diagram.

Block-diagram for the considered processes of catalytic cracking, hydrocracking, visbreaking, coking, differing by the fact that the values a different technological parameter, other of the error signal and change discrete is fed to unified module of adaptation unit.

5. Summary

A method has been developed for calculating the tuning parameters of adaptive control systems by the magnitude of temperature error of catalytic cracking, hydrocracking, coking, and visbreaking processes.

The use of an adaptive system for the error allows for the stabilization of technological parameters on the basis of the existing element base, based on a software algorithm without complicating the control circuit of additional sensors.

The application of the concept of software adaptive control parameter by the error in the electrodynamic microwave installation will allow you to maintain the deviation of the heating temperature of an oily substance within the specified limits. With this control, the optimum power consumption is carried out.

Thus, it can be noted that the proposed solutions for managing the process of heating oil-containing substance can significantly reduce energy costs and increase the reliability of high-intensity technology in an electrodynamic furnace.

References

[1] Bikbulatov I Kh, Kadyrov R R, Charikov P N and Bukharov V R 2011 Technology for producing benzine from residues of methyl-tret-butyl ether production using microwave radiation and in a special production building *Bash. Chem. J.* **18(2)** 169-72

[2] Aseltine J A, Mancini A R and Sarture C W 1958 A survey of adaptive control systems *IRE Trans. on Automatic Control* **AC-6(12)** 102-8
[3] Bastin G and Gevers M 1988 Stable adaptive observers for nonlinear timevarying systems *IEEE Trans. on Automatic Control* **33(7)** 650-8
[4] Bellman R and Kalaba R 1960 Dynamic programming and adaptive control processes: Mathematical foundations *IRE Trans. on Automatic Control* **AC-5** 5-10
[5] Bastin G, Bitmead R., Campion G and Gevers M 1992 Identification of linearly overparametrized nonlinear systems *IEEE Trans. on Automatic Control* **37(7)** 1073-8
[6] Krstic M, Kanellakopoulos I and Kokotovic P 1995 Nonlinear and Adaptive Control Design *Wiley and Sons Inc.*
[7] Krstic M and Wang H 2000 Stability of extremum seeking feedback for general nonlinear dynamic systems *Automatica* **36** 595-601
[8] Morse A S, Mayne D Q and Goodwin G C 1988 Applications of hysteresis switching in parameter adaptive control *IEEE Trans. on Automatic Control* **37(9)** 1343-54
[9] Kadyrov R R and Charikov P N 2014 High technologies in engineering industry *Proc. of All-Russia research and practice Conf.* (Ufa: USATU) p 25-6
[10] Grigoryev E S, Kadyrov R R, Charikov P N and Pryanichnikova V V 2017 Simulation of Honing of a Processed Workpiece on CNC Machine *K. E. M.* **743** 236-40

Mutual influence of the supports stiffness and the first natural frequency at bending vibrations of a spring-hinged beam

A E Mityaev[1], I V Kudryavtsev[1], M P Khomutov[2], M V Brungardt[1] and O I Rabetskaya[1]

[1] Siberian Federal University, 79 Svobodny Pr., Krasnoyarsk, 660041, Russia
[2] Kuban State Technological University, 2 Moskovskaya st., Kuban, 350072, Russia

E-mail: amityayev@sfu-kras.ru

Abstract. This paper discusses the analytical finding of the required stiffness values of the supports, which would provide the given first natural frequency of bending vibrations for a spring-hinged beam. The solution is obtained through the support coefficient, the value of which, according to the theory of beam vibrations, connects the stiffness of the support of the beam and the value of its first vibration frequency. For this purpose, preliminary approximation of dependence of support coefficient values on a support stiffness is performed using quadratic analytical functions. In this case, the problem of determining the stiffness of the supports becomes inverse and is reduced to finding combinations of the supports stiffness values, at which there is a solution of the quadratic analytical functions for the required support coefficient and, accordingly, the required first beam frequency. Normalization of the obtained support coefficients is proposed for a more convenient estimation of their influence on the first frequency of beam vibration.

1. Introduction

In the theory of vibrations, free bending vibrations of a simple hinged beam are well studied [1-21]. However, in practice, a purely hinged joint is unattainable and elastic forces arise in the hinges, which we take into account in the form of spring stiffness values k_1 and k_2 . These support stiffness values will affect the dynamic behavior of the beam, including its first natural bending frequency (figure 1).

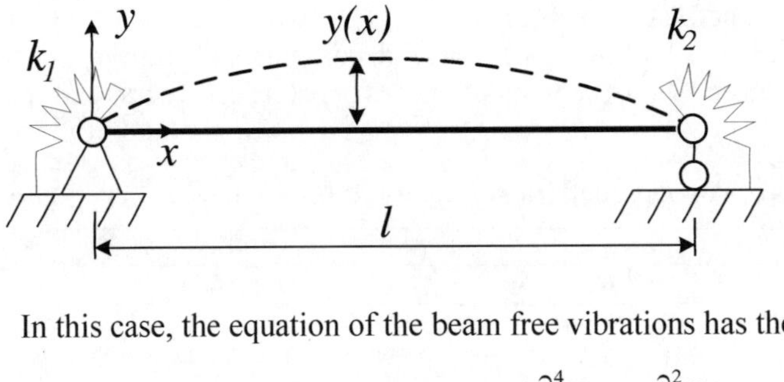

Figure 1. Free vibrations of a spring-hinged beam.

In this case, the equation of the beam free vibrations has the form [1-10]:

$$EJ_{\min}\frac{\partial^4 y}{\partial x^4}+m\frac{\partial^2 y}{\partial t^2}=0\,. \tag{1}$$

where $y=y(x)$ is the beam deflection; E is the elastic modulus; J_{min} is the minimum moment of inertia of the beam cross-section; m is the mass per beam length.

The solution for the first frequency of the beam free vibration according to [1-10] is defined as:

$$f_1 = \frac{\alpha^2}{2\pi l^2}\sqrt{\frac{EJ_{\min}}{m}}. \tag{2}$$

where α is the natural frequency dimensionless parameter, which depends on the boundary conditions; l is the length of the beam.

Reference literature [1-21] gives the values of support coefficients for common simple types of supports: hinge, fixed support, and its combinations. If the beam is supported in hinges with some given stiffness (figure 1), the support coefficient will be their function:

$$\alpha = \alpha(k_1, k_2). \tag{3}$$

Express the stiffness of the supports through the stiffness of the beam and get the relative stiffness:

$$C_1 = k_1 \frac{l}{EJ_{\min}};\quad C_2 = k_2 \frac{l}{EJ_{\min}}. \tag{4}$$

Then the dependence of the support coefficient will be a function of their relative stiffness:

$$\alpha = \alpha(C_1, C_2). \tag{5}$$

The function (5) is very nonlinear and there are no exact analytical expressions for it, so the determination of the support coefficient in these cases is carried out by any numerical method. Tables with different relative stiffness values are given in known reference books, from which the desired support factor can be found. However, there are several problems. Firstly, these tables contain a very limited set of possible combinations of relative stiffness of supports, which makes their use difficult. Further, if intermediate values of relative stiffness are used, then high nonlinearity of dependence (5) leads to large errors when trying to linear interpolate table values. Finally, these tables sometimes contain erroneous values that are not always obvious to inexperienced designers.

The purpose of this work is the inverse problem for function (5), namely in determining the stiffness of supports C_i, which will provide the required value of the support coefficient α. This corresponds to the function:

$$f_1(C_1, C_2) = \alpha(C_1, C_2)^2 \cdot \frac{1}{2\pi l^2}\sqrt{\frac{EJ_{\min}}{m}}. \tag{6}$$

Solution is performed by analytical approach using quadratic functions.

2. Quadratic approximation of support coefficients values

A numerical solution [22,23] of equation (5) was made for more than 500 combinations of beam support stiffness values. Table 1 shows a reduced set of combinations of relative stiffness and their corresponding support factor values α.

Table 1. Some values of support coefficient function α (C_1, C_2).

C_2	C_1										
	0	0.01	0.1	1	5	10	50	100	500	1000	∞
0	3.142	3.143	3.157	3.273	3.534	3.665	3.855	3.889	3.919	3.923	3.926
0.01		3.145	3.159	3.275	3.534	3.666	3.856	3.890	3.920	3.924	3.927
0.1			3.173	3.288	3.548	3.678	3.868	3.902	3.932	3.936	3.940
1				3.399	3.652	3.781	3.970	4.004	4.034	4.038	4.041
5					3.897	4.026	4.219	4.254	4.241	4.245	4.249

10						4.156	4.354	4.390	4.422	4.426	4.430
50			Symmetrical				4.563	4.602	4.636	4.640	4.645
100								4.641	4.676	4.681	4.685
500									4.711	4.716	4.720
1000										4.721	4.725
∞											4.730

Inside each range, we approximate the obtained numerical values of the support coefficients α with quadratic functions using the least-squares method [23-25]. As a result, we obtain a solution for each zone in the form:

$$\alpha_i\left(C_1,C_2\right)=a_{i0}+a_{i1}\cdot C_1+b_{i1}\cdot C_2+a_{i2}\cdot C_1^2+b_{i2}\cdot C_2^2. \quad (7)$$

where i is the zone number.

Using the least-squares method, we define the coefficients of approximating quadratic functions for each zone and summarize the results in the table 2.

Table 2. Coefficients of approximating quadratic functions $\alpha(C_1, C_2)$.

Zone	I	II		III		
Coefficient	C_1, C_2=0…10	C_1=0...10, C_2=10…100	C_1, C_2=10…100	C_1=0...10, C_2=100…1000	C_1=10...100, C_2=100…1000	C_1, C_2= 100…1000
a_0	3.1415	3.579	4.03	3.882	4.332	4.62
a_1	0.1008	0.1008	0.0075	0.1008	0.0075	0.000125
a_2	-0.005	-0.005	-0.0000435	-0.005	-0.0000435	-0.000000074
b_1	0.1008	0.0075	0.0075	0.000125	0.000125	0.000125
b_2	-0.005	-0.0000435	-0.0000435	-0.000000074	-0.000000074	-0.000000074

The obtained quadratic approximating functions (7) allow calculating the values of the support coefficient for equation (2) with an error of not more than 3.3% in the entire considered range of support stiffness. This is sufficient for the theory of beam vibrations, since the assumptions used in it lead to even greater deviations in the solution.

3. Squares of support coefficient values

In the equation of the frequency of free oscillations (6) of the hinge beam, the factor of supports is squared, so it makes sense to approximate the squares of their values. This will eliminate the accumulation of errors as a result of exponential operations. Using the analytical solution, we get the squares of the support coefficients and summarize them in the table 3.

Table 3. Square values of support coefficients $\alpha^2(C_1, C_2)$.

C_2	C_1										
	0	0.01	0.1	1	5	10	50	100	500	1000	∞
0	9.870	9.880	9.968	10.71	12.49	13.43	14.86	15.13	15.36	15.39	15.41
0.01		9.890	9.977	10.72	12.49	13.44	14.87	15.14	15.37	15.40	15.43
0.1			10.07	10.81	12.59	13.53	14.96	15.23	15.46	15.49	15.52
1				11.55	13.34	14.29	15.76	16.03	16.27	16.30	16.33
5					15.19	16.21	17.80	18.10	18.36	18.39	18.43
10						17.27	18.95	19.27	19.55	19.59	19.62
50			Symmetrical				20.82	21.18	21.49	21.53	21.57
100								21.54	21.86	21.91	21.95
500									22.20	22.24	22.28
1000										22.28	22.32
∞											22.36

The obtained values of the squares of the support coefficients are approximated using quadratic functions in the form (7). The coefficients of these functions are also determined by the least-squares method for each of the three stiffness zones and their values are shown in the table 4.

Table 4. Coefficients of approximating quadratic functions $\alpha^2(C_1, C_2)$.

Coef-ficient	Zone					
	I	II		III		
	C_1, C_2=0…10	C_1=0...10, C_2=10…100	C_1, C_2=10…100	C_1=0...10, C_2=100…1000	C_1=10...100, C_2=100…1000	C_1, C_2= 100…1000
a_0	9.87	12.972	16.6	15.243	18.872	21.3
a_1	0.736	0.736	0.0638	0.736	0.0638	0.00115
a_2	-0.0366	-0.0366	-0.0004	-0.0366	-0.0004	-0.00000065
b_1	0.736	0.0638	0.0638	0.00115	0.00115	0.00115
b_2	-0.0366	-0.0004	-0.0004	-0.00000065	-0.00000065	-0.00000065

The obtained approximating functions allow calculating the squares of the support factor α^2 for equation (2) with an error of not more than 3.5% in the entire considered range of support stiffness.

4. Determination of required supports stiffness

The use of quadratic approximating functions together with the division of the stiffness of supports into zones made it possible to analytically describe a complex nonlinear dependence (5) with sufficient accuracy for engineering calculations. At the same time, quadratic approximating functions allow solving the inverse problem of determining the stiffness of supports that provide the required support coefficients α or its quadratic value α^2.

The analytical solution of square equation (7) is complicated by the presence of two unknowns: C_1 and C_2. For certainty, link these unknowns through a coefficient n:

$$C = C_1 = \frac{C_2}{n}. \tag{8}$$

Then equation (7) is reduced to solving the square equation at a given value of α', which, depending on the problem to be solved, can be the factor of supports α or its square value α^2:

$$\left(a_{i2} + b_{i2} \cdot n^2\right) C^2 + \left(a_{i1} + b_{i1} \cdot n\right) C + \left(a_{i0} - \alpha'\right) = 0. \tag{9}$$

The root of this equation, taking into account the signs, is the expression:

$$C = \frac{-\left(a_{i1} + b_{i1} \cdot n\right) - \sqrt{\left(a_{i1} + b_{i1} \cdot n\right)^2 - 4\left(a_{i2} + b_{i2} \cdot n^2\right)\left(a_{i0} - \alpha'\right)}}{2 \cdot \left(a_{i2} + b_{i2} \cdot n^2\right)}. \tag{10}$$

The solution (10) is valid for the problem of determining the supports stiffness that provides the required factor of supports α or its quadratic value α^2.

Depending on the problem to be solved, the coefficients from table 2 or table 4 are used, respectively. After determination of stiffness value C, stiffness of supports according to equation (8) is defined as

$$k_1 = C \cdot \frac{EJ_{\min}}{l}, \quad k_2 = C \cdot \frac{EJ_{\min}}{l} n. \tag{11}$$

Stiffness of supports (11) will provide the specified value of support coefficient and, therefore, the first frequency of natural vibrations of the beam.

5. Discussion

Approximating functions (7) with calculated coefficients allow obtaining support coefficients and their squares in the almost full range of stiffness. The error in determining the values is not more than 3.5% compared to the exact values [21, 22], which is quite permissible for beam structures. However, when solving the inverse problem of equation (10), the error of determining the required stiffness of the supports can be significantly higher due to the high nonlinearity of the dependence (5) and the accumulation of errors from the rounding of irrational numbers. Therefore, after determining the required stiffness of the supports, it is recommended to perform a check calculation of the value of the first natural frequency of the beam bending vibrations.

The proposed solution for a single-span hinge beam is also valid for a multi-span hinge beam, provided that all spans and stiffness of all spring-hinged supports are equal.

6. Conclusion

In this paper, the values of support coefficients for the spring-hinged beam were calculated at different combinations of the supports relative stiffness. According to the obtained data, the least-squares method was approximated with quadratic functions over three zones. The obtained analytical functions make it possible to determine the beam support coefficient at any combination of support stiffness with an error of not more than 3.5%. These analytical quadratic approximations also allow solving the inverse problem of finding the necessary stiffness of supports at the known required value of the support coefficient.

Thus, the obtained results make it possible to analytically determine the first frequency of natural vibration of a single-span beam at any combination of supports stiffness values and solve the inverse problem of finding the required supports stiffness values, which will provide the beam with the required frequency of vibration.

Acknowledgements

The research was funded by RFBR, Krasnoyarsk Territory and Krasnoyarsk Regional Fund of Science, project number 20-48-242922.

References

[1] Lin Y K 1962 *Int Journal of Mechanical Sciences* **4** 409-23
[2] Balachandran B 2009 *Vibrations* (Cengage Learning, Toronto, Canada)
[3] Benaroya H, Nagurka M and Han S 2017 *Mechanical vibration* (CRC Press, Taylor & Francis Group)
[4] Bottega W J 2006 *Engineering vibrations* (CRC Press, Taylor & Francis Group)
[5] Clough R E 1995 *Dynamics of Structures* (McGraw-Hill College, USA)
[6] Geradin M and Rixen D J 2015 *Mechanical vibrations* (John Wiley & Sons Ltd, UK)
[7] Hartog J P 1985 *Mechanical vibrations* (McGraw-Hill, New York)
[8] Hagedorn P 2007 *Vibrations and waves in continuous mechanical systems* (John Wiley & Sons Inc., NJ, USA)
[9] Inman D J 2014 *Engineering vibration* (Pearson Education, NJ, USA)
[10] Kelly S G 2007 *Advanced vibration analysis* (CRC Press, Taylor & Francis Group, LLC, USA)
[11] Jazar R N 2013 *Advanced vibrations. A modern approach* (Springer, NY, USA)
[12] Kelly S G 2012 *Mechanical vibrations. Theory and applications* (Cengage Learning, USA)
[13] Leissa A W 2011 *Vibration of continuous systems* (McGraw-Hill, New York)
[14] Meirovitch L 1967 *Analytical methods in vibrations* (The Macmillan company, New York, USA)
[15] Meirovitch L 2001 *Fundamentals of vibrations* (McGraw-Hill,Book Co, New York, USA)
[16] Rades M 2010 *Mechanical vibrations II. Structural dynamic modeling* (Printech Publisher, Turin)
[17] Rao S 2018 *Mechanical vibrations* (Pearson Education Limited, UK)
[18] Shabana A S 2019 *Theory of vibration* (Springer-Verlag New York, USA)
[19] Blekhman I I 2004 *Selected topics in vibrational mechanics* (World Scientific Publishing Co Pte

Ltd, Singapore)
[20] Magnus K 1965 *Vibrations* (Blackie & Son, UK)
[21] Blevins R D 2016 *Formulas for dynamics, acoustics and vibration* (John Wiley & Sons, Ltd)
[22] Hibbeler R C 1975 J.Appl Mech **42** 501-502
[23] Hamming R W 1987 *Numerical Methods for Scientists and Engineers* (Dover Publications)
[24] Chapra S and Canale R 2014 *Numerical Methods for Engineers* (McGraw-Hill Education
[25] Burden R L 2015 *Numerical Analysis* (Cengage Learning)

Modern web application development technologies

A I Dzhangarov[1,3], K K Pakhaev[1] and N V Potapova[2]

[1] Faculty of information technology, Chechen State University, 32 Sheripov Street, Grozny, 364024, Russia
[2] Department of Information Educational Technologies, Kuban State University, Stavropol`skaya street 149, Krasnodar, 350001, Russia

[3] E-mail: dzhangarov1995@gmail.com

Abstract. One of the key areas of modern information technology is the area of programming. Programming is the creation of various programs. If earlier it was, for the most part, about writing programs for the interaction of a user and a computer, today the applied field of programming is no longer limited to anything. These are programs for portable gadgets, medical equipment, space nanotechnology, and so on. Within the framework of this research, the work will focus on web development and specifically - on the development of a web application. This direction is quite popular, since the Internet segment has played a key role in human life. Whether it's a government-owned company, a private enterprise, or a personal blog, having a website or application is a kind of business card. A web application is a client-server interaction model using browser software. In terms of their structure and logic of work, web applications often resemble the most common desktop solutions, but with one condition - such applications are always launched and executed in the browser.

1. Introduction

If we briefly consider modern trends in web development, we can highlight several areas at once:

1.1. Progressive Web Applications

The so-called Progressive Web Application architecture sets itself the task of integrating a full-fledged adaptive user interface for both web applications launched in a browser and web applications launched on mobile devices.

1.2. Implementation of artificial intelligence tools and voice assistants

The field of DataScience is densely bursting into all areas of human life. And, of course, such technologies need to be supported by applications that claim to be a serious and wide audience and strive to keep up with the times. Therefore, when developing web applications, developers try to include these options as well.

1.3. Serverless technologies

Perhaps the most important trend in modern web development is the use of cloud solutions as servers. In fact, this can be the same appeal to classic servers, but using cloud providers. The most important benefit is that the necessary server calculations are carried out directly in the cloud, and the ready generated request is already sent to the server [1]. Such an implementation can significantly speed up the processing requests and, accordingly, receiving responses.

Summing up the review of the key areas of 2021 in web development, it is significant to notice the importance of choosing the right development tools. First of all, it is a popular programming language that supports web development and has a variety of multifunctional frameworks [2].

2. Selection of development tools

Today there are many different ways of web development. In addition to choosing a specific programming language, IDE, there are already entire platforms that allow you to build an application from scratch without using programming tools and principles.

And at the same time, the decision to use such generators, without having practical programming skills, web technologies, can be considered reckless, since if any problems or failures occur, it will be almost impossible to resolve them on your own or to improve various components of the application yourself.

As a programming language, it was decided to choose Java. Despite the rather controversial public opinion regarding web development in the Java language, it has a very wide range of different tools and frameworks that set it apart from its competitors. The source of this opinion is that the development of web-applications on it is a more time-consuming and complex process than existing analogues.

A large IT company TIOBE has carried out a large-scale study to identify the most popular programming languages. The survey involved both professional developers from all over the world, large corporations, and novice programmers and startups [3]. This rating is updated every month and for April 2021 the list of the most popular languages is as follows:

Apr 2021	Apr 2020	Change	Programming Language
1	2	▲	JavaScript
2	1	▼	Java
3	3		Python
4	4		C++
5	5		C#

Figure 1. Rating of programming languages.

Development environment - IntelliJ IDEA Ultimate Edition. JetBrains software is the leader among the Java programming language IDEs due to its very user-friendly, intuitive interface and support for many different libraries and frameworks.

IntelliJ IDEA comes in two flavors: the free Community version and the paid Ultimate Edition. The essential difference between them is that Ultimate Edition (built on top of Community) is a full-fledged commercial IDE and, in addition to many additional functions, allows for corporate and web development. What, in fact, is required to achieve this goal.

3. The structure and properties of the developed application

As part of this research work, to demonstrate the capabilities of the selected tools, it was decided to develop a reference application that would have the following functions and properties:

- having your own dedicated server;
- data storage using a MySQL database;
- intuitive and user-friendly user interface.

Apache Tomcat was used as a dedicated server. Tomcat is an open source servlet container developed by the Apache Software Foundation. It implements the Servlet Specification, the JavaServer Pages (JSP) Specification, and the JavaServer Faces (JSF) Specification [4]. Written in Java language. Tomcat allows one to run web applications and contains a number of self-configuring programs. Tomcat is also used as a standalone web server, as a content server in conjunction with the Apache HTTP Server web server.

It is worth highlighting that a servlet or Java Servlet API is a standardized API designed to be implemented on the server and work with the client according to a request-response scheme.

A servlet is a class that can receive requests from a client and return responses to it. Thus, Java servlets are the building blocks of a client-server architecture.

As already mentioned, web development is divided into two main components: the client side and the server side. For client-side development, the Ext JS, Spring and Apache Maven framework files were additionally connected and the app.js file was created:

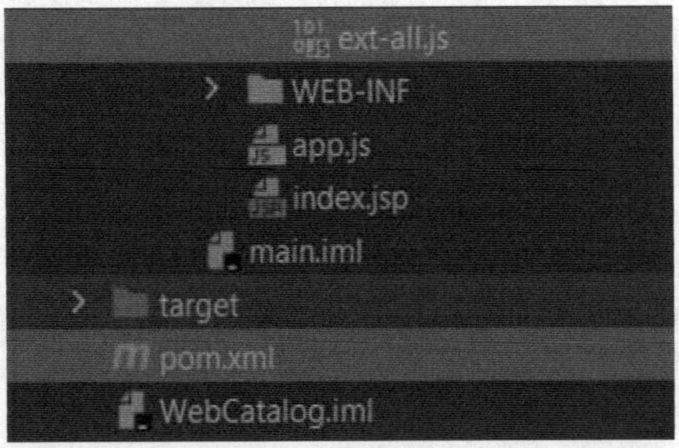

Figure 2. Connecting additional development tools.

The necessary Ext JS styles were included in the index.jsp file.The next step was to develop four views of the elements (which, after creation, must also be initialized in the app.js file) - this is the SearchWebView.js search view, the WebGridView.js table view, the AddWebFormView.js data-adding form view, and the WebCatalogView.js framework view, in which all views of the form are placed:

Figure 3. View elements.

Further, using the MVC programming pattern, the Controller Model and Store directories were additionally created. Controller - designed to initialize component handlers such as buttons, clickable fields, and so on. The model and storage in this case are required to process requests from the client side. And all of these files are registered in app.js as well [5]:

```
Ext.application( a: {
    name: 'WebCatalog',
    views: [
        'AddWebFormView',
        'WebCatalogView',
        'WebGridView',
        'SearchWebView'
    ],
    controllers : [
        'WebCatalogController'
    ],
    stores : [
        'WebCatalogStore'
    ],
    launch: function () {
        Ext.create('Ext.container.Viewport', {
            layout: 'fit',
            items: {
                xtype : 'WebCatalogView'
            }
        });
    }
});
```

Figure 4. Registering data in the app.js file.

Server side development. In the case of the development of the server side, the scheme for dividing the control logic into separate components was also used - controllers, models, service:

```
java
  > controllers
  > model
  > service
```

Figure 5. Control logic.

Of course, first of all, the main application model was developed, in which such boolean fields as id, name, position were initialized:

```
public class Web implements Serializable {
    @Id
    @Column(name = "ID")
    @GeneratedValue(strategy = GenerationType.IDENTITY)
    private Long id;
    @Column(name = "name")
    private String name;
    @Column(name = "position")
    private String position;
    public Web() {
    }
    public Long getId() { return id; }
    public void setId(Long id) { this.id = id; }
    public String getName() { return name; }
    public void setName(String name) { this.name = name; }
    public String getPosition() { return position; }
    public void setPosition(String price) { this.position = position; }
}
```

Figure 6. Example model.

After creating the models, controllers were developed, the main task of which is to process requests on the server side using the GET, POST, PUT, DELETE methods. As far as service is concerned, the internal file checks for duplicates and has built-in methods for handling such situations (deleting duplicate records, updating the list) [6].

After the management logic settings have been made, there are two steps left - connecting the database using Spring tools (entity manager):

Figure 7. Creating a database.

In the created document my-context, the data hosting facilities of the Spring framework were connected, settings for processing requests, authorization, session duration were set, a login and password were set to protect data. Data that will come from the WebService class.

Finally, the last step is to configure the servlet, which will be the final phase of processing requests:

Figure 8. Servlet mvc-dispatcher.

Dispatcher-servlet has the following main attributes:

- context: component-scan identification and initialization of components in a spring container;
- mvc: view-controller home page of the web application;
- mvc: resources processing requests for receiving static data;

In addition, additional methods for handling the functions of graphic items, such as: onSave, onAdd, onDel, onValidation, and so on [7].

As a result, as a result, the user gets the opportunity to work comfortably with the web application, which has an intuitive interface and contains handling of possible errors in the interaction process:

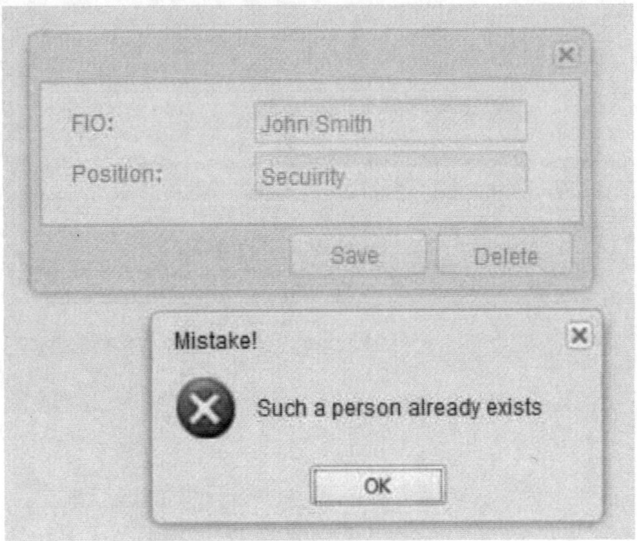

Figure 9. Duplicate error handling

4. Conclusion

The web development industry is one of the fastest growing and most active in the IT industry. Every day there are new libraries, frameworks and various constructors that allow you to automate the process of creating such applications or websites as much as possible [8].

Despite the fact that such technologies are very useful and reduce the time required for the development, it is necessary to understand that only knowledge and understanding of the fundamental functions of the interface and internal development makes a specialist out of a person, a way not only to improve work in short functional applications, but also to maintain it, to be able to deal with possible errors and problems [9].

In this research, the stages of design, development on the client and application, connecting several frameworks, and organizing data were mainly considered. A functioning web application was developed and the main goal of the work was achieved.

References

[1] Dzhangarov A I, Suleymanova M A and Zolkin A L 2020 Face recognition methods *IOP Conf. Ser.: Mater. Sci. Eng.* **862** 042046
[2] Ethan Marcotte, Responsive Web Design, Published June 7th 2011 by A Book Apart, 150 p.
[3] Dzhangarov A I and Suleymanova M A 2020 Electronic Digital Signature *IOP Conf. Ser.: Mater. Sci. Eng.* **862** 052054
[4] Crockford D 2008 *JavaScript: The Good Parts* (Yahoo Press)
[5] Wroblewski L 2011 *Mobile First* (A Book Apart)
[6] Simpson K 2015 *You Don't Know JS: Up & Going* (O'Reilly Media)
[7] Hunt A and Thomas D 1999 *The Pragmatic Programmer: From Journeyman to Master* (Addison-Wesley Professional) 321 p.
[8] Thomas D 2005 *Agile Web Development with Rails: A Pragmatic Guide* (Pragmatic Bookshelf)
[9] Magomedov I A and Sebaeva Z S 2020 Comparative Study of Finite Element Analysis Software Packages *Journal of Physics: Conference Series* **1515** 032073

IOP Conf. Series: Materials Science and Engineering 1155 (2021) 012101 doi:10.1088/1757-899X/1155/1/012101

Estimation of the minimum beam length for the static, dynamic, and stability problems

A E Mityaev[1], I V Kudryavtsev[1], M P Khomutov[2], M V Brungardt[1] and A V Kolotov[1]

[1] Siberian Federal University, 79 Svobodny Pr., Krasnoyarsk, 660041, Russia
[2] Kuban State Technological University, 2 Moskovskaya st., Kuban, 350072, Russia

E-mail: amityayev@sfu-kras.ru

Abstract. The calculation results of extended elements of structures according to the theory of beam will be correct if the condition for their length is met, which, as is known, should be much more transverse dimensions. The absence of an exact relationship between the transverse dimensions and the length of the beam is caused by the characteristics of the behavior under a load of various types and shapes of cross-sections: open, closed, thin-walled, etc. In this paper, the ratio of the length of the beam to its transverse dimensions is considered, which provides correct results during its static and dynamic loading, as well as during the loss of stability. To do this, two forms of thin-walled cross-sections are selected: circular and rectangular. The problem solution was carried out both analytically and by the finite element method. The results showed that the most stringent requirements are introduced by the theory of stability, according to which the length should be 15 or more times the largest transverse dimension.

1. Introduction

Beams are the most common structural elements and their calculation methods are the most developed. Beams are used both alone and create light and strong frames of spatial large structures. Beam calculation methods were probably the first in mechanics to be brought to the simplest linear ratios to allow manual strength calculations.

It would seem that today the calculation of beams no longer contains any uncertainties, but one controversial issue remains. It concerns the beam definition itself and its effect on the accuracy of the results. According to the known literature, a beam is a geometric object, one of the dimensions of which (length) is much larger than the other two. The exact ratio of length to transverse dimensions is not given in any scientific source for many reasons. Firstly, this ratio is highly dependent on the type and shape of the cross-section: open, closed, thin-walled, solid, etc. Further, this dimension ratio also depends on the method of restraining and loading the beam, the characteristics of the distribution of stress fields. вносят свои погрешности, которые будут сочетаться между собой по сложной зависимости Finally, the accuracy of the beam calculation is based on several assumptions that also introduce their errors, which will be combined by a complex dependence.

In this paper, an attempt was made to determine the necessary ratio of the length of the beam to its maximum transverse dimension, which provides correct results at static and dynamic loading, as well as during a loss of stability. To do this, in this work we will consider two common forms of thin-walled cross-section (figure 1): rectangular and circular.

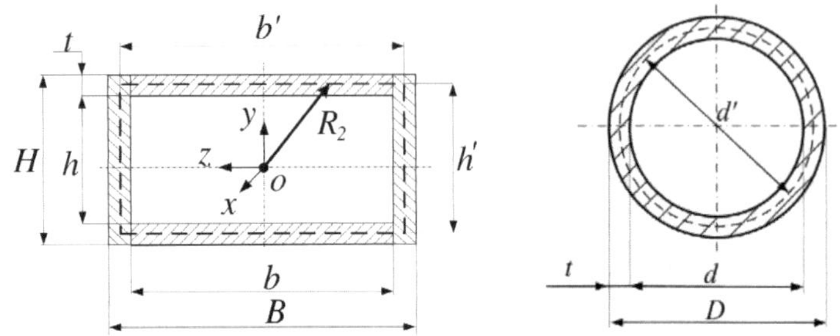

a) rectangular b) circular

Figure 1. Cross-sectional shapes of the beam.

The solution of the problem will be carried out using both known information from the literature and new data obtained as a result of analytical transformations and numerical calculations.

2. Beam length at static load

At static loading, the recommended ratio of beam length to its maximum transverse dimension can be found in many sources, for example in [1-5]. Length constraints depend on the type of cross-section, whether it is solid or thin-walled. For beams of a solid or thick-walled section, the ratio of their length to the maximum transverse dimension shall be:

$$\frac{l}{B_{max}} > 5, \tag{1}$$

where l is the beam length; B_{max} is the maximum cross-sectional dimension (B in figure 1, a and D in figure 1, b).

Condition (1) requires that the length of the beam ensure the fulfillment of the Saint-Venant principle, as well as the Bernoulli's plane section hypothesis. The latter requirement is most essential and is fulfilled by minimizing possible cross-sectional shifts, that is, the predominance of normal stresses over shear stresses. This is always valid, for example, when the long beams are transversely bent, at which the bending stresses will exceed the tangent stresses. If the beam is only subjected to stretching or compression, then condition (1) may well be weakened.

Thin-walled beams, especially open profiles, have a more complex stress distribution and require a more stringent condition for the applicability of beam theory [6-8]. Note here that wall thickness condition is added to length condition:

$$\frac{l}{B_{max}} > 10 \quad \text{and} \quad \frac{B_{max}}{t} > 10, \tag{2}$$

where t is the wall thickness (figure 1).

More stringent conditions (2) are caused by possible warping of the thin-walled cross-section and its effect on the calculation results.

3. Beam length at dynamic load

No information was found in the known literature on the applicability of beam theory in dynamic loading, it is possible that conditions (1) and (2) are preserved [9-10]. To check the length limits in the beam theory at dynamic load, numerical calculations were made using the finite element method [11,12] for beams with a circular and rectangular cross-section (figure 1). The finite element beam model was created in the Ansys software and consisted of approximately 20,000 finite elements of the type Shell128. A straight beam with two support types at the edges was studied: hinge and fixed. The first natural frequency of beam vibration was estimated by varying its length and wall thickness. The results were compared with reference values calculated from the beam vibration theory [10]. The difference in analytical and numerical results determined the error introduced by the length of the beam. The results of calculations in the form of plots are shown in figure 2.

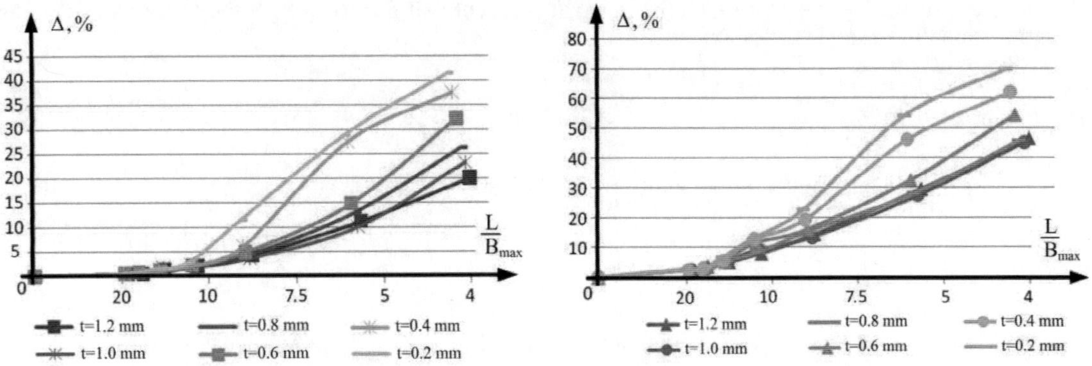

a) hinged beam b) fixed beam

Figure 2. Influence of beam length on error of calculation of the first natural frequency of vibration.

Plots show that for short beams, the calculation error is highly dependent on the wall thickness t, but even with a large wall thickness, condition (1) is incorrect. Condition (2) provides an error of 10-15%, which is not always permissible, although quite correct for beam theory.

4. Beam length in case of stability loss

The most often flexural stability of beams calculates by a formula Euler [13-23] which, as we know, is fair provided that the arising stresses are less than the limit of proportionality σ_{prop} and flexibility of a beam λ is more limit value λ_{min}.:

$$\lambda \geq \lambda_{\min} = \pi \sqrt{\frac{E}{\sigma_{prop}}}, \quad (3)$$

where E is the elastic modulus.

The flexibility of a beam is defined through its geometry and support conditions as:

$$\lambda = \mu \frac{l}{i_{\min}}, \quad (4)$$

where i_{min} is the minimum radius of gyration of beam cross-section:

$$i_{\min} = \sqrt{\frac{I_{\min}}{S}}. \quad (5)$$

Consider the two cross-section shapes (figure 1) and derive length conditions (1) for it.

4.1 Rectangular cross-section

Let us express the condition (3) through the ratio (1), for this, we output the flexibility of the waveguide with a rectangular cross-section through its dimensions, we get:

$$S = BH - bh \qquad I_{\min} = I_x = \frac{BH^3 - bh^3}{12}, \quad (6)$$

$$i_{\min} = i_x = \sqrt{\frac{BH^3 - bh^3}{12(BH - bh)}} \approx 0.289H \cdot \sqrt{\frac{3B + H}{B + H}} . \quad (7)$$

Let's substitute the expression (6) in (5) and accept that $H=2B$, we get the flexibility (4) in the form:

$$\lambda = \mu \frac{l}{i_{\min}} = \frac{\mu l}{0.373H} . \quad (8)$$

By substituting equation (8) into condition (3), we get:

$$\frac{\mu l}{0.373H} \geq \pi \sqrt{\frac{E}{\sigma_{\Pi}}} ,. \quad (9)$$

At last, taking into account the fact that actually μ=0.5,...,1, the condition (3) take a form:

$$\frac{l}{B_{\max}} \geq (1.17,....,2.34) \cdot \sqrt{\frac{E}{\sigma_{prop}}} ,. \quad (10)$$

The condition (10) is now defined only by the material properties that are discussed later. For other ratios of width B and height H of the cross-section, the coefficients in parentheses vary slightly.

4.2 Circular cross section

Perform the same transformations for the circular cross-sectional shape. The area and moment of inertia of the section in this case are equal:

$$S = \frac{\pi}{4}(D^2 - d^2) \qquad I_{\min} = I_x = I_y = 0.05 \cdot (D^4 - d^4), \quad (11)$$

$$i_{\min} = i_x = i_y = \frac{1}{4}\sqrt{D^2 + d^2} . \quad (12)$$

Flexibility is:

$$\lambda = \mu \frac{l}{i_{\min}} = \frac{4\mu l}{\sqrt{D^2 + d^2}} . \quad (13)$$

Take for uniqueness $D = d = d'$ (figure 1,b), then we get:

$$i_{\min} = i_x = i_y = 0.354 \cdot d' . \quad (14)$$

$$\lambda = 2.83 \cdot \frac{\mu l}{d'} . \quad (15)$$

By combining the resulting expressions, we obtain the condition of the ratio of the length of the beam to its transverse dimension for a circular cross-section in the form:

$$\frac{l}{d'} \geq (1.11,...,2.22) \sqrt{\frac{E}{\sigma_{prop}}} . \quad (16)$$

The coefficients in parentheses of expression (16) differ slightly from the coefficients in condition (10) and do not significantly affect the result. The properties of the material used have a much greater influence.

4.3 Calculation of limits for different materials

We get the conditions (10) and (16) in the form (1), for this, we substitute the values of mechanical parameters of various materials in them. Results are in table 1.

Table 1. Ratio of length to maximum cross-section dimension.

Material	E, GPa	σ_{prop}, MPa	rectangular		circular	
			from	to	from	to
Aluminum	70	70	37	74	35.1	70.2
Copper	132	400	21.3	42.5	20.2	40.3
AD31	71	210	21.5	43	20.4	40.8
D16T	72	320	17.5	35.1	16.6	33.3
Steel 45	210	500	24	48	22.7	45.5
Steel 35HM	218	600	22.3	44.6	21.2	42.3

The results of the calculations show that the application of the Euler formula is correct for most materials, if at least double the condition (2), that is, with

$$\frac{l}{B_{max}} > 20. \quad (17)$$

Condition (17) is obtained in the case of thin-walled cross-section; in the case of solid cross-section this condition will be slightly weaker:

$$\frac{l}{B_{max}} > 15. \quad (18)$$

Thus, the stability loss length condition is the strictest condition considered.

5. Discussion

The whole study shows that during dynamic loading the length of the beam should be noticeably longer than when static according to the conditions (17, 18). This beam length may in many cases be unacceptable due to a corresponding decrease in their first natural frequency due to a quadratic inverse relationship.

The dependence of the conditions (17, 18) on the properties of the material can somewhat weaken the coefficients. Since a limited number of crystal lattice structures exist for steels, it is possible to reduce the coefficient by increasing the proportionality limit, for example by alloying or quenching.

The use of shorter beams than the conditions recommend (17, 18) will result in the Euler formula giving somewhat overstated critical loads. This can lead to an accident, so other stability calculation methods need to be applied for short beams.

Thus, under static loading, conditions (1) and (2) are valid. Conditions (17, 18) should be applied for dynamic loading and stability loss.

6. Conclusion

In the work, studies were carried out on the limits of the correctness of the beam length to meet the criteria for calculating static, dynamic loading, as well as the loss of stability. The results obtained are well consistent with known values and refine them with respect to beam dynamics. The proposed approach can be used with beams with any cross-sectional shape.

Acknowledgements
The research was funded by RFBR, Krasnoyarsk Territory and Krasnoyarsk Regional Fund of Science, project number 20-48-242922.

References

[1] Carrera E 2011 *Beam Structures. Classical and Advanced Theories* (John Wiley & Sons, Ltd)
[2] Gere J M 2004 *Mechanics of materials* (Cengage Learning, Toronto, Canada)
[3] Hibbeler R C 2016 *Statics and Dynamics* (Pearson Education, NJ, USA)
[4] Timoshenko S 1940 *Strength of materials. Part I. Elementary theory and problems* (D. Van Nostrad Company, NY, USA)
[5] Ugural A C 2010 *Stresses in beams, plates and shells* (CRC Press, Taylor & Francis Group)
[6] Vlasov V Z 1961 *Thin-walled Elastic Beam* (English translation. Published for NSF and Department of Commerce by the Israel Program of Scientific Translations, Jerusalem)
[7] Meek J L and Lin W J 1990 *J Struct Eng-ASCE* **116(6)** 1473-89
[8] Djugash A C R 1994 *J. Construct. Steel Research* **31** 289-304
[9] Biedermann V L 1980 *Applied theory of mechanical vibrations* (VSH, Moscow, Russia)
[10] Blevins R D 2016 *Formulas for dynamics, acoustics and vibration* (John Wiley & Sons, Ltd)
[11] Shrikhande M 2008 *FEM and computational structural dynamics* (PHI Learning Private Limited, New Delhi)
[12] Moaveni S 2015 *Finite element analysis* (Pearson, USA)
[13] Timoshenko S P and Gere J M 2009 *Theory of Elastic Stability* (Dover Publications, USA)
[14] Thomsen J J 2003 *Vibrations and stability* (Springer-Verlag New York, USA)
[15] Farshad M 1994 *Stability of structures* (Elsevier Science Ltd)
[16] Ingerle K 2018 *Non-Conservative Systems: New Static and Dynamic Stability Criteria* (CRC Press)
[17] Jerath S 2020 *Structural Stability Theory and Practice: Buckling of Columns, Beams, Plates,* and Shells (John Wiley & Sons, Ltd)
[18] Abramovich H 2017 *Stability and Vibrations of Thin-Walled Composite Structures* (Woodhead Publishing)
[19] Gambhir M L 2004 *Stability Analysis and Design of Structures* (Springer-Verlag New York, USA)
[20] Wiggers S L and Pedersen P 2018 *Structural Stability and Vibration: An integrated introduction by analytical and numerical methods* (Springer-Verlag New York, USA)
[21] Vinson J R 2012 *The Behavior of Thin Walled Structures: Beams, Plates, and Shells: Beams,* Plates and Shells (Springer-Verlag New York, USA)
[22] Pignataro M 2013 *Stability, Bifurcation and Postcritical Behaviour of Elastic Structures* (Elsevier Science)
[23] Narayanan R 2007 *Beams and Beam Columns: Stability and Strength* (Applied science publisher)

AUTHOR INDEX